Solid Oxide Electrolysis: Fuels and Feedstocks from Water and Air

University of York, United Kingdom

13–15 July 2015

FARADAY DISCUSSIONS

Volume 182, 2015

ROYAL SOCIETY OF CHEMISTRY

Faraday Discussions (Print ISSN 1359-6640, Electronic ISSN 1364-5498) is published 9 times a year by the Royal Society of Chemistry, Thomas Graham House, Science Park, Milton Road, Cambridge, UK CB4 0WF.

Volume 182 ISBN-13: 978-1-78262-182-9

2015 annual subscription price: print+electronic £877, US $1635; electronic only £833, US $1553. Customers in Canada will be subject to a surcharge to cover GST. Customers in the EU subscribing to the electronic version only will be charged VAT. All orders, with cheques made payable to the Royal Society of Chemistry, should be sent to RSC Order Department, Royal Society of Chemistry, Thomas Graham House, Science Park, Milton Road, Cambridge, CB4 0WF, UK. Tel +44 (0) 1223 432398; E-mail **www.orders@rsc.org**

If you take an institutional subscription to any RSC journal you are entitled to free, site-wide web access to that journal. You can arrange access via Internet Protocol (IP) address at **www.rsc.org/ip**. Customers should make payments by cheque in sterling payable on a UK clearing bank or in US dollars payable on a US clearing bank.

Printed in the UK

Faraday Discussions are unique international discussion meetings that focus on rapidly developing areas of chemistry and its interfaces with other scientific disciplines.

Solid Oxide Electrolysis: Fuels and Feedstocks from Water and Air

Faraday Discussions

www.rsc.org/faraday_d

A General Discussion on Solid Oxide Electrolysis: Fuels and Feedstocks from Water and Air was held in York, UK on the 13th, 14th and 15th of July 2015.

CONTENTS

ISSN 1359-6640; ISBN 978-1-78262-182-9

Cover
Front cover image courtesy of Shutterstock.

INTRODUCTORY LECTURE

PAPERS AND DISCUSSIONS

SPONSORS

cogent | Skills for Science Based Industries

Metrohm Autolab U.K.

SPONSORING ORGANISATION

IChemE ADVANCING CHEMICAL ENGINEERING WORLDWIDE

CONCLUDING REMARKS

ADDITIONAL INFORMATION

Faraday Discussions

PAPER

Solid oxide electrolysis – a key enabling technology for sustainable energy scenarios

John Bøgild Hansen

Received 19th August 2015, Accepted 19th August 2015
DOI: 10.1039/c5fd90071a

Production of fuels and chemicals from steam and/or CO_2 with solid oxide electrolysis cells (SOEC) and electricity have attracted considerable interest recently. This paper is an extended version of the introductory lecture presented at the first Faraday Discussions meeting on the subject. The focus is on the state of the art of cells, stacks and systems. Thermodynamics, performance and degradation are addressed. Remaining challenges and potential application of the technology are discussed from an industrial perspective.

Introduction

The use of solid oxide cells to electrolyze steam or CO_2 is not a new development. It was suggested already in the 1980s and quite advanced technology developments were carried out in Germany and USA. Due to the low price of fossil fuels and the high cost associated with the SOEC (Solid Oxide Electrolysis Cells) technology at that time combined with challenges on material degradation the interest, however, waned.

During the last decade the field has attracted increasing interest and the number of research projects and indeed scientific papers has shown exponential growth.

This revival can be ascribed to mainly three factors:

• The progress in developing materials, cells, stacks and system for application of solid oxide cells for fuel cell applications.

• The impressive growth of renewable power generation from mainly wind and solar power which by nature are intermittent.

• The threat of climate change, which necessitates development of carbon neutral energy scenarios.

This paper is not intended to be a comprehensive review paper on the subject. Excellent recent reviews are available.[1,2] It is basically an extended version of the introductory lecture presented at the first Faraday Discussion dealing with the

Haldor Topsøe A/S, Nymøllevej 55, DK-2800, Kongens Lyngby, Denmark. E-mail: jbh@topsoe.dk

fundamental scientific details of the subject. The focus will be on the state of the art, the challenges remaining and the potential applications of the technology as seen from an industrial perspective.

The concept and basic principle

Solid Oxide Electrolyzer Cells (SOEC) operate in the reverse mode of Solid Oxide Fuel Cells (SOFC) and their operating principles are illustrated on Fig. 1.

In SOEC cells steam or CO_2 at the cathode (negative or fuel electrode) are reacting with electrons provided by an external power source producing H_2 and CO_2, respectively, and oxygen ions, which are transported through a ion-conducting, gas-tight membrane due to the electric field to the anode (positive or oxygen electrode) where the oxygen ions combine and liberate the electrons again. In the SOFC the reverse reactions are taking place so electricity is produced and sent to an external consumer.

Early developments

The decomposition of water by electricity was first accomplished by William Nicholson and Anthony Carlisle, who, in 1800 in London, sent a current from the newly developed Volta pile through water and to their astonishment saw hydrogen and oxygen evolve at the electrodes "several inches apart".

Michael Faraday coined the term electrolysis in 1833 and developed his two laws to describe these phenomena on a firm scientific basis.

The first use of SOEC was for space exploration, where interest was focused on oxygen generation for life support and the CO generated from CO_2 electrolysis was reacted to carbon and CO_2 (recycled back to the SOEC) over an iron catalyst.[4–6]

Isenberg at Westinghouse Electric[6] used tubular cells at high temperature and demonstrated convincingly that both steam and CO_2 electrolysis was technically feasible as seen in Fig. 2. He also foresaw that utilization of high temperatures may reduce the electrical input required. Furthermore the ability to produce synthesis gas and thus synthetic fuel was also obvious to him: "It is easily seen, that the ability to produce carbon monoxide from CO_2 electrolysis could be the basis for a host of organic syntheses *via* the Fischer–Tropsch process".

Quite intense development efforts were carried out by Dönitz and Erdle at Dornier also in the 1980s to develop commercial SOEC stacks for efficient hydrogen production by a concept called "Hot Elly".[7–9] Detailed engineering studies were performed and it was the intention to supply hydrogen for low

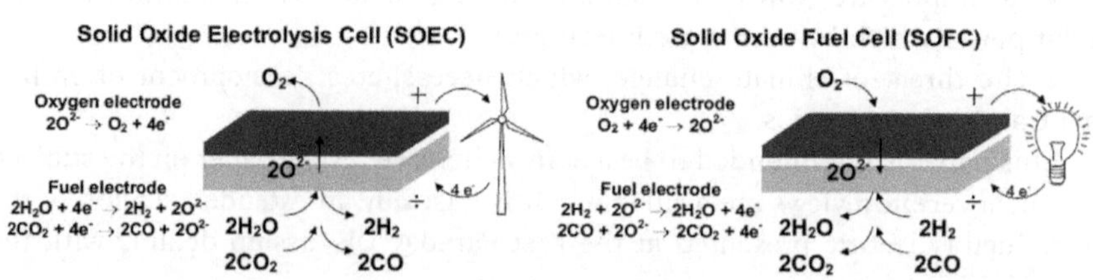

Fig. 1 Basic operating principles of SOEC and SOFC cells. Reprinted with permission from *ECS Trans.*, 2012, **50**, 167. Copyright 2012, The Electrochemical Society.[3]

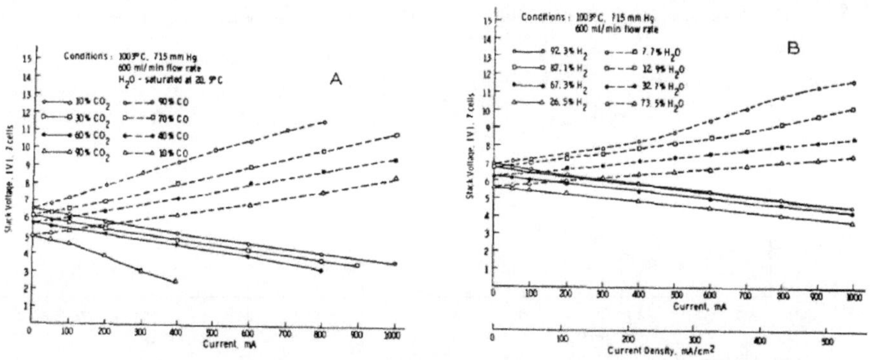

Fig. 2 *I*–*V* or polarization curves for CO_2 electrolysis (A) and steam electrolysis (B). Reprinted from *Solid State Ionics*, 1981, **3–4**, 431–437 with permission from Elsevier.[6]

quality oil and coal upgrading, an option which is overlooked today. Furthermore integration with coal gasification was studied and the possibility of doubling the output from the same amount of coal pointed out. The synergy for gasification with the oxygen produced from the SOEC instead from an air separation unit was already then seen as a bonus. The current density was -0.3 A cm^{-2} at 1.07 V and 100% Faradaic efficiency was demonstrated, but the operating temperature of around 1000 °C and the tubular design of course made the balance of the plant, as well as the SOEC unit, quite expensive.

Thermodynamics

It is important to understand the basic thermodynamics of water and CO_2 electrolysis because this is one of the main explanations for the high conversion efficiency achievable with SOEC.[1,10–13]

Both water decomposition

$$H_2O = H_2 + 0.5O_2\ (-\Delta H_{1023\ K} = 248.1\ \text{kJ mol}^{-1}) \tag{1}$$

and CO_2 electrolysis

$$CO_2 = CO + 0.5O_2\ (-\Delta H_{1023\ K} = 282.5\ \text{kJ mol}^{-1}) \tag{2}$$

are highly endothermal reactions requiring 3.07 and 3.50 kWh per Nm^3 of hydrogen and CO, respectively, produced at 750 °C. The reaction enthalpies, for both reactions are only weakly dependent on temperature, as illustrated on Fig. 3.

The reaction enthalpy consists of two terms

$$\Delta H = \Delta G + T \times \Delta S \tag{3}$$

where ΔG is the Gibbs free energy change which has to be provided in the form of electrical energy while the entropy part $T \times \Delta S$ can be supplied as heat. As ΔS are positive for both reaction (1) and (2), ΔG decreases with temperature. At 80 °C ΔG is 93% of ΔH but at 750 °C it is only 77%.

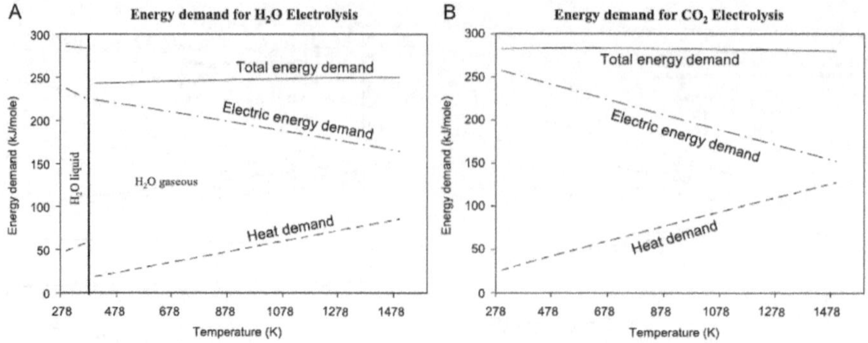

Fig. 3 ΔH, ΔG and $T \times \Delta S$ as a function of temperature for water splitting and CO_2 electrolysis. Reprinted from *Int. J. Hydrogen Energy*, 2007, **32**, 3253–3257 with permission from Elsevier.[14]

This heat could be provided by an outside source (such as, for instance, high temperature heat from a nuclear reactor, from solar heat or from exothermal chemical reaction heat in the SOEC stack). These are intriguing possibilities but the heat can simply be provided by the Joule heat generated due to the loss mechanism within the stack.

These loss mechanisms are normally lumped together in the so called area specific resistance (ASR) composed of ohmic resistance and activation overpotentials at both the anode and the cathode. The activation overpotentials are in turn composed of kinetic overpotentials due kinetic activation energies and overpotentials due to mass transport limitations.

The minimum operating voltage to accomplish the reactions is defined by:

$$V_O = \frac{\Delta G}{n \times F} \tag{4}$$

where n is the number of electrons involved in the reaction (2 for both reaction (1) and (2)) and F is Faraday's constant 96 485 C mol^{-1}. V_O is the open circuit voltage at standard conditions. At the actual operating conditions this minimum is defined by the Nernst equation, which for water decomposition takes the form

$$V_n = V_O - \frac{RT}{2F} \ln\left[\left(\frac{y_{H_2O}}{y_{H_2}\sqrt{y_{O_2}}}\right)\sqrt{\frac{P}{P_{std}}}\right] \tag{5}$$

Now the Joule heat generation is given by

$$Q_{rest} = i^2 \times \text{ASR} = i \times (V_{op} - V_n) \tag{6}$$

When there is no net heat flux to the SOEC cells the operating voltage is the so called thermoneutral voltage (or enthalpy voltage)

$$V_{tn} = \frac{\Delta H}{nF} \tag{7}$$

At this voltage the inlet and outlet temperature from a stack are equal. Although the local current densities across the cells are not identical, operation at this voltage will minimize local temperature differences and thus mechanical stresses. This voltage at 750 °C is 1.285 V for steam electrolysis and 1.464 V for CO_2 electrolysis.

The electrolysis of steam or carbon dioxide requires a large amount of energy compared to the heat capacity of the reactants or products. This can be illustrated by the fact that small deviation from the thermoneutral voltage will give rise to large adiabatic temperature increases or decreases. Operating without a sweep gas on the oxygen side and normal steam conversion at 33 mV will result in a delta temperature of 100 °C.

The efficiency of the electrolyzer is

$$\eta = \frac{\text{LHV}}{W} = \frac{V_{\text{tn}}}{V_{\text{op}}} \tag{8}$$

where LHV is the lower heating value of the fuel produced and W is the total electrical energy input. As LHV by definition is equal to the enthalpy change for reaction (1) and (2) it is seen that the efficiency of the electrolyzer stack is 100% if operated at the thermoneutral voltage. In practice it will normally be slightly lower due to heat losses, gas leakages in the stack and electronic leakage through the electrolytes in some cases, but close to 100% real efficiency has been demonstrated in numerous experiments.

In fact the coupling of exothermal loss mechanisms with endothermal reactions provides an elegant way of upgrading "waste heat" to chemical energy. The same principle is invoked when performing internal steam reforming of methane in SOFC, as discussed extensively in ref. 15. In these SOC cells the heat is furthermore transferred with a negligible temperature difference leading to minimum exergy loss. Nevertheless SOEC operation at the thermoneutral voltage still entails an exergy loss of around 5% even when accounting for the exergy content of the oxygen produced. This is an often overlooked fact, but transforming the pure exergy of electricity into chemical energy in this way comes at minor loss.

It is expedient to use the lower heating value in the efficiency calculation because this normally what is paid for in the fuel industry. In order to achieve 100% efficiency it is thus necessary to electrolyze steam and not start with liquid water. Otherwise the heat needed to evaporate the water needs to be added, which will increase the minimum amount of energy required by approx. 0.5 kWh per Nm^3 hydrogen.

When the hydrogen produced in the SOEC stack is used for a downstream chemical synthesis the reaction heat from the synthesis is normally sufficient to provide enough steam for the SOEC. This, together with the ability to

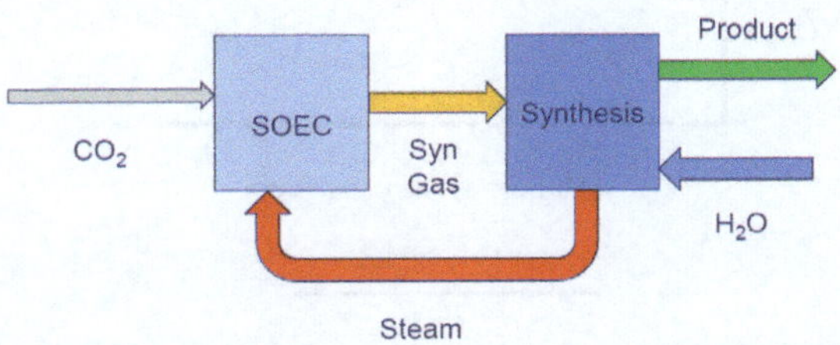

Fig. 4 Synergy between steam production in synfuels synthesis and SOEC.

electrolyze CO_2, creates a strong synergy between SOEC- and synthesis-gas- (mixtures of CO, H_2 and CO_2) based fuel syntheses, as illustrated on Fig. 4 and discussed later.

When performing co-electrolysis (simultaneous electrolysis of steam and CO_2) it is necessary to also consider the reverse water gas shift (RWGS) reaction

$$H_2 + CO_2 = H_2O + CO \; (-\Delta H_{1023\ K} = 248.1\ kJ\ mol^{-1}) \tag{9}$$

as well as the methanation reaction

$$CO + 3H_2 = CH_4 + H_2O \; (-\Delta H_{1023\ K} = 248.1\ kJ\ mol^{-1}) \tag{10}$$

The RWGS reaction is slightly endothermal whereas the methanation is quite exothermal. Both reactions will be fast at normal SOEC operating temperatures, at least with the ubiquitously used nickel containing cathodes, and both reactions can be assumed to proceed to equilibrium. The RWGS is independent of pressure whereas the conversion to methane increases rapidly according to the Le Chatelier principle with operating pressure.

The role of the RWGS in co-electrolysis is still under debate, with some being inclined to believe that for the co-electrolysis case only steam electrolysis is occurring, while CO_2 is only being converted to CO *via* the RWGS because CO_2 electrolysis is slower than steam electrolysis.

From above it can be seen that operation at high temperature allows for a higher current density for the stack than low temperature operation given equal ASR. A high temperature is also beneficial for reaction kinetics – even without noble metal electrodes – so ASR will decrease with temperature. Both factors allow for lower stack areas and thus costs.

Compared to the traditional low temperature alkaline electrolyzers as well as PEM based electrolyzers, SOEC thus has the advantages of efficiency and compactness, as well as the ability to produce synthesis gas directly. A comparison of the three electrolysis technologies is provided in Fig. 5.

On the other hand there are multiple challenges related to high temperature operation with respect to materials selection and degradation issues.

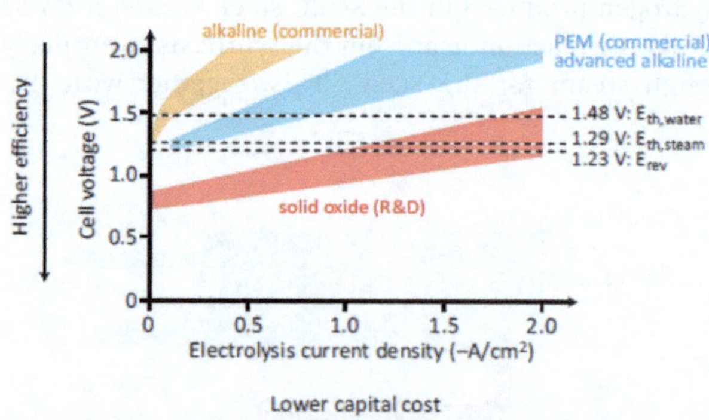

Fig. 5 Typical polarization curves for state of the art electrolyzer technologies. Reprinted from *Renewable Sustainable Energy Rev.*, 2011, **15**, 1–23 with permission from Elsevier.[16,94]

Materials for SOEC cells

The requirements to the materials used for SOEC are many and demanding:

• Chemical stability at the very reducing/oxidizing environment prevailing at the operating temperature.

• Stable with respect to sintering/agglomeration.

• Matching thermal expansion coefficients to avoid undue mechanical stresses.

• Chemical compatibility between adjacent materials.

• Mechanical strength and a good Weibull modulus as the materials are normally ceramic-like.

The electrolyte needs to be absolutely dense and free of pinholes and the electrodes needs to have:

• Adequate porosity to prevent undue diffusion restrictions of reactants and products.

• High electronic conductivity (for the widely used Ni–YSZ cermets this requires percolation of the nickel network).

• High ionic conductivity in order to extend the TPB into the active electrode structure.

• Excellent electrochemical activity.

These properties have now been optimized for decades for SOFC application and the materials used for SOEC have, accordingly, so far been SOFC materials although the operating conditions differ not only in obvious but also in more subtle ways between SOFC and SOEC operation, as will be discussed in the following.

Electrolyte. The most widely used electrolyte is zirconia doped with especially yttrium, which was used in the original experiments at both Westinghouse and Dornier and later at most of the research institute like DTU energy conversion (previously Risø National Lab), ECN, Forschungszentrum Jülich as well as most of the commercial companies.[6,7,9,17–21] Stabilizing of the zirconia with 8 mol% Y_2O_3 (YSZ) is the most commonly used due to its good ionic conductivity and good mechanical properties. Less yttrium (3YSZ) makes the electrolyte more robust but at the cost of lower ionic conductivity.[22]

Another good candidate for electrolyte is scandia doped yttrium (ScZ),[23] which has higher ionic conductivity and also lowers the polarization of the cathode. ScZ has also been used by Ceramatec for their electrolyte supported cells tested at INL (Idaho National Lab) with good results.[24]

Samarium-doped ceria (SDC) also shows very good performance, but due to the high electronic conductivity of ceria under the very reducing conditions prevalent in SOEC operation internal short circuiting occurs, which makes this material unsuitable for SOEC cells.[25] Bi-layered gadolinium doped ceria (GDC) and YSZ showed very promising performance, but decayed fast in steam electrolysis experiments.[26]

Lanthanum gallate doped with strontium and magnesium (LSGM) has been tried as an intermediate temperature electrolyte, but there are problems with formations of lanthanum nitrate by reaction with the fuel electrode.[27]

Cathodes (fuel electrode). Nickel has very high electro-catalytic activity for steam and CO_2 electrolysis, but has only electronic conductivity. Hydrogen/protons have high mobility in the metal, but low solubility, which limits the extent of the TPB to approximately 100 nm. Electrons, oxide ions and gas species have to be transported from and to the TPB. This is the reason why nickel is mixed with an ionic conducting material like YSZ in order to maximize the TPB length. Such electrodes are called cermets for ceramic–metal composites. This situation is illustrated in Fig. 6.

Nickel–YSZ cermet is the predominant choice for commercial cells but despite its good properties there some disadvantages. Nickel metal will oxidize if exposed to steam or CO_2 at SOEC relevant temperatures, so there is a need on a system level for recycling of part of the product gas from the stacks. According to Hauch it seem that 1% reducing gas should be sufficient.[29] Nickel is also very prone to carbon laydown and susceptible to poisoning by sulfur. The same would be the case for iron or cobalt containing electrodes.

There is thus an incentive to develop all-ceramic electrodes and this has been done successfully, especially at the Pacific Northwest National Lab[30] and at St. Andrews University.[31–35]

Lanthanum-doped strontium vanadate (LSV), strontium-doped lanthanum manganite partially substituted with chromium (LSCM), niobium-doped strontium titanates (STN) and lanthanum-doped strontium titanates with ceria (LST–ceria)[30] have all been used. LSCM with 0.5% Pd have been shown to be very effective CO_2 electrolysis electrodes.[36] Double perovskite Sr_2FeNbO_6 (SFN) have also been claimed to be better than Ni–YSZ.[37]

Very few experiments have been done with the very promising metal supported cathodes and the longest tests showed significant oxidation of the metal support structure[38] and a degradation rate of around 3.2% per 1000 h in the 2000 h durability experiment. Metal supported cells have been used successfully in SOFC operation where the steam partial pressure at the exit is comparable to SOEC operation. The metal supported cells are promising because of materials savings compared to Ni–YSZ and could furthermore be upscaled more easily with respect to their geometric area compared to ceramic supports.

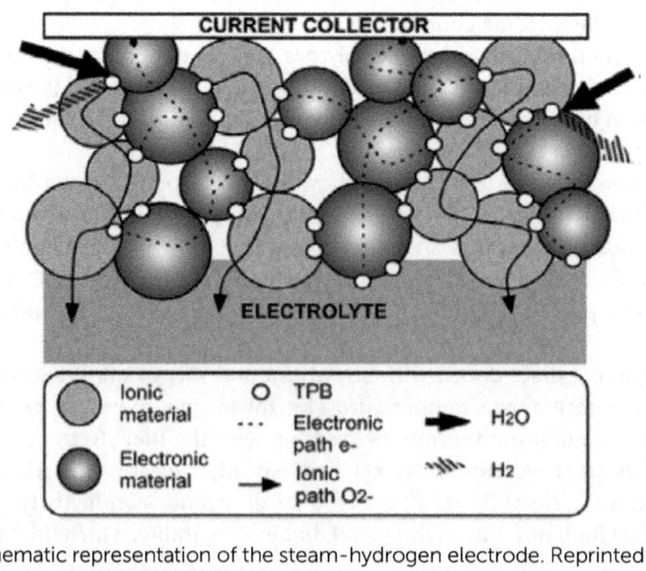

Fig. 6 Schematic representation of the steam-hydrogen electrode. Reprinted from *Int. J. Hydrogen Energy*, 2013, **38**, 15887–15902 with permission from Elsevier.[28]

Anodes (oxygen electrode). Most anodes are mixed conductor perovskite oxides, ABO_3, where A is usually a lanthanide ion, B small tri- or divalent 3d transition metals. The most widely used was initially a LSM–YSZ composite, but recently a host of mixed ionic and electronic conductors (MIEC) have been employed such as strontium doped lanthanum cobaltite (LSC),[17,39,40] strontium doped lanthanum ferrite partially cobalt doped LSCF[24,30,41–43] or nickelate (LSCN).[44] Most of these electrodes require a barrier layer (typically CGO) to prevent reaction with the electrolyte. They provide better performance and seem to have alleviated problems with delamination at the electrolyte/anode interface. The traditional LSM–YSZ performance is compared with LSC and LSCF anodes on Fig. 7.

Stacks. Apart from the early development of the Hot Elly concept by Dornier, tubular cells are rarely used in commercial stacks, which mostly use cathode-supported cells for SOFC applications. Some developers, however, prefer the use of electrolyte supported cells (ESC) despite the requirement for higher operating temperatures. The electrolyte supported cells have the advantage of having very thin electrodes so that polarization issues due to diffusion do not impede performance. ESC cells are also mechanically rugged.

Performance degradation

The initial results from long term testing of SOEC cells were rather discouraging with degradation rates exceeding 3% in applied voltage per 1000 hours for constant current. As the lifetime of SOEC stacks should for most applications be at least 5 years in order for the technology to become economically viable, this rate of degradation is clearly unacceptable as the operating voltage would long before exceed the tolerable. High voltages are unacceptable both with respect to efficiency and thermal management, which becomes critical on a system level if the operating voltage is much higher than the thermo-neutral (say 1.45 V maximum for steam electrolysis). Reference is made to the remark made under the thermodynamics section, that 33 mV would give rise to a 100 °C adiabatic temperature rise without large amounts of sweep gas and low steam conversions.

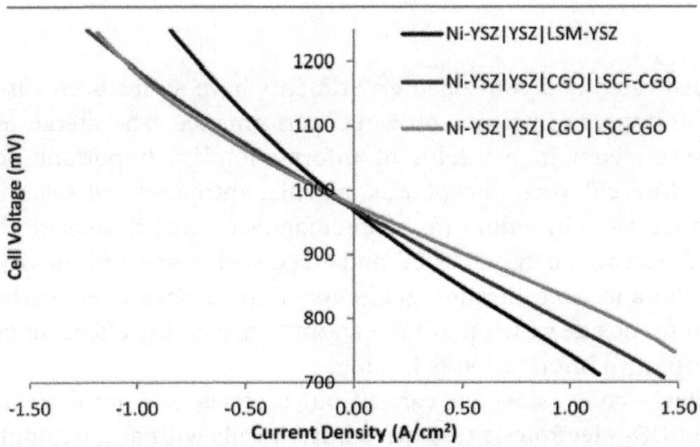

Fig. 7 Performance of different anodes with DTU cells at 800 °C with 50% H_2/50% H_2O. Reprinted from *Chem. Rev.*, 2014, **114**, 10697–10734 with permission from American Chemical Society.[1]

Unfortunately most tests are by tradition carried out in the galvanostatic mode, *i.e.* the current (and thus conversion of feed stock for constant flow) is kept constant whereas the voltage is allowed to increase. It is often claimed that this presents no problem when testing small button cells, and the temperatures have been carefully monitored and isothermal operation have been measured. It is however difficult to measure the exact temperature at the TPB and the results can thus be obfuscated severely. Starting the test far below the thermoneutral voltage and ending up far above would tend to mask an even more pronounced degradation than reported.

DTU energy conversion[3] has done some very careful experiments where it was shown that the temperature measured 1 mm from the cell deviated from the oven temperature by 0.6 °C, when operating at the voltage where maximum deviation from isothermal condition are found, namely halfway between V_{tn} and V_n. Impedance spectroscopy did, however, show a variation in pure ohmic resistance (which only depends on temperature) corresponding to a true temperature deviation from isothermal of 2.7 °C.

Another problem with small scale tests on button cells is the large surface area of tubing, seals *etc.* compared with the active cell area tested. Different poisons in these materials, along with assorted impurities in the feed gases, which are rarely subjected to dedicated cleanup, has thus probably had an unnoticed but major influence on the degradation results.

Finally it should be noted that it is experimentally very challenging to ensure a pulsation free and uninterrupted supply of steam with small-scale test facilities.

The earlier long-term tests did, however, serve to map more safe operating regimes with respect to especially current density and did also reveal by means of post-mortem analysis, impedance spectroscopy and other *in situ* techniques some important degradation mechanisms.

The database of degradation mechanisms for SOEC at the Haldor Topsøe company has more than 100 entries and it is beyond the scope of the present paper to discuss each of them, especially because there exist many interactions between them. Some of the more important mechanisms are, however, briefly discussed below.

Poisoning

Very few dedicated and controlled experiments have so far been carried out to evaluate the impact of poisons on SOEC performance. The literature on SOFC does, however, contain a wealth of information.[15,45] Important poisons are probably sulfur, chlorine, phosphorus, arsenic, antimony and selenium for the SOEC cathode and chromium (from interconnects, *etc.*) deactivate the anodes. There are differences between SOEC and SOFC with respect to the oxygen transport direction and concentration gradients of steam, hydrogen, carbon dioxide and carbon monoxide which may have an influence on the effects of poisons but so far quantitative information is lacking.

DTU energy conversion has carried out a series of careful steam, co-electrolysis and CO_2 electrolysis tests with button cells with and without gas clean up.[46] They also applied in-plane voltage measurements, which are an ingenious way of detecting poisoning progressing as a front through the cells, as illustrated on Fig. 8.

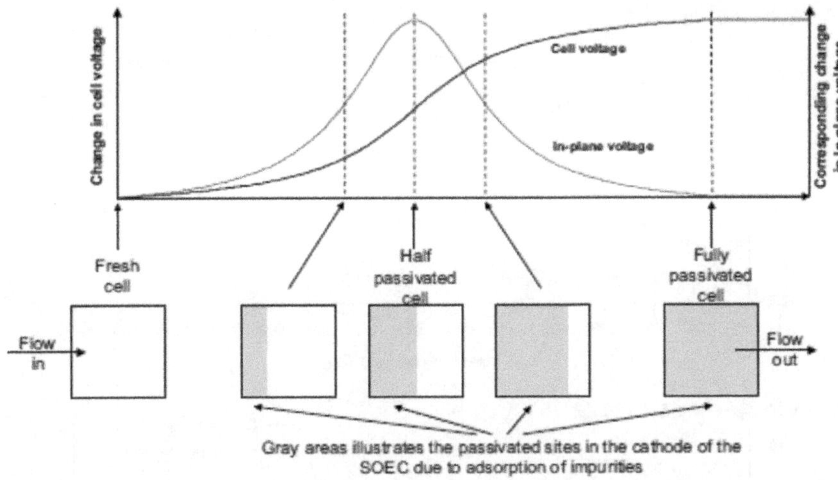

Fig. 8 Evolution of in-plane and cell voltage as a poison front moves through the cell. Reprinted with permission from *J. Electrochem. Soc.*, 2010, **157**, B1419–B1429. Copyright 2010, The Electrochemical Society.[46]

This technique can also reveal if the poisoning effect is temporary and reversible as the in plane voltage difference will show a similar picture when the poison(s) are removed from the cell again, like for instance silica being carried out of the cell by steam as silica hydroxide.

The results are quite revealing as shown on Fig. 9 for experiments with gases as received and on Fig. 10 with clean gases using a proprietary cleaning mass.

It was observed that when using the gases without cleaning the degradation rate was between 450 and 700 mV kh^{-1} for the first few hundred hours and then between 3 and 32 mV kh^{-1} for all three electrolysis modes.

With cleaned gases no degradation was observed applying current densities up to −0.5 A cm^{-2}. Some of the cells even activated slightly which was explained by removal by steam of impurities stemming from the raw materials, notably silica (33 ppm) used to produce the cells. DTU energy conversion has by SEM identified silica at the TPB.[18,47] They also note that for SOEC operation transport of impurities by steam is towards the TPB so that poisons have a more severe impact than in SOFC mode. Impurities should be removed to the ppb level.

The impedance studies showed that poisoning effects only affected the nickel cathodes, whereas the ohmic resistance was constant. A very slight degradation of the LSM/YSZ electrode (with a characteristic frequency of 1 to 3 kHz) was, however, observed even with clean gases.

DTU energy conversion identified traces of sulfur in the gases as received by mass spectroscopy but not in the cleaned gases.

Sulfur is probably the most commonly encountered poison. For SOFC operation Hansen[48] found the impact of sulfur could be correlated with a Temkin isotherm:

$$\Theta_s = 1.45 - 9.53 \times 10^{-5}T + 4.17 \times 10^{-5}T \ln Y \quad (11)$$

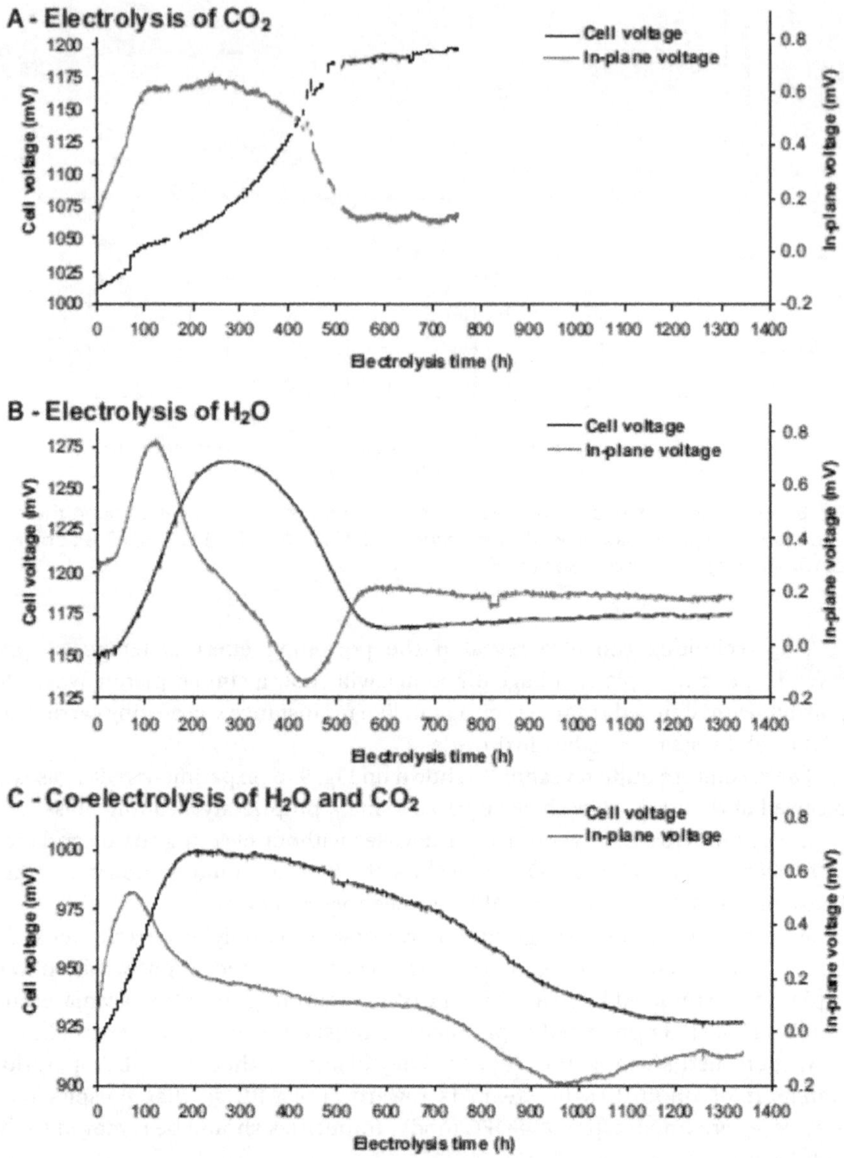

Fig. 9 Cell and in plane voltage as function of time at 850 °C with gases as received. Reprinted with permission from *J. Electrochem. Soc.*, 2010, **157**, B1419–B1429. Copyright 2010, The Electrochemical Society.[46]

where Y is the ratio between the partial pressures of hydrogen sulfide and hydrogen, Θ_s the sulfur coverage of the nickel surface. The impact of sulfur could then be correlated with

$$\mathrm{Pl} = k \times (\Theta_s - \Theta_{\min}) \quad (12)$$

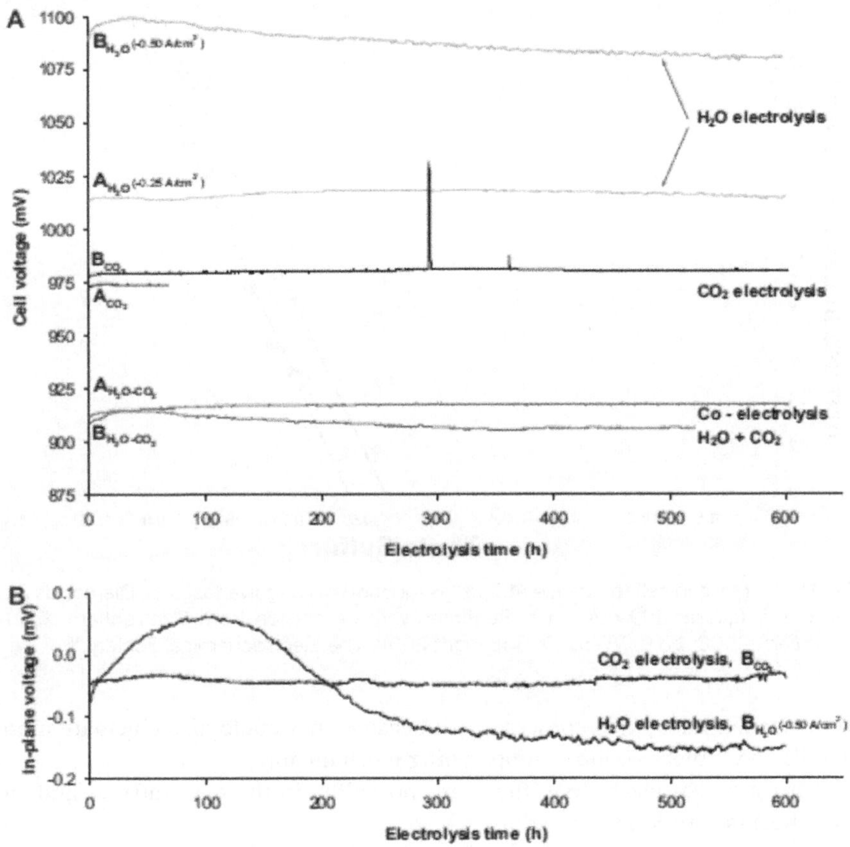

Fig. 10 Cell and in plane voltage as function of time at 850 °C with cleaned gases. Reprinted with permission from *J. Electrochem. Soc.*, 2010, **157**, B1419–B1429. Copyright 2010, The Electrochemical Society.[46]

where $\Theta_{min} > 0$ depends on the current density and anode material. The isotherm gave a very good fit to data covering temperatures from 700 to 1000 °C and H_2S contents from 0.05 to 50 ppm, as indicated on Fig. 11.

The fact that the performance loss was reduced at higher current densities was hypothesized to be due to a mitigating effect of oxygen reacting with part of the sulfur according to

$$H_2S + 1.5O_2 = SO_2 + H_2O \quad (13)$$

As this mechanism obviously should not be at play in SOEC mode it is to be expected that sulfur poisoning should be more severe in SOEC.

In order to minimize steam reforming of methane when co-electrolyzing biogas (consisting of methane and carbon dioxide) without impacting the electrochemistry too much, experiments were carried out by adding minute amounts of H_2S to the gas in SOEC mode.[49] It was found that 5–6 ppm H_2S completely eliminated the steam reforming activity and the performance loss could surprisingly be described quite well by eqn (12). More systematic tests

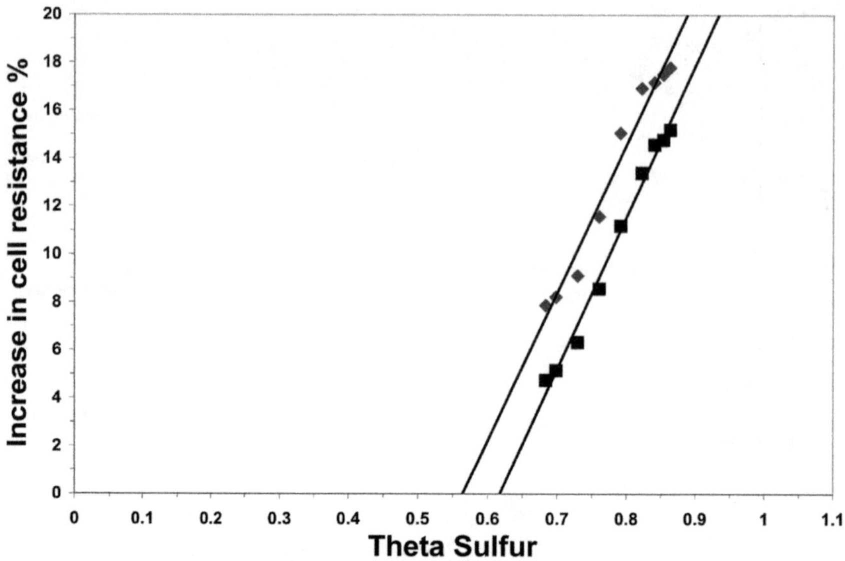

Fig. 11 Increase in cell resistance ASR (%) as function of Θ_s (galvanostatic). Diamonds 241 mA cm^{-2}, squares 409 mA cm^{-2}. Reprinted with permission from *Electrochem. Solid-State Lett.*, 2008, **11**, B178–B180. Copyright 2008, The Electrochemical Society.[46]

with cleaned inlet gases are warranted because they could also elucidate more details about the reaction and poisoning mechanism(s).

For pure CO_2 electrolysis there are no sulfur isotherms available and no experiments with SOEC operation.

Carbon formation

Carbon formation is set in by distinct mechanisms in the cathodes of high temperature electrolyzer cells (anodes in fuel cells) as well in the balance of plant equipment such as heat exchangers, which most often are steel based. Carbon nanotubes or whiskers may form notably on Ni, Fe, or Co. The limit where carbon is formed is influenced by the crystallite sizes. Carbon can also just be formed as graphite.

Graphite formation reduces the active area and blocks the pores, whereas whisker formation can lead to catastrophic destruction of the cathodes because the carbon nanotubes are very strong.

Graphite formation can occur from the following reactions:

$$2CO = C + CO_2 \tag{14}$$

$$CH_4 = C + 2H_2 \tag{15}$$

$$CO + H_2 = C + H_2O \tag{16}$$

When calculating the potential for carbon formation it is normally assumed that all steam reforming reactions, methanation and the shift reaction are in equilibrium. The potential for carbon then depends only on the C/H/O ratios, the pressure, and the temperature. The carbon limits found are shown in Fig. 12. It seen that the carbon deposition region increases at lower temperatures.

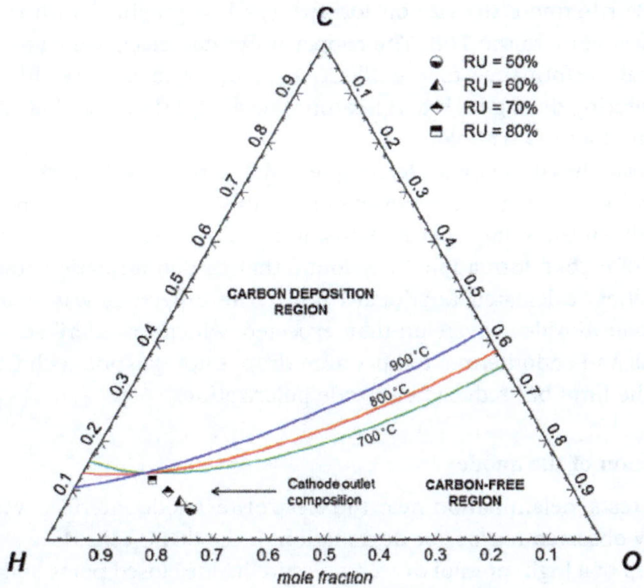

Fig. 12 Carbon deposition limit lines in the C–H–O diagram at 33 bars with different reactant utilizations. Reprinted from *J. Energy Storage*, 2015, 1, 22–37 with permission from Elsevier.[50]

In SOEC cells carbon formation is also challenging when performing CO_2 electrolysis especially at low temperature and if operated at higher pressures, which would otherwise be very advantageous as discussed later. DTU energy conversion has observed an interesting phenomenon where the carbon whiskers were actually formed at the three-phase boundary in co-electrolysis experiments at high current densities (>−1.0 A cm^{-2}), despite the fact that the actual H_2O + CO_2 conversion of 67% was below the critical conversion, calculated to be 99%. This discrepancy was explained by the concentration gradients and overpotential at high current densities, which are more critical in the SOEC mode than in the fuel-cell mode of operation.[51] The microstructure, including the porosity of the active layer thus plays a crucial role. The carbon whiskers actually formed on the zirconia nanoparticles from the YSZ components of the cermet.

Carbon formation has been studied by Raman and EDS spectroscopy on patterned 100 μm wide nickel stripes[52] with a CO : CO_2 gas mixture at 750 °C. The carbon deposited preferentially at the TPB in SOEC mode but was actually consumed in SOFC (being used as fuel most likely). The Raman spectroscopy showed that the carbon was mainly graphitic.

Researchers at Imperial College[53,54] have demonstrated with *ex situ* Raman and SEM that a 10 μm CGO interlayer between Ni/CGO electrodes and the YSZ electrolyte helps prevent carbon formation and delamination.

Using a copper infiltrated gadolinia doped ceria cathode[55] they also succeeded in demonstrating complete suppression of carbon formation in an aggressive simulated biogas mixture with 70% CH_4 and 30% CO_2. Without the copper the Ni–CGO cathode showed carbon formation across the whole electrode volume as opposed to the situation with CO/CO_2 where carbon was only observed close to the TPB. Furthermore the carbon formed was less graphitic with the methane present even close to the TPB. The copper infiltrated electrode had similar electrochemical performance to the Ni–CGO. In order to achieve this and avoid copper sintering during high temperature processing they used low temperature infiltration of a CGO scaffold.

Skafte *et al.* has developed a technique to detect the onset of carbon formation. The electrolysis current is gradually increased in steps of 0.1 A on a 16 cm^2 Ni–YSZ cathode. When the voltage became unstable and started increasing this defined the onset of carbon formation. They found that carbon formation started before the theoretical calculated Boudouard limit. The difference was at approx. 6% lower carbon dioxide conversion than expected, which was ascribed to diffusion limitations and endothermal temperature drop. Impregnation with CGO did not improve the limit but reduced electrode polarization.

Delamination of the anode

In earlier tests, delamination near the electrolyte–anode interface was the most commonly observed reason for degradation of the SOEC cells. It is explained by the buildup of a high internal oxygen pressure inside closed pores and cavities by the different Galvani and electromotive potentials in SOEC mode.[56,57] Use of more active anodes has alleviated the problem by lowering the over potential at the anode. Use of MIEC electrolyte has also been suggested although it is probably not the electronic conductivity but rather the lowered over potential which is responsible for the improvements.

Nickel sintering or agglomeration

Sintering of the nickel cathode has been observed at extreme operating conditions of -2.0 A cm^{-2} and 950 °C decreasing the TPB length.[18] The sintering has been ascribed to $Ni(OH)_2$ formation.[18,58,59]

Loss of reactant

By accident it was found by Schefold *et al.* that loss of steam to an SOEC lead to an increase in the cell voltage to 1.9 V at 810 °C, where it then stayed constant. The constant plateau was explained by emergence of electronic conduction in the YSZ electrolyte. The cell survived 64 hours without steam at -0.34 A cm^{-2}.

Durability tests

After many of these issues have been resolved recent results are much more promising and some of the most important ones will be discussed below. It was to be expected that degradation rates for complete stacks would be higher than for cell tests because additional poisoning sources (from, for instance, interconnects) are present and loss of contact, flow distribution *etc.* could play a role.

In a test with a 10-cell stack from Topsoe Fuel Cell A/S in co-electrolysis mode, no degradation was observed with 45% H_2O/45% CO_2/10% H_2 at 850 °C with a reactant conversion of 60% at a current density of -0.5 A cm^{-2} during the first 800 h and -0.75 A cm^{-2} during the last 400 h of the experiment.[60]

Researchers at Forschungszentrum Jülich[61] tested a two cell planar cathode supported Ni–YSZ/YSZ/LSCF stack for the first 4000 h in fuel cell mode at -0.5 A cm^{-2}, where the degradation rate was 0.6% per kh at 750 °C. The stack was then tested with steam electrolysis for 3450 h at 800 °C at −0.3 A cm^{-2} and finally in co-electrolysis mode with current densities varying between −0.3 and −0.875 A cm^{-2}. There was no degradation during the steam electrolysis mode for 2000 h. In the coelectrolysis mode the degradation increased to 2–4% per kh and especially when increasing the current density to −0.875 A cm^{-2} the degradation increased to 6% per kh. There is no mention of cleaning of the CO_2 used.

Schefold from EIFER reports on a 9000 h test of a Ni–YSZ/YSZ/LSCF single cell.[62] They operated at −1.0 A cm^{-2} at 780 °C. The initial voltage was very low (1.06 V) and the test was galvanostatic. There were many incidents with an unstable steam supply. Nevertheless the degradation was 3.8% per kh (40 mV kh^{-1}) during the entire test. During an incident-free period from 2000 to 5600 hours of operation time, the degradation rate was only 1.7% per kh. Tietz performed post mortem analyses of the cell and found severe pore formation along the grain boundaries and transport into the CGO barrier layer. The anode also showed signs of recrystallization.[63]

Tests at Idaho National lab have demonstrated the improvement in stack durability and the latest test of a Ceramatec stack even showed improvement in performance during 1900 h of operation, as illustrated in Fig. 13. The 10 cell stack consist of a nickel–ceria cathode, a scandia stabilized zirconia electrolyte (ScSz) and a LCF anode. It was tested at 800 °C at −0.25 A cm^{-2}. The improved performance compared to earlier tests was ascribed to the optimized electrodes, electrolyte and interconnecting coatings as well as interface microstructures.

The longest test so far has been done at EIFER with a electrolyte supported cell.[64] The results from the 11 000 hour test are shown on Fig. 14. The electrolyte used as support was scandia/ceria doped zirconia (6Sc1CeSZ) and an LSCF anode with a CGO barrier layer. The cathode was nickel based. Average voltage degradation amounted to 0.6% per kh (ASR increase 8 mΩ 7 kh). Interestingly enough, impedance spectroscopy revealed that the degradation was solely due to an ohmic

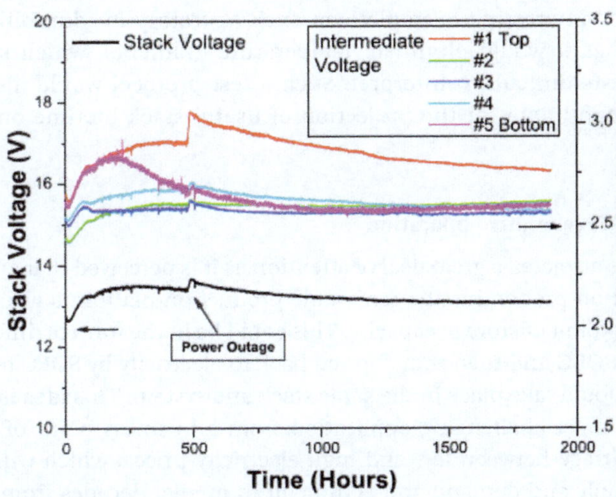

Fig. 13 Voltage *versus* time for a Ceramatec stack tested at INL. Reprinted from *Int. J. Hydrogen Energy*, 2013, **38**, 20–28 with permission from Elsevier.[24]

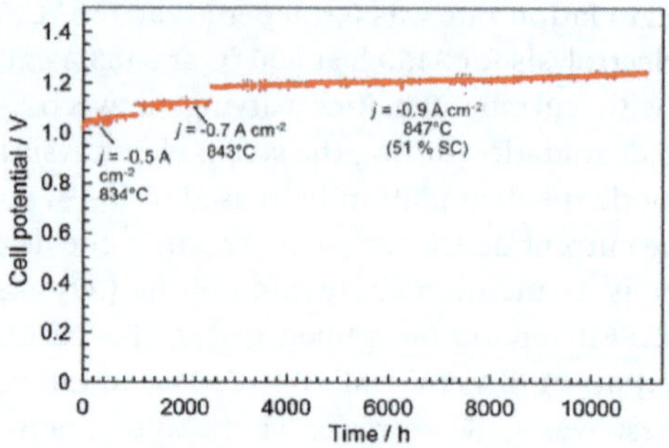

Fig. 14 Long term test of electrolyte supported cell at EIFER. Reprinted from *Electrochim. Acta*, 2015, DOI: 10.1016/j.electacta.2015.04.141 with permission from Elsevier.[64]

resistance increase and only after 9000 h did a small (<2 mV kh^{-1}) degradation in electrode overpotential became detectable.

Corre *et al.*[65] tested a 25 cell stack from Topsoe Fuel Cell A/S for steam electrolysis for 9000 h. They used two different current densities: −0.57 and −0.72 A cm^{-2}, a steam conversion of 50% and operated in the 750 to 780 °C range. An overall degradation of 2.3% per kh was observed, with cells near the bottom of the stack showing even lower degradation rates. The stack was subjected to several incidents during the tests without being affected. The temperature was adjusted to counter partially the degradation and it was calculated that this would prolong the useful lifetime by at least 7000 h.

Hansen has investigated strategies[66,67] to counteract degradation at the system level for both SOFC and SOEC plants and concluded that the optimum strategy for SOEC is to operate at thermoneutral conditions, constant current and then compensate for degradation by increasing the operating temperature. This incurs very little efficiency penalty as it only requires minute amounts of extra electrical preheating of the feed reactants. A doubling of the ASR can be counteracted by an increase in operating temperature of 98 °C.

It would be interesting to implement such a strategy for durability tests of SOEC stacks as it would eliminate temperature gradients, which sometimes makes the tests difficult to interpret. Such a test protocol would also provide more meaningful data for the projection of useful stack lifetime on a system level.

Reversible or regenerative operation

This topic has attracted a great deal of attention as it is perceived that an electrical system based on predominantly renewable production platforms would require a very large amount of storage capacity. This could be in the form of different fuels produced by SOEC and then transformed back to electricity by SOFC technology. Ideally this should take place in the same stack and system. This idea is, however, probably the most challenging application from a business point of view as it relies on arbitrage between low and high electricity prices, which will be determined by supply and demand and is difficult to predict decades from now. It is probably more profitable to produce transportation fuels especially for the heavy duty sector. The need for transportation fuels is of the same order of magnitude as electricity demand so in a future 100% renewable energy scenario the peaks in electricity production could be used for transport fuel production.

Nevertheless the R-SOFC concepts are interesting and the questions are now: can the SOEC/SOFC stacks withstand the transient operation required, which fuel is optimum for storage and round trip efficiency and how would the plant look and perform on a system level?

Operation in both the fuel cell and electrolysis mode is done routinely in the lab when testing SOEC cells and it was already proposed for energy storage by the early developer Isenberg as well as Dönitz and Erdle, but it is only recently that dedicated long term tests have been performed.

Versa Power has carried out an extensive development program for the DOE and tested more than 20 cell designs. They managed to bring down the degradation rate to 15 mV kh^{-1}, measured in the SOEC mode when operating with 0.25 A cm^{-2} and pure H_2 in the fuel cell mode and -0.5 A cm^{-2} and 50% H_2/50% H_2O in the SOEC mode. Both operation modes were at 750 °C. Fig. 15 shows result from this much extended test. The degradation rate in the reversible operation was not higher than in constant electrolysis operation.[68]

Wonsyld *et al.*[69] performed tests with a 10 cell Topsoe Fuel Cell AS/S stack with 12×12 cm^2 cells consisting of Ni/YSZ fuel electrodes, YSZ electrolyte and LSCF oxygen electrode with a CGO barrier layer. The interconnects are made of chrome rich steel. The tests were performed with 231 mA cm^{-2} in the fuel cell mode with a furnace temperature of 711 °C and 60% H_2 in N_2. SOEC operation was done with 55% steam/45% H_2 at 718 °C and at a current density of -656 mA cm^{-2}. The stack was operated for 2 h in each mode and the time for changing the cycles was approximately 10 min. In total 113 cycles were performed. The development in the cell voltages are shown in Fig. 16.

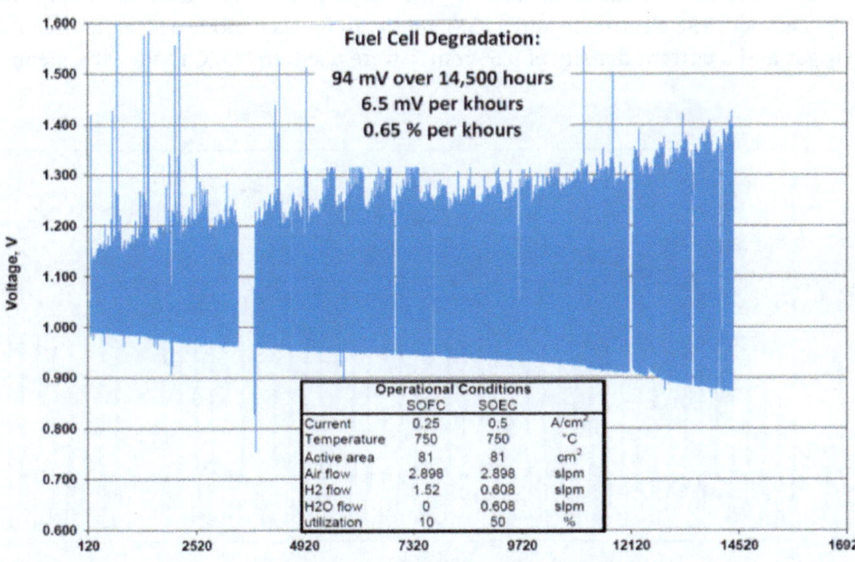

Fig. 15 Voltage development in tests of the Versa Power stack during reversible operation (daily cycles). Reprinted from *Chem. Rev.*, 2014, **114**, 10697–10734 with permission from American Chemical Society.[1]

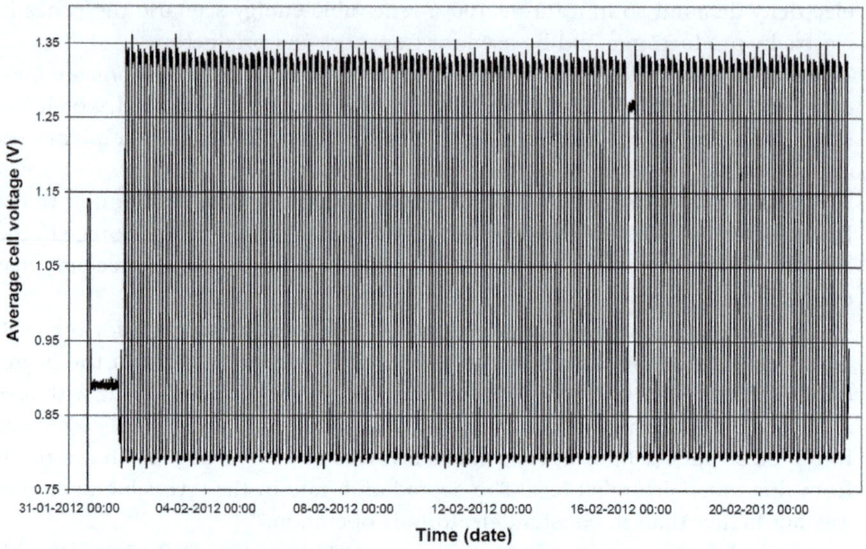

Fig. 16 Development in cell voltages of a 10 cell TOFC stack switching between SOFC and SOEC mode.

The small fluctuation in voltages in SOEC mode was due to pulsation of the water evaporator. It is seen that there is no detectable degradation in fuel cell mode and a slight activation in electrolysis mode. Such a test also demonstrates the robustness of the stack because the temperature profile changes quite dramatically when switching from SOEC to SOFC mode.

Graves *et al.*[70,71] undertook studies with single cells at 800 °C consisting of Ni/YSZ cathode, YSZ electrolyte and a LSM anode. In fuel cell mode a 50% steam/50% H_2 gas and a current density of 0.5 A cm^{-2} were used. In SOEC mode 90% steam/

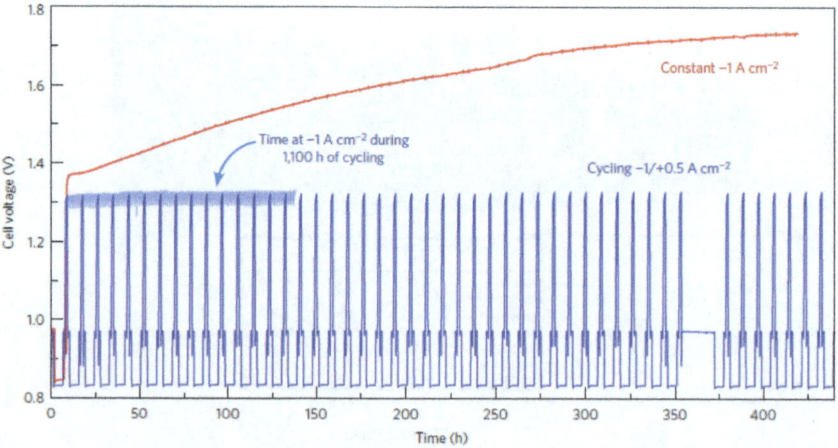

Fig. 17 Comparison of the SOEC stability during a constant-current electrolysis test and a reversible cycling test. Reprinted by permission from Macmillan Publishers Ltd: *Nat. Mater.*, 2015, **14**, 239–244. Copyright 2015.[77]

10% H_2 gas and a current density of -1.0 A cm^{-2} were used. The cycles were initially 1 h at SOEC conditions and 5 h at SOEC conditions. The results are shown in Fig. 17.

Also shown on the plot in Fig. 17 are data from an experiment with the same type of cell under constant electrolysis operation. This resulted in rapid degradation with an increase of impedance of a factor of 2.8. In the reversible cycling experiment no degradation could be observed.

Longer operation times in SOEC mode between switches were also tested and it was found that it was when 5 h periods (similar to the FC mode time) were applied that the ohmic resistance first gradually increased, albeit at a 20 times lower rate than with operation at constant electrolysis conditions.

Switching between SOEC mode and open circuit voltage but keeping the temperature did not result in the same elimination of degradation. The authors therefore concluded that the "healing" effect of fuel cell operation must be ascribed to reversal of oxygen bubble formation at the electrolyte/cathode interface by the low internal oxygen pressure in fuel cell mode and by the active electrochemical pumping of oxygen from the cavities.

The Barnett group at Northwestern University[72,73] has suggested an innovative concept for energy storage SOEC using methane rich synthesis gas as an intermediate product.

The concept is illustrated in Fig. 18.

Note that the system is closed and that the feedstock and fuel storage has to be kept hot to avoid steam condensation.

The round-trip efficiency η, neglecting system associated losses and assuming 100% faradaic efficiency will be

$$\eta = \frac{V_{FC}}{V_{EC}}$$

where V_{FC} and V_{EC} is the operating voltage in fuel cell and electrolysis mode, respectively. In a stand-alone energy storage plant the SOEC stack could not operate below the thermoneutral voltage unless significant heat was available from the outside, so the round-trip efficiency with hydrogen as intermediate

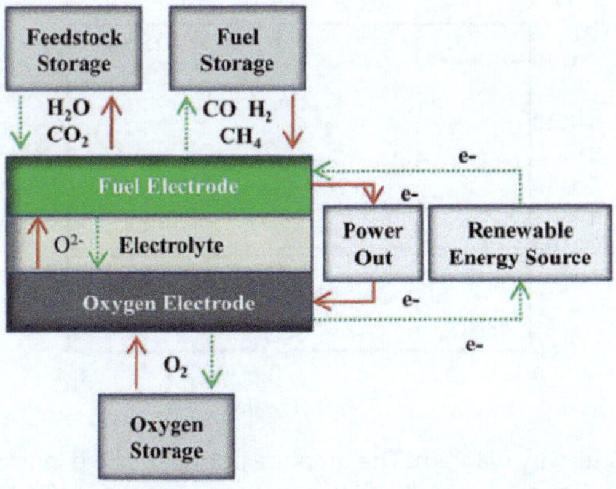

Fig. 18 Simplified reversible SOC system using H_2–CH_4 rich fuel. Reproduced from *Energy Environ. Sci.*, 2011, 4, 944–951 with permission from Royal Society of Chemistry.[72]

product could at maximum be less than 67%, which does not compare favorably with compressed-air storage or Li-ion batteries which can achieve $\eta_{\text{round trip}} \geq$ 80%.

If, however, the operating conditions for the SOEC mode are chosen so that significant amounts of methane are produced in the nickel cathode this will lower the V_{TN} due to the exothermic nature of the methanation reaction (4). In order to have appreciable amounts of methane due to equilibrium constraints the SOEC operating temperature needs to be lowered to 600 °C and/or the pressure increased to 10 bar. The conversion in SOEC also has to be limited so that the gas does not become prone to carbon decomposition, see Fig. 19.

The authors demonstrated by experiment that the nickel cathode was capable of equilibrating the gas with respect to methanation. Their papers also discuss the challenges involved and were of the opinion that pressurized operation at 750 °C will be easiest to realize in practice. Pressurised operation and use of pure oxygen in fuel cell mode will also help. Further internal reforming will be beneficial for FC efficiency and will lower parasitic losses.

One demanding challenge is, however, not addressed and that is thermal management of the methanation reaction, which probably will set in rather abruptly in the SOEC cathode channels.

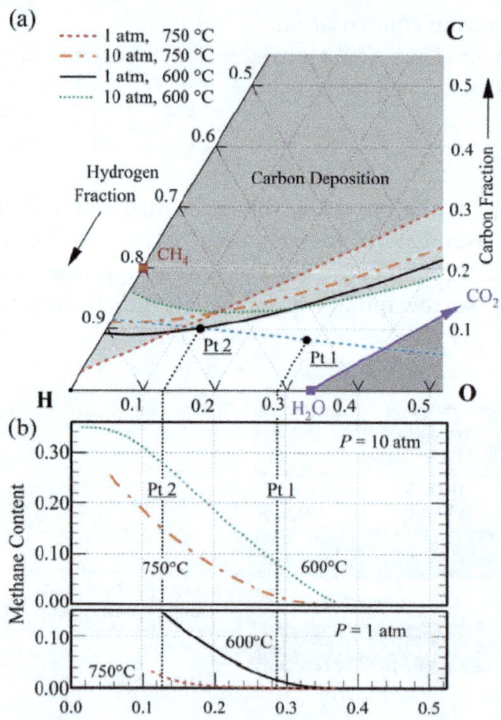

Fig. 19 (a) C–H–O ternary diagram. The graphite boundaries are indicated for $T = 600$ °C and 750 °C for P 0, 1 and 10 atm. (b) CH_4 content as function of oxygen mole fraction. Reproduced from *Energy Environ. Sci.*, 2011, 4, 944–951 with permission from Royal Society of Chemistry.[72]

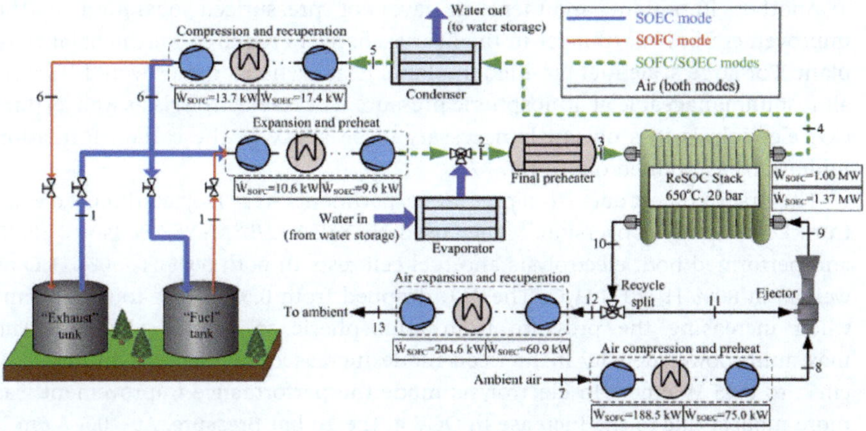

Fig. 20 Energy storage system with SOFC = 1 MW and SOEC = 1.37 MW. Reprinted from *J. Power Sources*, 2015, **276**, 133–144 with permission from Elsevier.[75]

The system analysis team at Colorado School of Mines has done careful analysis of the concept of the total system and varying the operating parameters.[74–76] They have concluded that considering parasitic losses from balance of plant components and heat integration issues the optimum operating conditions are stack temperatures of 680 °C and an operating pressure of 20 bar.

The plant would schematically look like Fig. 20.

Please note that the feed stock and fuel are stored at room temperature and 160 bar in the tanks, so that the steam is condensed and re-evaporated. Furthermore air is used as sweep gas in the SOEC and as coolant for the SOFC operation so oxygen storage is eliminated. The energy expended for compression is partly recuperated by expansion.

The overall round-trip efficiency is calculated to be 73% including balance plant losses, which is very respectable also considering the high energy density compared to alternative energy storage options.

Pressurized operation

Operating stacks under pressure would be beneficial from a systems level perspective because, the product being hydrogen, synthesis gas or carbon monoxide is normally used at pressure. Mechanical compression of gases is carried out in multi-step polytropic compressors with intercooling and a typical efficiency range from 70–80% depending on the application. The compression of liquid water or CO_2 requires very little energy. On the other hand the OCV voltage increases with the natural logarithm to the square root of the pressure according to the Nernst equation (5) but "electrochemical" compression is still more efficient. Furthermore, at higher operating voltages, such as for instance the most relevant thermoneutral voltage, the increase in pressure lowers the overpotential due to improved electrode kinetics and reduced diffusion restrictions as has been found both by modeling and the relatively few experiments carried out so far. This means that at a low current density the performance is worse at high pressure (higher voltage and thus power consumption) but at higher current density the pressurized operation is better.

Another important argument in favor of pressurized operation is the improvement in heat transfer in the heat exchangers required for the balance of plant. For large scale fuel production plants heat exchanger sizes would become almost unmanageable at atmospheric pressure. For co-electrolysis as well as pure CO_2 electrolysis it is obviously necessary to be aware of the carbon formation limitations as pointed out in ref. 77.

The first to report data from pressure experiments were Højgaard Jensen *et al.* from DTU energy conversion.[78] They used Ni–YSZ/YSZ/LSM–YSZ cells at 750 °C and performed both electrolysis and fuel cell tests in both 50% H_2/50% H_2O as well as in 80% H_2/20% H_2O. The ASR dropped from 0.52 Ω cm^2 to 0.42 Ω cm^2 when increasing the pressure from atmospheric to 10 bar absolute. The maximum power density in fuel cell mode increased significantly from 0.6 W cm^{-2} to 0.95 W cm^{-2}. In electrolysis mode the performance improvement was more modest due to the increase in OCV at the 10 bar pressure. At −0.5 A cm^{-2} the voltage was 1.29 at 0.4 bar absolute, whereas it was 1.26 V at 10 bar.

O'Brien *et al.* performed experiments with a two-cell stack at 800 °C and up to −0.4 A cm^{-2} below the thermoneutral voltage and found a decrease in ASR from 0.6 Ω cm^2 at atmospheric pressure to 0.46 Ω cm^2 at 6.9 bar.[79]

Bernadet *et al.*[80,81] carried out tests on two types of single cells. Both were cathode-supported with Ni–8YSZ/YSZ/CGO cells. The difference was that the type 1 cell had a 500 μm cathode and a LSC anode whereas type 2 had a 260 μm cathode and a LSCF anode. The crossover point on the *I–V* curve occurred at 1.3 V for type 1 and around 1.18 V for type 2. The improvement in hydrogen production was 11% for type 1 but 18% for type 2 at the thermoneutral voltage. The difference in the cell performances was mainly ascribed to microstructural differences. From the shape of the *I–V* curves it could also be seen that non-linearity indicating steam starvation set in at higher current densities with pressurized operation.

Applications

Hydrogen. The simplest application would obviously be hydrogen production. Many system studies have been performed on SOEC based hydrogen

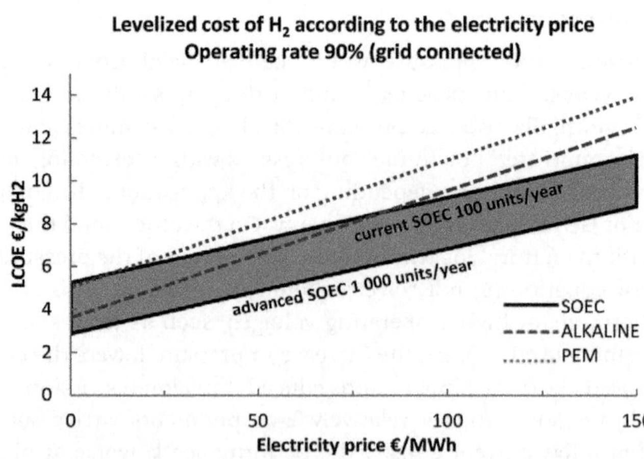

Fig. 21 Cost of hydrogen *versus* electricity price. Reprinted from *Int. J. Hydrogen Energy*, 2015, DOI: 10.1016/j.ijhydene.2015.04.085 with permission from Elsevier.[82]

production. One of the latest[82] also involves an economic evaluation based on experimental data and bottom-up cost estimation of plant components. The calculations have been carried out for a 100 kg per day H_2 plant (46 $Nm^3\ h^{-1}$) capacity. The SOEC stack is operated at 700 °C and 13 bar. It was assumed that steam would be available from an outside source. The results are displayed on Fig. 21. It is seen that the production, as always, depends strongly on electricity price, but that SOEC is more competitive than alkaline and PEM based electrolyzers.

Biogas upgrading. This is an interesting application because the CO_2 feedstock actually has a negative price in the sense that it costs on the order of 15 euro cents per Nm^3 to remove it by physical means in order to be able to upgrade the biogas to pipeline quality.

Biogas produced by anaerobic digestion consists of typically 60% methane, the rest being CO_2 and up to several thousand ppm of sulfur compounds, predominantly in the form of hydrogen sulfide. Biogas originating from landfill can also contain higher hydrocarbons, halides and siloxanes and varying amounts of nitrogen, which can be difficult to deal with.

Biogas can be upgraded by means of SOEC technology. Steam added to biogas and the CO_2 content can be co-electrolyzed generating a CO rich synthesis gas. Alternatively hydrogen from steam electrolysis can be added to the biogas. In both cases the synthesis gas is methanated.[83]

Production of "synthetic" natural gas by upgrading biogas by means of SOEC and wind power can also act to store renewable energy and may also provide various balancing services to the power grid (up and down regulation of either electricity consumption or production).

If the CO_2 in the biogas is co-electrolyzed with steam to produce CO and H_2 the synthesis gas can be converted to methane at pipeline quality at relatively low pressure. The present SOEC electrodes, based on nickel are, however, active for steam reforming. This will result in an inefficient plant, because the electrical energy will be used to drive the steam reforming in the SOEC and will be downgraded to steam generation in the methanator. If, however, a small amount of sulphur in the biogas feed to the SOEC is not removed, the steam reforming

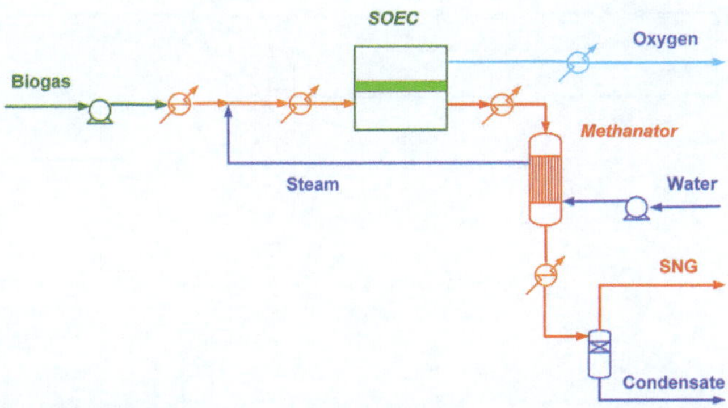

Fig. 22 Biogas upgrading by co-electrolysis of steam and biogas CO_2 route. Reprinted with permission from *ECS. Trans.*, 2013, **57**, 3089–3097. Copyright 2013, The Electrochemical Society.[83]

activity will be reduced to almost zero without sacrificing too much of the electrochemical electrolysis activity.[49,84]

As mentioned above the electrochemical activity is decreased by $(1 - \theta_s)$ where θ_s is the sulfur coverage of the nickel whereas the reforming activity is reduced proportional to $(1 - \theta_s)^3$. The experiments at Risø showed that steam reforming of the methane could be completely eliminated by having ppm levels of H_2S in the feed.

The plant layout is shown in Fig. 22. The overall energy balances are, however, such that the steam production from CO methanation is more than is needed to cover the need for the SOEC, also because it is only required to evaporate 3 moles of water per mole of methane synthesised, whereas for CO_2 methanation 4 moles of water need to be evaporated.

The net result is a surplus production of superheated steam. In a large plant this could be utilised to generate power in an expansion turbine (of course with losses incurred compared to the SOEC electricity input), but in a relatively small biogas plant this will most likely prove uneconomical due to the high investment and low availability of such small turbines.

The other possibility is to electrolyze steam to hydrogen separately and mix it with the cleaned biogas and then convert the CO_2 to methane.

The plant layout is shown in Fig. 23. The exergy streams for the steam electrolysis route are shown in Fig. 24.

Lanzini *et al.* has later studied the impact of sulfur for biogas upgrading both at atmospheric pressure and at 33 bar where there would be no driving force for steam reforming.[85] They use 3 adiabatic methanation reactors in series with recycling around the first in order to limit the temperature rise to 600 °C. They only looked at the co-electrolysis cases and found that at atmospheric pressure the sulfur passivation discussed above was the most efficient. Operating the stack at a pressure of 30 bar eliminated the need for sulfur passivation and turned to be the most energy efficient option.

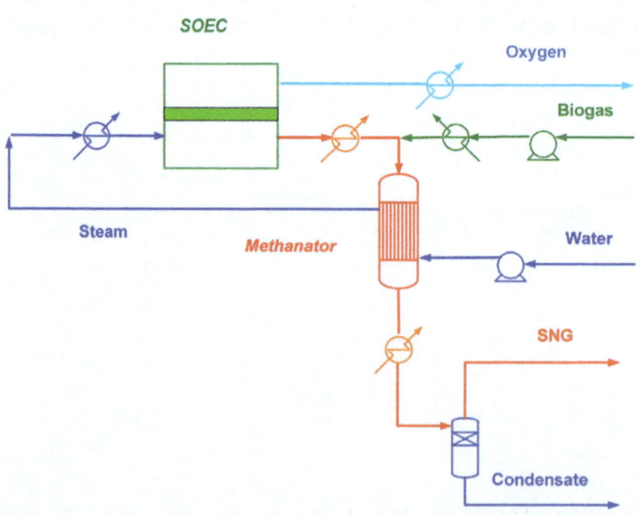

Fig. 23 Biogas upgrading by electrolysis of steam route. Reprinted with permission from *ECS. Trans.*, 2013, **57**, 3089–3097. Copyright 2013, The Electrochemical Society.[83]

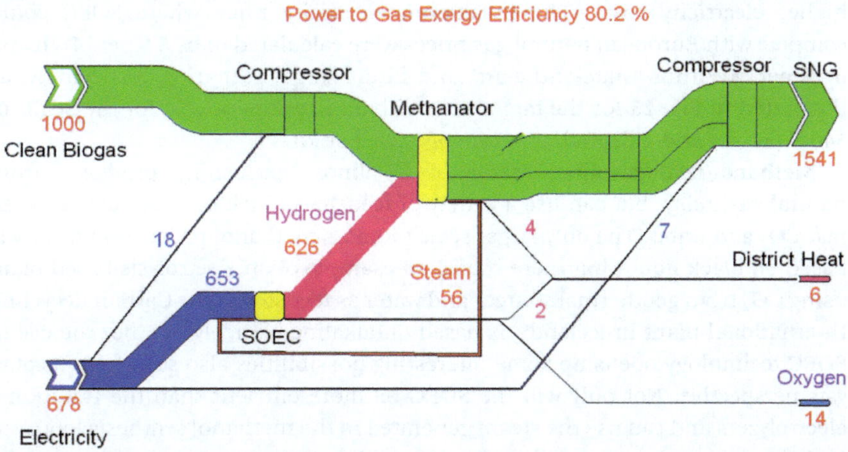

Fig. 24 Exergy streams in biogas upgrading plant. Exergy in incoming biogas stream set to 1000. Reprinted with permission from *ECS. Trans.*, 2013, **57**, 3089–3097. Copyright 2013, The Electrochemical Society.[83]

Methane from CO_2. Methane can also be produced *via* SOEC from pure CO_2, steam and electricity as discussed in ref. 77. In that case several adiabatic reactors as in the Topsøe TREMP scheme would be necessary and co-electrolysis could be beneficial because it would reduce the required methanation catalyst volume and there would a need for the extra steam produced.

Hansen *et al.* studied both pressurized and atmospheric operation of the SOEC and calculated LHV efficiencies of 74.8 and 77.3%, respectively. The efficiency calculations are based on the DC electricity supply, *e.g.* potential inverter losses are not included.

Giglio *et al.*[50,86] have also investigated either CO_2 or CO methanation *e.g.* steam or the co-electrolysis route and found that the latter has the highest efficiency but also the highest investment and operating cost. It is thus most competitive at

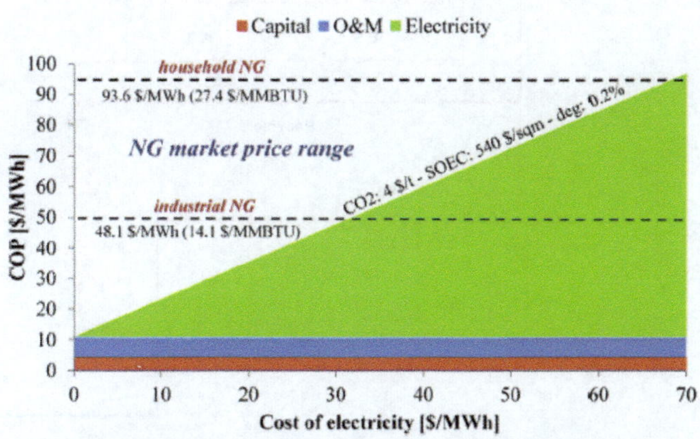

Fig. 25 Cost of production of methane from CO_2 with SOEC compared to European NG market price. Reprinted from *J. Energy Storage*, 2015, **2**, 64–79 with permission from Elsevier.[86]

higher electricity costs. The break-even electricity price where SOEC could compete with European natural gas prices were calculated to be 8 $ per MWh and 67 $ per MWh for "state-of-the-art" and future target scenarios, respectively, as illustrated in Fig. 25 for the target case with an investment cost for the SOEC of 540 $ per m^2 and a degradation rate of 0.2% per kh.

Methanol/DME/gasoline. Methanol is almost exclusively produced from natural gas today, but can use a variety of feedstocks including biomass or even just CO_2 and water. The only larger scale biomass methanol production facility is based on black liquor but there is also an example of an electrolysis based plant using CO_2 from geothermal sources and water as feedstock. The Carbon Recycling International plant in Iceland[87] is based on alkaline electrolyzers, but the use of SOEC technology opens up some interesting possibilities also seen from a catalysis perspective. Not only will the SOEC be more efficient than the traditional electrolyzers and can use the steam generated in the methanol synthesis loop, but it also provides almost complete freedom with respect to the composition of the synthesis gas, which can be virtually free of inert components.

A study has been undertaken to investigate the optimum amount of co-electrolysis in an SOEC coupled to a methanol synthesis loop.[88] The layout is shown in Fig. 26.

The CO_2 feedstock is either sent through the SOEC unit together with the steam or directly to the synthesis loop. It is well known that a maximum for the methanol synthesis rate exists at 2–4% CO_2 in an stoichiometric synthesis gas.[89] Results from laboratory tests as well as industrial feedback has shown that the deactivation rate in CO_2 rich gases is faster compared to more CO rich gases, probably due to steam-induced ageing caused by the water formed from the synthesis reaction. On the other hand a more CO-rich gas will lead to increased byproduct formation. Using proprietary Topsøe in-house kinetics including all

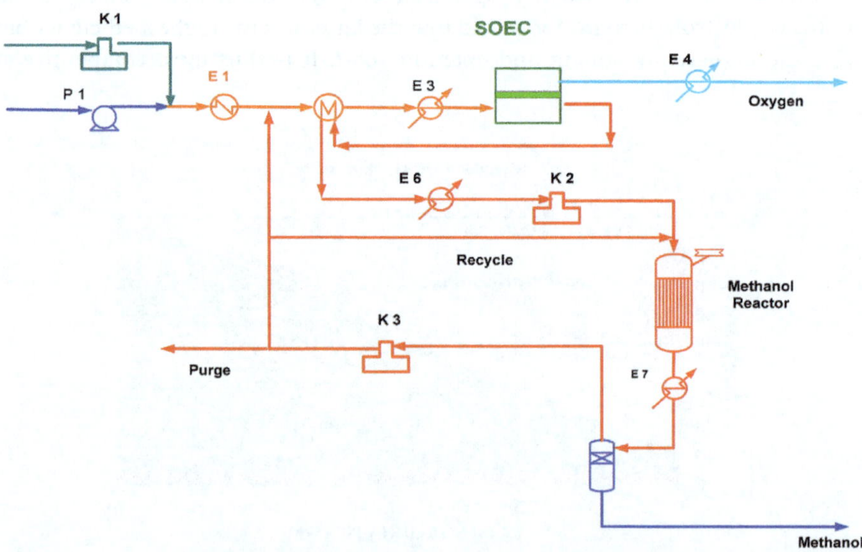

Fig. 26 Methanol plant based on co-electrolysis of CO_2 and steam. Reprinted with permission from *ECS Trans.*, 2011, **35**, 2941–2948. Copyright 2011, The Electrochemical Society.[77]

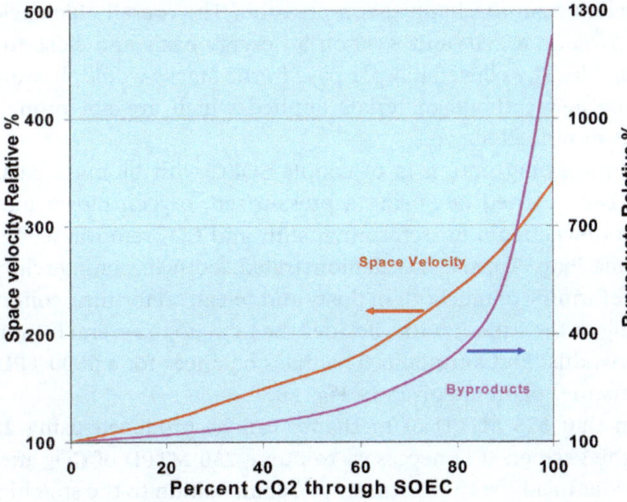

Fig. 27 Space velocity and byproduct formation as function of percentage co-electrolysis in a SOEC based methanol plant.

these factors the catalyst volumes, byproduct formation *etc.* have been calculated for end-of-run conditions (typically 3–4 years) as a function of the fraction of co-electrolysis. The results are shown in Fig. 27.

An optimum tradeoff between catalyst volume and byproduct formation is around 80% co-electrolysis. The needed amount of catalyst is reduced by a factor of 3–4 compared with a synthesis starting with only CO_2 in the make-up to the loop. The consequences of operating the SOEC stacks at a synthesis pressure of around 50 bar has also been calculated. The conversion efficiency is enhanced by 4% absolute due to the fact that "electrochemical" compression of the hydrogen

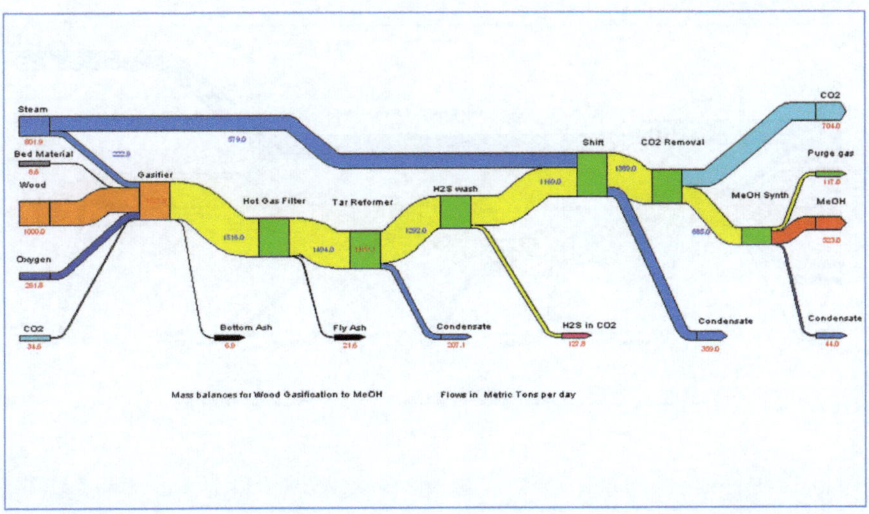

Fig. 28 Mass flows in a methanol plant based on gasification of 1000 MTPD of dried wood (20 wt H_2O).

is more efficient than mechanical compression. The overall efficiencies are also quite high, 76% on a LHV and 83% on an exergy basis and close to that theoretically possible. The conversion per pass in the stacks would, however, have to be limited or new cathode materials applied which are not prone to carbon formation as described above.

Another interesting option is to couple SOEC with biomass gasification. A study[90] has been carried out using a pressurized, oxygen blown bubbling bed gasifier with downstream tar reforming, shift and CO_2 removal feeding a methanol synthesis loop. Topsøe has demonstrated, both in commercial as well as pilot scale, reforming of tars both in dusty and "clean" reforming conditions. This is challenging as the gas contains, besides the tars, up to several hundred ppm of sulfur, olefins, dust and ammonia. The mass balances for a 1000 TPD dry wood (20 wt% moisture) plant are given in Fig. 28.

It is seen that 525 MTPD of methanol can be produced using 234 TPD of oxygen for the gasifier. It is necessary to purge 750 MTPD of CO_2 after the shift section in order to adjust the synthesis gas composition to the stoichiometric for methanol synthesis. This is a consequence of the fact that the hydrogen-to-carbon molar ratio in woody biomass is generally in the range of 1.4 to 1.6.

Liquid oxygenates and hydrocarbons are very convenient, energy-dense energy or hydrogen carriers and in fossil-free energy scenarios it will be biogenic, renewable carbon which will become the scarce and limiting resource, as already evidenced by the intense ongoing feed-or-fuel discussion. SOEC can then produce with high efficiency the "missing" hydrogen from physical energy sources and water.[91] This is illustrated in Fig. 29. The SOEC unit produces 112 MTPD of hydrogen using steam generated in the gasification and synthesis plant, which will require 144 MW of electric power. More than 3 times the needed oxygen for the gasification is at the same time produced as a byproduct, which will make it possible to eliminate the costly air separation unit.

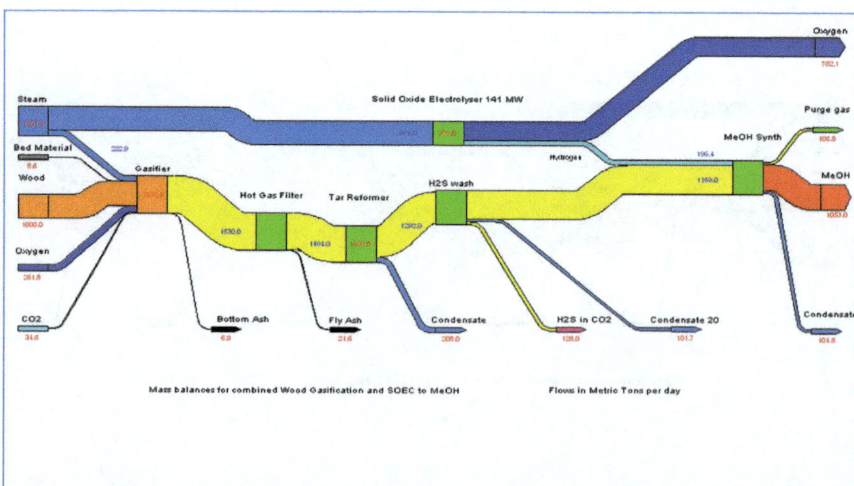

Fig. 29 Mass flow in a methanol plant based on gasification of 1000 MTPD of dried wood (20 wt H_2O) combined with a steam SOEC electrolyzer.

Economic analysis of the concept is difficult due to uncertainties about the temporal price structure in a future scenario with a high penetration of, for instance. wind turbines. An attempt was, however, made in the study and it turned out to be beneficial to operate the SOEC all the time even with high electricity prices at the minimum, where the gasifier could be supplied with oxygen. The gasifier will have to be operated in steady state at full load. It has been demonstrated that the SOEC can be modulated, when hot from 0 to 100% repeatedly within seconds but it remains to be demonstrated that the shift section, carbon dioxide removal and methanol synthesis proper can operated in such a flexible fashion. The catalyst no doubt will be able to sustain such dynamic operation, but balance of plant components like compressors may have other constraints. The economic evaluation showed that the production price for the combined SOEC plus gasification plant was lower than the stand-alone gasification based methanol plant.

Fischer–Tropsch. Small scale production of FT fuel *via* SOEC has already been demonstrated in the INL laboratories[92] but has also been studied by system modelling.

Fu *et al.* from EIFER[93] did a detailed study including sensitivity analyses on key parameters and calculated the production cost of Fischer–Tropsch diesel as function of CO_2 and electricity cost as shown on Fig. 30. Note that they included a small credit for oxygen production. They used \$20 per MT CO_2 as the cost from a nearby ammonia plant. They concluded that cathode supported SOEC cells was best and that the required lifetime should exceed 10 000 h. They also concluded that the diesel from SOEC was competitive compared to biomass to liquid production.

Graves *et al.*[94] also studied Fischer–Tropsch production using some overall estimated efficiencies for the different blocks in the overall plant and they came up with graphs showing production price as function of electricity price and on-stream capacity utilization, as shown on Fig. 31

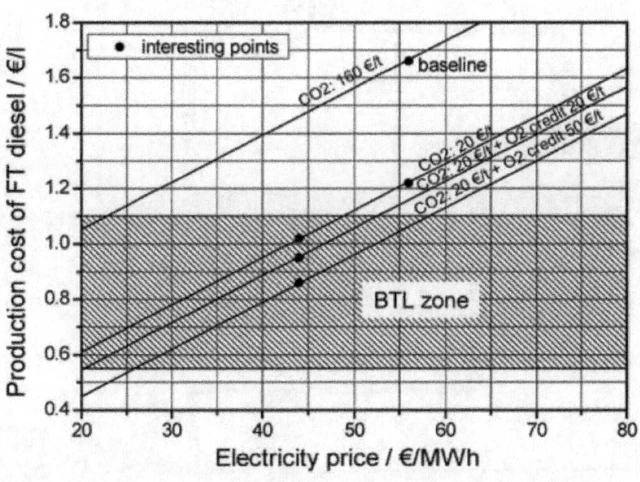

Fig. 30 Cost of Fischer–Tropsch diesel *versus* electricity price according to ref. 93. Reproduced from *Energy Environ. Sci.*, 2010, 3, 1382–1397 with permission from Royal Society of Chemistry.

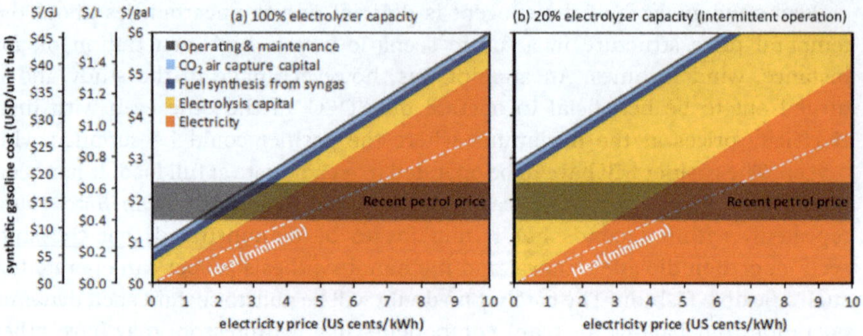

Fig. 31 Synthetic Fischer–Tropsch gasoline as function of electricity price and capacity factor. Reprinted from *Renewable Sustainable Energy Rev.*, 2011, **15**, 1–23 with permission from Elsevier.[94]

Becker *et al.*[95] from the Colorado School of Mines and NREL performed a very detailed study of Fischer–Tropsch production from CO_2, electricity and steam. They compared two different operating pressure levels: 1.6 bar and 5 bar and included methane production in the SOEC cathode in their model. The LPG produced in the FT unit was used to produce hydrogen for product upgrading and recycling to the SOEC as well as providing heat for the SOEC unit by being burned. The baseline study used a stack operating temperature of 800 °C, 90% reactant conversion and produced a H_2/CO ratio of 2.1 : 1, suitable for a cobalt based FT catalyst. This operation is very close to the carbon limits.[77] They did, however, also conclude that the low pressure case was preferred because the savings in compression was more than negated by

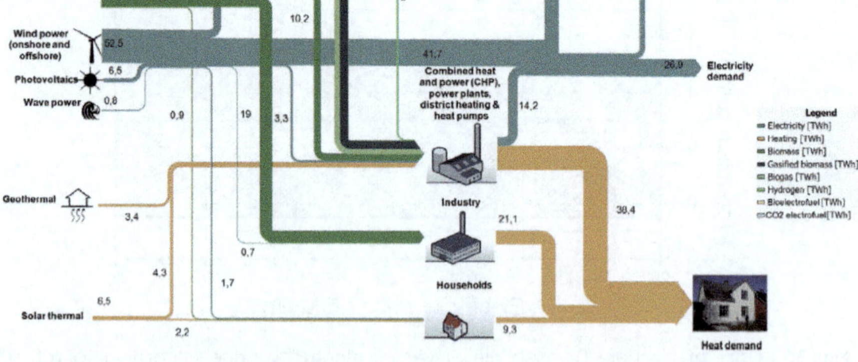

Fig. 32 Energy streams for Denmark in a 100% renewable energy scenario. Reprinted from *Appl. Energy*, 2015, **145**, 139–154 with permission from Elsevier.[100]

the increased inert molecules (CH_4 and CO_2) in the make-up gas stream to the FT unit.

They calculated an overall efficiency of electricity to FT liquid fuel of 54.8% HHV based (51.0% LHV). Operation at 5 bar lowered the efficiency by 2.6% absolute.

They cost-estimated all the equipment and calculated the production price on a $ per gge basis (gge being gasoline gallon equivalent which is approximately 120 MJ) as function of electricity price and plant capacity utilization percentage. The results ranged from 4.4 $ per gge to 15.0 $ per gge for electricity prices of 0.02 to 0.14 $ per kWh and capacity factors of 90 to 40% respectively.

They ascribed the lower efficiency numbers compared to the studies by Fu (59–62%) and Graves (70%) to the higher fidelity of their model and mainly to lower, realistic CO conversion in the FT reactor (80 *vs.* 95%) as well as greater LPG production in their upgrading unit.

Sunfire GMbH in Dresden has constructed a 1 bbl per day FT unit. The will operate a pressurized SOEC stack at 15 bar. They produce hydrogen by steam electrolysis in electrolyte-supported cells with Ni/GDC cathodes, 6ScCeSZ electrolyte and LSCF anodes from Kerafol at around 850 °C. Synthesis gas is then generated by reacting the part hydrogen with CO_2 in a separate RWGS reactor before entering the FT reactor. A conversion efficiency of 70% from electricity to liquid fuel is claimed. Their first product was produced in 2015.

Conclusions

Isenberg stated in one of the first papers on SOEC in 1981:[6] "As far as HTSOE electrolysis is concerned, one faces the immediately problem of finding the ecological niche for such a scheme".

There is no doubt that the progress in understanding the basic scientific details of SOEC, together with progress in materials for and engineering of SOEC stacks, as discussed above have expanded the niche considerably. It is, however, still very difficult to compete with fossil fuels with the present low prices in the absence of mechanisms invoking the external costs of such fuels with respect to local as well as global environment.

SOEC could, however, be competitive for small scale hydrogen production as they are more energy efficient than alkaline as well as PEM-based electrolyzers, which are already commercial today. The synergy with downstream chemical synthesis would also make SOEC attractive for biogas upgrading where the CO_2 actually has a negative price, because it incurs both investment as well as operating costs to remove it to produce pipeline quality methane.

The coupling of SOEC with biomass conversion is indeed an attractive option. Biomass is deficient in hydrogen with respect to the carbon based fuels used today. Adding hydrogen produced by SOEC could thus increase the biomass potential substantially and existing infrastructure could be used.

Finally there are geographical areas already today where there is an abundance of cheap, renewable electricity with a limited local demand. In such cases even production of liquid fuels like methanol, DME, gasoline or Fischer–Tropsch diesel may be viable.

If the vision of a fossil-free energy scenario with a large input of renewable electricity and limited biomass potential shall be realized there is no doubt that SOEC will be a key enabling technology. This is exemplified by studies[96–100] of the Danish energy system in 2050, where it is foreseen that 30 TWh worth of transport fuel is produced *via* SOEC per year, corresponding to almost 5 MWh per capita per year (Fig. 32).

References

1 S. D. Ebbesen, S. H. Jensen, A. Hauch and M. B. Mogensen, High temperature electrolysis in alkaline cells, solid proton conducting cells, and solid oxide cells, *Chem. Rev.*, 2014, **114**, 10697–10734.

2 M. A. Laguna-Bercero, Recent advances in high temperature electrolysis using solid oxide fuel cells: a review, *J. Power Sources*, 2012, **203**, 4–16.

3 S. D. Ebbesen and M. Mogensen, Kinetics of oxidation of H_2 and reduction of H_2O in Ni–YSZ based solid oxide cells, *ECS Trans.*, 2012, **50**, 167–182.

4 L. M. Elikan and J. P. Morris, Solid electrolyte system for oxygen generation, *NASA Contract. Rep.*, 1969, CR-185612.

5 L. M. Elikan, J. P. Wu and C. K. Wu, Development of a solid electrolyte carbon dioxide and water reduction system for oxygen recovery, *NASA Contract. Rep.*, 1972, CR-2014.

6 A. O. Isenberg, Energy conversion *via* solid oxide electrolyte electrochemical cells at high temperatures, *Solid State Ionics*, 1981, **3–4**, 431–437.

7 W. Dönitz, G. Dietrich, E. Erdle and R. Streicher, Electrochemical high temperature technology for hydrogen production or direct electricity generation, *Int. J. Hydrogen Energy*, 1988, **13**, 283–287.

8 W. Dönitz and E. Erdle, High-temperature electrolysis of water vapor-status of development and perspectives for application, *Int. J. Hydrogen Energy*, 1985, **10**, 291–295.

9 E. Erdle, W. Dönitz, R. Schamm and A. Koch, Reversibility and polarization behaviour of high temperature solid oxide electrochemical cells, *Int. J. Hydrogen Energy*, 1992, **17**, 817–819.

10 J. E. O'Brien, Thermodynamic consideration for thermal water splitting processes and high temperature electrolysis, in *2008 International Mechanical Engineering Congress and Exposition*, IMECE 2008, Boston, Massachusetts, USA, 2008.

11 J. E. O'Brien, Thermodynamics and transport phenomena in high temperature steam electrolysis cells, *J. Heat Transfer*, 2012, **134**, 031017.

12 M. Ni, M. K. H. Leung and D. Y. C. Leung, Energy and exergy analysis of hydrogen production by solid oxide steam electrolyzer plant, *Int. J. Hydrogen Energy*, 2007, **32**, 4648–4660.

13 X. Sun, M. Chen, S. H. Jensen, S. D. Ebbesen, C. Graves and M. Mogensen, Thermodynamic analysis of synthetic hydrocarbon fuel production in pressurized solid oxide electrolysis cells, *Int. J. Hydrogen Energy*, 2012, **37**, 17101–17110.

14 S. H. Jensen, P. H. Larsen and M. Mogensen, Hydrogen and synthetic fuel production from renewable energy sources, *Int. J. Hydrogen Energy*, 2007, **32**, 3253–3257.

15 J. B. Hansen, Direct reforming fuel cells, in *Fuel Cells: Technologies for Fuel Processing*, Elsevier, 2011, pp. 409–450.

16 M. Mogensen, S. H. Jensen, S. D. Ebbesen, A. Hauch, C. Graves, J. V. T. Høgh, X. Sun, S. Das, P. V. Hendriksen, J. U. Nielsen, A. H. Pedersen, N. Christiansen and J. B. Hansen, Production of "green natural gas" using solid oxide electrolysis cells (SOEC): status of technology and costs, in *25th World Gas Conference, WGC 2012*, International Gas Union, Kuala Lumpur, 2012, pp. 2314–2320.

17 A. Momma, T. Kato, Y. Kaga and S. Nagata, Polarization behavior of high temperature solid oxide electrolysis cells (SOEC), *J. Ceram. Soc. Jpn.*, 1997, **105**, 369–373.

18 A. Hauch, S. D. Ebbesen, S. H. Jensen and M. Mogensen, Solid oxide electrolysis cells: Microstructure and degradation of the Ni/yttria-stabilized zirconia electrode, *J. Electrochem. Soc.*, 2008, **155**, B1184–B1193.

19 A. Hauch, S. D. Ebbesen, S. H. Jensen and M. Mogensen, Highly efficient high temperature electrolysis, *J. Mater. Chem.*, 2008, **18**, 2331–2340.

20 A. Hauch, S. H. Jensen, S. Ramousse and M. Mogensen, Performance and durability of solid oxideelectrolysis cells, *J. Electrochem. Soc.*, 2006, **153**, A1741–A1747.

21 R. Knibbe, M. L. Traulsen, A. Hauch, S. D. Ebbesen and M. Mogensen, Solid oxide electrolysis cells: degradation at high current densities, *J. Electrochem. Soc.*, 2010, **157**, B1209–B1217.

22 S. C. K. Singhal, *High Temperature SOFCs: Fundamentals, Design and Applications*, Elsevier, 2003.

23 N. Osada, H. Uchida and M. Watanabe, Polarization behavior of SDC cathode with highly dispersed Ni catalysts for solid oxide electrolysis cells, *J. Electrochem. Soc.*, 2006, **153**, A816–A820.

24 X. Zhang, J. E. O'Brien, R. C. O'Brien, J. J. Hartvigsen, G. Tao and G. K. Housley, Improved durability of SOEC stacks for high temperature electrolysis, *Int. J. Hydrogen Energy*, 2013, **38**, 20–28.

25 K. Eguchi, T. Hatagishi and H. Arai, Power generation and steam electrolysis characteristics of an electrochemical cell with a zirconia- or ceria-based electrolyte, *Solid State Ionics*, 1996, **86–88**, 1245–1249.

26 P. Kim-Lohsoontorn, N. Laosiripojana and J. Bae, Performance of solid oxide electrolysis cell having bi-layered electrolyte during steam electrolysis and carbon dioxide electrolysis, *Curr. Appl. Phys.*, 2011, **11**, S223–S228.

27 S. Elangovan, J. J. Hartvigsen and L. J. Frost, Intermediate temperature reversible fuel cells, *Int. J. Appl. Ceram. Technol.*, 2007, **4**, 109–118.

28 P. Moçoteguy and A. Brisse, A review and comprehensive analysis of degradation mechanisms of solid oxide electrolysis cells, *Int. J. Hydrogen Energy*, 2013, **38**, 15887–15902.

29 A. Hauch, Solid oxide electrolysis cells – performance and durability, in *Risø National Laboratory & Department of Chemistry*, Technical University of Denmark, Roskilde, Denmark, 2007.

30 O. A. Marina, L. R. Pederson, M. C. Williams, G. W. Coffey, K. D. Meinhardt, C. D. Nguyen and E. C. Thomsen, Electrode performance in reversible solid oxide fuel cells, *J. Electrochem. Soc.*, 2007, **154**, B452–B459.

31 J. C. Ruiz-Morales, D. Marrero-López, J. Canales-Vázquez and J. T. S. Irvine, Symmetric and reversible solid oxide fuel cells, *RSC Adv.*, 2011, **1**, 1403–1414.

32 G. Tsekouras and J. T. S. Irvine, The role of defect chemistry in strontium titanates utilised for high temperature steam electrolysis, *J. Mater. Chem.*, 2011, **21**, 9367–9376.

33 G. Tsekouras, D. Neagu and J. T. S. Irvine, Step-change in high temperature steam electrolysis performance of perovskite oxide cathodes with exsolution of B-site dopants, *Energy Environ. Sci.*, 2013, **6**, 256–266.

34 X. Yang and J. T. S. Irvine, $(La_{0.75}Sr_{0.25})_{0.95}Mn_{0.5}Cr_{0.5}O_3$ as the cathode of solid oxide electrolysis cells for high temperature hydrogen production from steam, *J. Mater. Chem.*, 2008, **18**, 2349–2354.

35 X. Yue and J. T. S. Irvine, $(La,Sr)(Cr,Mn)O_3$/GDC cathode for high temperature steam electrolysis and steam-carbon dioxide co-electrolysis, *Solid State Ionics*, 2012, **225**, 131–135.

36 F. Bidrawn, G. Kim, G. Corre, J. T. S. Irvine, J. M. Vohs and R. J. Gorte, Efficient reduction of CO_2 in a solid oxide electrolyzer, *Electrochem. Solid-State Lett.*, 2008, **11**, B167–B170.

37 B. Ge, J. T. Ma, D. Ai, C. Deng, X. Lin and J. Xu, Sr_2FeNbO_6 applied in solid oxide electrolysis cell as the hydrogen electrode: kinetic studies by comparison with Ni–YSZ, *Electrochim. Acta*, 2015, **151**, 437–446.

38 G. Schiller, A. Ansar, M. Lang and O. Patz, High temperature water electrolysis using metal supported solid oxide electrolyser cells (SOEC), *J. Appl. Electrochem.*, 2009, **39**, 293–301.

39 W. Wang, Y. Huang, S. Jung, J. M. Vohs and R. J. Gorte, A comparison of LSM, LSF, and LSCo for solid oxide electrolyzeranodes, *J. Electrochem. Soc.*, 2006, **153**, A2066–A2070.

40 Y. Bo, Z. Wenqiang, X. Jingming and C. Jing, Status and research of highly efficient hydrogen production through high temperature steam electrolysis at INET, *Int. J. Hydrogen Energy*, 2010, **35**, 2829–2835.

41 P. Kim-Lohsoontorn and J. Bae, Electrochemical performance of solid oxide electrolysis cell electrodes under high-temperature coelectrolysis of steam and carbon dioxide, *J. Power Sources*, 2011, **196**, 7161–7168.

42 P. Hjalmarsson, X. Sun, Y. L. Liu and M. Chen, Influence of the oxygen electrode and inter-diffusion barrier on the degradation of solid oxide electrolysis cells, *J. Power Sources*, 2013, **223**, 349–357.

43 J. P. Ouweltjes, M. M. A. van Tuel, F. P. F. van Berkel and G. Rietveld, Solid oxide electrolyzers for efficient hydrogen production, in *10th International Symposium on Solid Oxide Fuel Cells, SOFC-X*, 2007, pp. 933–940.

44 M. A. Laguna-Bercero, N. Kinadjan, R. Sayers, H. El Shinawi, C. Greaves and S. J. Skinner, Performance of $La_{2-x}Sr_xCo_{0.5}Ni_{0.5}O_{4\pm\delta}$ as an oxygen electrode for solid oxide reversible cells, *Fuel Cells*, 2011, **11**, 102–107.

45 J. B. Hansen and J. Rostrup-Nielsen, Sulfur poisoning on Ni catalyst and anodes, in *Handbook of Fuel Cells. Fundamental Technology and Applications*, ed. W. Vielstich, H. Yokokawa and H. A. Gasteiger, Wiley, Chichester, 2009, vol. 6, pp. 957–969.

46 S. D. Ebbesen, C. Graves, A. Hauch, S. H. Jensen and M. Mogensen, Poisoning of solid oxide electrolysis cells by impurities, *J. Electrochem. Soc.*, 2010, **157**, B1419–B1429.

47 A. Hauch, S. H. Jensen, J. B. Bilde-Sørensen and M. Mogensen, Silica segregation in the Ni/YSZ electrode, *J. Electrochem. Soc.*, 2007, **154**, A619–A626.

48 J. B. Hansen, Correlating sulfur poisoning of SOFC nickel anodes by a Temkin isotherm, *Electrochem. Solid-State Lett.*, 2008, **11**, B178–B180.
49 S. D. Ebbesen, J. B. Hansen and M. B. Mogensen, Biogas upgrading using SOEC with a Ni–ScYSZ electrode, in *ECS Transactions*, Electrochemical Society Inc., 2013, pp. 3217–3227.
50 E. Giglio, A. Lanzini, M. Santarelli and P. Leone, Synthetic natural gas *via* integrated high-temperature electrolysis and methanation: part I-energy performance, *Journal of Energy Storage*, 2015, **1**, 22–37.
51 Y. Tao, S. D. Ebbesen, W. Zhang and M. B. Mogensen, Carbon nanotube growth on nanozirconia under strong cathodic polarization in steam and carbon dioxide, *ChemCatChem*, 2014, **6**, 1220–1224.
52 W. Li, Y. Shi, Y. Luo, Y. Wang and N. Cai, Carbon deposition on patterned nickel/yttria stabilized zirconia electrodes for solid oxide fuel cell/solid oxide electrolysis cell modes, *J. Power Sources*, 2015, **276**, 26–31.
53 V. Duboviks, R. C. Maher, M. Kishimoto, L. F. Cohen, N. P. Brandon and G. J. Offer, A Raman spectroscopic study of the carbon deposition mechanism on Ni/CGO electrodes during CO/CO_2 electrolysis, *Phys. Chem. Chem. Phys.*, 2014, **16**, 13063–13068.
54 V. Duboviks, R. C. Maher, G. Offer, L. F. Cohen and N. P. Brandon, In-*Operando* Raman spectroscopy study of passivation effects on Ni–CGO electrodes in CO_2 electrolysis conditions, in *ECS Transactions, Electrochemical Society Inc.*, 2013, pp. 3111–3117.
55 V. Duboviks, M. Lomberg, R. C. Maher, L. F. Cohen, N. P. Brandon and G. J. Offer, Carbon deposition behaviour in metal-infiltrated gadolinia doped ceria electrodes for simulated biogas upgrading in solid oxide electrolysis cells, *J. Power Sources*, 2015, **293**, 912–921.
56 T. Jacobsen and M. Mogensen, The course of oxygen partial pressure and electric potentials across an oxide electrolyte cell, in *Ionic and Mixed Conducting Ceramics 6-213th ECS Meeting*, Phoenix, AZ, 2008, pp. 259–273.
57 A. V. Virkar, A model for solid oxide fuel cell (SOFC) stack degradation, in *10th International Symposium on Solid Oxide Fuel Cells, SOFC-X*, 2007, pp. 453–454.
58 R. Knibbe, A. Hauch, J. Hjelm, S. D. Ebbesen and M. Mogensen, Durability of solid oxide cells, *Green*, 2011, **1**, 141–169.
59 T. Matsui, A. Ozaki, R. Kikuchi and K. Eguchi, Influence of steam concentration on the degradation behavior of reversible solid oxide fuel cells, in *9th International Symposium on Solid Oxide Fuel Cells, SOFC IX*, ed. S. C. Singhal and J. Mizusaki, Quebec, 2005, pp. 621–626.
60 S. D. Ebbesen, J. Høgh, K. A. Nielsen, J. U. Nielsen and M. Mogensen, Durable SOC stacks for production of hydrogen and synthesis gas by high temperature electrolysis, *Int. J. Hydrogen Energy*, 2011, **36**, 7363–7373.
61 V. N. Nguyen, Q. Fang, U. Packbier and L. Blum, Long-term tests of a Jülich planar short stack with reversible solid oxide cells in both fuel cell and electrolysis modes, *Int. J. Hydrogen Energy*, 2013, **38**, 4281–4290.
62 J. Schefold, A. Brisse and F. Tietz, Nine thousand hours of operation of a solid oxide cell in steam electrolysis mode, *J. Electrochem. Soc.*, 2012, **159**, A137–A144.
63 F. Tietz, D. Sebold, A. Brisse and J. Schefold, Degradation phenomena in a solid oxide electrolysis cell after 9000 h of operation, *J. Power Sources*, 2013, **223**, 129–135.

64 J. Schefold, A. Brisse and H. Poepke, Long-term Steam Electrolysis with Electrolyte-Supported Solid Oxide Cells, *Electrochim. Acta*, 2015, DOI: 10.1016/j.electacta.2015.04.141.
65 G. Corre and A. Brisse, 9000 hours operation of a 25 solid oxide cells stack in steam electrolysis mode, in *14th International Symposium on Solid Oxide Fuel Cells, SOFC 2015; held as part of the Electrochemical Society, ECS Conference on Electrochemical Energy Conversion and Storage*, ed. K. Eguchi and S. C. Singhal, Electrochemical Society Inc., 2015, pp. 3481–3490.
66 J. B. Hansen, J. Pålsson, J. U. Nielsen, E. Fontell, T. Kivisaari, P. Jumppanen and P. V. Hendriksen, Design aspects of a 250 kW NG fuelled SOFC system – strategies to counteract stack performance degradation, in *Fuel Cell Seminar*, Miami Beach, FL, USA, 2003, pp. 790–793.
67 J. B. Hansen, System level strategies to counteract ageing in SOEC plants, in *European Fuel Cell Conference*, Rome, 2013.
68 E. Tang, T. Wood, S. Benhaddad, C. Brown, H. He, J. Nelson, O. Grande, B. Nuttal, M. Richards and B. Petri, *Advanced materials for RSOFC Dual Operation with Low Degradation, Report for United States Department of Energy*, 2012.
69 K. Wonsyld, L. Bech, J. U. Nielsen and C. F. Pedersen, *Operational Robustness of Solid Oxide Electrolysis Stacks, in 11 European SOFC & SOE Forum*, Luzern, Switzerland, 2014, pp. B13–52/67.
70 C. Graves, Reversing and repairing microstructure degradation in solid oxide cells during operation, in *ECS Transactions*, Electrochemical Society Inc., 2013, pp. 3127–3136.
71 C. Graves, S. D. Ebbesen, S. H. Jensen, S. B. Simonsen and M. B. Mogensen, Eliminating degradation in solid oxide electrochemical cells by reversible operation, *Nat. Mater.*, 2014, **14**, 239–244.
72 D. M. Bierschenk, J. R. Wilson and S. A. Barnett, High efficiency electrical energy storage using a methane-oxygen solid oxide cell, *Energy Environ. Sci.*, 2011, **4**, 944–951.
73 D. M. Bierschenk, J. R. Wilson, E. Miller, E. Dutton and S. A. Barnett, *A proposed method for high efficiency electrical energy storage using solid oxide cells*, 2011, pp. 2969–2978.
74 P. Kazempoor, C. H. Wendel and R. J. Braun, Pressurized regenerative solid oxide cells for electrical energy storage, in *ECS Transactions*, Electrochemical Society Inc., 2013, pp. 45–54.
75 C. H. Wendel, P. Kazempoor and R. J. Braun, Novel electrical energy storage system based on reversible solid oxide cells: System design and operating conditions, *J. Power Sources*, 2015, **276**, 133–144.
76 C. H. Wendel, P. Kazempoor and R. J. Braun, Operating conditions and system considerations for a reversible solid oxide cell system for Electrical energy storage, in *ASME 2014 12th International Conference on Fuel Cell Science, Engineering and Technology, FUELCELL 2014 Collocated with the ASME 2014 8th International Conference on Energy Sustainability*, Web Portal ASME (American Society of Mechanical Engineers), 2014.
77 J. B. Hansen, N. Christiansen and J. U. Nielsen, Production of sustainable fuels by means of solid oxide electrolysis, in *12th International Symposium on Solid Oxide Fuel Cells, SOFC-XII – 219th ECS Meeting*, Montreal, QC, 2011, pp. 2941–2948.

78 S. H. Jensen, X. Sun, S. D. Ebbesen, R. Knibbe and M. Mogensen, Hydrogen and synthetic fuel production using pressurized solid oxide electrolysis cells, *Int. J. Hydrogen Energy*, 2010, **35**, 9544–9549.
79 J. E. O'Brien, X. Zhang, C. K. Housley, K. Dewall, L. Moore-McAteer and C. Tao, *High temperature electrolysis pressurized experiment design, operation and results*, DE-AC07-05ID14517, DOE, 2012.
80 L. Bernadet, G. Gousseau, A. Chatroux, J. Laurencin, F. Mauvy and M. Reytier, Influence of pressure on solid oxide electrolysis cells investigated by experimental and modeling approach, *Int. J. Hydrogen Energy*, 2015, **40**, 12918–12928.
81 L. Bernadet, G. Gousseau, A. Chatroux, J. Laurencin, F. Mauvy and M. Reytier, Assessment of pressure effects on high temperature steam electrolysis based on solid oxide technology, in *14th International Symposium on Solid Oxide Fuel Cells, SOFC 2015; held as part of the Electrochemical Society, ECS Conference on Electrochemical Energy Conversion and Storage*, ed. K. Eguchi and S. C. Singhal, Electrochemical Society Inc., 2015, pp. 3369–3378.
82 M. Reytier, S. Di Iorio, A. Chatroux, M. Petitjean, J. Cren, M. de Saint Jean, J. Aicart and J. Mougin, Stack performances in high temperature steam electrolysis and co-electrolysis, *Int. J. Hydrogen Energy*, 2015, DOI: 10.1016/j.ijhydene.2015.04.085.
83 J. B. Hansen, F. Fock and H. H. Lindboe, Biogas upgrading: by steam electrolysis or co-electrolysis of biogas and steam?, in *ECS Transactions*, Electrochemical Society Inc., 2013, pp. 3089–3097.
84 J. B. Hansen, *Process for converting biogas to a gas rich in methane*, WO2012003849, 2010.
85 G. Lorenzi, A. Lanzini and M. Santarelli, Digester gas upgrading to synthetic natural gas in solid oxide electrolysis cells, *Energy Fuels*, 2015, **29**, 1641–1652.
86 E. Giglio, A. Lanzini, M. Santarelli and P. Leone, Synthetic natural gas *via* integrated high-temperature electrolysis and methanation: part II-economic analysis, *Journal of Energy Storage*, 2015, **2**, 64–79.
87 K. C. Tran and O. Sigurbjornsson, That's why we should all go to Iceland, *Chem. Eng.*, 2011, 28–31.
88 J. B. Hansen, N. Christiansen and J. U. Nielsen, Production of sustainable fuels by means of solid oxide electrolysis, *ECS Trans.*, 2011, 2941–2948.
89 J. B. Hansen and P. E. Højlund-Nielsen, Methanol Synthesis Technology, in *Handbook of Heterogeneous Catalysis*, ed. G. Ertl, H. Knözinger, F. Schüth and J. Wetkamp, Wiley-VCH, Weinheim, 2008, pp. 2920–2949.
90 J. B. Hansen, *Green SynFuels, Final Project Report March 2011*, 2011.
91 H. Wenzel, *Breaking the biomass bottleneck of the fossil free society*, Concito, 2010.
92 J. Hartvigsen, S. Elangovan, L. Frost, C. Stoots, J. O'Brien, J. S. Herring, M. Sonai and G. Hawkes, Synthetic fuel production utilizing CO_2 recycling as an alternative to sequestration, in *Energy Technology 2010: Conservation, Greenhouse Gas Reduction and Management, Alternative Energy Sources – TMS 2010 Annual Meeting and Exhibition*, Seattle, WA, 2010, pp. 15–26.
93 Q. Fu, C. Mabilat, M. Zahid, A. Brisse and L. Gautier, Syngas production *via* high-temperature steam/CO_2 co-electrolysis: an economic assessment, *Energy Environ. Sci.*, 2010, **3**, 1382–1397.

94 C. Graves, S. D. Ebbesen, M. Mogensen and K. S. Lackner, Sustainable hydrocarbon fuels by recycling CO_2 and H_2O with renewable or nuclear energy, *Renewable Sustainable Energy Rev.*, 2011, **15**, 1–23.
95 W. L. Becker, R. J. Braun, M. Penev and M. Melaina, Production of Fischer–Tropsch liquid fuels from high temperature solid oxide co-electrolysis units, *Energy*, 2012, **47**, 99–115.
96 D. Connolly, B. V. Mathiesen and I. Ridjan, A comparison between renewable transport fuels that can supplement or replace biofuels in a 100% renewable energy system, *Energy*, 2014, **73**, 110–125.
97 F. Hvelplund, P. Østergaard, B. Möller, B. V. Mathiesen, D. Connolly and A. N. Andersen, Analysis: smart energy systems and infrastructures, in *Renewable Energy Systems: A Smart Energy Systems Approach to the Choice and Modeling of 100% Renewable Solutions*, Elsevier Inc., 2nd edn, 2014, pp. 131–184.
98 H. Lund, A. N. Andersen, P. A. Østergaard, B. V. Mathiesen and D. Connolly, From electricity smart grids to smart energy systems – a market operation based approach and understanding, *Energy*, 2012, **42**, 96–102.
99 B. V. Mathiesen, W. Liu, X. Zhang and W. W. Clark, Analysis: 100 percent renewable energy systems, in *Renewable Energy Systems: A Smart Energy Systems Approach to the Choice and Modeling of 100% Renewable Solutions*, Elsevier Inc., 2nd edn, 2014, pp. 185–238.
100 B. V. Mathiesen, H. Lund, D. Connolly, H. Wenzel, P. A. Ostergaard, B. Möller, S. Nielsen, I. Ridjan, P. KarnOe, K. Sperling and F. K. Hvelplund, Smart Energy Systems for coherent 100% renewable energy and transport solutions, *Appl. Energy*, 2015, **145**, 139–154.

Faraday Discussions

PAPER

Interpreting equilibrium-conductivity and conductivity-relaxation measurements to establish thermodynamic and transport properties for multiple charged defect conducting ceramics

Huayang Zhu,[a] Sandrine Ricote,[a] W. Grover Coors[b] and Robert J. Kee*[a]

Received 2nd February 2015, Accepted 16th March 2015
DOI: 10.1039/c5fd00012b

A model-based interpretation of measured equilibrium conductivity and conductivity relaxation is developed to establish thermodynamic, transport, and kinetics parameters for multiple charged defect conducting (MCDC) ceramic materials. The present study focuses on 10% yttrium-doped barium zirconate (BZY10). In principle, using the Nernst–Einstein relationship, equilibrium conductivity measurements are sufficient to establish thermodynamic and transport properties. However, in practice it is difficult to establish unique sets of properties using equilibrium conductivity alone. Combining equilibrium and conductivity-relaxation measurements serves to significantly improve the quantitative fidelity of the derived material properties. The models are developed using a Nernst–Planck–Poisson (NPP) formulation, which enables the quantitative representation of conductivity relaxations caused by very large changes in oxygen partial pressure.

1 Introduction

Multiple charged defect conducting (MCDC) ceramics play important roles in mature and emerging technologies. For example, solid-oxide fuel cell (SOFC) cathodes are often fabricated using mixed ionic-electronic conducting (MIEC) materials such as lanthanum strontium cobalt ferrite (LSCF). The present effort is concerned primarily with yttrium-doped barium zirconates (BZY), which have interesting proton-conducting properties. These materials can be used at intermediate-temperature (*i.e.*, around 500 °C) protonic-ceramic fuel cells (PCFCs) or electrolyzers.[1–4] Other emerging applications include hydrogen separations and catalytic membrane reactors.[5–7]

[a]*Mechanical Engineering, Colorado School of Mines, Golden, CO, USA. E-mail: rjkee@mines.edu; Fax: +1-303-273-3602; Tel: +1-303-273-3379*
[b]*CoorsTek, Inc., Golden, CO, USA*

Fig. 1 illustrates a PCFC configured as a button cell. The fuel compartment is on the lower anode side and air is on the upper cathode side. The ion transport through the MCDC electrolyte membrane is fundamentally different from a traditional SOFC with an oxide-ion-conducting electrolyte membrane. As illustrated in the balloon inset of Fig. 1, protonic defects ($OH_O^{\bullet}$) are conducted *via* a Grotthuss mechanism from the fuel side toward the oxidizer side. In a mixed conductor such as BZY, there is a counter flux of oxide-ion vacancies and "holes". As discussed subsequently, the holes are assumed to be small polarons that are associated with oxygen sites on the doped perovskite crystal lattice. To make clear that the holes are polarons, the O-site polarons are designated as $O_O^{\bullet}$. In BZY, the oxide-ion vacancies are typically minority charge carriers.

Unlike in an SOFC, the oxidation product H_2O is produced primarily in the cathode chamber of a PCFC, potentially leading to improved efficiency. Nevertheless, because the MCDC electrolyte permits some oxide-ion transport, low levels of H_2O can be formed in the anode chamber. Moreover, because there is some electronic conductivity in the mixed-conducting electrolyte membrane, the leakage current contributes to efficiency loss. The MCDC materials tend to have reasonable proton conductivity at relatively intermediate temperatures (*i.e.*, around 500–600 °C). The intermediate-temperature performance potentially leads to system-level improvements in terms of material selections, catalyst degradation, sealing strategies, *etc.*

The present paper draws a clear distinction between MIEC and MCDC materials. As discussed by Riess,[8] MIEC materials permit electronic conduction *via* band-state holes or electrons. Certain ceramic materials (*e.g.*, doped ceria, doped barium zirconate) generally do not have metallic-type electronic conduction. Nevertheless, previous literature frequently refers to "hole" mobility and MIEC behavior. Such nomenclature is used, even in cases where the "holes" are stated to be small polarons. The "holes" are understood to be small polarons that are associated with the lattice oxygen. Thus, it is reasonable to introduce a new nomenclature (MCDC), making clear that none of the charge-carrying defects have band-state behavior.

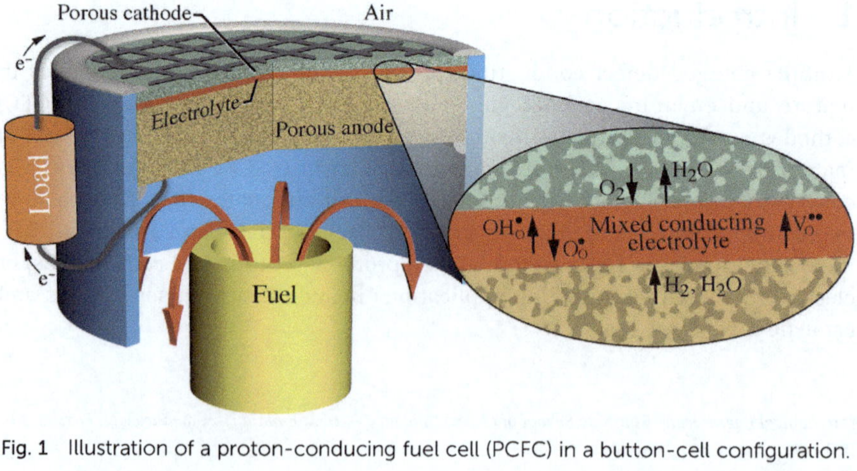

Fig. 1 Illustration of a proton-conducing fuel cell (PCFC) in a button-cell configuration.

Despite the current and potential importance of the MCDC ceramics, there are surprisingly few quantitatively predictive modeling capabilities that can be applied in technology design and process optimization. Because the MCDC materials typically have three or more charged-defects (*e.g.*, protonic defects, oxide ions, small polarons), widely-practiced models such as those based on ambipolar diffusion are not directly applicable. The Nernst–Planck–Poisson formulation is applicable, but computationally more complex than ambipolar diffusion.[9–12]

The present experimental and modeling approaches generally follow the one developed recently to establish thermodynamic and transport properties in doped ceria.[13,14] However, ceria has only two charge-carrying defects (oxide ions and reduced-cerium polarons). Thus, for MCDC materials such as BZY the task is significantly more complex because three charge-carrying defects must be considered.

The overarching objective of the present investigation is to develop experimental and modeling approaches that can be applied to establish thermodynamic and transport properties for MCDC ceramics. The required thermodynamic properties are the changes in enthalpy ΔH° and entropy ΔS° for each defect-incorporation reaction at the ceramic–gas interface. The required transport properties are diffusion coefficients for each of the defects within the MCDC ceramic. The transport properties are represented as pre-exponential factors and an activation energy.

The conductivity-relaxation experiments reveal kinetic limitations associated with the defect reactions at the MCDC–gas interfaces. Thus, in addition to establishing thermodynamic and transport properties, the model-based experimental interpretation leads to quantitative insight about the surface kinetics. The surface kinetics are represented as microscopically reversible reactions, with the forward and reverse rates being constrained by the thermodynamic equilibrium constants.

2 Experimental approach

2.1 Rod preparation

The extruded BZY10 rods were fabricated by solid state reactive sintering (SSRS) using precursor oxides and sulfates: barium sulfate – Blanc Fixe (Solvay Chemicals, Inc, $BaSO_4$), zirconium oxide (Neo Performance Materials (AMR Ltd), ≥99.5% ZrO_2), yttrium oxide (HJD Intl. 99.99+% ultra-fine Y_2O_3). The appropriate amounts of precursors were mixed with 1.0 wt% of black nickel oxide (Novomet, Grade F NiO†) and ball milled in deionized water for 24 hours. The mixture was subsequently dried in air and sieved through a 40 mesh screen. The resulting powder was blended with water-soluble acrylic and a cellulosic-ether plasticizer and pelletized for wet-extrusion processing. Green rods were extruded through an encapsulated die set, producing unfired rods with a diameter of 6.3 mm. The rods were dried and then hang-fired in air at 1600 °C for 6 hours. The fired shrinkage was about 31%, producing a fired diameter of 4.3 mm.

Alumina was used as a getter for NiO extraction from dense fired monolithic pellets of SSRS BZY10. The sintered dense rods were buried in Almatis calcined

† D50 of 1–2 microns. Weight-percent impurities: Co < 0.0015; Fe < 0.015; Cu < 0.001; S < 0.001.

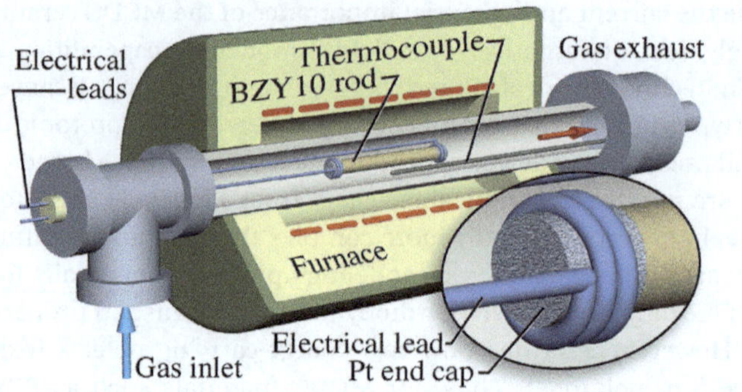

Fig. 2 Experimental configuration of measuring equilibrium conductivity and conductivity relaxation.

alumina (CT3000LS SG) inside an alumina crucible and re-fired at 1600 °C for 20 hours. Following NiO extraction, the rods were removed from the packed powder and surface ground to remove any surface-reaction layers that may have developed by reaction with the alumina powder during the Ni-extraction process. The grinding and polishing process reduced the rod diameter to 3 mm.

Finally, the samples were subjected to a 24 hour annealing process. For the annealing process, samples were placed on ceramic foam filter of SELEE® structure in a sealed alumina tube and heated in air to 800 °C at 5 °C min^{-1}. Once the furnace temperature reached 800 °C, hydrogen forming gas (4% H_2, balance argon) was added to the chamber at a rate of 1000 mL min^{-1} and exhausted to atmosphere for 24 hours. Platinum leads were attached at the ends of the rod, painted with Pt paste and fired at 1100 °C for 2 h.

2.2 Testing protocol

Because the experimental procedures have been previously reported,[13] only a brief summary is included here. Fig. 2 illustrates the experimental setup in which solid slender rods of BZY10 are placed within a furnace under controlled temperature and gas composition. The rods are approximately 3 mm in diameter and 20 mm long. This somewhat unusual configuration offers some important advantages compared to other sample configurations. Under conductivity-relaxation conditions (*i.e.*, rapidly switching the gas environment) the defect transport is effectively one-dimensional in the radial direction while the conductivity measurement is essentially one-dimensional in the axial direction. As reported previously,[14] one-dimensional radial models can be developed to assist quantitative data interpretation. Such models are relatively straightforward to solve computationally, even with very large variations in the gas-phase atmospheres.

The gas flow rates used in the experiments were sufficient to ensure defect equilibrium with the rod surfaces, but not so high as to introduce an axial temperature gradient in the specimen. Equilibrium conductivity measurements were made after the rod achieved equilibrium in a particular gas atmosphere. All the gases were either dried through Drierite ($CaSO_4$) or humidified through bubblers maintained at room temperature (approximately 3.8% humidification or $p_{H_2O} \approx 3154$ Pa, assuming laboratory atmospheric pressure of 0.83 atm).

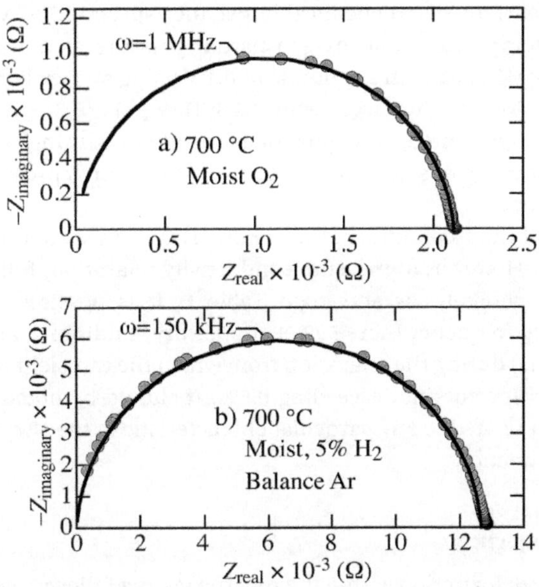

Fig. 3 Nyquist representation of typical electrochemical impedance measurements and RQ fits.

Following the Drierite, the H_2O partial pressure is estimated to be 10 ppm. To avoid water condensation, all the lines were heated to about 50 °C. The oxygen partial pressure was measured in the exhaust gas using a zirconia p_{O_2} sensor fabricated by CoorsTek. The p_{O_2} sensor was maintained at the same temperature as the sample chamber.

Table 1 Comparison of steady-state resistances based on full EIS sweeps and at a single frequency at different temperatures and gas compositions. The "error" columns represent the differences between the resistances measured by the full EIS sweep and those measured at single specified frequencies

Temperature (°C)	R (Ω) (RQ fit)	R (Ω) (500 Hz)	Error (%)	Temperature (°C)	R (Ω) (RQ fit)	R (Ω) (5 kHz)	Error (%)
Dry, O_2				**Moist 3.8% H_2O, O_2**			
600	3602	3605	0.08	600	6447	6415	0.49
700	1469	1469	0.00	700	2113	2108	0.23
800	794	789	0.60	800	980	979	0.10
Dry, 1% O_2, balance Ar				**Moist 3.8% H_2O, 1% O_2, balance Ar**			
600	7961	7969	0.10	600	9568	9575	0.07
700	3461	3463	0.06	700	4527	4510	0.37
800	1804	1805	0.05	800	2096	2093	0.14
Dry, 5% H_2, balance Ar				**Moist 3.8% H_2O, 5% H_2, balance Ar**			
600	199 000	198 000	0.50	600	15 140	15 160	0.13
700	97 030	97 040	0.01	700	12 660	12 675	0.11
800	107 000	106 600	0.37	800	11 730	11 705	0.20

Alternating current electrochemical impedance spectra (EIS) were recorded as a function of temperature and oxygen partial pressure at frequencies ranging from 20 Hz to 1 MHz, with an amplitude of 0.1 V using an Hewlett Packard 4284A Precision LCR Meter interfaced with LABVIEW.‡ Fig. 3 shows typical EIS measurements represented as Nyquist plots together with single-arc RQ fits. The ZSIMPWIN software§ was used to fit equivalent-circuit representations of the measured AC spectra.

Under equilibrium conditions, the conductivity is measured using RQ fits to full EIS sweeps. However, to measure conductivity relaxation, full EIS sweeps are impractical. Fortunately, as shown in Table 1, it is possible to identify and monitor a single frequency (here 500 Hz under dry conditions, and 5 kHz under moist conditions) during the relaxation from which the transient conductivity can be measured with errors not exceeding 0.5%. It should be noted that the single frequency is not related to any particular characteristic of the Nyquist plot such as the summit frequency.

2.3 Testing sequence

Because of slow dehydration within the rod, the testing follows a certain sequence from dry to moist conditions. The prepared rod (Section 2.1) was placed into the testing apparatus (Fig. 2), whereupon it was heated in dry (through Drierite) 100 mL min^{-1} of Argon to 800 °C, while recording the resistance. When the resistance reached a constant value (after about 3 days, slow dehydration), AC spectra were collected in dry atmospheres (both oxidizing and reducing) as a function of temperature. After having determined the particular frequency for conductivity-relaxation measurements (in this case 500 Hz), the resistance was recorded during the relaxation transient when switching from dry 100% O_2 to dry 1% O_2 (balance Ar) and *vice versa*, at 600, 700 and 800 °C.

Following experiments in dry environments, the atmosphere was humidified with a bubbler maintained at room temperature (approximately 3.8% H_2O) and a similar procedure was followed: first AC spectra at equilibrium were collected in both oxidizing and reducing atmospheres as a function of temperature and then conductivity relaxations (at 5 kHz) were recorded at 600, 700 and 800 °C when switching back and forth from moist O_2 to moist 1% O_2 and finally from moist O_2 to moist 5% H_2.

3 Defect chemistry

In our foregoing BZY models[10–12] we assumed four possible charge-carrying defects, protons $OH_O^\bullet$, oxide vacancies $V_O^{\bullet\bullet}$, electron holes $h^\bullet$, and electrons e'. Using Kröger–Vink notation, these defects at the surface are assumed to interact with the gas environment according to four defect reactions as

$$\frac{1}{2}H_2 + O_O^\times \rightleftharpoons OH_O^\bullet + e' \tag{1}$$

‡ National Instruments, Inc.

§ Princeton Applied Research.

$$\frac{1}{2}O_2 + V_O^{\bullet\bullet} \rightleftharpoons O_O^\times + 2h^\bullet \tag{2}$$

$$H_2O + V_O^{\bullet\bullet} + O_O^\times \rightleftharpoons 2OH_O^\bullet \tag{3}$$

$$\text{null} \rightleftharpoons h^\bullet + e' \tag{4}$$

The holes and electrons are understood to be small polarons, not delocalized or band-state electronic charge carriers. In practice, the models show that although electrons are included, they are essentially negligible for BZY materials.

3.1 Alternative defect reactions

Stokes and Islam[15] proposed that the "holes" are small polarons localized at the oxygen sites, which may be modeled as O^-. The present study assumes that a delocalized hole $h^\bullet$ associated with an oxygen site can be represented as

$$O_O^\times + h^\bullet \rightleftharpoons O_O^\bullet. \tag{5}$$

The (perhaps unusual) nomenclature $O_O^\bullet$ designates a hole (polaron) localized at an oxygen site. The nomenclature is intended to be analogous with $OH_O^\bullet$, meaning a protonic defect associated with an oxygen site. The combination of reactions (2) and (5) yields an alternative form of the oxygen-incorporation reaction as

$$\frac{1}{2}O_2 + V_O^{\bullet\bullet} + O_O^\times \rightleftharpoons 2O_O^\bullet. \tag{6}$$

Thus, reaction (6) differs from reaction (2) by twice the association energy of the oxygen-bound polaron relative to an (unbound) valance-band hole.

Stokes and Islam also suggested that electrons would be localized on a metal site. In this case, reaction (1) might be rewritten as

$$\frac{1}{2}H_2 + O_O^\times + Zr_{Zr}^\times \rightleftharpoons OH_O^\bullet + Zr_{Zr}'. \tag{7}$$

Then, reaction (4) can be rewritten as,

$$O_O^\times + Zr_{Zr}^\times \rightleftharpoons O_O^\bullet + Zr_{Zr}'. \tag{8}$$

For BZY10, the electrons are not favored thermodynamically even in reducing environments, and can be assumed to be immediately annihilated in favor of a hole. By combining reactions (7) and (8), the resulting hydrogen-incorporation reaction would be rewritten as

$$\frac{1}{2}H_2 + O_O^\bullet \rightleftharpoons OH_O^\bullet. \tag{9}$$

which would eliminate the need for reaction (4) or (8).

3.2 Defect-reaction equilibrium and kinetics

The reformulated defect reactions for BZY10 consider only three charge-carrying defects – protons $OH_O^\bullet$, oxygen vacancies $V_O^{\bullet\bullet}$, and O-site polarons $O_O^\bullet$. The three defect reactions at the ceramic–gas interface can be summarized as

$$\frac{1}{2}\mathrm{H_2} + \mathrm{O_O^{\bullet}} \underset{k_{\mathrm{b,H_2}}}{\overset{k_{\mathrm{f,H_2}}}{\rightleftharpoons}} \mathrm{OH_O^{\bullet}}, \tag{10}$$

$$\frac{1}{2}\mathrm{O_2} + \mathrm{O_O^{\times}} + \mathrm{V_O^{\bullet\bullet}} \underset{k_{\mathrm{b,O_2}}}{\overset{k_{\mathrm{f,O_2}}}{\rightleftharpoons}} 2\mathrm{O_O^{\bullet}}, \tag{11}$$

$$\mathrm{H_2O} + \mathrm{V_O^{\bullet\bullet}} + \mathrm{O_O^{\times}} \underset{k_{\mathrm{b,H_2O}}}{\overset{k_{\mathrm{f,H_2O}}}{\rightleftharpoons}} 2\mathrm{OH_O^{\bullet}}. \tag{12}$$

The present study considers that these reactions may be equilibrated or may be kinetically rate controlled. If the reactions proceed at finite rates, microscopic reversibility is enforced (*i.e.*, forward and backward rates are related through the equilibrium constants). If the reactions are maintained at equilibrium, then, assuming gas-phase equilibrium, one of the reactions is redundant.

The forward and backward rate expressions are denoted as $k_{\mathrm{f},i}$ and $k_{\mathrm{b},i}$. The defect production rates at the gas–surface interface can be represented as

$$\dot{s}_{\mathrm{O_O^{\bullet}}} = 2\dot{q}_{\mathrm{O_2}} - \dot{q}_{\mathrm{H_2}}, \tag{13}$$

$$\dot{s}_{\mathrm{V_O^{\bullet\bullet}}} = -\dot{q}_{\mathrm{O_2}} - \dot{q}_{\mathrm{H_2O}}, \tag{14}$$

$$\dot{s}_{\mathrm{OH_O^{\bullet}}} = 2\dot{q}_{\mathrm{H_2O}} + \dot{q}_{\mathrm{H_2}}, \tag{15}$$

$$\dot{s}_{\mathrm{O_O^{\times}}} = -\dot{q}_{\mathrm{O_2}} - \dot{q}_{\mathrm{H_2O}}, \tag{16}$$

where the gas-phase reaction rates of progress can be reprinted as

$$\dot{q}_{\mathrm{O_2}} = k_{\mathrm{f,O_2}}\left[\mathrm{V_O^{\bullet\bullet}}\right]\left[\mathrm{O_O^{\times}}\right][\mathrm{O_2}]^{1/2} - k_{\mathrm{b,O_2}}\left[\mathrm{O_O^{\bullet}}\right]^2, \tag{17}$$

$$\dot{q}_{\mathrm{H_2}} = k_{\mathrm{f,H_2}}\left[\mathrm{O_O^{\bullet}}\right][\mathrm{H_2}]^{1/2} - k_{\mathrm{b,H_2}}\left[\mathrm{OH_O^{\bullet}}\right], \tag{18}$$

$$\dot{q}_{\mathrm{H_2O}} = k_{\mathrm{f,H_2O}}\left[\mathrm{V_O^{\bullet\bullet}}\right]\left[\mathrm{O_O^{\times}}\right][\mathrm{H_2O}] - k_{\mathrm{b,H_2O}}\left[\mathrm{OH_O^{\bullet}}\right]^2. \tag{19}$$

The equilibrium constants, written in terms of molar concentrations, can be stated as

$$K_{c,\mathrm{O_2}} = \frac{k_{\mathrm{f,O_2}}}{k_{\mathrm{b,O_2}}} = \frac{\left[\mathrm{O_O^{\bullet}}\right]^2}{\left[\mathrm{V_O^{\bullet\bullet}}\right]\left[\mathrm{O_O^{\times}}\right][\mathrm{O_2}]^{1/2}} = K_{p,\mathrm{O_2}}\left[\frac{p_{\mathrm{atm}}}{RT}\right]^{-1/2}, \tag{20}$$

$$K_{c,\mathrm{H_2}} = \frac{k_{\mathrm{f,H_2}}}{k_{\mathrm{b,H_2}}} = \frac{\left[\mathrm{OH_O^{\bullet}}\right]}{\left[\mathrm{O_O^{\bullet}}\right][\mathrm{H_2}]^{1/2}} = K_{p,\mathrm{H_2}}\left[\frac{p_{\mathrm{atm}}}{RT}\right]^{-1/2}, \tag{21}$$

$$K_{c,\mathrm{H_2O}} = \frac{k_{\mathrm{f,H_2O}}}{k_{\mathrm{b,H_2O}}} = \frac{\left[\mathrm{OH_O^{\bullet}}\right]^2}{\left[\mathrm{V_O^{\bullet\bullet}}\right]\left[\mathrm{O_O^{\times}}\right][\mathrm{H_2O}]} = K_{p,\mathrm{H_2O}}\left[\frac{p_{\mathrm{atm}}}{RT}\right]^{-1}. \tag{22}$$

The equilibrium constants K_p based on the gas-phase partial pressures and lattice formula-unit concentrations are defined as

$$K_{p,\mathrm{H_2}} = \frac{\left[\mathrm{OH_O^{\bullet}}\right]_{\mathrm{L}}}{{p_{\mathrm{H_2}}}^{1/2}\left[\mathrm{O_O^{\bullet}}\right]_{\mathrm{L}}}, \tag{23}$$

$$K_{p,O_2} = \frac{[O_O^{\bullet}]_L^{\ 2}}{p_{O_2}^{\ 1/2}[O_O^{\times}]_L[V_O^{\bullet\bullet}]_L}, \tag{24}$$

$$K_{p,H_2O} = \frac{[OH_O^{\bullet}]_L^{\ 2}}{p_{H_2O}[O_O^{\times}]_L[V_O^{\bullet\bullet}]_L}. \tag{25}$$

The formula-unit concentrations $[X_k]_L$ are related to the molar concentrations $[X_k]$ *via* the molar volume V_m as $[X_k]_L = [X_k]V_m$. For BZY10, the lattice molar volume is $V_m = 44.792 \times 10^{-6}\ m^3\ mol^{-1}$. The three equilibrium constants K_{p,H_2}, K_{p,O_2} and K_{p,H_2O} are not independent, but are coupled through the gas-phase equilibrium constant $K_{p,G}$ as

$$K_{p,H_2}^{\ 2}K_{p,O_2} = K_{p,H_2O}K_{p,G}. \tag{26}$$

The present model assumes that the gas phase is maintained in equilibrium according to

$$H_2 + \frac{1}{2}O_2 \rightleftharpoons H_2O, \tag{27}$$

with $K_{p,G}$ being the equilibrium constant.

All the equilibrium constants can be evaluated from the Gibbs free energy change for the reactions $\Delta G^\circ = \Delta H^\circ - T\Delta S^\circ$ as

$$K_p = \exp\left(-\frac{\Delta G^\circ}{RT}\right) = \exp\left(\frac{\Delta S^\circ}{R}\right)\exp\left(-\frac{\Delta H^\circ}{RT}\right). \tag{28}$$

The gas-phase thermodynamics are well known; for reaction (27) $\Delta H^\circ = -241.11$ kJ mol^{-1} and $\Delta S^\circ = -55.48$ J mol^{-1} K^{-1}. Thermodynamic parameters for the defect reactions will be determined using the conductivity measurements.

3.3 Site and electroneutrality constraints

On the surface and within the ceramic, site constraints and electroneutrality must be preserved. The number of oxygen sites in a perovskite lattice (per formula unit) must be maintained at 3. Thus,

$$[V_O^{\bullet\bullet}]_L + [OH_O^{\bullet}]_L + [O_O^{\bullet}]_L + [O_O^{\times}]_L = 3. \tag{29}$$

In this formulation, eqn (29) is written to include the O-site polaron $[O_O^{\bullet}]_L$ in the site constraint. However, it may be questioned as to whether the electronic defect (*i.e.*, polaron) "occupies" the site in the same sense as an ionic defect (*i.e.*, an atom or vacancy) does. If the defect were a delocalized band hole, then it would not enter the oxygen site balance.

Charge neutrality requires that

$$2[V_O^{\bullet\bullet}]_L + [OH_O^{\bullet}]_L + [O_O^{\bullet}]_L - [Y'_{Zr}]_L = 0. \tag{30}$$

Considering BZY10 ($BaZr_{0.9}Y_{0.1}O_{3-\delta}$), the doping concentration is 10%, in which case $[Y'_{Zr}]_L = 0.1$.

3.4 Defect equilibrium concentrations

The defect equilibrium concentrations ($[V_O^{\bullet\bullet}]_L$, $[OH_O^{\bullet}]_L$, $[O_O^{\bullet}]_L$, and $[O_O^{\times}]_L$) on the surface are functions of the gas-phase partial pressures (p_{O_2}, p_{H_2}, and p_{H_2O}) and can be established by solving iteratively eqn (24), (25), (29) and (30). Note that eqn (23) is not used because eqn (23)–(25) are not independent of each other, but are coupled through the gas-phase equilibrium (eqn (26)).

The iteration process is initiated by first estimating a value of $[V_O^{\bullet\bullet}]_L$. Based on eqn (24), (25) and (30), the defect concentrations $[O_O^{\bullet}]_L$ and $[OH_O^{\bullet}]_L$ can be evaluated in terms of $[V_O^{\bullet\bullet}]_L$ as

$$[O_O^{\bullet}]_L = \frac{[Y'_{Zr}]_L - 2[V_O^{\bullet\bullet}]_L}{1+\Theta}, \tag{31}$$

$$[OH_O^{\bullet}]_L = \Theta[O_O^{\bullet}]_L, \tag{32}$$

where

$$\Theta = \frac{[OH_O^{\bullet}]_L}{[O_O^{\bullet}]_L} = \left[\frac{K_{p,H_2O}}{K_{p,O_2}}\frac{p_{H_2O}}{{p_{O_2}}^{1/2}}\right]^{1/2}. \tag{33}$$

Using eqn (24), $[O_O^{\times}]_L$ can be represented as

$$[O_O^{\times}]_L = \frac{{[O_O^{\bullet}]_L}^2}{K_{p,O_2}{p_{O_2}}^{1/2}[V_O^{\bullet\bullet}]_L}. \tag{34}$$

Finally, the oxygen site constraint (eqn (29)) is imposed. If all the equations above were to be explicitly substituted, the result would be one nonlinear equation with one unknown, $[V_O^{\bullet\bullet}]_L$. Once $[V_O^{\bullet\bullet}]_L$ is determined, all the other defect concentrations follow easily from eqn (31), (32) and (34).

4 Nernst–Planck–Poisson model

To model conductivity relaxations quantitatively, the spatial and temporal variations in defect concentrations are governed by conservation equations as

$$\frac{\partial[X_k]}{\partial t} + \nabla\cdot\mathbf{J}_k = 0, \tag{35}$$

where the Nernst–Planck fluxes can be written as

$$\mathbf{J}_k = -D_k\left(\nabla[X_k] + \frac{z_k F}{RT}[X_k]\nabla\Phi\right). \tag{36}$$

The electrostatic-potential profiles Φ are determined by solving the Gauss equation,

$$\nabla\cdot(\varepsilon_r\varepsilon_o\nabla\Phi) = -\rho_e = -F\sum_{k=1}^{K} z_k[X_k]. \tag{37}$$

The relative and vacuum permittivities are represented as ε_r and ε_o, respectively. The local charge density is represented as ρ_e. Boundary conditions for the defect concentrations are established in terms of the surface reactions. The boundary condition for the electrostatic potential is established in terms of a zero net charge transfer at the ceramic surface. Details of the Nernst–Planck–Poisson (NPP) model and the computational solution can be found elsewhere.[10]

As discussed previously,[13] the experimental configuration (Fig. 2) has been designed such that the NPP problem can be solved in one-dimensional radial coordinates. Although there are inevitably some end effects near the end electrodes, two-dimensional axisymmetric models show that the one-dimensional approximation is valid.

5 Parameter fitting algorithm

Altogether, twelve (ten independent) parameters are required to characterize the BZY10 material. Specifically, ΔH° and ΔS° are required for each of the three defect reactions (eqn (10)–(12)). Because the thermodynamics of the defect reactions are correlated through the gas-phase equilibrium, only four of these thermodynamic parameters are independent. The defect diffusion coefficients are represented in Arrhenius form as $D_k = D_k^\circ \exp(-E_k/RT)$. Thus, two parameters ($D_k^\circ$ and E_k) are needed to describe the diffusion coefficients for each of the three mobile charged defects ($V_O^{\bullet\bullet}$, $OH_O^\bullet$, and $O_O^\bullet$).

The fitting process begins with a least-squares fit of measured equilibrium conductivities, using measurements from a wide range of oxygen partial pressures and temperatures. The Nernst–Einstein equation (eqn (38)) forms the basis for this fitting process. Very good fits can be obtained in the sense that a set of parameters can be found that matches the data well. The problem is one of non-uniqueness. That is, alternative sets of different parameters do an equally good job of matching the equilibrium data.

The consideration of conductivity-relaxation data provides additional information that is valuable in resolving ambiguities associated with considering equilibrium conductivities alone. Using the parameter set from an initial fit to equilibrium conductivity measurements, the NPP model is used to predict relaxation profiles. In the first instance, the defect-incorporation reactions are assumed to be in equilibrium with the gas-phase environment, providing the boundary conditions for the NPP model. However, the initial set of parameters from the equilibrium fits usually do not lead to good representations of the measured relaxations. Often the predicted relaxation times (based on surface equilibrium) are slower than the measured relaxation, which cannot be correct. In other words, surface equilibration should produce relaxation times that are shorter than the measured relaxations. Thus, parameter adjustments are needed to assure that the modeled relaxations under surface equilibrium are no slower than the data. With adjusted parameters, the equilibrium fits are repeated, making sure that the equilibrium data is still represented accurately.

It is usually the case that after the thermodynamic and transport parameters are established, the model-predicted relaxation times are faster than the measurements for the reducing-to-oxidizing relaxations. This indicates a surface-reaction kinetic limitation. Thus, the next step is to establish forward rate expressions for the surface kinetics (*i.e.*, reactions (10)–(12)). The reverse rates are constrained by microscopic reversibility, involving the thermodynamic equilibrium constant and hence the thermodynamic fitting parameters. Thus, the kinetic limitations also provide information that is important to establishing the thermodynamic and transport parameters.

5.1 Equilibrium fits

Assuming a dilute limit, measurable conductivity σ is related to defect concentrations $[X_k]$ and diffusion coefficients D_k *via* the Nernst–Einstein relationship as

$$\sigma = \sum \sigma_k = \frac{F^2}{RT} \sum_{k=1}^{K} z_k{}^2 [X_k] D_k, \tag{38}$$

where σ_k are partial conductivities of the participating defects. In this expression F and R are the Faraday and gas constants, respectively, and z_k are the charges associated with the mobile defects. At equilibrium, the defect concentrations are established by reactions with the gas phase at the ceramic surface, thus incorporating the thermodynamic Gibbs free energy associated each reaction. The diffusion coefficients are represented in Arrhenius form as $D_k = D_k^{\circ} \exp(-E_k/RT)$.

The experimentally measured conductivities σ_i^* over a range of oxygen partial pressures p_{O_2} and temperature T can be used to determine ten physical parameters (*i.e.*, four thermodynamic parameters: $\Delta H^{\circ}_{H_2O}$, $\Delta H^{\circ}_{O_2}$, $\Delta S^{\circ}_{H_2O}$, and $\Delta S^{\circ}_{O_2}$, and six transport parameters: $D^{\circ}_{OH_O^{\bullet}}$, $D^{\circ}_{V_O^{\bullet\bullet}}$, $D^{\circ}_{O_O^{\bullet}}$, $E_{OH_O^{\bullet}}$, $E_{V_O^{\bullet\bullet}}$, and $E_{O_O^{\bullet}}$) by minimizing the following residual function:

$$\min_{\mathbf{S}} \sum_{i=1}^{N} (\ln \sigma_i - \ln \sigma_i^*)^2. \tag{39}$$

In this expression the vector **S** represents all the fitting parameters, and N is the total number of measured conductivities. The nonlinear least-squares fitting problem is implemented here using the "lsqnonlin" function in MATLAB.

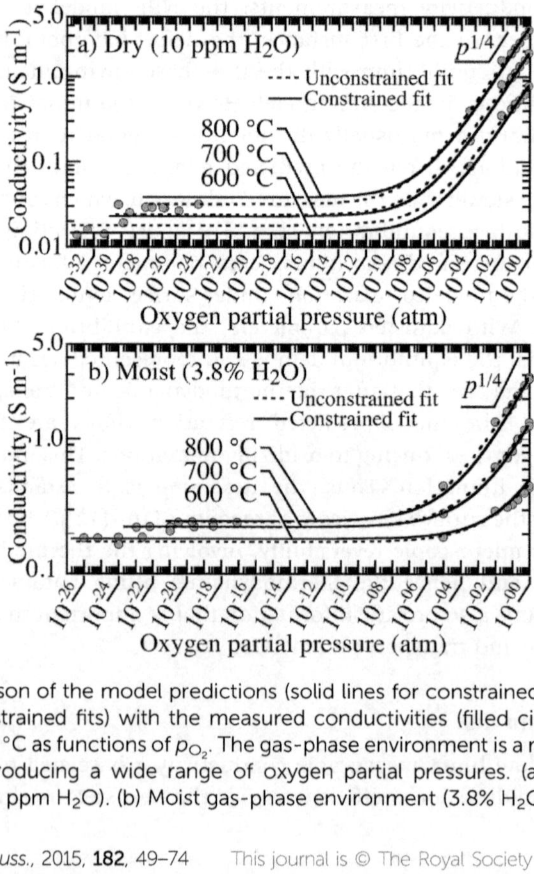

Fig. 4 Comparison of the model predictions (solid lines for constrained fits, and dashed lines for unconstrained fits) with the measured conductivities (filled circles) at 600 °C, 700 °C, and 800 °C as functions of p_{O_2}. The gas-phase environment is a mixture of H_2, O_2, H_2O, and Ar, producing a wide range of oxygen partial pressures. (a) Dry gas-phase environment (10 ppm H_2O). (b) Moist gas-phase environment (3.8% H_2O).

Using eqn (26), the enthalpy and entropy changes associated with the hydrogen-incorporation reaction (eqn (1)) can be obtained from the fitting thermodynamic parameters as

$$\Delta H^\circ_{H_2} = (\Delta H^\circ_G + \Delta H^\circ_{H_2O} - \Delta H^\circ_{O_2})/2, \tag{40}$$

$$\Delta S^\circ_{H_2} = (\Delta S^\circ_G + \Delta S^\circ_{H_2O} - \Delta S^\circ_{O_2})/2. \tag{41}$$

In these expressions, ΔH°_G and ΔS°_G refer to the gas-phase equilibrium reaction (*i.e.*, eqn (27)).

In principle, the equilibrium conductivity measurements and eqn (38) are sufficient to determine all the fit parameters. However, the "best" fits are usually found to predict some parameter values that conflict with independently established values. Also, there are significant ambiguities in the sense that many different combinations of parameters produce nearly equivalent representations of the measured conductivities. The following section develops a set of "unconstrained" fits. Then, certain parameters are fixed based on independent information, producing a set of "constrained" fits.

5.2 Unconstrained parameter fits

Assuming that there are no constraints for the fitting parameters, Fig. 4 compares the measured (filled circles) and modeled (dashed lines) equilibrium conductivities as functions of p_{O_2} at the temperatures of 600, 700, and 800 °C and 0.829 atm total pressure. The oxygen partial pressure p_{O_2} is evaluated assuming an equilibrium gas-phase mixture of H_2, O_2, H_2O, and Ar. By comparing Fig. 4a and b, it is clear that moisture significantly increases conductivity in the reducing regions, but tends to decrease the conductivity in the oxidizing regions. As illustrated in Fig. 4, the measured equilibrium conductivities can be reasonably well represented by the model over wide ranges of temperature and p_{O_2}. Table 2 shows

Table 2 Thermodynamic and transport parameters that result from constrained and unconstrained fits

Thermodynamic parameters						
	$\Delta H^\circ_{O_2}$	$\Delta H^\circ_{H_2O}$	$\Delta H^\circ_{H_2}$	$\Delta S^\circ_{O_2}$	$\Delta S^\circ_{H_2O}$	$\Delta S^\circ_{H_2}$
	(kJ mol^{-1})			(J mol^{-1} K^{-1})		
Unconstrained	−41.56	−96.65	−151.60	−73.45	−90.30	−36.17
Constrained	−36.08	−80.00	−146.01	−125.48	−71.78	−0.89

Transport parameters						
	$E_{O^\bullet_O}$	$E_{OH^\bullet_O}$	$E_{V^{\bullet\bullet}_O}$	$D^\circ_{O^\bullet_O}$	$D^\circ_{OH^\bullet_O}$	$D^\circ_{V^{\bullet\bullet}_O}$
	(kJ mol^{-1})			(m^2 s^{-1})		
Unconstrained	85.05	41.79	43.38	5.95×10^{-05}	2.22×10^{-08}	7.51×10^{-10}
Constrained	86.00	35.09	67.00	1.70×10^{-03}	9.38×10^{-09}	1.30×10^{-08}

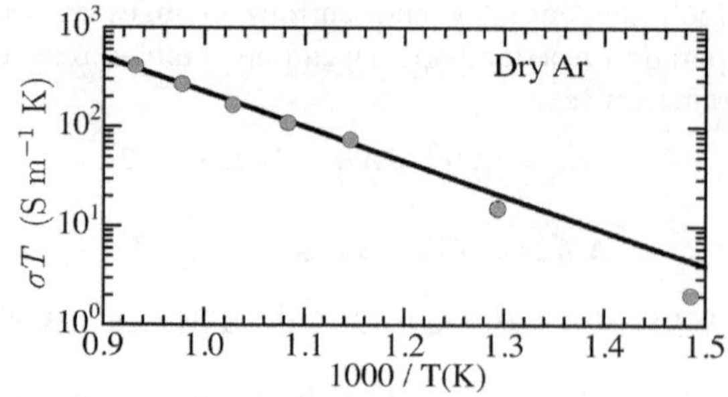

Fig. 5 Comparison of the model prediction (solid lines) with the experimentally measured total conductivity (filled circles) at functions of temperature. The gas-phase mixture contains Ar with 10 ppm H_2O and 50 ppm O_2.

thermodynamic and transport parameters for the unconstrained and constrained fits, with the constrained fitting being discussed in the following section.

5.3 Constrained parameter fits

Assuming that the gas-phase environment is pure Ar, then $[O_O^{\bullet}]_L = 0$ and $[OH_O^{\bullet}]_L = 0$. The charge-neutrality constraint (eqn (30)) leads to $[V_O^{\bullet\bullet}]_L = [Y'_{Zr}]_L/2$. In the Ar environment, the total conductivity can be represented as

$$\sigma = \frac{z_{V_O^{\bullet\bullet}}{}^2 F^2}{RT}\frac{\left[X_{V_O^{\bullet\bullet}}\right]_L}{V_m}D_{V_O^{\bullet\bullet}} = \frac{z_{V_O^{\bullet\bullet}}{}^2 F^2}{RT}\frac{\left[Y'_{Zr}\right]_L}{2V_m}D^{\circ}_{V_O^{\bullet\bullet}}\exp\left(-\frac{E_{V_O^{\bullet\bullet}}}{RT}\right). \quad (42)$$

Fig. 5 shows the measured temperature-dependent conductivity in the Ar environment, which can be fit using $D^{\circ}_{V_O^{\bullet\bullet}} = 1.43 \times 10^{-7}$ m^2 s^{-1} and $E_{V_O^{\bullet\bullet}} \approx 67$ kJ mol^{-1}. However, the experimental Ar gas mixture actually contains impurities of approximately 10 ppm H_2O and 50 ppm O_2. Consequently, the oxygen-vacancy diffusion coefficient $D^{\circ}_{V_O^{\bullet\bullet}}$ cannot be accurately evaluated, and is retained as a parameter in the fitting algorithm. But, $E_{V_O^{\bullet\bullet}} = 67$ kJ mol^{-1} is fixed.

There is general consensus in the literature[16–18] that $\Delta H^{\circ}_{H_2O} \approx 80$ kJ mol^{-1}. Therefore, in the fitting process $\Delta H^{\circ}_{H_2O}$ is constrained to be 80 kJ mol^{-1}. Additionally, based on the relaxation analysis discussed in later sections, $E_{O_O^{\bullet}} \approx 86$ kJ mol^{-1} and $D^{\circ}_{O_O^{\bullet}} \approx 1.70 \times 10^{-3}$ m^2 s^{-1} are also fixed. Fig. 4 shows that the measured equilibrium conductivity can also be fit reasonably well using these constrained parameters. All the fitting parameters are listed in Table 2. Compared to the unconstrained fitting parameters, the O-site polaron diffusivity $D_{O_O^{\bullet}}$ resulting from the constrained fitting is about 25 times larger than it is in the unconstrained fit, which is critical for the model to capture the conductivity relaxation processes.

5.4 Conductivity relaxation

As a practical matter, all the thermodynamic and transport parameters cannot be fit uniquely based on the measured equilibrium conductivity alone. Fortunately, the conductivity-relaxation measurements can provide additional data that assist establishing the required physical parameters.

The conductivity-relaxation measurements begin from an equilibrium state under a particular gas-phase composition and temperature. The relaxation is

Table 3 End-point gas-phase mole-fraction compositions for the conductivity relaxations

	Reducing				Oxidizing			
	H_2	O_2	H_2O	Ar	H_2	O_2	H_2O	Ar
DryOx	0	0.01000	0.00001	0.98999	0	0.99999	0.00001	0
MoistOx	0	0.00962	0.03774	0.95264	0	0.96226	0.03774	0
MoistRedox	0.04811	0.00005	0.03774	0.91410	0	0.96226	0.03774	0

Table 4 Comparison of measured and modeled relaxation times τ (seconds)

		Reducing to oxidizing			Oxidizing to reducing		
	T (°C)	Expr	Equil	Kinetic	Expr	Equil	Kinetic
DryOx	600	7224	4250	5440	6179	2760	5940
	700	1824	1640	1940	2233	1150	2320
	800	545	660	780	980	480	910
MoistOx	600	492	130	380	486	125	850
	700	462	90	220	449	75	520
	800	308	70	140	350	54	310
MoistRedox	600	21462	145	5490	69	118	202
	700	3500	100	2580	185	56	132
	800	1823	80	1480	92	30	99

initiated by switching rapidly from one gas-phase environment (*e.g.*, low p_{O_2}) to another gas-phase environment (*e.g.*, high p_{O_2}), and *vice versa*.[13] The relaxation is complete when the rod's conductivity reaches equilibrium with the final gas-phase environment. As listed in Table 3, the present study considers three initial and final gas-phase compositions. The relaxation sets are named "DryOx", "MoistOx", and "MoistRedox". Each relaxation experiment is repeated for three temperatures – 600, 700 and 800 °C. Thus, altogether eighteen relaxation experiments are considered. In all cases the total pressure is 0.829 atm (atmospheric pressure in Denver).

To be quantitative, the characteristic conductivity-relaxation time τ is taken to be the time required for the conductivity to achieve 95% of the final steady-state conductivity after the sudden switching between gas-phase environments. Table 4 lists the measured and modeled relaxation times. As shown in Table 4, the relaxation time for all the cases decreases as the temperature increases. The relaxation time from 1% to 100% O_2 is shorter than it is from 100% to 1% O_2 under both the dry and moist conditions, but the times are comparable at each temperature. However, introducing moisture into the oxidizing environment can significantly shorten the relaxation time. The moist H_2 reducing-to-oxidizing relaxation times are significantly longer than the characteristic times in the oxidizing-to-reducing direction.

5.5 Approximate ambipolar diffusion

When applicable, ambipolar diffusion is practically useful. An important aspect of the utility is that the important effects of internal electrostatic-potential fields are subsumed into the ambipolar diffusion coefficients, and thus do not need to be explicitly considered. Consequently, the mathematics appear as a pure diffusion problem.[12] However, the resulting ambipolar diffusion coefficient depends nonlinearly on the defect concentrations, the ordinary diffusion coefficients, and the defect charges.

Ambipolar-diffusion concepts are strictly limited to only two mobile charge carriers, but the ambipolar concepts can be used to assess qualitatively, and even quantitatively, the relative importance of individual defect diffusion coefficients. For instance, by neglecting the transport contribution from $V_O^{\bullet\bullet}$, the ambipolar diffusion coefficient $\mathcal{D}_{O_O^\bullet-OH_O^\bullet}$ based on the ionic pair $O_O^\bullet$ and $OH_O^\bullet$ can be defined as[12]

$$\mathcal{D}_{O_O^\bullet-OH_O^\bullet} \equiv \frac{\left(z_{OH_O^\bullet}{}^2\left[OH_O^\bullet\right] + z_{O_O^\bullet}{}^2\left[O_O^\bullet\right]\right) D_{OH_O^\bullet} D_{O_O^\bullet}}{z_{OH_O^\bullet}{}^2 D_{OH_O^\bullet}\left[OH_O^\bullet\right] + z_{O_O^\bullet}{}^2 D_{O_O^\bullet}\left[O_O^\bullet\right]}. \tag{43}$$

Although ambipolar diffusion coefficients vary spatially and nonlinearly during transients, relaxation times can be estimated as $\tau \approx R_{out}{}^2/\mathcal{D}_{O_O^\bullet-OH_O^\bullet}$, where the ambipolar diffusion coefficient is evaluated at the rod's surface ($r = R_{out}$). For example, assuming a relaxation time of $\tau \approx 100$ s from the MoistRedox oxidizing-to-reducing environment at 800 °C, $\mathcal{D}_{O_O^\bullet-OH_O^\bullet} \approx 2.2 \times 10^{-8}$ m^2 s^{-1}. Similarly, considering a relaxation time of $\tau \approx 550$ s at 800 °C from 1% to 100% O_2 under dry conditions, $\mathcal{D}_{O_O^\bullet-V_O^{\bullet\bullet}} \approx 4.1 \times 10^{-9}$ m^2 s^{-1}.

It is interesting to understand why the ambipolar diffusion coefficient evaluated at the rod surface is particularly relevant for estimating the relaxation time. Beginning with the entire rod being in equilibrium with a gas-phase environment, the gas environment is suddenly changed. Consequently, the defect concentrations at the rod surface rapidly come into equilibrium with the new gas-phase environment. The new defect concentrations must "diffuse" into the rod interior, ultimately bringing the entire rod into equilibrium with the new gas-phase environment. Thus, it is the ambipolar diffusion coefficients associated with the new gas-phase environment that control the characteristic transport, and hence relaxation, times.

Following some algebraic manipulation, the ambipolar diffusion coefficient as stated in eqn (43) can be rewritten as

$$\frac{1}{\mathcal{D}_{O_O^\bullet-OH_O^\bullet}} = \frac{1}{D_{OH_O^\bullet}\left(1 + \dfrac{z_{OH_O^\bullet}{}^2}{z_{O_O^\bullet}{}^2}\dfrac{\left[OH_O^\bullet\right]}{\left[O_O^\bullet\right]}\right)} + \frac{1}{D_{O_O^\bullet}\left(1 + \dfrac{z_{O_O^\bullet}{}^2}{z_{OH_O^\bullet}{}^2}\dfrac{\left[O_O^\bullet\right]}{\left[OH_O^\bullet\right]}\right)}, \tag{44}$$

which reveals that

$$\mathcal{D}_{O_O^\bullet-OH_O^\bullet} \leq \min\left\{ D_{OH_O^\bullet}\left(1 + \frac{\left[OH_O^\bullet\right]}{\left[O_O^\bullet\right]}\right),\ D_{O_O^\bullet}\left(1 + \frac{\left[O_O^\bullet\right]}{\left[OH_O^\bullet\right]}\right)\right\}. \tag{45}$$

In highly reducing environments, $[O_O^\bullet]/[OH_O^\bullet] \ll 1$, leading to

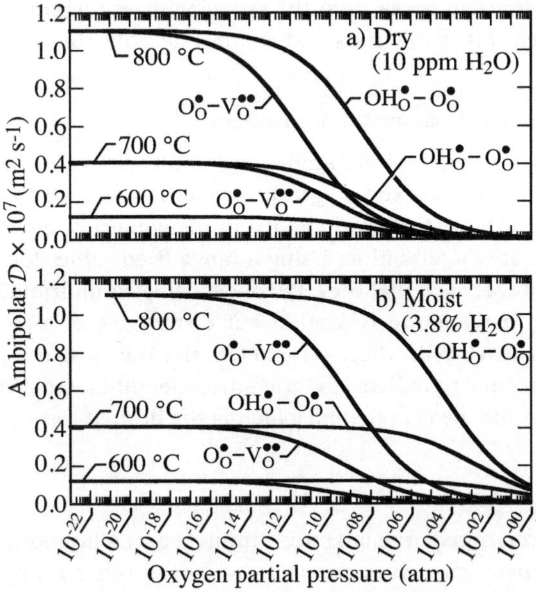

Fig. 6 Ambipolar diffusion coefficients at 600 °C, 700 °C, and 800 °C as functions of p_{O_2} for defect pairs $O_O^\bullet - OH_O^\bullet$ and $O_O^\bullet - V_O^{\bullet\bullet}$ under dry and moist environments.

$$\mathscr{D}_{O_O^\bullet-OH_O^\bullet} \leq \min\left\{D_{O_O^\bullet},\ D_{OH_O^\bullet}\frac{[OH_O^\bullet]}{[O_O^\bullet]}\right\}. \tag{46}$$

To achieve the $\mathscr{D}_{O_O^\bullet-OH_O^\bullet} \approx 2.2 \times 10^{-8}\ m^2\ s^{-1}$, which was estimated for the MoistRedox oxidizing-to-reducing relaxation at 800 °C, the ordinary diffusion coefficient for the O site polaron must be $D_{O_O^\bullet} \gtrsim 2.2 \times 10^{-8}\ m^2\ s^{-1}$. Using parameters from the unconstrained fits (Table 2), $D_{O_O^\bullet} \approx 4.31 \times 10^{-9}\ m^2\ s^{-1}$ at 800 °C. Thus, it is unlikely that the measured relaxation behaviors can be captured by the unconstrained fitting parameters.

Fig. 6 illustrates the behavior of the ambipolar diffusion coefficients for BZY10 over a wide range of oxygen partial pressures and temperatures for moist and dry conditions. The ambipolar diffusion coefficient $\mathscr{D}_{OH_O^\bullet-V_O^{\bullet\bullet}}$ is not shown here because $\mathscr{D}_{OH_O^\bullet-V_O^{\bullet\bullet}}$ is two orders of magnitude smaller than $\mathscr{D}_{O_O^\bullet-OH_O^\bullet}$ and $\mathscr{D}_{O_O^\bullet-V_O^{\bullet\bullet}}$, and nearly independent of p_{O_2}. It is evident that the functional forms of $\mathscr{D}_{O_O^\bullet-OH_O^\bullet}$ and $\mathscr{D}_{O_O^\bullet-V_O^{\bullet\bullet}}$ are qualitatively and quantitatively similar. Higher temperatures contribute to higher ambipolar diffusion coefficients, thus suggesting faster relaxations at higher temperatures. At a particular temperature, the values of $\mathscr{D}_{O_O^\bullet-OH_O^\bullet}$ and $\mathscr{D}_{O_O^\bullet-OH_O^\bullet}$ approach each other under both very reducing and very oxidizing conditions, but are separated in between. In oxidizing regions and at a given oxygen partial pressure, higher moisture levels tend to increase the ambipolar diffusion coefficients.

The binary-pair ambipolar diffusion coefficients provide substantial insight about the relaxation processes. One observation is that the relatively mobile polaron $O_O^\bullet$ plays an important role in relaxation processes. Because the ambipolar diffusion coefficients in the reducing environments are higher than those in the oxidizing environments, the relaxation times from the oxidizing environment should be shorter than those from the reducing environment to oxidizing environment. In fact, this is the measured result (*cf.*, Table 4).

5.6 Relaxation-based parameter adjustment

The O-site polaron transport is found to be closely coupled with the thermodynamics ($\Delta H^{\circ}_{O_2}$ and $\Delta S^{\circ}_{O_2}$) of the oxygen-incorporation reaction (eqn (11)). Thus, it is interesting to explore the relationships in the context of equilibrium conductivity and conductivity relaxation. Using a range fixed values for $E_{O_O^{\bullet}}$, lower-limit values for $D^{\circ}_{O_O^{\bullet}}$ are set such that the NPP model using equilibrium defect reactions on the surface matches the relaxation curve from 1% to 100% O_2 under dry conditions at 800 °C. With these parameters fixed, it is also found that all the other relaxation times predicted by the surface-equilibrated NPP model are at least faster than the measurements, which is the desired result.

5.7 Influence of $E_{O_O^{\bullet}}$

The O-site polaron transport has a strong influence on relaxation times. Thus, it is interesting to investigate the effects of $D_{O_O^{\bullet}}$ on all other fitting parameters by systematically varying the diffusion activation energy $E_{O_O^{\bullet}}$. Specifically, as listed in Table 5, six values of $E_{O_O^{\bullet}}$ are considered, and the $D^{\circ}_{O_O^{\bullet}}$ values are readjusted such that the model based on the equilibrium defect reactions achieve essentially the same relaxation time as with $E_{O_O^{\bullet}} = 86$ kJ mol^{-1}. In this study, two other

Table 5 Thermodynamic and transport parameters as functions of $E_{O_O^{\bullet}}$

Thermodynamic parameters

$E_{O_O^{\bullet}}$ (kJ mol^{-1})	$\Delta H^{\circ}_{O_2}$ (kJ mol^{-1})	$\Delta H^{\circ}_{H_2O}$	$\Delta H^{\circ}_{H_2}$	$\Delta S^{\circ}_{O_2}$ (J mol^{-1} K^{-1})	$\Delta S^{\circ}_{H_2O}$	$\Delta S^{\circ}_{H_2}$
65	6.31	−80.00	−167.21	−85.98	−71.78	−20.64
70	−3.76	−80.00	−162.17	−95.37	−71.78	−15.95
75	−13.85	−80.00	−157.13	−104.77	−71.78	−11.25
80	−23.94	−80.00	−152.08	−114.17	−71.78	−6.55
85	−34.05	−80.00	−147.03	−123.60	−71.78	−1.84
86	−36.08	−80.00	−146.01	−125.48	−71.78	−0.89
90	−44.18	−80.00	−141.96	−133.03	−71.78	2.88

Transport parameters

$E_{O_O^{\bullet}}$ (kJ mol^{-1})	$E_{OH_O^{\bullet}}$	$E_{V_O^{\bullet\bullet}}$	$D^{\circ}_{O_O^{\bullet}}$ (m^2 s^{-1})	$D^{\circ}_{OH_O^{\bullet}}$	$D^{\circ}_{V_O^{\bullet\bullet}}$
65	35.09	67.00	1.62×10^{-04}	9.38×10^{-09}	1.30×10^{-08}
70	35.09	67.00	2.83×10^{-04}	9.38×10^{-09}	1.30×10^{-08}
75	35.09	67.00	4.96×10^{-04}	9.38×10^{-09}	1.30×10^{-08}
80	35.09	67.00	8.68×10^{-04}	9.38×10^{-09}	1.30×10^{-08}
85	35.09	67.00	1.52×10^{-03}	9.38×10^{-09}	1.30×10^{-08}
86	35.09	67.00	1.70×10^{-03}	9.38×10^{-09}	1.30×10^{-08}
90	35.09	67.00	2.66×10^{-03}	9.38×10^{-09}	1.30×10^{-08}

parameters are fixed as $\Delta H^\circ_{H_2O} = 80$ kJ mol^{-1} and $E_{V_O^{\bullet\bullet}} = 67$ kJ mol^{-1}. Table 5 compares the resulting thermodynamic and transport parameters.

Table 5 shows that some parameters depend only weakly on $E_{O_O^\bullet}$, but others depend strongly. The transport parameters ($D^\circ_{OH_O^\bullet}$, $D^\circ_{V_O^{\bullet\bullet}}$, and $E_{OH_O^\bullet}$), as well as $\Delta S^\circ_{H_2O}$ have only weak dependence. Two parameters show strong dependence, which provides insight for estimating the value of the polaron transport activation energy, $E_{O_O^\bullet}$. In the range $65 \leq E_{O_O^\bullet} \leq 90$ kJ mol^{-1}, the oxygen-incorporation enthalpy $\Delta H^\circ_{O_2}$ decreases from nearly zero to approximately −44 kJ mol^{-1}. In the same range, the entropy associated with the hydrogen incorporation $\Delta S^\circ_{H_2}$ increases from approximately −20 J mol^{-1} K^{-1} to being slightly positive. On physical grounds, the oxygen incorporation is expected to be exothermic[19,20] and the H_2 incorporation entropy should remain negative. These considerations bound the acceptable range of $E_{O_O^\bullet}$. Although certainly not entirely definitive, the present study uses $E_{O_O^\bullet} = 86$ kJ mol^{-1}. This choice makes the oxygen incorporation as exothermic as possible, but does not permit the hydrogen-incorporation entropy to become positive. The other thermodynamic and transport parameters that result from the choice of $E_{O_O^\bullet}$ are listed in the corresponding rows of Table 5. These parameters are at least consistent with prior literature, satisfy physical constraints, and provide a good representation of all the measurements reported herein.

6 Recommended thermodynamic and transport parameters

Table 6 lists the recommended reaction enthalpies and entropies based on constrained fits of the equilibrium conductivity measurements. Table 6 also lists the corresponding defect diffusion coefficients. These parameters respect all the constraints as discussed previously, including those suggested by the relaxation

Table 6 Recommended thermodynamic and transport parameters

Defect reaction thermodynamics

Reactions	ΔH° (kJ mol^{-1})	ΔS° (J mol^{-1} K^{-1})	K_p (750 °C)
$\frac{1}{2}H_2 + O_O^\bullet \rightleftharpoons OH_O^\bullet$	−146.01	−0.89	2.56 × 10^{+07}
$\frac{1}{2}O_2 + O_O^\times + V_O^{\bullet\bullet} \rightleftharpoons 2O_O^\bullet$	−36.08	−125.48	1.94 × 10^{-05}
$H_2O + V_O^{\bullet\bullet} + O_O^\times \rightleftharpoons 2OH_O^\bullet$	−80.00	−71.78	2.16 × 10^{+00}
$H_2 + \frac{1}{2}O_2 \rightleftharpoons H_2O$	−248.11	−55.48	5.87 × 10^{+09}

Charged-defect diffusion coefficients, $D_k = D_k^\circ \exp(-E_k/RT)$

	D_k° (m^2 s^{-1})	E_k (kJ mol^{-1})	D_k (600 °C) (m^2 s^{-1})	D_k (800 °C) (m^2 s^{-1})
$OH_O^\bullet$	9.39 × 10^{-9}	35.09	7.46 × 10^{-11}	1.84 × 10^{-10}
$V_O^{\bullet\bullet}$	1.30 × 10^{-8}	67.00	1.28 × 10^{-12}	7.12 × 10^{-12}
$O_O^\bullet$	1.70 × 10^{-3}	86.00	1.22 × 10^{-08}	1.11 × 10^{-07}

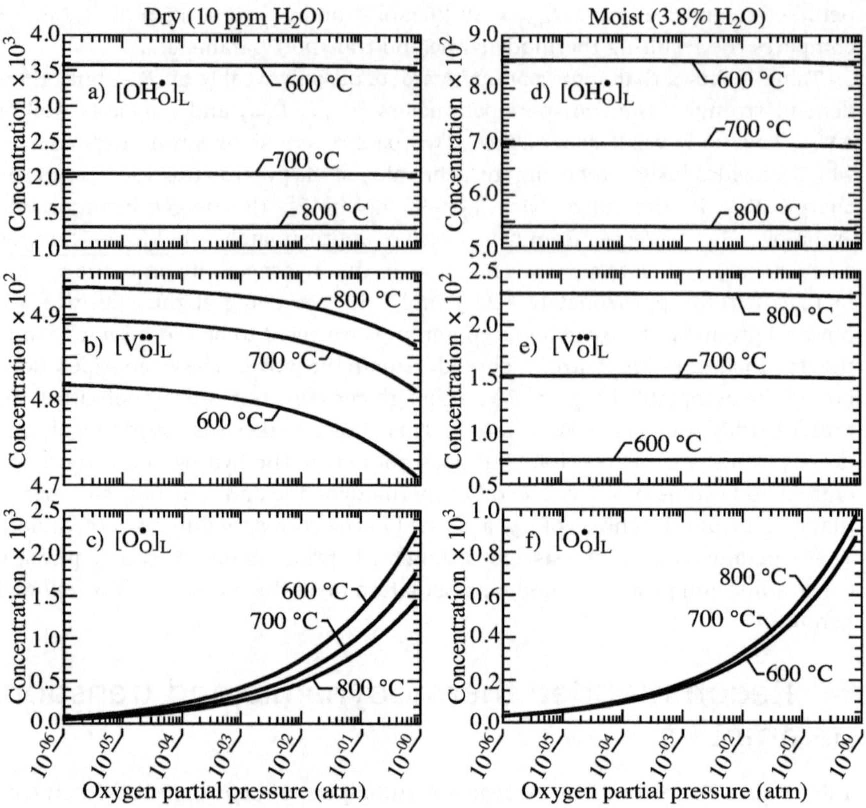

Fig. 7 Equilibrium defect concentrations (formula unit) as functions of oxygen partial pressure and temperature under dry and moist conditions.

measurements. Note, for example, that the $O_O^{\bullet}$ diffusion coefficient at 800 °C is $D_{O_O^{\bullet}} = 1.11 \times 10^{-7}$ m^2 s^{-1}, which satisfies the constraint suggested from the ambipolar-diffusion analysis. As noted earlier, Fig. 4 compares the results of the constrained fits (solid lines) with the equilibrium conductivity measurements. Although the conductivity measurements are represented well over wide ranges of conditions and certain constraints are respected, the parameter set cannot be claimed to be unique. In fact, where possible, further experiments or theories are required to unambiguously establish constraints.

6.1 Equilibrium defect concentrations

Using the recommended parameters, Fig. 7 shows equilibrium formula-unit concentrations of the three defects under dry and moist gas-phase environments. It is evident that $[V_O^{\bullet\bullet}]_L$ and $[OH_O^{\bullet}]_L$ depend only weakly on p_{O_2}, but depend strongly on temperature. By contrast, $[O_O^{\bullet}]_L$ depends strongly on p_{O_2} in the oxidizing region, but nearly vanishes in the reducing environment.

6.2 Dry oxygen conductivity

Although Fig. 4 reveals temperature dependencies over wide ranges of gas-phase environments, it can be difficult to read precise conditions from the typical log–

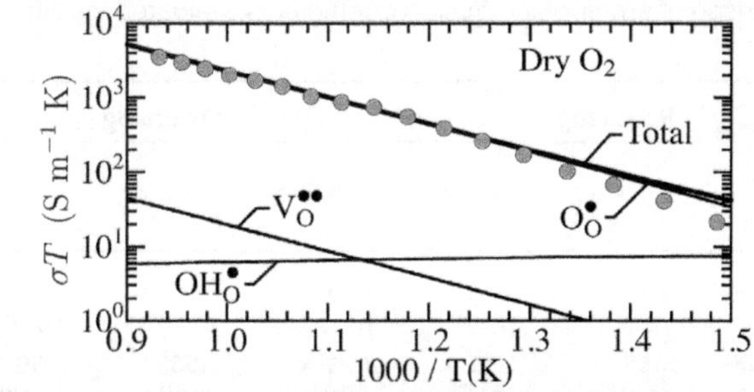

Fig. 8 Partial and total defect conductivities as functions of temperature in dry O_2 environment (10 ppm of H_2O). Filled circles are measurements and lines are model predictions.

log graphical representation. Fig. 8 compares the measured and modeled conductivity as a function of reciprocal temperature in the dry O_2 environment (10 ppm of H_2O). The figure also shows model-predicted partial conductivities of the mobile-defect charge carriers as well as the total conductivity as functions of temperature. In this highly oxidizing environment, the total conductivity is dominated by $O_O^{\bullet}$. However, even though the H_2O concentration is very low, there is still some proton $OH_O^{\bullet}$ contribution at low temperature. The model-predicted $OH_O^{\bullet}$ partial conductivity is greater than the $V_O^{\bullet\bullet}$ partial conductivity at low temperatures. However, above 600 °C, $\sigma_{V_O^{\bullet\bullet}}$ is larger than $\sigma_{OH_O^{\bullet}}$.

7 Surface kinetics and relaxation

As can be seen from Table 4, the predicted relaxation processes based on the equilibrium defect reactions at the rod surface are generally faster than the measured relaxation times. This strongly suggests that one or more surface reactions are kinetically rate-limiting. Using the thermodynamic and transport parameters shown in Table 6, the forward rate coefficients for the defect reactions (k_{f,O_2}, k_{f,H_2}, and k_{f,H_2O}) are adjusted such that the model matches the measured relaxation histories. The three reverse rate expressions (k_{b,O_2}, k_{b,H_2}, and k_{b,H_2O}) are evaluated based on eqn (20)–(22) to maintain microscopic reversibility. Thus, the constrained-fit thermodynamic parameters are directly involved in the rate equations. Table 7 lists the reaction rate coefficients based on fitting the relaxation experiments.

Table 7 Modified Arrhenius rate parameters for the defect-incorporation reactions, $k = AT^{\beta}\exp(-E_a/RT)$

	A (kmol, m, K)	β	E_a (kJ mol^{-1})
k_{f,O_2}	2.7087×10^{-05}	0.0	62.70
k_{f,H_2}	7.3995×10^{-02}	0.0	29.21
k_{f,H_2O}	1.8000×10^{-04}	0.0	0.00
k_{b,O_2}	$3.3922 \times 10^{+02}$	−0.5	98.77
k_{b,H_2}	2.8759×10^{-01}	−0.5	175.22
k_{b,H_2O}	$1.2327 \times 10^{+01}$	−1.0	80.00

Table 8 Comparison of ambipolar diffusion coefficients evaluated for different gas-phase environments

		Reducing			Oxidizing		
	T (°C)	$\mathscr{D}_{O_O^\bullet-OH_O^\bullet}$ ($m^2\ s^{-1}$)	$\mathscr{D}_{O_O^\bullet-V_O^{\bullet\bullet}}$	$\mathscr{D}_{V_O^{\bullet\bullet}-OH_O^\bullet}$	$\mathscr{D}_{O_O^\bullet-OH_O^\bullet}$ ($m^2\ s^{-1}$)	$\mathscr{D}_{O_O^\bullet-V_O^{\bullet\bullet}}$	$\mathscr{D}_{V_O^{\bullet\bullet}-OH_O^\bullet}$
DryOx	600	4.32×10^{-10}	3.32×10^{-10}	3.64×10^{-11}	1.90×10^{-10}	1.07×10^{-10}	3.62×10^{-11}
	700	5.63×10^{-10}	1.11×10^{-09}	8.91×10^{-11}	2.63×10^{-10}	3.59×10^{-10}	8.90×10^{-11}
	800	6.99×10^{-10}	2.98×10^{-09}	1.58×10^{-10}	3.47×10^{-10}	9.60×10^{-10}	1.58×10^{-10}
MoistOx	600	7.99×10^{-09}	1.35×10^{-10}	1.70×10^{-12}	4.60×10^{-09}	4.37×10^{-11}	1.70×10^{-12}
	700	1.66×10^{-08}	6.40×10^{-10}	6.02×10^{-12}	7.32×10^{-09}	2.06×10^{-10}	6.00×10^{-12}
	800	2.51×10^{-08}	2.08×10^{-09}	1.80×10^{-11}	9.49×10^{-09}	6.69×10^{-10}	1.80×10^{-11}
MoistRedox	600	1.22×10^{-08}	1.22×10^{-08}	1.70×10^{-12}	4.60×10^{-09}	4.37×10^{-11}	1.70×10^{-12}
	700	4.12×10^{-08}	4.11×10^{-08}	6.02×10^{-12}	7.32×10^{-09}	2.06×10^{-10}	6.00×10^{-12}
	800	1.11×10^{-07}	1.10×10^{-07}	1.80×10^{-11}	9.49×10^{-09}	6.69×10^{-10}	1.80×10^{-11}

Table 4 summarizes the measured and modeled characteristic relaxation times. Table 3 lists the relaxation end-point gas-phase mole-fraction compositions for the conductivity relaxations. Table 8 shows the corresponding three-pair ambipolar diffusivities, which assists in interpreting the relaxation behavior.

The following sections include figures of the measured and modeled relaxation histories. Two factors are important in adjusting model parameters to achieve fits to the relaxation profiles. First is the characteristic relaxation time. Second is the qualitative functional form of the profiles. In some cases, although the characteristic time and shapes match well, it is evident from the plots that the end-point conductivities do not always fully align. Beginning with the equilibrium conductivity measurements, the full set of experiments took over two months to complete. During this time the sample rod was repeatedly oxidized and reduced, and heated and cooled. Inevitably, this repeated cycling caused some material degradation. Thus, there are small variations between the equilibrium end states for the relaxations and the initial equilibrium conductivity measurements for the same gas-phase environments (*i.e.*, Fig. 4). Moreover, as is conventionally done, Fig. 4 shows conductivities on a logarithmic scale, which tends to diminish any scatter in the data.

7.1 Dry O_2

Fig. 9a compares modeled and measured conductivity-relaxation histories from 1% O_2 to 100% O_2 under the dry environment at three temperatures and pressure of 0.829 atm (Table 3, DryOx). Fig. 9b shows the corresponding reverse conductivity relaxation histories. It is apparent that the measured relaxation histories can

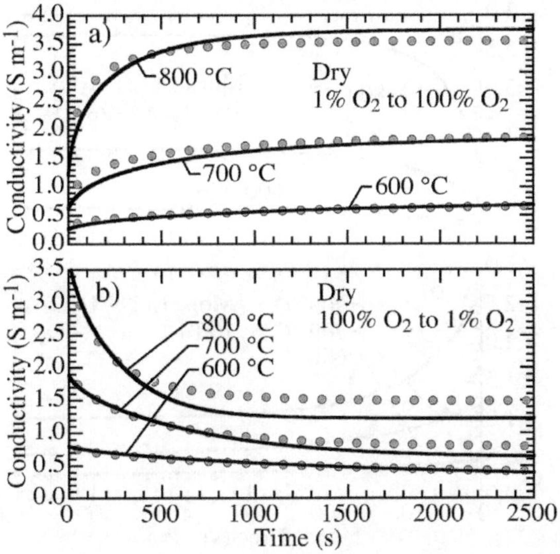

Fig. 9 Comparison of modeled (lines) and measured (points) conductivity relaxation histories under dry (10 ppm H_2O) conditions at three temperatures and $p = 0.829$ atm. (a) Relaxation from 1% to 100% O_2. (b) Relaxation from 100% to 1% O_2.

be reasonably well predicted with the kinetic-limited relaxation model. The magnitudes of the effective ambipolar diffusion coefficients can contribute additional insight.

The ambipolar diffusion coefficient $\mathscr{D}_{O_O^{\bullet}-V_O^{\bullet\bullet}}$ is larger than $\mathscr{D}_{O_O^{\bullet}-OH_O^{\bullet}}$ at the 700 and 800 °C, but $\mathscr{D}_{O_O^{\bullet}-V_O^{\bullet\bullet}}$ becomes smaller than $\mathscr{D}_{O_O^{\bullet}-OH_O^{\bullet}}$ at 600 °C, even though there is only about 10 ppm of H_2O (*cf.*, Table 8). Therefore, the relaxation time under the equilibrated-surface assumption is determined by $\mathscr{D}_{O_O^{\bullet}-V_O^{\bullet\bullet}}$ at 700 and 800 °C, but by $\mathscr{D}_{O_O^{\bullet}-OH_O^{\bullet}}$ at 600 °C. Because $\mathscr{D}_{O_O^{\bullet}-V_O^{\bullet\bullet}}$ and $\mathscr{D}_{O_O^{\bullet}-OH_O^{\bullet}}$ at 1% O_2 are larger than they are at 100% O_2, it is reasonable that the relaxation time with equilibrium surface reactions from 100% O_2 to 1% O_2 is shorter than it is at 1% O_2 to 100% O_2 (*cf.*, Table 4).

Table 4 and also Fig. 9 show that the relaxation times from 1% O_2 to 100% O_2 are significantly shorter than they are from 100% O_2 to 1% O_2 at the same temperatures. The relatively fast relaxation is the result of relatively faster defect transport. However, the actual relaxation time is limited by the kinetics associated with rate-limiting surface kinetics. In the dry O_2 environment, the dominant rate-limiting step is the O_2 incorporation. The relaxation time predicted by the kinetic model is consistent with the experimental observations.

7.2 Moist O_2

Fig. 10a shows that the measured conductivity-relaxation histories between 1% O_2 and 100% O_2 under a saturated moist environment (Table 3, MoistOx) can also be predicted reasonably well by the kinetic model. In the moist environment, Table 8 shows that the ambipolar diffusion coefficient $\mathscr{D}_{O_O^{\bullet}-OH_O^{\bullet}}$ is much larger than $\mathscr{D}_{O_O^{\bullet}-V_O^{\bullet\bullet}}$ at all the three temperatures. Therefore, the defect transport is associated primarily with $O_O^{\bullet}$ and $OH_O^{\bullet}$. Compared with the dry O_2 case, $\mathscr{D}_{O_O^{\bullet}-OH_O^{\bullet}}$ is also much

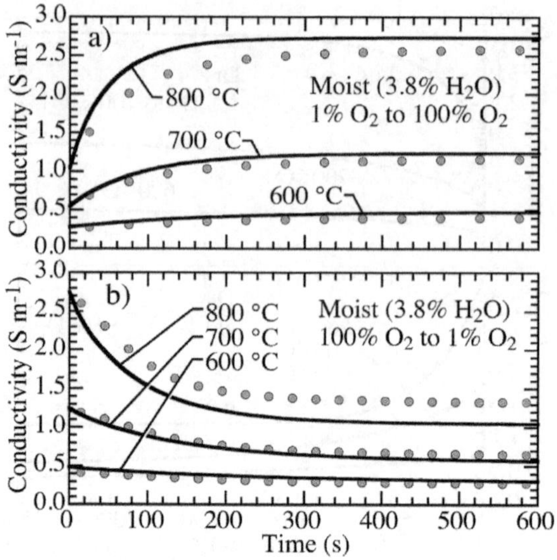

Fig. 10 Comparison of modeled (lines) and measured (points) conductivity relaxation histories under moist (3.8% H_2O) conditions at three temperatures and $p = 0.829$ atm. (a) Relaxation from 1% to 100% O_2. (b) Relaxation from 100% to 1% O_2.

larger in the moist environment. Thus, the relaxation time for the moist O_2 is much shorter than it is under dry conditions. However, the dominant rate-limiting step remains the O_2 incorporation.

7.3 Moist H_2–O_2

Fig. 11 illustrates that the measured conductivity relaxation from 100% O_2 to 5% H_2 in a moist environment (Table 3, MoistRedox) can be represented well by the NPP model considering the surface kinetics. In the highly reduced environment, the ambipolar diffusivity $\mathscr{D}_{O_O^\bullet-OH_O^\bullet}$ is almost equal to $\mathscr{D}_{O_O^\bullet-V_O^{\bullet\bullet}}$, but much larger than it is under the dry and moist oxidizing environment shown in Table 8. Therefore, the relaxation time from the oxidizing to reducing environment is very short (*cf.*, Table 4). In the reducing environment, the surface kinetics is dominated by the H_2 incorporation rate.

8 Summary and conclusions

A combination of equilibrium-conductivity and conductivity-relaxation measurements, together with model-based interpretation, is used to establish thermodynamic and transport properties for 10% yttrium-doped barium zirconate (BZY10). Specifically, enthalpy and entropy changes (ΔH° and ΔS°) are reported for reactions representing the incorporation of H_2, O_2, and H_2O. The transport properties are reported in terms of pre-exponential factors and activation energies for the diffusion coefficients of three mobile defects (oxygen vacancies $V_O^{\bullet\bullet}$, protons $OH_O^\bullet$, and O-site polarons $O_O^\bullet$). The equilibrium conductivities are measured using alternating-current impedance spectroscopy. The relaxation measurements consider switching the gas-phase environments rapidly

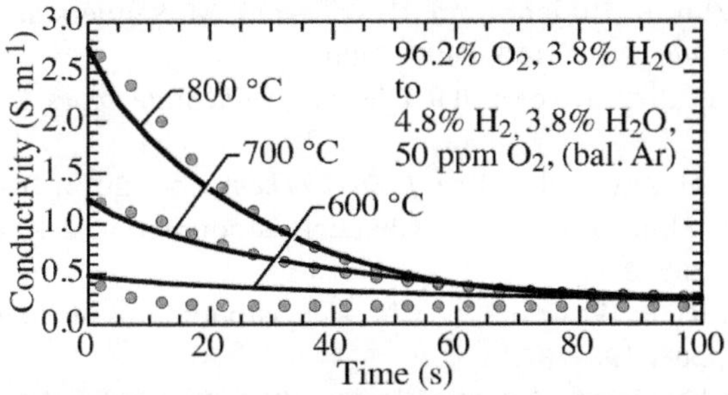

Fig. 11 Comparison of model-predicted conductivity relaxation histories (lines) and the experimental measurements (filled circles) from 100% O_2 to 5% H_2 under moist conditions at three temperatures and pressure of 0.829 atm.

from high to low oxygen partial pressures, and *vice versa*. The conductivity during relaxation is measured using impedance at a single frequency, which was selected based on measuring the same resistance as the full EIS sweeps at the asymptotic equilibrium end states. To model the relaxation experiments and considering three mobile charge carriers, the relaxation problem is solved computationally using a Nernst–Planck–Poisson formulation. The relaxation modeling incorporates the possibility of kinetic rate limitations of the surface reactions, but requires microscopic reversibility that is consistent with the equilibrium thermodynamics. Although not claimed to be entirely unique, the properties reported here are consistent with prior literature, the equilibrium and relaxation experiments, and physical constraints. Thus, complete and self-consistent set of properties is probably the best that is currently available for BZY10.

Acknowledgements

This research was supported by the Office of Naval Research *via* grant N00014-08-1-0539 and the Colorado School of Mines Foundation, Protonic Capital Funds. We gratefully acknowledge insightful and helpful discussions with Profs. Ryan O'Hayre, Greg Jackson, and Neal Sullivan (CSM). We are also grateful for the efforts of Dr. Anthony Manerbino (CoorsTek), who prepared the BZY10 rods used in the experiments.

References

1 W. Coors, *J. Power Sources*, 2003, **118**, 150–156.
2 M. Ni, D. Leung and M. Leung, *J. Power Sources*, 2008, **183**, 133–142.
3 C. Zamfirescu and I. Dincer, *Thermochim. Acta*, 2009, **486**, 32–40.
4 A. Arpornwichanop and Y. Patcharavorachot, *Chem. Eng. Res. Des.*, 2013, **91**, 1508–1516.
5 W. Li, R. Borry and E. Iglesia, *Chem. Eng. Sci.*, 2002, **57**, 4595–4604.
6 W. Coors, *J. Electrochem. Soc.*, 2004, **151**, A994–A997.
7 W. Coors, *Solid State Ionics*, 2007, **178**, 481–485.
8 I. Riess, *Solid State Ionics*, 2003, **157**, 1–17.
9 W. Lai and S. Haile, *Phys. Chem. Chem. Phys.*, 2008, **10**, 865–883.

10 R. Kee, H. Zhu, B. Hildenbrand, E. Vøllestad, M. Sanders and R. O'Hayre, *J. Electrochem. Soc.*, 2013, **160**, F290–F300.

11 M. Sanders, H. Zhu, R. Kee and R. O'Hayre, *Solid State Ionics*, 2013, **249–250**, 6–16.

12 E. Vøllestad, H. Zhu and R. Kee, *J. Electrochem. Soc.*, 2014, **161**, F114–F124.

13 S. Ricote, H. Zhu, W. Coors, C. Chatzichristodoulou and R. Kee, *Solid State Ionics*, 2014, **265**, 22–28.

14 H. Zhu, S. Ricote, W. Coors, C. Chatzichristodoulou and R. Kee, *Solid State Ionics*, 2014, **268**, 198–207.

15 S. Stokes and M. Islam, *J. Mater. Chem.*, 2010, **20**, 6258–6264.

16 K. Kreuer, S. Adams, W. Munch, A. Fuchs, U. Klock and J. Maier, *Solid State Ionics*, 2001, **145**, 295–306.

17 K. Kreuer, *Annu. Rev. Mater. Res.*, 2003, **33**, 333–359.

18 C. Kjolseth, L. Wang, R. Haugsrud and T. Norby, *Solid State Ionics*, 2010, **181**, 1740–1745.

19 E. Bévillon, G. Dezanneau and G. Geneste, *Phys. Rev. B: Condens. Matter Mater. Phys.*, 2011, **83**, 174101.

20 J. Hermet, F. Bottin, G. Dezanneau and G. Geneste, *Phys. Rev. B: Condens. Matter Mater. Phys.*, 2012, **85**, 205137.

Faraday Discussions

ROYAL SOCIETY OF CHEMISTRY

PAPER

Kinetics of CO/CO_2 and H_2/H_2O reactions at Ni-based and ceria-based solid-oxide-cell electrodes

Christopher Graves,* Christodoulos Chatzichristodoulou and Mogens B. Mogensen

Received 1st May 2015, Accepted 6th May 2015

DOI: 10.1039/c5fd00048c

The solid oxide electrochemical cell (SOC) is an energy conversion technology that can be operated reversibly, to efficiently convert chemical fuels to electricity (fuel cell mode) as well as to store electricity as chemical fuels (electrolysis mode). The SOC fuel-electrode carries out the electrochemical reactions $CO_2 + 2e^- \leftrightarrow CO + O^{2-}$ and $H_2O + 2e^- \leftrightarrow H_2 + O^{2-}$, for which the electrocatalytic activities of different electrodes differ considerably. The relative activities in CO/CO_2 and H_2/H_2O and the nature of the differences are not well studied, even for the most common fuel-electrode material, a composite of nickel and yttria/scandia stabilized zirconia (Ni–SZ). Ni–SZ is known to be more active for H_2/H_2O than for CO/CO_2 reactions, but the reported relative activity varies widely. Here we compare AC impedance and DC current–overpotential data measured in the two gas environments for several different electrodes comprised of Ni–SZ, Gd-doped CeO_2 (CGO), and CGO nanoparticles coating Nb-doped $SrTiO_3$ backbones (CGOn/STN). 2D model and 3D porous electrode geometries are employed to investigate the influence of microstructure, gas diffusion and impurities.
Comparing model and porous Ni–SZ electrodes, the ratio of electrode polarization resistance in CO/CO_2 *vs.* H_2/H_2O decreases from 33 to 2. Experiments and modelling suggest that the ratio decreases due to a lower concentration of impurities blocking the three phase boundary and due to the nature of the reaction zone extension into the porous electrode thickness. Besides showing higher activity for H_2/H_2O reactions than CO/CO_2 reactions, the Ni/SZ interface is more active for oxidation than reduction. On the other hand, we find the opposite behaviour in both cases for CGOn/STN model electrodes, reporting for the first time a higher electrocatalytic activity of CGO nanoparticles for CO/CO_2 than for H_2/H_2O reactions in the absence of gas diffusion limitations. We propose that enhanced surface reduction at the CGOn/gas two phase boundary in CO/CO_2 and in cathodic polarization can explain why the highest reaction rate is obtained for CO_2 electrolysis. Large differences observed between model electrode kinetics and porous electrode kinetics are discussed.

Department of Energy Conversion and Storage, Technical University of Denmark, Risø Campus, Frederiksborgvej 399, DK-4000 Roskilde, Denmark. E-mail: cgra@dtu.dk

Introduction

The growing need for efficient and affordable production and storage of electricity currently drives the development of electrochemical energy conversion technologies. The solid oxide electrochemical cell (SOC) is a promising candidate that could play a key role in the transition to sustainable energy because SOCs can both store renewable electricity as fuels (electrolysis mode) as well as convert fuels to electricity (fuel cell mode), with high efficiency and high reaction rates in both modes.[1–4] Furthermore, SOCs can operate with carbonaceous feedstocks (hydrocarbons, CO, CO_2, *etc.*)[5–8]

A good physical and chemical understanding of the electrochemical reactions (ERs) that occur in the fuel-electrode, $CO_2 + 2e^- \leftrightarrow CO + O^{2-}$ and $H_2O + 2e^- \leftrightarrow H_2 + O^{2-}$, is crucial for addressing performance and lifetime limiting factors as well as for designing new improved electrodes. The reactions take place either at the 3-phase boundary (3PB) between the gas, O^{2-} conductor and e^- conductor, or at the 2-phase boundary (2PB) between the gas and the surface of a mixed ionic and electronic conducting (MIEC) material, depending on the electrode materials. The conventional fuel-electrode is a 3PB-based, three-dimensional porous composite of Ni and yttria or scandia-stabilized zirconia (Ni–SZ, referring to either one) that exhibits high electrochemical activity but is often a major source of performance loss during long-term operation in both fuel cell and electrolysis modes due to 3PB poisoning, Ni mobility, carbon deposition, and other degradation mechanisms.[9–14]

Alternative electrodes comprised of MIEC materials show great promise with respect to avoiding these issues.[15,16] Acceptor-doped ceria is a well-known MIEC material when exposed to a reducing environment and it exhibits high activity for the H_2/H_2O reaction when applied as an electrode.[17–20] Considerable effort has gone into understanding and modelling the reaction kinetics and mechanisms for the H_2/H_2O and CO/CO_2 reactions, especially for Ni–SZ electrodes,[21–25] but also more recently for ceria-based electrodes.[16,26–28]

However, relative activities for the CO/CO_2 and H_2/H_2O reactions are not well studied, even for conventional Ni–SZ electrodes. Ni–SZ is known to be more active for H_2/H_2O than for CO/CO_2 reactions, but the relative activity varies widely,[23] apparently depending on the electrode microstructure, geometry, and purity of materials and gas supplies. Here we compare AC impedance and DC current–overpotential data measured in the two gas environments for several different electrodes comprised of Ni/SZ, Gd-doped CeO_2 (CGO), and CGO nanoparticles coating Nb-doped $SrTiO_3$ backbones (CGOn/STN). The aim is to better understand the nature of the electrocatalytic activities observed for different electrode materials and geometries, which vary considerably for these two reactions.

Model electrodes are advantageous in that they provide a more well-defined, usually 2-dimensional, interface and corresponding ER zone compared with complex 3-dimensional porous electrode structures. However, the small interface in a large system is more susceptible to contamination by segregation of material from the electrode itself or from other solid or gaseous sources, which can make it a poor representation of the actual interface that the model electrode aims to investigate in a realistic technological electrode. It is therefore worthwhile studying both model and porous electrodes together to obtain the most useful information, which we attempt here in order to investigate the influence of microstructure, gas diffusion and impurities present at reaction sites.

Experimental

Several types of electrochemical cells were fabricated and subjected to electrochemical characterization, as illustrated in Fig. 1. Regarding notation, we adopt Ni–SZ to indicate a composite porous electrode and Ni/SZ to indicate an interface of Ni and SZ (as in a model electrode).

Cells with a Ni metal wire electrode contacting the polished surface of a $Y_{0.16}Zr_{0.84}O_2$ (YSZ) electrolyte pellet with a porous Pt counter electrode (prepared by sintering Pt paste) were assembled and tested in a single-atmosphere custom test rig, as described in ref. 29 and 30. A pseudo-three-electrode set-up was used, with the relatively large counter electrode expected to contribute negligible resistance and therefore serving as reference electrode. These types of model

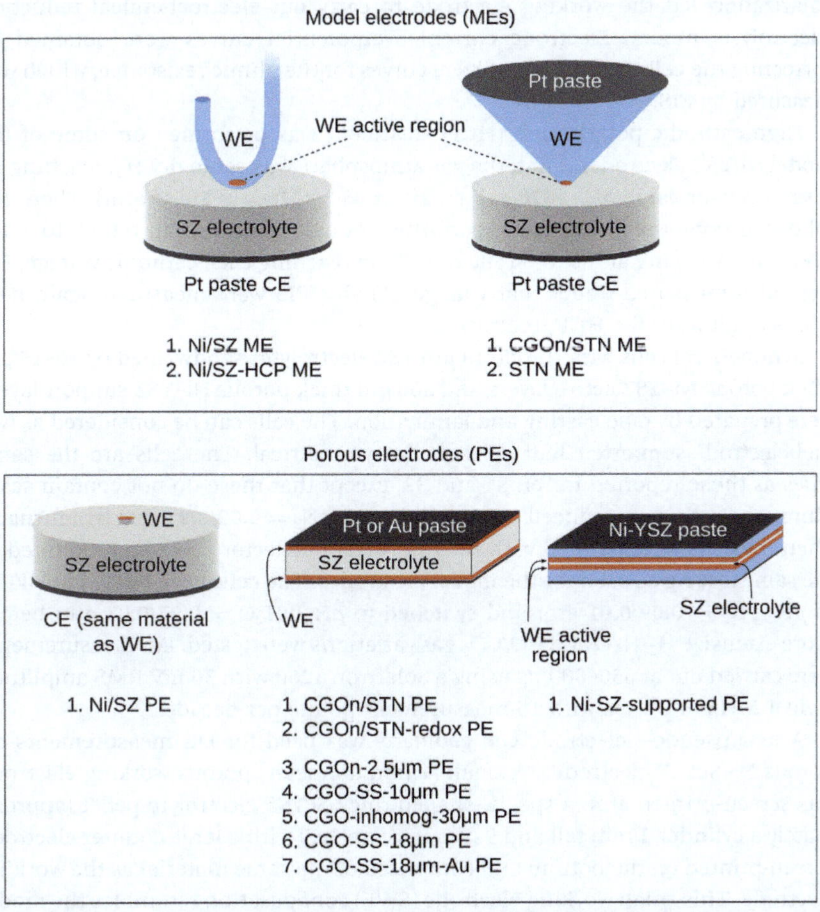

Fig. 1 Illustration of the different electrode geometries employed (WE = working electrode; CE = counter electrode) and list of electrode materials characterized (abbreviations are defined in the "Results and discussions" section).

electrodes have a well-defined 3PB which can be examined by microscopy before and after testing. The wires were 0.5 mm in diameter and 99.999% purity. The wires were cut and bent approximately 1 cm from the end, to form a "J" shape, with the bottom of the "J" contacting the YSZ surface. The wires were annealed in a reducing atmosphere ($pH_2/pN_2 = 0.09/0.91$ atm). Each wire electrode was inserted into an alumina tube which held it in place and applied pressure downwards from a weight of 200 g attached above. The test rig was heated to an initial temperature of 1000 °C in $pH_2/pN_2 = 0.09/0.91$ atm which was then switched to $pH_2/pH_2O = 0.97/0.03$ atm, and electrochemical impedance spectra (EIS) were measured near open-circuit continuously for at least 10 hours until the interface stabilized. More extensive H_2/H_2O and CO/CO_2 gas variations were then tested. EIS were measured near open-circuit, from ~1 MHz to 0.1 Hz with 12 measurement points per decade, with an AC RMS amplitude of 30 mV, using a Gamry Reference 600 potentiostat, which was also used for DC polarization measurements at 1–10 mV s^{-1} between −300 mV and +300 mV overvoltage. The cell was wired such that anodic polarization corresponded to electrochemical oxidation of H_2 or CO (fuel-cell mode) at the working electrode, whereas cathodic polarization led the working electrode to carry out electrochemical reduction (electrolysis mode). Electrode current–overpotential curves were obtained by correcting the cell current–overvoltage curves for the ohmic resistance, which was measured by EIS.

High cathodic polarization (HCP) treatment was performed on some of the model Ni/YSZ electrodes. First, the gas atmosphere was set to dry H_2, reaching an oxygen sensor value of −1370 mV *vs.* air at 850 °C ($pO_2 < 10^{-24}$ atm). Then, the following potential sweeps were performed at a 20 mV s^{-1} scan rate: 0 to −500 mV, 0 to −1000 mV and 0 to −1500 mV. Upon reaching each cathodic voltage, EIS were also measured under that voltage. Finally, EIS were measured again near open-circuit after the HCP treatment.

Symmetrical cells with thin 5–10 μm YSZ electrolytes sandwiched by 10–15 μm thick porous Ni–YSZ active layers and 300 μm thick porous Ni–YSZ support layers were prepared by tape-casting and lamination. The cells can be considered as two fuel-electrode-supported half cells made symmetrical. The cells are the same types as those reported in ref. 31 and 32, except that these do not contain scandium. The cells were reduced at 1000 °C in $pH_2/pN_2 = 0.09/0.91$ atm in a furnace. Then the cells were painted with Ni–YSZ current collector paste and mounted in the same test rig as used for the model electrodes. The cells were heated to 800 °C in $pH_2/pN_2 = 0.09/0.91$ atm and switched to $pH_2/pH_2O = 0.97/0.03$ atm before more extensive H_2/H_2O and CO/CO_2 gas variations were tested. EIS measurements were carried out at 650–800 °C using a Solartron 1260 with 30 mV RMS amplitude from 1 MHz to 0.1 Hz with 16 measurement points per decade.

A new pseudo-3-electrode cell geometry was used for DC measurements on porous Ni–ScYSZ electrodes. A small (1 mm diameter) porous working electrode was screen-printed atop a special-shaped thick ScYSZ electrolyte pellet (approximately a cylinder 3 mm tall and 9 mm in diameter) with a large counter electrode screen-printed on the bottom and comprised of the same material as the working electrode. This small working electrode (SWE) configuration resembles the model wire electrode configuration, but with a 3D porous working electrode. Like the model wire electrode, this geometry minimizes ohmic heating and gas diffusion resistance. Details of the cell design will be reported in a future publication. The electrode was characterized with a similar procedure as that used for the model wire electrodes, using a Gamry Reference 600 potentiostat.

Symmetrical cells were prepared with electrodes comprised of porous $Sr_{0.94}Ti_{0.9}Nb_{0.1}O_3$ (STN) backbones coated with $Ce_{0.8}Gd_{0.2}O_{1.9}$ by surfactant-assisted nitrate solution infiltration, on both sides of a YSZ electrolyte, as described in ref. 33. Pt paste was applied to both electrodes for current collection. The cells were mounted in between Pt meshes with weight applied, in the same test rig as above. The cells were heated to an initial temperature of 800 °C in $pH_2/pN_2 = 0.09/0.91$ atm and switched to $pH_2/pH_2O = 0.97/0.03$ atm before more extensive H_2/H_2O and CO/CO_2 gas variations were tested. EIS measurements were performed using a Solartron 1260 with similar settings as for the Ni–YSZ symmetrical cells. During redox cycles, N_2 was supplied for 30 minutes before supplying air, and N_2 was again supplied before supplying the reducing atmosphere again.

Cone-shaped electrodes comprised of STN were prepared by machining dense cylindrical STN pellets that had been sintered at 1400 °C in $pH_2/pN_2 = 0.09/0.91$ atm. The cone point (working electrode) was contacted to a smooth polished YSZ electrolyte pellet surface, with a much larger porous Pt counter electrode acting as a reference electrode (pseudo 3-electrode setup), as described previously for other oxide electrode materials.[34] The cell geometry was similar in principle to the model wire electrode cell described above. For some electrodes, a drop of the same $Ce_{0.8}Gd_{0.2}O_{1.9}$ solution that was used to infiltrate the porous STN backbones was placed at the tip of the cone before contacting the cone to the YSZ surface. The cells were heated to 850 °C in $pH_2/pN_2 = 0.09/0.91$ atm and switched to $pH_2/pH_2O = 0.97/0.03$ atm, and EIS were measured continuously for at least 10 hours until the interface stabilized. The electrodes were characterized with a similar procedure as that used for the model wire electrodes using a Gamry Reference 600 potentiostat.

Symmetrical cells with porous $Ce_{0.9}Gd_{0.1}O_{1.95}$ (CGO) electrodes were prepared by slurry spraying (SS) or screen-printing onto a sintered 200 μm thick YSZ electrolyte and sintering at 1200 °C, similar to in ref. 35 which used CGO electrolyte. For the screen-printed electrodes, due to poor dispersion in the ink, the CGO particles agglomerated, resulting in an inhomogeneous microstructure after sintering at 1200 °C. The electrodes were characterized with a similar procedure as that used for the STN-backbone symmetrical cells described above. For one SS cell, Au paste was used instead of Pt paste.

Symmetrical cells with nano-porous CGO electrodes were prepared by spin-coating a nitrate based precursor solution. The electrodes were only sintered *in situ* at 800 °C, which resulted in the nano-porous structure. More detail about the structure and electrochemistry of these highly active electrodes will be reported in a future publication. The electrodes were characterized with a similar procedure as that used for the STN-backbone symmetrical cells described above.

All EIS reported here were measured near open-circuit. The gas compositions of $pH_2/pH_2O = 0.5/0.5$ atm and $pCO/pCO_2 = 0.5/0.5$ atm (hereafter simply referred to as H_2/H_2O and CO/CO_2) were chosen as the basis for comparison for several reasons:

- In current–overpotential curves, they provide an equal concentration of reactants for both anodic and cathodic polarization.

• Gas diffusion impedance is minimized.[36] For porous electrodes, several important gas diffusion impedance contributions have been identified for electrodes measured in this test set-up: diffusion in a stagnant gas layer above the porous electrode that arises solely from the test set-up,[36,32] diffusion through thick porous layers,[32] and Knudsen diffusion through nanoparticle-loaded structures with fine porosity.[37] The stagnant gas layer diffusion resistance has been found to contribute 5 and 30 mΩ cm^2 in $pH_2/pH_2O = 0.5/0.5$ atm and $pCO/pCO_2 = 0.5/0.5$ atm, respectively, in the present study and in previous reports.[32]

• The 50–50 gas compositions give a very similar pO_2 in the measurement temperature range 750–850 °C.

Measurements were carried out in H_2/H_2O and CO/CO_2 as close in time as possible while also allowing enough time to stabilize, to avoid interference from relaxation and degradation, especially for the 3PB-based Ni/SZ model electrodes.[38] The ranges of the current–overvoltage curves were kept between +300 mV and −300 mV, since Ni oxidizes to NiO near +300 mV at the temperature and gas composition used – going to too high an anodic overvoltage led to hysteresis in the current–overvoltage curve due to Ni oxidation. Each symmetrical cell has an area of about 30 mm^2. At least two electrodes of each type were tested for reproducibility, except that only one cell was tested with Au paste instead of Pt paste for current collection, which will be discussed later.

All data analysis was carried out using the python-based Ravdav software.[39] EIS measurements were corrected for stray inductance using an automatic Kramers–Kronig procedure. The model electrodes were not normalized to 3PB length or contact area in this study because: in the Ni wire electrode case, after the HCP treatment the 3PB length was not well defined and due to coarsening during subsequent measurements it was not possible to take post-test measurements of the *in situ* 3PB formed by the HCP treatment; in the CGOn/STN model electrode case, the CGO nanoparticles at the interface formed *in situ* and were similarly not well enough defined for sensible normalization.

Results and discussions

Ni/stabilized-zirconia 3PB-based electrodes

To investigate the effects of electrode microstructure and 3PB morphology on the ER kinetics of the Ni/SZ/gas interface, several different Ni/SZ electrodes were employed, with either 2D or 3D 3PB geometries (illustrated in Fig. 1):

1. A Ni wire pressed down onto a polished YSZ surface, forming a 2D 3PB at the perimeter of the oval-shaped contact (hereafter referred to as the "Ni/SZ model electrode" or "Ni/SZ ME").

2. The same Ni wire contacting the YSZ surface after it had been subjected to HCP treatment, which forms a new nano-structured 3PB ("Ni/SZ-HCP model electrode").

3. A porous Ni–YSZ composite symmetrical cell with 300 μm thick porous Ni–YSZ support layers sandwiching the thin porous Ni–YSZ active layer and thin dense electrolyte ("Ni–SZ-supported porous electrode").

4. A small-sized porous Ni-ScYSZ composite working electrode in the pseudo-3-electrode "SWE" cell configuration ("Ni–SZ porous electrode").

Impedance spectra measured on each of the electrodes are shown in Fig. 2a–d. All cells except the 3rd cell (Ni–SZ-supported porous electrode; symmetrical cell) could be polarized to obtain DC current–overpotential measurements, which are shown in Fig. 3a–c.

For all of these electrodes, the electrochemical activity for the H_2/H_2O reaction was higher than for the CO/CO_2 reaction, but the relative activity near open-circuit varied from a factor of 33 (Fig. 2a and 3c) to a factor of 2 (Fig. 2c and d and 3i) depending on the electrode. These ratios are in a similar range to those previously reported for various Ni/SZ electrodes.[23] Here we address the reasons for the large differences in relative activity.

Impedance spectroscopy

The EIS measured on the model Ni/YSZ electrode exhibited a skewed semicircle that becomes smaller with decreasing frequency in the complex–complex plane (Fig. 2a), which matches previously reported measurements in H_2/H_2O on similar Ni wires contacting YSZ or ScYSZ surfaces.[29,40–42] The ratio of R_p for CO/CO_2 *vs.* H_2/H_2O (hereafter "C : H R_p ratio") is 33. After the HCP treatment, the impedance has changed dramatically (Fig. 2b), similar to the changes observed in measurements in H_2/H_2O for a model Ni/ScYSZ electrode subjected to HCP treatment at lower temperature.[42] The H_2/H_2O impedance is dominated by a single near-perfect

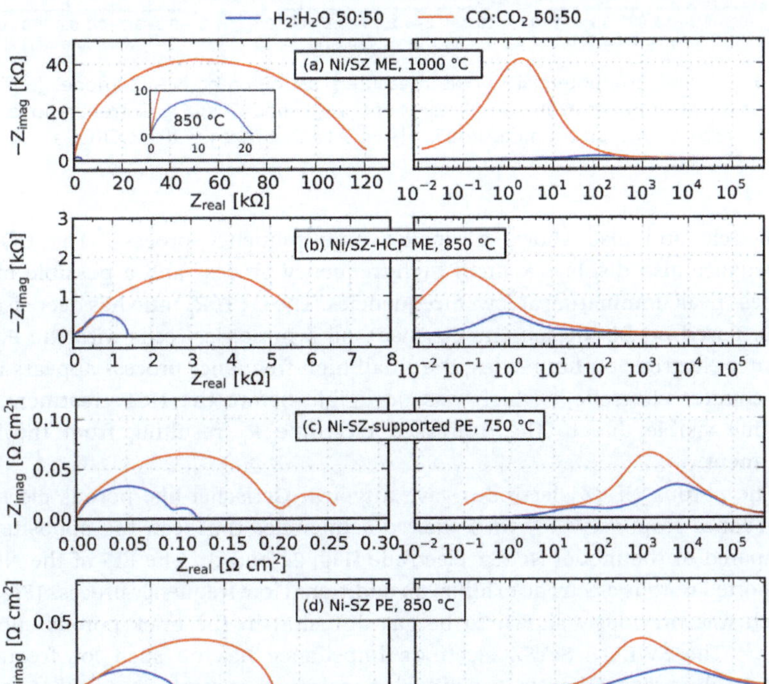

Fig. 2 Single-electrode impedance data measured on different Ni/SZ model (a and b) and porous (c and d) electrodes in 50 : 50 mixtures of H_2/H_2O and CO/CO_2. To obtain the single-electrode impedance, the ohmic resistance is subtracted for all data, and for (c) and (d) the symmetrical cell impedance is divided by 2.

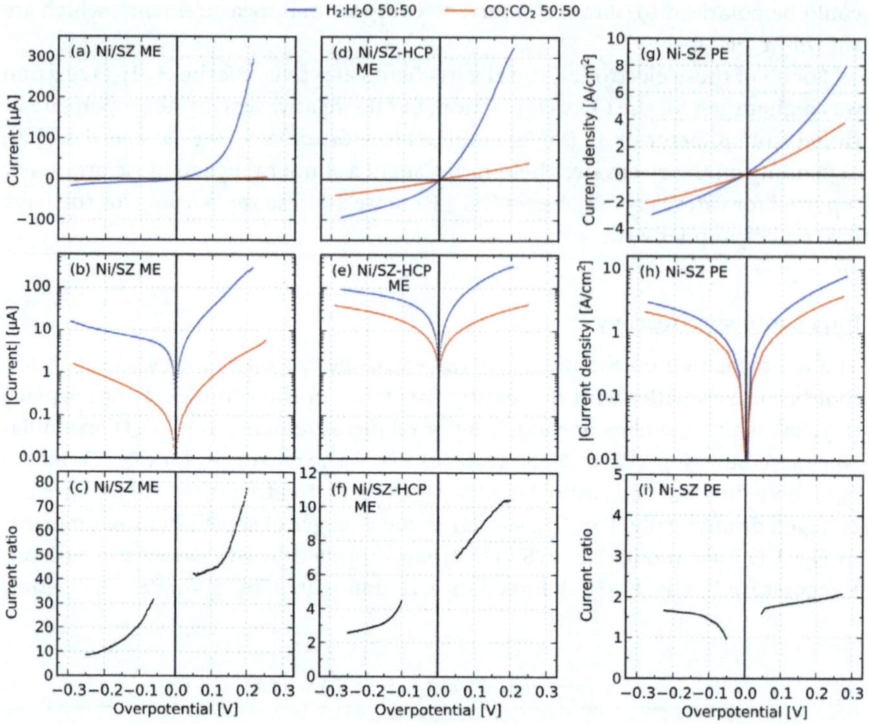

Fig. 3 Current–overpotential curves measured on different Ni–SZ model (a–f) and porous (g–i) electrodes in 50 : 50 mixtures of H_2/H_2O and CO/CO_2. Current ratio (c, f and i) is the ratio of the current measured in H_2/H_2O to that measured in CO/CO_2.

semicircle and also shows a smaller high-frequency process. The CO/CO_2 impedance also displays a small high-frequency process and a possible multi-process peak dominating at lower frequencies. The C : H R_p ratio has decreased to ~4.5. Based on EIS measurements made on symmetrical cells with the Pt/YSZ counter electrode on both sides, the small high-frequency process appears to be the counter electrode,[29] which was negligible before the HCP treatment and became visible due to the decreased electrode R_p resulting from the HCP treatment.

The porous Ni–SZ electrodes gave a typical Gerischer-like porous electrode impedance response,[32,43,44] with skewed semicircles that lean the opposite way compared to the model Ni/YSZ electrode (Fig. 2c and d). The EIS of the Ni–SZ-supported porous electrode exhibits an additional low-frequency process (Fig. 2c), which was previously shown to be gas diffusion in the thick porous support layer.[32] The SWE Ni–ScYSZ electrode impedance has no such low-frequency process, illustrating that the gas diffusion resistance is minimized in this porous electrode geometry. For both electrodes, the C : H R_p ratio is 2, suggesting a confirmation of previous reports that the use of ScYSZ does not change the 3PB reaction mechanism compared with YSZ.[42]

It should be noted that although the presented EIS data was measured at different temperatures, the C : H R_p ratios are approximately the same throughout the measurement range due to the similar activation energies of the H_2/H_2O and

CO/CO_2 reactions, as is evident from comparison with the current–overpotential curves which were all measured at 850 °C. The general shapes of the spectra are also very similar throughout the measurement range 750–1000 °C. The presented spectra were chosen for illustrative purposes. For example, the 1000 °C measurements are shown for the model Ni/SZ electrode because at lower temperature the CO/CO_2 impedance process goes to very low frequencies and therefore most measurements did not include the full arc. The inset shows the H_2/H_2O measurement at 850 °C, which has the same shape. Table 1 gives a summary of the EIS results for all of the model and porous Ni-based electrodes.

Current–overpotential curves

In agreement with the EIS results, at low current the current–overpotential curves show the same C : H R_p ratios of ~30, 5 and 2 for the model Ni/YSZ, model Ni/YSZ-HCP, and porous Ni–SZ electrodes, respectively (Fig. 3). For the two model electrodes, the ratio increases with anodic polarization and decreases with cathodic polarization to about double and half, respectively, at ±0.2 V compared with the value near open-circuit. However, the ratio for the porous electrode remains nearly constant in the range of 1.5 to 2. All three electrodes have considerably faster anodic than cathodic kinetics, with the largest difference for the model Ni/SZ (factors of 26 and 4.1 higher current at ±0.2 V for H_2/H_2O and CO/CO_2, respectively), a smaller difference for the model Ni/SZ-HCP (factors of 4.0 and 1.2), and the smallest difference for the porous Ni–SZ electrode (factors of 2.1 and 1.5).

Table 1 Summary of the EIS results for all of the electrodes tested

Electrode abbreviation	Materials|electrolyte	Geometry	Rxn. site	T [°C]	R_p Units	R_p H_2/H_2O	R_p CO/CO_2	R_p Ratio
Ni/SZ ME	Ni|YSZ	Model	3PB	1000	kΩ	3.68	125	34
Ni/SZ ME	Ni|YSZ	Model	3PB	850	kΩ	20.8	701	34
Ni/SZ ME	Ni|YSZ	Model	3PB	750	kΩ	98.0	2860	29
Ni/SZ-HCP ME	Ni|YSZ	Model	3PB	860	kΩ	1.41	5.90	4.2
Ni/SZ-redox ME	Ni|YSZ	Model	3PB	850	kΩ	7.55	25.8	3.4
Ni–SZ-supported PE	Ni-YSZ|YSZ	Porous	3PB	750	Ω cm^2	0.099	0.20	2.0
Ni–SZ PE	Ni-ScYSZ|ScYSZ	Porous	3PB	850	Ω cm^2	0.067	0.14	2.0
CGOn/STN ME	CGOn/STN|YSZ	Model	2PB	850	MΩ	2.50	0.40	0.16
CGOn/STN PE	CGOn/STN|YSZ	Porous	2PB	800	Ω cm^2	0.31	0.49	1.6
CGOn/STN-redox PE	CGOn/STN|YSZ	Porous	2PB	800	Ω cm^2	0.12	0.66	5.7
CGOn-2.5 μm PE	CGOn|YSZ	Porous	2PB	800	Ω cm^2	0.010	0.076	7.9
CGO-SS-10 μm PE	CGO|ScYSZ	Porous	2PB	800	Ω cm^2	0.066	0.17	2.6
CGO-SS-18 μm PE	CGO|YSZ	Porous	2PB	750	Ω cm^2	0.078	0.21	2.7
CGO-inhomog-30 μm PE	CGO|YSZ	Porous	2PB	750	Ω cm^2	0.43	0.67	1.6
CGO-SS-18 μm-Au PE	CGO|YSZ	Porous	2PB	750	Ω cm^2	1.35	0.86	0.64

For these 3 cells, the ohmic resistance, and corresponding heating during DC measurements, is minimal due to the cell geometries, as the current spreads outwards from the small working electrode into the large electrolyte pellet to the large counter electrode.

Mechanisms

One might consider that the higher activity in H_2/H_2O than CO/CO_2 has the same explanation as the higher activity in anodic than cathodic polarization. Indeed, gas diffusion has been shown to be partly responsible for both effects in the case of the Ni–YSZ composites with a thick porous support layer,[25] but here the largest gas composition difference and anodic/cathodic asymmetry is observed for the model electrodes which have negligible gas diffusion resistance.

We propose that impurities and the 3D microstructure are responsible for the decreasing difference in activity in H_2/H_2O *vs.* CO/CO_2 in the progression from the model electrode to the model electrode with HCP to the porous electrode. It is well established that in Ni/SZ model electrodes, the 3PB is blocked by glassy impurity phases that segregate from the raw materials,[14] unless extremely pure raw materials were used in preparation of the cell,[38,23] which was not the case here. The large activation of Ni/SZ during HCP treatment has been reported to result from nano-structuring of the 3PB which occurs when YSZ is partially reduced and interacts with Ni during HCP treatment[45,46] and/or from reduction and mobilization of impurities that were accumulated at the 3PB.[42] Both effects would lead to a fresh 3PB with a lower concentration of impurities. In this case, microscopy revealed that the brief treatment rearranged the Ni/YSZ interface and a new, considerably longer 3PB was formed. No analysis of impurities has been carried out. We observed that subjecting the model Ni/SZ electrodes to a redox cycle (oxidizing to NiO and then reducing again), which also rearranges the 3PB,[30] led to a similar decrease in the C : H R_p ratio down to 3.4 (see Table 1). We focus on the HCP results here because current–overpotential curves were not measured after the redox cycle. The segregated impurities apparently inhibit the CO/CO_2 reaction more so than the H_2/H_2O reaction, as formation of a fresh 3PB decreases the C : H R_p ratio by nearly an order of magnitude. This might be because H_2/H_2O related species are transported more rapidly through or on the surface of the impurity phases or maybe partly through the Ni.[47] The especially large activation for the Ni/SZ model electrode in anodic polarization in H_2/H_2O (Fig. 2a and b), which generates H_2O at the 3PB, might be due to the same effect and/or due to a temporary rearrangement of the impurities by their reaction with H_2O, *e.g.* vaporization of SiO_2 by forming the hydroxide, and subsequent re-deposition at the 3PB.[48,49] The well-known anodic activation by 3PB pH_2O dependence, exothermicity of the anodic electrochemical reaction (and cathodic endothermicity), and possibly enhanced kinetics of the actual charge-transfer reaction in anodic polarization[25,50] cannot account for the factor of 26 higher current at +0.2 V *vs.* −0.2 V overpotential in H_2/H_2O. Regardless of the impurity content and the exact value of the C : H R_p ratio, the consistently observed higher activity for the H_2/H_2O *vs.* CO/CO_2 reactions at Ni/SZ may be due to the high mobility of H on Ni,[51] or it could simply be due to a more beneficial energetic landscape for the overall reaction scheme.

The 3PB of the model Ni/SZ electrodes has a high impurity content due to the relatively large volumes of Ni and YSZ compared with the 3PB reaction zone. The Ni and YSZ components act as large reservoirs of impurities. Therefore, the fresh 3PB of the model Ni/SZ-HCP electrode should more closely match the 3PBs found in porous electrodes. However, the C : H R_p ratio decreases from 4–5 for the model Ni/SZ-HCP electrode to 2 for the porous Ni–SZ electrode. This difference can be explained by the 3D microstructure of the porous composite electrode, which extends the ER zone into the electrode thickness due to the ionic conductivity in the electrode. The lower CO/CO_2 reaction rate leads to extension of the reaction zone farther into the electrode, which partly compensates for the lower 3PB reaction rate. This effect is illustrated in Fig. 4a, which shows modelling results of the total porous electrode R_p *vs.* electrode thickness for two different ER rates that differ by a factor of 4 – the approximate C : H R_p ratio for the model Ni/SZ-HCP electrode. As the electrode becomes thick enough, the R_p ratio decreases to a factor of 2, in close agreement with the experimental results for the porous Ni–SZ electrode.

The porous electrode model is a 1-D transmission line model that represents the porous electrode as a continuum of ER resistances coupled with ionic and

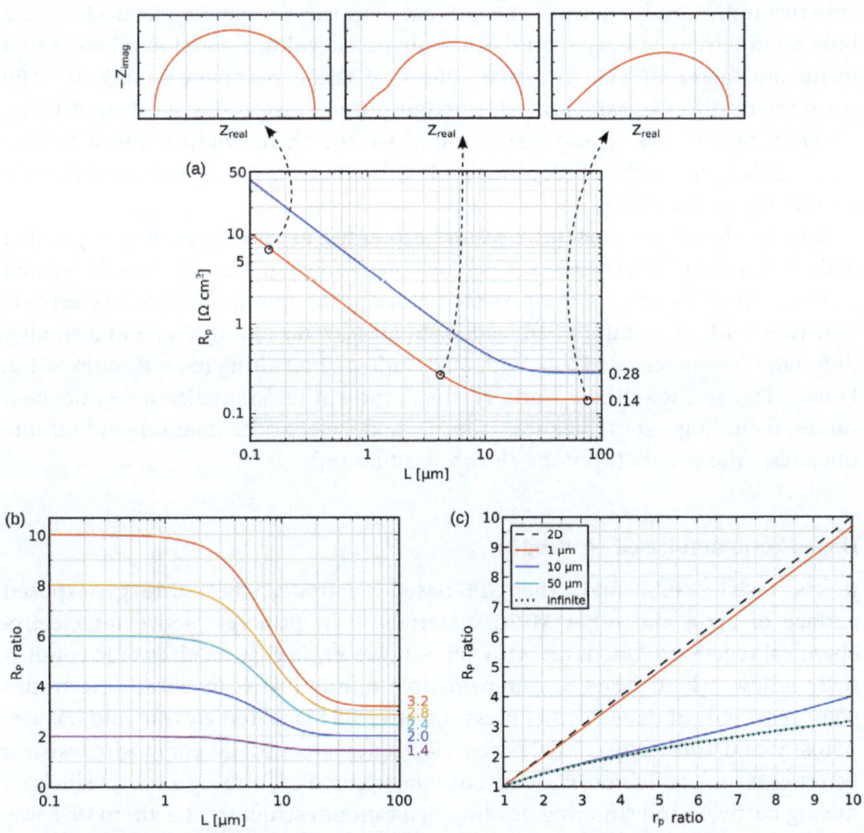

Fig. 4 Porous electrode modelling, showing the total electrode R_p as a function of electrode thickness L and ER site r_p. (a) Electrode R_p *vs.* thickness for 2 different r_p values that have a ratio of 4. (b) Electrode R_p ratio *vs.* thickness for 5 different r_p ratios. (c) Electrode R_p ratio *vs.* r_p ratio for 5 different electrode thickness values.

electronic currents which extend through the electrode thickness, away from the electrolyte and towards it, respectively:

$$R_p = \sqrt{r_p r_i}\,\coth\left(L\sqrt{\frac{r_p}{r_i}}\right)$$

where R_p is the total porous electrode polarization resistance, r_p is the polarization resistance of the ER sites (Ω cm), r_i is the ionic transport resistance (Ω cm^{-1}), and L is the electrode thickness (cm). This general analytical model has been shown to successfully describe the impedance of porous SOC composite and mixed-conducting electrodes that are macroscopically homogeneous.[43,44] It is valid for low current (near open-circuit) and neglects gradients in gas composition, temperature, *etc.*

For 3PB-based electrodes, r_p is the 3PB length-specific polarization resistance (LSR_p) divided by the 3PB density (3PB length per electrode volume). Realistic values of 4 kΩ cm for LSR_p and 3 μm μm^{-3} for the 3PB density were used for the modelling results. Changing the ionic resistivity r_i does not change the R_p ratios; it only changes the thickness value at which the minimum porous electrode R_p is reached and the value of that minimum R_p value. A realistic r_i value of 200 Ω cm was used for these simulations, which corresponds to an effective ionic conductivity of 0.005 S cm^{-1} in the composite electrode, *e.g.* a bulk conductivity of 0.04 S cm^{-1}, an ionic phase volume fraction of 0.3, and a tortuosity factor of 2.5. The real values of these parameters and the 3PB parameters differ considerably for actual porous electrodes fabricated using different procedures. These assumptions are simply the right order of magnitude and they are sufficient for illustrating the C : H R_p ratios because they have no impact on the ratios.

Fig. 4b shows the modelling results expressed as ratios (the line beginning with an R_p ratio of 4 is the ratio of the two lines shown in Fig. 4a). This shows that a larger difference in r_p is compensated more by the porous composite electrode structure, with an r_p ratio of 10 leading to an R_p ratio of only 3.2, and a smaller difference is compensated less, with an r_p ratio of 2 resulting in an R_p ratio of 1.4. Finally, Fig. 4c shows the R_p ratio *vs.* the r_p ratio at different electrode thickness values, including ~zero thickness (the 2D model electrode situation) and infinite thickness (the usual 3D porous electrode situation).

Doped ceria 2PB-based electrodes

In contrast to Ni/SZ and other 3PB-based electrodes, the entire gas-exposed surface of ceria and other MIEC materials is in principle active for electrochemical reactions. One issue with ceria, however, is that its electronic conductivity is low, which leads to non-negligible ohmic losses in porous electrodes comprised only of doped ceria unless the electrode is extremely thin and has low porosity and tortuosity. Ceria based electrodes are usually composites with a better electron conductor. The conventional method of forming a composite is by mixing particles and sintering, leading to a random structure. An alternative way is by coating the ceria onto an electronic backbone, which both ensures better percolation of the two phases and enables low temperature processing of the ceria, maintaining the ceria as nanoparticles. We use the coating approach here, with STN as the electronic conducting phase.

Similar to the Ni/SZ experiments, several different types of CGO-based electrodes were characterized to study the influence of nano/microstructure (illustrated in Fig. 1). The first set of electrodes employed STN as the electronic conducting backbone. The second set of electrodes included no additional electronic current collector in the porous structure, depending only on porous Pt or Au (by sintering the respective pastes) on top.

Electrodes with STN electron-conducting backbone:

1. CGO nanoparticles placed at the 2D interface of the electronic conductor STN pressed down onto a polished YSZ surface (hereafter referred to as the "CGOn/STN model electrode") – as well as the same type of electrode without the CGOn, *e.g.* only the STN/SZ interface, for comparison.

2. CGO nanoparticles coating the surfaces inside a 3D porous STN backbone ("CGOn-STN porous electrode").

3. The same type of electrode as #2, after a redox cycle ("CGOn-STN-redox porous electrode").

Electrodes comprised solely of doped ceria:

1. A 2.5 μm thick porous electrode comprised of CGO nanoparticles, prepared by spin-coating and sintering at 800 °C ("CGOn-2.5 μm porous electrode").

2. A 10 μm thick porous electrode comprised of larger (200–400 nm) particles, prepared by slurry spraying and sintering at 1200 °C ("CGO-SS-10 μm porous electrode").

3. A 30 μm thick porous electrode comprised of 200–400 nm size particles with a highly inhomogeneous, regionally agglomerated microstructure, prepared by screen-printing and sintering at 1200 °C ("CGO-inhomog-30 μm porous electrode").

4. An 18 μm thick porous electrode similar to #2, only thicker ("CGO-SS-18 μm porous electrode").

5. The same type of electrode as #4, but with a porous Au current collector layer (prepared by sintering Au paste; "CGO-SS-18 μm-Au porous electrode") rather than porous Pt as used for the other ceria-based porous electrodes.

Model electrodes

EIS and current–overpotential curves measured on the CGOn/STN model electrode are shown in Fig. 5 and 6, respectively. The electrode configuration is illustrated in Fig. 6a.

The most remarkable observation about the impedance of this electrode is that the activity in CO/CO_2 is much higher than in H_2/H_2O; the C : H R_p ratio is 0.16

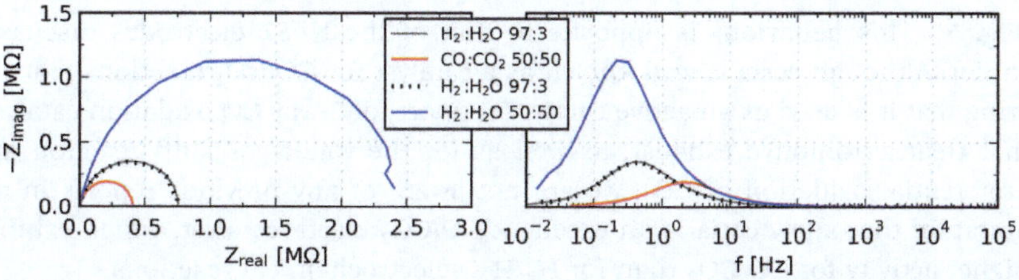

Fig. 5 EIS measured on a CGOn/STN model electrode at 850 °C in 3 different gas compositions.

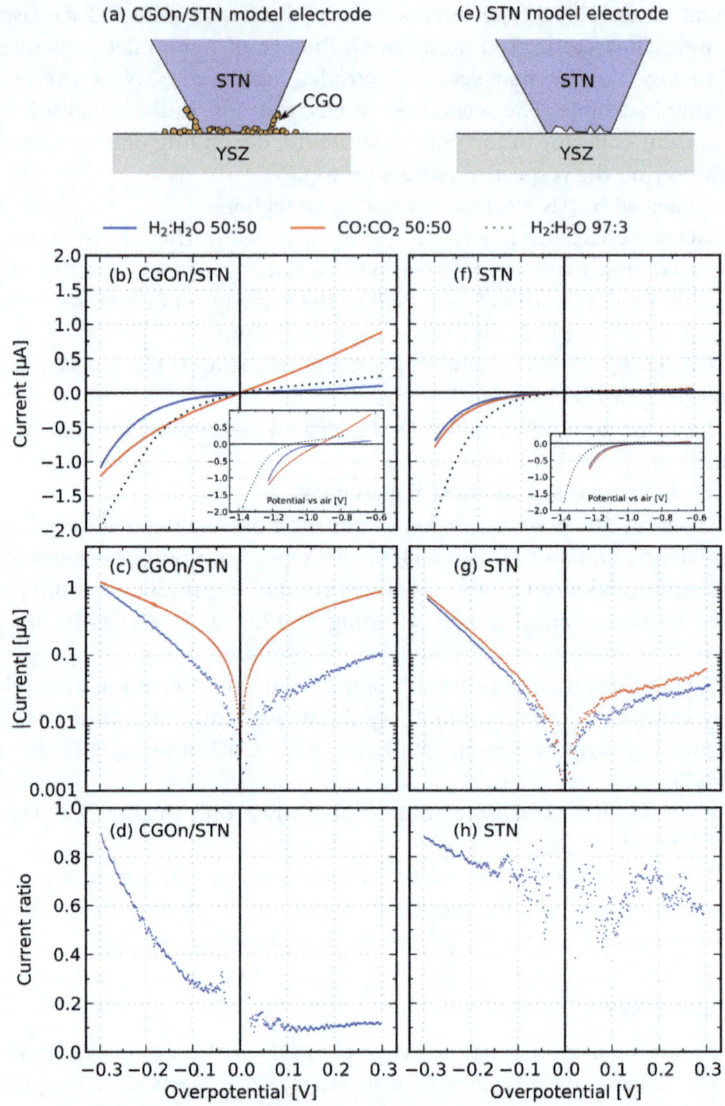

Fig. 6 Current–overpotential curves measured on CGOn/STN (b–d) and STN (f–h) model electrodes. The electrode configurations are illustrated in (a) and (b). Current ratio (d and h) is the ratio of the current measured in H_2/H_2O to that measured in CO/CO_2.

(Fig. 5). This behaviour is opposite to that of the Ni/SZ electrodes discussed earlier. Although ceria is well known as a catalyst for CO/CO_2 reactions (considering that it is used as an active catalyst support for 3-way CO oxidation catalysts that treat automotive exhaust, as well as for the water gas shift reaction and preferential oxidation of CO), we are not aware of any previous reports in the literature that show ceria is an excellent CO/CO_2 electrocatalyst, which exhibits higher activity for CO/CO_2 than for H_2/H_2O electrochemical reactions.

The electrode activity was stable throughout the measurement sequence, as can be seen in the EIS measurements in $pH_2/pH_2O = 0.97/0.03$ atm during the sequence (Fig. 5). At low current, the current–overpotential curves show the same relative activity (Fig. 6d). A major difference in the shapes of the current–overpotential curves can be observed for the H_2/H_2O and CO/CO_2 reactions – the CO/CO_2 curve is nearly linear, whereas the H_2/H_2O curve shows tremendous cathodic activation. The cathodic activation is so strong that the H_2O electrolysis current approximately "catches up" to the CO_2 electrolysis current at −0.3 V overpotential. Higher activity in H_2O electrolysis than in H_2 oxidation has been observed on various ceria electrodes and interpreted as being due to the increased surface reduction of ceria.[52–54] This observation is consistent with the reported increase in oxygen surface exchange rate with decreasing pO_2.[55–57] Only very slight cathodic activation is present in CO_2 electrolysis. The higher activity for the electrochemical reduction of H_2O and CO_2 than for oxidation of H_2 and CO is another example of opposite behaviour to the Ni/SZ electrodes.

We propose that a higher surface concentration of reaction sites on the ceria nanoparticles in CO/CO_2 than in H_2/H_2O underlies the higher activity. The nature of the rate limiting reaction step on Sm-doped ceria surfaces in H_2/H_2O and CO/CO_2 has been recently suggested to relate to electron transfer between Ce^{3+} and surface OH^- and $CO_3{}^{2-}$ respectively (speculated to be so at ambient pressure).[27,28] DFT modelling has shown that the ceria surface is more easily reduced by CO than by H_2.[58] Surface reduction leads to increased surface concentrations of oxygen vacancies and polarons (Ce^{3+}), which directly or indirectly provide ER sites.

The impedance data supports this explanation. All of the impedance measurements appear as single near-perfect semicircles in the complex plane. The measurements in $pH_2/pH_2O = 0.5/0.5$ atm and $pH_2/pH_2O = 0.97/0.03$ atm show a similar summit frequency (f_s) of 0.3–0.4 Hz, whereas the measurement in $pCO/pCO_2 = 0.5/0.5$ atm shows a considerably higher f_s value of 2.1. The summit frequency relates to the resistance R and capacitance C by $f_s = 1/(2\pi RC)$. For doped ceria and other mixed conductors, the capacitance is dominated by bulk chemical capacitance.[26,59,60] The chemical capacitance results from the presence of mixed valence Ce^{3+}/Ce^{4+} and increases with decreasing pO_2, within the pO_2 range explored here. The resistance is dominated by the 2PB ER resistance at the ceria|gas interface.[60,16] In the two H_2/H_2O environments, the unchanged summit frequency means the product of R and C is unchanged – in the more reducing H_2/H_2O environment, the reduction of the bulk ceria increases the chemical capacitance and the reduction of surface ceria decreases the surface reaction resistance. However, the $pH_2/pH_2O = 0.5/0.5$ atm and $pCO/pCO_2 = 0.5/0.5$ atm environments yield almost the same pO_2 (5.3×10^{-18} and 6.1×10^{-18} atm respectively) at the measurement temperature of 850 °C, so the chemical capacitance is the same in both environments. Yet the R_p has clearly decreased (causing the increase in f_s), suggesting that the ceria surface is more reduced (or provides a higher ER rate by another mechanism) in CO/CO_2 compared to H_2/H_2O when both environments provide the same pO_2.

The nearly linear current–overpotential curve in CO/CO_2 might be due to the presence of a high concentration of reactive surface species that is essentially independent of polarization and corresponding oxygen activity.

The higher activity in cathodic *vs.* anodic polarization can be similarly explained by reduction of the oxides during cathodic polarization, leading to enhanced surface concentration of reaction sites. This is not a special feature of ceria or nanoceria, evidenced by the similar current–overpotential behaviour of the STN model electrode without the ceria nanoparticles measured during the same test sequence (Fig. 6f and g). Cathodic activation has also been observed previously in the literature on oxide electrodes comprised of STN,[61] Sm-doped ceria,[27] a composite of CGO and La-doped $SrTiO_3$,[62] and various molybdenum oxide containing perovskites.[34]

Three important questions that arise are (i) whether the unique behaviour observed with the model electrodes can translate to a technologically relevant porous electrode, (ii) whether the high activity for CO/CO_2 reactions is specific to nanoparticulate ceria, and (iii) the possible role of the STN support. Here we briefly explore these considerations.

Porous electrodes

EIS measured on the CGOn/STN porous electrodes, which are comprised of the same materials (CGO nanoparticles on STN backbones) as the model electrodes, did not show a C : H R_p ratio far below 1 (Fig. 7), as was observed for the model electrodes. One CGO-infiltrated STN electrode exhibited a ratio of 1.6 (Fig. 7a) – still a lower value than observed for any of the Ni/SZ electrodes. The same type of electrode was subjected to a redox cycle, which causes a decrease in R_p, possibly due to formation of nano-sized cracks in the CGO coating, thereby re-gaining nanostructured features that were lost during particle coarsening.[30] The C : H R_p ratio increased to 5.7 (Fig. 7b). These two electrodes exhibit typical Gerischer-like porous electrode impedance, similar to the porous Ni–SZ electrodes. The presence of the Gerischer-like shapes is expected because the ionic conductivity is low in these electrodes, as it is provided only by the thin coatings of CGO nanoparticles, and therefore the ER zone is not expected to extend to the full 20 μm electrode thickness.

Electrodes made of only CGO were also studied (Fig. 8). A thin nano-structured electrode (CGOn-2.5 μm porous electrode) exhibited an exceptionally high activity ($R_p = 0.01\ \Omega\ cm^2$ in H_2/H_2O, including SGL diffusion). However the C : H R_p ratio was high at 6.9 (Fig. 8a). For a thicker electrode made of larger 200–400 nm sized particles (CGO-SS-10 μm porous electrode), the C : H R_p ratio was 2.6 (Fig. 8b). For

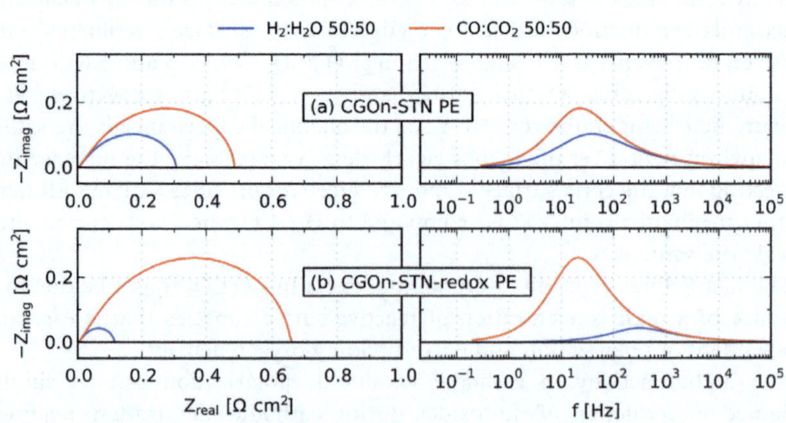

Fig. 7 EIS measured on porous electrodes with CGO nanoparticles coated on STN backbones in 50 : 50 mixtures of H_2/H_2O and CO/CO_2 at 800 °C.

a yet thicker electrode, which also had an inhomogeneous microstructure due to particle agglomeration during fabrication (CGO-inhomog-30 μm porous electrode), the ratio was 1.6 (Fig. 8c). Another thick electrode (CGO-SS-18 μm porous electrode) showed a ratio of 2.7. Finally, the same type of 18 μm thick electrode tested with Au paste as the current collector was the only one to show an inverse ratio of 0.6 — higher activity for CO/CO_2 than for H_2/H_2O, as observed for the model electrode. It should be noted that the impedance was relatively large for this last electrode, as well as the cell ohmic resistance which was more than double the ohmic resistance of the same type of cell with Pt paste (CGO-SS-18 μm porous electrode), indicating a poor coating of the Au paste. Therefore, the area-normalized impedance of the electrode with Au paste, which is calculated with the assumption of even current collection across the cell area, should not be compared with that measured on the Pt paste cell. The impedance response of these CGO electrodes varied from a somewhat skewed semicircle for thinner electrodes to a Gerischer-like porous electrode response for thicker electrodes, which indicates that the ER zone extends to a thickness of between 10–18 μm at these test conditions. As the ionic conductivity is much higher in these electrodes

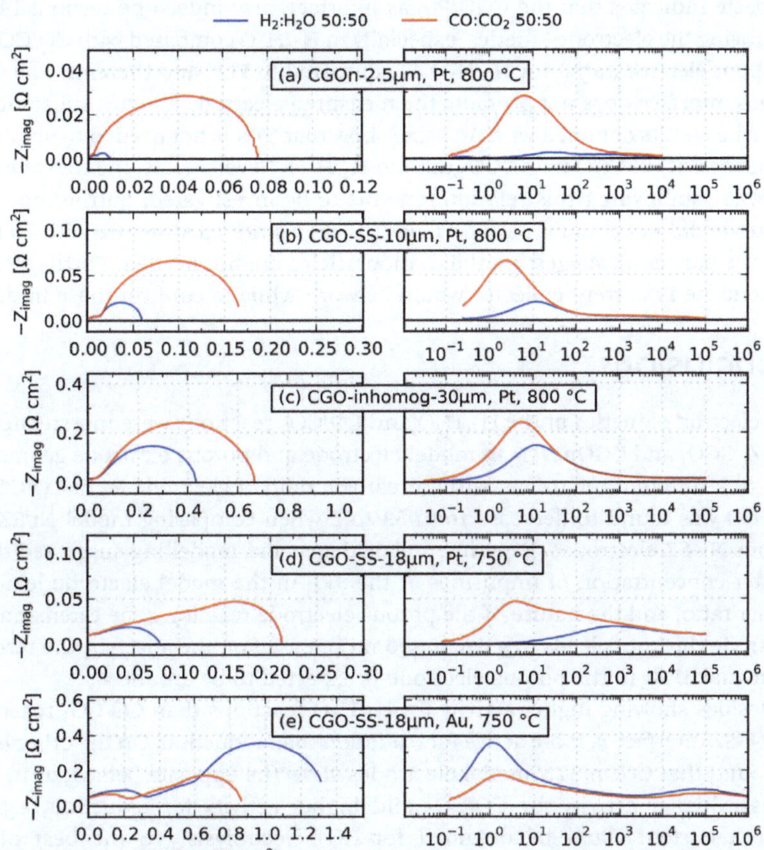

Fig. 8 EIS measured on various porous CGO electrodes with either Pt or Au paste used for electronic current collection, in 50 : 50 mixtures of H_2/H_2O and CO/CO_2.

due to the larger volume fraction of the ion conductor, CGO, than in the case of CGO-infiltrated STN, the ER zone extends farther into the thickness, assuming the same surface reaction rates. Table 1 gives a summary of the EIS results for all of the model and porous ceria-based electrodes.

The relatively high C : H R_p ratios for the nanostructured electrodes might be due to gas diffusion limitations in the porous structures. In the case of the CGOn-STN porous electrode, the nano-cracks may provide additional surface area that is only accessible to the smaller H_2/H_2O molecules, providing no additional reaction sites for CO/CO_2. Similarly, the CGOn-2.5 μm porous electrode has nano-pores which may not be accessible. Furthermore, Knudsen diffusion is present in such electrodes.[37]

The relatively low C : H R_p ratios for thicker electrodes might relate to the porous electrode thickness effect described earlier (Fig. 4). However, another possibility is that in the thinner electrodes the ER zone extends to the end of the electrode thickness, thereby reaching the CGO/Pt/gas interface at the current collector, which may provide enhanced activity. Sintered Pt paste consists of large particles and by itself shows extremely low activity ($R_p > 100\ \Omega\ cm^2$ for porous Pt when applied directly to the electrolyte), but the nanostructured CGO provides a high density of interfaces in contact with the Pt particles. The data measured with Au paste indicates that the CGO/Pt/gas interface may indeed be responsible for improving the electrode kinetics, especially in H_2/H_2O compared with CO/CO_2. Au is a poor electrocatalyst for H_2/H_2O as compared to Pt,[63] and therefore the CGO/Au/gas interface does not promote the measured electrode kinetics substantially. Also, the results obtained with Au paste show that STN is not needed to obtain the higher activity of CGO in CO/CO_2 than in H_2/H_2O. Nevertheless, the role of the Au paste in establishing this relation remains to be investigated. Furthermore, the electrode did not contain nanoparticles (it was a sintered structure with 200–400 nm size particles), suggesting that nanoparticles are not needed. Further testing without the Pt current collector would be worthwhile to confirm these findings.

Conclusions

The electrode kinetics of the H_2/H_2O and CO/CO_2 reactions were investigated for Ni/SZ, CGO, and CGOn/STN in model electrode and porous electrode geometries and with different micro/nanostructures. The ratio of electrode R_p in CO/CO_2 *vs.* H_2/H_2O was found to decrease from 33 to 2 when comparing model Ni/SZ and porous Ni–SZ electrodes. Experimental evidence and modelling suggested that a greater concentration of impurities at the 3PB in the model electrode lead to a higher ratio, and the nature of the porous electrode reaction zone extension into electrode thickness leads to a decreased ratio. Based on the analysis, the ratio for the actual 3PBs in the porous electrode is expected to be around 4.

Besides showing higher activity for H_2/H_2O reactions than CO/CO_2 reactions, the Ni/SZ interface is more active for oxidation than reduction. On the other hand, we found that CGOn/STN model electrodes show the opposite behaviour in both cases – higher activity for CO/CO_2 and higher activity for reduction, *e.g.* the highest current could be obtained for CO_2 electrolysis. To the best of our knowledge, this is the first report of a higher electrocatalytic activity for CO/CO_2 than for H_2/H_2O reactions on CGO nanoparticles in the absence of gas diffusion limitations. We proposed that enhanced surface reduction of the doped ceria in

CO/CO_2, at the same pO_2, underlies this behaviour. This effect could not be observed in CGO based porous electrodes, except in one case where Au instead of Pt was used as the electronic current collector, suggesting that the CGO/Pt interfaces may be selectively enhancing the activity of the H_2/H_2O reaction for those porous electrodes. Greater gas diffusion limitations for CO/CO_2 than for H_2/H_2O, especially with nanostructured features involved, is another possible explanation for the different relative activities of the porous electrodes and the model electrodes in CO/CO_2 and H_2/H_2O.

Acknowledgements

We would like to thank Peter Blennow for valuable discussions; Tânia Ramos for providing Ni–YSZ symmetrical cells; Bhaskar Reddy Sudireddy, Theis Løye Skafte, Jean-Claude Njodzefon, Vahid Vakili, Lene Knudsen, Annelise Mikkelsen and other colleagues and laboratory staff at DTU Energy for assistance with sample fabrication and experiments. We gratefully acknowledge financial support from Energinet.dk through the ForskEL programme "Solid Oxide Fuel Cells for the Renewable Energy Transition" contract no 2014-1-12231 and from the Program Commission on Sustainable Energy and Environment, The Danish Council for Strategic Research, *via* the SERC project (http://www.serc.dk), contract no. 2104-06-0011.

References

1 S. H. Jensen, P. H. Larsen and M. Mogensen, *Int. J. Hydrogen Energy*, 2007, **32**, 3253–3257.

2 N. Q. Minh and M. B. Mogensen, *Electrochem. Soc. Interface*, 2013, 55.

3 C. Graves, S. D. Ebbesen, M. Mogensen and K. S. Lackner, *Renewable Sustainable Energy Rev.*, 2011, **15**, 1–23.

4 C. Graves, S. D. Ebbesen, S. H. Jensen, S. B. Simonsen and M. B. Mogensen, *Nat. Mater.*, 2015, **14**, 239–244.

5 D. M. Bierschenk, J. R. Wilson and S. A. Barnett, *Energy Environ. Sci.*, 2011, **4**, 944–951.

6 S. D. Ebbesen and M. Mogensen, *J. Power Sources*, 2009, **193**, 349–358.

7 Z. Zhan and S. A. Barnett, *Science*, 2005, **308**, 844–847.

8 C. Graves, S. D. Ebbesen and M. Mogensen, *Solid State Ionics*, 2011, **192**, 398–403.

9 R. Knibbe, A. Hauch, J. Hjelm, S. D. Ebbesen and M. Mogensen, *Green*, 2011, **1**, 141–169.

10 A. Weber, in *Encyclopedia of Electrochemical Power Sources*, Elsevier, Amsterdam, 2009, pp. 120–134.

11 M. Chen, Y.-L. Liu, J. J. Bentzen, W. Zhang, X. Sun, A. Hauch, Y. Tao, J. R. Bowen and P. V. Hendriksen, *J. Electrochem. Soc.*, 2013, **160**, F883–F891.

12 A. Hauch, A. Hagen, J. Hjelm and T. Ramos, *J. Electrochem. Soc.*, 2014, **161**, F734–F743.

13 S. D. Ebbesen, C. Graves, A. Hauch, S. H. Jensen and M. Mogensen, *J. Electrochem. Soc.*, 2010, **157**, B1419–B1429.

14 M. Mogensen and K. V. Hansen, in *Handbook of Fuel Cells*, John Wiley & Sons, Ltd, 2010.

15 J. B. Goodenough and Y.-H. Huang, *J. Power Sources*, 2007, **173**, 1–10.
16 W. C. Chueh and S. M. Haile, *Annu. Rev. Chem. Biomol. Eng.*, 2012, **3**, 313–341.
17 M. Mogensen, N. M. Sammes and G. A. Tompsett, *Solid State Ionics*, 2000, **129**, 63–94.
18 C. Chatzichristodoulou, P. T. Blennow, M. Søgaard, P. V. Hendriksen and M. B. Mogensen, in *Catalysis by Ceria and Related Materials*, Imperial College Press, 2013, vol. 12, pp. 623–782.
19 W. C. Chueh, Y. Hao, W. Jung and S. M. Haile, *Nat. Mater.*, 2012, **11**, 155–161.
20 W. Jung, J. O. Dereux, W. C. Chueh, Y. Hao and S. M. Haile, *Energy Environ. Sci.*, 2012, **5**, 8682–8689.
21 M. Mogensen, J. Høgh, K. V. Hansen and T. Jacobsen, *ECS Trans.*, 2007, **7**, 1329–1338.
22 V. Yurkiv, A. Utz, A. Weber, E. Ivers-Tiffée, H.-R. Volpp and W. G. Bessler, *Electrochim. Acta*, 2012, **59**, 573–580.
23 A. Utz, A. Leonide, A. Weber and E. Ivers-Tiffée, *J. Power Sources*, 2011, **196**, 7217–7224.
24 W. G. Bessler, M. Vogler, H. Störmer, D. Gerthsen, A. Utz, A. Weber and E. Ivers-Tiffée, *Phys. Chem. Chem. Phys.*, 2010, **12**, 13888.
25 J.-C. Njodzefon, D. Klotz, A. Kromp, A. Weber and E. Ivers-Tiffée, *J. Electrochem. Soc.*, 2013, **160**, F313–F323.
26 R. D. Green, C.-C. Liu and S. B. Adler, *Solid State Ionics*, 2008, **179**, 647–660.
27 Z. A. Feng, M. L. Machala and W. C. Chueh, *Phys. Chem. Chem. Phys.*, 2015, **17**, 12273–12281.
28 Z. A. Feng, F. El Gabaly, X. Ye, Z.-X. Shen and W. C. Chueh, *Nat. Commun.*, 2014, **5**, 1–9.
29 C. R. Graves, S. D. Ebbesen and M. Mogensen, *ECS Trans.*, 2009, **25**, 1945–1955.
30 C. Graves, *ECS Trans.*, 2013, **57**, 3127–3136.
31 T. Ramos, K. Thydén and M. Mogensen, *ECS Trans.*, 2010, **28**, 123–139.
32 T. Ramos, M. Søgaard and M. B. Mogensen, *J. Electrochem. Soc.*, 2014, **161**, F434–F444.
33 P. Blennow, K. K. Hansen, L. R. Wallenberg and M. Mogensen, *ECS Trans.*, 2008, **13**, 181–194.
34 C. Graves, B. R. Sudireddy and M. Mogensen, *ECS Trans.*, 2010, **28**, 173–192.
35 A. Samson, M. Søgaard, R. Knibbe and N. Bonanos, *J. Electrochem. Soc.*, 2011, **158**, B650–B659.
36 S. Primdahl and M. Mogensen, *J. Electrochem. Soc.*, 1999, **146**, 2827–2833.
37 A. M. Hussain, J. V. T. Høgh, T. Jacobsen and N. Bonanos, *Int. J. Hydrogen Energy*, 2012, **37**, 4309–4318.
38 A. Utz, H. Stormer, A. Leonide, A. Weber and E. Ivers-Tiffee, *J. Electrochem. Soc.*, 2010, **157**, B920–B930.
39 C. Graves, *RAVDAV data analysis software, version 0.9.8*, 2014.
40 M. S. Schmidt, K. V. Hansen, K. Norrman and M. Mogensen, *Solid State Ionics*, 2008, **179**, 2290–2298.
41 M. S. Schmidt, K. V. Hansen, K. Norrman and M. Mogensen, *Solid State Ionics*, 2008, **179**, 1436–1441.
42 J. Høgh, K. V. Hansen, K. Norrman, I. Chorkendorff, T. Jacobsen and M. Mogensen, *Solid State Ionics*, 2013, **234**, 11–18.
43 V. Sonn, A. Leonide and E. Ivers-Tiffee, *J. Electrochem. Soc.*, 2008, **155**, B675–B679.

44 J. Nielsen and J. Hjelm, *Electrochim. Acta*, 2014, **115**, 31–45.
45 D. Klotz, B. Butz, A. Leonide, J. Hayd, D. Gerthsen and E. Ivers-Tiffee, *J. Electrochem. Soc.*, 2011, **158**, B587–B595.
46 D. Klotz, J. Szasz, A. Weber and E. Ivers-Tiffée, *ECS Trans.*, 2012, **45**, 241–249.
47 K. V. Jensen, R. Wallenberg, I. Chorkendorff and M. Mogensen, *Solid State Ionics*, 2003, **160**, 27–37.
48 M. S. Schmidt, K. Vels Hansen, K. Norrman and M. Mogensen, *Solid State Ionics*, 2009, **180**, 431–438.
49 A. Hauch, S. D. Ebbesen, S. H. Jensen and M. Mogensen, *J. Electrochem. Soc.*, 2008, **155**, B1184–B1193.
50 S. D. Ebbesen and M. Mogensen, *ECS Trans.*, 2013, **50**, 167–182.
51 M. Mogensen and S. Skaarup, *Solid State Ionics*, 1996, **86–88**(2), 1151–1160.
52 C. Zhang, M. E. Grass, A. H. McDaniel, S. C. DeCaluwe, F. E. Gabaly, Z. Liu, K. F. McCarty, R. L. Farrow, M. A. Linne, Z. Hussain, G. S. Jackson, H. Bluhm and B. W. Eichhorn, *Nat. Mater.*, 2010, **9**, 944–949.
53 L. Wang, C. Zhang, M. Grass, Y. Yu, K. Gaskell, Z. Hussain, Z. Liu, B. Eichhorn and G. Jackson, *ECS Trans.*, 2011, **35**, 1435–1444.
54 S. C. DeCaluwe and G. Jackson, *ECS Trans.*, 2011, **35**, 2883–2895.
55 A. Karthikeyan and S. Ramanathan, *Appl. Phys. Lett.*, 2008, **92**, 243109–243109.
56 T. Horita, K. Yamaji, N. Sakai, M. Ishikawa, H. Yokokawa, T. Kawada, and M. Dokiya, *Electrochem. Solid-State Lett.*, 1998, **1**, 4–6.
57 M. Katsuki, S. Wang, K. Yasumoto, and M. Dokiya, *Solid State Ionics*, 2002, **154–155**, 589–595.
58 D. Knapp and T. Ziegler, *J. Phys. Chem. C*, 2008, **112**, 17311–17318.
59 S. B. Adler, J. A. Lane and B. C. H. Steele, *J. Electrochem. Soc.*, 1996, **143**, 3554–3564.
60 W. C. Chueh, W. Lai and S. M. Haile, *Solid State Ionics*, 2008, **179**, 1036–1041.
61 P. Blennow, K. K. Hansen, L. R. Wallenberg and M. Mogensen, *Solid State Ionics*, 2009, **180**, 63–70.
62 O. A. Marina, L. R. Pederson, M. C. Williams, G. W. Coffey, K. D. Meinhardt, C. D. Nguyen and E. C. Thomsen, *J. Electrochem. Soc.*, 2007, **154**, B452–B459.
63 R. Baker, J. Guindet and M. Kleitz, *J. Electrochem. Soc.*, 1997, **144**, 2427–2432.

Faraday Discussions

PAPER

Development of a diffuse reflectance infrared fourier transform spectroscopy (DRIFTS) cell for the *in situ* analysis of co-electrolysis in a solid oxide cell

Denis J. Cumming,*[a] Christopher Tumilson,[ab] S. F. Rebecca Taylor,[b] Sarayute Chansai,[b] Ann V. Call,[a] Johan Jacquemin,[b] Christopher Hardacre*[b] and Rachael H. Elder*[a]

Received 2nd March 2015, Accepted 12th March 2015
DOI: 10.1039/c5fd00030k

Co-electrolysis of carbon dioxide and steam has been shown to be an efficient way to produce syngas, however further optimisation requires detailed understanding of the complex reactions, transport processes and degradation mechanisms occurring in the solid oxide cell (SOC) during operation. Whilst electrochemical measurements are currently conducted *in situ*, many analytical techniques can only be used *ex situ* and may even be destructive to the cell (*e.g.* SEM imaging of the microstructure). In order to fully understand and characterise co-electrolysis, *in situ* monitoring of the reactants, products and SOC is necessary. Diffuse Reflectance Infrared Fourier Transform Spectroscopy (DRIFTS) is ideal for *in situ* monitoring of co-electrolysis as both gaseous and adsorbed CO and CO_2 species can be detected, however it has previously not been used for this purpose. The challenges of designing an experimental rig which allows optical access alongside electrochemical measurements at high temperature and operates in a dual atmosphere are discussed. The rig developed has thus far been used for symmetric cell testing at temperatures from 450 °C to 600 °C. Under a CO atmosphere, significant changes in spectra were observed even over a simple Au|10Sc1CeSZ|Au SOC. The changes relate to a combination of CO oxidation, the water gas shift reaction, carbonate formation and decomposition processes, with the dominant process being both potential and temperature dependent.

1 Introduction

Electrochemical devices based on solid-state ceramic electrolytes have a wide range of applications from sensors to membrane reactors; batteries to fuel cells.

[a]*Department of Chemical and Biological Engineering, University of Sheffield, Mappin St, Sheffield, S1 3JD, UK. E-mail: d.cumming @sheffield.ac.uk; r.elder@sheffield.ac.uk*

[b]*School of Chemistry and Chemical Engineering, Queen's University Belfast, Belfast, BT9 5AG, N. Ireland. E-mail: c.hardacre@qub.ac.uk*

One of the attractive features of these devices is their flexibility and efficiency when operated at high temperature. Solid oxide fuel cells have received much attention over the last four decades, resulting in many advances in the understanding of the technology. It was recognised very early on that solid oxide cells (SOC) were equally capable of reverse operation in electrolysis mode, for example the pioneering work of Doenitz *et al.*[1,2] on steam electrolysis.

Recent research interest has focused on the co-electrolysis of steam and carbon dioxide for the single-step production of syngas. This process is attractive because it represents an efficient route to producing synthetic liquid hydrocarbon fuels with all of the associated benefits of high energy density, transport infrastructure and ease of handling. In cases where the carbon dioxide used during electrolysis is recycled from the air or another point source, a closed carbon cycle is possible. Furthermore, the efficient conversion of electricity into an energy-dense chemical offers the possibility for commercially viable large-scale energy storage.

There are compelling reasons to pursue high temperature co-electrolysis, but as it stands the technology remains by-and-large in the laboratory. One of the key challenges facing co-electrolysis is the lack of understanding of fundamental reaction mechanisms. Knowledge of the rate-limiting reactions within a cell would aid rational design of materials and microstructures, leading to increased device efficiency and longevity. Comprehensive understanding of the relationship between the electrolysis, surface catalytic and water gas shift reactions would enable advancement of electrolysis technology. To achieve the level of detailed understanding required, the development of *in situ* analysis techniques is essential.

Typical *in situ* or *operando* testing of fuel cells and electrolysers has been limited to electrochemical methods such as impedance spectroscopy and direct current techniques. *In situ* spectroscopy could provide invaluable information about the surfaces and species that play a crucial role in the overall macroscopic reaction, along with information on the complex interplay of reactions occurring during co-electrolysis. Progress in the development and application of *in situ* spectroscopy methods to high temperature electrochemical devices has accelerated recently and techniques such as Raman[3–11] and X-ray photoelectron spectroscopy[7,12–15] now deliver valuable information from operational devices. Typically these techniques give potential-dependent structural or chemical information, but do not always yield direct information about the chemistry of the reactions, intermediates or surface interactions.

Infrared absorption spectroscopy can bridge the gap between the surface electronic/crystal structure and link to the chemistry of the gas-phase and absorbed species. It is particularly suited to this application because most molecular vibrational modes fall into the energy range of mid-infrared light (2.5–50 μm, 4000–200 cm^{-1}). Fourier transform infrared (FTIR) spectroscopy has become routine in most analytical laboratories throughout the world and commercially available instruments tend to be highly adaptable, allowing numerous combinations of gas–solid interactions to be probed. This has made *in situ* IR techniques highly popular for heterogeneous catalysis research.[16–18] The majority of early catalysis studies were carried out in transmission mode (TIR) whereby the incident beam passes through the sample/gases of interest. This requires that samples are self-supporting and not completely opaque to the incident beam. Whilst this method has been applied to the study of fuel cell materials,[19,20] *operando* studies are difficult and not representative of working conditions due to the requirement of transparent electrodes for cell polarization. An alternative is to use diffuse reflectance infrared Fourier transform spectroscopy (DRIFTS), another technique commonly employed in the catalysis community.[16] The experimental configuration has the advantage of allowing the typical powder-basket arrangement to be replaced by a small solid oxide cell operated in a scenario very similar to most lab-scale electrochemical tests.

There are few reported studies using FTIR to probe SOCs and those that exist mainly focus on fuel cell operation. Lu *et al.* used IR emission spectroscopy to study the oxygen electrode material in SOCs, finding that two to three different oxygen ion species were present on the electrode surface. Murai *et al.*[21] simplified the experiment by using symmetric Pt|YSZ|Pt cells to study the oxygen reduction reaction, finding electrode microstructure and silica contamination caused significant variation in both electrochemical performance and spectroscopic response. There have been very few IR studies to examine the fuel electrode in detail, with the exception of Kirtley *et al.*[22] It is rather surprising that there are so few examples of *in situ* or *operando* research on the fuel electrode given the large number of examples of application of IR spectroscopy to low and high temperature water–gas shift reactions[23–32] using metal catalyst particles on active catalyst supports (such as zirconia and ceria), which bear close resemblance to the reactions and materials systems in a SOC.

In this paper we report the development of a modified commercial DRIFTS unit for the study of SOCs under operating conditions. Herein, we describe the technical design of this apparatus, and demonstrate its viability and potential to provide surface specific information on the electrochemistry of polarized electrodes and ionic surfaces. This is illustrated by preliminary experiments using symmetric gold electrodes on scandia-stabilized zirconia substrates under a range of technologically relevant conditions.

2 Technical design of the DRIFTS apparatus

A schematic of the DRIFTS setup is shown in Fig. 1. The Spectra-Tech system fits an environmental chamber (from here on referred to as the base unit) into a demountable mirror assembly (Fig. 1D), which decouples the base unit from the mirror assembly allowing modifications to be made to the base unit design. Previous catalysis studies in our laboratory[33] required modifications to the original factory base unit design (Fig. 1A) to improve sealing around the crucible and reduce the pressure drop through the catalyst bed (Fig. 1B). Further improvements to this design were incorporated to include electrical feed-throughs and gas ports (Fig 1C and detail in Fig. 2). There are several design requirements for the modified apparatus to operate SOCs effectively, which can be broken down into four main categories:

- Temperature;
- Geometrical considerations;
- Electrical and spectroscopic measurement;
- Materials.

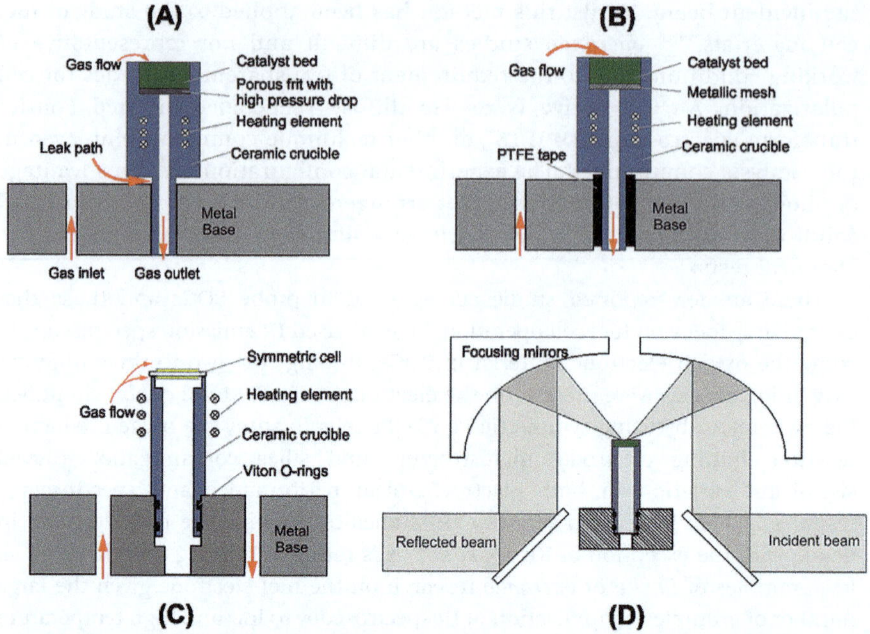

Fig. 1 Schematic drawing the evolution of the base unit design and the optical path through the DRIFTS apparatus. (A) As purchased Spectra-Tech configuration. (B) Modification to reduce gas leak in chamber. (C) SOC operation configuration. (D) Full schematic view of DRIFTS rig showing mirror assembly and beam path.

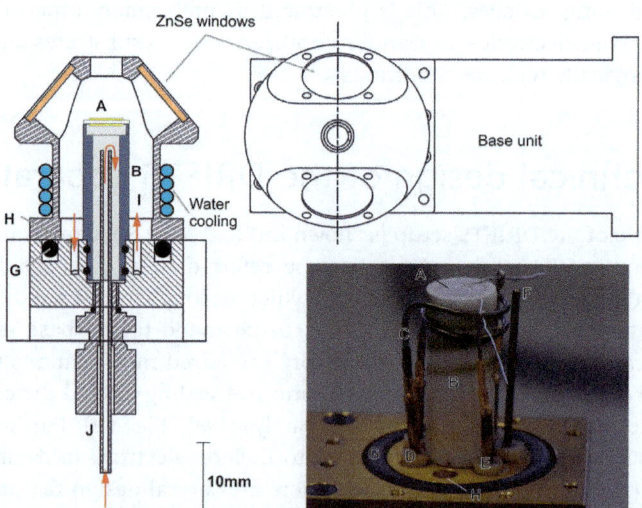

Fig. 2 Detailed cross section of base and dome and a photograph of apparatus showing electrochemical cell (A), crucible (B) heating element (C), gold crimp connections for the heating wire (D) and the electrical cell connections (E), K-type thermocouple (F), dome Viton O-ring seal (G) cell, gas feed-throughs (H and I) and underside inlet port (J) (see text for more detail).

2.1 Overview of design criteria

2.1.1 Temperature. In many commercially available base units, temperature is limited to around 500 °C and the space available for crucible heating elements is restricted by the small geometrical volume (*i.e.* ~5 cm^3) within the chamber. The operating temperature is also limited by the optical windows, composed of ZnSe, which must be kept below 200 °C to prevent oxidation. In order to reduce the temperature of the optical windows, cooling water is flowed around a closed circuit in the base of the dome, keeping the temperature of the optical windows close to room temperature (*i.e.* <60 °C).

To gather data relevant to solid oxide electrolysis cells, operation in the range of 500–800 °C is required. The temperature needs to be distributed evenly across the cell to ensure reproducible electrochemical performance and to correlate the spectra with the correct operational temperature. Heating is achieved by placing an electrical resistance wire as near to the sample as possible. This is commonly achieved by utilizing a wire-wrapped crucible[33] or by installing a cartridge heater close the sample.[34]

2.1.2 Geometrical considerations. The base unit must retain the original dimensions of the factory unit to fit inside the existing optical mirror assembly. In this setup, the environmental compartment where gas and electrical feed through components pass is only 18 mm in diameter and at least four electrical feed-throughs (two each for the heater and cell electrical connections) are necessary in addition to inlet and outlet gas ports. For full cell testing with two atmospheres, provision must also be made for gas flow to the counter electrode.

The cell configuration design is only limited by the cell diameter, which has to be a minimum of 8.5 mm to fit the top of the cell and be supported by the crucible used in this rig (Fig. 2A). As with any electrochemical cell it must be self-supporting, but can be either electrode or electrolyte supported.

The previous version of this DRIFTS apparatus[33] used a crucible with a narrow neck which was susceptible to cracking during sample loading or high temperature operation. More recent iterations of the design widened the crucible neck to accommodate an internal gas feed and which also reduced the likelihood of cell failure due to cracking at the crucible neck.

2.1.3 Electrical and spectroscopic measurements. Achieving both optical and electrical access to the cell whilst maintaining a leak tight chamber is challenging. The use of IR transparent ZnSe windows enables spectroscopic access to the cell. Careful sealing is required around electrical feed-throughs for heating and measurement. Electrodes and current collector geometries were carefully considered to collect accurate spectroscopic measurements from a known area as near to the electrode–electrolyte interface as possible. At this stage materials with low reflectivity (such as Ni–YSZ and many oxygen electrodes) present a real challenge. Gold electrodes, therefore, make an excellent model electrode for preliminary work.

2.1.4 Materials. Due to the high temperatures, small volume, and therefore very high thermal gradients present, rig materials near to the hot zone must have a small thermal expansion coefficient to withstand multiple, fast thermal cycles without damage. The base unit and dome should have high thermal conductivity to maximize the effect of water cooling. Materials must not degrade significantly under the range of highly oxidising and highly reducing gas compositions used and must not produce volatile species that may poison the electrochemical cell or form layers that interfere with spectroscopic measurement. Version 1 of the base unit was built from brass for ease of construction, whilst version 2 is made from stainless steel. PEEK, ceramics, silver and gold were selected as materials suitable for components within the rig.

2.2 Detailed design of the cell used in these studies

Fig. 2 shows the base unit in detail: the electrochemical cell (Fig. 2A) is mounted on the crucible (Fig. 2B) which is machined from a commercial glass ceramic (Macor, RS Components). Previous crucible designs (Fig. 1B) had a very narrow neck which was susceptible to mechanical failure, causing loss of the gas-tight seal. In the design iteration reported herein, the wider base adds space for a gas tube to feed gas to the internal electrode of the cell. Introduction of a double Viton O-ring seal (Fig. 2G) prevents gas leakage between the crucible and the base. In another departure from previous reported designs,[33] an exposed heater wire (Fig. 2C) was used which can be detached independently from the crucible. The heater consisted of a spiral-wound length of nichrome furnace wire, 0.8 mm in diameter and 16 cm in length with a total resistance of $\sim$0.7 Ω. It was attached to the feed-through pins (Fig. 2D) using gold plated crimp connectors (RS components). Temperature control was achieved within an accuracy of $\pm$1 °C by means of a custom built 24 V system with a furnace controller (2216e, Eurotherm) with the input from a type K thermocouple (Fig. 2F) mounted as close to the cell as practically possible. Hot zone temperatures were limited to 600 °C as the low heater wire resistance means a very large current is drawn to deliver the power needed for operation above 600 °C. This heater configuration was retained as it allows significant flexibility in the chamber configuration. Future iterations are planned to increase the heater resistance to allow higher temperature operation.

For the testing reported in this paper, a symmetric cell was operated in a single atmosphere. To achieve this, the outlet below the crucible (Fig. 2J) was closed and the cell placed on top of the crucible suspended by a few dots of silver paste to form a small gap below the cell, allowing gas flow across both electrodes. Electrode wires were connected to the same type of pins and crimp connectors (Fig. 2E) used for the heater wire. Gas feed-throughs consist of gold plated pins embedded in a barbed PEEK collar (Fig. 2H). Within the collar, the pin is soldered to a length of wire. To achieve a hermetic seal, the PEEK collar, along with the pin, is press fit into the base unit. Both the feed-though and the O-rings are all capable of withstanding temperatures up to 200 °C.

Water cooling of the base unit is required when the cell operating temperature is above 200 °C and is supplied *via* internal channels in the base unit, fed (*via* a peristaltic pump) into a tube coil wrapped around the neck of the dome (shown in Fig. 2). The base unit remains below 50 °C when the water-cooling is in operation up to hot zone temperatures of 600 °C. Addition of the gas feed-throughs, extra gas port and compression fitting on the underside of the base unit provide the capability to perform full cell testing while still fitting into the existing mirror assembly.

3 Experimental

3.1 Sample preparation

SOCs were prepared by brushing 6 mm diameter symmetric Au electrodes (C5756, Heraeus) onto commercial 10Sc1CeSZ electrolytes (Kerafol) before being dried at 120 °C for 1 h. Cells were then sintered in air at 600 °C for 1 h. Typical electrolyte thicknesses were 130 μm with electrode thicknesses of 90 μm. The cells were then placed on top of the crucible and Ag flying lead current collectors were attached to the electrodes using Ag paste (Gwent Electronic Materials), which were subsequently fired *in situ* at 600 °C for 1 h.

3.2 Sample characterisation

The reactant gases, Ar (99.998% Ar), Kr (5% Kr in Ar), and CO (10% CO in Ar) supplied by BOC, were individually regulated by Aera™ FC-7700C mass flow controllers at a total flow rate of 40 cm^3 min^{-1}. The cell was heated to temperature in Ar at a rate of 3 °C min^{-1} and pre-cleaned by two potentiodynamic scans in Ar between ±2.0 V prior to testing. The experimental atmosphere was then applied for a minimum of 30 min to achieve steady state conditions. Infrared spectra were obtained using a Bruker Vertex 70 FTIR spectrometer equipped with a liquid N_2 cooled detector. The *in situ* DRIFTS spectra were recorded twice at each voltage with a resolution of 4 cm^{-1}. A background spectrum of the cell (56 scans averaged over 30 s) was taken in the reactant atmosphere at temperature, at open circuit, and has been subtracted from the results presented. All DRIFT spectra are displayed as log $1/R$ where R is the relative reflectance (R = reflectance of the sample recorded in the gas feed at temperature and potential/reflectance of the sample recorded under gas feed at temperature and open circuit potential). During measurements potential was applied, starting at 0 V, in steps of 200 mV. The cell was held at each step voltage for a duration of 1 min to enable the data for two spectra to be recorded. Spectra were then analyzed using proprietary OPUS software. Electrochemical measurements were performed using a potentiostat (Ecochemie B.V., Autolab PGSTAT302) with the error in the applied potentials of ±10 mV. The reacted gas phase species were analyzed using an online Hiden Analytical HPR20 Mass Spectrometer (MS). Hermetic sealing of the system was verified by monitoring the levels of H_2O (m/z = 18), O_2 (m/z = 32) and CO_2 (m/z = 44) using the MS.

4 Results

The effect of potential on surface species during polarization (−2 V to +2 V) of the symmetric Au|10Sc1CeSZ|Au SOC as detected by *in situ* infrared spectroscopy under a gas feed of 5% CO in Ar is depicted in Fig. 3. For clarity, only steps at 0 V, ±0.8 V and ±1.6 V are shown.

DRIFTS data recorded at 450 °C (selected potentials shown in Fig. 3A) show distinct peaks at 3245, 2360 and 1320–1896 cm^{-1}. An increase in the intensity of peaks at 3245 and 1320–1896 cm^{-1}, attributed to surface hydroxyl species,[19,20] is observed when a negative potential is applied. Interestingly, these hydroxyl species appear to accumulate with applied potential over time and the surface response does not return to base levels even at zero potential, indicating a

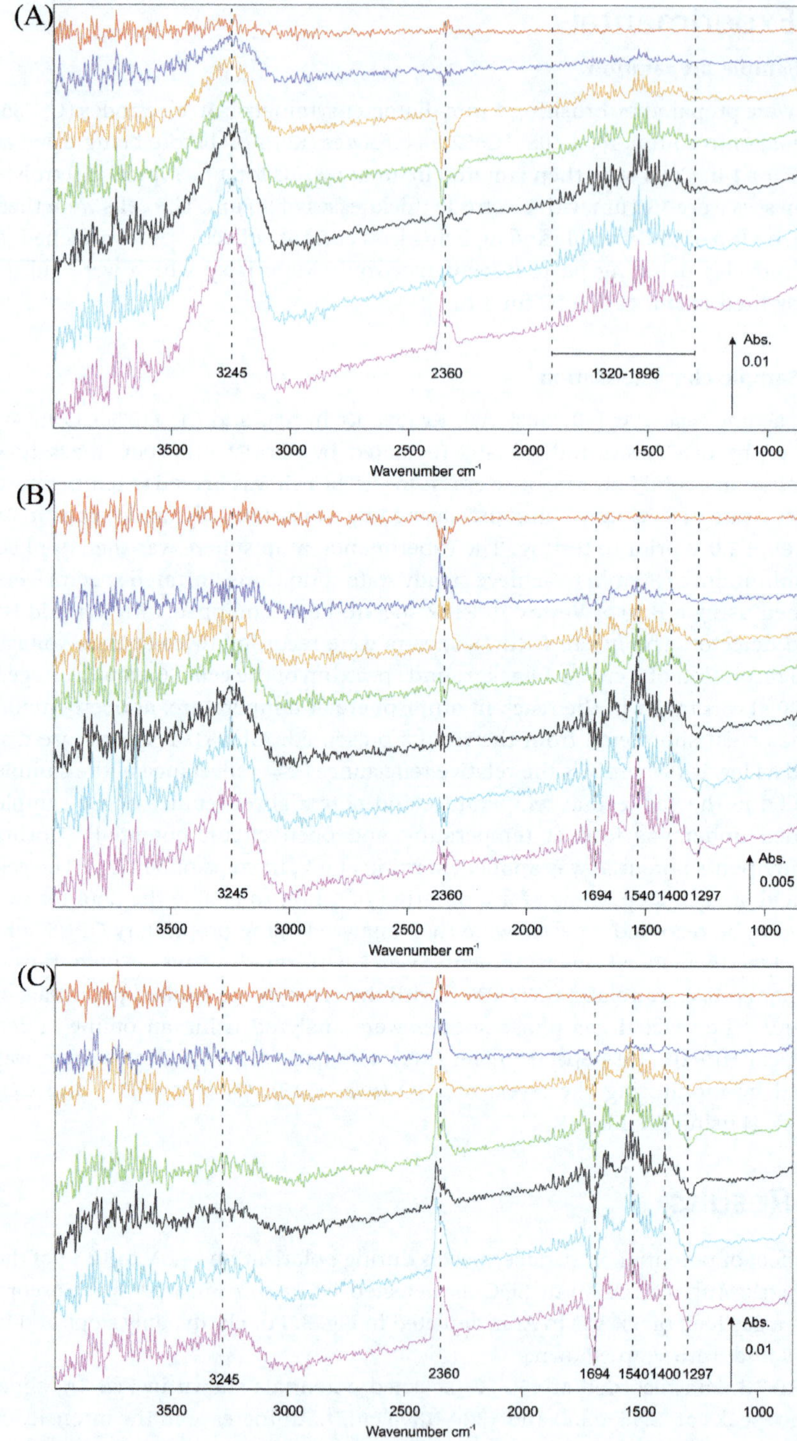

Fig. 3 Comparison of DRIFTS spectra obtained for symmetric SOC utilizing a doped zirconia electrolyte and Au electrodes, operated in a 5% CO in Ar atmosphere under D.C. polarization at (A) 450 °C, (B) 550 °C and (C) 600 °C.

non-reversible accumulation. In contrast, the strong signal associated with gas-phase CO_2 (2360 cm^{-1}) is clearly potential dependent. This band shows a negative feature when positive potential is applied, which is associated with a loss of CO_2 from the gas phase, *i.e.* a reduction in the rate of CO_2 formation compared with that formed at open circuit potential at 450 °C. This loss, still present on returning to 0 V, is then reversed under negative applied potential, resulting in an increase in the production rate of gas phase CO_2 compared with the initial state of the catalyst.

When increasing the temperature from 450 °C to 550 °C (Fig. 3A and B) or to 600 °C (Fig. 3C), the peaks attributed to surface hydroxyls[20] (3245 cm^{-1}) and CO_2 (2360 cm^{-1}) persist. The intensity of the hydroxyl signal shows a distinct decrease in intensity with increasing temperature across the three experiments. A noticeable change in the potential dependence of the band associated with CO_2 is observed by comparing spectra at 450 °C and 550 °C. At 550 °C (Fig. 3B) the intensity of the CO_2 signal increases and then decreases under positive and negative potentials, respectively. Inspection of the 1200–1800 cm^{-1} region reveals several possible species with signals observed at 1694, 1540, 1400 and 1297 cm^{-1}. These peaks are tentatively assigned to carbonate and bicarbonate[19,20] species on the electrode surface with the positive peaks indicating formation and negative peaks removal from the surface. The positive signal detected at 1540 cm^{-1}, assigned to bidentate carbonate species formation, shows the largest change in intensity under positive potentials, while bicarbonate species (1694, 1400 and 1297 cm^{-1}) exhibited significant response under applied negative potential. In the latter case, both formation and removal are observed.

In contrast to observations at 450 °C (Fig. 3A) and 550 °C (Fig. 3B), the CO_2 peak observed at 600 °C (Fig. 3C) remains positive throughout the experiment, showing an increased rate of gas phase CO_2 production with respect to the initial open circuit potential state of the catalyst over time when a potential is applied. When comparing data collected at 550 °C and at 600 °C (Fig. 3B *vs.* 3C), the intensity of the signals associated with carbonate and bicarbonate species (1200–1800 cm^{-1}) also increase with operating temperature. This indicates that the formation of carbonate and bicarbonate species is both potential and temperature dependent. Increased operating temperature induces the formation of carbonate and bicarbonate species while more complex effects such as formation/adsorption and removal are observed when a potential is applied.

5 Discussion

A rig for *operando* DRIFTS studies of SOCs has been designed and constructed and used successfully to analyse the surface and gas species under both fuel cell and electrolysis operating potentials.

DRIFTS spectra are collected from the top electrode of the cell with the option to switch polarization between positive and negative potential, resulting in electrode dynamics which reflects either electrolysis (negative) or fuel cell (positive) operation. Since the atmosphere is swept continuously by the feed gas, it is unlikely that products from electrode reactions occurring at the bottom electrode are detected by the spectroscopy and that changes observed in the DRIFTS are only associated with the top electrode.

The results show three clear processes occurring during cell operation: (1) the potential dependent production of CO_2, (2) the formation of complex carbonate species,[19,20] and (3) the role of water in the form of an associated surface hydroxide species[35] which are surface species that interact *via* hydrogen bonding giving rise to the broad hydroxyl peak in the 3500–3000 cm^{-1} range. There are frequent spectroscopic observations in the literature of carbonate formation in both the ceramic-only zircona system[19,20] and the phenomena is more widely observed in the supported catalysis literature, particularly with reference to the water–gas shift reaction (WGS).[36] Additionally, there is some debate in the literature as to the precise assignment of the carbonate species due to the possibility of the development of formate species.[32] However, it should be noted that as well as the features between 1200–1800 cm^{-1}, additional bands, just below 3000 cm^{-1}, are also typically observed[24] when formates are present. In the present study, these features were not present and thus it is thought that carbonate rather than formate is the predominant surface species.

For carbonates to form from CO, either water or a hydroxylated species must be present.[35] Gas phase CO_2 can then subsequently be formed by decomposition of the carbonate. The surface and gas phase species observed are not thought to originate from outgassing of organics from the electrode paste due to the combination of high temperature preprocessing of the electrodes in air (600 °C), long equilibration time in each atmosphere, and evidence in the literature suggesting that preparation methods using pastes were not considered an origin of carbonate and hydration products.[37]

It is possible that the hydroxyl species observed, which are prevalent throughout the experiments and continuously increase, may indicate some water ingress in the experimental apparatus. This is, however, unlikely due to the high purity of the feed gases, the very low water concentration detected by the mass spectrometer during the measurements and the gas-tight seal of the chamber. These species are, therefore, thought to arise from the fact that the surface and sub-surface of the zirconia electrolyte was partially hydroxylated[35,38] before testing and the electrochemical cleaning procedure used was not sufficient to remove surface species. This is in agreement with previous studies which showed that pre-treatment of yttria-doped zirconia at 600 °C in 20% O_2/Ar for 1 h did not completely remove surface hydroxyl species.[20]

Fig. 3A shows a large band at 450 °C associated with OH species (3245 cm^{-1}). As the temperature increases this band became smaller in keeping with higher protonic (or hydroxylates species) surface diffusion observed in zirconia.[39] At 450 °C, as the cell is polarized in the positive direction, there is a marked decrease in CO_2(g). This is consistent with the formation of surface carbonate according to Fig. 4.

Electrochemically, a negative bias corresponds to pumping oxygen away from the spectroscopically probed surface (electrolysis mode). The changes observed can be attributed to both the thermal hydroxide/CO reaction, *i.e.* the water gas shift reaction, and electrochemical processes; however, it is not clear at this point which dominates. When the cell polarity is reversed, we see an increase in CO_2 production. Once again this could be due to carbonate decomposition or the electrochemical oxidation of CO with oxygen ion transport. However, given the low temperature the oxygen ion conductivity is expected to be small, therefore, it is more likely that the heterogeneous catalytic reaction is controlled by the diffusion of protonic species and which is modified by the applied bias. At 550 °C (Fig. 3B), the OH bands are noticeably weaker. Carbonate species are removed throughout the period of the experiment, but the potential dependent change in the CO_2 concentration has switched. The fuel cell mode produces CO_2, presumably through the decomposition of remaining surface carbonates. There is no longer a sufficient amount of protonic species to allow carbonate formation during this process. During polarisation in electrolysis mode, very little CO_2 production is observed, but carbonate removal continues. At 600 °C (Fig. 3C), the trend continues with the OH band decreasing, carbonate removal increasing, resulting in an increase in CO_2 production, both from carbonate decomposition and CO oxidation.

There is some difficulty accounting for the presence of some of these species on the gold surface. Hydroxides and carbonates are often observed on the surfaces of metal oxides, but the mechanism for their existence on a pure metal surface is less well understood. Some studies have suggested that the penetration depth of an IR beam in a reflectance geometry is a few hundred microns[33] suggesting subsurface detection would be possible. However, at 90 μm thick it is unlikely that the beam is passing through the electrode but further work is required to confirm this. It should also be noted that the observations from IR studies and the electrochemical behaviour of the cell do not necessarily match. This is because the majority of the electrochemistry occurring at the triple-phase boundaries is located at the Au|SSZ interface. Species and chemical responses that are observed in the IR spectra are, therefore, not necessarily dominated by intermediates or products of the electrochemical reactions. This is also why the only potential dependent species observed in this study is CO_2, a by-product of the catalysis and electrochemical reactions.

One would also expect that, particularly in fuel cell mode when oxygen in pumped towards the spectroscopically active electrode, that a marked increase in CO oxidation would be observed. If the presence of hydroxides and carbonates preferentially react with electrochemically supplied oxygen or simply adsorb to sites active to CO oxidation at the phase boundary between gas|Au|SSZ, the absence of CO peaks at (2200–1950 cm^{-1}) would be justified.

In summary, under negative bias the water gas shift reaction occurs at 450 °C but its importance decreases with temperature due to the fact that it is an exothermic equilibrium reaction. The changes with temperature under this bias are, therefore, likely to be due to the thermal/proton mediated decomposition of carbonates which increases in importance with temperature. Under positive bias, little CO oxidation occurs at 450 °C due to slow oxygen ion conduction; however, this process becomes more important as the temperature increases due to the increase in oxygen ion conduction. Hence at 450 °C, CO_2 production only occurs at negative bias whereas at 600 °C it occurs under positive and negative bias.

6 Outlook

One of the main challenges facing any *in situ* measurement is to characterise the sample as close to realistic operational conditions as possible. Many of the limitations associated with this challenge have been overcome with this design of experimental apparatus, however, there remain some key aspects that need further development. Achieving the desired cell operating temperature and

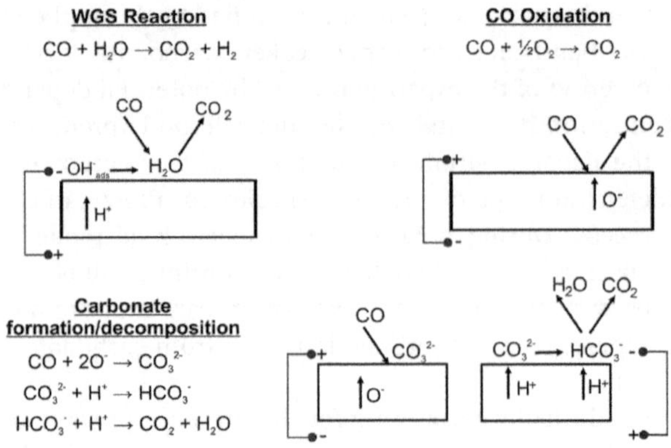

Fig. 4 Proposed reaction mechanisms at the electrode surface.

distribution remains a challenge as operating closer to 800 °C will induce even higher thermal gradients across the base unit components. The location of the heating element within the chamber atmosphere presents challenges to the integrity of the furnace wire but also introduces the risk of contamination of SOCs from volatile elements (such as chromium) in the heater wire. Future work will focus on a more improved thermal design to minimize these issues. Another limitation of the reflectance technique is difficulty in analysing the surface of electrodes that absorb strongly – which includes most state-of-the-art materials. Efforts are being focused on improved collection efficiency for the optical path and IR spectrometer.

Despite careful control of gas inlet composition and design of a hermetically sealed chamber, the spectra showed that surface hydroxyl and complex carbonate species play a subtle but crucial role in the surface chemistry in a SOC. Although regularly discussed in the catalysis literature, the origin and nature of these species is not documented in the solid-state electrochemistry literature. From what has been observed in this study and those in literature, these species influence the water–gas shift reaction, an important reaction during co-electrolysis, and may also influence rate limiting intermediates during electrochemical reaction. These observations also highlight that IR spectroscopy will see all species present on the electrode surface meaning it is not only important to determine the region of the cell or electrode the IR signal originates from, but it is crucial to interpret which species are participating in the electrochemical reaction, which are influenced by those reactions and which are spectating. Once this is understood the key rate-limiting reactions may be identified and steps taken to improve cell chemistry and microstructure. This will be developed using analysis of the temporal behaviour of the electrochemistry and spectroscopic species by inducing a step change in potential or gas composition, for example.[40,41]

The above observations also highlight the importance of cell cleanliness. What is the origin of the observed surface species? Have they formed recently or is their occurrence based of sample history? Factors such as sample processing (*e.g.* electrode pastes) and the previous testing history are very relevant to cell cleanliness. Based on the results presented here, and from other IR studies in the literature, it is essential to monitor and tightly control the moisture in the system.

A suitable cleaning routine, which could include electrochemical pre-treatment, is necessary to form a benchmark during future testing.

In an effort to better understand the role of some of the species observed in this work there is significant scope for use of alternative materials, such as doped ceria instead of zirconia for some electrode components. There is a large body of catalysis literature that shows the mechanism of carbonate formation and decomposition is highly dependent on the how easily the cations can be reduced in the ionic conductor support material. Comparison with the zirconia system and the influence of electrochemical bias will give further insights into the carbonate cycle.

The inability to determine the exact depth profile of the DRIFTS beam has meant there is some uncertainty surrounding the precise location of the observed species. By switching from pure metal electrodes to composite electrodes containing an ionic conductor the electrochemically active region will be extended away from the electrode–electrolyte interface to make it more optically accessible. Other *in situ* experiments have overcome this problem by using well defined patterned electrodes, however this reduces the cell performance to a level that is difficult to reconcile with a real cell. Another issue with this approach is the low concentration of surface species which will decrease the available spectroscopic information.

Finally, the main thrust of future work should be focused on the operation of full SOCs *in situ*. The apparatus we present here has the capability to do this and future publications will cover this area. There are important reactions on both the fuel and oxygen electrode which will benefit from full cell testing. This paper demonstrates the sensitivity of DRIFTS and it is anticipated that full cell testing will enable detailed *operando* study of degradation and poisoning mechanisms.

7 Conclusions

A new diffuse reflectance spectroscopic setup has been developed to enable the surface reactions within a SOC under bias and at temperature to be probed. Under a CO atmosphere, even over a simple Au|10Sc1CeSZ|Au significant changes were observed related to a combination of CO oxidation, the water–gas shift reaction and carbonate decomposition processes. The dominant process was found to be both potential and temperature dependent. This is a powerful technique with which to understand SOC in electrolysis and fuel cell mode and is currently being developed to allow the use of asymmetric cells in the presence of different atmospheres.

Acknowledgements

This work was carried out as part of the "4CU" programme grant, aimed at sustainable conversion of carbon dioxide into fuels, led by The University of Sheffield and carried out in collaboration with The University of Manchester, Queens University Belfast and University College London. The authors acknowledge gratefully the Engineering and Physical Sciences Research Council (EPSRC) for supporting this work financially (Grant no. EP/K001329/1).

References

1 W. Doenitz and R. Schmidberger, *Int. J. Hydrogen Energy*, 1982, **7**, 321–330.
2 W. Doenitz, R. Schmidberger, E. Steinheil and R. Streicher, *Int. J. Hydrogen Energy*, 1980, **5**, 55–63.
3 M. B. Pomfret, J. C. Owrutsky and R. A. Walker, *J. Phys. Chem. B*, 2006, **110**, 17305–17308.
4 R. C. Maher, L. F. Cohen, P. Lohsoontorn, D. J. Brett and N. P. Brandon, *J. Phys. Chem. A*, 2008, **112**, 1497–1501.
5 J. Malzbender, R. W. Steinbrech and L. Singheiser, *Fuel Cells*, 2009, **9**, 785–793.
6 D. J. L. Brett, A. R. Kucernak, P. Aguiar, S. C. Atkins, N. P. Brandon, R. Clague, L. F. Cohen, G. Hinds, C. Kalyvas, G. J. Offer, B. Ladewig, R. Maher, A. Marquis, P. Shearing, N. Vasileiadis and V. Vesovic, *ChemPhysChem*, 2010, **11**, 2714–2731.
7 M. B. Pomfret, J. C. Owrutsky and R. A. Walker, in *Annual Review of Analytical Chemistry*, 2010, vol. 3, pp. 151–174.
8 R. C. Maher, G. Offer, N. P. Brandon and L. F. Cohen, *2011 MRS Fall Meeting*, Boston, MA, 2012.
9 J. E. A. Saunders and M. H. Davy, *Int. J. Hydrogen Energy*, 2012, **37**, 3403–3414.
10 B. C. Eigenbrodt and R. A. Walker, *Spectroscopy*, 2014, **29**, 24–30.
11 X. Li, M. Liu, S. Y. Lai, D. Ding, M. Gong, J. P. Lee, K. S. Blinn, Y. Bu, Z. Wang, L. A. Bottomley, F. M. Alamgir and M. Liu, *Chem. Mater.*, 2015, **27**, 822–828.
12 S. C. DeCaluwe, G. S. Jackson, R. L. Farrow, A. H. McDaniel, F. El Gabaly, K. F. McCarty, S. Nie, M. A. Linne, H. Bluhm, J. T. Newberg, Z. Liu and Z. Hussain, *Solid State Ionic Devices 6-Nanoionics - 214th ECS Meeting*, Honolulu, HI, 2008.
13 M. Yang, E. Bucher and W. Sitte, *J. Power Sources*, 2011, **196**, 7313–7317.
14 B. Bozzini, M. Amati, L. Gregoratti, C. Mele, M. K. Abyaneh, M. Prasciolu and M. Kiskinova, *Electrochem. Commun.*, 2012, **24**, 104–107.
15 Y. Yu, B. Mao, A. Geller, R. Chang, K. Gaskell, Z. Liu and B. W. Eichhorn, *Phys. Chem. Chem. Phys.*, 2014, **16**, 11633–11639.
16 F. Zaera, *Chem. Soc. Rev.*, 2014, **43**, 7624–7663.
17 C. Lamberti, A. Zecchina, E. Groppo and S. Bordiga, *Chem. Soc. Rev.*, 2010, **39**, 4951–5001.
18 J. Ryczkowski, *Catal. Today*, 2001, **68**, 263–381.
19 E.-M. Köck, M. Kogler, R. Pramsoler, B. Klötzer and S. Penner, *Rev. Sci. Instrum.*, 2014, **85**, 084102.
20 E. M. Kock, M. Kogler, T. Bielz, B. Klotzer and S. Penner, *J. Phys. Chem. C*, 2013, **117**, 17666–17673.
21 T. Murai, K. Yashiro, A. Kaimai, T. Otake, H. Matsumoto, T. Kawada and J. Mizusaki, *30th Symposium on Solid State Ionics in Japan*, 2005, **176**, 2399–2403.
22 J. D. Kirtley, D. A. Steinhurst, J. C. Owrutsky, M. B. Pomfret and R. A. Walker, *Phys. Chem. Chem. Phys.*, 2014, **16**, 227–236.
23 F. C. Meunier, A. Goguet, C. Hardacre, R. Burch and D. Thompsett, *J. Catal.*, 2007, **252**, 18–22.
24 F. C. Meunier, D. Reid, A. Goguet, S. Shekhtman, C. Hardacre, R. Burch, W. Deng and M. Flytzani-Stephanopoulos, *J. Catal.*, 2007, **247**, 277–287.

25 D. Tibiletti, F. C. Meunier, A. Goguet, D. Reid, R. Burch, M. Boaro, M. Vicario and A. Trovarelli, *J. Catal.*, 2006, **244**, 183–191.

26 F. C. Meunier, D. Tibiletti, A. Goguet and R. Burch, *Oil Gas Sci. Tech.*, 2006, **61**, 497–502.

27 G. Jacobs, P. M. Patterson, L. Williams, E. Chenu, D. Sparks, G. Thomas and B. H. Davis, *Appl. Catal., A*, 2004, **262**, 177–187.

28 E. Chenu, G. Jacobs, A. C. Crawford, R. A. Keogh, P. M. Patterson, D. E. Sparks and B. H. Davis, *Appl. Catal., B*, 2005, **59**, 45–56.

29 G. Jacobs, U. M. Graham, E. Chenu, P. M. Patterson, A. Dozier and B. H. Davis, *J. Catal.*, 2005, **229**, 499–512.

30 T. Tabakova, F. Boccuzzi, M. Manzoli and D. Andreeva, *Appl. Catal., A*, 2003, **252**, 385–397.

31 T. Shido and Y. Iwasawa, *J. Catal.*, 1993, **141**, 71–81.

32 G. Jacobs, E. Chenu, P. M. Patterson, L. Williams, D. Sparks, G. Thomas and B. H. Davis, *Appl. Catal., A*, 2004, **258**, 203–214.

33 F. C. Meunier, A. Goguet, S. Shekhtman, D. Rooney and H. Daly, *Appl. Catal., A*, 2008, **340**, 196–202.

34 X. Lu, P. W. Faguy and M. Liu, *J. Electrochem. Soc.*, 2002, **149**, A1293–A1298.

35 M. Kogler, E. M. Kock, T. Bielz, K. Pfaller, B. Klotzer, D. Schmidmair, L. Perfler and S. Penner, *J. Phys. Chem. C*, 2014, **118**, 8435–8444.

36 S. Hilaire, X. Wang, T. Luo, R. J. Gorte and J. Wagner, *Appl. Catal., A*, 2001, **215**, 271–278.

37 P. Vernoux, L. Lizarraga, M. N. Tsampas, F. M. Sapountzi, A. De Lucas-Consuegra, J. L. Valverde, S. Souentie, C. G. Vayenas, D. Tsiplakides, S. Balomenou and E. A. Baranova, *Chem. Rev.*, 2013, **113**, 8192–8260.

38 D. L. Hoang and H. Lieske, *Catal. Lett.*, 1994, **27**, 33–42.

39 B. Scherrer, M. V. F. Schlupp, D. Stender, J. Martynczuk, J. G. Grolig, H. Ma, P. Kocher, T. Lippert, M. Prestat and L. J. Gauckler, *Adv. Funct. Mater.*, 2013, **23**, 1957–1964.

40 S. Chansai, R. Burch, C. Hardacre, J. Breen and F. Meunier, *J. Catal.*, 2011, **281**, 98–105.

41 S. Chansai, R. Burch and C. Hardacre, *J. Catal.*, 2012, **295**, 223–231.

Faraday Discussions

PAPER

Is the surface oxygen exchange rate linked to bulk ion diffusivity in mixed conducting Ruddlesden–Popper phases?

Alex C. Tomkiewicz,[a] Mazin A. Tamimi,[a] Ashfia Huq[b] and Steven McIntosh*[a]

Received 4th February 2015, Accepted 27th February 2015

DOI: 10.1039/c5fd00014a

The possible link between oxygen surface exchange rate and bulk oxygen anion diffusivity in mixed ionic and electronic conducting oxides is a topic of great interest and debate. While a large body of experimental evidence and theoretical analyses support a link, observed differences between bulk and surface composition of these materials are hard to reconcile with this observation. This is further compounded by potential problems with simultaneous measurement of both parameters. Here we utilize separate techniques, *in situ* neutron diffraction and pulsed isotopic surface exchange, to examine bulk ion mobility and surface oxygen exchange rates of three Ruddlesden–Popper phases, general form $A_{n-1}A'_2B_nO_{3n+1}$, $A_{n-1}A'_2B_nX_{3n+1}$; $LaSrCo_{0.5}Fe_{0.5}O_{4-\delta}$ ($n = 1$), $La_{0.3}Sr_{2.7}CoFeO_{7-\delta}$ ($n = 2$) and $LaSr_3Co_{1.5}Fe_{1.5}O_{10-\delta}$ ($n = 3$). These measurements are complemented by surface composition determination *via* high sensitivity-low energy ion scattering. We observe a correlation between bulk ion mobility and surface exchange rate between materials. The surface exchange rates vary by more than one order of magnitude with high anion mobility in the bulk of an oxygen vacancy-rich $n = 2$ Ruddlesden–Popper material correlating with rapid oxygen exchange. This is in contrast with the similar surface exchange rates which we may expect due to similar surface compositions across all three samples. We conclude that experimental limitations lead to inherent convolution of surface and bulk rates, and that surface exchange steps are not likely to be rate limiting in oxygen incorporation.

Introduction

Mixed ionic-electronic conducting (MIEC) materials for oxygen anion transport find potential applications as dense membranes for oxygen separation and oxidation reactions and as electrode materials in solid oxide fuel cells (SOFCs). The performances of these devices are dictated by the rate of oxygen anion

[a]*Department of Chemical and Biomolecular Engineering, Lehigh University, Bethlehem, PA, 18015, USA. E-mail: mcintosh@lehigh.edu; Fax: +1-610-758-5057; Tel: +1-610-758-6835*

[b]*Spallation Neutron Source, Oak Ridge National Laboratory, Oak Ridge, TN, 37830, USA*

formation and incorporation from oxygen gas at the MIEC surface, and the subsequent rate of oxygen anion transport through the MIEC bulk. These parameters are typically measured and denoted as surface oxygen exchange rates, *k*,[1] and bulk anion diffusivity, *D*.

An intriguing observation in the literature is a proposed correlation between these two parameters across a wide range of MIEC materials.[2–4] Direct correlations between these parameters have been derived,[3] with De Souza suggesting a 'universal' equation for the surface exchange coefficient.[2] The most common interpretation[1,4] for 'electron-rich' materials is that the two processes are linked through the oxygen vacancy concentration. Where bulk anion transport *via* vacancy hopping depends directly on vacancy concentration, the relationship between the surface exchange coefficient and the vacancy concentration depends on the assumed surface reaction mechanism and the rate limiting step. For example, Merkle *et al.*[5] derived a direct correlation between these bulk and surface parameters by assuming a two-step surface mechanism; a molecular oxygen rapidly dissociates to form O^-, followed by a rate-determining step of oxygen incorporation to fill a vacancy.

An underlying assumption in all of these derived correlations is that the bulk composition and oxygen vacancy concentration extend up to the outer MIEC surface. This assumption is in stark contrast to the observation by multiple authors that the surface compositions of these materials vary dramatically from the bulk. This is most commonly observed as surface enrichment of A-site cations, notably Sr.[6–10] With such segregation it is difficult to reconcile mechanisms that link the outermost surface reaction rate to the bulk composition, structure or oxygen vacancy concentration. If the outermost surface compositions of these materials are similar, we would expect similar surface reaction rates irrespective of the bulk composition or structure. Reconciling these contradictions remains an important challenge to the field.

We have previously suggested that the observed correlation may in part be due to the inherent difficulty in accurately extracting two independent parameters from a single exponential curve;[9] a technique applied in most simultaneous measurements of *k* and *D*. This suggestion motivated our group to adopt two independent measurements of these parameters. We utilize *in situ* neutron diffraction[11–15] to examine bulk ion transport and an isotopic pulse exchange technique to characterize surface exchange rates.[16–19] This information is then complemented by determination of the surface composition *via* X-ray photoelectron spectroscopy (XPS) or high sensitivity low energy ion scattering (HS-LEIS).

High temperature, controlled atmosphere *in situ* powder neutron diffraction is utilized to extract details of crystal symmetry, lattice constants and atomic positions, occupancies, and displacement parameters for both the cations and anions in MIEC materials.[15] While this does not yield a direct measurement of ionic conductivity or a *D* parameter, this crystallographic information allows the ionic transport pathways in the materials to be discerned and enables direct comparison of ion motion between materials of the same crystal structure. Oxygen anions are transported through the bulk structure *via* a vacancy hopping mechanism, and thus insights into the rate of oxygen transport for a given material can be gained by simultaneous consideration of the concentration of oxygen vacancies, the hopping distance between neighboring oxygen sites, and the magnitude and orientation of the anisotropic atomic displacement parameters for the anions.

We have previously performed these combined studies on both cubic[19] and layered perovskites,[18] and in both cases we have found the same correlation between high bulk ion mobility and rapid surface exchange, even noting changes in surface exchange rate upon changes in bulk crystal symmetry. This is despite the surface compositions of these materials varying dramatically from the average bulk compositions. Herein we present a related body of work, studying layered Ruddlesden–Popper structures and suggesting a possible explanation for the observed correlation. The Ruddlesden–Popper phases are layered materials of general form $A_{n-1}A'_2B_nO_{3n+1}$, $A_{n-1}A'_2B_nX_{3n+1}$, consisting of n perovskite layers alternating with a rock salt layer. Fig. 1 illustrates the structures of the $n = 1$, $n = 2$ and $n = 3$ materials considered in this study.

Experimental

Material synthesis

Formation of $n = 1$, $n = 2$ and $n = 3$ Ruddlesden–Popper phases with identical La/Sr and Co/Fe ratios was not possible due to sample decomposition into a mixture of the desired phase along with perovskite, simple oxide and other Ruddlesden–Popper phases. Sharma and Singh developed a tolerance factor for Ruddlesden–Popper $n = 1$ phases.[20] This approach was used to vary the La/Sr ratio at a constant Co/Fe ratio to synthesize three Ruddlesden–Popper materials, $LaSrCo_{0.5}Fe_{0.5}O_{4-\delta}$ (RPn1), $La_{0.3}Sr_{2.7}CoFeO_{7-\delta}$ (RPn2), and $LaSr_3Co_{1.5}Fe_{1.5}O_{10-\delta}$ (RPn3). Each material has a tolerance factor t_f of approximately 1.08, corresponding to a tetragonal phase K_2NiF_4-type structure. The La/Sr ratios vary from 1 : 1 to 1 : 9 and 1 : 3 for RPn1, RPn2, and RPn3 materials, respectively.

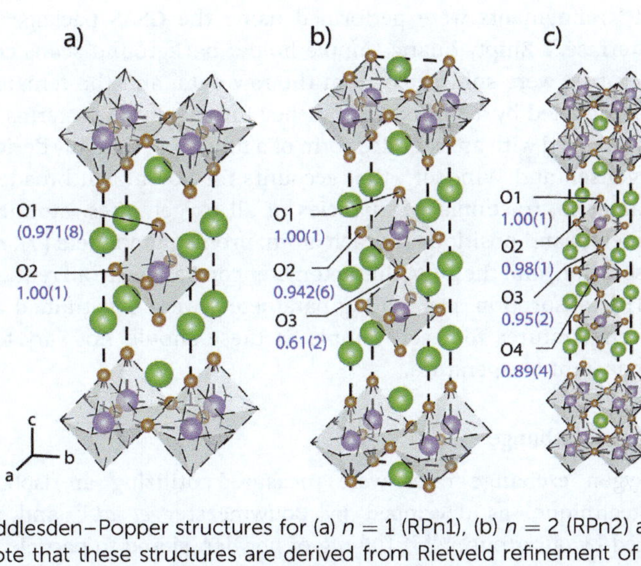

Fig. 1 Ruddlesden–Popper structures for (a) $n = 1$ (RPn1), (b) $n = 2$ (RPn2) and (c) $n = 3$ (RPn3). Note that these structures are derived from Rietveld refinement of the neutron diffraction data collected at 967 K with a pO_2 of 0.1 atm. The numbers in blue denote the refined oxygen site occupancies at these conditions.

All materials were synthesized using stoichiometric ratios of titrated metal nitrate precursor solutions. $La(NO_3)_3$ [99.9%, Alfa Aesar, Ward Hill, MA, USA], $Sr(NO_3)_2$ [99.9%, Alfa Aesar], $Co(NO_3)_2$ [98.0–102.0%, Alfa Aesar] and $Fe(NO_3)_3$ [98.0–102.0%, Alfa Aesar] powders were dissolved in distilled water and the concentration was verified using an EDTA titrant [0.100 M, Ricca Chemical, Arlington, TX, USA]. The metal nitrate solutions were combined with citric acid and EDTA as complexing agents in a modified Pechini method, as described previously.[21] The resulting mixtures were heated to remove excess water, resulting in the formation of a gel which was then combusted in air at 573 K for 12 hours. The cake resulting from this combustion was then ground into powder and calcined in air at 1623 K for 10 hours. The final powder samples were characterized by X-ray diffraction [Rigaku Miniflex II, Texas, USA] as an initial verification of phase purity.

In situ determination of bulk crystal structure

Neutron diffraction measurements were conducted at the POWGEN beamline, Spallation Neutron Source (SNS), Oak Ridge Laboratory, Oak Ridge, TN, USA. Powder samples were suspended within the neutron beam using an *in situ* measurement cell.[22] The sample temperature was calibrated using a ZnO reference sample.[23] The sample environment was created using a continuous flow of an ultra-high purity mixture of O_2 and N_2 with a pO_2 of 10^{-1} atm at 500 sccm, with each sample allowed to equilibrate for at least one hour prior to each measurement. Diffraction patterns were collected at 967 K in the *d*-spacing range of 0.45–5.35 Å until a total accelerator source proton charge of 9.0×10^{12} pC was reached (~2 hour measuring time). Several patterns with shortened collection times (7.50×10^{11} pC) were collected and examined prior to the 2 hour measurement to ensure that the sample had reached equilibrium. Samples were declared equilibrated after there was no observed change in the primary peak positions for at least 3 collection periods.

Rietveld[24] refinements were performed using the GSAS package[25] with the EXPGUI interface.[26] Empty quartz sample holder background scans collected at each temperature were subtracted from the raw data, and the remaining background was modeled by a shifted Chebyschev function with 14 terms. The peak profile was modeled with an extended form of a function originally derived by Von Dreele, Jorgensen and Windsor[27] that accounts for Lorentzian broadening. The lattice parameters, fractional occupancies of all oxygen sites, atomic displacement parameters and positions for each atom, profile parameters (γ_1, σ_1 and σ_2), sample absorption, and the DIFA diffractometer constant were all refined for each sample. DIFA, absorption and scaling parameters were constrained across the measured temperatures for each sample as these should not vary for a given sample at constant temperature.

Surface oxygen exchange

Surface oxygen exchange rates were measured utilizing an isotopic pulse exchange technique, as developed by Bouwmeester *et al.*[17] and previously implemented by our group.[16,18,28] 100 mg of powder, sieved to particle diameters between 106 μm and 150 μm, was placed within a 4 mm id. quartz tube and held in place by plugs of quartz wool. Measurements of reaction rates at higher temperatures were performed using samples diluted with $SrTiO_3$ (STO) to avoid exchange of the entire isotopic pulse. STO is inert relative to the materials of interest for the oxygen exchange reaction.[18]

The reactor tube was placed in a tube furnace and heated to 1073 K in an atmosphere of 21% O_2/N_2 for one hour prior to measurement to clean the surface of any carbonates.[29] The sample was then cooled to measurement temperature and allowed to equilibrate in the 21% O_2/N_2 stream for 2 hours prior to rate measurements.

A computer controlled 6-way valve with a pneumatic actuator [Swagelok model 131 SR] was utilized in combination with a 500 μL sample loop to inject a pulse of 21% $^{18}O_2/N_2$ (97 atom%, Sigma-Aldrich, Miamisburg, Oh, USA] into a stream of 21% O_2/N_2 with a total flow rate of 150 mL min^{-1}. The pO_2 within the natural abundance stream and the pulse loop were identical in order to avoid any driving force for the overall transport of oxygen into or out of the sample. Pulses were measured in sets of five at each temperature, with the standard deviation between measurements represented as error bars in the data. Each pulse within a set was separated from the others by a 90 s delay to allow sufficient time for the measured isotopic fraction to return to baseline values.

The outlet gas stream was continually analyzed (sampling frequency: 1.67 s^{-1}) using a residual gas analyzer (RGA) fitted with a quadrupole analyzer and associated Faraday and secondary electron multiplier detectors (Cirrus 2, MKS Instruments UK Ltd). The $^{16}O_2$ ($m/z = 32$) signal did not vary significantly during the pulse experiments due to the short duration of the pulse compared to the residence time in the RGA chamber. Considering this, only the $^{16}O^{18}O$ ($m/z = 34$) and $^{18}O_2$ ($m/z = 36$) signals were measured. The reduction in the $^{18}O_2$ pulse response and the corresponding increase in the $^{16}O^{18}O$ pulse response directly correlate to the rate of reaction of oxygen on the surface of the sample. The surface exchange rate, R_O (mol ^{18}O m^{-2} s^{-1}), is calculated based on the RGA response for the $^{16}O^{18}O$ and $^{18}O_2$ pulse signals assuming first order kinetics.[17]

Surface composition

The surface compositions were measured by high sensitivity-low energy ion scattering (HS-LEIS, ION-TOF Qtac-100, Münster, Germany). Powder samples were treated in air at 1173 K for 12 hours and pressed into pellets at 2000 psi. An initial scan was performed using 3 keV He^+ to identify the presence of lighter surface contaminants such as surface carbonates.[19] The samples were then exposed to an oxygen plasma to remove any surface carbon species without altering the distribution or composition of the outer surface. Measurements of the surface composition were then performed using 5 keV Ne^+ to further distinguish between the La and Sr peaks, as the two peaks partially overlap when utilizing a He^+ ion source. The ion beam impacted normal to the surface and was rastered over a 1000 μm × 1000 μm area. Repeated sputtering of the sample surface with 2.1 × 10^{14} cm^{-2} of a 0.5 keV Ar^+ beam followed by subsequent measurement with the 5 keV Ne^+ ion source allowed for the development of a depth profile to examine the transition from surface to bulk compositions within the samples. To enhance the signal to background ratio for a given measurement time, measurements were focused on three regions: (1) 1100–1400 eV with 7.2 ×

10^{13} Ne^+ cm^{-2} for Fe and Co, (2) 1900–2300 eV with 1.3 × 10^{13} Ne^+ cm^{-2} for Sr; and (3) 2650–3000 eV with 1.3 × 10^{13} Ne^+ cm^{-2} for La.

Results

In situ neutron diffraction

At 967 K and $pO_2 = 0.1$ atm, the structures of all three samples were fit satisfactorily to a tetragonal structure, space group *I4/mmm*, as depicted in Fig. 1. Fig. 2 shows the measured, fit, and difference patterns, with structural details and quality of fit parameters listed in Table 1. Less than 4 wt% of a cubic perovskite impurity phase, space group $Pm\bar{3}m$, was present in each sample, with positions denoted by an asterisk in Fig. 2. No evidence for A-site (La/Sr) or B-site (Co/Fe) ordering nor deviations from the proposed material stoichiometries were observed from the refinement. As such the occupancies of A-sites and B-sites by La/Sr and Co/Fe were fixed at the proposed overall material compositions and the atomic displacement parameters for the individual A- and B-sites were identical irrespective of cation.

Of primary interest to oxygen anion transport is the relative occupancies, Fig. 1 and Table 1, anisotropic atomic displacement parameters, Table 1, and direct site-to-site hopping distance, Table 2, of the oxygen sites in the structure. These represent the oxygen vacancy concentrations, the average motion of the oxygen anions, and the hopping distance for oxygen transport, respectively.

The oxygen sites in the RPn1 material were almost completely occupied under the measurement conditions, with only a small vacancy concentration, 97.1(8)%

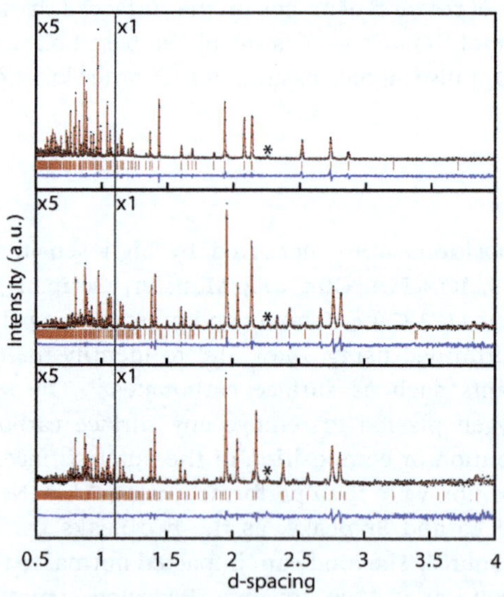

Fig. 2 Neutron diffraction patterns collected at 967 K with a pO_2 of 0.1 atm for (a) $LaSrCo_{0.5}Fe_{0.5}O_{4-\delta}$ (RPn1), (b) $La_{0.3}Sr_{2.7}CoFeO_{7-\delta}$ (RPn2), and (c) $LaSr_3Co_{1.5}Fe_{1.5}O_{10-\delta}$ (RPn3) samples. All three materials were fitted to structures with the *I4/mmm* space group. A perovskite ($Pm\bar{3}m$) impurity was identified in each sample, with the primary diffraction peak marked by an asterisk. Measured data (black dots), calculated diffraction pattern (top red line), difference pattern (bottom blue line) and expected peak positions are shown.

Table 1 Tetragonal space group *I4/mmm*, Rietveld fit parameters for RPn1, RPn2 and RPn3 structures at 967 K with sites: RPn1: La/Sr1: 4e (0,0,*Z*), Co/Fe1: 2a (0,0,0), O1: 4e (0,0,*Z*), O2: 4c (0.5,0,0); RPn2: La/Sr1: 4e (0,0,*Z*), La/Sr2: 2b (0,0,0.5), Co/Fe1: 4e (0,0,*Z*), O1: 4e (0,0,*Z*), O2: 8g (0,0.5,*Z*), O3: 2a (0,0,0); RPn3: La/Sr1: 4e(0,0,*Z*), La/Sr2: 4e (0,0,*Z*), Co/Fe1: 4e (0,0,*Z*), Co/Fe2: 2a (0,0,0), O1: 4e (0,0,*Z*), O2: 8g (0,0.5,*Z*), O3: 4e (0,0,*Z*), O4: 4c (0,0.5,0)

$LaSrCo_{0.5}Fe_{0.5}O_{3.942}$ (RPn1), *I4/mmm*, 967 K					
a, *b* (Å)	3.868(1)		χ^2		4.522
c (Å)	12.840(1)		R_P%		0.1055
V (Å^3)	192.08(1)		wR_P%		0.0423
ρ_c (g cm^{-3})	5.999				

	Atomic displacement parameters (Å^2) × 100				
	U_{11}/U_{iso}	U_{22}	U_{33}	*Z*	Fractional occupancy
La/Sr1	1.90(2)			0.359(1)	
Co/Fe1	1.53(3)				
O1	3.52(7)	3.52(7)	1.8(1)	0.167(1)	0.971(8)
O2	1.0(1)	2.7(1)	3.3(1)		1.0

$La_{0.3}Sr_{2.7}CoFeO_{6.378}$ (RPn2), *I4/mmm*, 967 K					
a, *b* (Å)	3.912(1)		χ^2		4.529
c (Å)	20.409(1)		R_P%		0.1459
V (Å^3)	312.28(1)		wR_P%		0.0587
ρ_c (g cm^{-3})	5.265				

	Atomic displacement parameters (Å^2) × 100				
	U_{11}/U_{iso}	U_{22}	U_{33}	*Z*	Fractional occupancy
La/Sr1	2.87(6)			0.317(1)	
La/Sr2	2.52(7)				
Co/Fe1	1.76(5)			0.101(1)	
O1	3.65(7)			0.196(2)	1.0
O2	3.4(1)	1.8(1)	4.4(2)	0.091(1)	0.942(6)
O3	4.7(3)				0.61(2)

$LaSr_3Co_{1.5}Fe_{1.5}O_{9.582}$ (RPn3), *I4/mmm*, 967 K					
a, *b* (Å)	3.900(1)		χ^2		2.786
c (Å)	28.523(1)		R_P%		0.1577
V (Å^3)	433.78(2)		wR_P%		0.0587
ρ_c (g cm^{-3})	5.568				

	Atomic displacement parameters (Å^2) × 100				
	U_{11}/U_{iso}	U_{22}	U_{33}	*Z*	Fractional occupancy
La/Sr1	2.5(1)			0.2013(1)	
La/Sr2	2.8(1)			0.0710(2)	
Co/Fe1	1.5(1)			0.1409(1)	

Table 1 (*Contd.*)

	Atomic displacement parameters (Å^2) × 100				
	U_{11}/U_{iso}	U_{22}	U_{33}	Z	Fractional occupancy
Co/Fe2	2.7(1)				
O1	3.8(2)			0.2114(2)	1
O2	2.5(1)			0.1380(2)	0.98(1)
O3	4.7(3)			0.0654(3)	0.95(2)
O4	12(1)	2.2(4)	11.4(9)		0.89(4)

occupied, for the apical oxygen of the perovskite layer. The RPn2 and RPn3 both show increased vacancy concentrations and a degree of vacancy localization within the perovskite layers, with the oxygen sites in the rock-salt layer refined to full occupancy. This localization trend is analogous to previously published data on both Ruddlesden–Popper[30–34] and other layered materials.[13,22,35] The RPn2 structure was found to have the lowest oxygen stoichiometry, with an occupancy of 0.61(2) at the centermost apical oxygen site (O3) and an occupancy of 0.942(6) at the neighboring equatorial (O2) site. The oxygen vacancies in the RPn3 structure were more diffuse, although still centered within the central octahedron, with the central equatorial site (O3) showing the lowest occupancy and with occupancy increasing outward from the center through O3, O2 and O1.

The compositions thus calculated are $LaSrCo_{0.5}Fe_{0.5}O_{3.94}$ (RPn1), $La_{0.3}Sr_{2.7}CoFeO_{6.38}$ (RPn2) and $LaSr_3Co_{1.5}Fe_{1.5}O_{9.58}$ (RPn3); that is, the oxygen vacancy concentration follows the trend RPn2 > RPn3 > RPn1. This trend is expected *via* charge compensation from the increasing Sr^{2+} *vs.* La^{3+} content at a fixed Co/Fe ratio in the same trend of RPn2 > RPn3 > RPn1.

Ideally, anisotropic atomic displacement parameters would be utilized for each oxygen site in each structure. However, the addition of these fit parameters is only justified if it leads to significant improvement in refinement quality (measured by χ^2, R_P and wR_P parameters). This was the case for only a limited number of sites in this study; for all other cases, the atomic displacement

Table 2 Direct oxygen transport distances through least occupied sites at 967 K. Transport distances between neighboring equatorial oxygen sites as well as between axial and equatorial sites are considered

Direct oxygen transport distances at 967 K			
	RPn1	RPn2	RPn3
O_1–O_2	2.885(2)	2.905(4)	2.860(6)
O_2–O_2	2.734(1)	2.766(1)	2.758(1)
O_2–O_3		2.690(2)	2.844(7)
O_3–O_4			2.699(5)
O_4–O_4			2.758(1)

parameters were refined isotropically. The isotropic atomic displacements for the least occupied equatorial sites of each material are shown as density maps in Fig. 3, illustrating their relative anisotropy.

The refined anisotropic oxygen displacement parameters for the two oxygen sites in the RPn1 structure indicate that the apical (O1) site moves primarily within the *a*–*b* plane towards the nearest neighbor equatorial oxygen site (O2). In turn, this neighboring site shows preferential displacement along the *c*-axis towards the apical site.

Unfortunately, any anisotropy of the apical sites (O1 and O3) in the RPn2 phase could not be differentiated from the Rietveld refinement. However, we can observe a significant increase in atomic displacement for the lowest occupancy central apical site (O3) when compared to the site in the rock salt layer (O1). We were able to differentiate anisotropic displacement for the equatorial site (O2) which showed statistically significant preferential displacement along the *c*-axis towards the apical sites and along the face of the unit cell towards the equatorial neighbor, when compared with motion towards the central B-cation.

Similar limitations were observed in determining the anisotropy of the oxygen sites in the RPn3 material. Only the O4 equatorial site showed significant anisotropic displacement, again along the *c*-axis towards the apical site and along the face of the unit cell towards the neighboring equatorial site.

The neutron diffraction data also provides information on the relative direct oxygen transport distances between materials, Table 2. Note that we only consider hops between nearest neighbors and that these direct distance comparisons do not account for the real curved path that the oxygen anions must take around the cations. The shortest hop across all materials is between the least occupied apical O3 site and the neighboring equatorial O2 site in the RPn2 structure. The second shortest is between the similar lowest occupied equatorial O4 site to the apical O3 site in the RPn3 structure. The differences in direct hopping distances are due to atomic shifts from the ideal lattice sites towards the lower oxygen occupancy planes in the structure, Table 1.

In situ neutron diffraction data was also collected at 765 K, 865 K and 1070 K and will be presented in detail elsewhere. The crystal structure and, in general terms, all of the trends between materials observed at 967 K are conserved across the temperature range.

When taken in combination, the locations and densities of oxygen vacancies, the atomic displacement magnitudes and directionalities, and the distances for vacancy hopping all suggest an oxygen anion transport pathway through the

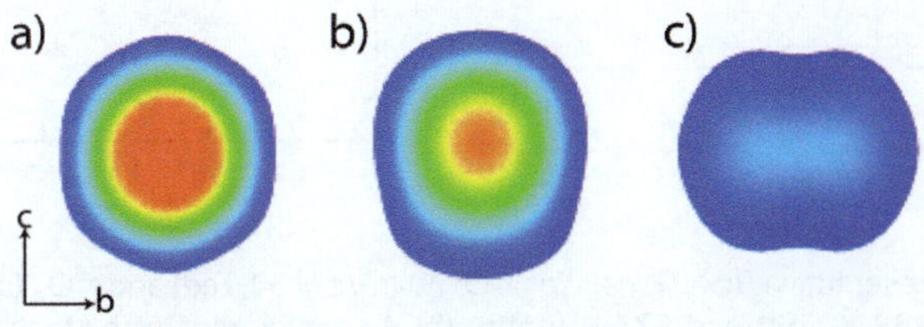

Fig. 3 Observed nuclear density cut along the *b*–*c* plane of the least occupied equatorial oxygen for (a) the O2 site of RPn1, (b) the O2 site of RPn2 and (c) the O4 site of RPn3.

central perovskite layers of all of these materials. Rapid ion transport should be favored by high local vacancy concentration (low oxygen occupancy) and short hopping distances. Based on these considerations, we conclude that ion transport through the perovskite layers decreases in the order RPn2 > RPn3 > RPn1.

In particular, the O2 to O3 hop in RPn2 is favored by having the shortest distance and highest vacancy concentration. The relatively diffuse and highly oriented atomic displacement between these two sites is further confirmation that transport between these sites is facile. We suggest that this is the most facile pathway observed in all three of these materials. Hops between other sites in RPn2 are less favorable due to greater distances and lower vacancy concentrations, also reflected in the more localized atomic displacement parameters. The RPn1 structure is the least favorable due to high oxygen occupancy in the perovskite layer and relatively long hopping distances between nearest neighboring sites. The RPn3 structure is a middle case, with vacancies distributed across the sites in the perovskite layers and some short hops comparable to those in the RPn2 structure.

Surface exchange and composition

Representative data for the isotopic pulse experiments are shown in Fig. 4a. At room temperature, there is no detectable exchange of $^{18}O_2$, resulting in a single peak corresponding to the $^{18}O_2$ signal ($m/z = 36$). At operating temperatures, there is significant exchange of ^{18}O from the gas phase with ^{16}O species in the material, yielding a large $^{16}O^{18}O$ ($m/z = 34$) signal with corresponding decrease in the $^{18}O_2$ signal.

The overall surface oxygen exchange rate, R_O, is shown as a function of inverse temperature for all three materials in Fig. 4b. The reaction rate of the RPn2 phase was found to be the highest, followed reasonably closely by RPn3, with the rate over RPn1 being more than one order of magnitude lower. The experiment was repeated with multiple pure samples and samples diluted with STO in order to

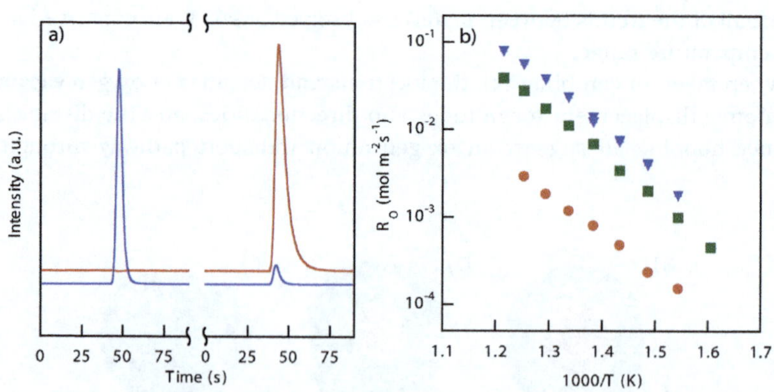

Fig. 4 (a) Representative RGA signals for $^{16}O^{18}O$ ($m/z = 34$, red) and $^{18}O_2$ ($m/z = 36$, blue) for pulses at 298 K (left) and 973 K (right). (b) Arrhenius plot of surface reaction rates derived from $^{18}O_2$ isotopic pulse experiments for RPn1 (red circle), RPn2 (blue inverted triangle) and RPn3 (green square).

measure reaction rates at the highest temperatures. No deviation was observed upon repetition or dilution. The activation energies calculated for each of the Ruddlesden–Popper samples were 85.5 kJ mol^{-1}, 95.7 kJ mol^{-1} and 95.4 kJ mol^{-1} for the RPn1, RPn2 and RPn3 phases, respectively.

Initial scans of the as-prepared surfaces of each material using the 3 keV He^+ beam (selected to observe light elements on the surface) showed well defined O, C (likely as carbonate) and Sr peaks with a small shoulder to the Sr peak hinting at the presence of La when using He^+ as the probe ion. The carbon contaminants were removed by treatment of the surface with oxygen plasma.

Switching to 5 keV Ne^+ probe ions to differentiate between Sr and La showed each surface to be dominated by Sr, with small amounts of surface La present (Fig. 5 and 6). Co and Fe were absent from the surface. Repeated sputtering with 0.5 keV Ar^+ enabled determination of a depth profile. Oxygen is not measured due to the use of the Ar^+ ion. Peaks attributed to Co and Fe were identified after the initial surface was removed. The measurement of a depth profile continued until the Sr, La and Co/Fe peak areas were constant, considered to be equal to the bulk ratios. These bulk values were used to normalize the surface spectra and calculate the composition profiles as functions of depth, Fig. 6.

All three materials show a significant surface enrichment of Sr with a corresponding absence of B-site cations Co and Fe. Fig. 6 shows that the surfaces of the materials have 68%, 96% and 70% surface coverage of Sr cations for the RPn1, RPn2 and RPn3 materials, respectively, with the remainder comprised of La cations. This follows the trend of the Sr/La ratios in the materials and is in line with previous reports.[6–10,36,37,39] The similarities of the surfaces of these three Ruddlesden–Popper materials, despite the vastly differing bulk structures, would hint that the three should have comparable surface reaction rates. The approximate sputtering depth was estimated from previously published work.[10] The compositions reach the bulk values within ~10 nm of the surfaces.

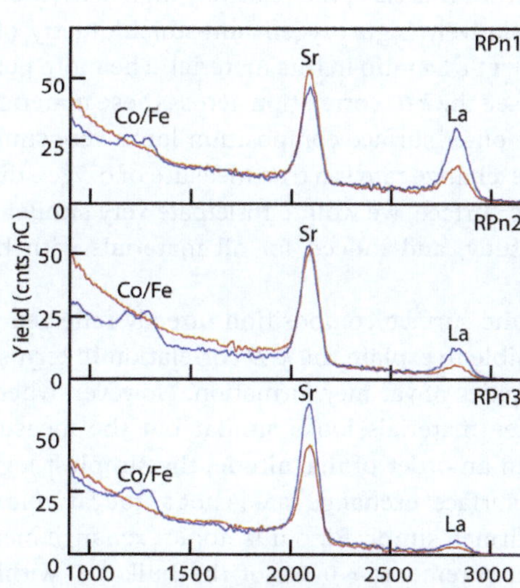

Fig. 5 Example LEIS spectra for RPn1, RPn2 and RPn3 samples, showing initial scans of the surfaces after treatment with O atoms (red) and after achieving bulk compositions (blue).

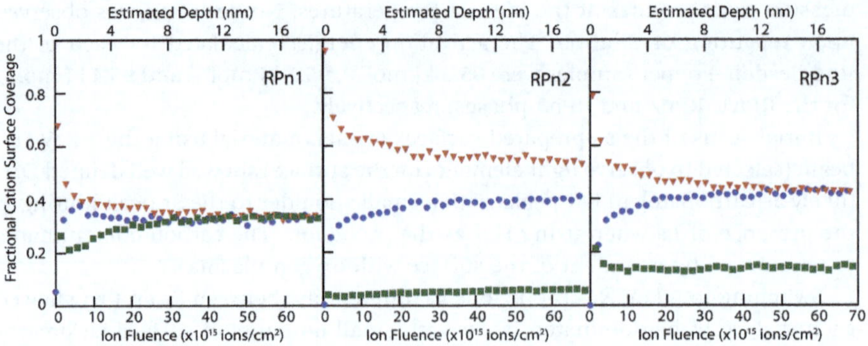

Fig. 6 Depth profiles generated by LEIS, showing the fractional surface coverages of Co/Fe (blue circle), Sr (red inverted triangle) and La (green square) for RPn1, RPn2 and RPn3 structures.

Discussion

Numerous research groups have identified correlations between surface and bulk oxygen transport rates. For example, the trends proposed by De Souza show a direct correlation between surface exchange and bulk diffusivity for a wide range of materials.[2] The existence of this trend over such a wide range of materials despite that fact that surface structures can vary greatly depending on synthesis route, and more importantly differ significantly from the bulk structure due to cation segregation[36,37] or the presence of surface impurities,[38] shows a fundamental issue with the separation between "surface" and "bulk" properties.

Our work on these Ruddlesden–Popper phases appears to confirm the previously reported correlation between rapid oxygen surface exchange and high ion mobility. The ionic mobility follows the same trend as the surface exchange rates: RPn2 > RPn3 > RPn1. In this case, the relatively high activity of the RPn2 phase is likely due to the relatively high oxygen non-stoichiometry of the RPn2 phase, induced by the higher Sr/La ratio in this material. The more puzzling aspect is the question of why we see the *k*–*D* correlation across these materials, especially when we observe that the outer surface composition looks very similar between materials. If the surface exchange rate is a true measure of oxygen dissociation kinetics on the outer atomic surface, we would anticipate very similar rates for all of the materials in this study, and indeed for all materials with heavily Sr-enriched surfaces.

If the outer atomic surface composition directly reflected the bulk composition, it may be possible to explain the *k*–*D* correlation in terms of oxygen vacancy availability or energetics of vacancy formation. However, when the outer surface composition of three materials looks similar but the measured exchange rate differs by more than an order of magnitude, the simplest logical explanation is that the measured 'surface' exchange rate is not a true 'atomic outer surface' rate.

This can be confirmed simply for our isotopic exchange measurements. In the course of these measurements ~0.4% of the bulk ^{16}O within each particle is exchanged for ^{18}O. Assuming that the ^{18}O is completely exchanged with ^{16}O as a moving front, the ^{18}O penetrates ~200 nm into the ~150 μm particles. The actual value is likely to be higher due to bulk mixing. In contrast, the surface Sr enrichment reaches the bulk composition within ~10 nm of the outer surface. Thus, the isotopic pulse experiments exchange oxygen across a depth ~20 times greater than the thickness of the outer "surface" composition of the material, resulting in a significant contribution of "bulk" transport rates within the measurement of the surface exchange rate. We suggest that a similar 'surface sensitivity' analysis should be conducted for all of the approaches to measuring surface exchange rates and bulk diffusivity. Certainly the observed consistent k–D correlation across materials and, to the best of our knowledge, universal variation of surface composition from the bulk suggests that measured values for k are influenced by penetration into the outer layers of the material.

Despite this, materials exhibiting rapid surface exchange also typically yield high performance electrodes. Any oxygen incorporated into these materials must be transferred from the outer atomic surface into the bulk, and we propose that the surface exchange rates in the literature simply reflect the incorporation rate into the first few hundred nanometers of the material. Thus, the measured parameters are still very useful, and we would suggest, still applicable for some modelling efforts.

The concept that we are inherently measuring a bulk property when we attempt to measure surface exchange rates has implications for our understanding of surface incorporation rates. It certainly suggests that oxygen dissociation and incorporation in the outermost atomic layers is facile when compared with ionic transport across a few hundred nanometers. If these outermost steps, which must be determined by the outermost atomic composition, were rate limiting, we would measure the same surface exchange rates across materials in this study. Instead, we find a correlation between the surface exchange rate and the bulk ion mobility. It appears that the bulk transport is rate limiting.

An alternative explanation is that all samples are surface limited and that the Sr content on the outer surface determines the rate. The determined surface exchange rates do increase as the Sr surface concentration increases, however it is difficult to correlate a greater than one order of magnitude difference in the surface exchange rate between RPn1 and RPn3 with the relatively small difference (2%) in surface composition. It seems more likely that the large difference in vacancy concentration and ion mobility is responsible.

The challenge to the field then remains to measure the rates of these outermost surface steps. In our case, it is difficult to increase the surface sensitivity of the isotopic pulse technique as we are limited by limits of isotope detection in the RGA. We can become more surface sensitive by utilizing smaller pulses or more rapid gas flow rates but the measurements simultaneously become much more difficult to make due to the reduced area of the resulting peaks generated by the RGA. It is not clear how surface sensitive the other commonly used techniques are.

Conclusion

Three Ruddlesden–Popper phases of general form $A_{n-1}A'_2B_nX_{3n+1}$, $LaSrCo_{0.5}Fe_{0.5}O_{4-\delta}$ (n = 1, RPn1), $La_{0.3}Sr_{2.7}CoFeO_{7-\delta}$ (n = 2, RPn2) and $LaSr_3Co_{1.5}Fe_{1.5}O_{10-\delta}$ (n = 3, RPn3), were synthesized and found to have a tetragonal structure, space group $I4/mmm$, at 967 K and $pO_2 = 0.1$ atm. The oxygen vacancies in these materials were localized within the perovskite layers, with vacancy concentration following the expected trend, concurrent with Sr/La ratios of RPn2 > RPn3 > RPn1. Analysis of the crystal structure indicates that vacancy mobility follows the same trend. A similar trend in oxygen surface exchange rate is observed despite the surface compositions of the materials being quite similar. All surfaces showed significant surface segregation of Sr. It is suggested that the observed correlation between surface and bulk properties is due to experimental limitations that inherently convolute the two properties. Furthermore, our results indicate that surface oxygen incorporation is not rate limiting in these materials. The challenge remains to experimentally determine the true outer surface oxygen exchange rate.

Acknowledgements

Partial support for this work was provided by Lehigh University through the Faculty Innovation Grant program. We thank Dr Henry Luftman for his assistance with the HS-LEIS experiments. Experimentation at POWGEN was performed as part of the User Program of the Spallation Neutron Source at Oak Ridge National Laboratory, funded by the Office of Basic Energy Sciences of the U.S. Department of Energy. Mazin Tamimi is a sponsored student supported by the Saudi Arabian Oil Company, Saudi Aramco.

References

1 J. Maier, *Solid State Ionics*, 1998, **112**, 197–228.
2 R. A. De Souza, *Phys. Chem. Chem. Phys.*, 2006, **8**, 890–897.
3 R. A. De Souza and J. A. Kilner, *Solid State Ionics*, 1999, **126**, 153–161.
4 J. A. Kilner, R. A. De Souza and I. C. Fullarton, *Solid State Ionics*, 1996, **86–88**(2), 703–709.
5 R. Merkle, J. Maier and H. J. M. Bouwmeester, *Angew. Chem., Int. Ed.*, 2004, **43**, 5069–5073.
6 F. S. Baumann, J. Fleig, M. Konuma, U. Starke, H. Habermeier and J. Maier, *J. Electrochem. Soc.*, 2005, **15210**, A2074–A2079.
7 S. P. Jiang and J. G. Love, *Solid State Ionics*, 2001, **138**, 183–190.
8 W. Wang and S. P. Jiang, *Solid State Ionics*, 2006, **177**, 1361–1369.
9 R. A. Cox-Galhotra and S. McIntosh, *Solid State Ionics*, 2010, **181**, 1429–1436.
10 J. Druce, H. Tellez, M. Burriel, M. D. Sharp, L. J. Fawcett, S. N. Cook, D. S. McPhail, T. Ishihara, H. H. Brongersma and J. A. Kilner, *Energy Environ. Sci.*, 2014, **7**, 3593–3599.
11 S. McIntosh, J. F. Vente, W. G. Haije, D. H. A. Blank and H. J. M. Bouwmeester, *Solid State Ionics*, 2006, **177**, 833–842.
12 S. McIntosh, J. F. Vente, W. G. Haije, D. H. A. Blank and H. J. M. Bouwmeester, *Chem. Mater.*, 2006, **18**, 2187–2193.
13 R. A. Cox-Galhotra, A. Huq, J. P. Hodges, C. Yu, X. Wang, W. Gong, A. J. Jacobson and S. McIntosh, *Solid State Ionics*, 2013, **249–250**, 34–40.
14 A. C. Tomkiewicz, M. A. Tamimi, A. Huq and S. McIntosh, *Solid State Ionics*, 2013, **253**, 27–31.
15 M. A. Tamimi and S. McIntosh, *J. Mater. Chem. A*, 2014, **2**, 6015–6026.
16 H. H. Shin and S. McIntosh, *J. Mater. Chem. A*, 2013, **26**, 7639.

17 H. J. M. Bouwmeester, C. Song, J. Zhu, J. Yi, M. van Sint Annaland and B. A. Boukamp, *Phys. Chem. Chem. Phys.*, 2009, 9640–9643.
18 A. C. Tomkiewicz, M. Meloni and S. McIntosh, *Solid State Ionics*, 2014, **260**, 55–59.
19 M. A. Tamimi, A. C. Tomkiewicz, A. Huq and S. McIntosh, *J. Mater. Chem. A*, 2014, **2**, 18838–18847.
20 I. Sharma and D. Singh, *Bull. Mater. Sci.*, 1998, **21**, 363–374.
21 R. H. E. van Doorn, H. Kruidhof, A. Nijmeijer, L. Winnubst and A. J. Burggraaf, *J. Mater. Chem.*, 1998, **8**, 2109–2112.
22 R. Cox-Galhotra, A. Huq, J. P. Hodges, J. Kim, C. Yu, X. Wang, A. J. Jacobson and S. McIntosh, *J. Mater. Chem. A*, 2013, **1**, 3091–3100.
23 Y. S. Touloukian, R. K. Kriby, R. E. Taylor and P. D. Desai, *Thermophysical Properties of Matter*, Purdue Research Foundation, 1977.
24 H. M. Rietveld, *J. Appl. Crystallogr.*, 1969, **2**, 65–71.
25 A. C. Larson and R. B. von Dreele, *General Structure Analysis System (GSAS)*, Los Alamos National Laboratory, Los Alamos, New Mexico, 2004.
26 E. B. H. Toby, *J. Appl. Crystallogr.*, 2001, **34**, 210–213.
27 R. B. Von Dreele, J. D. Jorgensen and C. G. Windsor, *J. Appl. Crystallogr.*, 1982, **15**, 581–589.
28 H. H. Shin and S. McIntosh, *ECS Electrochem. Lett.*, 2013, **2**, F88–F91.
29 M. van den Bossche and S. McIntosh, *J. Catal.*, 2008, **255**, 313–323.
30 T. Armstrong, F. Prado and A. Manthiram, *Solid State Ionics*, 2001, **140**, 89–96.
31 J. Y. Lee, J. S. Swinnea, H. Steinfink, W. M. Reiff, S. Pei and J. D. Jorgensen, *J. Solid State Chem.*, 1993, **103**, 1–15.
32 S. E. Dann, M. T. Weller and D. B. Currie, *J. Solid State Chem.*, 1992, **97**, 179–185.
33 F. Prado, L. Mogni, G. J. Cuello and A. Caneiro, *Solid State Ionics*, 2007, **178**, 77–82.
34 A. Manthiram, F. Prado and T. Armstrong, *Solid State Ionics*, 2002, **152–153**, 647–655.
35 C. Frontera, A. Caneiro, O. J. Carrillo and J. L. Garcia-Munoz, *Chem. Mater.*, 2005, **17**, 5439–5445.
36 E. J. Crumlin, E. Mutoro, Z. Liu, M. E. Grass, M. D. Biegalski, Y. Lee, D. Morgan, H. M. Christen, H. Bluhm and Y. Shao-Horn, *Energy Environ. Sci.*, 2012, **5**, 6081–6088.
37 P. Decorse, G. Caboche and L. Dufour, *Solid State Ionics*, 1999, **117**, 161–169.
38 K. V. Hansen, K. Norrman and M. Mogensen, *Surf. Interface Anal.*, 2006, **38**, 911–916.
39 J. Druce, T. Ishihara and J. Kilner, *Solid State Ionics*, 2014, **262**, 893–896.

Faraday Discussions

PAPER

Proton uptake in the H^+-SOFC cathode material $Ba_{0.5}Sr_{0.5}Fe_{0.8}Zn_{0.2}O_{3-\delta}$: transition from hydration to hydrogenation with increasing oxygen partial pressure

Daniel Poetzsch, Rotraut Merkle* and Joachim Maier

Received 3rd February 2015, Accepted 16th March 2015
DOI: 10.1039/c5fd00013k

Thermogravimetric investigations on the perovskite $Ba_{0.5}Sr_{0.5}Fe_{0.8}Zn_{0.2}O_{3-\delta}$ (BSFZ, with mixed hole, oxygen vacancy and proton conductivity) from water vapor can occur by acid–base reaction (hydration) or redox reaction (hydrogen uptake), depending on the oxygen partial pressure, *i.e.* on the material's defect concentrations. In parallel, the effective diffusion coefficient of the stoichiometry relaxation kinetics also changes. These striking observations can be rationalized in terms of a defect chemical model and transport equations for materials with three mobile carriers. Implications for the search of cathode materials with mixed electronic and protonic conductivity for application on proton conducting oxide electrolytes are discussed.

1 Introduction

Amongst oxide-based fuel cells, cells based on proton-conducting electrolytes (abbreviated H^+-SOFC or PCFC) have great potential to reduce the operating temperature to below 700 °C. This is essentially due to a higher ionic conductivity of the electrolyte compared to oxide ion conductors at a given temperature,[1,2] and offers advantages such as faster startup, less degradation, and last but not least reduced cost (less expensive materials, *e.g.* less effort for sealing, bipolar plates). Furthermore, formation of water occurs at the cathode without diluting the fuel. However, reduced temperatures increase the overpotentials, in particular the cathode process which typically shows the highest activation energy. To achieve a high cathode performance it is feasible to use materials with mixed ionic and electronic conductivity because this increases the catalytically active zone beyond the triple-phase-boundary (TPB, where the electrode, electrolyte and gas phase are in contact).[3] For the cathodes of H^+-SOFCs, materials with sufficiently high hole (p-type conducting in oxidizing atmosphere) and proton conductivity are desired. In the literature, as first attempt perovskites which show mixed oxygen vacancy and hole conductivity (employed already as cathodes on YSZ) are used in the form

Max Planck Institute for Solid State Research, Stuttgart, Germany. E-mail: r.merkle@fkf.mpg.de

of porous cathodes or porous composites mixed with the proton-conducting electrolyte,[4–10] and show good performances (the maximum power reported so far exceeds 2 W cm^{-2} at 700 °C [11]). However, these investigations do not provide direct evidence for proton conductivity in cathode materials (while the bulk proton conductivity of typical acceptor-doped $Ba(Zr,Ce)O_3$ electrolyte materials is in the range of 10^{-3} to 2×10^{-2} S cm^{-1} at 300–600 °C and 30 mbar H_2O [12]). Already the determination of the proton concentration in such materials is difficult, because – in contrast to the proton conducting electrolyte – more than one reaction may contribute to proton uptake as discussed in detail in Sections 2 and 3. Proton contents extracted from thermogravimetric measurements without considering these complications, *e.g.* in ref. 4 and 5, have to be considered with great care. A chemical method of determining proton concentrations such as Karl–Fischer titration is much more reliable, but limited to measuring quenched samples under *ex situ* conditions.[13,14] In a recent publication we extracted proton concentrations in $Ba_{0.5}Sr_{0.5}Fe_{0.8}Zn_{0.2}O_{3-\delta}$ (BSFZ) from *in situ* thermogravimetric measurements and full defect chemical modeling under low p_{O_2},[15] and could also determine a lower bound for the proton mobility. In the present article, we extend this thermogravimetric investigation to conditions of lower and higher p_{O_2} and rationalize the observed changes in the defect chemical behavior. A detailed discussion if the corresponding proton conductivity is expected to suffice for the "bulk path", as well as experimental results on dense BSFZ microelectrodes will be given elsewhere.[16]

2 Theory

In this section the defect chemistry of a material with three mobile carriers such as oxygen vacancies $V_O^{\bullet\bullet}$, protonic defects $OH_O^{\bullet}$ and electron holes $h^{\bullet}$ is briefly laid out. A more quantitative analysis together with numerical simulations of scenarios covering various types of materials can be found in ref. 17.

Thermodynamics – The equilibration of the perovskite with the gas phase is described by two reactions: eqn (1) is the oxidation reaction where oxygen vacancies are filled by oxygen from the gas phase forming holes.

$$1/2O_2 + V_O^{\bullet\bullet} \rightleftharpoons O_O^{\times} + 2h^{\bullet} \quad K_O = \frac{[O_O^{\times}][h^{\bullet}]^2}{p_{O_2}{}^{1/2}[V_O^{\bullet\bullet}]} \tag{1}$$

Instead of activities, concentrations denoted by brackets are used, assuming ideally dilute defects as a first approximation for this general defect model. In a material without perceptible proton concentration this reaction (together with the electroneutrality condition) suffices to describe the defect chemistry. But for a material with the ability of proton incorporation in a humid atmosphere a second reaction such as the hydration reaction (2) is required.

$$H_2O + V_O^{\bullet\bullet} + O_O^{\times} \rightleftharpoons 2OH_O^{\bullet} \quad K_W = \frac{[OH_O^{\bullet}]^2}{p_{H_2O}[O_O^{\times}]^2[V_O^{\bullet\bullet}]} \tag{2}$$

Here, the oxygen vacancies are filled by dissociative absorption of water in the gas phase, and the transfer of the second proton to a regular lattice oxygen creates a second protonic defect. This reaction is a pure acid–base reaction not involving electronic defects. In proton conducting electrolytes with low redox activity (large bandgap, low electronic defect concentration) this is the only way for proton incorporation. But in redox-active cathode materials the combination of reactions (1) and (2) corresponds to

$$H_2O + 2O_O^\times + 2h^\bullet \rightleftharpoons 2OH_O^\bullet + 1/2O_2 \quad K_{W/O} = \frac{p_{O_2}{}^{1/2}[OH_O^\bullet]^2}{p_{H_2O}[O_O^\times][h^\bullet]^2} = \frac{K_W}{K_O} \tag{3}$$

This reaction describes proton uptake by a redox reaction at the expense of holes and can thus be named "hydrogenation" (uptake of hydrogen, in contrast to water uptake by reaction (2) = hydration). By this reaction a hole-conducting cathode material could incorporate protons even if it had no oxygen vacancies. However, one of equilibria (1)–(3) is mathematically redundant for the description of the defect chemistry. Furthermore, for the defect-chemical calculations (neglecting defect interactions such as trapping) the electroneutrality condition is required ([Acc′] = concentration of acceptor dopants):

$$[Acc'] = 2[V_O^{\bullet\bullet}] + [OH_O^\bullet] + [h^\bullet] \tag{4}$$

Using eqn (1), (2) and (4) the concentrations of oxygen vacancies, protons and holes are calculated numerically as a function of $K_O p_{O_2}{}^{1/2}$ ("redox activity", *i.e.* potential to change the valence of the A or B cation) and $K_W p_{H_2O}$ (potential to take up protons) assuming an acceptor dopant concentration of unity and shown in a triple-logarithmic plot in Fig. 1(a). This plot represents a "materials map": depending on $K_O p_{O_2}{}^{1/2}$ and $K_W p_{H_2O}$ values and the resulting major defect (and corresponding conductivity), one can recognize different material classes: (1)

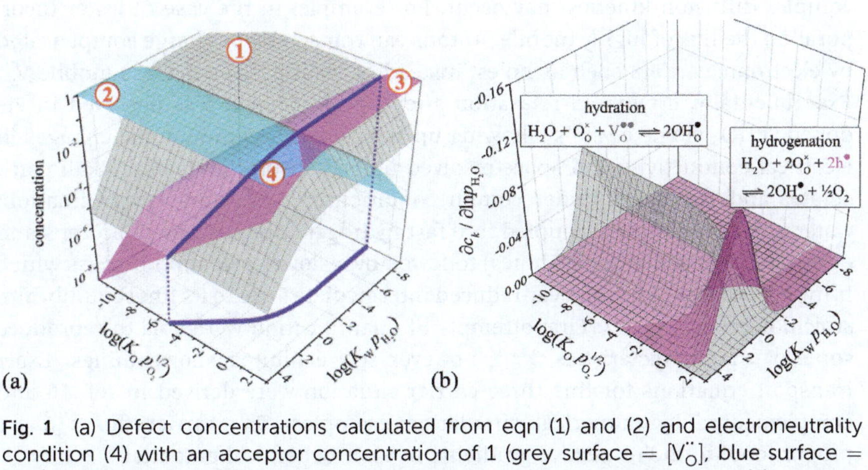

Fig. 1 (a) Defect concentrations calculated from eqn (1) and (2) and electroneutrality condition (4) with an acceptor concentration of 1 (grey surface = $[V_O^{\bullet\bullet}]$, blue surface = $[OH_O^\bullet]$, pink surface = $[h^\bullet]$). Different material classes (1) = oxide electrolytes, (2) = proton electrolytes, (3) = mixed $V_O^{\bullet\bullet}/h^\bullet$ conductors and (4) = mixed $OH_O^\bullet/V_O^{\bullet\bullet}/h^\bullet$ conductors can be recognized in this "materials map". The blue line indicates the intersection of the grey and pink surface, *i.e.* the condition $V_O^{\bullet\bullet} = h^\bullet$. (b) Changes of $V_O^{\bullet\bullet}$ (grey surface) and $h^\bullet$ concentrations (pink surface) upon p_{H_2O} increase as calculated from the general defect model. Reproduced with permission from ref. 17, copyright Wiley-VCH (2015).

oxide ion conducting electrolytes such as (La,Sr) (Ga,Mg)O_{3-x}, (2) proton conducting electrolytes such as (Ba,Sr)(Zr,Ce,Y)O_{3-x}, (3) mixed oxide ion/hole conducting electrodes such as (La,Sr)(Co,Fe)$O_{3-\delta}$, and finally (4) mixed proton/hole conducting electrodes such as BSFZ, the desired case for cathodes on proton conducting electrolytes.

As discussed above, in $OH_O^{\bullet}/V_O^{\bullet\bullet}/h^{\bullet}$ mixed conductors, protons can be incorporated by hydration (eqn (2), proton uptake compensated by annihilation of $V_O^{\bullet\bullet}$) and/or hydrogenation (eqn (3), proton uptake compensated by annihilation of $h^{\bullet}$). From the general defect chemical model the concentration changes of $V_O^{\bullet\bullet}$ and $h^{\bullet}$ upon increasing p_{H_2O} can be calculated. In Fig. 1(b) one can recognize that in the region of the electrolyte materials proton uptake is mainly compensated by $V_O^{\bullet\bullet}$ annihilation (grey surface corresponding to $\partial \ln[V_O^{\bullet\bullet}]/\partial \ln p_{H_2O}$ is higher than the pink surface $\partial \ln[h^{\bullet}]/\partial \ln p_{H_2O}$), *i.e.* proton uptake occurs by the hydration reaction (2). For materials with high redox-activity located close to the right front edge of the materials map, proton uptake is mainly compensated by hole consumption, *i.e.* according to reaction (3). The border between both regimes approximately follows the line where the hole equals the oxygen vacancy concentration marked as the bold blue line in Fig. 1(a) (strictly speaking the condition is $2[V_O^{\bullet\bullet}] = [h^{\bullet}]$;[18] additional deviations are to be expected when activities differ from concentrations).

The presence of a third mobile carrier does not only lead to a more complex equilibrium defect chemistry, but even more to complications in stoichiometry relaxation kinetics (assumed here to be diffusion controlled, *i.e.* fast surface reaction). In the case of two carriers, local electroneutrality directly couples their fluxes, leading to a chemical diffusion coefficient of the neutral component which takes a value in-between the two respective defect diffusivities (in the absence of trapping effects[19]). In the presence of an additional mobile carrier, only all three defects together have to fulfill electroneutrality, with the consequence that the defect fluxes are not directly coupled by simple prefactors any more, and quite complex diffusion kinetics may occur. For example, in the case of water incorporation the flux of highly mobile protons can transiently be charge compensated by electronic carriers such as holes, instead of "waiting" for the less mobile $V_O^{\bullet\bullet}$. Experimentally, a two-fold relaxation after a p_{H_2O} increase was observed in Fe-doped $SrTiO_3$ single crystals (showing up as peculiar non-monotonic changes in electrical conductivity and space-resolved optical absorption[20]) and alkali earth cerates and zirconates under certain conditions.[21,22] In a first approximation, water uptake becomes decoupled into fast hydrogenation (uptake of protons and electrons, or annihilation of holes) followed by a slower oxidation process which brings the sample from an over-reduced intermediate state to its final equilibrium stoichiometry, *cf.* Fig. 2. First attempts of quantification were able to reproduce some of the characteristics,[19,23–25] however still exhibited shortcomings. Exact transport equations for this three carrier situation were derived in ref. 16 and complemented by numerical simulation examples. The carrier fluxes J_i are proportional to the respective gradients in electrochemical potential[26]

$$\begin{pmatrix} J_{V_O^{\bullet\bullet}} \\ J_{OH_O^{\bullet}} \\ J_{h^{\bullet}} \end{pmatrix} = -\frac{1}{F^2} \begin{pmatrix} \sigma_{V_O^{\bullet\bullet}}/4 & 0 & 0 \\ 0 & \sigma_{OH_O^{\bullet}} & 0 \\ 0 & 0 & \sigma_{h^{\bullet}} \end{pmatrix} \begin{pmatrix} \nabla\tilde{\mu}_{V_O^{\bullet\bullet}} \\ \nabla\tilde{\mu}_{OH_O^{\bullet}} \\ \nabla\tilde{\mu}_{h^{\bullet}} \end{pmatrix}. \quad (5)$$

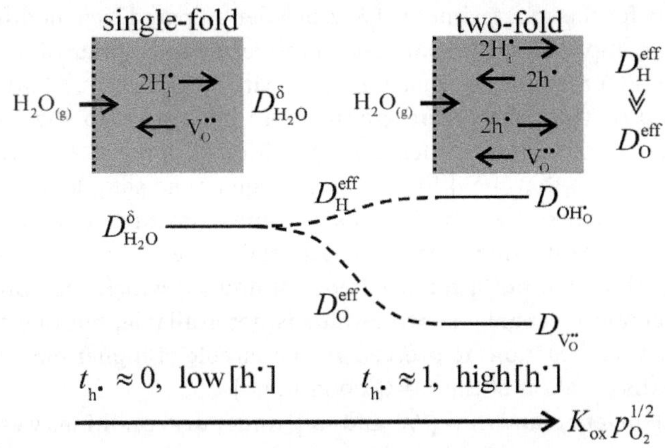

Fig. 2 Change of stoichiometry relaxation process after increase of p_{H_2O} at constant p_{O_2}. Left: the situation with only two carriers (low hole concentration) can be described by a single chemical diffusion coefficient of water. Right: materials with perceptible concentration of mobile holes show two-fold relaxation of fast hydrogen uptake followed by slower re-oxidation. Reproduced with permission from ref. 17, copyright Wiley-VCH (2015).

Using the electroneutrality condition for three carriers $J_{V_O^{\cdot\cdot}} + J_{OH_O^{\cdot}} + J_{h^{\cdot}} = 0$ and expressing the chemical potentials by concentrations *via* $\mu_k = \mu_k^0 + RT \ln c_k$ (in the ideally dilute limit) finally yields expressions in which the defect fluxes are unambiguously related to the concentration gradients ($D_{V_O^{\cdot\cdot}}, D_{OH_O^{\cdot}}, D_{h^{\cdot}}$ denote the defect diffusivities, $t_{V_O^{\cdot\cdot}}, t_{OH_O^{\cdot}}, t_{h^{\cdot}}$ are the respective transference numbers):

$$J_{V_O^{\cdot\cdot}} = -\left[\left(1 - t_{V_O^{\cdot\cdot}}\right)D_{V_O^{\cdot\cdot}} + t_{V_O^{\cdot\cdot}}D_{h^{\cdot}}\right]\nabla c_{V_O^{\cdot\cdot}} - 1/2\left[t_{V_O^{\cdot\cdot}}\left(D_{h^{\cdot}} - D_{OH_O^{\cdot}}\right)\right]\nabla c_{OH_O^{\cdot}} \quad (6)$$

$$J_{OH_O^{\cdot}} = -2\left[t_{OH_O^{\cdot}}\left(D_{h^{\cdot}} - D_{V_O^{\cdot\cdot}}\right)\right]\nabla c_{V_O^{\cdot\cdot}} - \left[\left(1 - t_{OH_O^{\cdot}}\right)D_{OH_O^{\cdot}} + t_{OH_O^{\cdot}}D_{h^{\cdot}}\right]\nabla c_{OH_O^{\cdot}} \quad (7)$$

$$J_{h^{\cdot}} = -(-2)\left[t_{h^{\cdot}}D_{V_O^{\cdot\cdot}} + (1 - t_{h^{\cdot}})D_{h^{\cdot}}\right]\nabla c_{V_O^{\cdot\cdot}} - (-1)\left[t_{h^{\cdot}}D_{OH_O^{\cdot}} + (1 - t_{h^{\cdot}})D_{h^{\cdot}}\right]\nabla c_{OH_O^{\cdot}} \quad (8)$$

The third equation is redundant and given for completeness only. These expressions show that the flux of one carrier (*e.g.* $V_O^{\cdot\cdot}$ given by (6)) depends on its "own" concentration gradient as well as on a second gradient ($\nabla c_{OH_O^{\cdot}}$ in (6)) which is not coupled to the "own" gradient by a simple prefactor ($\nabla c_{V_O^{\cdot\cdot}} = -1/2\nabla c_{OH_O^{\cdot}}$ as for a system containing only $V_O^{\cdot\cdot}$ and $OH_O^{\cdot}$ is not valid here any more, instead only $2\nabla c_{V_O^{\cdot\cdot}} + \nabla c_{OH_O^{\cdot}} + \nabla c_{h^{\cdot}} = 0$ holds). It is important to note that the peculiar kinetic behaviors possible in such three carrier systems (two-fold non-monotonic relaxation) occur already at small driving forces. They arise from the additional degree of freedom given by the fact that only all three carriers together have to fulfill electroneutrality, not from concentration-dependent diffusivities under high driving forces.

3 Experimental

To avoid complications from water surface adsorption, and to be able to extract diffusion coefficients after partial pressure changes, BFSZ was used in the form of dense pellets for thermogravimetry. BSFZ powder prepared by a modified Pechini method was compacted using spark plasma sintering in graphite dies under an Ar atmosphere (FCT Systeme, Germany, 1050 °C with 300 K min^{-1}, 50 MPa pressure, holding 2 min). The pellets were ground from both sides to slices of 1.5 mm thickness, 5–10 mm lateral dimension. Densities of about 97% were obtained, sufficient to ensure that the diffusion length equals the sample thickness.[27] For the oxygen exchange reaction (1) this thickness ensures that stoichiometry relaxation is in the diffusion controlled (critical thickness of related perovskites is much lower than 1 mm [28]); for reactions (2) and (3) which may proceed by a different reaction mechanism such a value is not available, but one can reasonably assume there reactions to proceed at comparable or higher rate (*e.g.* (2) does not require dissociation of the O=O bond in O_2).

Oxygen deficiency (and thus $[V_O^{\bullet\bullet}]$ and $[h^{\bullet}]$) under dry conditions were extracted from thermogravimetry (STA449C Jupiter TG, Netzsch, Germany) by stepwise changes of temperature and p_{O_2}. The absolute O stoichiometry of a quenched sample was determined using iodometric titration.[24] The proton concentration and the relaxation behavior were analyzed after stepwise changes in p_{H_2O} from 5.7 to 9.8 and 14.5 mbar (by changing the temperature of the humidifier, monitored by a HygroClip sensor (Rotronic, Germany)) under different oxygen partial pressures. Since the relative weight changes of the p_{H_2O} steps are very small, ≈4 g of BSFZ pellets were used in these measurements to obtain sufficient precision. For p_{H_2O} changes, buoyancy effects were found to be negligible, for T and p_{O_2} changes, they were corrected based on a measurement with an empty crucible (even there the corrections are small compared to sample weight changes).

For an independent determination of the proton content, BSFZ pellets which already had numerous cracks (and thus equilibrated much faster than the samples used for stoichiometry relaxation in thermogravimetry) were equilibrated at given T, p_{O_2} and p_{H_2O} and rapidly quenched. The water content was measured by rapidly heating the sample to 600 °C in a gas stream of dry argon and analyzing the output gas by Karl–Fischer titration.[24] Total conductivity (≈hole conductivity) was measured by 4-point-DC (2400 Digital Sourcemeter, Keithley, USA) on a bar-shaped SPS sintered sample in a dry, oxidizing atmosphere.

Details of the finite difference numerical simulations of the stoichiometry relaxation kinetics based on eqn (6)–(8) can be found in ref. 16. The input parameters are the defect diffusivities and the initial and final defect concentrations.

4 Results and discussion

4.1 Dry conditions

BSFZ is a perovskite with predominant p-type electronic conduction as illustrated in Fig. 3(a). In such a perovskite with a high concentration of redox-active transition metals, the holes can be identified with Fe^{4+} on Fe^{3+} sites ($Fe_{Fe}^{\bullet}$). Typically, Fe-based perovskites exhibit polaron hopping.[30] In agreement with this expectation, the hole mobility (calculated from σ_{tot} and the hole concentration in Fig. 4(a)) is found to be thermally activated. Thus the decrease of σ_{tot} with increasing temperature is caused by the decrease of the hole concentration with increasing T.

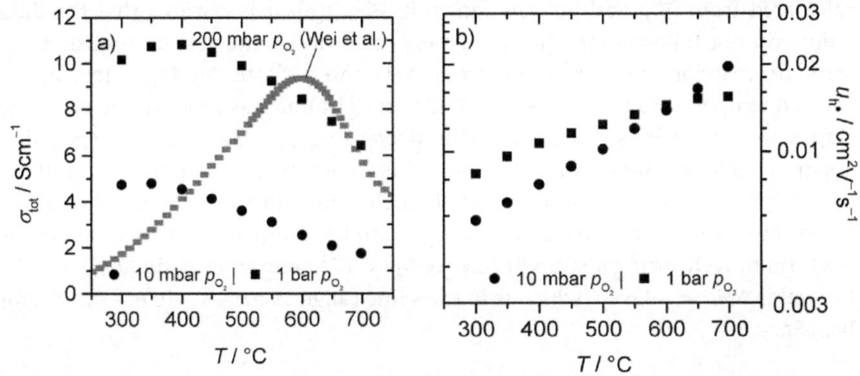

Fig. 3 (a) Total conductivity ≈ hole conductivity of BSFZ in a dry atmosphere. The black symbols are measured under conditions equilibrated with $p_{O_2} = 0.01$ or 1 bar, while the grey squares taken from ref. 29 were apparently obtained without equilibration of the sample. (b) Hole mobility calculated from σ_{tot} and $[h^{\cdot}]$ in Fig. 4(a).

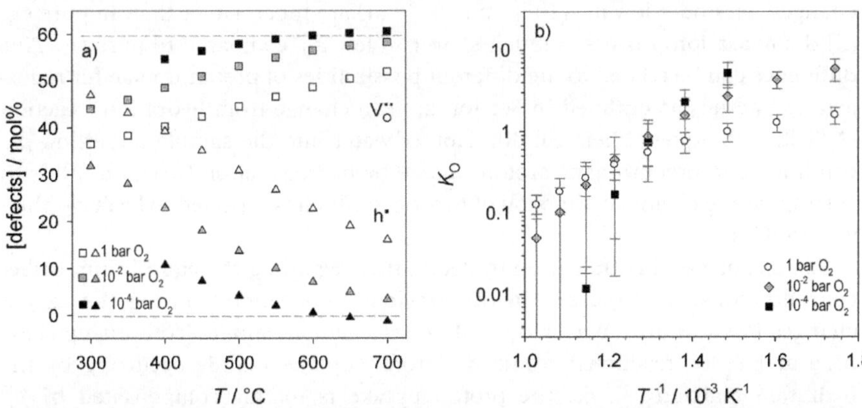

Fig. 4 (a) $V_O^{\cdot\cdot}$ (squares) and $h^{\cdot}$ concentrations (triangles) in BSFZ in dry $p_{O_2} = 10^{-4}, 10^{-2}$ or 1 bar as determined from thermogravimetry (error bar ≈ symbol size). Data for $p_{O_2} = 10^{-2}$ bar previously published in ref. 14. (b) Mass action constant K_O of oxygen exchange reaction (1) calculated from $[V_O^{\cdot\cdot}]$ and $[h^{\cdot}]$ under dry conditions.

The $[V_O^{\cdot\cdot}]$ and $[h^{\cdot}]$ concentrations determined from thermogravimetry in dry 10^{-4}, 10^{-2} and 1 bar O_2 are shown in Fig. 4(a). Under dry conditions the hole concentration is obtained from the measured $[V_O^{\cdot\cdot}]$ and the electroneutrality condition $[Acc'] = 2[V_O^{\cdot\cdot}] + [h^{\cdot}]$ with $[Acc'] = 1.2$ (the perovskite's A-site is fully acceptor doped by Ba and Sr, and additionally 20% Zn^{2+} on replacing Fe^{3+} on the B site further increases the acceptor concentration). These oxygen non-stoichiometry values are in good agreement with measurements in 0.2 bar O_2.[29] The dashed lines in Fig. 4(a) indicate the $[V_O^{\cdot\cdot}]$ and $[h^{\cdot}]$ concentrations for BSFZ when all iron is in the 3+ oxidation state. Obviously, this plateau is reached in 10^{-4} bar O_2 at temperatures ≥600 °C (*cf.* the plateaus in $V_O^{\cdot\cdot}$ concentration observed for $La_{1-x}Sr_xO_{3-\delta}$ perovskites, indicating that a reduction to Fe^{2+} is not favorable[31]). The mass action constants K_O for the oxygen exchange reaction (1) calculated from $[V_O^{\bullet\bullet}]$ and $[h^{\bullet}]$ are shown in Fig. 4(b). It is obvious that the data points do not follow a straight line as expected under the assumption that the standard reaction enthalpy and entropy ΔH_O^0 and ΔS_O^0 are *T*-independent in the covered temperature range. Also, even if a straight line was fitted to the data, the slopes would obviously increase with decreasing p_{O_2} (*i.e.* the corresponding apparent ΔH_O^0 becomes more negative). This indicates that $[V_O^{\bullet\bullet}]$ and $[h^{\bullet}]$ in BSFZ do not behave as ideally dilute defects (which is not surprising given that one of these defect concentrations is always high, and the temperatures are comparably low). Tentatively fitting a straight line to K_O *vs.* T^{-1} at an intermediate $p_{O_2} = 10^{-2}$ bar yields $\Delta H_O^0 = -0.6$ eV, which is in the same range as $\Delta H_O^0 = -(0.6 - 0.7 \text{ eV})$ for $Ba_{0.5}Sr_{0.5}FeO_{3-\delta}$.[32]

4.2 Humid conditions

Proton uptake of BSFZ was investigated thermogravimetrically by stepwise changes of p_{H_2O} while temperature and p_{O_2} were kept constant. Under all investigated p_{O_2}, a p_{H_2O} increase leads to a weight increase of the sample. However, there are two striking differences as depicted in Fig. 5(a) and (b): (i) the weight changes Δm under low p_{O_2} (10^{-2} and 10^{-4} bar) are much larger than in pure O_2, (ii) the relaxation process is much slower in low p_{O_2} compared to pure O_2. This difference can be related to the different possibilities of proton uptake for redox-active materials as outlined in Section 2, *i.e.* a change from hydration (reaction (2)) with subsequent chemical diffusion of water into the sample under low p_{O_2} conditions, to predominant proton uptake by hydrogenation (reaction (3)) with corresponding chemical diffusion of hydrogen which is expected to be faster than water diffusion.

Let us analyze this transition in detail, first regarding the equilibrium defect concentrations. As discussed in the previous publication,[14] fitting the weight changes in $p_{O_2} = 10^{-2}$ bar by the full defect chemical model (comprising acid–base and redox reactions) results in proton uptake mainly occurring by the hydration reaction (2), *i.e.* the proton uptake is mainly compensated by $V_O^{\bullet\bullet}$ annihilation. The resulting proton concentrations are shown in Fig. 6(a). Since in $p_{O_2} = 10^{-4}$ bar $[V_O^{\bullet\bullet}]$ further increases compared to 10^{-2} bar while $[h^{\bullet}]$ decreases

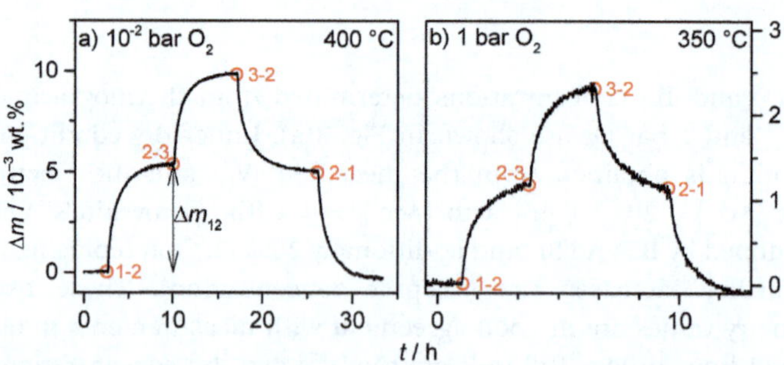

Fig. 5 Weight relaxation of 1.5 mm thick BSFZ pellets after p_{H_2O} steps from 5.7 to 9.8 and 14.5 mbar. The red circles indicate the time of switching between different p_{H_2O}. (a) In $p_{O_2} = 10^{-2}$ bar at 400 °C (data taken from ref. 14). (b) In $p_{O_2} = 1$ bar at 350 °C. Please note the smaller range on the *x* and *y*-axis; the smaller Δm also reflects the noise becoming more visible.

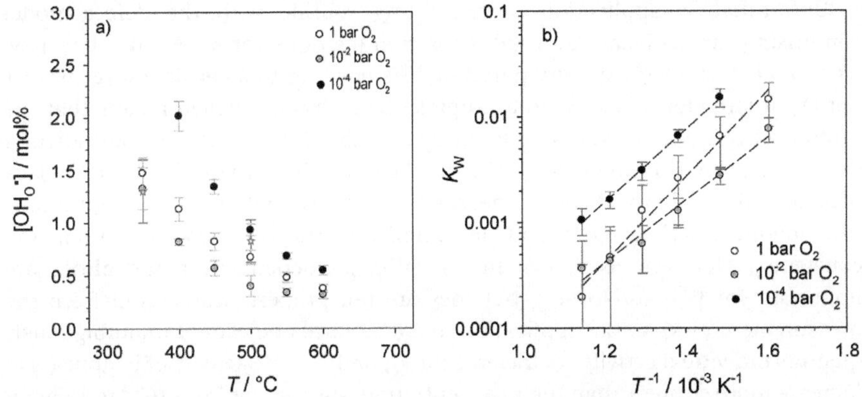

Fig. 6 (a) Proton concentrations of BSFZ in $p_{H_2O} = 20$ mbar and different p_{O_2}, obtained from the analysis of p_{H_2O} changes in the thermogravimetry. The red stars are obtained from Karl–Fischer titration on samples quenched from $p_{H_2O} = 20$ mbar and $p_{O_2} = 1$ bar. Data for $p_{O_2} = 10^{-2}$ taken from ref. 15. (b) Hydration mass action constant K_W calculated from the defect concentrations in (a), plotted *versus* inverse temperature.

(Fig. 4(a)), the general defect chemical considerations shown in Fig. 1(b) indicate a negligible contribution from the hydrogenation reaction (3). Thus these weight changes were directly fitted with the hydration reaction (2) only. The black symbols in Fig. 6(a) show that this very low p_{O_2} leads to moderately increased proton concentrations compared to $p_{O_2} = 10^{-2}$ bar.

In pure oxygen, the situation differs strongly. The general defect model (Fig. 1) predicts that with increasing p_{O_2} at some point the proton uptake will change from predominant hydration (2) to hydrogenation reaction (3), and the drastic decrease of Δm in combination with shorter relaxation times indicates this change of proton uptake reaction occurred between 10^{-2} bar and 1 bar O_2. Furthermore, if the sample is exposed to D_2O instead of H_2O, the weight changes between corresponding p_{D_2O} are significantly larger (typically by a factor of 1.5–1.8) than for the respective p_{H_2O} changes. This comes close to the factor of two expected in case of pure hydrogenation (under the assumption of $[OH_O^{\bullet}] = [OD_O^{\bullet}]$), and is clearly different from p_{D_2O} and p_{H_2O} changes in $p_{O_2} = 10^{-2} \le 10^{-2}$ bar where Δm_{D_2O} exceeds Δm_{H_2O} only slightly by a factor of 1.05 (which is essentially still within the error bar because of the very small weight changes in pure O_2; expected ratio for water uptake is $20/18 = 1.1$). However, while this argument qualitatively supports predominant hydrogenation, it should not be attempted to quantitatively extract a ratio between hydration and hydrogenation reaction from Δm_{D_2O} *versus* Δm_{H_2O} (ΔH_W^0 and ΔS_W^0 could differ between H and D, leading to differences in equilibrium proton, deuterium concentrations even for identical partial pressures).

The proton concentrations in 1 bar O_2 were extracted from the TG results based on the hypothesis of exclusive hydrogenation by reaction (3). They are shown in Fig. 6(a) as open symbols, and are comparable to $[OH_O^{\bullet}]$ in $p_{O_2} = 10^{-2}$ bar. This interpretation is also supported by the comparison of these *in situ* measured $[OH_O^{\bullet}]$ with results from Karl–Fischer titration on samples quenched from equilibration with 20 mbar H_2O and 1 bar O_2 (red stars in Fig. 6(a)) which agree well with the TG data.

Unfortunately, application of the fitting routine with the defect model comprising the acid–base and redox reaction (as done for $p_{O_2} = 10^{-2}$ bar, now using K_O determined from dry pure O_2) does not lead to a consistent result in 1 bar O_2 (it still yields largely proton uptake by hydration, which means that the proton concentrations would be lower by a factor of 9 – obviously contradicting the Karl–Fischer titration result). This can be traced back to the fact that – in lack of actual values – in the defect model the activities are replaced by concentrations (the non-unity activity coefficients are implicitly absorbed into the mass action constants). However, as shown in Fig. 4(b), p_{O_2}-dependent nonidealities are important for $[V_O^{\cdot\cdot}]$ and/or $[h^{\cdot}]$, but they are not properly transferred from the measurement of K_O to the application in K_W because one cannot unambiguously quantify individual activity coefficients for $V_O^{\cdot\cdot}$ and $h^{\cdot}$. The activity of $h^{\cdot}$ appears to increase more strongly than their concentration when going from 10^{-2} to 1 bar O_2 (but the apparent K_O which reflects only the concentration change is lower in 1 bar O_2 compared to 10^{-2} bar) which leads to the change of the dominant proton uptake reaction. In view of these considerations, we note that the proton concentration determined in $p_{O_2} = 10^{-2}$ bar might slightly be underestimated. But with a potential underestimation of $[OH_O^{\cdot}]$ we stay on the safe side in answering the question if the proton conductivity of BSFZ as a cathode material suffices for oxygen reduction by the bulk path.

The values of the hydration mass action constant K_W calculated from the defect concentrations are plotted in Fig. 6(b) *versus* inverse temperature (even if proton uptake in pure O_2 actually proceeds by hydrogenation reaction (3), K_W values can still be calculated). In contrast to K_O in Fig. 4(b), for a given p_{O_2} the K_W values follow a straight line, *i.e.* the protons behave as dilute defects. The non-idealities of $V_O^{\cdot\cdot}$ and $h^{\cdot}$ can be absorbed in activity coefficients deviating from one, which depend on p_{O_2} but are, however, independent of p_{H_2O} (*i.e.* independent of the low $[OH_O^{\cdot}]$). The standard hydration enthalpy ΔH_W^0 extracted from the slopes amounts to -0.7 ± 0.15 eV. Fig. 7 compares ΔH_W^0 and ΔS_W^0 values for BSFZ with data measured for barium zirconate and cerate electrolyte materials.[12] While ΔH_W^0 and ΔS_W^0 individually fall within the range of the electrolyte perovskites, the combination of ΔH_W^0 and ΔS_W^0 for BSFZ results in a less negative free enthalpy ΔG_W^0, corresponding to the lower proton content of BSFZ. Further investigations (experimental as well as *ab initio* calculations) are required to understand which material parameters are decisive for the proton concentration in such mixed conducting perovskites (and responsible for the lower $[OH_O^{\cdot}]$ compared to the electrolytes). A decreased basicity of the oxygen related to a more covalent Fe–O bonding (compared to Zr–O and Ce–O in electrolytes) might be one reason. Nonetheless, as discussed in ref. 14 and experimentally demonstrated on dense BSFZ microelectrodes[15] the proton concentration in BSFZ gives rise to a proton conductivity which suffices for oxygen reduction by the "bulk path".

4.3 Stoichiometry relaxation kinetics

Let us now analyze the diffusion kinetics of stoichiometry relaxation in more detail. Fig. 8 shows exemplary fits of normalized weight changes after p_{H_2O} increase. These curves can reasonably well be fitted with a single, constant effective diffusion coefficient D_m^{eff}.[33] The resulting D_m^{eff} and overall weight changes Δm for a p_{H_2O} step are collected in Fig. 9, showing that D_m^{eff} as well as Δm strongly

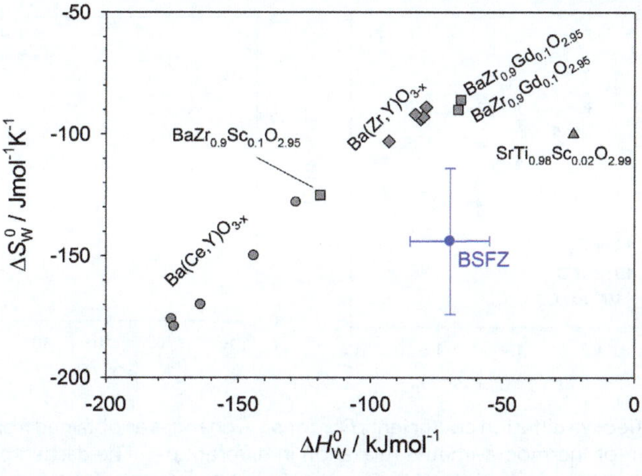

Fig. 7 Standard enthalpies and entropies of the hydration reaction (2) for several barium zirconate and barium cerate electrolyte materials (data taken from ref. 12; for material families comprising different acceptor concentrations, the oxygen deficiency amounts to $[V_O^{\cdot\cdot}] = 0.5[A']$). The data obtained for BSFZ in the present study are overlaid in blue. So far BSFZ is the only cathode material for which ΔH^0_W and ΔS^0_W were determined.

vary with p_{O_2}. As discussed in the previous section, the decrease of Δm at higher p_{O_2} can be traced back to the change from predominant hydration reaction to hydrogenation reaction with increasing hole activity. It is interesting to note that at high p_{O_2} the extracted diffusion coefficient is close to the proton diffusivity in $Ba(Zr_{0.9}Y_{0.1})O_{2.95}$ (blue dashed line in Fig. 9(a), $E_a = 0.44$ eV), while at low p_{O_2} it is in the range of $D_{V_O^{\cdot\cdot}}$ (grey dashed line, $E_a = 0.62$ eV). Overall, the diffusion kinetics is in line with the interpretation in terms of changes from hydrogenation to hydration, but the detailed analysis becomes somewhat more involved.

The first point to be discussed is that based on the general considerations in Section 2, one should expect to observe two relaxation processes with different

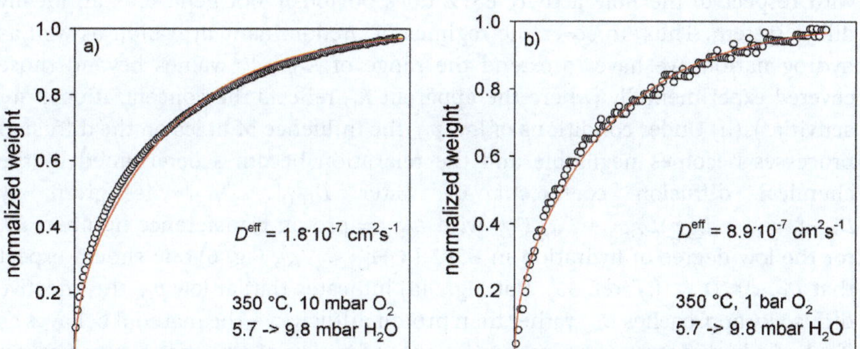

Fig. 8 Relaxation of normalized weight of the 1.5 mm thick BSFZ pellets (symbols) after p_{H_2O} increase from 5.7 to 9.8 mbar at 350 °C. (a) In $p_{O_2} = 10^{-2}$ bar, (b) in $p_{O_2} = 1$ bar. The solid lines represent the fit curve.

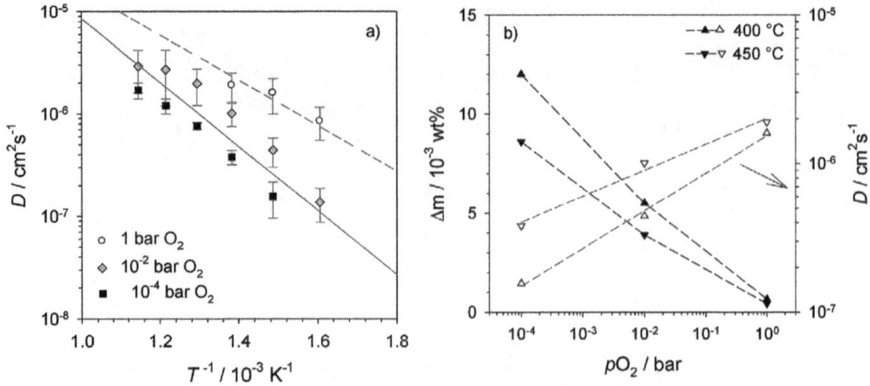

Fig. 9 (a) Effective diffusion coefficients D_m^{eff} for p_{H_2O} changes as obtained from the single-parameter fit of thermogravimetric relaxation in different p_{O_2}. The dashed blue line indicates the bulk proton diffusivity of $Ba(Zr_{0.9}Y_{0.1})O_{2.95}$ taken from ref. 30, the solid grey line indicated the $V_O^{\cdot\cdot}$ diffusivity calculated from O permeation measurements on BSFZ.[34] (b) Weight changes Δm for p_{H_2O} increase from 5.7 to 9.8 mbar (solid symbols), and effective diffusivities D_m^{eff} (open symbols) plotted as function of p_{O_2} (lines are guide to the eye only).

time scales (at least under some p_{O_2} conditions). Mathematically, the relaxation curves could be fitted with a linear combination of two diffusion processes (D_O^{eff} and D_H^{eff}), but owing to the low weight contribution of proton concentration changes relative to O, the errors of the resulting D_H^{eff} are so large that these values are not meaningful in the present case. Only if the total weight increase is indeed dominated by H uptake at sufficiently high p_{O_2}, *i.e.* the p_{H_2O} increase mainly leads to chemical diffusion of hydrogen, a reliable D_H^{eff} value is obtained from fitting the relaxation curve with a single diffusivity. This behavior will now be illustrated by numerical simulations.

In ref. 17, the exact transport equations for the case of three mobile carriers were derived, and some numerical examples given. Here we use the same approach. However, we face two complications: (i) as discussed in Section 4.2, with respect to the hole activity BSFZ does obviously not behave as an ideally dilute system. Thus, to cover the regimes of predominant hydration as well as hydrogenation, we have to extend the range of $K_O p_{O_2}{}^{1/2}$ values beyond those covered experimentally (where the apparent K_O reflects the concentrations, not activities). (ii) Under conditions of low p_{O_2} the influence of holes on the diffusion processes becomes negligible and the relaxation becomes determined by the chemical diffusion coefficient of water $D_{H_2O}{}^{\delta}$. $D_{H_2O}{}^{\delta}$ is given by $D_{H_2O}{}^{\delta} = (1 - t_{OH_O^{\cdot}})D_{OH_O^{\cdot}} + t_{OH_O^{\cdot}} D_{V_O^{\cdot\cdot}}$ with $t_{OH_O^{\cdot}}$ = proton transference number, *i.e.* for the low degree of hydration in BSFZ ($[OH_O^{\cdot}] < [V_O^{\cdot\cdot}]$, Fig. 6) one should expect that $D_{H_2O}{}^{\delta} \approx D_{OH_O^{\cdot}}$ (*cf.* ref. 35). But Fig. 9(a) indicates that at low p_{O_2} the effective diffusivity approaches $D_{V_O^{\cdot\cdot}}$ rather than proton diffusivity – the material behaves as if it had a high degree of hydration (*i.e.* only a fraction of all lattice oxygens might be preferred sites for protonation. An increase of $[OH_O^{\cdot}]$ by a constant factor at all T would make ΔS_W^0 less negative but leave ΔH_W^0 unchanged). Thus, the proton concentration and/or mobility also seems affected by defect interactions.

Therefore, the input parameters for calculation of defect concentrations (ideally dilute model) have to be somewhat modified compared to the K_O and K_W

values in Fig. 4 and 6 (higher degree of hydration *i.e.* higher $K_W p_{H_2O}$ value, larger $K_O p_{O_2}{}^{1/2}$ range). The defect diffusivities correspond to values for BSFZ ($D_{V_O^{\cdot\cdot}}, D_{h^{\cdot}}$) and $Ba(Zr_{0.9}Y_{0.1})O_{2.95}$ ($D_{OH_O^{\cdot}}$) at 450 °C. Fig. 10(a) gives examples of the defect concentration changes after p_{H_2O} increase for conditions of low (top panel) and high (bottom panel) p_{O_2}, the corresponding weight changes are shown in Fig. 10(b). At low p_{O_2}, the p_{H_2O} increase mainly leads to proton uptake by the hydration reaction (the fraction of protons taken up by redox reaction (3) as indicated by the red arrow is small), and the proton and $V_O^{\cdot\cdot}$ concentrations have almost the same time behavior (the hole concentration is still too low for perceptible decoupling of H and O transport). The weight increase is dominated by the contribution from O, and the effective diffusion coefficient D_m^{eff} fitted from the normalized weight change (Fig. 10(b)) is close to $D_{H_2O}{}^{\delta}$ ($\approx D_{V_O^{\cdot\cdot}}$ for the defect concentrations and mobilities used in the simulation). The situation strongly changes at high p_{O_2}, where proton uptake happens mainly by hydrogenation reaction (3) (*cf.* the large redox fraction of 82% as indicated in the figure), which drastically decreases the overall weight increase. There, the diffusion process is mainly ambipolar diffusion of protons and holes (= chemical diffusion of hydrogen, for which under the given conditions of high hole conductivity $D_H{}^{\delta} \approx D_{OH_O^{\cdot}}$). In such a case the H uptake constitutes a large share of the overall weight increase, and proceeds faster than the incorporation of oxygen (which, given the small weight changes and the even smaller contribution from O, is difficult to be distinguished at all from baseline drift). Thus the weight relaxation can still be fitted with a single effective diffusion coefficient, which now corresponds to D_H^{eff} and comes close to $D_{OH_O^{\cdot}}$, as shown in Fig. 10(c).

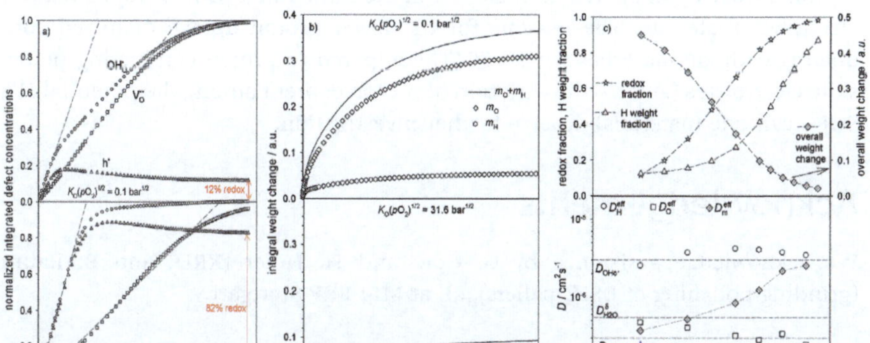

Fig. 10 Results from numerical simulations of p_{H_2O} changes (10 to 20 mbar) at 450 °C for different $K_O p_{O_2}{}^{1/2}$ values calculated with $D_{V_O^{\cdot\cdot}} = 3 \times 10^{-7}$ cm² s⁻¹, $D_{OH_O^{\cdot}} = 3 \times 10^{-4}$ cm² s⁻¹, $D_{h^{\cdot}} = 5 \times 10^{-6}$ cm² s⁻¹. (a) Changes of normalized defect concentrations, the red arrows indicate the fraction of proton uptake which occurs by redox reaction (3). With the applied $t^{1/2}$ scale, for small time diffusion processes appear as a straight line. (b) Total weight increase (line) and individual O and H contributions (symbols) *versus* time. (c) Top: fraction of proton uptake by redox reaction, overall weight change Δm and H contribution to Δm upon increased p_{H_2O}. Bottom: effective diffusivities D_O^{eff}, D_H^{eff}, D_m^{eff} extracted from the slopes of $[V_O^{\cdot\cdot}]$, $[OH_O^{\cdot}]$ and normalized weight *versus* $t^{1/2}$ (the blue arrows indicate the conditions of (a and b)); the dashed lines give defect diffusivities $D_{V_O^{\cdot\cdot}}$, $D_{OH_O^{\cdot}}$ and the value of the water chemical diffusion coefficient $D_{H_2O}{}^{\delta}$ would take in the absence of holes.

The alternative hypothesis that the lower weight changes upon p_{H_2O} increase in high p_{O_2} are completely caused by decreased proton uptake (but still occurring by the hydration reaction) can be discarded as it could not explain the increased effective diffusivities at high p_{O_2}. The combination of decreasing Δm and increasing D_m^{eff} with increasing p_{O_2} gives strong evidence for the gradual transition from hydration to hydrogenation.

So we can summarize that overall the numerical simulations reproduce the experimentally observed trends of overall weight changes and effective diffusion coefficients with p_{O_2}, and thus support the given interpretation of the stoichiometry relaxation kinetics although the nonideal defect chemical behavior of BSFZ (not surprising given the high defect concentrations and comparably low measurement temperatures) requires some adjustments in the input parameters.

5 Conclusions

The present thermogravimetric investigation shows that BSFZ incorporates a nonnegligible amount of protons (although less than barium zirconate cerate electrolyte materials). The mode of proton incorporation changes from hydration (at low p_{O_2}) to hydrogenation (at high p_{O_2}), indicating that the presence of oxygen vacancies is not necessarily required for proton uptake. The modification of the overall proton incorporation reaction is accompanied by pronounced changes in the diffusion kinetics. These findings clearly show that in such systems with more than two carriers the *complete* defect chemistry and transport behavior *must* be analyzed (the weight change after p_{H_2O} increase is not necessarily uptake of water; the corresponding diffusion coefficient may differ from $D_{H_2O}{}^{\delta}$). Interestingly, the proton diffusivity in BSFZ is found to be in the same range as in Y-doped $BaZrO_3$ electrolyte materials. The reasons for the lower proton uptake of mixed conducting cathode materials for H^+-SOFC compared to proton conducting perovskite electrolytes (and for the variation of proton content among the potential H^+-SOFC cathode materials) deserve further investigation.

Acknowledgements

We acknowledge assistance by G. Götz and H. Hoier (XRD) and B. Baum (grinding/polishing of BSFZ pellets), all at MPI FKF Stuttgart.

References

1 T. Norby, *Solid State Ionics*, 1999, **125**, 1.
2 K. D. Kreuer, *Solid State Ionics*, 2001, **145**, 295.
3 J. Fleig, *Annu. Rev. Mater. Res.*, 2003, **33**, 361.
4 A. Grimaud, F. Mauvy, J. M. Bassat, S. Fourcade, M. Marrony and J. C. Grenier, *J. Mater. Chem.*, 2012, **22**, 16017.
5 A. Grimaud, F. Mauvy, J. M. Bassat, S. Fourcade, L. Rocheron, M. Marrony and J. C. Grenier, *J. Electrochem. Soc.*, 2012, **159**, B683.
6 F. He, T. Wu, R. Peng and C. Xia, *J. Power Sources*, 2009, **194**, 263.
7 R. Peng, T. Wu, W. Liu, X. Liu and G. Meng, *J. Mater. Chem.*, 2010, **20**, 6218.
8 E. Fabbri, L. Bi, D. Pergolesi and E. Traversa, *Adv. Mater.*, 2012, **24**, 195.

9 L. Zhao, B. B. He, Y. H. Ling, Z. Q. Xun, R. R. Peng, G. Y. Meng and X. Q. Liu, *Int. J. Hydrogen Energy*, 2010, **35**, 3769.

10 S. Ricote, N. Bonanos, F. Lenrick and R. Wallenberg, *J. Power Sources*, 2012, **218**, 313.

11 R. O'Hayre, *oral presentation at SSPC17*, Seoul, Korea, June 15-19 2014.

12 K. D. Kreuer, *Annu. Rev. Mater. Res.*, 2003, **33**, 333.

13 D. Han, Y. Okumura, Y. Nose and T. Uda, *Solid State Ionics*, 2010, **181**, 1601.

14 D. Hashimoto, D. Han and T. Uda, *Solid State Ionics*, 2014, **262**, 687.

15 D. Poetzsch, R. Merkle and J. Maier, *Phys. Chem. Chem. Phys.*, 2014, **16**, 16446.

16 D. Poetzsch, R. Merkle, J. Maier, 2015, to be submitted.

17 D. Poetzsch, R. Merkle and J. Maier, *Adv. Funct. Mater.*, 2015, **25**, 1542.

18 The exact boundary between predominant hydration and hydrogenation reaction in Fig. 1 is given by the condition $2[V_O^{\cdot\cdot}] = [h^{\cdot}]$, which can be derived as follows: solving eqn (2) for $[V_O^{\cdot\cdot}]$ and eqn (3) for $[h^{\cdot}]$ and taking the derivative with respect to p_{H_2O}, one obtains $\frac{\partial \ln[V_O^{\cdot\cdot}]}{\partial \ln p_{H_2O}} = 2\frac{\partial \ln[OH_O^{\cdot}]}{\partial \ln p_{H_2O}} - 1$ and $\frac{\partial \ln[h^{\cdot}]}{\partial \ln p_{H_2O}} = \frac{\partial \ln[OH_O^{\cdot}]}{\partial \ln p_{H_2O}} - \frac{1}{2} = \frac{1}{2}\frac{\partial \ln[V_O^{\cdot\cdot}]}{\partial \ln p_{H_2O}}$, and correspondingly $\frac{\partial[h^{\cdot}]}{[h^{\cdot}]} = \frac{1}{2}\frac{\partial[V_O^{\cdot\cdot}]}{[V_O^{\cdot\cdot}]}$. This yields the condition $2[V_O^{\cdot\cdot}] = [h^{\cdot}]$ for the boundary where $\partial[h^{\cdot}] = \partial[V_O^{\cdot\cdot}]$.

19 J. Maier, *J. Am. Ceram. Soc.*, 1993, **76**, 1212–1223.

20 J. H. Yu, J.-S. Lee and J. Maier, *Angew. Chem., Int. Ed.*, 2007, **46**, 8992.

21 H.-I. Yoo, J.-Y. Yoon, J.-S. Ha and C.-E. Lee, *Phys. Chem. Chem. Phys.*, 2008, **10**, 974.

22 H.-I. Yoo, J. I. Yeon and J.-K. Kim, *Solid State Ionics*, 2009, **180**, 1443.

23 J. H. Yu, J.-S. Lee and J. Maier, *Solid State Ionics*, 2010, **181**, 154.

24 E. Kim and H.-I. Yoo, *Solid State Ionics*, 2013, **252**, 132.

25 G.-R. Kim, H.-H. Seo, J.-M. Jo, E.-C. Shin, J. H. Yu and J. S. Lee, *Solid State Ionics*, 2015, **272**, 60.

26 While the proton flux is designated by $J_{V_O^{\cdot\cdot}} + J_{OH_O^{\cdot}} + J_{h^{\cdot}} = 0$, it is important to keep in mindthat in proton and mixed conducting perovskites only the proton migrates, not the whole hydroxide ion.

27 D. Poetzsch, PhD thesis, University of Stuttgart, Germany, 2014.

28 H. J. M. Bouwmeester, H. Kruidhof and A. J. Burggraaf, *Solid State Ionics*, 1994, **72**, 185.

29 B. Wei, Z. Lu, X. Q. Huang, M. L. Liu, N. Li and W. H. Su, *J. Power Sources*, 2008, **176**, 1.

30 J. Mizusaki, T. Sasamoto, W. R. Cannon and H. K. Bowen, *J. Am. Ceram. Soc.*, 1983, **66**, 247.

31 J. Mizusaki, M. Yoshihiro, S. Yamauchi and K. Fueki, *J. Solid State Chem.*, 1985, **58**, 257.

32 L. Wang, R. Merkle and J. Maier, *ECS Trans.*, 2009, **25**(2), 2497.

33 J. Crank, Solution for the 1-dimensional diffusion problem, in *The Mathematics of Diffusion*, Oxford University Press, New York, 1975.

34 J. Martynczuk, K. Efimov, L. Robben and A. Feldhoff, *J. Membr. Sci.*, 2009, **344**, 62.

35 K. D. Kreuer, E. Schönherr and J. Maier, *Solid State Ionics*, 1994, **70**, 278.

Faraday Discussions

PAPER

Relating surface chemistry and oxygen surface exchange in $LnBaCo_2O_{5+\delta}$ air electrodes

Helena Téllez,*[a] John Druce,[a] John A. Kilner[ac] and Tatsumi Ishihara[ab]

Received 23rd February 2015, Accepted 7th April 2015
DOI: 10.1039/c5fd00027k

The surface and near-surface chemical composition of electroceramic materials often shows significant deviations from that of the bulk. In particular, layered materials, such as cation-ordered $LnBaCo_2O_{5+\delta}$ perovskites (Ln = lanthanide), undergo surface and sub-surface restructuring due to the segregation of the divalent alkaline-earth cation. These processes can take place during synthesis and processing steps (*e.g.* deposition, sintering or annealing), as well as at temperatures relevant for the operation of these materials as air electrodes in solid oxide fuel cells and electrolysers. Furthermore, the surface segregation in these double perovskites shows fast kinetics, starting at temperatures as low as 400 °C over short periods of time and leading to a decrease in the transition metal surface coverage exposed to the gas phase. In this work, we use a combination of stable isotope tracer labeling and surface-sensitive ion beam techniques to study the oxygen transport properties and their relationship with the surface chemistry in ordered $LnBaCo_2O_{5+\delta}$ perovskites. Time-of-Flight Secondary-Ion Mass Spectrometry (ToF-SIMS) combined with ^{18}O isotope exchange was used to determine the oxygen tracer diffusion (D^*) and surface exchange (k^*) coefficients. Furthermore, Low Energy Ion Scattering (LEIS) was used for the analysis of the surface and near surface chemistry as it provides information from the first mono-atomic layer of the materials. In this way, we could relate the compositional modifications (*e.g.* cation segregation) taking place at the electrochemically-active surface during the exchange at high temperatures and the oxygen transport properties in double perovskite electrode materials to further our understanding of the mechanism of the surface exchange process.

1 Introduction

Many theoretical studies on the performance of perovskites and perovskite-related materials as oxygen electrodes for electrochemical energy conversion

[a]*International Institute for Carbon Neutral Energy Research (wpi-I2CNER), Kyushu University, 744 Motooka, Nishi-ku, Fukuoka, 819-0395, Japan. E-mail: htellez@i2cner.kyushu-u.ac.jp; Tel: +81 92-802-6738*
[b]*Department of Applied Chemistry, Kyushu University, 744 Motooka, Nishi-ku, Fukuoka, 819-0395, Japan*
[c]*Department of Materials, Imperial College London, Prince Consort Road, South Kensington, London, SW7 2BP, UK*

applications, such as solid oxide fuel cells (SOFCs) and solid oxide electrolysers (SOECs) are based on the assumption that their chemical composition and structure are homogeneous from the immediate surface (*i.e.* gas phase–electrode interface) to the bulk material. Nevertheless, quite often there are significant compositional and structural deviations at the surface and near-surface of these materials due to segregation of matrix species,[1–5] atomic rearrangements[5–7] and other extrinsic processes (*e.g.* surface poisoning).[8–10]

In the case of mixed-ionic electronic conductors (MIECs), these processes might have a dramatic impact on the mechanism of the oxygen exchange between the gas phase and the electrode surface. Unlike traditional electrode materials, such as $La_{1-x}Sr_xMnO_3$, the surface exchange mechanism is not limited to the triple-phase boundary (TPB), but extended along the gas–electrode interface.[11] The incorporation of oxygen from the gas atmosphere into the air electrode lattice (or oxygen evolution from the electrode to the surrounding gas in SOEC mode) involves a series of intermediate steps and species, although the specific pathway is not yet known.[12] The overall reaction can be written in Kröger–Vink notation as follows:

$$O_2 + 4e' + 2V_O^{\cdot\cdot} \rightleftarrows 2O_O^{\times}$$

The incorporation (or evolution, in electrolysis) of oxygen then requires the adsorption (desorption, in electrolysis) of the O_2 gas molecule, dissociation by redox reaction and incorporation into an oxygen vacancy.[13] This surface exchange is believed to be the limiting factor in the performance of MIECs as air electrodes, and hence will determine, to a large extent, the efficiency and durability of the electrochemical device.

In order to improve the performance of these devices by tailoring the properties of the MIEC electrodes, there is an essential need to comprehend the fundamental mechanisms of the surface exchange and bulk diffusion. Although the diffusion properties are generally well understood in terms of the bulk defect chemistry, the surface exchange process cannot be precisely explained on the basis of the bulk composition.

One of the main challenges to face in order to establish the relationship between the surface chemistry and the oxygen exchange process is characterizing the relevant electrocatalytically-active surface. However, most surface analysis techniques provide information from several atomic layers, and hence, integrated over the near-surface region. In order to obtain information *at the atomic scale* of the immediate gas–electrode interface and to unequivocally link this composition to the surface exchange kinetics, extremely-surface sensitive techniques, such as low-energy ion scattering (LEIS) and Time-of-Flight secondary ion mass spectrometry (ToF-SIMS), are being recently applied to study MIEC materials.[14–16]

In LEIS, the elemental composition of the outermost surface is determined by analysing the kinetic energy distribution of noble gas ions (typically He^+, Ne^+ or Ar^+ accelerated to energies between 1 and 8 keV) scattered by the surface.[17] Due to its extreme surface specificity (*i.e.* the information is limited to the first atomic layer of the surface), LEIS is becoming a very popular technique to investigate the composition of the immediate surface in MIEC materials and ionic conductors.[1,2,18–23] On the other hand, ToF-SIMS enables measurement of ^{18}O diffusion profiles to extract diffusion and surface exchange kinetic parameters (D^* and k^*)

which characterise the ionic transport properties in MIECs.[24,25] In addition, the technique allows isotopic chemical mapping with high lateral resolution and high sensitivity to obtain 3 dimensional information of the sample composition.[26,27]

In MIEC materials, it is widely accepted that the transition metal cation at the gas–electrode interface plays a major role in order to facilitate the electron transfer for O_2 reduction reaction (*e.g.* many theoretical studies assumed that the electrochemically-active surface is typically BO_2-terminated or mixed BO_2- and AO-terminated).[28–30] However, most LEIS studies on surface segregation in MIECs have shown a predominantly A′O-terminated surface after being subjected to relatively high temperatures[1,21,22] or for long periods of time[1,31] (with $A' =$ divalent cation, for the general case of an acceptor-substituted (3,3) $A_{1-x}A'_xBO_3$ perovskite). Moreover, recent works have shown that the segregation of the acceptor-substituent cation segregation is activated at low temperature. For instance, Druce *et al.* showed that the segregation of Sr in $La_{0.6}Sr_{0.4}Co_{0.2}Fe_{0.8}O_{3-\delta}$ was observed after annealing at 400 °C for 8 h, leading to a significant decrease of the transition metals exposed to the gas–electrode interface.[3] Furthermore, this segregation was shown to take place in a very short-time, as observed in the case of $GdBaCo_2O_{5+\delta}$ double perovskites with a significant Ba surface coverage after 15 minutes of thermal treatment at 400 °C.[2]

In this work, we study the effects of the acceptor-substituent segregation on the surface exchange process by using a combination of ^{18}O isotope exchange and analysis by ToF-SIMS and LEIS. For this purpose, we determined the surface exchange coefficient (k^*) in $GdBaCo_2O_{5+\delta}$ double perovskites as a function of the Ba coverage at the outer surface. Given the fast kinetics of this segregation, we paid particular attention to the evolution of the surface and near-surface chemistry as a function of the time subjected to the thermal treatments required for the isotope exchange experiments.

2 Experimental

Sample preparation

Dense (>95% of theoretical) ceramic pellets of $GdBaCo_2O_{5+\delta}$ were prepared by a sol–gel based Pechini method. A detailed description of the synthesis method can be found elsewhere.[2] The powders were calcined at 800 °C for 6 h and pressed into pellets before sintering in air at 1000 °C for 8 h. SEM/EDX analysis showed neither the presence of any impurities, nor any deviations from the nominal compositions for the double perovskites. The single phase purity of the materials was checked by X-ray diffraction, confirming the cation ordering for both materials by extra low angle reflections in the XRD spectrum. To prepare the samples for the isotope exchange measurements, they were ground with SiC paper and polished to a mirror finish using an aqueous-based diamond suspension (final particle size 1/4 μm). Any residue from the polishing was removed by ultrasonic cleaning in acetone, ethanol and de-ionised water before performing the exchanges.

Stable isotope tracer exchange methodology

^{18}O exchange experiments were performed in order to extract surface exchange and diffusion coefficients, k^* and D^*, respectively. The exchange experiment involves two different annealing steps in different atmospheres: an equilibration stage in normal oxygen and a subsequent isotopic exchange stage in an ^{18}O-enriched atmosphere. Firstly, the samples were introduced into a custom-built set-up which was evacuated by means of a turbomolecular pump to $<2 \times 10^{-7}$ mbar. Then, in the first annealing step, the samples were equilibrated at the temperature and oxygen partial pressure of interest (in this case, we used $pO_2 =$ 200 mbar and 400 °C). Several equilibration times were used for the different samples in order to lead to different Ba coverages on the surface and to investigate the effect of the surface segregation on the exchange kinetics. The equilibration times used were $t = 0$, 1 h, 8 h and 24 h. After the equilibration stage, the sample tube was again evacuated and re-filled with 200 mbar O_2 enriched in ^{18}O (^{18}O isotopic fraction of 91%) to perform the exchange at 400 °C for 15 minutes. It should be noted that given the low temperature and exchange time used, the influence of chemical diffusion can considered to be minimal for the sample that was not equilibrated before the ^{18}O exchange.[32] In fact, the oxygen stoichiometry in GBCO is just slightly decreased upon heating in air from room temperature up 400 °C.[33] Oxygen isotopic fractions were measured prior to and after the exchange using a residual gas analyzer (M-070QA-TDF, Canon Anelva, Japan) to assure there was no significant ^{18}O depletion in the gas phase during the experiment.[32]

Measurement of the ^{18}O diffusion profiles by ToF-SIMS

Isotope exchange depth profiling was performed using a ToF-SIMS V instrument (IonToF GbmH, Germany) in order to obtain the oxygen surface exchange and diffusion coefficients, as described elsewhere.[14,34] The depth profiles were measured in dual beam mode using a 2 keV Cs^+ beam for fast sputtering (rastered over an area of 250 × 250 μm²) and a 30 keV Bi^+ analytical beam (70 μs cycle time; 150 × 150 μm² raster), both at an incident angle of 45° degrees to the sample surface normal. The depth profiles were acquired using a high current Bi^+ beam in order to obtain good accuracy and precision along the whole profile, while detector dead time effects and signal saturation were avoided by selective attenuation of secondary ions (SASI), as described in ref. 25.

Surface and near-surface characterisation by low-energy ion scattering (LEIS) spectroscopy

The outermost surface chemistry of the annealed samples was analyzed by LEIS (Qtac^{100} spectrometer, IonToF GbmH, Germany). The cation surface coverage was analysed using 6 keV Ne^+ ions as the primary ion beam. In our specific set-up, the primary ions are directed towards the surface at normal incidence, and back-scattered ions are collected at a 145° scattering angle from the total azimuth angle. Due to the surface sensitivity of the LEIS technique, the samples require *in situ* cleaning by oxidation with reactive atomic oxygen to remove any organic contamination from air exposure that would hinder the detection of any matrix species at the first atomic layer of the surface.[35] Since this oxidation procedure is performed at room temperature, it is not expected to produce further composition changes (*i.e.* cation segregation). Additionally, the near-surface composition can be also measured by combining the analysis with low-energy Ar^+ sputtering at shallow angles (typically, 500 eV at 60°). Further details of the instrument set-up can be found elsewhere.[2] In order to minimize surface damage during the analysis, the ion fluence was kept below the static dose by rastering the analysis beam over rather large areas (typically 1 mm^2), while the sputtering areas were 1.4 × 1.4 mm^2 to avoid crater-edge effects during depth profiling.

3 Results and discussion

Segregation of matrix species and sub-surface atomic rearrangements

As discussed in the introduction, recent LEIS work has shown that after subjecting MIEC materials to temperatures relevant for materials processing, the outer atomic surfaces are predominantly A′O-terminated (with A′ = acceptor-substituent cation in (3,3) $A_{1-x}A'_xBO_3$ perovskites). For instance, the thermal treatment of a ceramic $GdBaCo_2O_{5+\delta}$ pellet at 1000 °C for 12 h led to a BaO-terminated surface, with no transition metal detected at the outermost surface (Fig. 1e).[1] The segregation of the acceptor-substituent Ba cation is produced to achieve surface neutrality, avoiding polar catastrophe that would otherwise take place in (3,3) perovskites.[36,37]

Moreover, a later work showed that the segregation process is initiated very fast even at low annealing temperatures, as observed in Fig. 1 that shows the kinetics of the surface segregation for a GBCO sample annealed in O_2 at 400 °C.[2] The outer surface of a polished GBCO pellet analysed by 6 keV Ne^+ scattering showed all the matrix cationic species, as expected from the bulk composition (Fig. 1a). After the thermal treatments in oxygen, the cation surface coverage is significantly modified even after 15 minutes (Fig. 1b), showing an increase in the Ba coverage at the expense of Gd and Co at the outer surface. The Ba segregation became more significant as the annealing time was increased (Fig. 1b–d). The segregation of the divalent cation is thus activated at temperatures relevant for SOFC and SOEC operation, with much faster kinetics than anticipated.

Although the previous studies in ref. 1 and 2 focussed only on qualitative changes in the composition of the very outer surface, in the present work we build

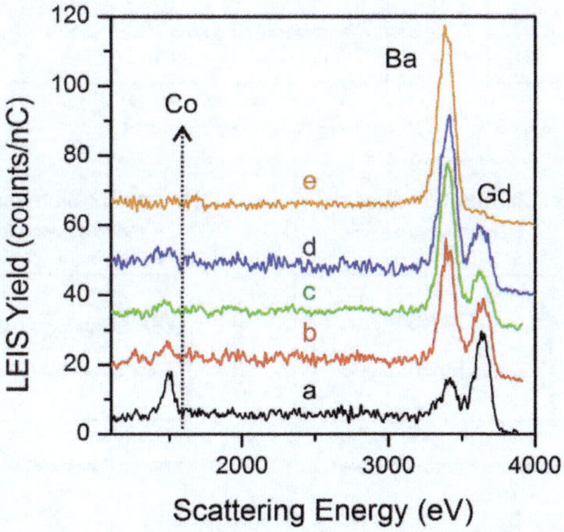

Fig. 1 LEIS surface spectra obtained using a 6 keV Ne^+ analysis beam for a GBCO ceramic pellet (a) as polished, annealed at 400 °C in 200 mbar O_2 for (b) 15 minutes, (c) 1 h, (d) 8 h, and annealed at 1000 °C for 12 h, and (e) annealed at 1000 °C for 12 h. Adapted from ref. 1 and 2.

on this by studying the extent of the atomic rearrangements produced at low temperatures by the segregation of the alkaline-earth cations and the evolution of the near-surface composition with the annealing time. This has been accomplished by LEIS depth profiling analysis on a new set of samples on which the different exchange experiments described in the experimental section were performed. The LEIS depth profiles were quantified by assuming that the plateau signals at the end of the analysis correspond to the bulk stoichiometry in order to obtain the cation ratios plotted in Fig. 2.

In Fig. 2, a clear evolution of the near-surface region can be observed as a function of the equilibration time for the ^{18}O-exchanged GBCO samples. The polished GBCO sample already shows some compositional deviations at the near-surface region (0–4 nm), compared to the bulk composition (Fig. 2a). This result was quite surprising, as any segregation produced during sintering was expected to be removed during the polishing step. In principle, the compositional deviation could be related to an analysis artefact due to ion sputtering. However, this possibility was ruled out as both Gd and Ba show similar sputter yields (*i.e.* 0.25 atoms of Gd and Ba/Ar^+ incident ion, as estimated by SRIM simulations).[38] The different profiles observed for the A cation species should be then attributed to real compositional deviations from the bulk stoichiometry.

One possible explanation for the depth profile obtained for the sample as-polished (Fig. 2a) could be related to insufficient polishing that did not remove the whole segregated layer produced during the sintering process. However, the samples were polished using a final particle size of 1/4 μm, so the polishing should remove any segregation occurring during sintering (*e.g.* a GBCO sample

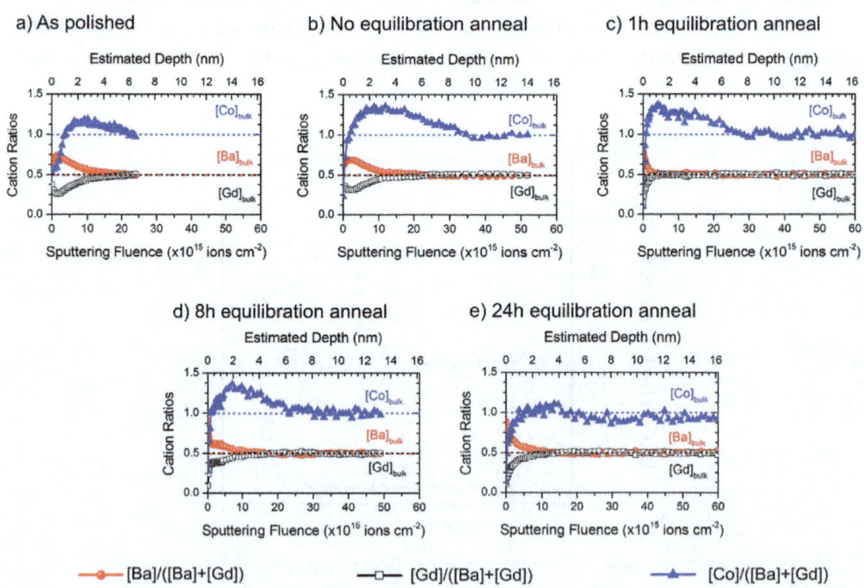

Fig. 2 LEIS depth profiles on GBCO (a) after polishing and (b–e) as a function of the equilibration time at 400 °C. Samples (b–e) were equilibrated in normal oxygen for the different times indicated in the captions (from 0 to 24 h equilibration) and then exchanged for 15 min in ^{18}O-enriched atmosphere.

annealed at 1000 °C for 12 h showed a segregation profile that extended up to 8 nm, and hence, even the segregation produced by long annealing would be removed by the polishing step).[1]

It should be noted that LEIS requires the analysis of rather large areas in order to avoid surface damage during the analysis, and hence, a large number of grain boundaries are also probed in the analysis region. Although the Ba segregation at the grain surfaces would be removed by the polishing, the contribution of any segregation profile along the grain boundaries might be still noticeable. In addition, other phases might be formed at the surface by reaction of the segregated Ba with atmospheric species (*e.g.* sulphates or carbonates).[2,39] Nevertheless, the outer surface composition shows a cation ratio similar to the bulk composition due to a decrease in the Ba coverage at the outer surface (first point of the depth profile in Fig. 2a). This might be explained by the use of aqueous solvents to eliminate any polishing residues that will remove any segregated water-soluble species at the same time.[40]

Upon annealing, the Co surface coverage decreases drastically with almost no Co is detected at the outer surface, as observed in Fig. 2b–e. The segregation of the acceptor-substituent Ba cations towards the surface gives rise to a B-enriched subsurface, as previously observed for high temperature treated GBCO.[1] The same behaviour is observed for the exchanges reported in this work, although the thermal treatments were performed at lower temperature in this case (400 °C). The Co-enriched sub-surface region spreads in depth and the compositional deviations from the bulk stoichiometry are less significant as the thermal treatments last for a longer time (including the two steps for the exchange experiments, *i.e.* equilibration and ^{18}O-enriched annealing) (Fig. 2, profiles b–e). For instance, the [Co]/([Ba] + [Gd]) ratios reached the bulk values very quickly after an equilibration annealing of 24 h (plus 15 minutes of ^{18}O exchange). Conversely, the segregation of the divalent species at the outer surface is also significantly larger with 90% of the A-sites occupied by Ba atoms.

The dynamic nature of the surface in these MIEC materials is evident given the fast atomic rearrangements occurring at low temperatures, yielding significant deviations in composition and microstructure from the bulk material. This suggests that cation segregation might not be the only phenomenon responsible for the long-term degradation observed in MIEC oxygen electrodes, as this process is already present after a few hours subjected to low temperatures.

The Ba segregation in GBCO might have important implications not only for the oxygen surface exchange between the gas phase and the electrode surface, but also in terms of the structural deviations at the near-surface region. The ordering of the lanthanide (Ln) cation and the aliovalent Ba cation on alternating AO planes is responsible for the high ionic conductivity in $LnBaCo_2O_{5+\delta}$ in comparison with their disordered analogous $Ln_{1-x}Ba_xCoO_{3-\delta}$ single perovskites.[41–43] Fast oxygen diffusion takes place along the LnO planes, where the oxygen vacancies are located, following an interstitialcy mechanism with the co-participation of oxygen sites at the CoO planes[44] and providing the anisotropic transport properties characteristic of this sort of material.[45] From the "static snapshots" provided by LEIS, it seems the GBCO material should not be regarded as a single phase at the near-surface region, but as a multiphase material with a compositional gradient due to the disruption of the A-cation sub-lattice due to cation segregation at the near-surface region to ordered $GdBaCo_2O_{5+\delta}$ double perovskite in the bulk. This lack of ordering produced by the Ba segregation towards the outer surface might affect the diffusion properties at the near-surface.

From the first points of the depth profiles shown in Fig. 2, we could compare the cation ratios at the outer surface as a function of the total time that the samples were subjected to the thermal treatments. The total annealing time includes both the equilibration step corresponding to the annealing in $^{16}O_2$ (which we varied to modify the Ba coverage) and the exchange time in $^{18}O_2$ (which was fixed at 15 minutes).

As observed in Fig. 3, the fraction of the A-sites occupied by the alkaline-earth cations increases with the total annealing time, showing a predominantly BaO-terminated surface after annealing for just 1 h and with no significant changes observed at the outer composition after long equilibrations for 8 and 24 h. For short equilibration times ($t < 1$ h), mixed AO- and BO_2-terminated surfaces with different [Ba]/([Ba] + [Gd]) ratios were observed. Although the surface ratios were very similar for long annealing times, it should be noted that the near-surface composition shows a clear evolution depending on the total annealing time. Fig. 3 also shows the surface cation ratios obtained after annealing at the sintering temperature (1000 °C for 12 h) (dashed lines), which led to almost 100% BaO-coverage, with no Co detected at the gas–solid interface, as a reference for the surface cation ratios observed for the GBCO samples after a thermal treatment similar to the conditions used during the sintering of the samples.

Effects of surface segregation on surface exchange and diffusion

In order to understand the influence of these changes in surface and near-surface composition on the oxygen exchange and diffusion properties, we performed isotope exchange depth profiling (IEDP) measurements after different equilibration treatments for 0, 1 h and 8 h. It should be kept in mind that even the sample that was not subjected to any equilibration treatment would, however, show a

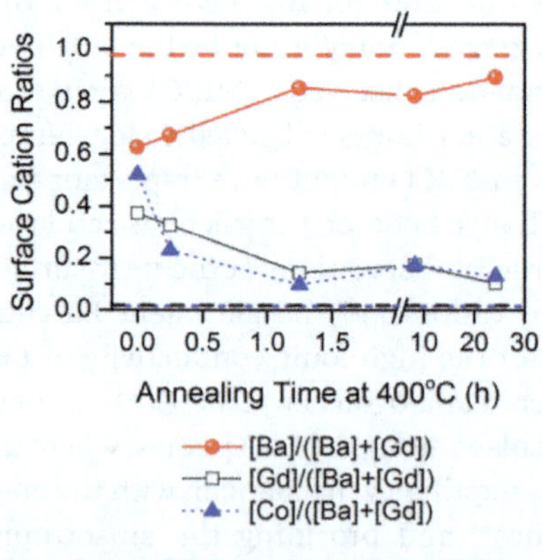

Fig. 3 Surface cation ratios of the GBCO samples after polishing (annealing time = 0) and after the ^{18}O exchanges using different equilibration times. Note that the surface cation ratios are plotted as a function of the total annealing time at 400 °C, including the equilibration time (0, 1, 8 and 24 hours) and the exchange time (15 min for all the experiments). The dotted lines correspond to the surface cation ratios obtained for the GBCO sample annealed at 1000 °C for 12 h (ref. 1).

certain degree of Ba segregation during the exchange experiment itself at 400 °C for 15 min in an ^{18}O-enriched atmosphere. Fig. 4 shows the ^{18}O isotope diffusion profiles (a) and derived oxygen diffusion (D^*) and surface exchange (k^*) kinetic parameters (b). Note the D^* and k^* plot has a linear *y*-axis, rather than the more typical logarithmic scale, indicating that the dependence on equilibration time (and hence surface composition as seen in the previous sections) is not very strong.

The ionic transport properties of GBCO double perovskites have been widely studied using both experimental and theoretical approaches,[43–47] and their anisotropic behaviour is well known. Nevertheless, in most of these studies, the near-surface compositional gradient due to the observed cation segregation was not considered. A recent study by Tomkiewicz *et al.* called attention to the importance of thermal history on the extent of the surface segregation.[48] Based on analysis of $LnBaCo_2O_{5+\delta}$ (Ln = Nd, Pr and Gd) double perovskite powders using XPS, which integrates over several atomic layers, they assumed that the Ba segregation was less significant than what we observe by LEIS analysis of the single outer atomic layer. Therefore, these authors suggested that the surface exchange rate is still directly influenced by the bulk composition and structure, as Co and Ln species would still be present at the surface.

However, our study shows that both surface and near-surface undergo rapid restructuring during the thermal treatments necessary for the measurement of the surface exchange rate itself by either isotopic or electrochemical methods. The extent of the Ba surface coverage does not seem to have such a direct influence on the oxygen exchange rate as initially expected. In fact, the comparison of the surface exchange and diffusion coefficient obtained for different Ba-surface coverages with previously reported values obtained by IEDP on thin films and polycrystalline GBCO revealed no significant differences for the k^* values (Fig. 5). For the standard IEDP measurements, the equilibration time used is normally between 4 and 10 times the exchange duration,[32] and hence, we expect the GBCO surfaces in ref. 45 and 46 to be close to a 100% BaO-termination.

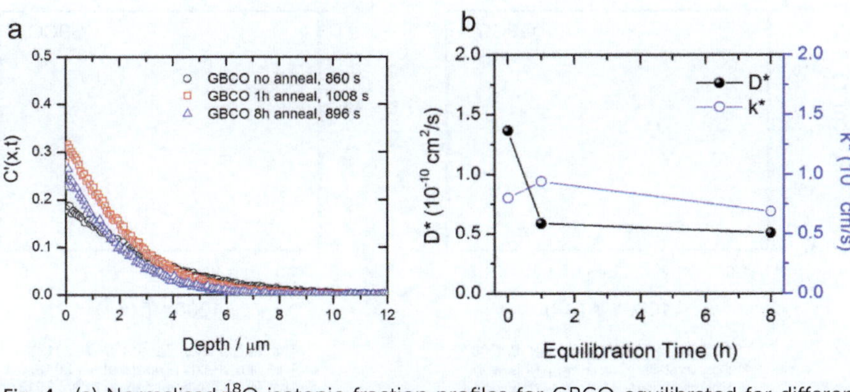

Fig. 4 (a) Normalised ^{18}O isotopic fraction profiles for GBCO equilibrated for different times before exchange in ^{18}O-enriched atmosphere at 400 °C for 15 min. Solid lines correspond to the fittings to Crank's solution for the diffusion profile. (b) Surface exchange and diffusion coefficients obtained from the fitting as a function of the equilibration time.

From these results, it seems clear that the surface termination might not be the most important factor playing a role for the surface exchange kinetics, as already suggested by Tomkiewicz *et al.*[48] The near-surface disorder introduced by the cation segregation might have a more dramatic influence on the observed surface exchange kinetics. On the other hand, this Ba segregation could also increase the oxygen vacancy concentration at the surface that could also facilitate the access for the oxygen in the gas phase to the transition metal at the subsurface.

4 Conclusions

In this work, we have used a combination of surface sensitive compositional analysis by Low Energy Ion Scattering (LEIS) spectroscopy and ^{18}O tracer exchange coupled with Secondary Ion mass Spectrometry (SIMS) to probe the relationship between surface composition and surface exchange coefficient in a promising solid oxide electrode material, namely double perovskite $GdBaCo_2O_{5+\delta}$ (GBCO). The LEIS results highlight that the surface and near-surface composition is highly dynamic, and significant compositional changes can occur at relatively low temperatures and short periods of time (*e.g.* 15 minutes at 400 °C), giving rise to structural rearrangements. In particular, the GBCO shows rapid segregation of the divalent Ba cation to the surface to form a BaO terminated layer, with a compositional profile which suggests the near-surface region should be regarded more as a multiphase material due to the formation of other secondary phases and the disruption of the ordering of the A cation sub-lattice produced by Ba segregation.

Oxygen isotope exchange experiments were performed on samples with different degrees of Ba surface segregation, achieved by using different equilibration times. Despite the different surface compositions and the extension of the Ba segregation profile, which would conventionally be expected to passivate the

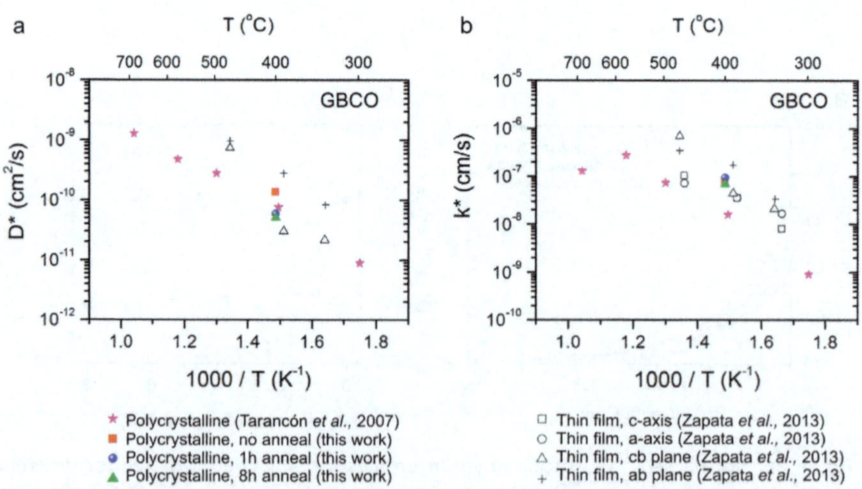

Fig. 5 Surface exchange (k^*) and diffusion (D^*) coefficient reported in the literature for GBCO.

surface by decreasing the electrocatalytically-active surface (*i.e.* CoO_2-terminated surface, assuming the transition metal is the active species for the oxygen reduction reaction), the obtained surface exchange coefficients (k^*) seemed less sensitive to the surface composition than anticipated, even though the cation segregation led to a significantly different surface in terms of the Ba surface coverage.

Overall, our results suggest that the cation segregation might not be the only phenomenon responsible for the long term degradation observed in MIEC oxygen electrodes, as the rapid kinetics of cation segregation ensures a divalent cation terminated surface is already present after even a few hours at low temperatures. Since all measurements of surface exchange kinetics require some equilibration time at temperature during which segregation will occur, it seems impossible to measure the intrinsic properties of a "fresh" unsegregated surface. The lack of observed dependence of k^* on this surface composition may indicate the importance of point defects at the surface, such as oxygen vacancies which would facilitate access to transition metal cations in subsurface layers. The presence of the restructured sub-surface layer also raises the possibility that surface exchange activity may be limited not by the incorporation reaction, but by diffusion of oxygen through the near surface compositional gradient.

Acknowledgements

The authors acknowledge the support of the International Institute for Carbon Neutral Energy Research (wpi-I2CNER) funded by the World Premier Research Centre initiative of the Ministry of Education, Culture, Sports, Science and Technology (MEXT) of the Japanese Government. H.T. also thanks the financial support of the Japanese Society for Promotion of Science for her JSPS postdoctoral fellowship and the Kakenhi Grant-in-Aid project, 25-03770.

References

1 J. Druce, H. Téllez, M. Burriel, M. Sharp, L. Fawcett, S. N. Cook, D. McPhail, T. Ishihara, H. H. Brongersma and J. A. Kilner, *Energy Environ. Sci.*, 2014, **7**, 3593–3599.

2 H. Téllez, J. Druce, Y.-W. Ju, J. Kilner and T. Ishihara, *Int. J. Hydrogen Energy*, 2014, **39**, 20856–20863.

3 J. Druce, T. Ishihara and J. Kilner, *Solid State Ionics*, 2014, **262**, 893–896.

4 E. J. Crumlin, E. Mutoro, Z. Liu, M. E. Grass, M. D. Biegalski, Y.-L. Lee, D. Morgan, H. M. Christen, H. Bluhm and Y. Shao-Horn, *Energy Environ. Sci.*, 2012, **5**, 6081–6088.

5 H. Dulli, P. A. Dowben, S. H. Liou and E. W. Plummer, *Phys. Rev. B: Condens. Matter Mater. Phys.*, 2000, **62**, R14629–R14632.

6 D. E. E. Deacon-Smith, D. O. Scanlon, C. R. A. Catlow, A. A. Sokol and S. M. Woodley, *Adv. Mater.*, 2014, **26**, 7252–7256.

7 J. H. Lee, G. Luo, I. C. Tung, S. H. Chang, Z. Luo, M. Malshe, M. Gadre, A. Bhattacharya, S. M. Nakhmanson, J. A. Eastman, H. Hong, J. Jellinek, D. Morgan, D. D. Fong and J. W. Freeland, *Nat. Mater.*, 2014, **13**, 879–883.

8 S.-N. Lee, A. Atkinson and J. A. Kilner, *J. Electrochem. Soc.*, 2013, **160**, F629–F635.

9 E. Bucher, C. Gspan, F. Hofer and W. Sitte, *Solid State Ionics*, 2013, **238**, 15–23.
10 J. Druce, H. Téllez and J. Hyodo, *MRS Bull.*, 2014, **39**, 810–815.
11 S. B. Adler, J. A. Lane and B. C. H. Steele, *J. Electrochem. Soc.*, 1996, **143**, 3554–3564.
12 R. Merkle and J. Maier, *Angew. Chem., Int. Ed.*, 2008, **47**, 3874–3894.
13 R. A. De Souza, *Phys. Chem. Chem. Phys.*, 2006, **8**, 890–897.
14 J. A. Kilner, S. J. Skinner and H. H. Brongersma, *J. Solid State Electrochem.*, 2011, **15**, 861–876.
15 J. A. Kilner, H. Tellez-Lozano, M. Burriel, S. Cook and J. Druce, *ECS Trans.*, 2013, **57**, 1701–1708.
16 H. Tellez, A. Aguadero, J. Druce, M. Burriel, S. Fearn, T. Ishihara, D. S. McPhail and J. A. Kilner, *J. Anal. At. Spectrom.*, 2014, **29**, 1361.
17 H. Brongersma, M. Draxler, M. Deridder and P. Bauer, *Surf. Sci. Rep.*, 2007, **62**, 63–109.
18 J. Druce, T. Ishihara and J. Kilner, *Solid State Ionics*, 2014, **262**, 893–896.
19 H. H. Shin and S. McIntosh, *J. Mater. Chem. A*, 2013, **1**, 7639–7647.
20 J. Druce, H. Téllez, N. Simrick, T. Ishihara and J. Kilner, *Int. J. Hydrogen Energy*, 2014, **39**, 20850–20855.
21 M. M. Viitanen, R. G. v. Welzenis, H. H. Brongersma and F. P. F. van Berkel, *Solid State Ionics*, 2002, **150**, 223–228.
22 I. C. Fullarton, J.-P. Jacobs, H. E. van Benthem, J. A. Kilner, H. H. Brongersma, P. J. Scanlon and B. C. H. Steele, *Ionics*, 1995, **1**, 51–58.
23 M. de Ridder, A. G. J. Vervoort, R. G. van Welzenis and H. H. Brongersma, *Solid State Ionics*, 2003, **156**, 255–262.
24 R. A. De Souza, J. Zehnpfenning, M. Martin and J. Maier, *Solid State Ionics*, 2005, **176**, 1465–1471.
25 H. Tellez, J. Druce, J. E. Hong, T. Ishihara and J. A. Kilner, *Anal. Chem.*, 2015, **87**, 2907–2915.
26 M. Kubicek, G. M. Rupp, S. Huber, A. Penn, A. K. Opitz, J. Bernardi, M. Stoger-Pollach, H. Hutter and J. Fleig, *Phys. Chem. Chem. Phys.*, 2014, **16**, 2715–2726.
27 G. Holzlechner, D. Kastner, C. Slouka, H. Hutter and J. Fleig, *Solid State Ionics*, 2014, **262**, 625–629.
28 Y. A. Mastrikov, R. Merkle, E. Heifets, E. A. Kotomin and J. Maier, *J. Phys. Chem. C*, 2010, **114**, 3017–3027.
29 Y.-L. Lee, J. Kleis, J. Rossmeisl, Y. Shao-Horn and D. Morgan, *Energy Environ. Sci.*, 2011, **4**, 3966.
30 J. W. Han and B. Yildiz, *Energy Environ. Sci.*, 2012, **5**, 8598–8607.
31 M. Burriel, S. Wilkins, J. P. Hill, M. A. Munoz-Marquez, H. H. Brongersma, J. A. Kilner, M. P. Ryan and S. J. Skinner, *Energy Environ. Sci.*, 2014, **7**, 311–316.
32 R. A. De Souza and R. J. Chater, *Solid State Ionics*, 2005, **176**, 1915–1920.
33 D. Munoz-Gil, D. Perez-Coll, J. Pena-Martinez and S. Garcia-Martin, *J. Power Sources*, 2014, **263**, 90–97.
34 J. A. Kilner, B. C. H. Steele and L. Ilkov, *Solid State Ionics*, 1984, **12**, 89–97.
35 M. de Ridder, R. G. van Welzenis and H. H. Brongersma, *Surf. Interface Anal.*, 2002, **33**, 309–317.
36 P. W. Tasker, *J. Phys. C: Solid State Phys.*, 1979, **12**, 4977.
37 W. A. Harrison, *Phys. Rev. B: Condens. Matter Mater. Phys.*, 2011, **83**, 155437.
38 J. F. Ziegler, M. D. Ziegler and J. P. Biersack, *Nucl. Instrum. Methods Phys. Res., Sect. B*, 2010, **268**, 1818–1823.

39 A. Berenov, A. Atkinson, J. Kilner, M. Ananyev, V. Eremin, N. Porotnikova, A. Farlenkov, E. Kurumchin, H. J. M. Bouwmeester, E. Bucher and W. Sitte, *Solid State Ionics*, 2014, **268**, 102–109.
40 G. M. Rupp, A. Limbeck, M. Kubicek, A. Penn, M. Stoger-Pollach, G. Friedbacher and J. Fleig, *J. Mater. Chem. A*, 2014, **2**, 7099–7108.
41 G. Kim, S. Wang, A. J. Jacobson, Z. Yuan, W. Donner, C. L. Chen, L. Reimus, P. Brodersen and C. A. Mims, *Appl. Phys. Lett.*, 2006, **88**, 024103.
42 A. A. Taskin, A. N. Lavrov and Y. Ando, *Appl. Phys. Lett.*, 2005, **86**, 091910.
43 D. Parfitt, A. Chroneos, A. Tarancon and J. A. Kilner, *J. Mater. Chem.*, 2011, **21**, 2183–2186.
44 H. Shiiba, M. Nakayama, T. Kasuga, R. W. Grimes and J. A. Kilner, *Phys. Chem. Chem. Phys.*, 2013, **15**, 10494–10499.
45 J. Zapata, M. Burriel, P. Garcia, J. A. Kilner and J. Santiso, *J. Mater. Chem. A*, 2013, **1**, 7408–7414.
46 A. Tarancon, A. Morata, G. Dezanneau, S. J. Skinner, J. A. Kilner, S. Estrade, F. Hernandez-Ramirez, F. Peiro and J. R. Morante, *J. Power Sources*, 2007, **174**, 255–263.
47 R. Moreno, J. Zapata, J. Roqueta, N. Bagues and J. Santiso, *J. Electrochem. Soc.*, 2014, **161**, F3046–F3051.
48 A. C. Tomkiewicz, M. Meloni and S. McIntosh, *Solid State Ionics*, 2014, **260**, 55–59.

Faraday Discussions

PAPER

Electrochemistry of $La_{0.3}Sr_{0.7}Fe_{0.7}Cr_{0.3}O_{3-\delta}$ as an oxygen and fuel electrode for RSOFCs

Beatriz Molero-Sánchez, Paul Addo, Aligul Buyukaksoy, Scott Paulson and Viola Birss*

Received 26th February 2015, Accepted 30th March 2015
DOI: 10.1039/c5fd00029g

The use of a single porous mixed ion-electron conducting (MIEC) material as both the oxygen and fuel electrodes in reversible solid oxide cells is of increasing interest, primarily due to the resulting simplified cell design and lower manufacturing costs. In this work, $La_{0.3}Sr_{0.7}Fe_{0.7}Cr_{0.3}O_{3-\delta}$ (LSFCr-3) was studied in a 3-electrode half-cell configuration in air, pure CO_2 and in a 1 : 1 CO_2 : CO mixture, over a temperature range of 650–800 °C. A detailed analysis of the impedance (EIS) data, under both open circuit and polarized conditions, as well as the cyclic voltammetry response of LSFCr-3 has shown that it is very active in all of these environments, but with oxygen evolution being somewhat more facile that oxygen reduction, and CO_2 reduction more active than CO oxidation. Evidence for a chemical capacitance, associated with the $Fe^{3+/4+}$ redox process in LSFCr-3, was also obtained from the EIS and CV data in all gas environments.

1. Introduction

As part of the world-wide efforts to identify new energy conversion and storage systems, there is great interest in the development of reversible fuel cells, particularly reversible solid oxide fuel cells (RSOFCs). RSOFCs can run in both the electrolysis mode (SOEC) to store energy by converting H_2O to H_2 and O_2, or co-electrolyze CO_2 and H_2O to syngas and O_2, when excess electricity is available, and then run in the fuel cell mode (SOFC) to convert H_2 or syngas, and oxygen, to electricity and heat when electricity is needed. The primary goal of our recent work has therefore been to develop high performance and stable electrode materials for use in RSOFCs, with our specific focus being on $La_{1-x}Sr_xFe_{1-y}Cr_yO_{3-\delta}$ (LSFCr) and $La_{1-x}Ca_xFe_{1-y}Cr_yO_{3-\delta}$ (LCFCr) mixed ionic-electronic conducting (MIEC) perovskite electrodes, operating at 650–800 °C.

Department of Chemistry, University of Calgary, Calgary, Alberta, Canada, T2N 1N4. E-mail: birss@ucalgary.ca

In addition to the excellent performance of this family of perovskites as RSOFC electrode materials,[1,2] a key concept behind our work is that these materials can be employed as both the fuel and air electrodes in both the SOFC and SOEC modes, thus significantly lowering cost by decreasing the number of materials used and also by simplifying the manufacturing process. It is worth noting that the latter concept is a relatively new topic of investigation and thus there are only a few studies focussed on developing symmetrical RSOFCs in the literature.[1,3,4]

Earlier work in our group has shown that the LSFCr perovskite, specifically the Cr-3 member of this family, gives the optimum trade-off between structural, electrical conduction, thermal expansion, and electrochemical properties.[1] LSFCr-3 is a very good air electrode, a promising H_2 electrode,[1] and also shows very good promise as a catalyst for CO_2 reduction and CO oxidation.[5] However, a detailed analysis of the electrochemistry of these reactions at this material has not yet been carried out. Our preliminary work showed that LSFCr-3 and LCFCr-3 are both promising as RSOFC electrodes, although LCFCr-3 has the better thermal expansion compatibility with YSZ (yttria-stabilized zirconia).[2] Even so, the focus in the present work is on LSFCr-3, as a broader set of results are available for this MIEC material.

In this paper, the electrochemical performance of LSFCr-3 has been studied in detail, both at open circuit and under polarization, between 650 and 800 °C in air, pure CO_2 and CO_2/CO mixtures. Importantly, all of this work has involved a 3-electrode configuration. Notably, in-depth studies of the performance and stability of oxygen and fuel electrodes using a 3 electrode setup, especially under polarization, are lacking in the literature, making it difficult to deduce reaction mechanisms. It is shown here that the performance of this material in all of the gas environments is very good, although CO oxidation is the most inhibited. Efforts are also made to understand the slow steps in these reactions, based primarily on the interpretation of the electrochemical impedance spectroscopy and cyclic voltammetry data collected for the LSFCr-3 electrode.

2. Experimental methods

2.1 Electrode material preparation

$La_{0.3}Sr_{0.7}Fe_{0.7}Cr_{0.3}O_{3-\delta}$ (LSFCr-3) powders were prepared by the combustion method starting with nitrate precursors. Details of the synthetic procedure have been published elsewhere.[1]

2.2 Half-cell construction

The $La_{0.3}Sr_{0.7}Fe_{0.7}Cr_{0.3}O_{3-\delta}$ powders, obtained from the regular combustion method,[1] were milled (high energy planetary ball mill, Pulverisette 5, Fritsch, Germany) in an isopropanol medium at a rotation speed of 300 rpm for 2 h using zirconia balls. A $Gd_{0.9}Ce_{0.1}O_{2-\delta}$ (GDC) buffer layer for the working, counter and reference electrodes was screen printed onto both sides of a 300 μm thick, dense YSZ electrolyte (Fuel Cell Materials), followed by firing at 1200 °C for 2 hours. The *ca.* 30 μm thick LSFCr-3 working (WE) and counter (CE) electrodes were then screen-printed symmetrically (over an area of 0.5 cm^2) onto both sides of the GDC buffer layer. LSFCr-3 was also used as the reference electrode (RE), screen-printed onto the same side of the YSZ, as the working electrode (WE), placed 2 to 4 mm from the WE to prevent electrode geometry artifacts. The LSFCr-3 electrodes were then fired at 1100 °C for 2 h. Au paste (C 5729, Heraeus Inc., Germany) was painted on all three electrode layers to serve as the current collectors.

2.3 Electrochemical studies

In all of this work, the electrochemical measurements to evaluate the cell performance were performed using 3 electrode methods in a range of gas atmospheres ($p_{O_2} = 1\text{–}10^{-18}$), with a total flow rate of 50 ml min^{-1}, and a range of temperatures. Electrochemical impedance spectra (EIS) were collected under open circuit and polarized conditions between 650 °C and 800 °C, using an amplitude of 50 mV in the frequency range of 0.01 to 65 kHz and using a Solartron 1287/1255 potentiostat/galvanostat/impedance analyzer. Zview software was used to fit and analyze the impedance data. CV measurements (sweep rates from 1–10 mV s^{-1}) between −0.7 V and 0.7 V *vs.* the RE were performed using a Solartron 1287 interface, with control and data collection handled by Corrware software.

3. Results and discussion

3.1 Oxygen reduction and evolution at LSFCr-3

The electrochemical activity towards both the oxygen reduction (ORR) and evolution (OER) reactions was examined by cyclic voltammetry (CV) and electrochemical impedance spectroscopy (EIS) at $p_{O_2} = 0.1\text{–}1$ atm and 650–800 °C. The EIS data were collected both at the open circuit potential (OCP) when the oxygen reaction is at equilibrium, and also under polarization (±0.3 V), conditions under which either the ORR or OER is the dominant reaction. While this approach is not commonly used in the literature in SOFC/SOEC studies, and especially not for mixed ion and electron conducting (MIEC) oxide electrodes, there have been a few efforts made in this direction, including determining the effect of polarization on the electrochemical behavior of La_2NiO_4, LSM, LSF and LSCF.[6–9]

3.1.1 Open circuit EIS studies. 3-Electrode AC impedance measurements were carried out initially at the open circuit potential (OCP) in air using the LSFCr-3/GDC/YSZ/GDC/LSFCr-3 cells, with LSFCr-3 also used as the reference electrode. Fig. 1a shows the typical impedance spectra obtained for LSFCr-3 at open circuit under an oxygen partial pressure of 0.21 atm at temperatures between 650 and 800 °C. Some high frequency noise and an inductance were sometimes observed during these measurements, generally attributed to the instrument and the leads, and thus ignored. In all cases, two time constants were clearly observed over the full temperature range at the OCP, revealing two different processes in the oxygen reaction at LSFCr-3. Thus, all of the spectra obtained in oxygen atmospheres were fitted using the equivalent circuit Rs(R2/CPE)(R3/CPE) shown in Fig. 1b, with the best-fit values of the circuit elements given in Table 1. Rs corresponds to the series resistance, arising from the contacts, leads and the resistance of the electrolyte between the WE and RE, while the sum of R2 (high frequency, (RHF)) and R3 (low frequency, (RLF)) is equivalent to the total polarization resistance (Rp).

The CPEs are constant phase elements and were converted to capacitors using eqn (1):[10]

$$C_i = (R_iQ_i)^{1/ni}/R_i \tag{1}$$

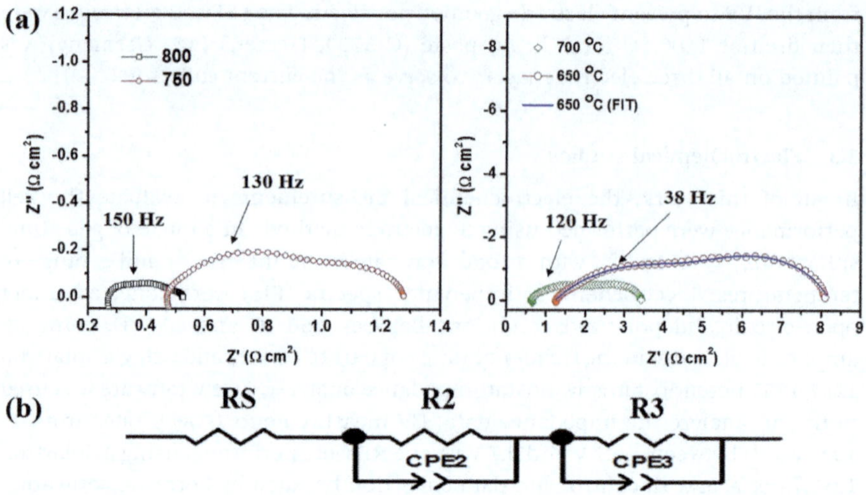

Fig. 1 (a) Typical impedance spectra of a LSFCr-3 working electrode in a 3-electrode half-cell configuration at open circuit under an oxygen partial pressure of 0.21 atm and variable temperatures. (b) Equivalent circuit used for data fitting.

where R is a resistance and Q is a constant phase element (CPE), representing time-dependent capacitive elements, and n is the associated parameter which indicates the similarity of Q to an ideal capacitor, for which $n = 1$.

Table 1 shows that the resistance associated with the process at high frequencies (RHF), often ascribed to the charge transfer reaction, increases with decreasing temperature, from 0.26 to 4.1 Ω cm^2, as expected for this type of process, while the associated capacitance (CHF) remains almost unchanged at *ca.* 2×10^{-3} F cm^{-2} (Table 1). The resistance of the low frequency arc (RLF) at the OCP (Table 1) decreases even more noticeably with increasing temperature (Table 1) and thus cannot be related to gas phase diffusion limitations, which are known to be almost temperature independent.[10] Its associated capacitance (CLF) is very large, typical of the chemical capacitance obtained when using MIEC electrodes[1,11] and related to their redox chemistry.

A series of experiments involving varying p_{O_2} (0.1–1 atm) was then conducted to elucidate the effect of the oxygen partial pressure on the EIS response at equilibrium (at the OCP). Fig. 2 shows a set of open circuit EIS data at 700 °C, as an

Table 1 Circuit elements[a] associated with the data in Fig. 1[b]

T (°C)	RHF (Ω cm^2)	RLF (Ω cm^2)	CHF (F cm^{-2})	CLF (F cm^{-2})	nHF	nLF
800	0.26	0.02	1.5×10^{-3}	2×10^{-1}	0.64	0.70
750	0.53	0.27	2.3×10^{-3}	2.0×10^{-1}	0.71	0.79
700	1.9	1.02	1.8×10^{-3}	1.8×10^{-1}	0.77	0.55
650	4.1	2.9	2.8×10^{-3}	1.6×10^{-1}	0.81	0.58

[a] (RHF, CHF, nHF) are obtained from the high frequency arc, while (RLF, CLF, nLF) are from the low frequency arc. [b] Three electrode LSFCr-3 half-cell configuration, $p_{O_2} = 0.21$ atm, 50 ml min^{-1} flow rate.

example, and Table 2 shows the resistance and capacitance values for the high (HF) and low frequencies (LF) processes, obtained by fitting the Fig. 2 data to the same equivalent circuit as shown in Fig. 1b and using eqn (1) to convert the CPEs to capacitance values. Rs and RHF are seen to be independent of p_{O_2}, as expected, (Table 2) and CHF remains almost unchanged at *ca.* 1.8×10^{-3} F cm^{-2} (Table 2).

In contrast, the resistance associated with the low frequency arc decreases quite significantly with increasing p_{O_2}, while CLF hardly changes, but remains large, typical of a chemical capacitance. As shown in Fig. 2a, the low frequency arc is absent in pure oxygen ($p_{O_2} = 1$). However, when p_{O_2} was decreased to 0.2 atm, the low frequency arc is easily discerned in both the Bode and Nyquist depictions, centered at *ca.* 3.6 Hz (Fig. 2b). Thus, the process associated with the low frequency arc is strongly affected by oxygen only when p_{O_2} is low (below 0.2 atm). As this arc is not diffusional (gas phase) in nature, this may point to surface reactions involving oxygen (*e.g.* oxygen adsorption or dissociation) as the slow step. Similar characteristics have been reported by several authors for the ORR/OER at the OCP at $La_{1-x}Sr_xMnO_3$ (LSM), $La_{0.6}Sr_{0.4}Co_{0.2}Fe_{0.8}O_{3-\delta}$ (LSCF) and La_2NiO_4 electrodes.[10,12–16]

The ORR (and the reverse OER) is known to be very complex, occurring in multiple elementary steps, with one mechanism suggested to be:[17] (1) diffusion of O_2 molecules from the gas phase to the active site, (2) oxygen dissociative adsorption on the cathode surface, (3) electron transfer step, (4) incorporation of O^{2-} into the electrode at the double phase boundary (DPB), (5) O^{2-} diffusion in the bulk of cathode, and (6) O^{2-} transfer from the cathode to electrolyte.[17] Any one of these steps can be rate-determining (the rds), and if microscopic reversibility applies, this step should also be slow during the OER. Using conventional electrochemical theory (Butler–Volmer[18]) and depending on which step is the slowest, a different Tafel slope and hence transfer coefficient (α value, inversely proportional to the Tafel slope[19]) should be obtained. Another parameter that helps in determining the rds of electrochemical reactions is the reaction order (m), in this case with respect to oxygen. For studies done at the OCP, m can be obtained from the dependence of Rp on the oxygen partial pressure, following eqn (2):[17]

$$\mathrm{Rp}^{-1} \propto p_{O_2}{}^{m} \tag{2}$$

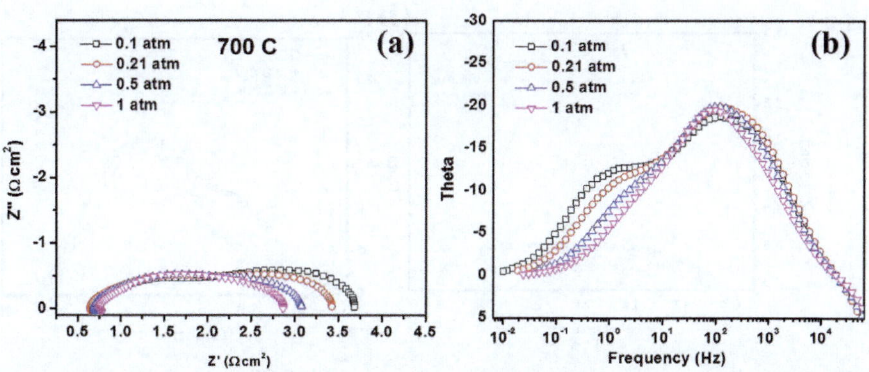

Fig. 2 (a) Nyquist and (b) Bode phase angle plots of a LSFCr-3 working electrode in a 3-electrode half-cell configuration, evaluated at a p_{O_2} of 0.1, 0.21, 0.5 and 1 atm, all at 700 °C.

Table 2 Circuit elements[a] associated with the impedance data in Fig. 2[b]

p_{O_2} (atm)	RHF (Ω cm²)	RLF (Ω cm²)	CHF[c] (F cm⁻²)	CLF[c] (F cm⁻²)	nHF	nLF
0.1	1.63	1.61	1.5×10^{-3}	1.6×10^{-1}	0.59	0.73
0.21	1.96	1.02	1.8×10^{-3}	1.9×10^{-1}	0.54	0.78
0.5	2.04	0.56	1.9×10^{-3}	2.0×10^{-1}	0.56	0.86
1	2.08	0.22	2.3×10^{-3}	3.4×10^{-1}	0.58	1.0

[a] RHF, CHF, and nHF are obtained from the high frequency arc, while RLF, CLF, and nLF are from the low frequency arc. [b] Three electrode LSFCr-3 half-cell configuration, 700 °C, 50 ml min^{-1} flow rate. [c] CHF and CLF represent the high and low frequency arcs, respectively, and were obtained by fitting to the equivalent circuit shown in Fig. 1b, and calculated by using $C = (R \times \text{CPE})^{1/n}/R$.[10]

Considering the mechanism above and assuming that the rate of each individual step is 10 times or more different from the previous or following steps so that they can be distinguished kinetically from each other, then for $m = 1$, molecular oxygen should be involved in the rate-determining step, when $m = 1/2$, atomic oxygen is involved in the rate-determining step, and when $m = 1/4$, then charge transfer is the rate-determining process.[17]

3.1.2 Polarized EIS studies. To further investigate the performance of LSFCr-3 towards both the ORR and OER, polarized EIS analysis was carried out in air, with the application of an overpotential of 300 mV in both the positive (SOEC) and negative (SOFC) directions. OCP and polarized Nyquist (Fig. 3a) and Bode (Fig. 3b) plots are shown at a p_{O_2} of 0.21 atm and 700 °C under these conditions, as an example. Under polarization, a decrease in the polarization resistance is seen from 2.98 Ω cm² at the OCP to 1.63 Ω cm² at −300 mV (ORR), while it is 1.03 Ω cm² at +300 mV (OER), as expected.

In general terms, under positive polarization, a somewhat less suppressed low frequency arc is observed than under negative polarization (Fig. 3a), consistent with what is seen in the Bode plots (Fig. 3b). Moreover, the arcs under either OCP or negative polarization remain at the same frequency, while under positive polarization, both arcs shift to higher frequencies and the high frequency

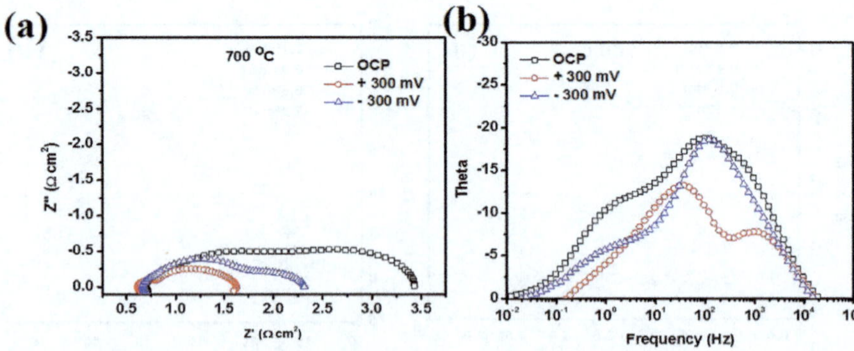

Fig. 3 (a) Nyquist and (b) phase angle Bode plots of the OCP and polarized impedance data of a LSFCr-3 working electrode in a 3-electrode half-cell configuration, evaluated under an oxygen partial pressure of 0.21 atm at 700 °C.

resistance diminishes greatly. These results show that there are some important differences between the ORR and the OER that could not have been detected at the OCP, but do need to be better understood. In relation to this, when anodically polarized EIS was performed in variable p_{O_2} atmospheres, *i.e.*, under active OER conditions (data not shown), Rp is independent of p_{O_2} from 0.1 atm to 1 atm. This suggests that the local p_{O_2} must be dominated by the oxygen generated electrochemically, rather than by the p_{O_2} of the gas stream passed through the cell. This phenomenon is also seen in the CV data, as will be shown below.

The resistance of the high frequency arc decreases under polarization (Table 3), especially at +300 mV, while the associated CHF value increases as the potential is changed from +300 to −300 mV. Thus the time constant changes quite significantly for the HF arc. In terms of RLF, this circuit element decreases slightly with polarization, which may again suggest that RLF is related to molecular oxygen dissociation/adsorption/desorption processes.[15,16] Further, the capacitance associated with the low frequency process changes equally with anodic and cathodic polarization, becoming *ca.* 10^{-2} F cm^{-2} in both cases (Table 3).

In order to better understand the OER and ORR at the LSFCr-3 materials, p_{O_2} dependence studies were also carried out under polarization to obtain the reaction orders in comparison with those at the OCP. This is not normally done in the literature, but is actually quite important, especially as it is not clear whether the OCP EIS data correctly predicts oxygen electrochemistry at MIEC electrodes under relevant polarization conditions. Fig. 4 displays the dependence of the total polarization resistance (ASRp) on the oxygen partial pressure at 650, 700, 750 and 800 °C at both the OCP, and at +300 and −300 mV *vs.* the OCP, with the *m* values given in Table 4.

From Table 4, it is seen that the reaction orders, obtained from Rp at the OCP and at negative overpotentials, are similarly p_{O_2} dependant at all temperatures. Furthermore, the reaction orders at positive overpotentials (OER) seem to be less dependent on p_{O_2} than those at the OCP and under negative polarization, as fully expected, as oxygen is the product of the OER.

To further investigate the p_{O_2} dependence of the high and low frequency processes, the p_{O_2} dependence of each of the resistances identified from the EIS studies at both the OCP and at polarization is shown in Fig. 5, for the example of 700 °C. A very strong p_{O_2} dependence ($m = 0.99$) of the low frequency process at

Table 3 Circuit elements obtained by fitting of impedance data[a] in Fig. 3[b]

Potential of WE	RHF (Ω cm^2)	RLF (Ω cm^2)	CHF[c] (F cm^{-2})	CLF[c] (F cm^{-2})	nHF	nLF
OCP	1.96	1.02	1.8×10^{-3}	1.9×10^{-1}	0.55	0.78
+300 mV (OER)	0.12	0.91	5×10^{-4}	1.1×10^{-2}	1.00	0.73
−300 mV (ORR)	0.91	0.72	4.2×10^{-3}	1.2×10^{-2}	0.89	0.44

[a] (RHF, CHF, nHF) are obtained from the high frequency arc, while (RLF, CLF, nLF) are from the low frequency arc, with the data fitted to the circuit shown in Fig. 1b. [b] Three electrode LSFCr-3 half-cell configuration, 700 °C, 0.21 atm p_{O_2}, 50 ml min^{-1} flow rate. [c] CHF and CLF represent the high and low frequency arcs, respectively, and were calculated by using $C = (R \times \mathrm{CPE})^{1/n}/R$.[10]

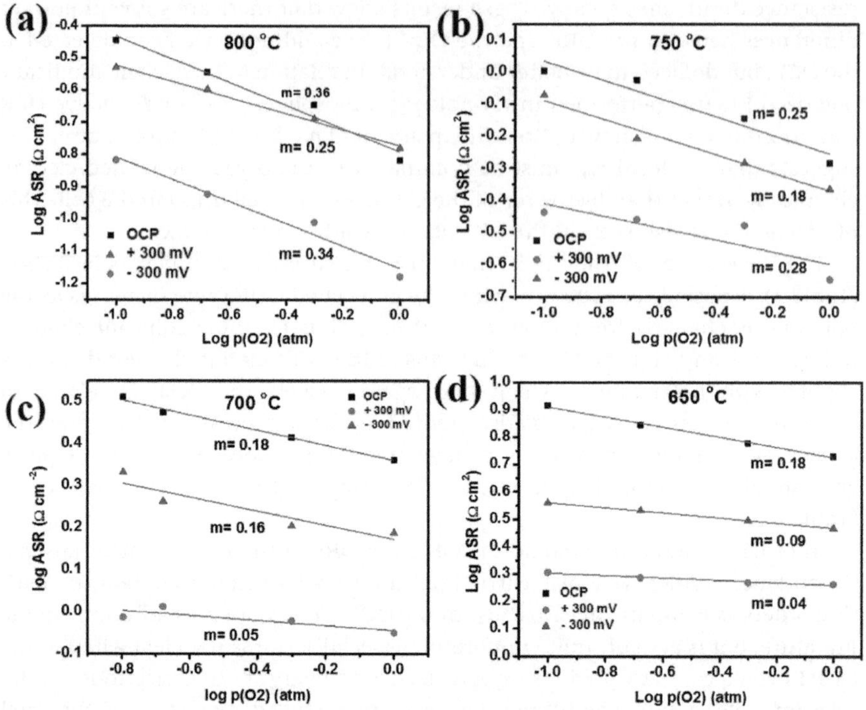

Fig. 4 $p(O_2)$ dependence of ASRp of the LSFCr-3 WE at the OCP and at both anodic (300 mV) and cathodic (−300 mV) overpotentials at (a) 800 °C, (b) 750 °C, (c) 700 °C, and (d) 650 °C, also showing the corresponding *m* values at different temperatures.

OCP conditions is observed in Table 5, which suggests that molecular oxygen is involved in the slowest step at the OCP, either adsorbing to the surface, or dissociating. This observation clearly supports our previous discussion about the low frequency process.

On the other hand, the high frequency process shows only a weak oxygen partial pressure dependence ($m = 0.1$) at the OCP. According to past studies,[10,15] this may indicate that a charge transfer step is the rate determining process. Moreover, the p_{O_2} dependence of both high and low frequency processes under positive and negative polarization, shown in Table 5, indicates a very weak p_{O_2}

Table 4 Reaction order (*m*) with respect to O_2 at various potentials and temperatures (from Fig. 4)[a]

	Temperature (°C)			
Potential (mV)	800	750	700	650
OCP	0.36	0.25	0.18	0.18
+300	0.25	0.18	0.05	0.04
−300	0.35	0.28	0.16	0.09

[a] Three electrode (LSFCr-3) half-cell configuration, 50 ml min^{-1} flow rate.

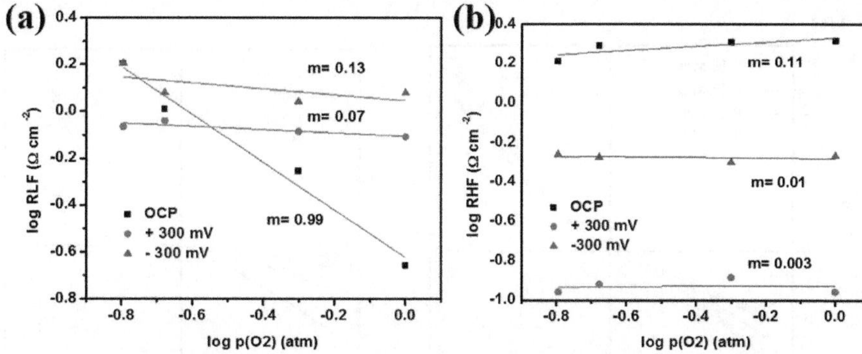

Fig. 5 p_{O_2} dependence of the (a) low and (b) high frequency processes observed at LCFCr-3 WE at the OCP and at anodic and cathodic overpotentials at 700 °C.

dependence for both the ORR and OER reactions, and very different dependencies with respect to what is seen at the OCP. These data confirm the earlier conclusions, namely that during the OER, the O_2 that is being produced generates a local p_{O_2} that makes the rate of the reaction independent of the external p_{O_2}. On the other hand, in the negative direction (ORR), we observe no p_{O_2} dependence for the high frequency feature, but a slightly stronger dependence at low frequencies.

3.1.3 Cyclic voltammetry studies. The cyclic voltammetry (CV) response of the ORR and OER at the LSFCr-3 electrodes was also investigated, revealing that it was independent of the potential sweep rate and gas flow rate. This indicates there are no bulk mass transport limitations associated with getting O_2 to or from the electrode surface. Fig. 6a shows the effect of p_{O_2} at constant temperature (800 °C), while Fig. 6b shows the effect of temperature on the CV behavior a constant oxygen partial pressure of 0.21 atm.

It is worth noting that LSFCr-3 appears to be a significantly better catalyst for the OER reaction as opposed to the ORR, as seen in the asymmetrical nature of all of the CVs, similar to the Ca analogue, LCFCr-3.[20] This is consistent with the polarized EIS data in Table 3, where Rp is 1.03 Ω cm^2 at +300 mV and 1.63 Ω cm^2 at −300 mV. As expected, the currents increase with increasing p_{O_2} in the ORR region, except at 1 atm (Fig. 6a), consistent with the effect of O_2 on the EIS results (Fig. 2). However, in the OER region, no obvious effect is observed when the p_{O_2} is increased from 0.16 to 1 atm, as O_2 is the reaction product in the OER and should thus not affect its kinetics under anodic polarization. On the other hand, the

Table 5 Reaction order (*m*) of O_2 at 700 °C at the OCP and under positive and negative polarization[a]

Potential (mV)	*m* (low frequencies)	*m* (high frequencies)
OCP	0.99	0.11
+300	0.07	0.003
−300	0.13	0.01

[a] Three electrode (LSFCr-3) half-cell configuration, 50 ml min^{-1} flow rate, based on data in Fig. 5.

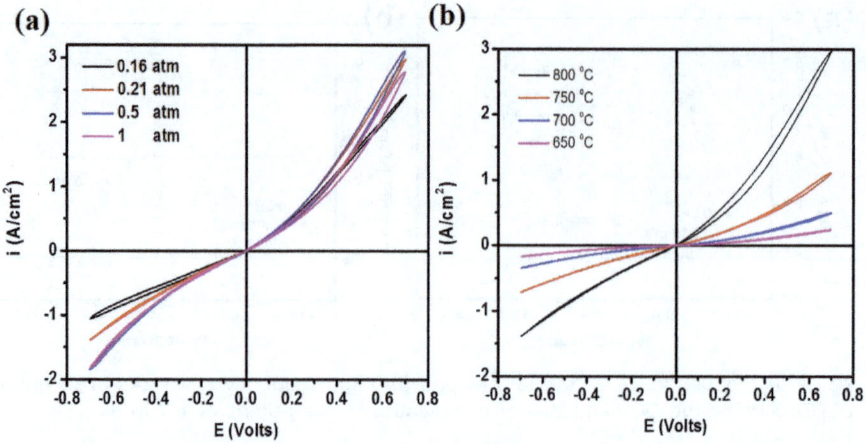

Fig. 6 Cyclic voltammograms (5 mV s^{-1}) for LSFCr-3 WE in a 3 electrode half-cell configuration evaluated (a) at various $p(O_2)$ at 800 °C and (b) at various temperatures at 0.21 atm p_{O_2}.

effect of temperature on the OER and ORR is very pronounced, as shown in Fig. 6b, confirming again that the slow step in either of these reaction is not mass transport in the gas phase.

A key kinetic parameter obtained from CV analysis is i_o, the exchange current density. The i_o values, obtained using low-field analysis (eqn (3))[19,21] of the CV data (Fig. 6b) are shown in Table 6 and were also compared to those obtained from Rp in the OCP EIS data (Fig. 1).

$$i_o = (RT\nu/nF)\text{slope} \tag{3}$$

where the slope is obtained from the i–V plots (Fig. 6b) within ±50 mV of the OCP, and assuming that $n = 4$ and $\nu = 1$, where n is the total number of electrons transferred in the oxygen reaction (*i.e.*, 4) and ν is the stoichiometric coefficient (ORR mechanisms proposed in the literature suggest a value of 1 (ref. 21)).

The exchange current density values (Table 6) show relatively good agreement for the results obtained using these two electrochemical techniques. Overall, this adds confidence that the open circuit and polarized 3-electrode experiments, and the associated data analysis methods used, have been carried out correctly in this work.

Table 6 Exchange current densities (i_o) obtained by CV[a] and EIS[b] measurements (Fig. 1 and 6b)

Temperature (°C)	CV i_o (A cm^{-2})	EIS i_o (A cm^{-2})
800	0.12	0.09
750	0.026	0.02
700	0.015	0.007
650	0.005	0.004

[a] Fitted parameters are based on the low field approximation in the potential range of ±50 mV *vs.* the OCP in Fig. 6b. [b] Based on Rp obtained from EIS experiments (Table 1).

3.2 Three electrode CO_2 reduction and CO oxidation studies

The electrochemical performance and stability of the LSFCr-3 material towards the reduction of CO_2 and oxidation of CO was also investigated here. A 3 electrode half-cell was again constructed, with LSFCr-3 pasted on both sides of the YSZ electrolyte, a LSFCr reference electrode attached on the WE side, and a GDC buffer layer present between YSZ and the LSFCr-3 electrodes.

3.2.1. 100% CO_2 studies. Fig. 7 shows both the open circuit and polarized impedance data (all at an overpotential of ±100 mV), obtained for LSFCr-3 in a pure CO_2 atmosphere, showing three well-defined time constants, with one process seen at high frequencies (R1/CPE1) and two at low frequencies (R2/CPE2 and R3/CPE3). CO_2 alone was investigated in an attempt to simplify data interpretation and help to identify any diffusion controlled processes, as CO will be present in very low concentrations under these conditions, generated only as the product of CO_2 reduction. As expected, the full polarization resistance (Rp = R1 + R2 + R3) is smaller at negative polarization during CO_2 reduction than at the OCP, and is the largest at +100 mV polarization in the absence of CO in the gas feed stream. To further understand the various processes occurring in this cell, the EIS data, obtained both at the OCP and under polarization, were fitted to the Rs(R1/CPE1)(R2/CPE2)(R3/CPE3) equivalent circuit, where the CPEs were again converted to capacitance (C) using eqn (1). Table 7 shows the values of the circuit elements and the calculated capacitances under all of these conditions.

The resistance associated with the high frequency arc (R1) is suggested to be due to a charge transfer process. This interpretation is strengthened by the relatively low value of the parallel capacitor (C1), 0.6–9 × 10^{-5} F cm^{-2}, which is typical for an interfacial capacitance. The weak dependence of R1 on overpotential is consistent with being in the 'low field' region, *i.e.*, at ±100 mV polarization, and thus R1 would be expected to be very similar to the value obtained at the OCP. The low frequency arc (R3/C3) is ascribed to a gas diffusion process, as the C3 values are very large, in the range of 0.67–0.85 F cm^{-2}, and are essentially independent of overpotential. Also, the n3 values are generally close to unity. R3 is much higher under anodic *vs.* cathodic polarization, presumably because there is no CO in the gas stream and thus only the small quantities of CO

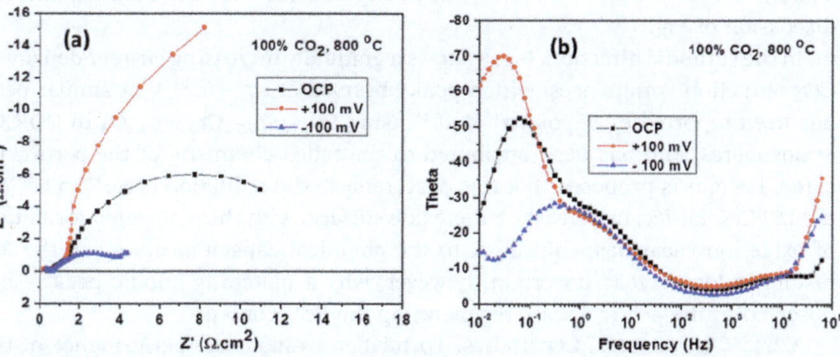

Fig. 7 OCP and polarized (±100 mV) EIS response of LSFCr-3 in a three electrode half-cell, operated at 800 °C in 100% CO_2, showing (a) the Nyquist plots and (b) the phase angle Bode plots.

Table 7 Fitted parameters[a] from Fig. 7a[b] using the Rs(R1/CPE1)(R2/CPE2)(R3/CPE3) equivalent circuit model

Potential (mV)	R1 (Ω cm^2)	C1[c] (F cm^{-2})	n1	R2 (Ω cm^2)	C2[c] (F cm^{-2})	n2	R3 (Ω cm^2)	C3[c] (F cm^{-2})	n3	Rp (Ω cm^2)
OCP	0.20	8.66×10^{-5}	0.64	1.09	0.131	0.67	12.6	0.68	0.97	13.9
+100	0.45	6.53×10^{-6}	0.38	0.91	0.143	0.71	40.7	0.85	1.00	42.1
−100	0.22	8.19×10^{-5}	0.64	0.87	0.172	0.67	2.77	0.67	0.82	3.85

[a] (R1, C1, n1) were obtained from the high frequency (10 kHz) arc, while (R2, C2, n2) and (R3, C3, n3) are from the low (1 Hz) and still lower (0.05 Hz) frequency arcs, respectively. [b] Three electrode LSFCr-3 half-cell, operated at 800 °C in 100% CO_2. [c] C1, C2 and C3 represent the high, low and lower frequency arcs, respectively, and were calculated using eqn (1).

generated during CO_2 reduction are available for the oxidation process under anodic polarization. In contrast, under cathodic polarization, R3 is quite small, since CO_2 is being rapidly reduced to CO.

The meaning of R2 and C2 is less clear. C2 may be reflective of the chemical capacitance (changes in stoichiometry) of the LSFCr material, similar to the interpretation used in the air study above. As seen in Table 7, C2 is in the range of 0.13–0.17 F cm^{-2}, observed at *ca.* 1 Hz, and is relatively independent of polarization. The fact that C2 is the largest at −100 mV (0.17 F cm^{-2}), *vs.* at the OCP and at +100 mV polarization, may indicate that more oxygen vacancies are created in this potential range due to the enhanced redox chemistry of Fe^{4+} to Fe^{3+} in LSFCr-3 under these conditions. R2 is only very weakly dependent on overpotential and thus it is possible that it reflects an interfacial process, such as oxygen ion vacancy formation/filling in the LSFCr-3 lattice, which probably influences the CO_2/CO adsorption/dissociation step.

Cyclic voltammetry (CV) was also carried out to obtain further insights into the electrochemical mechanism of CO_2 reduction and CO oxidation at the LSFCr-3 electrode. Fig. 8 shows the 3-electrode CV in pure CO_2 at 800 °C at 5 mV s^{-1}. The CV shows higher current densities in the cathodic scan than in the anodic direction, which is not surprising, since there is no CO in the gas stream. This is confirmed by the anodic limiting current density behavior (concentration polarization) observed from 0 to +0.7 V, in full agreement with the EIS data and the discussion of Fig. 7.

In the cathodic direction, Fig. 8 shows a gradually increasing current density as CO_2 reduction commences, with a peak observed at *ca.* −0.25 V. A similar peak has been reported by Sapountzi *et al.*[22] using $La_{0.75}Sr_{0.25}Cr_{0.9}Fe_{0.1}O_3$ in CO/CO_2 atmospheres, and has been attributed to the redox chemistry of the perovskite oxide. Here, it is proposed that this peak reflects the reduction of Fe^{4+} to Fe^{3+} in the LSFCr-3 lattice, or at least on the oxide surface, with the associated formation of oxide ion vacancies, equivalent to the chemical capacitance seen in the EIS results in Fig. 7. It is uncertain, however, why a matching anodic peak is not observed in the positive scan, regenerating the Fe^{4+} content.

3.2.2. 50 : 50 CO_2 : CO studies. To further evaluate the performance of the LSFCr-3 catalyst towards CO_2 reduction and also CO oxidation, a CO_2 : CO ratio of 1 : 1 was fed to the cell. Fig. 9 shows the OCP and polarized EIS data at 800 °C in 30% CO_2 : 30% CO : 40% N_2. The impedance data show again that the cell

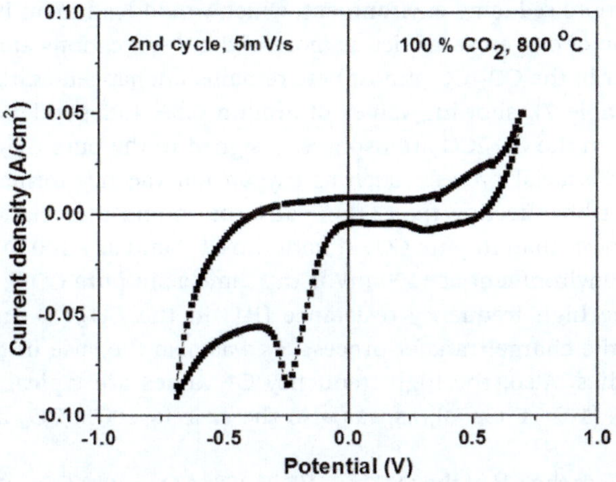

Fig. 8 Cyclic voltammetry data obtained for LSFCr-3 WE in three electrode half-cell, operated at 800 °C in pure CO_2 at 5 mV s^{-1}.

performs better in the cathodic mode (reduction of CO_2) than in the anodic mode (CO oxidation), suggesting that there are some real kinetic differences between these two reactions.

The EIS data were fitted to a Rs(R1/CPE1)(R2/CPE2) equivalent circuit model, with only two time constants appearing under these CO_2/CO conditions (Fig. 9). The (R3/CPE3) time constant, observed at very low frequencies (0.05 Hz) in the pure CO_2 environment in Fig. 7a and ascribed to gas phase diffusion limitations, is minimal or barely detectable in Fig. 9a and thus it was excluded from the fitting model. Table 8 shows the fitted parameters, demonstrating that the polarization resistance at the OCP is 0.87 Ω cm^2, while at +100 mV, it is 1.35 Ω cm^2, and at −100 mV, Rp is 0.67 Ω cm^2.

In the CO_2/CO system (Fig. 9), the low frequency C2 element (Table 8) is now slightly higher in value than in 100% CO_2 (Table 7). If it is still ascribed to the LSFCr-3 chemical capacitance, then this is perhaps not surprising because the cell

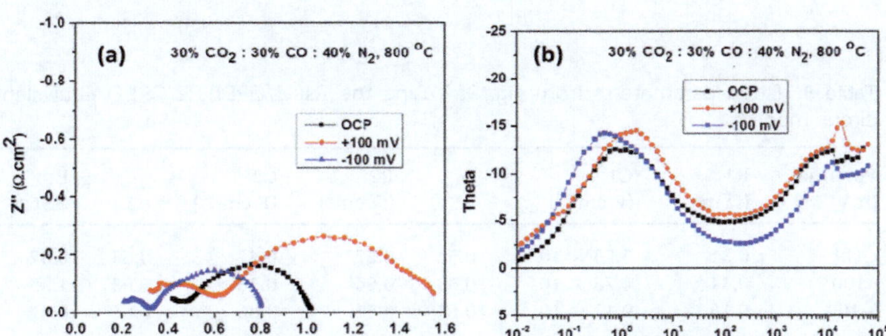

Fig. 9 OCP and polarized (±0.1 V) EIS response of the LSFCr-3 WE in a 3 electrode half cell experiment at 800 °C in 30% CO_2 : 30% CO : 40% N_2, showing (a) the Nyquist plots and (b) the phase angle Bode plots.

is now in a more reducing environment, which would lead to an increase in the concentration of oxygen vacancies as more of the B site cations are reduced. The n2 parameter in the CO_2/CO atmosphere remains unchanged *vs.* the situation in pure CO_2 (Table 7), showing values of around 0.65, independent of potential. However, R2 in the CO_2/CO atmosphere, assigned in the pure CO_2 studies to be due to an interfacial process, such as oxygen ion vacancy formation/filling in LSFCr-3, possibly affecting the CO_2/CO adsorption and dissociation process, is about 50% lower than in pure CO_2 at both the OCP and at −100 mV, while R2 in the CO_2/CO environment at +100 mV is the same as in pure CO_2.

Again, the high frequency resistance (R1) in the CO_2/CO environment is ascribed to the charge transfer process, as was also the case in the 100% CO_2 and air studies. Also, the high frequency C1 values are typical of interfacial capacitances (10^{-5} F cm^{-2}), as was also the case in 100% CO_2 and in the air studies.

Fig. 10 shows the CV of the LSFCr-3 WE in 30% CO_2 : 30% CO : 40% N_2 at 1 mV s^{-1} at 800 °C. It is interesting to note that much higher current densities are observed in the negative scan during the reduction of CO_2 to CO than in the positive scan during the oxidation of CO to CO_2, which may indicate that a CO adsorption step is the rds, consistent with the EIS data (Fig. 9, Table 8). Even so, the performance of the LSFCr-3 electrode towards CO oxidation is quite good, comparable to other types of perovskites, such as LSCM.[23]

It is also interesting that the CO_2 reduction currents are roughly 10 times higher in the CO_2/CO mixture used in Fig. 10 than in 100% CO_2, suggesting that the LSFCr-3 material is in a much different redox state in the two CO_2-containing environments. The potential of the CO_2/CO equilibrium is *ca.* 670 mV more negative (more reducing) than in pure CO_2, arguing that a higher concentration of Fe^{3+} should be present at the surface, leading to better catalytic activity. We expect Cr to remain in the +6 state under all conditions at the surface, in agreement with studies by Mizusaki[24] on $LaCrO_3$.

The excellent performance in the cathodic direction during CO_2 reduction could be due to preferential adsorption of CO_2 on the active sites and also an increase in oxygen vacancy concentration due to the reduction of Fe^{4+} to Fe^{3+}, which will favour the reduction of CO_2, according to reaction (4).

$$CO_2 + V_O^{\cdot\cdot} + 2e^- = O_O^{\times} + CO \quad (4)$$

Table 8 Fitted parameters[a] from Fig. 9a[b] using the Rs(R1/CPE1)(R2/CPE2) equivalent circuit model

Potential (mV)	R1 (Ω cm^2)	C1[c] (F cm^{-2})	n1	R2 (Ω cm^2)	C2[c] (F cm^{-2})	n2	Rp (Ω cm^2)
OCP	0.31	5.75×10^{-5}	0.52	0.57	0.47	0.65	0.87
+100	0.41	4.78×10^{-5}	0.53	0.94	0.26	0.63	1.35
−100	0.16	9.12×10^{-5}	0.65	0.51	0.86	0.66	0.67

[a] (R1, C1, n1) and (R2, C2, n2) obtained from the high (10 kHz) and low (1 Hz) frequency arcs, respectively. [b] Three electrode (LSFCr-3) half-cell, operated at 800 °C in 30% CO_2 : 30% CO : 40% N_2. [c] C1 and C2 represent the capacitance in the high and low frequencies arcs, respectively and were calculated using eqn (1).

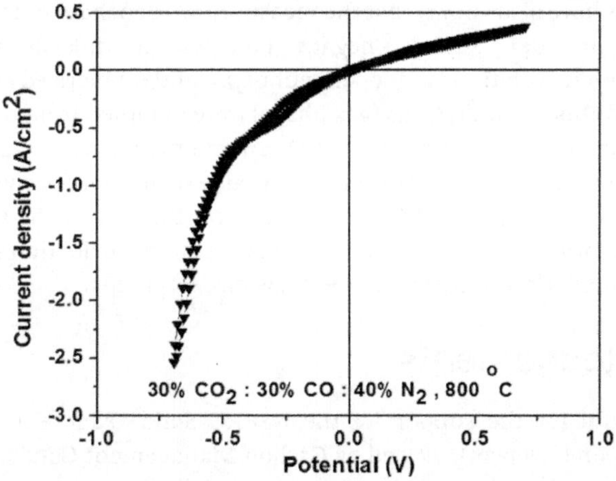

Fig. 10 Cyclic voltammetry data obtained for LSFCr-3 WE in a 3 electrode half-cell at 800 °C in 30% CO_2 : 30% CO : 40% N_2 at 1 mV s^{-1}.

This is in agreement with the explanation for R2 and C2 in Table 8, where a lower R2 and higher C2 is seen at −100 mV polarization *vs.* at the OCP.

To further analyze the CV data, the low field exchange current density (i_o) was obtained from the slope of the i–V plot (Fig. 10) within ±50 mV of the OCP, based on eqn (3). For the CO_2/CO reduction reaction, n and ν were assumed to be 2 and 1, respectively (as the total number of electrons passed per molecule of CO_2 reduced is 2 and the number of occurrences of the rds (ν) would probably be 1). The i_o values obtained from this CV analysis were then compared to the i_o obtained from the Rp of OCP EIS work (Table 8), giving similar values, *i.e.*, 41 mA cm^{-2} from the CV and 53 mA cm^{-2} from the EIS data. These high current densities are fully consistent with the excellent activity of the LSFCr-3 catalyst towards the CO/CO_2 redox process. Notably, the high field Butler–Volmer approximation was not used here, because it is questionable if Butler–Volmer formalisms can be applied to electrochemical data acquired at an electrode material that is changing its composition (and catalytic activity) over the potential range being analyzed.

4. Conclusions

The electrochemical performance of LSFCr-3 has been studied in detail, both at open circuit and under polarization, in 3-electrode half-cells between 650 and 800 °C, in air, pure CO_2 and CO_2/CO mixtures. It is shown here, using impedance spectroscopy (EIS) and cyclic voltammetry (CV), that the performance of this material in all of these gas environments is very good. The 3 electrode half-cell configuration allowed us to isolate the reactions occurring in the fuel cell and electrolysis modes and determine the catalytic activity of the material towards each reaction separately. This showed that LSFCr-3 is a better catalyst for the oxygen evolution and CO_2 reduction reactions than for oxygen reduction or CO oxidation, indicating an overall better performance in the electrolysis *vs.* fuel cell mode.

Our results have also shown that the measured p_{O_2} dependence is very different at the OCP than under positive or negative polarization, while also revealing that the rate of oxygen evolution is independent of p_{O_2} under polarized conditions. No evidence for diffusion limitations (gas phase) were obtained from our EIS and CV experiments in air. In both air and CO/CO_2 environments, there is good evidence for the presence of a chemical capacitance, as also seen from the significant effect of the reducing atmosphere (CO_2 *vs.* CO_2/CO) on the rate of the CO/CO_2 redox reactions. This was ascribed to significant changes in the oxidation state of the Fe component in LSFCr-3 in these different gas environments.

Acknowledgements

We are grateful for the support of the NSERC Solid Oxide Fuel Cell Canada Strategic Research Network as well as Carbon Management Canada, a Canadian National Centre of Excellence. The authors gratefully acknowledge the Eyes High PDF Program at the University of Calgary and Alberta Innovates – Technology Futures (AITF) for the support of AB. We also would like to thank AITF for their scholarship support of PA.

References

1 M. Chen, S. Paulson, V. Thangadurai and V. Birss, *J. Power Sources*, 2013, **236**, 68–79.
2 B. Molero-Sánchez, J. Prado-Gonjal, D. Ávila-Brande, M. Chen, E. Morán and V. Birss, *Int. J. Hydrogen Energy*, 2015, **40**, 1902–1910.
3 J. C. Ruiz-Morales, D. Marrero-Lopez, J. Canales-Vazquez and J. T. S. Irvine, *RSC Adv.*, 2011, **1**, 1403–1414.
4 S. Xu, S. Li, W. Yao, D. Dong and K. Xie, *J. Power Sources*, 2013, **230**, 115–121.
5 P. Addo, B. Molero-Sánchez, M. Chen, S. Paulson and V. Birss, *11th European SOFC and SOE forum*, Luzerne, Switzerland, 2014.
6 D. Pérez-Coll, A. Aguadero, M. J. Escudero and L. Daza, *J. Power Sources*, 2009, **192**, 2–13.
7 Z. Lu, J. Hardy, J. Templeton and J. Stevenson, *J. Power Sources*, 2011, **196**, 39–45.
8 S. McIntosh, S. B. Adler, J. M. Vohs and R. J. Gorte, *Electrochem. Solid-State Lett.*, 2004, **7**, A111–A114.
9 R. Küngas, A. S. Yu, J. Levine, J. M. Vohs and R. J. Gorte, *J. Electrochem. Soc.*, 2013, **160**, F205–F211.
10 M. J. Escudero, A. Aguadero, J. A. Alonso and L. Daza, *J. Electroanal. Chem.*, 2007, **611**, 107–116.
11 P. Hjalmarsson, M. Søgaard and M. Mogensen, *Solid State Ionics*, 2009, **180**, 1395–1405.
12 E. P. Murray, T. Tsai and S. A. Barnett, *Solid State Ionics*, 1998, **110**, 235–243.
13 S. P. Jiang, J. G. Love and Y. Ramprakash, *J. Power Sources*, 2002, **110**, 201–208.
14 D. Marinha, L. Dessemond and E. Djurado, *J. Power Sources*, 2012, **197**, 80–87.
15 F. S. Baumann, J. Fleig, H.-U. Habermeier and J. Maier, *Solid State Ionics*, 2006, **177**, 1071–1081.
16 F. S. Baumann, J. Maier and J. Fleig, *Solid State Ionics*, 2008, **179**, 1198–1204.
17 B. Liu, Y. Zhang and L. Zhang, *Int. J. Hydrogen Energy*, 2009, **34**, 1008–1014.

18 D. A. Noren and M. A. Hoffman, *J. Power Sources*, 2005, **152**, 175–181.
19 O. Bockris, *Modern Electrochemistry*, New York, NY, 1973, vol. 2.
20 B. Molero-Sánchez, P. Addo, M. Chen, S. Paulson and V. Birss, *11 th European SOFC & SOE FORUM*, Luzern, Switzerland, 2014.
21 J. Liu, A. C. Co, S. Paulson and V. I. Birss, *Solid State Ionics*, 2006, **177**, 377–387.
22 F. M. Sapountzi, S. Brosda, K. M. Papazisi, S. P. Balomenou and D. Tsiplakides, *J. Appl. Electrochem.*, 2012, **42**, 727–735.
23 X. Yue and J. T. S. Irvine, *J. Electrochem. Soc.*, 2012, **159**, F442–F448.
24 J. Mizusaki, *Solid State Ionics*, 1992, **52**, 79–91.

Faraday Discussions

DISCUSSIONS

Fundamental electrochemistry: general discussion

Mogens Mogensen, Raymond Gorte, John Vohs, Ian Metcalfe, Sune Dalgaard Ebbesen, Steven McIntosh, Scott Barnett, San Ping Jiang, Ming Li, Tatsumi Ishihara, Rotraut Merkle, John Bøgild Hansen, John Druce, Helena Téllez, Göran Lindbergh, Viola Birss, Xiangling Yue, Joseph Hartvigsen, Christopher Graves, John Irvine, Bilge Yildiz, Robert Kee and Denis Cumming

DOI: 10.1039/C5FD90073E

Ian Metcalfe opened the discussion of the paper by John Hansen: You have discussed the use of both solid oxide fuel cells and solid oxide electrolysers. What about the possibility of a solid oxide device for the partial oxidation of hydrocarbons? Do you think such a device would have utility and if so under what circumstances?

John Bøgild Hansen responded: There are certainly some interesting options to couple catalysis and solid oxide electrolysis. You can for instance apparently achieve better selectivity for oxidative coupling reactions (most recently described by Caravaca *et al.*[1]), but as the research on the topic is rather limited, in a very early phase and slightly off the topic of this discussion theme, I chose not to include it in my presentation.

1 A. Caravaca, A. de Lucas-Consuegra, V. J. Ferreira, J. L. Figueiredo, J. L. Faria, J. L. Valverde and F. Dorado, *Appl. Catal., B*, 2013, **142–143**, 298–306.

Joseph Hartvigsen said: Ian Metcalfe discussed with John Hansen the possibility of using a methane fuelled SOFC to power solid oxide electrolysis. John said that would not be practical from a systems perspective and that syngas from methane would be more efficiently created using steam reforming. I agree that John's explanation is obvious, but I think that instead perhaps Ian was suggesting an anode depolarization configuration where the oxygen moving through the electrode in steam or steam–CO_2 co-electrolysis mode was used to partially oxidize methane to syngas. I think that it could have some advantages from an exergy view but that reduced driving potential of high efficiency operation may also lead to low current density and high specific capital cost. Further there are serious materials challenges in anode depolarized cells related to having strongly reducing conditions on both sides of the cell.

Ian Metcalfe replied: Yes, this is one possibility. I was also thinking more generally of exploiting the possibilities for partial oxidation. So instead of using a fuel cell followed by electrolysis is there any scope for looking at *e.g.* methane partial oxidation in fuel cells? This would lead to energy generation and useful chemical products.

Joseph Hartvigsen commented: Does industry want to go beyond meeting electricity needs and consider also producing transportation fuels? The power required to produce transportation fuels and displace fossil fuels from transportation would push the electric production capacity so high that even in relatively low wind conditions there would be enough wind power to meet the non fuel production grid demand. Therefore using wind power to focus fuel synthesis to displace fossil fuels from the transportation sector would have the side effect of reducing the probability that available wind power was less than the conventional, non-dispatchable, grid loads. Having fuel synthesis as a large dispatchable load, with an associated gross over-capacity of wind generation with respect to conventional non-dispatchable grid loads would help eliminate much of the need for dispatchable backup electrical generation.

Mogens Mogensen stated: This discussion is causing confusion because partial oxidation is only reasonable in context of fossil fuels. The work we discuss here is about how we can promote renewable energy.

Mogens Mogensen opened the discussion of the paper by Robert Kee: I think that the oxygen partial pressure dependence (Fig. 4) can qualitative be understood in a simple way by reaction 11: half an oxygen molecule plus an oxide vacancy plus an oxide ion form two O^-, plus the fact that the mobility of an electron hole (O^-) is much higher than the mobility of a proton.

Robert Kee replied: This comment is a generally correct interpretation of the conductivity dependence shown in Fig. 4. Although we did not include a plot of the defect partial conductivities as a function of oxygen partial pressures, they can be recovered from the concentration plots shown in Fig. 7, the Nernst–Einstein relationship (eqn 38), and the defect diffusion coefficients (Table 2). Fig. 8 also shows the high partial conductivity of the O-site polaron contribution under highly oxidizing conditions.

Raymond Gorte asked: How can oxygen enter a non-reducible oxide such as BZY? Do you have any evidence, such as coulometric titration or TGA, to estimate the amount of oxygen that is entering or leaving?

Robert Kee replied: It is not necessary that one of the cations be reduced to enable oxygen incorporation. The heterovalent doping (here yttrium) produces vacancies that can be filled by oxygen incorporation. The O-site polarons (or holes) provide the needed charge compensation to maintain electroneutrality upon oxygen incorporation, and hence variations in the vacancy concentration. Although we think it to be unlikely, it is possible that the Zr could undergo reduction. This possibility is discussed in the paper in eqn 8 and 9.[1]

Mogens Mogensen queried: How do you describe a small polaron? You write that polaron conduction is different from a band conduction. Isn't it a matter of formalism? To the best of my knowledge physicists describe polarons as associated with very narrow conduction bands. The localisation of the electron is associated with a local distortion of the lattice, and the small polaron is the electron plus the local distortion of the lattice. It takes a certain thermal activation energy that must be enough to rearrange the lattice distortion in order to set the electron free to move.

Robert Kee responded: Our intent is to make clear that the small-polaron mobility occurs *via* an activated hopping mechanism that is different from metallic-like band conduction. A small polaron can be defined as a localized charge in a potential well with a Fermi surface that is smaller than the unit cell. As such, it has no band structure. As the question notes, the surrounding lattice distorts to accommodate the localized charge, with the electron residing in the potential well. Small-polaron mobility is typically thermally activated, which can usually be described with Arrhenius expressions. In contrast to small polarons (the present situation), large polarons are characterized by lattice distortions that exceed the lattice unit cell. In this case, Fermi surfaces associated with mobile electrons can overlap and band-like electronic conduction is possible.

Mogens Mogensen asked: So according to your definition, which sounds attractive, there is not a band in which the electrons are moving as described from the physics perspective. Other definitions define the small polaron as being localised due to a very narrow band gap. This causes a distortion of the lattice near the ion, where the electron (or hole) is localised, and the electron can only hop when this distortion is thermally "dissolved". Is your definition in disagreement with the usual physicist definition?

Robert Kee answered: We don't see the situation as being one of a "physicist's interpretation" or a "chemist's interpretation," *etc.* Rather, different materials behaving differently. For example, if we think of semiconducting materials such as doped silicon, the electronic conduction should be associated with band structure and overlapping wave functions. Such electronic conduction is only weakly activated. However, in the BZY (and other ceramic mixed conductors), there is evidence for highly activated electronic transport. This behaviour strongly suggests localized small polarons. That is, the localized polaron, which causes local lattice distortion, must hop over activation barriers associated with the local lattice distortions.

Although the material itself has a band structure, including a band gap, we don't see a band structure as being associated with individual localized charges (*i.e.* the polarons). Rather, the localized polaron resides at an energy level within the band gap. Based on your question, it appears that your minds-eye picture of a "very narrow band gap" leads to the same net behaviour as the small-polaron mobility that we have implemented in our models. Nevertheless, we think that the material's band structure should be associated with the overall lattice, not with individual defects.

John Irvine questioned: Why do you dismiss a band picture? The difference between a small polaron, a large polaron and a band carrier is a matter of degree. A hole will be located on an oxide ion in the small polaron picture, whereas in a band picture it is delocalised across the oxide orbitals; however these are not exclusive as there will be a component due to localisation in the band picture. In a band scheme, semiconductor conductivity will be thermally activated.

Robert Kee answered: Based on our results and on recent literature,[1] there is evidence for localized charge in the sense of small polarons. Delocalized band conduction is typically not activated. Certainly, the properties derived from our modelling indicates activated electronic conduction. All this having been said, we cannot definitively dismiss the possibility of some band-like contributions. Nevertheless, our results are best fit by a small-polaron hopping mechanism.

1 D. N. Mueller. M. L. Machala, H. Bluhm, W. C. Chueh, *Nature Comm.*, 2015, **6**, 6097.

Bilge Yildiz asked: Conductivity relaxation measurements are typically performed with smaller step changes in the partial pressure of gases. This is to ensure that the properties of the material is not changing drastically and spatially during the measurement. Why did you choose very large changes in gas composition during the measurements? How does this affect the uncertainties in the fitted kinetic parameters?

Robert Kee responded: As you note, essentially all conductivity–relaxation literature is based upon small changes in gas-phase partial pressures. An important reason for this approach is that the model-based analysis and interpretation is invariably based upon assuming ambipolar transport and analytic separation-of-variables solutions (*e.g.*, from the classic book by Crank[1]). The ambipolar diffusion coefficients (sometimes called chemical diffusion coefficients) necessarily depend nonlinearly upon local defect compositions (*cf.*, eqn 43 in the paper). So, any large variations would invalidate the analytic-solution analysis, which depends on assuming constant diffusion coefficients. Because they are more fundamental, and hence more generally useful, we are usually more interested in determining the intrinsic defect diffusion coefficients, not the ambipolar diffusion coefficients. Ambipolar diffusion is necessarily restricted to materials with two charge-carrying defects. Thus, for mixed conductors, such as BZY that have three mobile charge-carrying defects, the ambipolar theory does not apply, or applies in narrow circumstances. Thus, the analysis approach using analytic solutions and small perturbations is generally inapplicable. The Nernst–Planck–Poisson (NPP) approach developed herein has no restrictions associated with numbers of charge-carrying defects or the magnitude of gas-phase composition changes. It does, however, require computational solution. We find that measuring relaxation performance over large gas-phase variations provides more useful information with fewer experiments, compared to many more experiments with small perturbations. The NPP models incorporate the intrinsic diffusion coefficients directly. Thus, the model-based experimental interpretation comes more directly to the needed intrinsic diffusion coefficients, without the complications of deconvoluting the ambipolar diffusion coefficients.

1 J. Crank, *The Mathematics of Diffusion*, Oxford University Press, London, 1975.

Ian Metcalfe questioned: Could you comment on the uncertainty of the parameters that you estimate using such a model? Is it not possible that the model prediction is relatively insensitive to the values used for certain parameters and that consequently the values you return for such parameters have large uncertainty?

Robert Kee replied: We agree that the six thermodynamic and six transport parameters of BZY10 cannot be claimed to be entirely unique. Thus, some uncertainty must be acknowledged. With equilibrium conductivity measurements alone, the paper shows that parameter non-uniqueness, and hence uncertainty, is a major problem. In fact, as the paper shows, parameter uniqueness can be greatly improved by using the relaxation results. Where there is direct evidence and/or there is general consensus in the literature, we constrain parameters to be within known ranges. In any case, we think that there is great value in having a self-consistent set of properties that span very wide ranges of temperatures and oxidizing/reducing conditions. There are very few, if any, such complete parameter sets for mixed-conducting perovskites. Thus, as long as the operating ranges are within the ranges for which the properties were fit, the property sets can be used in models to predict behaviour of systems that use the materials, such as fuel cell, electrolyzers, membrane reactors, *etc.*

Steven McIntosh asked: Have you considered that some of the changes you observe in behaviour during the long duration required for the experiments may be due to surface segregation of species, particularly A-site cations?

Robert Kee replied: We have not explicitly considered the possibility that cation segregation and surface reconstructions affect long-term behaviour. However, the redox processes are reproducible, even after multiple cycles and over periods of weeks to months. So, if surface segregation and reconstruction were happening and relevant, then it appears that they are reversible.

Steven McIntosh asked: The model utilizes the bulk vacancy concentrations in the reaction kinetics at the surface of the material, but it seems likely that the surface vacancy concentration will vary significantly from this bulk value. Have you considered how to include this and do you think it would be a dramatic difference?

Robert Kee replied: It is correct that we assume the defect concentrations at the BZY10 surface to be the same as they are in the near-surface bulk material. Based on your own work (DOI: 10.1039/c5fd00014a and **ref. 1 and** 2 below), and that of others, there is some evidence that the surface stoichiometry can be different from that of the near-surface bulk material. At this point, we do not have a means to quantitatively evaluate the surface structure in terms of cation segregation, *etc.* However, although we cannot be certain, we do not expect the results to change dramatically by incorporating surface reconstruction behaviour.

1 A. C. Tomkiewicz, M. Meloni and S. McIntosh, *Solid State Ionics*, 2014, **260**, 55–59.

2 M. A. Tamimi, A. C. Tomkiewicz, A. Huq and S. McIntosh, *J. Mater. Chem. A*, 2014, **2**, 18838–18847.

Mogens Mogensen questioned: How important is your chemical *versus* physical interpretation? Will eqn 29 still be valid and have the same dependency on the concentration of localised electrons?

Robert Kee answered: As noted in the previous response, we don't see a meaningful distinction between chemical or physical interpretations. Rather, the important result is that the electronic-defect transport occurs by an activated hopping mechanism that is a characteristic of small polarons. Eqn 29 expresses the oxygen site balance. If the electronic defect is localized at the oxygen site, then it should contribute to the site balance. However, because the O-site polaron concentration is usually quite low, the model predictions are not greatly influenced by the polaron's contribution to the site balance.

Ming Li asked: The hole is located at O^{2-} giving rise to O^{-}. Is there any experimental evidence to prove the existence of O^{-}?

Robert Kee responded: Yes, there is now some experimental evidence to support the notion of an O-site polaron. Recently, based on X-ray adsorption measurements, Cheuh reported that "The importance of anion-redox chemistry in oxygen-deficient metal-oxide electrocatalysts, particularly that arising from the localized electronic states on oxygen, challenges the current paradigm of electrocatalyst design, which has been focused largely on the cation redox chemistry."

1 D. N. Mueller. M. L. Machala, H. Bluhm, W. C. Chueh, *Nature Comm.*, 2015, **6**, 6097.

Mogens Mogensen questioned: Do you assume no defect interaction? If so, do you think this is true?

Robert Kee answered: The Nernst–Planck formulation does indeed neglect defect–defect interactions (*i.e.* dilute limit). Although this assumption is widely practised in the solid-state ceramics literature, it certainly is an assumption that is not strictly correct. So, in part, our approach is based on pragmatism. In principle, there are no fundamental barriers to incorporating defect–defect interactions. However, doing so requires matrices of transport parameters that replace diffusion coefficients assigned to individual defects. Inasmuch as establishing unique sets of thermodynamic and transport properties is difficult on the Nernst–Planck setting, substantially increasing the number of unknown, and difficult-to-measure, parameters is difficult to justify. Although it is difficult to prove, we expect that the errors associated with dilute-limit approximations are relatively small, especially at high temperature.

John Vohs opened the discussion of the paper by Christopher Graves: Have you thought about how the different catalytic properties of Ni and CGO affect performance? Ni would be expected to be a much better hydrogen oxidation catalyst than CGO, it is less clear which would be better for CO oxidation. Is the catalytic activity what is important here or is the microstructure just as important?

Christopher Graves replied: I agree there is no reason to expect different activity of CGO for H_2–H_2O and CO–CO_2 reactions. In the case of porous electrodes, of course the microstructure is very important. One intention of the model electrodes is to remove microstructure effects and measure isolated electrocatalytic activity. This is basically successful with the Ni/YSZ model electrodes. However the model electrode with CGO nanoparticles (placed at the tip of a titanate cone current collector) has an important micro/nanostructure, it does not have just a simple well-defined interface. A CGO film could instead be used, but that may give a different response than the CGO nanoparticles. I would say both the electrochemical activity and the microstructure are important here and it's very difficult to isolate their effects. Model electrodes may also introduce other issues, as discussed in the paper.

Raymond Gorte asked: With regards to your results with model electrodes, do you have evidence that the metal is not moving at the point of contact with the electrolyte? My question comes up because of the observations of J. Nielsen and T. Jacobson who observed that metals from such a configuration can move out over the electrolyte.[1]

1 J. Nielsen and T. Jacobson, *Solid State Ionics*, 2008, **178**, 1769.

Christopher Graves replied: Yes, for the Ni wire model electrodes in this study, the Ni did not spread out onto the YSZ electrolyte surface. The paper you reference is for Ag and other metals as oxygen-electrodes, but I have seen similar behavior of Ni model electrodes under certain conditions that were avoided here. For example in high anodic polarization approaching the NiO threshold, Ni transports out and forms nanoparticles on the electrolyte surface, see M. S. Schmidt *et al.*[1]

But here we avoided these issues, and the interface stability was sufficient at least during sequences of measurements in H_2–H_2O and CO–CO_2, which are the measurements compared in this study. The stability was always checked by doing the same measurement again, *e.g.* H_2–H_2O then CO–CO_2 then H_2–H_2O, with less than 10% change in polarization resistance. These kinds of concerns are briefly addressed in the Experimental section of the paper.

1 M. S. Schmidt, K. Vels Hansen, K. Norrman, M. Mogensen, *Solid State Ionics*, 2009, **180**, 431–438.

Mogens Mogensen commented: It depends on the conditions. If we *e.g.* polarize hard in the anodic direction then volatile Ni hydroxide is formed due the steam formation. As soon as the steam is diluted a small distance (microns) away from the electrolyte–electrode interface (non-porous Ni-model electrodes) then the Ni is precipitated as a metal again.[1]

1 M. S. Schmidt, K. Vels Hansen, K. Norrman, M. Mogensen, *Solid State Ionics*, 2009, **180**, 431–438.

John Bøgild Hansen asked: You ascribe the difference between model and porous electrodes with respect to the ratio of the electrode polarisation resistance between CO_2 and steam electrolysis to accumulation of impurities in the model electrodes. Given that impurities are probably also present in the porous electrodes how can you state anything definite about the ratio as you do, indicating a value of 4 for the nickel based electrodes?

Christopher Graves responded: To clarify, in the paper we ascribed the different ratios for the model and porous electrodes (about 4 and 2, respectively) to the nature of the reaction zone extension into the porous electrode thickness, assuming the model and porous electrodes have the same "high purity" Ni/YSZ interfaces. This possible explanation does fit the values well. However, we did not intend to state anything definite. The much higher ratio of 33 obtained for the impure Ni/YSZ model electrode was likely due to a high level of accumulated impurities. We expect that the 3-phase-boundaries in the porous electrode have a lower impurity level than the interface of the model electrode – instead we expect them to be more similar to the interface of the "high purity" model electrodes, as mentioned in the paper. To confirm this, a detailed nano-scale chemical analysis of the 3-phase-boundaries in both electrodes would of course be needed.

Viola Birss questioned: In Table 1 of the paper, it is interesting that the ratios of the H_2/steam reaction to CO/CO_2 reaction rates are all roughly in the range of 2–4 for the porous Ni/SZ and CGO based electrode systems (the larger ratios seen for the model electrodes were explained as being due to impurity effects.) What is your feeling about why the rates of these very different reactions (very different bond energies, very different adsorption energies of intermediates, *etc.*) are so similar, especially compared to the many orders of magnitude differences in electrochemical reaction rates seen at lower temperatures? Does this imply that a good catalyst for H_2/steam can thus just be assumed to be a good catalyst for CO/CO_2 as well? And what is your general explanation for the ratio of 2–4 for the rates for H_2/steam *vs.* CO/CO_2?

Christopher Graves answered: I agree with Mogens, who has long said that in high temperature systems the rate-limiting steps are very different, things like competitive adsorption can dominate. Charge-transfer seems to not be rate-limiting. The ratios of the H_2/H_2O *vs.* CO/CO_2 reaction rates that we report are sometimes very different – going from ratios of 4 to 0.2 for different model electrodes is more than a factor of 10. But I know what you mean, for many of the different porous electrodes, the ratios are within a factor of 2 or so. One reason we did this study is because there is a considerable amount of literature that states that the reaction rate for CO_2 electrolysis is 10× slower than H_2O electrolysis, usually referring to Ni-based electrodes, but these results clearly show that is not the case except with very poor electrodes, and that it depends heavily on impurities and microstructure. I don't have a clear general explanation for a ratio of 2–4 for the porous electrodes, but gas diffusion limitations (absent in the model electrodes) certainly may play a role, as we mention in the paper. The binary diffusion coefficient for H_2/H_2O is about 5.5× higher than CO/CO_2 at the operating temperatures.

Viola Birss communicated: What was the volume or area difference of the counter/reference electrode (C/RE) *vs.* the working electrode? The ratio of C/RE : WE area/volume must be very large in order to prevent polarization of the C/RE, and if this occurred at any time, or differentially in the CO/CO_2 *vs.* H_2/H_2O environments, then the measured R_P values and their ratios would not be reliable.

Christopher Graves communicated in reply: Apologies that these dimensions were not reported in the paper. For the point-contact electrodes (the Ni wire electrode and the CGO/STN cone-shaped electrode), the ratios of the counter to working electrode areas were greater than 1000. For the small porous working electrode cell with working and counter electrodes made of the same Ni–ScYSZ composite, the ratio was about 100, so the counter electrode is expected to make up around 1% of the impedance and current–voltage measurements. Therefore, because the counter electrodes were much larger and the electrolyte thickness was sufficient to allow the current to distribute outwards to the whole counter electrode, the counter electrode response is expected to be negligible. The cells are known as "pseudo 3-electrode" cells with the counter electrode acting as the reference electrode. (A reference electrode could also be added to eliminate this minor contribution, but reference electrode placement must be done very carefully and can introduce other inaccuracies.)

Viola Birss communicated: Please comment on the reproducibility and error range of the results (ratios) in Table 1, in relation to the reporting of the ratios to two decimal places.

Christopher Graves communicated in response: The reproducibility was within 10% for two electrodes of the same type. Therefore, the ratios were reported with only two significant digits. More detailed error handling and analysis would be beneficial but would not significantly change the ratios.

Joseph Hartvigsen asked: The paper compared the kinetics of CO/CO_2 with H_2/H_2O electrolysis of Ni/YSZ (three-phase-boundary) and ceria (two phase surface). A Ni/ceria electrode is well known and has the high electronic conductivity of Ni/YSZ and the active two phase surface of a doped ceria electrode. What was the rationale that excluded Ni/ceria from consideration in this study?

Christopher Graves replied: That composite electrode would be interesting for a follow-up study. The composite was not specifically excluded in this paper, but since there are two types of reaction sites (3PB and 2PB) operating simultaneously in that composite, it would be more difficult to understand. A composite of ceria and an acceptor-doped strontium titanate was included in the paper (CGO nanoparticles coating the titanate), but that should be simpler and reflect only the electrochemistry of the ceria because the electron-conducting titanate has negligible electrocatalytic activity compared to the ceria.

Joseph Hartvigsen asked: The presented kinetics provide a very interesting level of detail on relative rates. Have these fundamental measurements been translated to practical validation in single cells? Are these electrode materials applicable for implementation in cells and stacks? Are there barriers for implementation such as in an electrode supported thin film cell?

Christopher Graves responded: There has been some work in our lab and others in making full sized cells with titanate/ceria electrodes. As far as I know, developing electrolyte-supported cells has been successful whereas there have been some difficulties in obtaining sufficient mechanical strength and electronic current collection in electrode-supported cells, although there was a report back in 2010 of a successfully tested large-sized titanate-electrode-supported cell.[1] A recent paper reviews some of the progress (also see other recent papers by these authors).[2]

1 Q. Ma, F. Tietz, A. Leonide and E. Ivers-Tiffée, *Electrochem. Commun.*, 2010, **10**, 1326–1328.
2 M. C. Verbraeken, T. Ramos, K. Agersted, Q. Ma, C. D. Savaniu, B. R. Sudireddy, J. T. S. Irvine, P. Holtappels and F. Tietz, *RSC Adv.*, 2015, **5**, 1168–1180.

San Ping Jiang said: Have you considered the effect of the interface of the model electrode and porous Ni–SZ electrodes on the electrochemical activity and kinetics of the CO/CO_2 and H_2/H_2O reactions? In the case of model electrodes, there is no pre-sintered interface between the electrode and electrolyte, while in porous Ni–SZ electrodes, an interface is formed during the high temperature sintering process.

Christopher Graves responded: That is a good point, the interfaces will be different because in the model electrode the Ni/YSZ interface is heat treated from room temperature to 1000 °C *in situ* in a reducing atmosphere (with Ni metal wire) whereas the porous Ni–SZ electrodes were sintered as NiO–SZ at around 1300 °C in air and then subsequently reduced at 1000 °C *in situ*. This relates to the concern we briefly mentioned in the paper that a model electrode can be a "poor representation of the actual interface that the model electrode aims to investigate in a realistic technological electrode". Nevertheless, the electrochemistry of both of these types of *in situ* formed Ni/SZ interfaces follows the same trends despite considerable variability, as also shown in other papers cited in the paper.[1,2]

1 A. Utz, A. Leonide, A. Weber and E. Ivers-Tiffée, *J. Power Sources*, 2011, **196**, 7217–7224.
2 A. Utz, H. Stormer, A. Leonide, A. Weber and E. Ivers-Tiffée, *J. Electrochem. Soc.*, 2010, 157, B920–B930.

Scott Barnett asked: What role did the Pt current collector play in the electrode characteristics of the oxide electrodes? Prior literature has shown that Pt current collectors can help to promote the hydrogen oxidation kinetics of oxide SOFC anodes, so it is important to provide some additional evidence that this is not the case here.

Christopher Graves responded: Exactly, the possible effect of the Pt current collectors on the pure CGO electrodes is a point of discussion in our paper. Even though the porous Pt (prepared by sintering the Pt paste) by itself is a terrible electrode, it may be that the fine CGO structure which makes many interfaces with the Pt, makes an active interface and enhances the performance. However, these electrodes should be thick enough that the reaction zone does not extend all the way to the current collector – especially for the electrodes made of titanate backbones with only a small volume fraction of CGO coating (which provides the ion conduction pathways). But this is what we are considering and may be why the gold current collector shows a different response in H_2/H_2O *vs.* CO/CO_2. Regarding the magnitudes of the electrode polarization resistances, I have tested pure CGO electrodes with Pt, Au, Ni and Cu current collectors and Pt gives the best results but it may be simply due to the quality of the contact area rather than the current collector material enhancing the electrochemical activity. We have a paper in preparation that addresses this.

Tatsumi Ishihara said: How is the stability of the Ni electrode during coke formation in comparison to the CeO_2 electrode, Ni easily forms carbon and is deactivated whereas CeO_2 is stable against coke formation.

Since Ni is soft compared to the ceramic electrode, I wonder if the contact area could quite easily change during the measurement, how stable and reproducible is the data?

Christopher Graves responded: In this study we did not go near the pCO/pCO_2 thermodynamic threshold that leads to carbon deposition by the Boudouard reaction. But in other studies we have indeed observed carbon deposition on Ni and a lack of carbon deposition on ceria.

As mentioned in the paper, at the start of the test the Ni wire creeps and the contact area changes, so an initial stabilization period was included in all the model electrode tests. After that initial period, the contact area changed very little. The stability and reproducibility were sufficient that the measurements were reproduced in the same H_2/H_2O gas mixture before and after a CO/CO_2 measurement (see *e.g.* Fig. 5 in the paper for an example).

Raymond Gorte opened the general discussion of the paper by Denis Cumming: Meilin Liu and his students published work on diffuse reflectance FTIR of SOFC cathodes that showed peaks when he polarized his electrodes.[1] I believe most of the peaks are associated with changes in the phonon modes of the electrode as he reduced or oxidized them. My question is, have you considered the possibility that artifacts like this could creep into your results?

1 X. Lu, P. W. Faguy and M. L. Liu, *J. Electrochem. Soc.*, 2002, **149**, A1293.

Denis Cumming replied: We are aware of this work and agree that artefacts could play a significant role. However through the careful background subtraction (*i.e.* to eliminate emission effects) and sample history considerations we think genuine electrochemical effects can be observed. Furthermore, this work examines the fuel electrode which contains many more IR active species (such as carbonates) that have been well characterised in the catalysis community so we will be able to draw out true electrochemical effects.

John Bøgild Hansen said: Have you noted any formate species in your spectra?

Denis Cumming replied: This was considered, particularly if there was a shift reaction occurring. Many of IR active bands from formates overlap with carbonates but there a characteristic formate band at 3000 cm^{-1} that is absent from our spectra.[1] It does not preclude their presence but there is no direct evidence so far.

1 F. C. Meunier, D. Reid, A. Goguet, S. Shekhtman, C. Hardacre, R. Burch, W. Deng and M. Flytzani-Stephanopoulos, *J. Catal.*, 2007, **247**, 277–287.

Sune Dalgaard Ebbesen continued the general discussion of the paper by Denis Cumming: In the figure the claimed surface OH groups look more like gas phase scissor bending of water. How sure are you, that no gas phase steam is present?

Denis Cumming responded: The spectra does indicate the presence of gas-phase water but the source of the water is unclear. Further experiments are planned which will control the gas atmosphere more tightly.

Mogens Mogensen said: You don't have a defined electrode potential – do you? Do you know where the water comes from? Would you try with CO/CO_2 in argon in order to avoid any hydrogen?

Denis Cumming responded: Cells were operated in a single atmosphere and both electrodes to experience the same atmosphere. The atmosphere for the experiments consisted of only carbon monoxide and argon so was not thermodynamically defined. The source of the water is unknown at this stage. Our apparatus has been used for many years to perform heterogeneous catalysis experiments and this problem has not been encountered. Experiments in CO/CO_2 mixtures are in progress and the results will be reported in a future publication.

Ian Metcalfe asked: Clearly some of your experiments are dynamic in nature as under open-circuit conditions you feed carbon monoxide and observe the production of gas phase carbon dioxide. Have you had the chance to do any time-resolved work yet?

Denis Cumming answered: Not yet. This links in some way to the observation where results appear to be sample history dependent. Time-resolved work will therefore form an important part of our future work.

Sune Dalgaard Ebbesen said: How do you contact the sample? And how sure are you that you are probing the electrochemical active area?

Denis Cumming answered: Samples were contacted using a silver wire and silver paste. The electrodes were gold. Lateral conductivity was high (tested using two point continuity) so there should have been no issue polarising the electrode across its entire surface. The observation of the electrochemically active region is still under investigation.

John Vohs questioned: How thick was your electrode and are you sampling the electrochemically active region?

Denis Cumming replied: Our electrodes were ~90 microns thick. We are still to determine the depth of penetration of the IR beam. Literature suggests it could be several tens of microns but even if that were the case we would still not be probing the electrochemically active region. It is our opinion that what we have presented in this paper shows other surface species that are the result of direct catalysis or electrochemically mediated secondary catalysis.

John Irvine opened the general discussion of the paper by Steven McIntosh: Could the majority of the surface be terminated in impurity phases, but a small portion might correspond to the parent perovskite. This would give same activation energy but limit exchange rates in accord with the available area?

Steven McIntosh answered: I do think that the Sr-rich impurity phases will influence the true outer surface rate, and while we see some possible correlation, the differences in measured surface rate don't track well with the changes in surface composition. We typically observe orders of magnitude differences in surface rate for relatively small (10% or so) changes in surface composition. The only real link we can find is with the inferred bulk oxygen anion mobility from the neutron diffraction studies.

John Irvine asked: If the surface exchange is limited by the coverage of impurity, Sr-rich, phases, could the removal of Sr improve surface exchange rates?

Steven McIntosh responded: I believe that this could. There are numerous papers in the literature reporting on the "activation" of LSM based electrodes.[1] In many of these studies, removal of Sr species is often linked to an improvement in electrode performance. We do have to be careful here though. As presented by Telléz and co-workers at this meeting, the surface segregation can be rapid, even at 400 °C.

1 S. P. Jiang, J. G. Love, *Solid State Ionics*, 2001, **138**, 183–190.

Helena Téllez asked: In your paper, you mentioned that the outer surface of the LSCF Ruddlesden–Popper phases are very similar in terms of the Sr coverage. However, in the LEIS depth profiles in Fig. 6, the Sr surface coverage varies from about 70% for the RPn1 up to about 95% for the RPn2. Interestingly, the surface exchange rates follow the same trend as the Sr surface coverage in these RP phases: the higher the Sr surface coverage, the higher the surface exchange rate (RPn2 > RPn3 > RPn3). Do you think that the different Sr coverages could explain the trend in the surface exchange rates? Does the segregated Sr make the RP surface more active?

Steven McIntosh replied: I think it's extremely likely that the difference in Sr coverage will influence the kinetics at the true surface *i.e.* the outer atomic layer. However, this trend is not exact. The RPn2 shows the highest rate and highest Sr coverage at ~ 95% (this would imply that Sr is an active surface species, against many literature reports from electrode studies), but presents a similar rate to RPn3 which shows ~80% Sr coverage. The RPn1 rate is more than one order of magnitude lower with a surface coverage of ~70%. I don't think it's feasible that the drop from 95% to 80% makes a relatively small change in observed rate, whereas the further drop from 80% to 70% Sr coverage leads to an order of magnitude drop.

The only clear correlation we can find with these materials and the other material systems we have studied is with an anticipated trend in bulk oxygen mobility determined by neutron diffraction.[1,2] This seems to be the simplest explanation – our technique is not surface sensitive enough to measure the rapid, true outer surface, oxygen exchange rate. We always end up with a bulk convoluted number. That is not to say that no technique can access the true rate, but we must be very careful when interpreting this data.

1 A. C. Tomkiewicz, M. Meloni, S. McIntosh, *Solid State Ionics*, 2014, **260**, 55–59.
2 M. A. Tamimi, A. C. Tomkiewicz, A. Huq, S. McIntosh, *J. Mater. Chem. A*, 2014, **2**, 18838–18847.

Helena Téllez enquired: In the case of the double perovskites that you showed in your slides, do you know if they lose the A-cation ordering due to the Ba segregation? Can you obtain any evidence of the lack of A-cation ordering at the surface by neutron diffraction?

Steven McIntosh answered: We haven't directly looked for this loss of ordering but it may be. We can't see this by powder neutron diffraction as the technique measures average bulk properties. We will try to look by HRTEM.

Bilge Yildiz asked: Surfaces of the samples are clearly not the same as the bulk, as you showed evidence from LEIS. However, LEIS alone is not enough to say what the surface structure is exactly. So we can't say if the segregated surface in your case is a good or a bad one. What other analysis have you performed or will perform to deduce what exactly is the surface and sub-surface structure?

Steven McIntosh answered: We have not performed any analysis of the surface or near-surface structure. I anticipate, based on your own work and that of others, that we would find Sr-rich islands of material across the surface. These would near 100% coverage for the RPn2 case. This is an area for further investigation.

Bilge Yildiz asked: Indeed the surface composition has a lot of Sr. If we assume that all the Sr is in a blocking phase that partially covers the surface (as was shown in other works on LSC and LSM), it would still be possible that one can correlate the reactivity (oxygen exchange kinetics) to the remaining Fe or Co exposed at the surface. Even if the amount of Fe and Co is small at the surface, the kinetics could be correlated to the presence of those active transition metals at the surface. Have you checked whether your kinetics data correlates to the available transition metals at the surface?

Steven McIntosh responded: A case can be made for some influence of relative coverage on the observed rate; however, we see dramatic (order or magnitude or more) changes in rate for relatively small changes in Sr coverage between materials. We have not been able to investigate multiple Sr coverages on the same material up to this point. The only clear correlation we find is with the inferred oxygen anion mobility from neutron diffraction. This is true for the RP phase presented here, and double perovskites and single perovskites in our previously published work. It seems to be a strong trend and leads us to conclude that our techniques may not be adequate to measure the rapid true outer surface rate.

Bilge Yildiz commented: A possible follow-up to check the correlation of the pristine (bulk-like) surface chemistry to the oxygen exchange kinetics is first etching the surface to get rid of the blocking Sr-rich layer and then performing the LEIS and the oxygen exchange measurements at lower temperatures (<400 °C). This might avoid or minimize the artifacts related to surface segregation in establishing correlations of bulk-like surface chemistry to surface reactivity.

Steven McIntosh responded: I think there would be great value in pursuing experiments as a function of Sr coverage – we will likely do this in the future to further test our hypothesis.

Mogens Mogensen asked: Does the measured exchange coefficient tell whether the oxygen exchange takes place at the very outer-most atomic layer or somewhere else within the first few nanometres from the surface? Can the bottleneck be some special feature within the surface region with much lower conductivity than the rest of the structure?

Steven McIntosh responded: My belief is that the exchange technique is probing oxygen isotopic exchange within the first few nanometers of the surface and not the outermost surface layer. This is likely fine if we use the resulting data with caution in, for example, an electrode model, but we have to be careful if we use it to imply some outer surface reaction mechanism. The bottleneck could very well be at the near surface. It's clear that the surface enrichment of one species will leave a depleted layer within the near surface region.

Raymond Gorte remarked: With regards to the comment that much of the surface might be covered and that adsorption only occurs at specific places, have you estimated a "sticking coefficient"? Kinetic theory of gases places limits on the flux of gas to a surface and one may be able to rule out certain models that imply only collisions with a small fraction of the surface have a probability of adsorbing.

Steven McIntosh replied: The kinetic theory of gases predicts ~1.75×10^{27} collisions m^{-2} s^{-1} or around 3000 moles of collision m^{-2} s^{-1} for O_2 at 725 K. Our measured rates of isotope exchange in this temperature range are of the order 10^{-2} moles m^{-2} s^{-1}, implying that very few of the collisions result in isotope exchange with the bulk. This would give a very low effective "sticking coefficient" for the reaction. The number of "successful" collisions is reduced from the theoretical maximum due to the measured high activation energy (around 90 kJ mol^{-1}), and the fact that we are only probing collisions which go through the number of necessary kinetic steps to exchange with the oxygen in the material. We can't get a clear picture of the effect of available active surface.

John Vohs asked: Will polarization affect Sr segregation and surface oxygen exchange?

Steven McIntosh replied: It appears as if it does have an impact. A number of groups have attributed LSM electrode activation to changes in surface composition upon electrode polarization. Some examples from the literature.[1–3]

1 M. A. Haider and S. McIntosh, *J. Electrochem. Soc.*, 2009, **156**, B1369.
2 J. Kim H.-I. Ji, H. Prasad Dasari, D. Shin, H. Song, J.-H. Lee, B.-K. Kim, H.-J. Je, H.-W. Lee, K. J. Yoon, *Int. J. Hydrogen Energy*, 2013, **38**, 1225.
3 S. P. Jiang, *J. Solid State Electrochem.*, 2007, **11**, 93–102.

John Druce remarked: I presume you quantified the cation coverages by an internal calibration from the depth profiles with Ne scattering. Of course, with Ne we miss out information on light contaminants such as Na, K and Ca, as well as oxygen. Did you see any evidence of these elements in the He survey spectra? If they are present, but not accounted for in the internal calibration, this could throw off the quantification.

Steven McIntosh answered: We have seen Na contamination in our work on some single perovskites, but we haven't observed any in this study.[1]

1 M. A. Tamimi, A. C. Tomkiewicz, A. Huq, S. McIntosh, *J. Mater. Chem. A*, 2014, **2**, 18838–18847.

John Druce asked: As a follow up to my previous question, one doubt I have about the internal calibration we often use is that it forces all of the surface coverages of the elements we consider to add up to one. I wonder if using external standards to give us absolute concentrations of the cations might allow us to glean more information about the presence of any point defects. I have a feeling that the defect chemistry (particularly oxygen vacancy concentration) at the surface, rather than the elemental composition, might play a role.

Steven McIntosh replied: This is a good point. We have previously used both external and internal standards and they do give differences in relative coverage (although the trends remain the same). This is certainly something to look into further, especially if we can pre-treat our samples carefully in an attempt to reproduce operating conditions.

Ian Metcalfe asked: Any individual instrument has a limited range of time constants that it can access. Thus if k (surface oxygen exchange rate) and D (bulk anion diffusivity) vary, over the temperature range of study, by a few orders of magnitude is it possible that the instrument is forcing a correlation on k and D as you are confined to the time constants it can access. And of course you don't study materials that are not suitable for study in your kit.

Steven McIntosh responded: This is correct. Any experimental system designed to measure both k and D is limited by the time constants of the system, and additionally the sample geometry. For example, thin films are probably the best way to isolate and measure the surface exchange coefficient without influence of the bulk under the caveat that film sample preparation may be very different from bulk sample conditions.

Rotraut Merkle said: The extraction of k^{δ} and D^{δ} from conductivity relaxation measurements can lead to larger uncertainties in both values because we analyze a signal which is integrated over the whole sample, and the curve shapes for pure k-limitation and pure D-limitation differ only slightly.

On the other hand, the extraction of k^* and D^* from isotope exchange and SIMS measurements is expected to be more reliable because their spatially resolved concentration profiles are measured. The conditions under which k^* and D^* can simultaneously be determined are analyzed *e.g.* in work by Fischer *et al.*[1], and for materials where k^* and D^* are outside this window, Fielitz *et al.*[2] suggest to extract k^* from one very short isotope exchange, and D^* from a much longer run (even if the surface degrades during this run, this should not affect D^* which is a bulk property). Thus, careful tracer experiments are a strong tool to extract reliable k and D.

1 E. Fischer, J. L. Hertz, *Solid State Ionics*, 2012, **281**, 18.
2 P. Fielitz, G. Borchardt, *Solid State Ionics*, 2001, **144**, 71.

Steven McIntosh replied: It may be that other techniques can more accurately determine these parameters; however, I believe that it is an outstanding challenge in the field to demonstrate that these k parameters are truly reflective of the kinetics on the outermost surface of the material where the oxygen dissociation occurs.

Rotraut Merkle commented: For $La_{0.8}Sr_{0.2}Mn_{1-y}Co_yO_3$ and $La_{0.5}Sr_{0.5}Mn_{1-y}Co_yO_3$ perovskites, the activation energies of k^* and D^* are clearly different.[1] Also, for $La_{0.6}Sr_{0.4}CoO_3$, D^* has a negative pO_2 dependence while k^* has a positive pO_2 dependence.[2] In particular this last example indicates that really two separate physical processes – surface reaction and bulk diffusion – were measured (D^* is expected to have a negative pO_2 dependence originating from the oxygen vacancy concentration, while k^* in generally has a positive pO_2 dependence because the positive oxygen reaction order typically dominates over defect contributions with potentially negative pO_2 dependence). Thus, I think that in these cases a possible correlation between k^* and D^* does not simply result from improper resolution of the relaxation kinetics into k and D.

1 R. A. De Souza, and J. A. Kilner, *Solid State Ionics*, 1999, **166**, 153.
2 A. V. Berenov *et al.*, *Solid State Ionics*, 2010, **191**, 891.

Steven McIntosh replied: I think it is not the case that we can *never* accurately measure a true surface rate but I do believe that we must be careful when we interpret the data. This is especially true when we continue to observe significant surface segregation/enrichment of some cations on these materials. Of course this enrichment, and thus its influence on a measurement, will vary with sample preparation and pre-treatment. I would maintain that a true outer surface value for k must correlate with the composition and structure of the outer material surface. I believe that it's a challenge to the field to obtain such a clear correlation.

Rotraut Merkle stated: While LEIS shows very similar surface cation compositions for the samples in Fig. 5, the method might not be very sensitive to defects present in small concentrations (*e.g.* oxygen vacancies, or particular adsorption sites) which are nevertheless decisive for the surface kinetics. Thus, the surface concentration of active sites for oxygen reduction may differ, leading to the different surface reaction rates.

Steven McIntosh responded: There are always limitations in the sensitivity of the instruments. The primary advantage of the HS–LEIS is that we can gain composition information on the outermost surface of the material; however, it is not an *in situ* measurement and it is likely that the outermost surface structure and surface concentration of oxygen will play important roles in dictating kinetics. It is unfortunate that we have no such *in situ* measurement due to the high temperature and atmospheric pressure of operation. These factors may dominate the true surface rate, but it would, in my opinion, be quite a significant coincidence if these factors correlate with the bulk transport properties of the materials. We continue to find this correlation, leading us to believe that the measured k involves some convolution of bulk properties for these materials.

Mogens Mogensen addressed Steven McIntosh :A k-value can be measured independently of the corresponding oxide ion diffusion coefficient using a thin layer of the mixed oxide under investigation, but unfortunately thin layers behave differently from bulk, so there will still be some uncertainty on k/D ratios.

John Druce said: I would like to add that because of the spatial resolution of SIMS, the isotopic fraction profile deeper in the sample also contains information about the surface exchange rate. In other words, the value of "k^*" that we obtain is not given entirely by the surface isotopic fraction, but by fitting the combination of D^* and k^* to match the shape of the whole measured profile.

John Druce asked: Your profiles seem to show some subsurface enrichment in the transition metal cations, similar to our recent report.[1] Do you agree that the apparent "surface segregation" of the transition metals reported by authors using other surface sensitive techniques could be because the information depth of those techniques integrates this subsurface layer rather than the outer atomic surface we see by LEIS?

1 J. Druce and H. Téllez, *Energy Enrivon. Sci.*, 2014, 7, 3593.

Steven McIntosh answered: I would agree entirely with that comment – the depth of the "surface" probe is of critical importance.

Tatsumi Ishihara asked: Measurement of the k-value was performed by isotopic exchange between the gas phase and lattice oxygen. In this kind of measurement, $^{16}O^{16}O$ is generally formed and how much of $^{16}O^{16}O$ is observed? I also think that the total amount of ^{18}O was decreased by diffusion into bulk. How about the change in total amount of ^{18}O? This may suggest the diffusivity of oxygen in the bulk of the three different materials.

Steven McIntosh replied: Unfortunately we cannot measure the amount of $^{16}O_2$ formed due to the time constant of the RGA chamber for gas mass analysis. We introduce a very short pulse of $^{18}O_2$ into a continuous flow of $^{16}O_2$. The RGA signal for $m/z = 32$ never drops below the level of the flowing gas before the pulse.

There is some background signal of ^{18}O species measured in the RGA chamber after the pulse decays; however, it is not clear if these small amounts are from the sample powder or from residual signal in the RGA chamber.

Tatsumi Ishihara said: Since surface activity may be related with the formation rate of surface oxygen vacancy, what is the relationship between diffusivity of oxygen vacancies in the bulk of the three materials? Diffusivity could be estimated with the change in total ^{18}O during isotopic measurement. I think the difference in surface activity is more strongly influenced by the difference in oxide ion conductivity of the materials but not the surface segregation property.

Steven McIntosh responded: This is the exact reported correlation between surface exchange rate and bulk ion conductivity that we have been trying to probe. Instead of the ^{18}O loss, we have been utilizing our *in situ* neutron diffraction data to infer the relative mobility of oxygen between samples. This inference is based on an assumption that the energetics of vacancy hopping will be similar between materials of similar structure and composition, and that we can thus rank ion conductivity between materials based on measured oxygen site occupancies and average atomic displacement. We have found that our measured surface exchange coefficients are correlated with this bulk inferred ionic conductivity for the materials presented in this Faraday Discussions paper and for double perovskite and single perovskite materials (with relevant citations given below). What is puzzling is that all of these materials show surface enrichment of A-site cations. We believe that if the reported surface exchange coefficient is truly reflective of the kinetics of oxygen dissociation and exchange on the outer atomic surface of the material, then the surface exchange coefficient should correlate directly with surface composition. We believe that the fact that this correlation does not hold, and that the surface exchange correlates with a bulk property, indicates that the measured surface exchange coefficient is actually probing a near-surface or bulk property. It comes down to how we define surface (do we mean the outermost atomic layer?) and what information we wish to extract from the measured constant. The reported exchange coefficients are likely sufficient in, for example, an electrode model, but may be insufficient to infer outermost surface kinetic mechanisms.[1,2]

1 A. C. Tomkiewicz, M. Meloni, S. McIntosh, *Solid State Ionics*, 2014, **260**, 55–59.
2 M. A. Tamimi, A. C. Tomkiewicz, A. Huq, S. McIntosh, *J. Mater. Chem. A*, 2014, 2, 18838–18847.

John Irvine opened the general discussion of the paper by Rotraut Merkle: It is quite surprising that "hydrogenation" is suggested for the reaction under air and "hydration" under reducing conditions, please can you clarify this?

Rotraut Merkle responded: In the present paper the term "hydrogenation" is used to describe hydrogen uptake by the sample (not exposure to hydrogen). Under conditions of low pO_2 (here: 100 ppm), BSFZ exhibits a large concentration of oxygen vacancies and only a small hole concentration (Fig. 4a). This naturally favours proton uptake by the hydration reaction (2) which requires oxygen vacancies as the reactant. But at high pO_2 with low vacancy and high hole concentration (Fig. 4a: more holes than oxygen vacancies in 1 bar O_2 at T below 450 °C), the hydrogen uptake reaction (3) from water leaving behind O_2 in the gas phase becomes preferable (it proceeds without O vacancies but needs a sample in a quite oxidized state). This change in the dominant proton uptake reaction also follows from the numerical solution of the defect model as shown in Fig. 1b: for high values of $K_O pO_2^{1/2}$ the change in hole concentration after pH_2O increases is larger than the change in O vacancy concentration (pink surface is higher than grey surface).

Mogens Mogensen asked: I think that "hydrogenation" means '"to combine or treat with molecular hydrogen, especially in organic chemistry to add hydrogen to a compound". Thus, it seems incorrect to use this term for the reaction described by eqn 3 as this is a probably a mainly entropy driven deoxygenation. No molecular hydrogen is involved in any way. Water is reacting with the oxidized BSFZ displacing some of the O^{-1} (electron holes) with protons (OH^-).

Rotraut Merkle replied: As a result of reaction (3) the sample overall has taken up neutral hydrogen (increase of proton concentration, decrease of hole concentration). The important point is that reaction (3) describes a mode of increasing the sample's proton concentration by a redox reaction which differs from the pure acid–base reaction (2), and which makes a significant contribution only for redox-active materials such as BSFZ but not for electrolyte materials such as Y-doped $BaZrO_3$. To avoid misunderstandings, we now denote reaction (3) in the manuscript as a hydrogen uptake reaction. Since $K_{W/O} = K_W/K_O$ and the reaction enthalpies of reactions (1) and (2) are of comparable magnitude, indeed the reaction enthalpy of the hydrogen uptake reaction (3) is close to zero.

John Irvine said: Are water and oxygen competing for vacancies? In p-type materials, oxygen fills vacancies under oxidising conditions, during hydration water competes for those sites and pushes oxygen out, this looks like a redox reaction, but the driving force is competition. This would fit with a lower amount of water uptake. The end reaction is exactly as you say. Lose oxygen from lattice creates vacancies, water can then go in these holes, is perhaps a two step reaction occurring rather than a one step reaction, adding them up gives apparent hydrogenation.

Rotraut Merkle answered: Yes, one can view the hydrogen uptake reaction (3) as water uptake (eqn (2)) followed by oxygen release (reaction (1) proceeding from right to left). In such a system with three carriers, we need two independent reactions (mass action constants) to describe the defect concentrations. If the vacancy concentration is decreased according to the hydration reaction (2) after increasing pH_2O, then the oxygen exchange equilibrium (1) will respond by releasing some O_2. This release will be the stronger the larger the relative change of vacancy concentration is (stronger at a low absolute vacancy concentration, *i.e.* at high pO_2). Under conditions of sufficiently high pO_2, the resulting overall concentration changes then correspond to eqn (3). Since we deal with equilibrium concentrations (thermodynamics) we cannot deduce from the data how the reaction proceeds mechanistically.

John Irvine asked: The amount of water uptake is much less in oxidising conditions, your model doesn't explain that.

Rotraut Merkle responded: The TG measurements show that indeed the weight changes after pH_2O increase are much smaller in high pO_2. However, we assign this weight change in high pO_2 predominantly to hydrogen uptake according to eqn (3) (not water uptake according to eqn (2)), and the corresponding increase in proton concentration is comparable to that from water uptake at low pO_2 (Fig. 6, *cf.* also the red data points from the Karl–Fischer titration of the samples quenched from 20 mbar H_2O in 1 bar O_2). This assignment is supported by the fact that the uptake – which is diffusion limited for the dense BSFZ pellets – is faster in high pO_2 (Fig. 5) as expected for hydrogen chemical diffusion compared to water chemical diffusion in low pO_2. Furthermore, the simulations in Fig. 10 show that for the defect model employed, an increase of pO_2 from 100 ppm to 1 bar suffices to switch from mainly hydration (eqn (2), corresponding to low redox fraction in Fig. 10c) to largely hydrogen uptake (eqn (3), large redox fraction).

John Irvine questioned: What colour are the samples? Are the colour changes dominated by Fe? In the past molecular oxygen uptake has been seen to form superoxide/peroxide species, have you looked for this?[1]

1 J. T. S. Irvine, D. J. D. Corcoran, A. Lashtabeg, J. C. Walton, *Solid State Ionics*, 2002, **154–155**, 447–453.

Rotraut Merkle responded: The samples are essentially black (similar to $SrFeO_3$), and do not visibly change in color after exposure to different pO_2 or to different pH_2O (note that $SrFeO_3$ also has to be exposed to dilute H_2 at elevated T to convert into $SrFeO_{2.5}$ to achieve the typical rust-like colour of an iron oxide not containing a trace of any higher oxidation state than Fe^{3+}). Regarding potential incorporation of peroxide ions: in principle it could be possible in a perovskite such as BSFZ with a large lattice constant and high basicity owing to the high Ba content. We have not checked for this – EPR measurements as performed in the given reference are difficult for materials with a high concentration of paramagnetic transition metal ions, and the present samples were not treated with pressurized oxygen. Thus, while traces of peroxide ions cannot completely be excluded, given the high concentration of other point defects we think they are of minor importance regarding the proton uptake.

Mogens Mogensen enquired: Iwahara showed that if you put cerium into hydrogen, then you get proton conductivity by reducing the oxygen and the cerium[1] — is this hydrogenation?

1 H. Iwahara, T. Esaka, H. Uchida and N. Maeda, *Solid State Ionics*, 1981, **3–4**, 359–363.

Rotraut Merkle answered: Yes, I would call this hydrogen uptake or equivalently hydrogenation. In this example it occurs under conditions where the sample is exposed to gaseous hydrogen because Ce^{4+} needs strongly reducing conditions to become reduced to Ce^{3+}. But as the present measurements and the defect chemical modeling show, such a combined reaction of proton uptake and reduction of a cation occurs also in a less reducing atmosphere when the material contains a redox-active cation such as $Fe^{3+/4+}$ in BSFZ under conditions where the low concentration of oxygen vacancies compared to holes (Fe^{4+}) favors it over the hydration reaction.

Steven McIntosh opened the discussion of Helena Téllez' paper: Your data shows that your determined surface exchange coefficient, k is independent (or nearly independent) of the surface composition of your sample. This is the same observation that we have made, and included in our paper for this Faraday Discussion and references therein. However, since oxygen adsorption, dissociation, and initial incorporation occur on the outer surface of the material, considerations from heterogeneous catalysis would suggest that the outer surface composition should dictate the nature and density of active catalytic sites, and hence the surface exchange coefficient. What is your current thinking as to why no correlation between surface composition and surface exchange rate is observed for these materials? We do observe a link between measured surface exchange rate and bulk ionic transport, as has been reported previously by Kilner and co-workers. The subsurface may play a role in determining the activity but should this always correlate with bulk transport rates?

One possibility that has been discussed is that surface coverage of the A-site cation changes in an exactly correlated way with high activity of the underlying material to coincidentally give the data your group and my group observe. I find this hard to reason and suggest that it may be that our current techniques are not sensitive enough to accurately probe the true outer surface exchange coefficient in the most active materials. Instead, we think we are measuring a convolution of the true outer surface exchange coefficient and bulk ionic transport in the outer few nanometers of the sample.

Helena Téllez replied: I think the observed lack of correlation between the surface exchange and the surface composition is due to the fast kinetics of the surface segregation. We have tried to study the relationship between the Ba-surface coverage and the surface exchange coefficient (k^*), but the segregation of Ba is so fast that after just 15 minutes at 400 °C, we already have about 70% of the outer surface covered by the segregated Ba. It is very difficult to perform any measurement of oxygen exchange (relaxation, isotope exchange depth profiling, impedance) in short time scales (*e.g.* 10 min), since there is always an annealing and equilibration time involved in these experiments. Then, during any equilibration or step that involves high temperature, the outer surface will already be mostly covered by the Ba cations, and hence, it is very difficult to unequivocally determine the influence of the outer composition and the surface exchange rate. Another possibility may be that point defects at the surface play a role. The argument made by Kilner *et al.*[1] to which you refer was that D^* and k^* are linked by their common dependence on oxygen vacancy concentration. We might expect the concentration of oxygen vacancies to be different at the surface, but I would assume that the surface concentration of vacancies will depend on that of the bulk, as well as depending on local electro-neutrality due to the surface cation segregation (*e.g.* Ba segregation at the near-surface in our GBCO). Unfortunately, what we are measuring by LEIS is the surface cation composition – we have little information on the oxygen coverage, and it will be quite a challenge to extract any information about surface oxygen vacancy concentrations.

As you mentioned, even if the segregation process were slow enough as to finely control the Ba coverage through the pre-equilibration step in the exchange experiment, there are other experimental limitations of the techniques that might make the measurement of independent k^* and D^* coefficients very challenging. In the case of pulse-isotope exchange methodologies, there are other considerations as described in your paper, since the exchange length involves several hundreds of nm, far deeper than a single atomic layer or even thicker than the altered composition at the near-surface due to cation segregation (less than 10 nm for Ba segregation in GBCO). In the case of the isotope-exchange SIMS depth profiling, a proper fitting of the diffusion profile might be even more complicated since, as we discussed with Prof. Metcalfe during this meeting, the measured k^* parameter could be indeed affected by a non-uniform D^* coefficient at the near-surface region, with a compositional gradient until the bulk composition is reached.

1 J. A. Kilner, R. A. De Souza and I. C. Fullarton, *Solid State Ionics*, 1996, **86–88**, 703–709.

John Druce replied: I think it is important to make a distinction between "bulk" ionic transport (*i.e.* the over all diffusivity or ionic conductivity probed by macroscopic techniques), and diffusion through a subsurface layer. Indeed, for our two-parameter fitting of SIMS data, any resistance due to the latter would result in an apparently lower value for "k", but I would expect the fitted D value to correspond to the true bulk diffusivity.

Steven McIntosh addressed Helena Téllez and John Druce: Can you estimate the error involved in fitting the surface exchange coefficient, k and bulk ion diffusivity, D, from the SIMS depth profiles? Is it possible to provide error bars based on quality of fit and fit sensitivity metrics? In particular, I'm interested in how sensitive the determined value of k is to the quality of fit of the first few data points.

Helena Téllez answered: At the moment I cannot provide an estimation of the uncertainty of the surface and diffusion coefficients from the fitting of the SIMS depth profiles in this particular work, as we have not conducted a full evaluation of our experimental error. As a rule of thumb, we would expect it to be of the order of 10% rather than, say a factor of 2. Some authors suggested (see work by De Souza *et al.*[1]) that the error in the estimation of the diffusion/exchange coefficient could be around 10%, taking into account experimental errors, such as the measurement of the crater depth. Recently, our colleague Richard Chater from Imperial College London, conducted a systematic study of many of these aspects and recently published these results.[2] We refer to this paper and to Dr Chater's thesis for a more detailed description of the parameters to consider in order to assess the errors in the fitting of the SIMS diffusion profiles.

1 R. A. De Souza and R. J. Chater, *Solid State Ionics*, 2005, **176**, 1915–1920.
2 R. Chater, *ECS Transactions*, 2015, **68**, 631–649.

John Druce responded: The fitting of the diffusion profiles requires two parameters to describe the entire profile. As well as increasing the isotope fraction at the very surface, a higher *k* value also results in a higher fraction in the bulk of the profile. In other words, the profile deeper within the sample still contains information on *k*. Our fitting uses all of the data points to fit the whole profile with equal weighting – in fact, I would say that all the data points up to half or even one diffusion length into the profile actually control the *k* value we get from the fitting. It is even possible to disregard the first few data points completely for the fitting, and obtain similar values, so I would not assume the values are particularly sensitive to errors in the first few points.

Mogens Mogensen continued a general discussion of the paper by Helena Téllez: As said in the paper, the 15 min "O^{18} exchange time" is quite long and a lot may happen during this period. It is generally known that polished/cut surfaces relax almost immediately at relative low temperature, and thus it is not possible to determine *k* for a well-defined surface composition and structure. It is in spite of this that I have the following questions: 1) is it possible to determine where in the near-surface layer/region that you have a bottleneck (if such exists there), *e.g.* a double- or multi-layer with space charge, or where you may point out a "bottleneck" *i.e.* a low vacancy concentration? 2) Do you have any information about how the space charge region looks? What prevents looking at the near-surface atom layer by layer using LEIS? What do the ^{18}O profiles look like after the different equilibration times?

Helena Téllez answered: 1. As observed in our LEIS depth profiles, there is a compositional gradient from the surface to the bulk of the pellet due to the segregation of the Ba cations towards the surface, leading to a B-enriched region at the near-surface. In principle, one could speculate that the concentration of vacancies at the surface could be higher than in the sub-surface, as the near-surface would be highly-doped with the acceptor-dopant. A counter to the concentration argument may also be that at high surface vacancy concentrations, association and clustering may occur which decrease the mobility. 2. No, we don't have specific information about the space charge layer, although we have measured the near-surface cation composition by dual-beam depth profiling (combining low energy Ar sputtering with analysis using a Ne ion beam). A layer-by-layer analysis is challenging, as we need to account both for artefacts induced by non-steady state sputtering (the build up of a mixed altered layer, preferential sputtering), as well as the fact that we are measuring a 2D projection of a 3D sample. The latter will come into play especially if we have precipitated secondary phases.

Ian Metcalfe said: We know that the surface is very different from the bulk. Let's imagine that the first layer of the solid exchanges rapidly with the gas phase. This is so fast that you cannot measure any rate for this step. Is it then possible that your technique is picking up what is in effect an exchange coefficient between the 'first layer' and the 'second layer' of the solid? Would it then be conceivable that such an exchange coefficient would have some relationship to the diffusion coefficient (and of course now the diffusion coefficient is no longer uniform)?

Helena Téllez responded: Yes, that is plausible, and would seem to agree with the results that Prof. McIntosh has shown for the Ruddelsden–Popper phases in his presentation and also in previous results on double perovskites (see Tomkiewicz *et al.*[1]), where they showed a link between the ionic mobility in the bulk and the surface exchange. Nevertheless, the situation is rather complex when you consider the atomic rearrangements produced at the surface and sub-surface region that leads to compositional gradients at the near-surface region. In previous studies we performed experiments at high temperature on $GdBaCo_2O_{5+\delta}$ double perovskites (GBCO) (*i.e.* annealing for 12 h at 1000 °C), we observed that the near-surface (about 8 nm in depth) region of the material shows a different chemical composition due to the segregation of the alivalent cations (Ba), with a Co-enriched subsurface in comparison with the bulk composition (see Druce *et al.*[2]). As you suggest, this region could well show a different, and non-uniform, diffusion coefficient for oxygen. In the present paper and other references therein, we have shown that such segregation and compositional deviations take place at much lower temperatures and in a very short period of time (*i.e.* after 15 min at 400 °C).

1 A. C. Tomkiewicz, M. Meloni, S. McIntosh, *Solid State Ionics*, 2014, **260**, 55–59.
2 J. Druce and H. Téllez, *Energy Enrivon. Sci.*, 2014, 7, 3593–3599.

Bilge Yildiz asked: The LEIS ratios indicate that the material is not retaining the original structure or composition of the bulk phase. This may be associated with a phase change or precipitation. Did you have a chance to look at the surface with AFM or SEM both before and after the annealing, to see whether such phase precipitation occurred? AFM may be difficult to implement on the polished surface at high resolution, but SEM may be more helpful for this.

Helena Téllez responded: We studied the evolution of the surface morphology in these double perovskites upon short-term annealing at relatively low temperature by AFM (see Téllez *et al.*[1]). In the case of the $GdBaCo_2O_{5+\delta}$, the segregation of the Ba cations did not seem to have a strong impact on the surface roughness due to the formation of secondary phases (*i.e.* there was a slight increase of the surface roughness after annealing at 400'°C, but the polished surface already showed some roughness). In the same work, we also studied the surface chemistry and morphology of $PrBaCo_2O_{5+\delta}$ double perovskite and we observed that the roughness was significantly increased by the formation of secondary precipitates which appeared to be associated with impurities (*e.g.* Na, Ca, K). We did not obtain any SEM images of these surfaces, but from the AFM images in GBCO, it seems there was no clear phase separation of the segregated species.

1 H. Téllez *et al.*, *Int. J. Hydrogen Energy*, 2014, **39**, 20856–20863.

Viola Birss asked: Have you or others been able to correlate the observed Sr surface enrichment with the electrochemical activity of these MIEC electrodes? Is there a direct correlation between the build-up of Sr and the depression of electrochemical activity?

Helena Téllez responded: Several groups have been trying to establish a relationship between Sr segregation and electrochemical performance, giving a general trend on the behaviour but not a direct correlation. In my opinion this is quite a complicated situation in which many factors, not all of them controlled completely, might have an impact on the conclusions obtained. For instance, many works linked the segregation of Sr with an increase of the polarisation resistance in $La_{0.6}Sr_{0.4}CoO_{3-\delta}$ thin films (see for instance work by Kubiceck *et al.*[1,2]). On the other hand, other works suggested that surface decoration with SrO can improve ORR kinetics due to a more catalytically active interface between the SrO precipitates and the $La_{0.8}Sr_{0.2}CoO_{3-\delta}$ thin film surface (see E. Mutoro *et al.*[3]). During this meeting, we have also seen the work of Prof. Yildiz in which the electrochemical performance seems to be more obviously affected by the formation of surface precipitates than elemental segregation in $La_{0.8}Sr_{0.2}CoO_{3-\delta}$ thin films. Coming back to the previous argument of the fast surface segregation, it might be very difficult to establish a direct, quantitative relationship between Sr segregation and electrochemical performance as electrochemical testing also involves *e.g.* equilibration time at high temperature during which the surface and near-surface composition may be modified. From our own work on LEIS analysis of MIEC materials, we think that not only the Sr surface termination influences the electrochemical behaviour, but also the formation of other additional phases might have even a larger effect. For instance, we recently performed some work in collaboration with Prof. Fleig's group in Vienna, which seems to suggest that it is a water-soluble secondary phase (possibly SrO or $Sr(OH)_2$ that starts to form during the equilibration at high temperatures (1 h at 600 °C) during impedance measurements that is responsible for a decrease in the surface exchange resistance of $La_{0.6}Sr_{0.4}CoO_{3-\delta}$ thin films. This work has been submitted and is currently under review.

1 M. Kubiceck *et al.*, *J. Electrochem. Soc.*, 2011, **58**, B727–B734.
2 M. Kubiceck *et al.*, *ACS Nano*, 2013, 7, 3276–3286.
3 E. Mutoro *et al.*, *Energy Environ. Sci.*, 2011, **4**, 3689–3696.

John Irvine asked: LEIS gives a lot of good information on the surface, but what is the surface? The surface terminates in different ways depending on the orientation, so you could have an Sr—O termination for example. When you get down to 1 nm the surface dimension becomes of similar order to the crystal unit cell. Therefore you would expect enrichment at the surface under these length scales. Generally the surface will reorganise to improve its stability as a cleaved crystal surface is unlikely to be stable at the atomic level.

Helena Téllez replied: In the particular case of LEIS analysis, the information provided comes from the first atomic layer of atoms at the surface, that is, the interface between the solid surface and the gas phase. This atomic layer resolution is due to the high neutralization probability of the noble gas ions used for probing the surface (*i.e.* if these ions penetrate into the second layer of atoms, they will be neutralized and then will not be detected if they are backscattered as neutrals). As you mentioned, in the case of a perfect cleaved crystal at sufficient temperature, the surface will reconstruct in order to stabilise and minimise its energy. A similar phenomenon is also observed in the polycrystalline materials for which most of our results are reported. For instance, the analysed area by LEIS is normally rather large (typically 1 mm^2) as we try to minimise the ion fluences during the analysis to avoid surface damage. In this large area, we expect to have grains of many different orientations exposed at the surface. As a general observation in LEIS analysis of mixed-ionic electronic materials and, in particular, for acceptor-doped (3,3) perovskites, the surface stabilizes by segregation of the divalent dopant, such as Sr or Ba, in order to achieve surface neutrality. In other words, our results are not sampling a single grain orientation, but reflect an overall tendency for a large number of reasonably randomly oriented grains.

Mogens Mogensen asked: What does the surface look like? Is it a compact layer of BaO without anything else, or are there several phases? What is the preferred crystallographic plane of the BaO surface? Can you observe any difference between the first and second atom layer?

Helena Téllez replied: The surface roughness is slightly increased after the thermal treatment. Although we don't have evidence of an additional phase formed in these particular samples, the segregation of the Ba might induce the formation of secondary phases, especially if the annealing is performed for longer times (*e.g.* we have seen this for LSC thin films after annealing for long time, see Y. Chen *et al.*[1]). I cannot say exactly what it is the preferred crystallographic plane of the BaO surface for sure, but I would assume it would be the (100) plane, as this plane is both close-packed and charge neutral. We observe significant differences between the outer surface and the second layer, as the transition metal is detected after gently sputtering around one monolayer of material, to expose the second atomic layer.

1 Y. Chen *at al.*, *Chem. Mater.*, 2015, **2**, 5436.

John Irvine commented: At the resolution of 0.1 nm, even with a perfect crystal the surface will be enriched *e.g.* see Fig. 2 in **ref. 1** below for a simple (100) perovskite termination.

1 D. Neagu, T-S. Oh, D. N. Miller, H. Menard, S. M. Bukhari, S. R. Gamble, R. J. Gorte, J. M. Vohs and J. T. S. Irvine, *Nat. Commun.*, 2015, **6**, 8120.

Helena Téllez replied: Yes, a perfect cleaved (100) surface could correspond to either 100% BO_2 termination or 100% AO-termination. However, we refer to a surface enrichment after annealing of these MIEC materials because the surface is not terminated in these AO planes as a consequence of cleavage, but due to segregation processes and sub-surface rearrangements produced to stabilise the outer surface. This is clearly observed in the chemical composition deviations observed in our LEIS depth profiles at the near-surface (around 10 nm) compared to the bulk composition of the material. These atomic re-arrangements can be produced in any process that involves intermediate or high temperatures, and even during the growth of thin films (atomic re-arrangements have been also predicted and experimentally confirmed by other techniques during growth of Sr_2TiO_4 thin films (see J.H. Lee *et al.*[1])).

1 J. H. Lee *et al.*, *Nat. Mater.* 2014, **13**, 879–883.

Mogens Mogensen said: Do you know how k^* (isotope exchange) compares to electrochemically measured k_q? Is k_q also independent of coverage by segregates at the surface?

Helena Téllez replied: We have not measured k_q yet, but it will be an interesting experiment to do. Nevertheless, the situation during the electrochemical measurements might be different because of the necessity to make electrical contacts, and to deposit the MIEC onto an electrolyte. This processing might introduce other cation species that change the composition of the GBCO surface. We have seen evidence for this on PLD microelectrodes after electrochemical testing, where, in addition to the segregation of Sr cations (in this case it was LSCF microelectrode on YSZ), there was also impurity (*e.g.* Si and Na) migration from the electrolyte (see Druce *et al.* and Téllez *et al.*[1,2]). These impurities might also have an impact on any measured *k*, by blocking the surface for the oxygen exchange. Therefore, one must take care to ensure the samples for the two experiments are as comparable as possible.

1 J. Druce *et al.*, *Int. J. Hydrogen Energy*, 2014, **39**, 20850–20855.
2 Téllez et *al.*, *ECS Trans.*, 2015, **66**, 69–77.

John Irvine commented: The sample is normally covered in water, what do you do to this sample before you do the LEIS measurement? If there is a hydroxide layer on the surface as seen by XPS, would this mean that you see more oxygen stuck on the surface, if you remove the water would this layer be the perovksite/layered phase or decomposed Sr oxide? When you take the sample and do a measurement you do it in the real world, not a vacuum, the surface you see in an experiment will be modified once you put it into a real atmosphere. You can get a huge amount of information from this technique, but the surface is modified once it is in the atmosphere.

Helena Téllez replied: Due to the surface sensitivity of LEIS, it is necessary to clean the surface *in situ* in order to remove any organic contamination due to exposure to atmospheric air. There are different methodologies in order to clean the surface (see different approaches described by De Ridder *et al.*[1]). Typically, in the case of oxide materials, we oxidise the sample surface by using atomic oxygen radicals (at room temperature) that react with any organic contamination at the surface forming volatile species that are pumped away in the vacuum chamber.

In the case of a surface covered in water, one should keep in mind that if there were any segregated species on the surface (such as SrO_2 or $Sr(OH)_2$) they might have been water-soluble phases that would be removed by the water (as we have observed previously in LSC thin films). If the hydroxide species are formed and attached to the surface, this outer layer will be the observed composition given by LEIS. And finally, I agree the surface is very different when it is exposed to the atmosphere at room temperature, since it will be covered by any adsorbed species and organic contamination from the air. At the elevated temperatures with which we are concerned, we expect the coverage of adsorbates to be lower. Of course, we could clean the sample surfaces by heating, but we try to avoid this as it may lead to further segregation. However, if we want to get information on the surface composition of the relevant surface exposed to the gas interface, the only way to achieve such monolayer surface-sensitivity is by removing any adventitious species that can be adsorbed on the surface that hinder the detection of the outer atomic layer of our material. In addition, it is possible to perform *in situ* annealing experiments in different atmospheres and then analyse the surface by LEIS without any exposure to air. We are now working in these *in situ* experiments in order to gain a fundamental understanding of the kinetics of the segregation processes and compositional changes occurring at intermediate and high temperature for different MIEC materials.

1 De Ridder *et al.*, *Surf. Interface Anal.*, 2002, **33**, 309–317.

Steven McIntosh commented: We are analyzing powder samples that are mechanically pressed into a loose pellet (that is, they are not sintered as a pellet). The powders are annealed as powders under atmospheres identical to those used to measure surface exchange coefficient, pressed into pellets, and then transferred to the HS-LEIS. We apply a gentle oxygen treatment in the instrument prior to measuring the surface composition. We see the same very high A-site coverage of the surface over a wide range of materials.

Helena Téllez answered: Since we typically work with dense ceramics, which are of course very different to the porous structures that these materials often have when they are applied as electrodes, it is useful to know that measurements of a different sample form (powders *vs.* the dense ceramics we tend to use) from a different group yield consistent trends. Of course the powders are probably more representative of the large surface areas when we have a porous electrode.

Raymond Gorte opened the general discussion of the paper by Viola Birss: Our experience with 3 electrode measurements is that they simply do not work (see McIntosh *et al.*[1]). The problems have been summarized very nicely by S. B. Adler.[2] The criterion you have used to argue that the reference electrodes properly sample electrode performance, that your reference electrode is several millimeters from the working electrodes, is not relevant, as discussed in some detail by Adler. Your measurements in air are probably fine but I am uncomfortable with your results at the fuel electrode.

1 S. McIntosh, J. M. Vohs, and R. J. Gorte, *J. Electrochem. Soc.*, 2003, **150**, A1305–1312.
2 B. Adler, J. Electrochem. Soc., 2002, **149**, E166.

Viola Birss responded: We understand that there is significant debate on this point, and yet many good papers published in the SOFC literature are using 3 electrode methods. In our work, we used a relatively thick YSZ electrolyte (300 μm) plus two GDC buffer layers (one at each electrode), giving 400 μm in total. This seems to be sufficiently thick so that 3 electrode data are more likely to be trustworthy, based on the work of various authors, including Jiang *et al.*,[1] (with the RE placed to the side of the WE, but at a sufficient distance, should be safe). In fact, we were very careful in this work with electrode placement, ensuring the symmetrical positioning and sizes of the WE and CE, both composed of LSFCr-3, and also depositing the CE and WE in such a way so that they had a very similar thickness, thus achieving very similar electrochemical characteristics. All of these factors play a role in making 3 electrode studies more reliable. We also carried out some 2-electrode half-cell and full cell experiments for comparison with our three-electrode half-cell results on the same cells (these comparisons were not included in our paper). In one experiment in air at 800 °C, Rp for the WE in the 3 electrode experiment at the OCP was *ca.* 0.25 Ω cm^2, while in the 2 electrode work, Rp was *ca.* 0.5 Ω cm^2 (dividing by 2, this gives a similar Rp value of 0.4 Ω cm^2). In a 1 : 1 CO:CO_2 environment, again using LSFCr-3 as both the WE and CE and making them as similar as possible in all respects, in the 3 electrode study, an Rp of about 0.9 Ω cm^2 was observed for the WE, while the Rp obtained for a 2 electrode symmetrical full cell was *ca.* 1 Ω cm^2. LSFCr-3 is a very good air electrode catalyst, giving an Rp of 0.1–0.25 Ω cm^2 at 800 °C. As such, the Rp obtained in the full cell study should be dominated by the fuel CO/CO_2 reactions at the fuel electrode. This suggests that the Rp of about 0.9 Ω cm^2 observed for the WE in the 3 electrode half cell experiment, is quite reasonable. Overall, careful experiments comparing the performance of electrodes, as well as their capacitance values, in 2 and 3 electrode experiments should be carried out. This is because there is a critical need for mechanistic and kinetic information about SOFC reactions in both the fuel cell and electrolysis modes in order to further improve and understand cell performance, with these data only obtainable through 3 electrode experiments. Finally, if the data collected in the 3 electrode experiments in air are considered likely to be reliable, then the data in CO/CO_2 environments should also be acceptable, considering that the performance of the cell in these two environments is quite similar.

1 S. P. Jiang, *J. Appl. Electrochem.*, 2004, **34**, 1045.

Mogens Mogensen asked: Regarding the reference electrode, even if you measure a resistance that is half of the electrolyte resistance, this is no proof that your reference electrode measures the potential that you should measure. You mention that the reference electrode is 2–4 mm from the working electrode, but there is no rule that it has to be this distance away. The main problem is that the reference electrode does not measure the relevant potential because of uneven current density distribution due to asymmetry (misalignment of the working and counter electrodes) or due to big edge effects.

Viola Birss responded: In our work, we used a relatively thick YSZ electrolyte (300 μm) plus two GDC buffer layers (one at each electrode), giving a 400 μm electrolyte thickness in total. This seems to be sufficiently thick so that 3 electrode data are more likely to be trustworthy, based on the work of various authors, including Jiang *et al.*[1] (with the RE placed to the side of the WE, but at a sufficient distance away). In fact, we were very careful in this work with electrode placement and other factors, ensuring the symmetrical positioning and sizes of the WE and CE, both composed of LSFCr-3, and also depositing the CE and WE in such a way so that they had a very similar thickness, thus achieving very similar electrochemical characteristics. All of these factors play a role in making 3 electrode studies more reliable. We also carried out some 2-electrode half-cell and full cell experiments for comparison with our three-electrode half-cell results on the same cells (these comparisons were not included in our paper). In one experiment in air at 800 °C, Rp for the WE in the 3 electrode experiment at the OCP was *ca.* 0.25 Ω

cm^2, while in the 2 electrode work, Rp was *ca.* 0.5 Ω cm^2 (dividing by 2, this gives a similar Rp value of 0.4 Ω cm^2). In a 1 : 1 CO : CO_2 environment, again using LSFCr-3 as both the WE and CE and making them as similar as possible in all respects, in the 3 electrode study, an Rp of about 0.9 Ω cm^2 was observed for the WE, while the Rp obtained for a 2 electrode symmetrical full cell was *ca.* 1 Ω cm^2. LSFCr-3 is a very good air electrode catalyst, giving an Rp of 0.1–0.25 Ω cm^2 at 800 °C. As such, the Rp obtained in the full cell study should be dominated by the CO/CO_2 reactions at the fuel electrode. This suggests that the Rp of about 0.9 Ω cm^2, observed for the WE in the 3 electrode half cell experiment, is quite reasonable. Even so, further work is being carried out currently in our lab to compare our 2 and 3 electrode results so that we can be more certain that our 3-electrode data can be trusted.

1 S. P. Jiang *et al.*, *J. App. Electrochem.*, 2004, **34**, 1045.

Mogens Mogensen said: The *m*-value in eqn 2 is not a reaction order. Furthermore eqn 2 is only “an often observed approximate relation”. It is not a theory-basic exact equation. As the polarisation resistance at OVC depends on both the anodic and the cathodic reactions, *m* cannot be a reaction order. The definition of an electrochemical reaction order with respect to a single reactant in a single electrochemical reaction has a form similar to equation 2. It is: rate ∝ (concentration-i)^m, but everything other than concentration-*i* has to be constant (including the electrode potential *vs.* a reference electrode). To measure the rate of an electrochemical reaction in only one direction, it is necessary to stay well away from OCV.

Viola Birss responded: We completely agree that the polarisation resistance at the OCP depends on the rate of both the anodic and cathodic reactions, and thus the measured “*m*” cannot be a reaction order, even though many authors in the literature do use this term for data of this kind collected at the OCP (which we disagree with). We also agree that, in order to determine the rate of an electrochemical reaction in only one direction, it is essential to polarize well away from the OCP. Demonstrating this was one of the motivations behind our work in carrying out the polarized impedance experiments at both the OCP and at various overpotentials up to +/− 300 mV *vs.* the OCP, for comparison. In fact, we have shown in our paper that the apparent *m*-values are quite different under these two conditions *vs.* what is seen at the OCP, as we had expected. Under polarization *vs.* a reliable reference electrode (which was kept in the same gas environment and is composed of the same material as our working electrode), we therefore do consider that we are determining the reaction order correctly.

Mogens Mogensen asked: The potential of your reference electrode is given by the composition of your gas, *i.e.* it is the oxygen potential of your gas, which is determined by the oxygen partial pressure. When you change from oxygen to CO/CO_2 then you change the potential of your reference electrode by almost 1 V. In case of pure CO_2 the potential is not defined at all. Try to write the electrochemical reaction that should determine the potential and look at the Nernst equation for this. You need a redox pair.

Viola Birss responded: Based on our full cell data in pure CO_2 at 800 °C, the experimental OCP is *ca.* −0.3 V *vs.* air. This is equivalent to a theoretical Nernst potential of 0.3 V, assuming a gas composition of ca. 99.9999% CO_2 : 0.0001% CO. Further, we observe that, as we electrochemically reduce CO_2 in this experiment, the OCP moves negatively as CO is produced, consistent with Nernst predictions. These results indicate that data collected in pure CO_2 environments are difficult to interpret, and some CO or steam should also be present.

Göran Lindbergh remarked: It turns out, while deriving an analytical expression of impedance at large overpotentials, *i.e.* at several hundred mV from the equilibrium potential, that the impedance related to the charge transfer reaction is only dependent on the slope of the polarisation curve and the current density, and that the dependence on the concentration is different compared to equilibrium conditions. So it is not surprising that you are not getting the same concentration dependence of the reaction far from equilibrium.

Viola Birss answered: Based on fundamental electrochemical theory, the dependence of the rate of the oxidation reaction (at positive overpotentials) and the reduction reaction (at negative overpotentials) on reactant concentration should only be determined when well away from the equilibrium potential, conditions under which the back reactions can be neglected. We agree completely that the concentration dependence is therefore different under polarization *vs.* at the OCP. However, there is a significant amount of literature where the "reaction order" or "concentration dependence" of SOFC reactions has been reported under OCP conditions, which we believe is incorrect or at least misleading.

Rotraut Merkle remarked: When discussing kinetics, we should be precise about terminology. The parameter "m" in eqn 2 gives the overall pO_2 dependence of the polarization resistance (*i.e.* surface reaction resistance). However, m is related to but not necessarily identical to the reaction order. The reaction order describes the molecularity of the rate determining step; the reaction order for oxygen is 1/2 (or 1) when the oxygen dissociation is (or is not) a fast preceding equilibrium. The overall pO_2 dependence also contains contributions from point defect concentrations of the mixed conducting electrode material appearing in the rate equation (because they are involved in preceding equilibria or in the rds), which themselves typically have a pO_2 dependence. Thus the resulting m values generally deviate from 1/2 or 1; typically they are lower than the oxygen reaction order because defects such as oxygen vacancies or electrons have a negative pO_2 dependence. This also means that for a measured $m < 1$, oxygen can still have a reaction order of one.

Viola Birss answered: This is an important point. For MIECs, changing the gas concentrations likely also changes the surface (and bulk) properties of the materials (*e.g.*, surface vacancy concentration), consistent with the presence of a chemical capacitance. [Note that there is uncertainty in the surface concentrations relative to the bulk, but even so, some changes in surface concentrations should certainly be expected with changing potential]. Thus, the concentration dependence of the current that we report, both under equilibrium and polarized conditions in air and CO/CO_2 environments, should probably be described as "apparent reaction orders" or "measured concentration dependencies". While the "*m* values" that we measured are still very relevant to what the gas concentration dependence of the cell performance will be under operation, it is more difficult to interpret these values from a fundamental perspective without knowing what the surface composition is as a function of gas environment and potential (or pO_2).

Tatsumi Ishihara asked: Although tetravalent Fe is a major species in the $SrFeO_3$ based oxide used, you doped Cr^{3+} for the Fe site. Why is Cr^{3+} chosen as the dopant and are there any effects on the stability of the materials by doping with +3? In general, a +4 cation like Sn^{4+} may be adopted for stablizing the structure.

Viola Birss replied: Doping the B site of (La,Sr) FeO_3 with Cr has been shown to significantly stabilize the perovskite structure over a wide pO_2 range, without any evidence for the formation of a brownmillerite phase, even in low pO_2 environments (see our published work in ref. 1 of our paper). The structure of LSFCr-3, with Cr shown in our still unpublished work in the +3 and +4 oxidation state, is stable down to a pO_2 of 10–21 (ref. 1), as also shown by Barnett *et al.*, therefore making it suitable for use as either an air or fuel electrode. While we are not aware of any reported effects of Sn doping of similar MIEC materials, this would certainly be worth trying.

Tatsumi Ishihara asked: What are the electronic properties of this material? Is this material a stable hole conductor across a wide pO_2 range with no electronic conductivity in a reducing atmosphere?

I ask this as in the perovskite structure, Fe^{4+} may be dominant and is quite unstable, therefore under applied potential, this material decomposes easily and the electronic conducting properties may be drastically changed. Did you check the change in crystal structure of the materials used?

Viola Birss answered: We have not yet determined the ionic and electronic conductivity of our LSFCr-3 material in either fuel or air environments, but have measured the total conductivity. In air, the total conductivity is 80 S cm^{-1}, while in 10% H_2, it is roughly 0.25 S cm^{-1}. However, this relatively low conductivity seems to be sufficient to still give a very good performance in CO/CO_2 environments. In terms of the stability of this material under reducing conditions, XRD analysis has shown it to be stable down to a pO_2 of at least 10–21 (10% H_2/bal. N_2), consistent with the lack of degradation seen in our H_2/steam and CO/CO_2 studies, even when run for over 200 h.

Scott Barnett asked: For the low frequency response of the oxide anodes, one possible mechanism that should be considered is adsorption, specifically hydrogen dissociative adsorption. We have seen this as a key mechanism with similar oxide anodes, showing a low frequency response similar to yours. This is distinguished from gas diffusion by noting that the response is strongly temperature dependent.

Viola Birss responded: We believe that the low frequency arc is due to the chemical capacitance of our MIEC material, arising from bulk changes in the oxygen vacancy concentration (and metal oxidation state) with changing potential (or pO_2). The interpretation of the low frequency capacitance as being due to an adsorption/desorption process does not seem likely for our system, as the measured capacitances are so large (up to 10 F cm^{-2}) and are seen consistently (at values of 0.1–10 F cm^{-2}) over a wide range of potentials and in both air and fuel environments. For an adsorption/desorption process, this could only involve up to a maximum of one monolayer of charge, and thus the measured capacitances should be smaller (*e.g.*, 0.001 to 0.1 F cm^{-2}), especially if the process occurs over a wide potential range, as we observe. Also, an approximate calculation has shown that the maximum chemical capacitance that our materials could generate is in the range of 10–20 F cm^{-2}, which is close to what we sometimes do observe. Finally, we observe this large low frequency capacitance in air, H_2/steam, and dry CO/CO_2 environments, and it seems unlikely that adsorption would be the slow step, with similar characteristics, in all three environments.

Xiangling Yue continued the general discussion of the paper by Viola Briss: The low frequency response, you said it was from some sort of chemical capacitance in the material, and it was related to oxygen adsorption/desorption processes, is this capacitance calculation reliable based on your equivalent circuit modelling of impedance? If you have the adsorption/desorption dominant processes, what values would you expect for the capacitance? Do you have any ideas about this?

Viola Birss responded: We believe that the low frequency arc yields the chemical capacitance of our MIEC material. Other processes that could generate a low frequency capacitor include gas phase mass transport and adsorption/desorption reactions involving charge transfer. The former process (mass transport) can be ruled out in our work, as we observe a significant temperature dependence of the low frequency resistance, which could be expected for both the chemical capacitance and adsorption/desorption processes, but not for gas phase diffusion. The interpretation of the low frequency capacitance as being due to an adsorption/desorption process does not seem likely, at least for our system, as the measured capacitances are so large (up to 10 F cm^{-2}) and seen consistently over a wide range of potentials and in both air and fuel environments. For an adsorption/desorption process, this could only involve up to a maximum of one monolayer of charge, and thus the measured capacitances should be quite a bit smaller (*e.g.*, 0.001 to 0.1 F cm^{-2}), especially if the process occurs over a wide potential range, as we observe for our case.

Scott Barnett asked: The relatively high capacitance associated with the low frequency arc could be the chemical capacitance associated with changing the oxygen content in the oxide, even though the rate limiting step for hydrogen oxidation is adsorption.

Viola Birss replied: It is possible that the resistance associated with the low frequency impedance arc, for which the capacitor has been assigned as being a "chemical capacitance of the MIEC", is related to a slow gas adsorption step. In air, the low frequency resistance has only a very weak dependence on potential but strong dependence on pO_2, consistent with this suggestion. However, in CO/CO_2, our low frequency resistance depends strongly on potential, which is quite different from the behaviour in air. These results, which do show the benefits of doing polarized EIS studies (and the need for 3 electrode geometries), seem to hold important clues to the reaction mechanisms, but further work is needed to examine a broader range of variables and then to interpret the data.

Mogens Mogensen asked: I agree with you that the low frequency arc probably is a redox process, not adsorption. It is possible to calculate, from the resistance and the maximum of this low frequency arc of the impedance spectrum, an equivalent capacitance of the process. If we assume that the redox process involves oxide ions then 2 electrons per oxide is involved, and a way to estimate the capacitance of the redox process is to find the difference in oxide vacancy concentration by changing the equilibrium potential as much as the peak-to-peak amplitude of the alternating current used in the impedance spectra. This gives a number of coulombs which, when divided by p-p amplitudes gives a capacitance that can be compared to the capacitance found from the impedance spectra. An experimental way of investigating the low frequency arc is to change the CO/CO_2 ratio some orders of magnitude. This should change the size of the low frequency arc in a manner equivalent to the effect of changing the overvoltage.

Viola Birss replied: The interpretation of the low frequency capacitance as being due to a chemical capacitance is indeed likely correct, partly as the measured capacitances are so large (sometimes up to 10 F cm^{-2}) and also as they are seen in both air and fuel environments and over a relatively wide range of overpotentials in each gas. An approximate calculation has shown that the maximum chemical capacitance that our MIEC electrodes could generate is in the range of 10–20 F cm^{-2}, which is close to what we sometimes do observe. Also, as we obtain this large low frequency capacitance in air, H_2/steam, and also dry CO/CO_2 environments, it seems unlikely that adsorption would be the slow step, with such similarly high capacitive characteristics, in all three environments.

In relation to the suggested approach involving altering the CO : CO_2 ratio, we did do this type of experiment (see Tables 7 and 8 in our paper), giving results similar to what the questioner has suggested. In other analysis of our data in relation to this point, the EIS results in two different gas environments (30% CO_2 : 30%, CO : 40%, bal N2 and 90% CO2 : 10% CO) were compared, but at the same potential (−0.94 V) *vs.* air. −0.94 V is the Nernst equilibrium potential in the first environment and this potential is reached at a −100 mV overpotential *vs.* the MIEC reference electrode in the second environment, and thus the MIEC oxide should be in the same state (compositionally) in these two cases. This gave two similar low frequency capacitance values, both in the range of 0.4 F cm^{-2}, suggesting a chemical capacitance, again as the adsorption characteristics should be quite different (and smaller) under these two conditions.

Christopher Graves communicated: This question is regarding the higher performance in electrolysis mode than in fuel-cell mode for the mixed ionic–electronic conducting oxide fuel-electrode that you report in this paper (see also my paper and my comment on the paper by Torrell *et al.*). In this paper you write that "the cell performs better in the cathodic mode (reduction of CO_2) than in the anodic mode (CO oxidation), suggesting that there are some real kinetic differences between these two reactions." and then later mention that the effect may be due to the variation of the oxygen vacancy concentration with the polarization. Indeed, isn't it possibly just the opening up of more reaction sites (surface oxygen vacancies) under cathodic polarization (decreasing oxygen partial pressure)? Related to this, there was just a lot of discussion about the low-frequency impedance process and the possibility that it may be an adsorption process. In the paper you ascribe the low frequency to a bulk chemical capacitance and report that the impedance of that process decreases in cathodic polarization and increases in anodic polarization. This sounds reasonable and may describe the process without a rate-limiting adsorption component.

Viola Birss communicated in reply: In answer to the first question, we completely agree that the higher performance observed in the cathodic mode during CO_2 reduction (*vs.* CO oxidation) could be due to an increase in the surface oxygen vacancy concentration. This would be related directly to changes in the bulk vacancy concentration (and thus the change in the chemical capacitance) with potential. In terms of whether the changed surface vacancy concentration also influences the rate of an adsorption step or not, this remains uncertain.

Faraday Discussions

PAPER

Surface modification of $La_{0.8}Sr_{0.2}CrO_{3-\delta}$–YSZ dual-phase membranes for syngas production

Anthony S. Yu, Tae-Sik Oh, Ran Zhu, Alexa Gallegos, Raymond J. Gorte* and John M. Vohs*

Received 16th January 2015, Accepted 10th April 2015
DOI: 10.1039/c5fd00007f

Oxygen permeation fluxes were studied in Mixed Ionic and Electronic Conducting (MIEC) membranes based on composites of 40 vol% $La_{0.8}Sr_{0.2}CrO_3$ (LSCr) and 60 vol% yttria-stabilized zirconia (YSZ), using ambient air and flowing CO to establish a $P(O_2)$ gradient. The ambipolar conductivity of the dense LSCr–YSZ composite was determined for membranes with dense layers that were 115 μm and 650 μm thick. Other parts of the investigation focused on how modifications to the surface on the CO side affected the fluxes. Using a porous LSCr–YSZ composite on the surface as the base case, oxygen fluxes were shown to increase dramatically upon addition of 5 wt% CeO_2 as a catalyst and an additional increase was observed with 1 wt% Pt. Changes in the structure of the porous composite LSCr–YSZ surface to improve connectivity of the YSZ phase also led to large increases in the oxygen fluxes.

Introduction

While the transport of ions and electrons in the ionically and electronically conductive phases of solid state electrochemical devices such as ion transfer membranes, solid oxide fuel cells (SOFCs), and solid oxide electrolysis cells (SOECs) are clearly important, in many cases the overall device performance is determined by the electrochemical reactions that occur on the surface of the membrane or electrodes. The intrinsic rates of these reactions can often be difficult to quantify and depend on both the catalytic activity of the surface and its microstructure. In the work reported here we have quantified the importance of these properties in determining the performance of the fuel side of a $La_{0.8}Sr_{0.2}CrO_{3-\delta}$ (LSCr)–YSZ dual-phase, mixed-conducting, oxygen ion transfer membrane that has been proposed for use in the production of syngas from methane.[1] While the study focuses on membranes with Mixed Ionic and Electronic Conductors (MIECs), the insight obtained are is applicable to the other

Chemical & Biomolecular Engineering, University of Pennsylvania, Philadelphia, PA 19104, USA. E-mail: vohs@seas.upenn.edu

aforementioned systems, especially the anode side of electrolysis cells where electrochemical oxidation of oxygen ions and desorption of O_2 can be rate limiting.

As indicated above, ceramic membranes based on MIECs show great promise for air separation and for production of syngas (a mixture of CO and H_2).[1] In both of these applications, MIEC membranes produce pure O_2 by allowing oxygen ions to diffuse through the ceramic lattice, with counter diffusion of electrons to balance charge. For syngas production, the O_2 produced by the membrane is added to a reactor where Steam Reforming of Methane (SRM) is being carried out in order to provide the energy required to run the endothermic SRM reaction under autothermal conditions. Because SRM must be carried out at high temperatures (>1073 K) in order to achieve high equilibrium conversions,[2] the temperatures required for operation of MIEC membranes and the *in situ* production of O_2 are significant advantages over other ways of producing O_2. Compared to conventional technologies, syngas production using MIEC membranes would improve energy efficiency, decrease the plant's footprint and capital costs.[1]

Two types of MIEC membranes are presently under development for syngas production. The first type is a single-phase perovskite (*e.g.* $La_{0.9}Ca_{0.1}FeO_{3-\delta}$ (LCF)) that has both ionic and electronic conductivity,[3–8] while the second type is a two-phase composite in which one phase is predominantly an electronic conductor (*e.g.* LSCr) and the other is an ionic conductor (*e.g.* scandia-stabilized zirconia (ScSZ) or yttria-stabilized zirconia (YSZ)).[9–12] In composite membranes, the volume fraction of each phase must be high enough to allow each to percolate the structure. One of the big challenges for syngas production using MIEC ceramics is that there is a very large $P(O_2)$ gradient across the membrane, with one side exposed to air and the other to syngas. With a few exceptions, most single-phase MIEC ceramics are not stable over this wide range of $P(O_2)$.[4,13–15] Composite membranes can be made from materials that are stable over a wider range of $P(O_2)$ but they are more difficult to fabricate. Furthermore, it seems likely that O_2 fluxes will be lower in composite membranes because oxygen ions can only pass through one phase.

Previous studies of LCF membranes have demonstrated that O_2 fluxes in this system are often limited by surface reactions and not by diffusion through the membrane itself.[3,11,16,17] For example, Miller, *et al.* showed that there was a ten-fold increase in the oxygen flux of an LCF membrane when porous layers of LCF were added to each side.[13] Yu, *et al.* found that additional large increases in the O_2 flux could be achieved with LCF membranes when catalysts were incorporated into the porous layers at the surface of the membranes.[14] Both of these studies demonstrated that the surfaces of the membranes are often not in equilibrium with the gas phases. The addition of porous layers, which increases the surface area for reaction, and the addition of catalysts both cause the chemical potential of the membrane surfaces to more closely approach that of the gas phase. Similar surface-reaction limitations almost certainly exist with two-phase membranes. Indeed, Christie, *et al.* reported that the addition of oxidation catalysts to the porous layer at the surface of their two-phase membranes also increased oxygen fluxes.[15]

In order to understand the performance of MIEC membranes, we believe it is useful to quantify the various rate processes in terms of the resistance to oxygen transfer across the membrane and the driving force for transfer which is the difference between the gas-phase oxygen chemical potentials on either side of the membrane.[16] The total resistance to oxygen transfer, R_{tot}, is the sum of a bulk diffusion resistance in the membrane itself, R_b, and interfacial loss components, R_s.[3,14] Here, R_s arises primarily from slow surface-exchange rates that cause the $P(O_2)$ at the surfaces of the membrane to differ from the $P(O_2)$ in the adjacent gas phases. Using this electrical circuit analogy, R_{tot} can be calculated from the Nernst potential across the membrane, V_N, and the O_2 flux, J_{O_2}, using eqn (1)–(3). Eqn (3) is simply a statement that the flow of charged species, i, across the membrane is proportional to the O_2 flux, with a proportionality constant of $4F$, where F is Faraday's constant.

$$V_N = \frac{RT}{4F}\ln\left(\frac{P(O_{2,\,air})}{P(O_{2,\,syngas})}\right) \tag{1}$$

$$R_{tot} = \frac{V_N}{i} \tag{2}$$

$$i = 4FJ_{O_2} \tag{3}$$

In the present study, we examined the effect of surface modification and catalyst addition on the performance of composite membranes made from LSCr and YSZ. By changing the thickness of the dense membrane, we were able to estimate the bulk and surface resistances. Changes in the surface structure and catalyst addition both had a large impact on the membrane performance.

Experimental

YSZ (8 mol% Y_2O_3-doped ZrO_2) was purchased from Tosoh. LSCr ($La_{0.8}Sr_{0.2}CrO_3$) was synthesized *via* sol–gel methods with an aqueous precursor containing the appropriate amounts of $La(NO_3)_3\cdot 6H_2O$ (Alfa Aesar, 99%), $Sr(NO_3)_2$ (Alfa Aesar, 99.9%), and $Cr(NO_3)_3\cdot 9H_2O$ (Alfa Aesar, 98.5%). To ensure formation of the LSCr perovskite phase, citric acid (Fisher Scientific) was added as a complexing agent and the pH of the precursor solution was adjusted to 7 with ammonium hydroxide (Fisher Scientific). The precursor solution was dried at 373 K, and the resulting powder was calcined in air at 1373 K to form the perovskite phase. To form the LSCr–YSZ composites, LSCr and YSZ were mixed in a 40 : 60 volume ratio (assuming bulk densities) and then attritor milled together for 9 h.

Slurries that were used for tape casting were produced by combining the attritor-milled LSCr–YSZ mixture together with the appropriate amounts of solvents (ethanol, Decon, 100% and xylene, Sigma-Aldrich, 98.5%), a dispersant (blown menhaden fish oil, Tape Casting Warehouse), a binder (polyvinyl butyl B-98, Tape Casting Warehouse), and plasticizers (benzyl butyl phthalate, Sigma-Aldrich, 98% and polyethylene glycol 400, Alfa Aesar). After ball milling the slurry for 48 h, green tapes were prepared using a Mistler Model TTC-1200 tape caster. Green tapes for porous catalyst support layers were fabricated in a similar fashion but with the addition of graphite (Sigma-Aldrich) as a sacrificial pore former. Porous catalyst support layers were prepared from both a LSCr–YSZ mixture and from pure YSZ, without LSCr. Circular wafers were punched from the green tapes, and then laminated into porous-dense-porous structures. After calcination to 1773 K for 8 h, the dense layer in all of the membranes was 115 μm thick and 1 cm in diameter, while the porous layers were 50 μm thick and 0.7 cm in diameter. The porosities of the catalyst-support layers were always between 60 and 65%.

Because it was difficult to tape cast very thick dense layers, membranes were also fabricated by pressing pellets of the same attritor-milled, LSCr–YSZ mixture used in tape casting. Porous layers of YSZ on one side and an LSCr–YSZ mixture on the other, were then painted onto the pellet with a slurry containing the oxide powders together with graphite (Sigma-Aldrich) and α-terpinol (Alfa Aesar, 95%). After firing, the dense layers in these membranes were approximately 650 μm in thickness.

For each membrane, the catalyst support layer to be used on the air side was made from pure YSZ. 30 wt% $La_{0.8}Sr_{0.2}FeO_3$ (LSF) was then infiltrated into this porous YSZ layer in order to maximize the oxygen-exchange rates on the air side. The LSF was formed in the YSZ scaffold by infiltration of nitrate salts, using methods identical to that for preparing LSF–YSZ cathodes in studies of Solid Oxide Fuel Cells (SOFCs).[17] After infiltration of La, Fe, and Sr nitrate salts, the membranes were calcined to 1123 K for 4 h to form the perovskite phase.

For the syngas side, the surface of the membranes was composed of either pure porous YSZ or porous 40 vol% : 60 vol%, LSCr–YSZ composites. For the membranes with pure YSZ on the syngas side, 38 wt% LSCr was added into the porous layer by infiltration of aqueous salt solutions. LSCr was added using a nitrate salt precursor solution similar to the one used to make the powder. Infiltration steps were repeated until the target loading was achieved, with each infiltration step being followed by heating to 723 K to decompose the nitrate ions. After infiltration, membranes containing infiltrated LSCr were calcined to 1473 K for 4 h (for those cells with infiltrated LSCr, LSF was added to the air side after addition of LSCr to avoid the higher calcination temperature). As a final step, Pt (using tetraammineplatinum(II) nitrate, Alfa Aesar) and/or CeO_2 catalysts (using $Ce(NO_3)_3\cdot 6H_2O$, Alfa Aesar, 99.5%) were added to the syngas side of some membranes by infiltration. The compositions of the porous layer in the various membranes that were used in this study are listed in Table 1.

For oxygen flux measurements, the membrane discs were attached to the end of a YSZ tube using a glass sealant (ElectroScience type 4461). The air side of the membrane was exposed to ambient air while pure CO was passed over the syngas

Table 1 Membrane compositions

Membrane #	Fuel-side porous layer	Catalysts
M1	LSCr–YSZ composite	—
M2	LSCr–YSZ composite	5 wt% CeO_2
M3	LSCr–YSZ composite	1 wt% Pt 5 wt% CeO_2
M4[a]	LSCr–YSZ composite	1 wt% Pt 5 wt% CeO_2
M5	YSZ + infil. LSCr	—
M6	YSZ + infil. LSCr	1 wt% Pt

[a] The dense layers were 115 μm thick for all membranes except M4, which was 650 μm thick.

side of the membrane at a flow rate of 10 mL min^{-1}. The CO conversion was then determined with an online gas chromatograph (GC) (Buck Scientific 910) equipped with a Carboxen 1000 column (Supelco).

The phase and structure of the dense layers of the sintered dual-phase membranes were analyzed using X-ray diffraction (XRD) and scanning electron microscopy (SEM, JEOL 7500F HRSEM). The total conductivity of the 40 : 60 LSCr–YSZ, dense composite was determined by 4-probe, DC measurements in ambient air using a dense rectangular slab (10 mm × 7.5 mm × 3 mm).

Results

Physical characterization of LSCr–YSZ composites

The physical characteristics of the dense 40 vol% LSCr–60 vol% YSZ rectangular slab sintered at 1773 K were determined by XRD and electrical conductivity measurements. Density calculations based on the weight and dimensions of the slab indicated that it was greater than 95% dense. The XRD pattern, shown in Fig. 1, confirms that the LSCr–YSZ composite remains as two separate phases. The total conductivity of the slab was determined as a function of temperature in ambient air from 4-probe measurements and the values are shown in Fig. 2. The conductibility was measured as the temperature of the slab was ramped from 873 K to 1073 K and then back to 873 K. While there is some variability in the data obtained while heating and cooling, the conductivity is nearly independent of temperature over this range, with a value of 1 S cm^{-1}.

It is interesting to compare the conductivity of the LSCr–YSZ composite to that of LSCr (37.3 S cm^{-1})[18] and YSZ (0.019 S cm^{-1})[19] at 973 K. First, the total conductivity of the composite is much higher than the ionic conductivity of YSZ, showing that electronic conductivity of the membrane will not limit performance.

Second, since the LSCr phase is responsible for most of the electronic conduction, it is reasonable to view the electronic conductivity of the composite as the product of the volume fraction of the LSCr phase and the effective relative path length that the charge carriers must travel in the composite. The inverse of the relative path length is the tortuosity, which is calculated to be 14 in this case

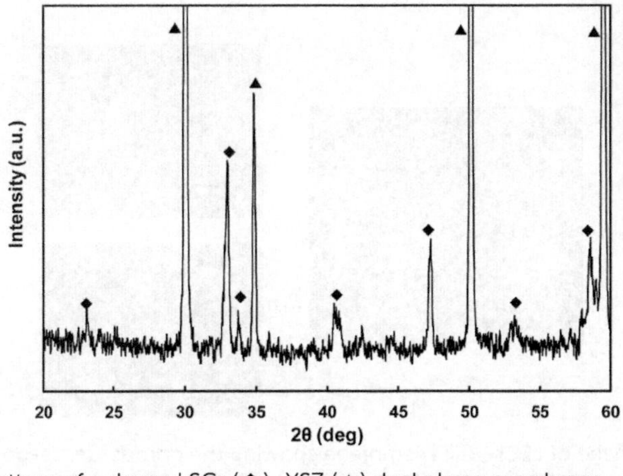

Fig. 1 XRD pattern of a dense LSCr (◆)–YSZ (▲) dual phase membrane.

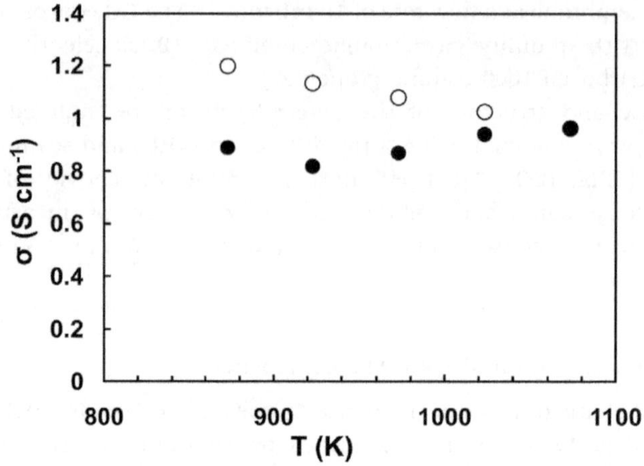

Fig. 2 Temperature dependence of the total conductivity of a dense LSCr–YSZ slab measured while heating (●) and cooling (○).

(bulk LSCr conductivity × 40 vol% ÷ composite conductivity). The tortuosity will be a strong function of the volume fraction of LSCr and the manner in which the LSCr is distributed within the composite, but a value of 14 is high for a composite that is 40 vol% conductive phase. For example, the tortuosity of 65% porous YSZ scaffold, which would be 35 vol% YSZ, has previously been determined to be 7 by similar conductivity measurements.[20] Therefore, the large tortuosity of the LSCr–YSZ composite suggests that the LSCr particles are not well connected within the composite.

A cross-sectional SEM image of one of the membranes (M1) prepared by tape casting is shown in Fig. 3. The low-magnification micrograph on the left side demonstrates that the membrane consists of a dense layer, 115 μm thick, surrounded by porous layers that are each 50 μm thick. A higher magnification image of the dense middle layer indicates that this layer contains very few occlusions. The YSZ and LSCr–YSZ porous layers on either side of the membrane were similar

Fig. 3 SEM image of LSCr–YSZ membrane showing the porous-dense-porous structure with insets showing higher magnification of the dense and porous layers.

to each other in structure (because micrographs of the porous YSZ have been reported previously,[21] only the porous LSCr–YSZ composite is shown here). The porous LSCr–YSZ layer is made up of pores that are approximately 5 μm in diameter, with a sponge-like structure.

Oxygen permeation fluxes

In previous O_2 flux studies in LCF membranes,[14] it was demonstrated that the addition of oxidation catalysts to the syngas side of the membrane had a large effect on the O_2 fluxes. To determine whether catalyst addition to the syngas side could cause similar increases with LSCr–YSZ composite membranes, O_2 flux measurements were performed on membranes M1, M2, and M3 as a function of temperature, with CO flowing over one surface. These three membranes differ only in that 5 wt% CeO_2 was added to the porous LSCr–YSZ composite on the syngas side in M2 and 1 wt% Pt and 5 wt% CeO_2 were added in M3. All three of these membranes had a porous composite of LSF–YSZ on the air side and identical LSCr–YSZ dense layers that were 115 μm thick. Results for these membranes are shown in Fig. 4.

The addition of the catalysts affected the performance dramatically. For example, at 1073 K, adding 5 wt% CeO_2 increased the oxygen permeation flux by more than a factor of 20 at 1073 K, from 2.9×10^{-8} mol cm^{-2} s^{-1} (M1) to 6.9×10^{-7} mol cm^{-2} s^{-1} (M2). Since CeO_2 is a mixed conductor under reducing conditions, some of this increase may be due to increased conductivity within the surface region; however, it should be noted that 5 wt% CeO_2 in a 60% porous LSCr–YSZ scaffold corresponds to only 2 vol%, an amount that seems too small to significantly change conductivity. CeO_2 is also known to have reasonably high activity for oxidation reactions, so that catalytic effects are a more likely explanation for the increased flux. Data from a membrane for which 1 wt% Pt was added along with 5 wt% CeO_2 (M3) provides further evidence that catalytic

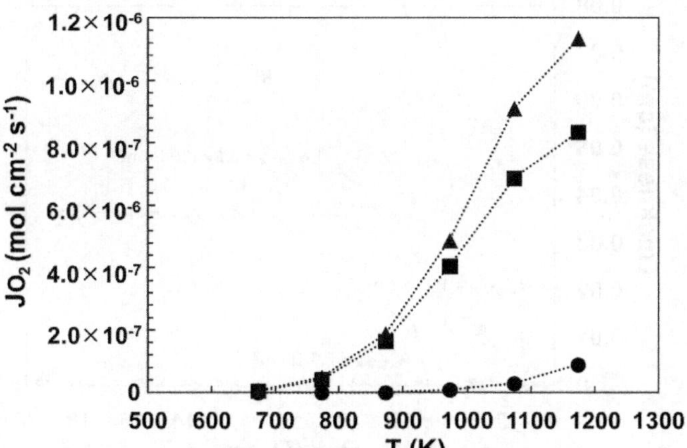

Fig. 4 Temperature dependence of the oxygen permeation flux of LSCr–YSZ membranes with LSCr–YSZ composite porous layer and various catalysts on the syngas side: unmodified (M1) (●), 5 wt% CeO_2 (M2) (■), and 5 wt% CeO_2 with 1 wt% Pt (M3) (▲).

properties are critical. The addition of Pt further increased the oxygen flux at 1073 K to 9.1×10^{-7} mol cm^{-2} s^{-1}, almost certainly due to the catalytic activity of Pt.

It is instructive to quantify these changes in the oxygen flux by calculating the membrane resistances in electrical terms. For CO oxidation, the Nernst potential (V_N) is close to 1 V ($\pm$0.1 V) at these temperatures for CO-to-CO_2 conversions between 7% and 80%.[22,23] Using this fact, together with eqn (1)–(3), R_{tot} can be calculated for each of the membranes. At 1073 K, the resistances of the three membranes, M1 to M3, were 88.9, 3.77, and 2.85 Ω cm^2, respectively. For comparison, the total cell resistance for a high-performance SOFC is typically targeted to be less than 0.5 Ω cm^2.[24] Clearly, the resistance of the membrane that did not have any catalyst must be limited by surface reactions at the syngas side. The effect of adding 5 wt% CeO_2 was dramatic but the addition of 1 wt% Pt also caused a non-trivial reduction in the surface resistance.

To determine how much of the membrane resistance was due to surface reactions and how much was due to bulk diffusion in the dense membrane, oxygen fluxes were measured for membrane M4, which was nearly identical to membrane M3 except for the thickness of the dense layer, 650 μm in M4 and 115 μm in M3. The surface of the syngas side of the membrane was again prepared from a porous composite of LSCr and YSZ, to which 5 wt% CeO_2 and 1 wt% Pt was added. Rather than report the fluxes directly, we have calculated R_{tot} at 973 and 1073 K and plotted these values in Fig. 5, together with the corresponding resistances from membrane M3. The plot of dense-layer thickness *versus* R_{tot} has a slope that is equal to the ambipolar conductivity of the dense layer, and the intercept with the abscissa is the resistance due to the surfaces. While only two membrane thicknesses were tested, the large difference in the thickness of the two membranes decreases the uncertainties in the calculated conductivities and surface resistances. From this plot, the ambipolar conductivities of the dense LSCr–YSZ composite were estimated to be 4.6×10^{-3} and

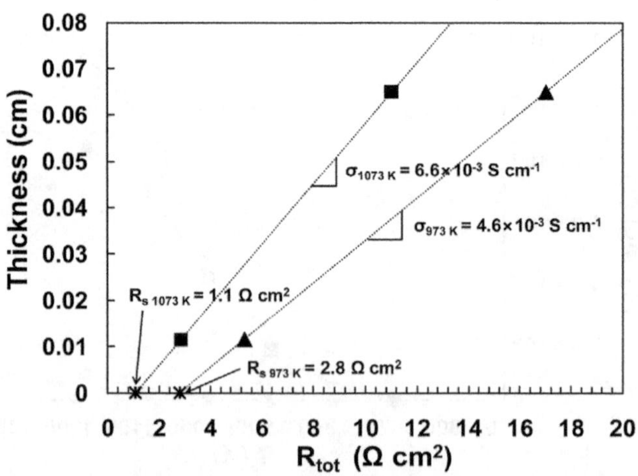

Fig. 5 R_{tot} of 115 μm and 650 μm thick LSCr–YSZ composite membranes at 973 K (▲) and 1073 K (■). The slope of a line through the two data points at each temperature was used to estimate the corresponding ionic conductivity.

6.6×10^{-3} S cm^{-1} at 973 and 1073 K. The surface resistances with 1 wt% Pt and 5 wt% CeO_2 catalysts were 2.8 and 1.1 Ω cm^2 at 973 and 1073 K, respectively.

Since the ambipolar conductivity is essentially equal to the ionic conductivity for materials that have high electronic conductivity, it is useful to compare the ambipolar conductivities of the 40 vol% LSCr–60 vol% YSZ composite to the ionic conductivities of bulk YSZ, which are 1.9×10^{-2} and 4.2×10^{-2} S cm^{-1} at 973 and 1073 K respectively.[19] Using the same arguments as those presented earlier for the electronic conductivity, the ambipolar conductivity of the composite is the ratio of the volume fraction YSZ and the tortuosity within the YSZ phase. From this, the tortuosity is calculated to be 1.7 from the data at 973 K and 2.5 from the data at 1073 K. Although tortuosity should not depend on temperature, a value of approximately 2 for a 60 vol% composite is reasonable.

The surface resistances can be compared to results from a recent investigation of an SOFC with electrode compositions that were similar to that of membrane M3.[25] That study used a similar LSF–YSZ cathode, again prepared by infiltration, and an LSCr–YSZ anode with 5 wt% CeO_2 and 0.5 wt% Pd. Two significant differences in the fuel-cell study were that the LSCr–YSZ composite anode was prepared by infiltration of LSCr into a porous YSZ scaffold and that humidified H_2 was used as the fuel. At 973 K, the non-ohmic resistance of the fuel cell was reported to be 0.39 Ω cm^2, which is much lower than the surface resistance of 2.8 Ω cm^2 determined for the membrane. This suggests that significant improvements can still be made in the performance of the membrane.

Effect of surface structure

In SOFCs, the surface structure is just as important as composition for influencing electrode performance. The best electrodes have ion-conducting channels to allow ions from the electrolyte to diffuse into the electrode, increasing the active region of the electrode. A potential problem with the porous LSCr–YSZ composites used on the syngas side of membranes M1 to M4 is that the LSCr phase within these channels will block conduction of the ions. A better structure would have channels that are pure YSZ to allow ion conduction, coated with LSCr for electron conduction. Fig. 6 shows a schematic of these two structures, with the one on the left representing the surfaces of membranes M1 to M4 and the one on the right representing the structure of the LSCr–YSZ anode in the earlier fuel-cell study.[25]

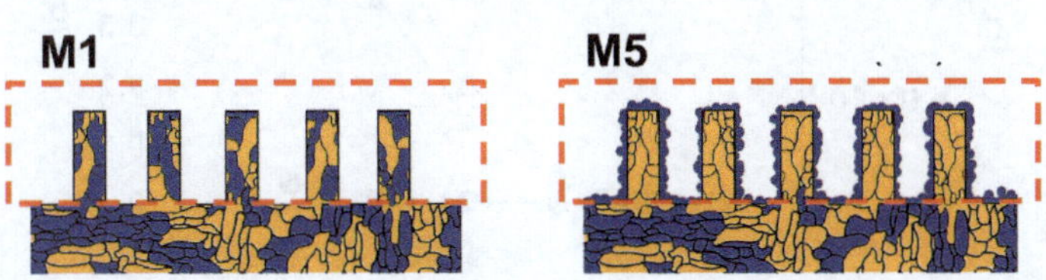

Fig. 6 Cartoon showing the different surface microstructures for the syngas side porous layers in membrane M1 and M5 with YSZ in yellow and LSCr in blue. Membrane M1 has a porous layer with a random mixture of LSCr–YSZ, while membrane M5 has a porous layer with LSCr infiltrated on top of YSZ. The dense layers in both membranes are LSCr–YSZ composite.

To test the concept that an improved surface structure could increase performance, membranes were prepared in which the porous scaffold on the syngas side was YSZ. 38 wt% LSCr (equivalent to 40 vol% of the solid phase) was then infiltrated into the porous scaffold using aqueous solutions of the nitrate salts, followed by calcination to produce the perovskite structure. The membrane with just the infiltrated LSCr is referred to as M5, while an identical cell with 1 wt% Pt added is M6. The oxygen permeation fluxes for these two membranes, together with results from the membrane prepared with the LSCr–YSZ composite scaffold, M1, are shown in Fig. 7. These data demonstrate that surface structure is equally important in the membranes as in SOFC. At 1073 K, the oxygen flux for the membrane with infiltrated LSCr was 8.9×10^{-7} mol cm^{-2}, compared to a flux of only 2.9×10^{-8} mol cm^{-2} s^{-1} when the membrane scaffold was a random composite. This 30-fold increase is nearly as large as the effect of adding 1 wt% Pt and 5 wt% CeO_2 to the random composite. Adding 1 wt% Pt to the infiltrated LSCr–YSZ composite (M6) further increased the flux to 1.2×10^{-6} mol cm^{-2} at 1073 K.

The calculated resistances for membranes M1, M5, and M6 are reported in Table 2. Clearly, changing the structure of the LSCr–YSZ composite from random to infiltrated had the largest effect on the oxygen fluxes; however, enhancement of the surface exchange rate by adding a catalyst like Pt was also important. This is easier to see when the contribution of the dense layer to the total membrane resistance is removed. Using the ambipolar conductivity of the dense LSCr–YSZ composite determined from Fig. 5, the contribution from the 115 μm dense layer is 1.75 Ω cm^2 at 1073 K. Since R_{tot} for membranes M5 and M6 at this temperature were 2.89 and 2.14 Ω cm^2, respectively, the addition of Pt decreased the surface resistance from approximately 1.1 Ω cm^2 to 0.39 Ω cm^2. Interestingly, this value for the surface resistance approaches values that are typical of SOFC electrodes.

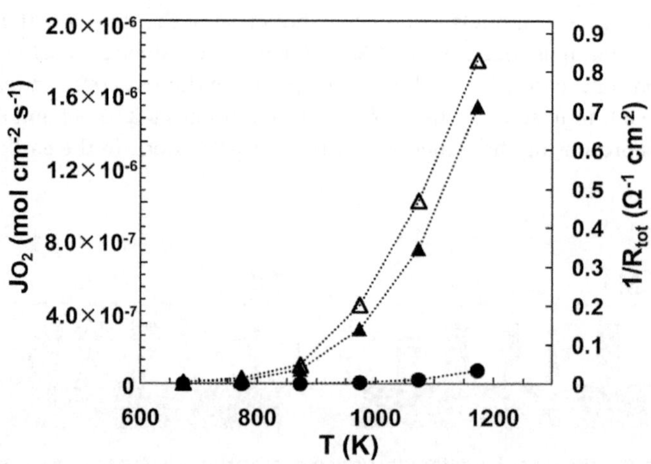

Fig. 7 Temperature dependence of the oxygen permeation flux of a membrane with an LSCr–YSZ composite porous layer on the syngas side (M1) (●), in comparison with that of a membrane with a pure YSZ porous layer on the syngas side infiltrated with LSCr (M5) (▲). Addition of Pt to M5 further improves oxygen flux (M6).

Table 2 Total resistances of LSCr–YSZ membranes M1, M5, and M6

T (K)	R_{tot} M1 (Ω cm^2)	R_{tot} M5 (Ω cm^2)	R_{tot} M6 (Ω cm^2)
973	293	7.15	4.94
1073	88.9	2.89	2.14
1173	29.1	1.41	1.21

Discussion

The results of the present study demonstrate that surface reactions can easily limit oxygen fluxes in MIEC membranes. Although it is sometimes assumed that oxygen fluxes are limited by bulk diffusion in a dense membrane when the membrane is thick enough, the surface resistances in the unmodified membranes of the present study were so large that they would limit the performance of any membrane with a reasonable thickness. It is therefore critically important to always consider what the surface limitations are.

Quantifying membrane losses in terms of resistances is very useful for showing where changes need to be made that will significantly impact performance. This is common practice in SOFC and SOEC research, where losses due to the cathode, the anode, and the electrolyte are each calculated; however, this idea is less commonly applied to research on MIEC membranes. For example, the oxygen fluxes in the membrane with the porous, random LSCr–YSZ composite on the syngas surface were very low (M1). Attempts to increase the fluxes by decreasing the thickness of dense layer would have been unproductive, since the resistance due to bulk diffusion was only 2% of the total resistance at 1073 K (1.75 Ω cm^2 out of 88.9 Ω cm^2). Clearly, the primary focus in this case should be on decreasing the surface resistances. The ability to quantify the various losses also allows one to determine how much improvement can be obtained by decreasing the thickness of the dense layer or by adding a catalyst. For example, although the addition of 1 wt% Pt to the membrane with infiltrated LSCr (M6) gave only modest improvements in the oxygen fluxes, the surface resistances were reduced by nearly a factor of 3. In that particular case, the resistance associated with the surface had been reduced to a point where further improvements in the oxygen permeation flux would require thinner membranes.

With regards to choice of catalysts, it should be noted that other transition-metal catalysts, including inexpensive metals like Ni and Co,[26] would likely provide a similar enhancement as Pt. In the present study, we chose to use Pt because it does not become volatile or undergo coking in the presence of CO. However, under typical syngas conditions, these other metals would almost certainly provide adequate catalytic activity.

Finally, it is worth emphasizing that concepts used for optimizing SOEC and SOFC electrodes were used extensively in the present study to reduce membrane resistances. The surface regions on the syngas and air sides of the MIEC membrane experience very similar conditions to those experienced by SOFC and SOEC anodes and cathodes. Much more is known about losses in these systems, in large part because electrical measurements are easier to make and losses from the various components are easier to separate. This body of work can and should be applied to the optimization of MIEC membranes.

Conclusions

Oxygen fluxes in MIEC membranes made from composites of LSCr and YSZ are often limited by surface reactions. Concepts from SOFC electrodes can be used to engineer the surface structures and compositions of the membranes so as to dramatically reduce surface losses.

Acknowledgements

The authors would like to thank the U.S. National Science Foundation for financial support for this research project (grant no. DMR-1210388).

Notes and references

1 P. N. Dyer, R. E. Richards, S. L. Russek and D. M. Taylor, *Solid State Ionics*, 2000, **134**, 21–33.
2 C. H. Bartholomew and R. J. Farrauto, *Fundamentals of Industrial Catalytic Processes*, Wiley, 2006.
3 C. S. Chen, Z. P. Zhang, G. S. Jiang, C. G. Fan, W. Liu and H. J. M. Bouwmeester, *Chem. Mater.*, 2001, **13**, 2797–2800.
4 T. Kida, D. Takauchi, K. Watanabe, M. Yuasa, K. Shimanoe, Y. Teraoka and N. Yamazoe, *J. Electrochem. Soc.*, 2009, **156**, E187–E191.
5 Z. P. Shao, G. X. Xiong, H. Dong, W. H. Yang and L. W. Lin, *Sep. Purif. Technol.*, 2001, **25**, 97–116.
6 X. L. Dong, W. Q. Jin, N. P. Xu and K. Li, *Chem. Commun.*, 2011, **47**, 10886–10902.
7 P. M. Geffroy, J. Fouletier, N. Richet and T. Chartier, *Chem. Eng. Sci.*, 2013, **87**, 408–433.
8 P. M. Geffroy, A. Vivet, J. Fouletier, C. Steil, E. Blond, N. Richet, P. Del Gallo and T. Chartier, *J. Electrochem. Soc.*, 2013, **160**, F60–F68.
9 J. H. Joo, K. S. Yun, Y. Lee, J. Jung, C. Y. Yoo and J. H. Yu, *Chem. Mater.*, 2014, **26**, 4387–4394.
10 T. Liu, Y. Wang, R. H. Yuan, J. F. Gao, C. S. Chen and H. J. M. Bouwmeester, *Acs Appl. Mater. Interfaces*, 2013, **5**, 9454–9460.
11 B. Wang, M. C. Zhan, D. C. Zhu, W. Liu and C. S. Chen, *J. Solid State Electrochem.*, 2006, **10**, 625–628.
12 J. Kim and Y. S. Lin, *AIChE J.*, 2000, **46**, 1521–1529.
13 C. F. Miller, J. Chen, M. F. Carolan and E. P. Foster, *Catal. Today*, 2014, **228**, 152–157.
14 A. S. Yu, J. Kim, T. S. Oh, G. Kim, R. J. Gorte and J. M. Vohs, *Appl. Catal., A*, 2014, **486**, 259–265.
15 G. M. Christie, J. R. Wilson and B. A. V. Hassel, *US Pat.*, 8,323,463, 2012.
16 A. S. Yu, J. M. Vohs and R. J. Gorte, *Energy Environ. Sci.*, 2014, 7, 944–953.
17 W. S. Wang, M. D. Gross, J. M. Vohs and R. J. Gorte, *J. Electrochem. Soc.*, 2007, **154**, B439–B445.

18 M. Mori, T. Yamamoto, H. Itoh and T. Watanabe, *J. Mater. Sci.*, 1997, **32**, 2423–2431.

19 K. Sasaki and J. Maier, *Solid State Ionics*, 2000, **134**, 303–321.

20 F. Bidrawn, R. Kungas, J. M. Vohs and R. J. Gorte, *J. Electrochem. Soc.*, 2011, **158**, B514–B525.

21 R. Kungas, J. S. Kim, J. M. Vohs and R. J. Gorte, *J. Am. Ceram. Soc.*, 2011, **94**, 2220–2224.

22 *Fuel Cell Handbook 7th Edition*, EG&G Technical Services, Inc., U.S. Department of Energy, Morgantown, WV, 2004.

23 S. McIntosh and R. J. Gorte, *Chem. Rev.*, 2004, **104**, 4845–4865.

24 B. C. H. Steele and A. Heinzel, *Nature*, 2001, **414**, 345–352.

25 T. S. Oh, A. S. Yu, L. Adijanto, R. J. Gorte and J. M. Vohs, *J. Power Sources*, 2014, **262**, 207–212.

26 J. S. Kim, V. V. Nair, J. M. Vohs and R. J. Gorte, *Scr. Mater.*, 2011, **65**, 90–95.

Faraday Discussions

PAPER

Oxygen deficient layered double perovskite as an active cathode for CO_2 electrolysis using a solid oxide conductor

Tae Ho Shin,*[ac] Jae-Ha Myung,[a] Maarten Verbraeken,[a] Guntae Kim[b] and John T. S. Irvine*[a]

Received 17th February 2015, Accepted 26th March 2015
DOI: 10.1039/c5fd00025d

A-site ordered $PrBaMn_2O_{5+\delta}$ was investigated as a potential cathode for CO_2 electrolysis using a $La_{0.9}Sr_{0.1}Ga_{0.8}Mg_{0.2}O_3$ (LSGM) electrolyte. The A-site ordered layered double perovskite, $PrBaMn_2O_{5+\delta}$, was found to enhance electrocatalytic activity for CO_2 reduction on the cathode side since it supports mixed valent transition metal cations such as Mn, which could provide high electrical conductivity and maintain a large oxygen vacancy content, contributing to fast oxygen ion diffusion. It was found that during the oxidation of the reduced $PrBaMn_2O_{5+\delta}$ (O5 phase) to $PrBaMn_2O_{6-\delta}$ (O6 phase), a reversible oxygen switchover in the lattice takes place. In addition, here the successful CO_2 electrolysis was measured in LSGM electrolyte with this novel oxide electrode. It was found that this $PrBaMn_2O_{5+\delta}$, layered perovskite cathode exhibits a performance with a current density of 0.85 A cm^{-2} at 1.5 V and 850 °C and the electrochemical properties were also evaluated by impedance spectroscopy.

1. Introduction

With growth of renewable energy technologies, it is necessary to store vast amounts of energy by, *e.g.*, converting electricity overcapacity to chemical energy in the form of compounds such as hydrogen, methane, or methanol. In particular, the variety of products from high temperature electrolysis can be used as feedstocks for chemical synthesis or converted into highly dense electric power; it would provide a means of storing renewable electricity in a convenient, high energy-density form. Recently, hydrogen has attracted attention as an alternative mediator for energy conversion and storage because of its high gravimetric energy density and zero carbon emission.[1-5] However, the utilization of hydrogen has been limited due to the practical engineering and economic limitations with

[a]*School of Chemistry, University of St Andrews, St Andrews, Fife, KY16 9ST, Scotland, UK. E-mail: ths@st-andrews.ac.uk; ceramist95@gmail.com; Tel: +44-(0)759-676-2424*
[b]*Dept. of Chemical and Energy Engineering, Ulsan National Institute of Science and Technology (UNIST), Korea*
[c]*Electronic Materials Convergence Division, Korea Institute of Ceramic Engineering & Technology (KICET), Korea*

respect to its generation and distribution.[6] On the other hand hydrocarbon fuels and carbon monoxide cost-effectively match the existing energy infrastructure well because of their similarity to current fossil fuels; they would be effective energy carriers in a period of transition toward zero carbon emission.[7–9] Recycling or reuse of CO_2 could be an effective approach for accessing a new form of energy carrier and enabling a carbon-neutral cycle. In particular, electrochemical reduction of CO_2 into CO fuel with renewable electricity has been proposed as an alternative method to store this energy, overcoming the inherent problems with intermittency and geographical distribution of renewable energy sources.[10–13]

Solid oxide electrolysis cells (SOECs) have recently attracted a great deal of interest as a highly efficient electrolyser and a green energy technology. Current developments are mainly based upon steam electrolysis for hydrogen production. However, reduction of CO_2 using a highly efficient solid oxide electrolyser is also feasible. This would be an alternative to carbon capture and advanced utilization of CO_2.[14,15] Solid oxide electrolysers with oxygen ion conducting electrolytes are able to separately produce CO and O_2 from CO_2 ($CO_2 \rightarrow 1/2O_2 + CO$) at high and intermediate temperature (500–900 °C) with high efficiencies (as they are not limited by the Carnot cycle). Both products, CO and O_2 may be useful in industrial processes. The electrolysis of CO_2 in a solid oxide device is analogous to high temperature steam electrolysis by SOECs, *i.e.* CO_2 molecules are electrochemically reduced into CO at the cathode (fuel electrode) and oxygen ions are transported through the oxygen-ion conducting electrolyte to the anode (oxygen electrode) where gaseous O_2 is produced and released. High temperature solid oxide electrolysis reduces electricity demand, and offers fast electrode kinetics, reducing the overall cell internal resistance as a consequence and increasing electrolysis efficiency. Due to the operating conditions in SOECs using CO_2, cathode materials require stability in a wide pO_2 and temperature range, and additionally need to be tolerant towards carbon coking. To date, SOECs predominantly comprise materials and techniques from solid oxide fuel cells (SOFCs), but operated in reverse modes. This means that electrode materials have not been tailored specifically for use in electrolysis conditions, which can differ substantially from fuel cell conditions. Long term stability of solid oxide electrode materials is still one of the important challenging issues for both SOECs and SOFCs for commercialisation into the energy market.

Currently, high temperature electrolysis of CO_2 has been mainly performed using Ni–YSZ cermet cathodes in oxygen ion conducting SOECs. However, Ni is easily deactivated by reoxidation or carbon formation when it is directly exposed to CO/CO_2 or a hydrocarbon atmosphere.[16–18] Therefore, a significant reduction atmosphere is required to flow over the Ni surface base electrode to avoid oxidation of Ni to NiO, which would cause a loss of electronic conductivity and the failure of the electrode. In the most serious circumstances, the electrode fails mechanically as a result of deposited coke or re-oxidation.[19] Consequently, redox stability and reversibility have recently been emphasised as important requirements for the cathode (fuel electrode) material in SOECs. So far, several conductive perovskite oxides have been reported as potential anode materials in SOFCs, which may also be useful as cathode materials in SOECs. For more sturdy and superior materials in durable SOFC/SOEC systems, various mixed ionic and electronic conducting (MIEC) oxides have been explored. Among these, researchers have mainly demonstrated that $La(Sr)MO_3$ (M = Cr, Mn, Fe, Ti) perovskite oxides[20–22] can, in principle, be good fuel electrode materials with sufficient electronic conductivity. For instance, the efficient electrolysis of CO_2 based on a ceramic cathode, La(Sr)Mn(Cr)O_3 (LSCM) perovskite oxide have been recently demonstrated.[20,23–25] However, electrochemical performance of the ceramic electrodes is still smaller than that using a Ni based electrode, due to insufficient electrical conductivity and low electrocatalytic activity. LSCM suffers from a drop in conductivity under the reducing SOEC cathode conditions.

Recently we have succeeded in synthesising the oxygen deficient layered phase, $PrBaMn_2O_{5+\delta}$ under reducing atmosphere. Whereas polycrystalline samples sintered in air adopt the cubic ABO_3 perovskite structure, the A-site ordered layered perovskite $PrBaMn_2O_{5+\delta}$, can be grown under reducing conditions. Furthermore, it was found that the double perovskite, $PrBaMn_2O_{5+\delta}$ exhibits high electrical conductivity and excellent redox and coking tolerance when used as an anode for an SOFC.[26] In this work, $PrBaMn_2O_{5+\delta}$, employed as a potential ceramic cathode for CO_2 electrolysis SOEC with a configuration of $PrBaMn_2O_{5+\delta}$|La doped ceria (LDC)|$La_{0.9}Sr_{0.1}Ga_{0.8}Mg_{0.2}O_3$ (LSGM)|La(Sr)Fe(Co)O_3. The performance of the $PrBaMn_2O_{5+\delta}$ cathode and the electrochemical process of high temperature electrolysis of CO_2 in a cell using LSGM as the electrolyte were investigated and evaluated.

2. Experimental

The $Pr_{0.5}Ba_{0.5}MnO_{3-\delta}$ was prepared using the conventional solid-state reaction method and then A-site ordered layered $PrBaMn_2O_{5+\delta}$ is obtained by *in situ* annealing of $Pr_{0.5}Ba_{0.5}MnO_{3-\delta}$ in reducing conditions using 5% H_2 at temperatures higher than 800 °C. The crystalline phase characterization of the oxide powder was performed by X-ray diffraction (PANalytical Empyrean, Mo Kα radiation and Cu Kα1 radiation). Thermal gravimetric analysis (TGA) was carried out on a NETZSCH TG 209 instrument (NETZSCH Geraetebau GmbH, Selb, Germany) with a heating & cooling rate of 2–5 °C min^{-1} to evaluate the oxygen loss and incorporation on redox cycling in air and 5% H_2 on the compositions synthesized and dilatometry was also performed at 100 °C to 800 °C with a heating/cooling rate of 2 °C min^{-1} in air and 5% H_2 to characterize the thermo-physical properties. Electrical conductivity was measured on the sintered pellets by the standard four terminal DC method or van der Pauw method up to 900 °C in ambient air. A current of 50 mA (model: Keithley 220, Keithley Instruments Inc., U.S.A.) was applied in both directions, and resistance was calculated as a gradient of potential *vs.* current. These conductivity details were previously reported.[26]

For solid oxide electrolyser experiments, a configuration of $PrBaMn_2O_{5+\delta}$|LDC|$La_{0.9}Sr_{0.1}Ga_{0.8}Mg_{0.2}O_3$ (LSGM)|La(Sr)Fe(Co)O_3 was fabricated using 200 μm thick LSGM electrolyte supports. Commercially available powder, $La_{0.9}Sr_{0.1}Ga_{0.8}Mg_{0.2}O_{3-\delta}$ (LSGM, 99.9% Kceracell co., Ltd, South Korea) was used for the preparation of electrolyte. The LSGM electrolyte supports were prepared by dry-pressing powder into a circular green body followed by high-temperature sintering in air at 1450 °C for 6 h and mechanical polishing until 2 cm diameter and 200 μm thick electrolyte supports were obtained. Electrodes were deposited onto the faces of the LSGM electrolyte disk by screen-printing inks of the respective powders, followed by sintering at 1100 °C for 30 min (thickness of screen printed electrode, <25 μm). The active area of the cells, equal to the area to the cathode area ($PrBaMn_2O_{5+\delta}$), was 0.5 cm^2 and the area of the anode was same.

To evaluate the electrochemical properties, *I–V* curves and impedance were measured using a four terminal configuration with silver mesh/wire as the electrode current collectors. In order to perform electrochemical testing in fuel cell mode, a dried CO_2/CO (70 : 30 vol) mixture was supplied into the cathode with a 100 ml min^{-1} flow rate while the anode side was exposed to constant oxygen flow as an oxidant gas (100 ml min^{-1}). The cell was maintained in each condition for at least one hour to reach steady state, before carrying out any electrochemical tests. AC impedance spectroscopy was recorded using an IM6 Electrochemical Workstation (Zahner, Germany) with frequency ranged from 0.1 Hz to 100 kHz with an amplitude of 10 mV.

3. Results and discussion

The disordered cubic perovskite $Pr_{0.5}Ba_{0.5}MnO_3$ was prepared by a typical solid state reaction. To avoid the complete oxidization of Mn^{3+} to Mn^{4+}, which would favour the formation of $BaMnO_3$, the powder was calcined in an atmosphere with lower oxygen partial pressure (using a mixture of Ar/air) below the synthesis temperature (<1300 °C). A-site ordering in the layered manganite perovskite would be partially hindered by the formation of the hexagonal related perovskite $BaMnO_3$. Fig. 1 shows the XRD pattern for the cubic disordered perovskite (Fig. 1a), prepared by solid state reaction at 1200 °C in air; reduction in 5% H_2 leads to A-site ordering and hence doubling of the unit cell, giving a layered phase with tetragonal symmetry (Fig. 1b). The XRD pattern of disordered cubic perovskite shown in Fig. 1a, suggests the presence of a small amount secondary hexagonal phase, as similarly observed in a previous study in which the powder was prepared by Pechini synthesis at lower temperature.[26] However, they are not serious impurities but a single phase of $Pr_{0.5}Ba_{0.5}MnO_3$ was mainly indicated. Furthermore, the A-site ordered layered perovskite $PrBaMn_2O_{5+\delta}$ could also be grown by annealing the disordered phase in reducing atmosphere in accordance

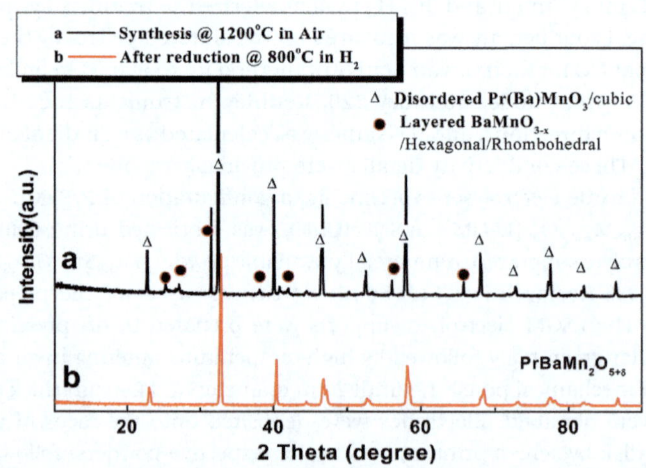

Fig. 1 X-ray diffraction patterns of (a) $Pr_{0.5}Ba_{0.5}MnO_3$ (O3) prepared by solid state reaction at 1200 °C in air, and (b) $PrBaMn_2O_{5+\delta}$ (O5) was obtained by reducing the as-prepared powder in 5% H_2 at 800 °C for 4 h (Cu Kα radiation).

with previous results.[26] To fully understand the phase changes during redox cycling, we checked the phase transition from disordered cubic $Pr_{0.5}Ba_{0.5}MnO_3$ phase (O3) to oxygen deficient layered $PrBaMn_2O_{5+\delta}$ (O5) by *in situ* variable temperature/atmosphere X-ray diffraction (Mo Kα1/Kα2), whilst performing reduction–oxidation cycles. Fig. 2a–c shows Rietveld refinement profiles of $Pr_{0.5}Ba_{0.5}MnO_3$ phase (O3), $PrBaMn_2O_{5+\delta}$ (O5) and $PrBaMn_2O_{6-\delta}$ (O6) from *in situ* XRD data at 800 °C in various atmospheres. The refined unit cell parameters and refinement statistics are summarized in Tables 1 and 2 and show selected bond distances and angles.

The *in situ* high temperature X-ray diffraction study confirms cubic symmetry for the disordered O3 phase at 800 °C. Upon reduction at 800 °C, the A-site ordered O5 phase emerges with tetragonal symmetry, causing a relative shrinkage of the *c*-axis as compared to the *a*-axis. Refinement is successful using the *P4/mmm* space group. Subsequent re-oxidation of the A-site ordered O5 phase at 800 °C results in the emergence of the double perovskite, $PrBaMn_2O_{6-\delta}$ (O6) phase. This O6 phase also has tetragonal symmetry with space group *P4/mmm*. Re-oxidation of the O5 phase causes a decrease in tetragonality as evidenced by the drop in $2a/c$ ratio from 1.018 to 1.005. Despite this, direct evidence of A-site ordering and thus doubling of unit cell in both O5 and O6 phased from the XRD data is the (001) reflection found at $2\theta \approx 5°$as shown in Fig. 2b and c. Despite the weak intensity of the (001) reflection due to small contrast between Pr^{3+} and Ba^{2+}, its emergence on redox cycling is significant.

The fully oxidised ordered O6 phase shows A-site ordering along the *c*-axis, with a stacking pattern of alternating BaO–MnO_2–PrO–MnO_2–BaO layers. As expected from perovskite materials, Mn is six coordinated, whereas both Ba and Pr are twelve coordinated. The MnO_6 octahedron is slightly distorted, causing the Mn and O2 sites to be closer to Pr than to Ba. The reduced O5 variant has an identical stacking sequence but additionally exhibits oxygen vacancy ordering, with the O3 site vacant, resulting in Pr layers without oxygen ions, shown in Fig. 3. In this structure Mn becomes square pyramidally coordinated, whereas Pr is now 8 coordinated to O2. Ba becomes less coordinated to O2, as the latter moves closer to Pr. The lattice volume is increased due to Mn reduction as shown in Tables 1 and 2.

Due to the reduction of Mn^{3+}/Mn^{4+} in the O3/O6 phases to Mn^{2+}/Mn^{3+} in O5 the Mn–O bond length at 800 °C increases from 1.89–2.03 to 2.04–2.07 Å. It is expected that the oxygen deficient PrO_x layer provides a channel for fast oxygen transport and therefore A-site ordering is in fact enhancing the oxygen mobility, albeit only in two dimensions.[27] The coexistence of Mn oxidation states is further expected to give rise to electronic conductivity, which is a requirement for any SOEC/SOFC electrode material. The mixed valence of transition metals such as Mn could give rise to different Mn–O bond lengths at various oxygen partial pressure conditions; the swollen Mn–O bonding length (*ca.* 2.046 Å) of the oxygen deficient phase could be expected from the reduced oxidation state of Mn (Mn^{3+} or Mn^{2+}). The vacancy would be mainly located between reduced Mn^{3+}/Mn^{2+} state cations, between pairs of square pyramidal MnO_5. Thus oxygen vacancy ordering provides a mechanism to reduce the coordination number of the smaller Pr^{3+} ion without reducing the coordination number of the larger Ba^{2+} ion. Under reducing conditions, oxygen atoms in the PrO_x plane can be partially or entirely removed,

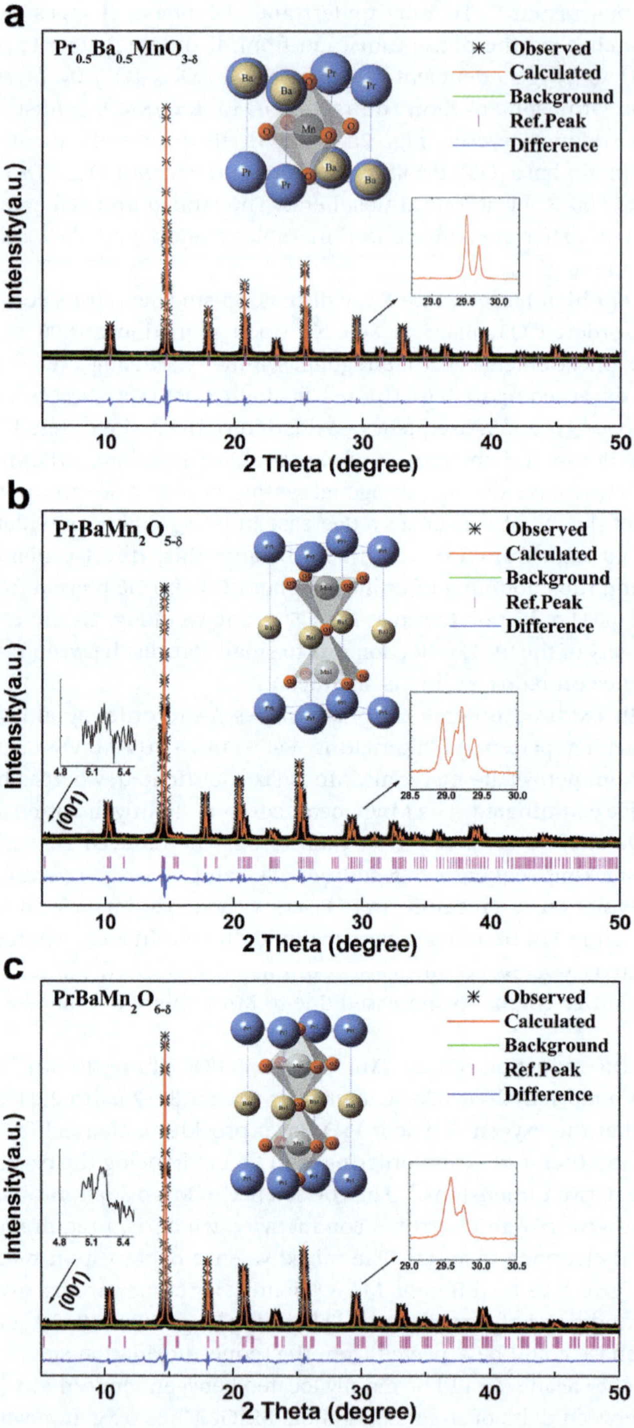

Fig. 2 Rietveld-refined XRD profiles of (a) $Pr_{0.5}Ba_{0.5}MnO_3$ (O3), (b) $PrBaMn_2O_{5+\delta}$ (O5) and (c) $PrBaMn_2O_{6-\delta}$ (O6) obtained by simultaneous fitting of *in situ* high temperature XRD data at 800 °C in air, 5% H_2 and air, respectively. (Mo Kα1, Kα2 radiation).

Table 1 Refined structural parameters for $Pr_{0.5}Ba_{0.5}MnO_3$ (O3), $PrBaMn_2O_{5+\delta}$ (O5) and $PrBaMn_2O_{6-\delta}$ (O6) obtained by simultaneous fitting of *in situ* high temperature XRD data at 800 °C

	$Pr_{0.5}Ba_{0.5}MnO_3$ (O3) in air	$PrBaMn_2O_{5+\delta}$ (O5) in H_2	$PrBaMn_2O_{6-\delta}$ (O6) in air
Space group	$Pm\bar{3}m$ (cubic)	$P4/nmm$ (tetra)	$P4/mmm$ (tetra)
$a = b$ (Å)	3.93593(5)	4.02405(9)	3.93985(9)
c (Å)	—	7.90618(24)	7.84136(24)
$2a/c$		1.018	1.005
V (Å^3)	60.9735(13)	128.024(6)	121.717(5)
λ (Å) Mo Kα1	0.0793	0.7093	0.7093
λ (Å) Mo Kα2	0.7136	0.7136	0.7136
R_p%	7.93	6.43	6.06
R_{wp}%	11.19	8.30	8.01

creating many oxygen vacant sites in the crystal sites. This might illustrate the close coupling between A-site ordering and oxygen vacancy ordering.[27]

What makes the layered perovskite-manganese oxide $PrBaMn_2O_{5+\delta}$ a promising candidate for high oxygen mobility is its remarkable variability in oxygen content under redox conditions. It was generally found that for $ReBaMn_2O_{5+\delta}$ (Re = Pr, Nd, La, Gd) the oxidation of the reduced (O5) phase $PrBaMn_2O_{6-\delta}$ (O6), in the case of $ReBaMn_2O_{5+\delta}$, a reversible oxygen switchover in the lattice takes place. The reversibility of the phase change during redox cycle was also established by thermal analysis. The fully oxygen charged $PrBaMn_2O_{6-\delta}$ (O6) had been reduced in 5% H_2/95% Ar at elevated temperature up to 800 °C and cooled in 5% H_2/95% Ar to room temperature (2.5 °C min^{-1}) and re-oxidized up to 700 °C in air. Fig. 4 shows the thermogravimetric analysis (TGA) in which a weight gain of *ca.* 3.58 wt % was observed on both reduction and re-oxidation corresponding to a loss/uptake of 1 formula unit of oxygen ($\delta = 1$). This symmetry between reduction and

Table 2 Selected interatomic distances and angles of $Pr_{0.5}Ba_{0.5}MnO_3$ (O3), $PrBaMn_2O_{5+\delta}$ (O5) and $PrBaMn_2O_{6-\delta}$ (O6) obtained by simultaneous fitting of *in situ* high temperature XRD data at 800 °C

	$Pr_{0.5}Ba_{0.5}MnO_3$ (O3) in air	$PrBaMn_2O_{5+\delta}$ (O5) in H_2	$PrBaMn_2O_{6-\delta}$ (O6) in air
Distance (Å)			
Mn–O1	1.96796(2) 6	2.067(5) × 1	2.03072(6) × 1
Mn–O2	—	2.0405(18) × 4	1.97022(4) × 4
Mn–O3	—	—	1.88996(6) × 1
Mn–O (average)	1.96796	2.0458	1.96693
Angles (°)			
Mn–O1–Mn	180	180	180
Mn–O2–Mn	—	160.83(57)	178.01680(10)
Mn–O3–Mn	—	—	180
O1–Mn–O3	180	—	180
O2–Mn–O2	180	160.83(57)	178.01680(10)

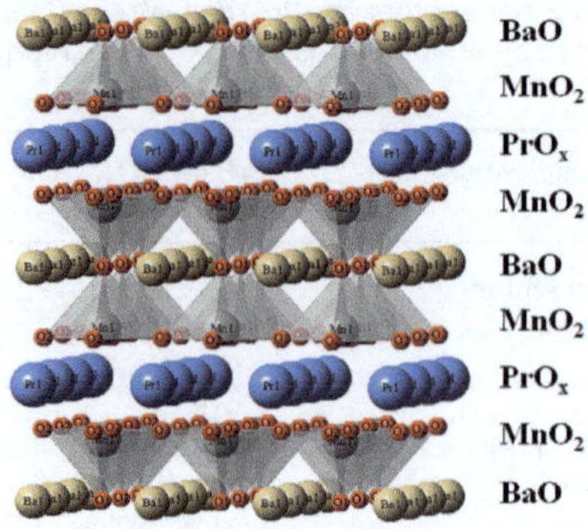

Fig. 3 A schematic drawing of crystal structure model determined for A-site ordering layered $PrBaMn_2O_{5+\delta}$ (O5) and oxygen vacancy ordering in oxygen deficient O5 phase by Rietveld-refined XRD profile.

oxidation cycles suggests excellent reversibility, which is in agreement with our previously reported dilatometry data.[26]

Since $PrBaMn_2O_{5+\delta}$ shows superior redox reversibility as well as potential electrochemical performance as oxide anode for SOFCs, application of $PrBaMn_2O_{5+\delta}$ for a potential cathode in CO_2 electrolysis using solid oxide ion conductor, LSGM, was further studied. Fig. 5a shows the temperature dependent *I–V* curves for CO_2 electrolysis between 750 °C and 850 °C under the following gas feed conditions: air on the anode side and a mixture of 70% CO_2/30% CO on the cathode side with 100 ml min^{-1}. It was found that $PrBaMn_2O_{5+\delta}$ is very effective for CO_2 electrolysis with a remarkable cathodic current density of 0.84 A cm^{-2} at 1.55 V and 850 °C. At lower temperature (750 °C), it seems that there was no significant current flow at lower voltages suggesting a slow electrolysis reaction, which might be explained by the large overpotential on the cathode side; the cell

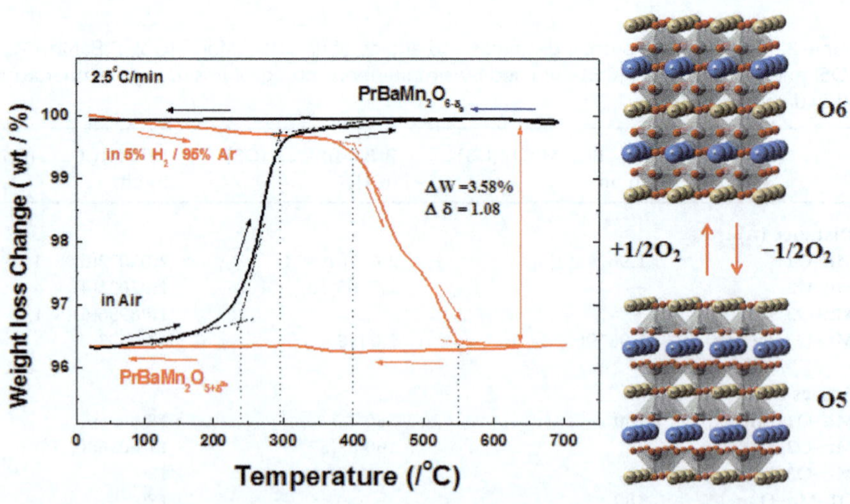

Fig. 4 TGA curves of $PrBaMn_2O_{5+\delta}$ in 5% H_2/Ar from room temperature to 700 °C to room temperature and in re-oxidation cycle from room temperature to 700 °C.

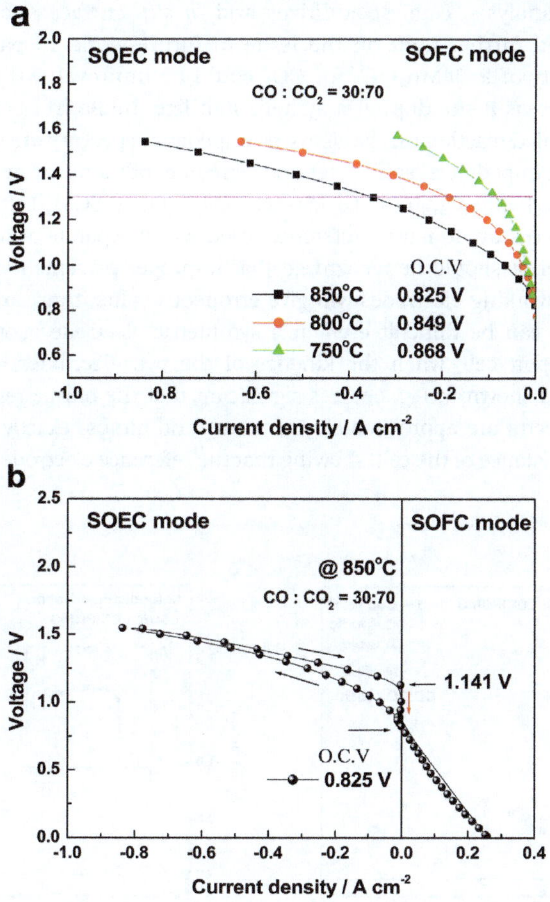

Fig. 5 (a) Temperature dependent *I*–*V* curves for CO_2 electrolysis under gas feed conditions with elevating temperature (750–850 °C) with the cell configuration of $PrBaMn_2O_{5+\delta}$|LDC|$La_{0.9}Sr_{0.1}Ga_{0.8}Mg_{0.2}O_3$ (LSGM)|La(Sr)Fe(Co)O_3 and (b) the *I*–*V* curves, voltages during steady current cycling at 850 °C.

voltage sharply increased with increasing current density at initial low current range. Fig. 5b shows the *I*–*V* curve at 850 °C, during current cycling, which was measured with increasing current up to the maximum current range and subsequent decreasing current, leaving long stabilisation times at every current step to attempt to attain equilibrium. However, it was observed that voltage recovery was slow and strongly time dependent. As shown in Fig. 5b, initial OCV (0.825 V) was slightly increased to 1.141 V after current cycling; the initial OCV was however recovered over time. The increased OCV after cycling could be related with momentary carbon deposition or carbon monoxide adsorption on the catalyst surface, as 1.14 V is very close to the theoretical OCV value for carbon fuel; slow surface kinetics of $PrBaMn_2O_{5+\delta}$ might explain such desorption/adsorption CO/CO_2 or carbon species. Alternatively, poor catalytic activity for carbon oxidation or a Ba carbonation reaction might play a role too. The slow chemical kinetics are not fully understood at this stage, but a more detailed analysis using *in situ* exhausts gas analysis, catalyst additives and *in situ* surface analysis would be helpful to shed further light on the issue in future work. In particular, stable catalytic activity of $PrBaMn_2O_{5+\delta}$ for CO_2 could be improved with other catalyst additives, whereas B-site doping may help stabilize the Ba from carbonisation.

For the detailed reaction mechanisms, ac impedance spectra were measured. Fig. 6 shows complex impedance plots under open circuit conditions for internal resistance of the cell $PrBaMn_2O_{5+\delta}$|LDC|$La_{0.9}Sr_{0.1}Ga_{0.8}Mg_{0.2}O_3$ (LSGM)|La(Sr)Fe(Co)O_3. In general, it is necessary to have a reference electrode to separate anode and cathode losses. However, it should be recognized that improper placement of the reference electrode and working electrodes will give erroneous value; thus, exactly separating each electrode can be difficult even in a symmetric electrode geometry in a thin electrolyte support cell, when the kinetics of the two electrodes are significantly different.[28–30] As shown in Fig. 6a, it is significant that the ohmic resistance in both impedance spectra are approximately the same, and almost exactly one-half of the total ohmic resistance of the cell, showing that the reference electrode is sampling the

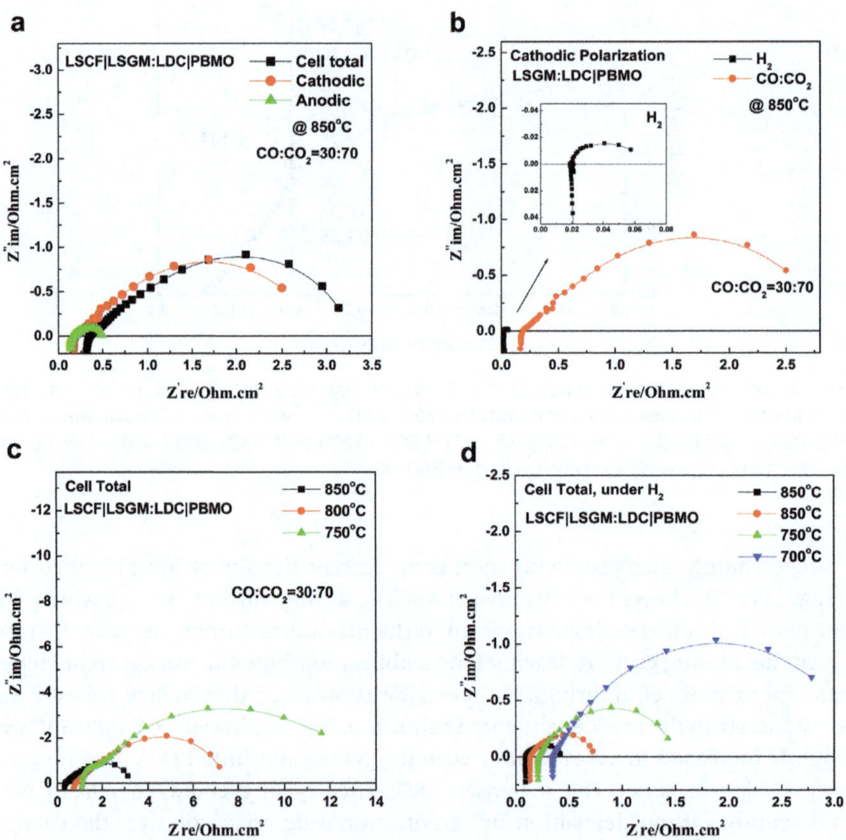

Fig. 6 Nyquist plot of the $PrBaMn_2O_{5+\delta}$|LDC|$La_{0.9}Sr_{0.1}Ga_{0.8}Mg_{0.2}O_3$ (LSGM)|La(Sr)Fe(Co)O_3 cell under open circuit: (a) impedance plot of cells, cathodic and anodic under 70% vol CO_2 gas fuel condition at 850 °C, (b) comparison of cathodic impedance plot under H_2 with 70% vol CO_2, (c) temperature dependant impedance plot of the total cell under 70% vol CO_2 gas fuel, (d) temperature dependant impedance plot of the total cell under H_2 gas fuel.

potential at the centre of the electrolyte. Thus, it can be roughly inferred that total polarization resistance was clearly dominated by cathodic electrode ($PrBaMn_2O_{5+\delta}$) because cathodic impedance spectra was similar to the total cell impedance. It is obvious that at least two semicircles are present in the impedance plot. Considering the response frequencies of the two parallel RQ circuits, the resistance at higher frequency (R_{p1}) could be assigned to a charge transfer process, whereas the lower frequency process (R_{p2}) seems related to a diffusion process of the cathodic reaction. The contribution of R_{p2} is much larger than that of the surface reaction step (R_{p1}), and so the observed impedance may indicate a slow CO_2 diffusion, which may be due to slow chemical surface adsorption/desorption and/or slow surface-diffusion processes for CO_2 reduction on the surface of $PrBaMn_2O_{5+\delta}$ at lower temperature. Although fast oxygen ion mobility would be expected due to a large concentration of ordered oxide ion vacancies in $PrBaMn_2O_{5+\delta}$, it shows lower surface activity for CO_2 reduction at lower temperature. However, fairly good performances were still achieved at temperatures higher than 800 °C and its surface activity for CO_2 may be improved if other catalytic additives are effectively employed. Due to improved surface activity, various metal catalyst additives are now under study, and the results will be reported in the future. Fig. 7 shows SEM micrographs of the cathode of the $PrBaMn_2O_{5+\delta}$|LDC|$La_{0.9}Sr_{0.1}Ga_{0.8}Mg_{0.2}O_3$ (LSGM)|La(Sr)Fe(Co)O_3 cell before and after CO_2 electrolysis. There appears to be some limited coking on the top surface that might be explained by Boudouard reaction at high temperature with CO_2/CO mixture or slow response of $PrBaMn_2O_{5+\delta}$ for the chemical surface adsorption/desorption of CO_2 molecules;

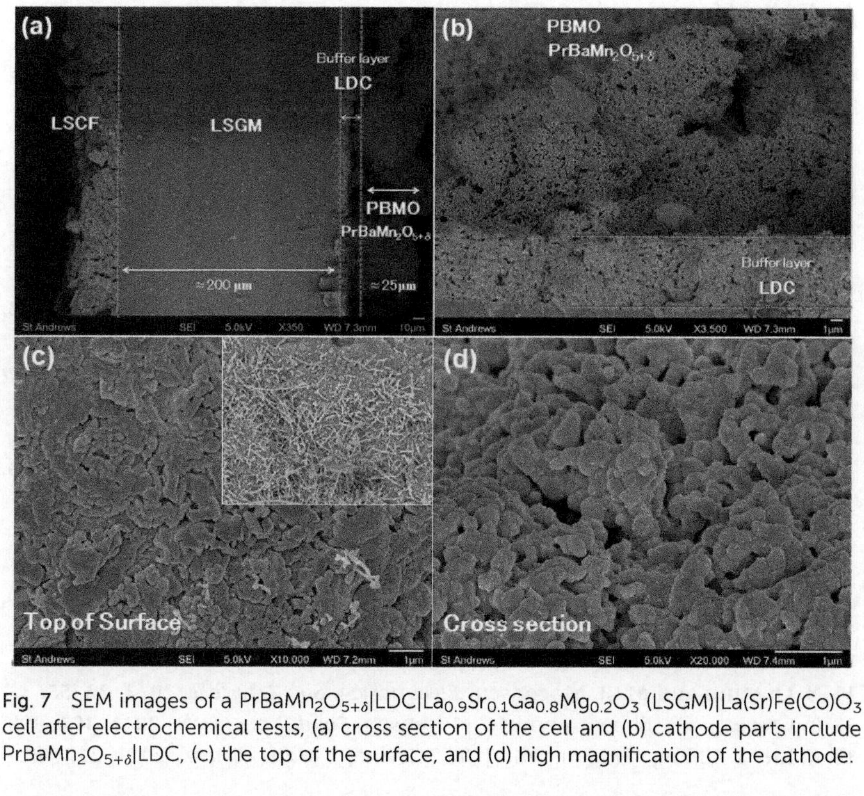

Fig. 7 SEM images of a $PrBaMn_2O_{5+\delta}$|LDC|$La_{0.9}Sr_{0.1}Ga_{0.8}Mg_{0.2}O_3$ (LSGM)|La(Sr)Fe(Co)O_3 cell after electrochemical tests, (a) cross section of the cell and (b) cathode parts include $PrBaMn_2O_{5+\delta}$|LDC, (c) the top of the surface, and (d) high magnification of the cathode.

however, there were no serious surface morphology changes. It was therefore concluded that A-site ordered layered perovskite, $PrBaMn_2O_{5+\delta}$, could be used as a potential cathode in solid oxide electrolysis with excellent redox tolerance.

4. Conclusion

In summary, $PrBaMn_2O_{5+\delta}$ was successfully prepared as a single phase cathode material from a disordered O3 perovskite phase after reduction treatment. The reduction to obtain layered $PrBaMn_2O_{5+\delta}$ can be performed *in situ* during SOEC/SOFC testing to limit processing steps. An *in situ* high temperature X-ray diffraction study confirmed the phase changes under reducing and oxidising conditions, showing that the A-site ordered layered perovskite can be formed from disordered cubic phase under reducing conditions at 800 °C with reversible phase change between $PrBaMn_2O_{5+\delta}$ (O5) and $PrBaMn_2O_{6-\delta}$ (O6) during redox cycles. The superior redox ability of $PrBaMn_2O_{5+\delta}$ (O5) was also observed by thermo-gravimetric methods, showing a reversible weight change during redox cycling corresponding to a change in oxygen stoichiometry of $\delta = 1$.

Moreover, the reduction of CO_2 *via* elevated-temperature (750–850 °C) electrolysis using an SOEC with LSGM as the electrolyte was studied using a $PrBaMn_2O_{5+\delta}$ (O5) cathode. High current density (0.845 A cm^{-2}) was obtained at 850 °C and electrolysis potential of 1.55 V for a cell using the $PrBaMn_2O_{5+\delta}$|LDC|$La_{0.9}Sr_{0.1}Ga_{0.8}Mg_{0.2}O_3$ (LSGM)|La(Sr)Fe(Co)O_3. As a result, this study revealed that $PrBaMn_2O_{5+\delta}$ (O5) is a potential cathode material for SOECs.

Acknowledgements

We thank EPSRC (EP/I022570/1, EP/K015540/1, EP/I038950/1) and the Royal Society (Wolfson Research Merit award) for support.

References

1 J. L. Bernal-Agustín and R. Dufo-López, *Int. J. Hydrogen Energy*, 2008, **33**, 6401.
2 J. O. M. Bockris and T. N. Veziroglu, *Int. J. Hydrogen Energy*, 2007, **32**, 1605.
3 S. H. Jensen, P. H. Larsen and M. Mogensen, *Int. J. Hydrogen Energy*, 2007, **32**, 3253.
4 W. Doenitz, R. Schmidberger, E. Steinheil and R. Streicher, *Int. J. Hydrogen Energy*, 1980, **5**, 55.
5 A. O. Isenberg, *Solid State Ionics*, 1981, **3–4**, 431.
6 E. H. Seymour, L. Murray and R. Fernandes, *Int. J. Hydrogen Energy*, 2008, **33**, 3015.
7 L. Schlapbach, *Nature*, 2009, **460**, 809.
8 L. Wang and R. T. Yang, *Energy Environ. Sci.*, 2008, **1**, 268.
9 B. Sakintuna, F. Lamari-Darkrim and M. Hirscher, *Int. J. Hydrogen Energy*, 2007, **32**, 1121.
10 S. D. Ebbesen and M. Mogensen, *J. Power Sources*, 2009, **193**, 349.
11 N. Q. Minh and M. B. Mogensen, *Electrochem. Soc. Interface*, 2013, **22**, 55.
12 C. Graves, S. D. Ebbesen and M. Mogensen, *Solid State Ionics*, 2011, **192**, 398.
13 K. Xie, Y. Zhang, G. Meng and J. T. S. Irvine, *J. Mater. Chem.*, 2011, **21**, 195.

14 C. Graves, S. D. Ebbesen, S. H. Jensen, S. B. Simonsen and M. B. Mogensen, *Nat. Mater.*, 2015, **14**, 239.

15 D. M. Bierschenk, J. R. Wilson and S. A. Barnett, *Energy Environ. Sci.*, 2011, **4**, 944.

16 E. P. Murray, T. Tsai and S. A. Barnett, *Nature*, 1999, **400**, 649.

17 P. Huang, A. Horky and A. Petric, *J. Am. Ceram. Soc.*, 1999, **82**, 2402.

18 B. C. H. Steele, I. Kelly, H. Middleton and R. Rudkin, *Solid State Ionics*, 1988, **28–30**, 1547.

19 S. McIntosh and R. J. Gorte, *Chem. Rev.*, 2004, **104**, 4845.

20 X. L. Yue and J. T. S. Irvine, *J. Electrochem. Soc.*, 2012, **159**, F442.

21 S. Wang, H. Tsuruta, M. Asanuma and T. Ishihara, *Adv. Energy Mater.*, 2015, **5**, 1401003.

22 Y. Li, J. Zhou, D. Dong, Y. Wang, J. Z. Jiang, H. Xiang and K. Xie, *Phys. Chem. Chem. Phys.*, 2012, **14**, 15547.

23 X. Yue and J. T. S. Irvine, *Electrochem. Solid-State Lett.*, 2012, **15**, B31.

24 X. L. Yue and J. T. S. Irvine, *Solid State Ionics*, 2012, **225**, 131.

25 F. Bidrawn, G. Kim, G. Corre, J. T. S. Irvine, J. M. Vohs and R. J. Gorte, *Electrochem. Solid-State Lett.*, 2008, **11**, B167.

26 S. Sengodan, S. Choi, A. Jun, T. H. Shin, Y.-W. Ju, H. Y. Jeong, J. Shin, J. T. S. Irvine and G. Kim, *Nat. Mater.*, 2015, **14**, 205.

27 M. T. Anderson, J. T. Vaughey and K. R. Poeppelmeier, *Chem. Mater.*, 1993, **5**, 151.

28 S. McIntosh and R. J. Gorte, *Chem. Rev.*, 2004, **104**, 4845.

29 S. B. Adler, *J. Electrochem. Soc.*, 2002, **149**, E166.

30 S. McIntosh, J. Vohs and R. Gorte, *J. Electrochem. Soc.*, 2003, **150**, A1305.

Faraday Discussions

PAPER

Co-electrolysis of steam and CO_2 in full-ceramic symmetrical SOECs: a strategy for avoiding the use of hydrogen as a safe gas†

M. Torrell,[a] S. García-Rodríguez,[b] A. Morata,[a] G. Penelas[b] and A. Tarancón*[a]

Received 9th February 2015, Accepted 12th March 2015

DOI: 10.1039/c5fd00018a

The use of cermets as fuel electrodes for solid oxide electrolysis cells requires permanent circulation of reducing gas, *e.g.* H_2 or CO, so called safe gas, in order to avoid oxidation of the metallic phase. Replacing metallic based electrodes by pure oxides is therefore proposed as an advantage for the industrial application of solid oxide electrolyzers. In this work, full-ceramic symmetrical solid oxide electrolysis cells have been investigated for steam/CO_2 co-electrolysis. Electrolyte supported cells with $La_{0.75}Sr_{0.25}Cr_{0.5}Mn_{0.5}O_{3-\delta}$ reversible electrodes have been fabricated and tested in co-electrolysis mode using different fuel compositions, from pure H_2O to pure CO_2, at temperatures between 850–900 °C. Electrochemical impedance spectroscopy and galvanostatic measurements have been carried out for the mechanistic understanding of the symmetrical cell performance. The content of H_2 and CO in the product gas has been measured by in-line gas micro-chromatography. The effect of employing H_2 as a safe gas has also been investigated. Maximum density currents of 750 mA cm^{-2} and 620 mA cm^{-2} have been applied at 1.7 V for pure H_2O and for $H_2O : CO_2$ ratios of 1 : 1, respectively. Remarkable results were obtained for hydrogen-free fuel compositions, which confirmed the interest of using ceramic oxides as a fuel electrode candidate to reduce or completely avoid the use of safe gas in operation minimizing the contribution of the reverse water shift reaction (RWSR) in the process. $H_2 : CO$ ratios close to two were obtained for hydrogen-free tests fulfilling the basic requirements for synthetic fuel production. An important increase in the operation voltage was detected under continuous operation leading to a dramatic failure by delaminating of the oxygen electrode.

Introduction

Massive emissions of anthropogenic carbon dioxide are destabilizing the natural carbon cycle. New strategies for the conversion and valorization of CO_2 are today

[a]*Catalonia Institute for Energy Research (IREC), Jardins de les Dones de Negre, 1, 08930-Sant Adrià de Besòs, Barcelona, Spain. E-mail: atarancon@irec.cat*

[b]*REPSOL Technology Center, Ctra de Extremadura A-5, km 18, 28935 Móstoles, Madrid, Spain*

† Electronic supplementary information (ESI) available. See DOI: 10.1039/c5fd00018a

considered with much attention since they can reduce the climate change impact while giving new opportunities for sustainable industrial development.[1] One of the main routes for CO_2 valorization is based on its chemical transformation for the generation of high-added value products. In particular, employing CO_2 as a carbon source allows the generation of hydrocarbon fuels by following Power-to-Fuel (P2F) schemes. Power to fuel is a set of technologies that allow the conversion of electricity into a fuel. The conversion of electricity surplus into easy-to-transport and -storage fuels might play a crucial role in future energy strategies based on large-scale energy storage and the interconnection of the two major energy infrastructures currently available, *i.e.* fuel and electricity networks.

High temperature electrolysis (HTE) is one of the most promising alternatives for the generation of gas fuel from electricity due to its high efficiency and flexibility.[2] In particular, solid oxide electrolysis cells (SOECs) seem to be a strong candidate because of their low internal resistances that allow high production rates at low voltages.[2] Moreover, besides typical steam electrolysis for direct hydrogen production, high temperature SOECs can efficiently electrolyze other molecules like CO_2 that, as previously mentioned, could be employed as the main carbon source for producing C-based chemicals. By introducing CO_2 and H_2O as a fuel, SOECs are able to generate CO and H_2, *i.e.* syngas. The fabrication of syngas is considered the first step for the production of synthetic hydrocarbons since it is the precursor of high added value hydrocarbons through well-known Fischer–Tröpsch catalytic routes.[3] When the electricity employed in the electrolysis comes from free carbon and renewable energy sources, this synthetic fuel production scheme becomes a sustainable route.

Favorable thermodynamic conditions permit the reduction of the electrical consumption in SOECs (compared to low T electrolysis). Part of the energy needed for the electrolysis reactions is provided in the form of heat, which can eventually come from waste heat in other industrial processes increasing the overall efficiency. The main advantage of simultaneous electrolysis of CO_2 and H_2O is the low polarization resistance (with values similar to the pure steam electrolysis) and the negligible reduction of CO to carbon.[4,6] This coke deposition takes place following the Boudouard reaction that is favored at low temperatures and high concentrations of CO, *i.e.* opposite to high temperature and low CO concentrations typically found in solid oxide electrolysis. After the evidence of co-electrolysis reported by A. O. Isenberg,[5] one of the first major descriptions of co-electrolysis using SOECs was published by Stoots *et al.* in 2008,[6] and has been followed by an increase in research activity during the last years.[3,6–17] However, the co-electrolysis in SOECs is still under study since it is a more complicated process than pure steam or carbon dioxide electrolysis, involving different reactions competing in the typical operation conditions. The current understanding of co-electrolysis is not complete since the controlling mechanism depends on the kinetics of multiple reactions, *i.e.* it is dependent on the catalysts involved. In this sense, it has not been shown yet if the syngas generation by co-electrolysis of steam and CO_2 actually proceeds by simultaneous electrolysis of both CO_2 and H_2O or by predominant steam electrolysis combined with reverse water shift reaction (RWSR) (hydrogen production by electrolysis that immediately reacts with CO_2 to produce CO).[18]

Since CO and CO_2 are similarly present in Solid Oxide Fuel Cells (SOFCs), even carbon monoxide being a typical fuel, the same materials employed in SOFCs and

SOECs have been used for the first prototypes of co-electrolysers of solid oxide cells (SOCs).[10] However, standard Ni-based fuel electrodes can cause problems in electrolysis because the poor redox stability and the possible carbon deposition at high CO_2 concentrations, especially in dry CO_2.[9,19] To avoid dramatic effects due to redox cycles, SOECs based on Ni-cermets usually employ hydrogen as a safe gas during operation. However, simple recirculation of hydrogen in co-electrolysis systems is unpractical and, therefore, it is of great interest to find more redox stable fuel electrodes than the classic Ni–YSZ. Only a few metal-free fuel electrodes have been successfully proven in solid oxide cells. In particular, Tao *et al.*[20] proposed a simple perovskite with composition $La_{0.75}Sr_{0.25}Cr_{0.5}Mn_{0.5}O_3$ (LSCM) that showed high stability and performance as a fuel electrode.[21,22] A reasonable catalytic activity of these materials as oxygen electrodes was also proved, resulting in a remarkable performance in symmetrical SOFCs and SOECs.[23,24]

The present paper studies an ytterbium scandium stabilized zirconia (YbScSZ) electrolyte-supported SOEC cell with symmetrical LSCM electrodes operating under different $H_2O : CO_2$ fuel compositions. Pure steam electrolysis, pure CO_2 electrolysis and different intermediate compositions of steam and CO_2 are evaluated. The paper compares the *I–V* curves obtained for this set of different fuels and compares them with the ones obtained by addition of safe gas (6% H_2). Only a few studies of co-electrolysis with low percentages of safe gas have been previously published[9,10,16,25] and, to the best of our knowledge, this is the first that totally removes hydrogen from the inlet fuel. Finally, $H_2 : CO$ production ratios are presented under different operation conditions.

Experimental method

Material synthesis, cell fabrication and characterization

$La_{0.75}Sr_{0.25}Cr_{0.5}Mn_{0.5}O_{3-\delta}$ (LSCM) powder was prepared by a solid state reaction method using La_2O_3 (99.999% puratronic Alfa Aesar), Cr_2O_3 (99.6% Alfa Aesar), $MnCO_3$ (99.9% Alfa Aesar) and $SrCO_3$ (puratronic 99.994%, Alfa Aesar) as precursors, they were mixed and heat treated in a single step at 1200 °C for 10 h in air. The electrolyte was a tape casted disc of scandium and ytterbium stabilized zircon defined as $(ZrO_2)_{0.9}(Yb_2O_3)_{0.06}(Sc_2O_3)_{0.04}$ ($Yb_{0.6}Sc_{0.4}$SZ or YbScSZ) with a diameter of 19.7 mm and thickness of 165 μm, sintered at 1490 °C for 3 h. The symmetrical LSCM electrodes were sintered at 1200 °C for 3 h after deposition by airbrushing of an ethanol based LSCM ink. Painted and cured gold current collectors were employed for avoiding cross effects in the electrochemical response.

X-ray diffraction (XRD) was performed using a Bruker D8 Advanced diffractometer, equipped with a primary 80 monochromator and a LynxEye detector. The scans were performed in high resolution mode from 20 to 80°, 2 Ω range, using copper Kα1 radiation. XRD was used to analyze the phase purity of the as-synthesized LSCM powder and to evaluate the reactivity of the LSCM powder under different concentrations of CO_2 after 75 h of exposition at 750 °C and 900 °C. Electrodes and electrolyte interfaces and morphology were studied by employing a Zeiss Auriga scanning electron microscope.

I–V polarization curves, galvanostatic aging test and Electrochemical Impedance Spectroscopy (EIS) measurements were carried out to evaluate the cell performance inside a Probostat test station (NorECS) at temperatures rainging

850–900 °C by using a galvanostat/potentiostat with EIS capabilities (Parstat 2273, PAR). Fig. 1 shows the experimental set up for the delivery of steam and CO_2 to the cell at the working temperature. The system permits the change of the gas composition and water content of the mixture. Ultrapure water is evaporated by a controlled evaporator and mixing liquid delivery system (Bronkhorst-CEM system) that regulates and monitors the steam flow by a liquid flow controller. 50% argon was used as a carrier gas on the test station and the total flow of fuel was balanced to 140 ml min^{-1}. Micro gas chromatography (Agilent 490-μGC) was used to evaluate the concentration of produced H_2 and CO gases and H_2/CO ratio by a MolSieve 5A with Ar and He as respective carriers.

Experimental results

Characterization of the elements of the cell

Fig. 2 shows an Arrhenius plot of the YbScSZ electrolyte conductivity. The values presented correspond to the deconvolution of the impedance spectra of symmetrical Pt/YbScSZ/Pt cells (not presented here). A remarkable improvement in the YbScSZ conductivity with respect to the state-of-the-art Yttrium stabilized zirconia defined as $(ZrO_2)_{0.92}(Y_2O_3)_{0.08}$ (8YSZ) electrolyte is observed (conductivity of 8YSZ is presented in the same figure for comparison). It is important to notice

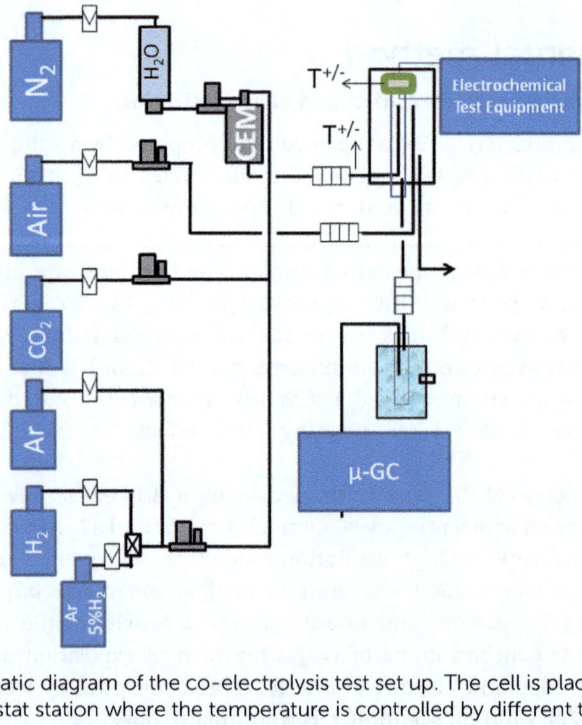

Fig. 1 Schematic diagram of the co-electrolysis test set up. The cell is placed to measure on the Probostat station where the temperature is controlled by different thermocouples in different places. The different gas flows are monitored by Bronkhorst mass-flow controllers. Steam is produced and controlled by a Bronkhorst CEM and liquid flow controller system. Galvanostat/potentiostat Parstat 2273, PAR is used for the electrochemical measurements and Agilent μ-GC for the characterization of the produced fuel.

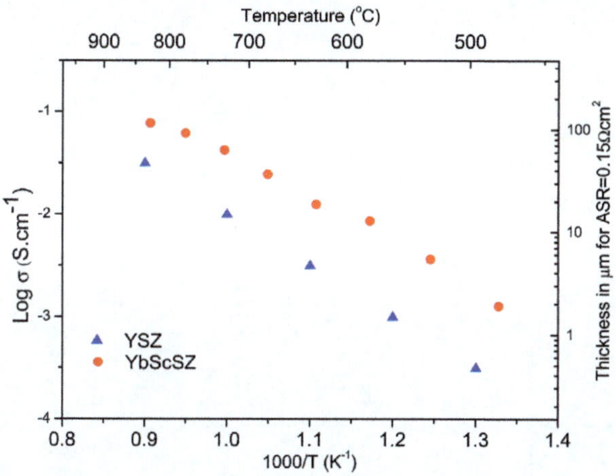

Fig. 2 Arrhenius plot of the electrolyte conductivity as a function of the inverse of the temperature for YbScSZ and YSZ. The *y*-axis on the right shows the required thickness of the electrolyte for reaching an Area Specific Resistance ASR = 0.15 Ω cm^2.

that working with YbScSZ allows the use of electrolyte supported cells (thickness >150 μm) at temperatures above 800 °C. Therefore, a low enough resistance contribution to the cell is expected for the 165 μm thick electrolyte employed in this work at 850 °C. The conductivity of the electrolyte is in agreement with previously reported values for similar compositions.[26,27]

Fig. 3 shows the XRD of the as-synthesized LSCM powders together with the diffraction patterns of the powder after being thermally treated for 75 hours in dry and wet CO_2 at temperatures between 750 °C and 900 °C. According to the XRD patterns, the as-synthesized LSCM presents a single phase, which remains stable to carbonation under dry and wet CO_2 atmospheres at operating temperatures. The stability of the LSCM under typical atmosphere and temperature co-electrolysis conditions shows a structural suitability of LSCM for this particular application.

An optimization study for the attachment temperature was carried out. Area specific resistance (ASR) of the cells was measured through the corresponding EIS spectra deconvolution. The optimal temperature to attach the LSCM on the YbScSZ electrolyte was determined to be 1200 °C, which leads to an ASR = 0.5 Ω cm^2 at 900 °C.

Cell performance under different fuel compositions

Fig. 4 shows the *I–V* curves for a single symmetrical cell tested at 850 °C employing five different fuels: pure steam, pure CO_2 and three intermediate compositions with different $H_2O : CO_2$ ratios. Low values of open circuit voltage (OCV) approaching zero were measured independently on the fuel. This is due to the absence of a significant gradient of oxygen partial pressure across the cell before starting the electrolysis reactions. A maximum current density value (780 mA cm^{-2}) was achieved when operating as a pure steam electrolyzer at 1.7 V. The maximum current density was progressively reduced for higher concentrations of CO_2 (inset Fig. 4). For pure CO_2 electrolysis, the maximum current density achieved was 700 mA cm^{-2} at 1.7 V. This decrease in the performance indicates that

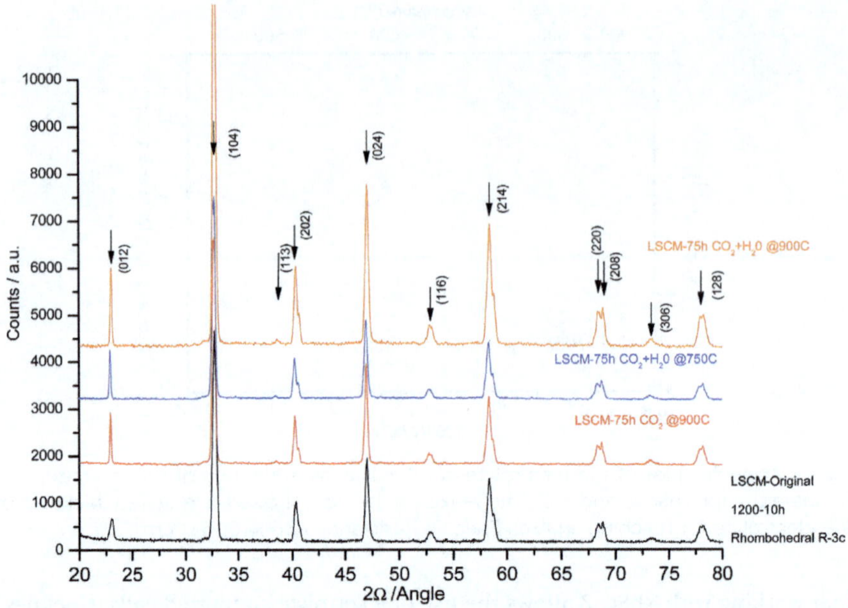

Fig. 3 XRD of the as-synthetized LSCM and after heat treatments under different dry and wet CO_2 atmospheres. The original and treated LSCM show a rhombohedral crystallographic phase (SG $R\bar{3}c$) without noticeable secondary phases.

the reduction reaction of the carbon dioxide becomes the limiting step of the co-electrolysis mechanism.[28] In all cases, initial high resistances were observed at low current densities due to required activation potential of the electrodes at very low voltages. Current densities above 100 mA cm^{-2} were required for reaching lower polarization resistances.

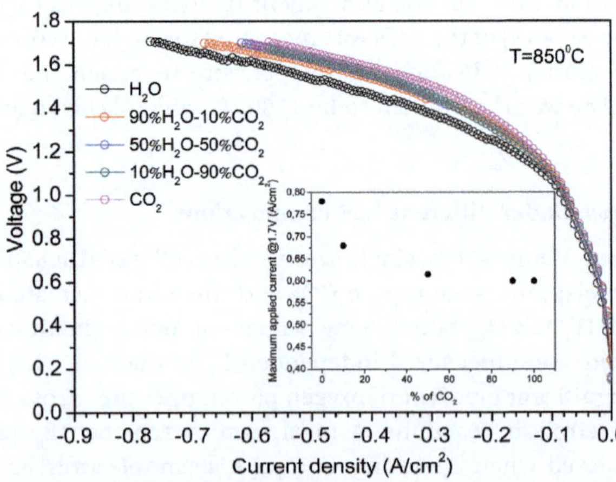

Fig. 4 *I–V* curves of a symmetrical SOEC measured at 850 °C for different hydrogen-free fuel compositions. The insert shows the maximum current densities obtained at 1.7 V as a function of the CO_2 concentration.

As previously mentioned, the whole set of measurements was repeated by adding safe gas (6% H_2) to the original fuel composition (Fig. 5). OCV values close to 0.8 V were observed independently on the fuel composition. These values are below the theoretical OCVs calculated by assuming a Nernst cell with a gradient of chemical potential based on the inlet fuel composition (OCV = 0.94 V) at the operation temperature and 6% of H_2. OCV cannot be accurately predicted by the standard calculations in the presence of water and CO_2 due to the effect of the reverse water shift reaction taking place all along the test station (RWSR: $CO_2 + H_2 \rightarrow H_2O + CO$).[29] The best performance of the cell was achieved under pure steam electrolysis conditions (96% H_2O–6% H_2), injecting a current density of 600 mA cm^{-2} at 1.7 V. Similar to hydrogen-free tests, the maximum ohmic resistance of the *I–V* curves was observed for the pure CO_2. However, in this case, there is not clear correlation between the evolution of the maximum power and the CO_2 concentration in the fuel (inset Fig. 5). This is probably due to the promotion of the RWSR, *i.e.* the CO_2 dissociation, by the addition of the hydrogen safe gas in the inlet.

Comparison of the *I–V* curves presented in Fig. 4 and 5 allows quantification of the effect of the addition of the safe gas in the cell performance. Cells operated with hydrogen-free fuels show even better performance. The lower performance of the cell in the presence of hydrogen can be ascribed to the reduction of the LSCM.[21] Maximum values of injected current densities are higher than previously published symmetrical cells and similar to reported values for classical cell configurations in co-electrolysis mode.[4,10,30–32] This indicates that the use of YbScSZ electrolyte supported cells with symmetrical electrodes of LSCM results in remarkable performances and allows the elimination of the use of safe gas. This makes the system simpler and intrinsically protected from gas supply interruptions or unexpected incidents during operation.

Products composition in co-electrolysis mode

The composition of gas produced by LSCM/YbScSZ/LSCM cell operating in co-electrolysis mode was analyzed by μ-GC and compared for different inlet fuel

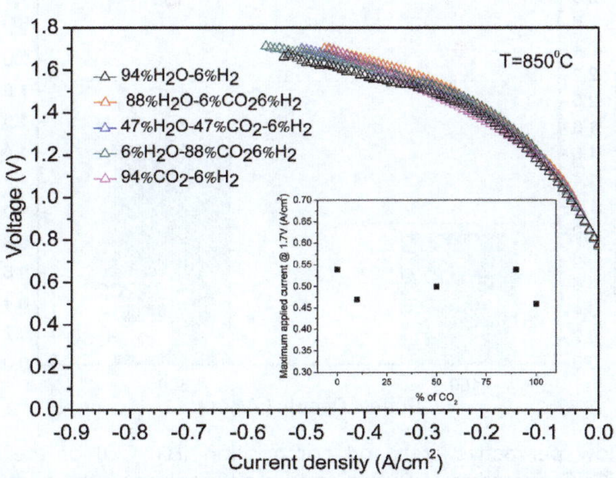

Fig. 5 *I–V* curves of a symmetrical SOEC measured at 850 °C for different fuel compositions containing safe gas (6% H_2). The insert shows the maximum current densities obtained at 1.7 V as a function of the CO_2 concentration.

compositions at 900 °C. Figures 6–9 show the gas flow produced per active area of the cell and (if applicable) the $H_2 : CO$ ratio at two applied current densities (150 mA cm^{-2} and 300 mA cm^{-2}) for different fuel compositions. In all the cases, the total produced gas is below the values predicted by Faraday's law (dashed lines in the plots). Since the thermodynamics do not predict other reactions than the RWSR and the electrolysis of H_2O and CO_2 at the operation temperature and initial fuel compositions, the total amount of gas produced should accomplish the predicted values. The observed disagreement can be associated to small gas or current leakages. Although not usually considered, the application of high voltages through the electrolyte generates a significant electronic current (even in wide electrolytic domain materials such as stabilized zirconia[33]). The phase stability and electronic transport of the employed YbScSZ electrolyte under high applied voltages is currently under study for perfectly matching expected and measured gas productions.

Fig. 6 shows the H_2 and CO generation for an atmosphere with the composition 50% H_2O–50% CO_2. Production rates per active area of 0.67 ml min^{-1} cm^{-2} of H_2 and 0.32 ml min^{-1} cm^{-2} of CO were detected when applying 150 mA cm^{-2}. Flows are almost twice these values when the current density is doubled (300 mA cm^{-2}), reaching production rates for H_2 and CO of 1.13 ml min^{-1} cm^{-2} and 0.54 ml min^{-1} cm^{-2}, respectively. In both cases, $H_2 : CO$ ratios between 2.1 and 2.3 are maintained. Ratios close to 2 are highly desired for the synthetic fuel production using syngas as a precursor.[10] However, although co-electrolysis at low temperatures reaches the 2 : 1 ratio under regular working conditions,[6,16,34] the operation at high temperatures requires low H_2 concentration in the inlet fuel to maintain high $H_2 : CO$ values. This is due to the fact that, at high temperatures, the RWSR can play an important role when a large amount of H_2 is added or high current densities are applied.

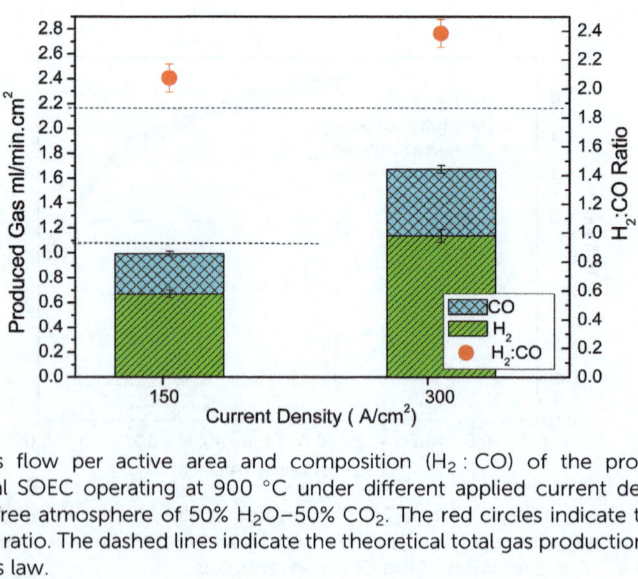

Fig. 6 Gas flow per active area and composition ($H_2 : CO$) of the products of a symmetrical SOEC operating at 900 °C under different applied current densities in a hydrogen-free atmosphere of 50% H_2O–50% CO_2. The red circles indicate the $H_2 : CO$ production ratio. The dashed lines indicate the theoretical total gas production according to Faraday's law.

The composition of the produced gas was drastically modified when H_2 was introduced as the safe gas in the inlet fuel (see Fig. 7). For injected density currents of 150 mA cm^{-2}, the production rates of H_2 and CO are 0.29 ml min^{-1} cm^{-2} and 0.83 ml min^{-1} cm^{-2}, respectively, achieving a very low $H_2 : CO$ ratio (0.35). Similarly, when doubling the current density to 300 mA cm^{-2}, a small ratio of $H_2 : CO = 0.67$ is observed (with production rates of 0.73 ml min^{-1} cm^{-2} of H_2 and 1.08 ml min^{-1} cm^{-2} of CO). This behavior can be explained by means of the reverse water shift reaction. The high amount of H_2 introduced as a safe gas, promotes the displacement of the RWSR to the production of CO. Higher production of CO makes the $H_2 : CO$ ratio decrease. RWSR is typically reported for Ni electrodes,[6,8,10,13,17] but it is clearly evidenced in the present work through the remarkable observed changes in the gas production selectivity. The RWSR effect becomes more evident at low current densities because the added H_2 represents a higher fraction of the total amount of H_2 produced by the steam electrolysis. The relative increase of CO in the produced gas composition is roughly in agreement with the thermodynamics equilibrium calculation at operation temperatures of 900 °C and 6% enriched H_2 fuel.[34] The role of the RWSR can be so important that it even questions the existence of real electrochemical co-electrolysis. Some authors suggest that a maximum of 10% of CO is produced by RWSR at 800 °C[35] while others conclude that only at high limiting current densities the CO_2 contributes to the electrochemical performance and the main production of CO is due to the thermo-chemical path.[36]

Fig. 8 shows the gas production rates for pure CO_2 electrolysis. CO flows of 0.85 ml min^{-1} cm^{-2} and 2.19 ml min^{-1} cm^{-2} were measured at current densities of 150 mA cm^{-2} and 300 mA cm^{-2}, respectively. CO production becomes remarkable due to the increasing interest of CO_2 recirculation and revalorization. CO_2 electrolysis or co-electrolysis are some of the main routes to minimize the emissions of the produced CO_2 obtaining a product such as CO with high added value.

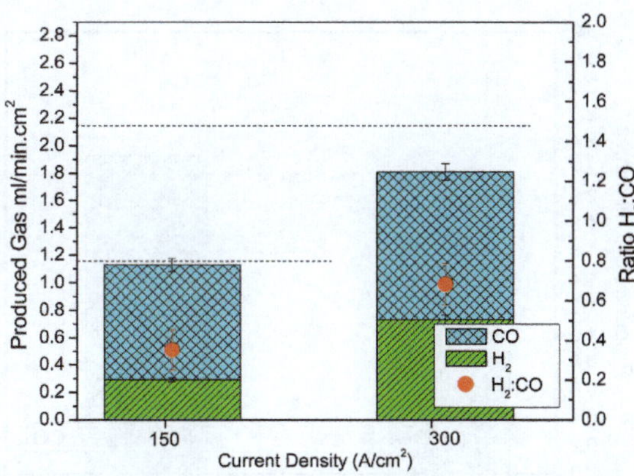

Fig. 7 Gas flow per active area and composition ($H_2 : CO$) of the products of a symmetrical SOEC operating at 900 °C under different applied current densities in an atmosphere of 47.5% H_2O–47.5% CO_2–6% H_2 atmosphere. The red circles indicate the $H_2 : CO$ production ratio. The dashed lines indicate the theoretical total gas production according to Faraday's law.

Degradation of the cell in co-electrolysis mode

Fig. 9 shows the voltage evolution with time for a LSCM/YbScSZ/LSCM cell measured at 850 °C under galvanostatic conditions (300 mA cm^{-2}) in a hydrogen-free atmosphere with composition 50% H_2O–50% CO_2. The voltage evolution observed in the galvanostatic test is in agreement with the observed evolution of the *I–V* curves (Fig. 10) obtained at different times of cell operation. From the starting *I–V* curve, it is possible to conclude a similar behavior to previously showed cells with current densities of 650 mA cm^{-2} at 1.6 V. An accelerated degradation is clearly observed after a few hours of operation. It has to be highlighted that the performance of the cell at the starting point is in accordance with the reported performance in co-electrolysis.[8,10,11]

Further analyses performed by EIS in the operation point (Fig. 11) show that the ohmic resistance is the main contribution of the dramatic cell degradation that takes place. Although the degradation during the first hour can be mainly attributed to an increase in the polarization resistance of the electrodes, the rapid increase of the ohmic resistance observed in EIS is clearly linked to the final failure. The initial ohmic resistance of 0.26 Ω cm^2 reaches values of 1.25 Ω cm^2 after 6.5 h of test (prior to collapse).

Post-mortem cross-section SEM images of the cell clearly point the focus of degradation in the electrolyte–oxygen electrode interface (Fig. 12). Fig. 12a shows the oxygen–electrode/electrolyte interface before testing, while Fig. 12b shows the equivalent interface after 7 h of operation. There is clear delaminating of the oxygen electrode likely due to the increase of oxygen pressure on the interface as previously described by other authors for the co-electrolysis mode[11,37–39] and widely discussed on pure steam electrolysis.[41–44] Electrolyzer cells under operation are forced to generate a given amount of gas proportional to the current density applied. When the cathode is not catalytically active enough to transform the O^{2-} ions on O_2, or there is a poor diffusion of generated O_2, an over-pressure of oxygen is created. This over-pressure starts to mechanically damage the interface leading to small pores or generating micro-cracks. This mechanism reduces the contact

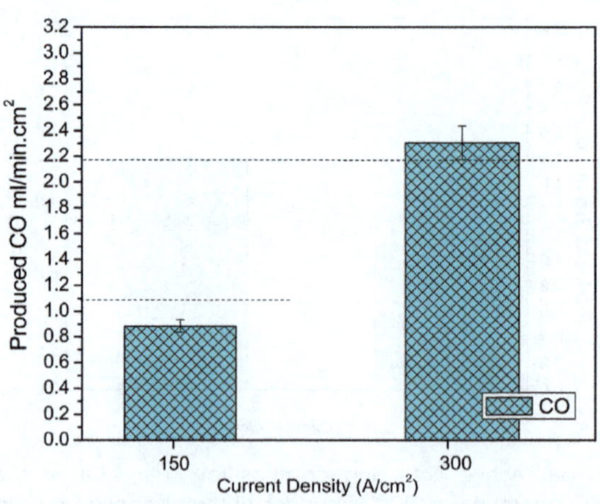

Fig. 8 CO flow per active area of a symmetrical SOEC operating at 900 °C under different applied current densities in a pure CO_2 atmosphere. The dashed lines indicate the theoretical total gas production according to Faraday's law.

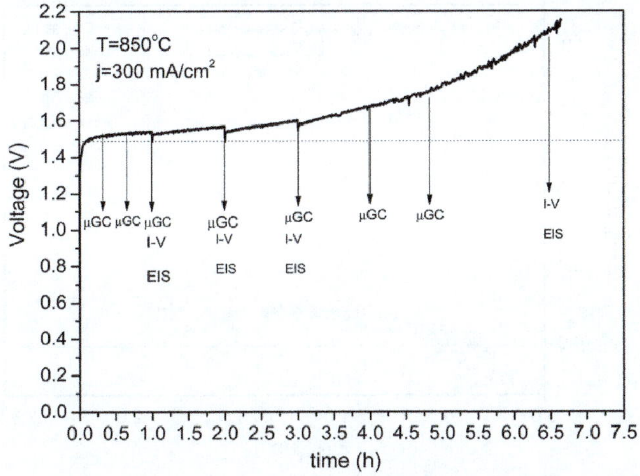

Fig. 9 Evolution of the voltage with time for a symmetrical cell operated under galvanostatic conditions (300 mA cm^{-2}) at 850 °C in 50% H_2O–50% CO_2 atmosphere. The arrows indicate the moment at which gas-chromatography analysis, *I*–*V* curves or EIS measurements were carried out. The solid line represents the thermoneutral potential for co-electrolysis.

area between electrode and electrolyte exponentially. Reduction of contacting area increases the current density on the remaining contacting paths between electrode and electrolyte, accelerating the degradation process. Eventually, the initial generated micro porosity ends as a macroscopic discontinuity between the electrode and the electrolyte. This mechanism is consistent with the evolution of the impedance spectra observed from the LSCM symmetrical cells since the initial polarization resistance increase could correspond to a decrease of the active

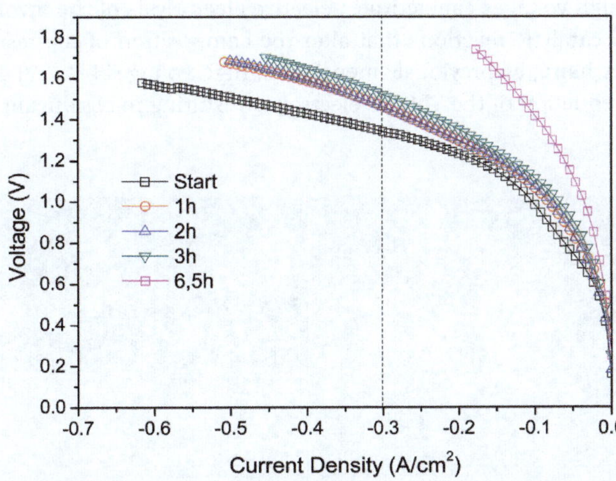

Fig. 10 *I*–*V* curves of the symmetrical cell at different times during the galvanostatic aging test (at a fixed current density of 300 mA cm^{-2}). The cell was operated at 850 °C and in 50% H_2O–50% CO_2 hydrogen-free atmosphere.

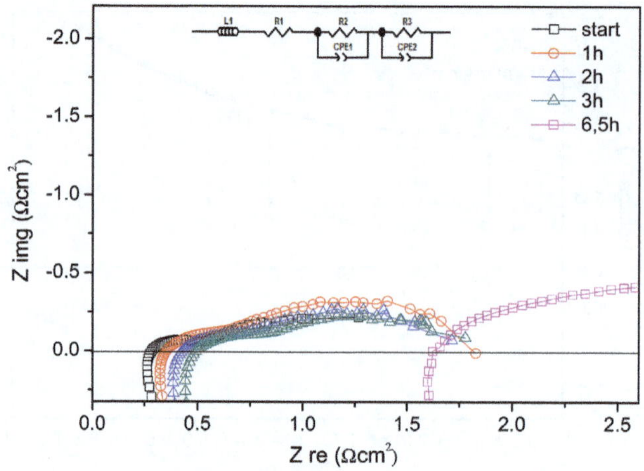

Fig. 11 Nyquist plot of the evolution with time of the impedance spectra obtained for the symmetrical cell during the galvanostatic test at a fixed current density of 300 mA cm^{-2} at 850 °C in 50% H_2O–50% CO_2 hydrogen-free atmosphere.

oxygen reduction points on the oxygen electrode.[40,43] The final sudden increase of the ohmic resistance can be ascribed to the located depletion of the oxygen electrode. The depleted interface shown in Fig. 12b is representative of the effect of the described mechanism but is not present along the entire interface and it is probably magnified by the dismantling of the cell from the measurement station.

A complementary study of the evolution of the composition of the produced gas with time confirms the values of H_2 and CO production and $H_2 : CO$ ratio presented in the previous section (Fig. 13). Two effects can be observed with the operation time: (i) a decrease of the hydrogen-to-carbon monoxide and (ii) a decrease of the gas production rate. Both phenomena could be explained by the continuous increase of the operation voltage due to the degradation process. On one hand, high voltages can activate selective electrolysis of the involved species and parallel catalytic reactions that alter the composition of the resulting gas.[30] On the other hand, as previously mentioned, high voltages (>1.7 V) can produce the partial reduction of the YbScSZ electrolyte resulting in significant leakages of

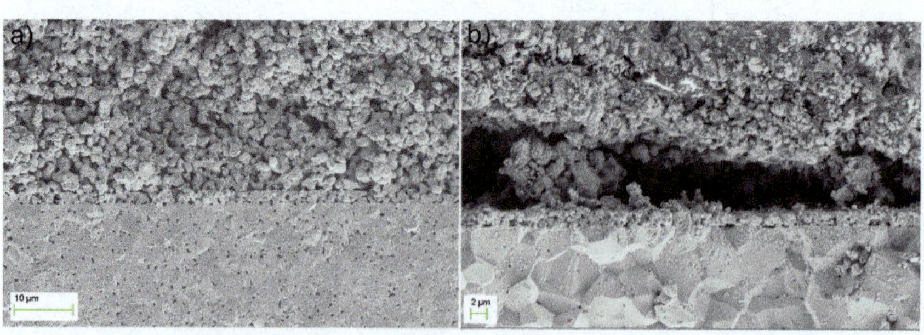

Fig. 12 SEM cross sections of the LSCM/YbScSZ interlayer (a) before and (b) after 6.5 h of galvanostatic test at a fixed density current of 300 mA cm^{-2} at 850 °C in 50% H_2O–50% CO_2 hydrogen-free atmosphere.

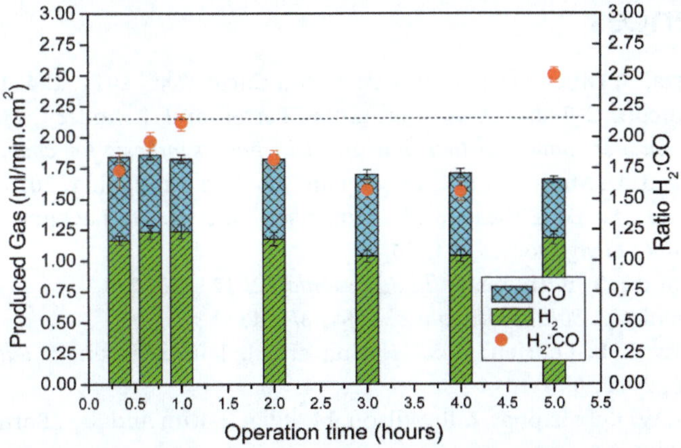

Fig. 13 Evolution of the H_2 and CO production, and their relative ratio, during the first 5 hours of the galvanostatic test at a fixed density current of 300 mA cm^{-2} at 850 °C in 50% H_2O–50% CO_2 hydrogen-free atmosphere.

the applied current[33] (electrons going through the electrolyte without promoting any reaction). In this sense, some evidences of the creation of a monoclinic phase have been found by XRD analysis (ESI Fig. S1†) of the YbScSZ electrolyte after operation at voltages higher than 1.9 V. This phase transformation has been related in the past with the electrolyte reduction and an increase of the electronic conductivity.[44]

The strong delaminating and poor durability observed in the galvanostatic tests presented here should encourage further investigation on (i) ways for improving the catalytic activity of the LSCM electrode by, for instance, infiltration of functional active layers or improved microstructures[10,45] and (ii) reversible operation as recently proposed by Graves *et al.*[38]

Conclusions

A good electrochemical performance of full-ceramic LSCM/YbScSZ/LSCM SOEC cells was demonstrated in co-electrolysis mode for different fuel compositions. It was shown that no safe gas is required when using symmetrical full ceramic SOEC cells. The addition of H_2 in the initial fuel composition produces undesirable effects on the final H_2 : CO ratios and could reduce the overall performance of the cell. Current densities above 700 mA cm^{-2} at 1.7 V and 850 °C were applied in symmetrical cells for co-electrolysis tests based on hydrogen-free fuels. Moreover, ratios of H_2 : CO about 2 were obtained, which are required for the synthesis of standard hydrocarbon fuels or alcohols. The study suggest that H_2 : CO ratio can be tuned by changing the operation temperature, initial fuel composition and operation voltages. Despite these good results, significant degradation issues have been observed under galvanostatic operation conditions caused by the delaminating of the oxygen–electrolyte interface. This could be associated to an increase of the internal oxygen pressure close to the surface of the oxygen electrode. The use of more stable and more catalytically active functional layers is proposed as a solution for minimizing the delaminating problem together with the easy-to-integrate in symmetrical cells reversible operation mode (SOFC/SOEC).

References

1 M. Aresta, A. Dibenedetto and A. Angelini, *Chem. Rev.*, 2014, **114**, 1709.

2 A. Tarancón, C. Fábrega, A. Morata, M. Torrell and T. Andreu, *Materials for energy, chapter: power to fuel and artificial photosynthesis for chemical energy storage*, ed. D. Muñoz and X. Moya, Pan Stanford Publishing, 2015.

3 C. Graves, S. D. Ebbesen, M. Mogensen and K. S. Lackner, *Renewable Sustainable Energy Rev.*, 2011, **15**, 1.

4 X. Yue and J. T. S. Irvine, *Solid State Ionics*, 2012, **225**, 131.

5 A. O. Isenberg, *Solid State Ionics*, 1981, **3/4**, 431.

6 C. Stoots, J. E. O'Brien, J. S. Herring and J. J. Hartvigsen, *J. Fuel Cell Sci. Technol.*, 2008, **6**, 011014.

7 Z. Zhan, W. Kobsiriphat, J. R. Wilson, M. Pillai, I. Kim and S. A. Barnett, *Energy Fuels*, 2009, **23**, 3089.

8 C. Stoots, Production of Synthesis Gas by H_2O and CO_2 (Coelectrolysis), *Sustainable Fuels from CO_2, H_2O and Carbon Free Energy*, Columbia University, New York, May 4, 2010.

9 S. E. Yoon, S. H. Song, J. Choi, J. Y. Ahn, B.-K. Kim and J. S. Park, *Int. J. Hydrogen Energy*, 2014, **39**, 5497.

10 C. Stoots, J. O'Brien and J. Hartvigsen, *Int. J. Hydrogen Energy*, 2009, **34**, 4208.

11 P. Kim-Lohsoontorn and J. Bae, *Proc. 2010 Eur. Solid Oxide Fuel Cell Forum*, 2011, vol. 196, p. 7161.

12 S. D. Ebbesen, C. Graves and M. Mogensen, *Int. J. Green Energy*, 2009, **6**, 646.

13 M. Ni, M. Leung and D. Leung, *Int. J. Hydrogen Energy*, 2008, **33**, 2337–2354.

14 C. R. Graves, *Recycling CO_2 into Sustainable Hydrocarbon Fuels: Electrolysis of CO_2 and H_2O*, Ph.D by, Columbia University, 2010.

15 J. P. Stempien, O. L. Ding, Q. Sun and S. H. Chan, *Int. J. Hydrogen Energy*, 2012, **37**, 14518.

16 Q. Fu, C. Mabilat, M. Zahid, A. Brisse and L. Gautier, *Energy Environ. Sci.*, 2010, **3**, 1382.

17 C. Graves, S. D. Ebbesen and M. Mogensen, *Proc. 17th Int. Conf. Solid State Ionics*, 2011, vol. 192, p. 398.

18 S. Uhm and Y. D. Kim, *Curr. Appl. Phys.*, 2014, **14**, 672.

19 F. Bidrawn, G. Kim, G. Corre, J. T. S. Irvine, J. M. Vohs and R. J. Gorte, *Electrochem. Solid-State Lett.*, 2008, **11**, B167.

20 S. Tao and J. T. S. Irvine, *Nat. Mater.*, 2003, **2**, 320.

21 X. Yang and J. T. S. Irvine, *J. Mater. Chem.*, 2008, **18**, 2349.

22 T. Delahaye, T. Jardiel, O. Joubert, R. Laucournet, G. Gauthier and M. T. Caldes, *Solid State Ionics*, 2011, **184**, 39.

23 D. M. Bastidas, S. Tao and J. T. S. Irvine, *J. Mater. Chem.*, 2006, **16**, 1603.

24 Q. Liu, X. Dong, G. Xiao, F. Zhao and F. Chen, *Adv. Mater*, 2010, **22**, 5478.

25 C. M. Stoots, J. E. O'Brien, K. G. Condie and J. J. Hartvigsen, *Int. J. Hydrogen Energy*, 2010, **35**, 4861.

26 O. Yamamoto, *Solid State Ionics*, 1995, **79**, 137.

27 V. Vijaya Lakshmi and R. Bauri, *Solid State Sci.*, 2011, **13**, 1520.

28 W. Li, H. Wang, Y. Shi and N. Cai, *Int. J. Hydrogen Energy*, 2013, **38**, 11104.

29 J. P. Stempien, Q. Liu, M. Ni, Q. Sun and S. H. Chan, *Electrochim. Acta*, 2014, **147**, 490.

30 Y. Tao, S. D. Ebbesen and M. B. Mogensen, *J. Electrochem. Soc.*, 2014, **161**(3), F337–F343.
31 X. Yue and J. T. S. Irvine, *J. Electrochem. Soc.*, 2012, **159**, F442.
32 S. D. Ebbesen and M. Mogensen, *Fuel Cell Syst.*, 2009, **193**, 349.
33 T. H. Etsell and S. N. Flengas, *J. Electrochem. Soc.*, 1972, **119**, 1.
34 W. L. Becker, R. J. Braun, M. Penev and M. Melaina, *Asia-Pacific Forum on Renewable Energy 2011*, 2012, **47**, 99.
35 J. Aicart, M. Petitjean, J. Laurencin, L. Tallobre and L. Dessemond, *Int. J. Hydrogen Energy*, 2015, **40**, 3134–3148.
36 S. W. Kim, H. Kim, K. J. Yoon, J. H. Lee, B. K. Kim, W. Choi, J.-H. Lee and J. Hong, *J. Power Sources*, 2015, **280**, 630.
37 A. V. Virkar, *Int. J. Hydrogen Energy*, 2010, **35**, 9527.
38 C. Graves, S. D. Ebbesen, S. H. Jensen, S. B. Simonsen and M. B. Mogensen, *Nat. Mater.*, 2015, **14**, 239–244.
39 R. Knibbe, M. L. Traulsen, A. Hauch, S. D. Ebbesen and M. Mogensen, *J. Electrochem. Soc.*, 2010, **157**, B1209.
40 P. Hjalmarsson, X. Sun, Y.-L. Liu and M. Chen, *J. Power Sources*, 2013, **223**, 349.
41 T. Jacobsen and M. Mogensen, in *ECS Trans.*, ECS, 2008, pp. 259–273.
42 F. Tietz, D. Sebold, A. Brisse and J. Schefold, *J. Power Sources*, 2013, **223**, 129.
43 J. I. Gazzarri and O. Kesler, *J. Power Sources*, 2007, **167**, 100.
44 A. Kumar, D. Rajdev and D. L. Douglas, *J. Am. Ceram. Soc.*, 1972, **117**, 378.
45 J. C. Ruiz-Morales, D. Marrero-López, J. Canales-Vázquez and J. T. S. Irvine, *RSC Adv.*, 2011, **1**, 1403.

Faraday Discussions

PAPER

Improved electrochemical stability at the surface of $La_{0.8}Sr_{0.2}CoO_3$ achieved by surface chemical modification

Nikolai Tsvetkov,[ab] Qiyang Lu[ac] and Bilge Yildiz*[abc]

Received 15th February 2015, Accepted 9th April 2015
DOI: 10.1039/c5fd00023h

The degradation of the surface chemistry on perovskite (ABO_3) oxides is a critical issue for their performance in energy conversion systems such as solid oxide fuel/electrolysis cells and in splitting of H_2O and CO_2 to produce fuels. This degradation is typically in the form of segregation and phase separation of dopant cations from the A-site, driven by elastic and electrostatic energy minimization and kinetic demixing. In this study, deposition of Ti at the surface was found to hinder the dopant segregation and the corresponding electrochemical degradation on a promising SOFC cathode material, $La_{0.8}Sr_{0.2}CoO_3$ (LSC). The surface of the LSC films was modified by Ti (denoted as LSC-T) deposited from a $TiCl_4$ solution. The LSC and LSC-T thin films were investigated by electrochemical impedance spectroscopy, nano-probe Auger electron spectroscopy, and X-ray photoelectron spectroscopy (XPS), upon annealing at 420–530 °C in air up to about 90 hours. The oxygen exchange coefficient, k^q, on LSC-T cathodes was found to be up to 8 times higher than that on LSC cathodes at 530 °C and retained its stability. Sr-rich insulating particles formed at the surface of the annealed LSC and LSC-T films, but with significantly less coverage of such particles on the LSC-T. From this result, it appears that modification of the LSC surface with Ti reduces the segregation of the blocking Sr-rich particles at the surface, and a larger area on LSC surface (with a higher Sr doping level in the lattice) is available for the oxygen reduction reaction. The stabilization of the LSC surface through Ti-deposition can open a new route for designing surface modifications on perovskite oxide electrodes for high temperature electro- and thermo-chemical applications.

Introduction

The surface chemical composition of functional oxides is an important factor that determines the surface reaction kinetics and affects the performance in multiple

[a]Laboratory for Electrochemical Interfaces, Massachusetts Institute of Technology, 77 Massachusetts Avenue, Cambridge, Massachusetts 02139, USA. E-mail: byildiz@mit.edu

[b]Department of Nuclear Science and Engineering, Massachusetts Institute of Technology, 77 Massachusetts Avenue, Cambridge, Massachusetts 02139, USA

[c]Department of Material Science and Engineering, Massachusetts Institute of Technology, 77 Massachusetts Avenue, Cambridge, Massachusetts 02139, USA

applications, including solid oxide fuel and electrolysis cells (SOFCs, SOECs),[1–11] thermochemical splitting of H_2O and CO_2,[12] oxygen permeation membranes,[13,14] batteries,[15–18] magnetic,[19–22] and catalytic[23,24] devices. For SOFC and SOEC applications, cation composition on perovskite oxide surfaces governs the reactivity and stability for oxygen reduction (ORR) and oxygen evolution (OER) reactions.[1–3,5,6,8–10,25,26] Thus, to achieve SOFCs and SOECs with high energy conversion efficiencies and long-term durability, it is important to design surfaces that are highly reactive to these reactions and have high stability. The mixed ionic and electronic conducting cobaltites, including $La_{1-x}Sr_xCoO_3$, $La_{1-x}Sr_xCo_{1-y}Fe_yO_3$ and $Ba_{1-x}Sr_xCo_{1-y}Fe_yO_3$ perovskite oxides, are state-of-the-art cathode materials which have been intensely studied and optimized to enhance the oxygen transport kinetics and surface exchange kinetics at intermediate temperatures.[27–29] One of the most promising materials is $La_{0.6}Sr_{0.4}CoO_3$, which has a very low area specific resistance (ASR) of 0.01 Ohm cm^2 at 600 °C.[27] However, recent work has shown that the surface of such perovskite oxides is not stable and undergoes segregation and phase separation of dopant oxides, typically at temperatures above 400 °C.[30–32] Similar chemical degradation of perovskite oxide surfaces was observed also for other applications of these materials, such as in magnetism,[33,34] superconductors[35] or memristors.[36,37] The consequence of such surface chemical degradation can be severe; for example, on $La_{0.6}Sr_{0.4}CoO_3$ thin film cathodes the polarization resistance of the surface was found to increase by up to 2 orders of magnitude within 50 hours at 550 °C,[9,29] correlated with the segregation and separation of SrO-related insulating phases at the surface.[7,34,38–40]

Several recent studies have attempted to prevent the surface segregation and electrochemical degradation in SOFC cathodes.[41–43] One of the proposed ways is the coating of $La_{0.8}Sr_{0.2}CoO_3$ (ref. 37) or $La_{0.6}Sr_{0.4}Co_{0.2}Fe_{0.8}O_3$ (ref. 41) electrodes with $La_{0.8}Sr_{0.2}MnO_3$. It was also shown that coating of the $La_{0.8}Sr_{0.2}CoO_3$ surface with a 5 nm thick ZrO_2 layer helps to stabilize the surface and improve electrochemical performance.[41] However the exact mechanisms on how these surface modifications improve the cathode stability have not yet been made clear, making it difficult to go beyond these empirical observations.

In our recent work, we have shown that the key driving forces behind dopant segregation at perovskite oxide surfaces are the elastic and electrostatic interactions of the dopant with the surrounding lattice.[32] The size mismatch between the dopant and host cations and the associated elastic energy minimization push the larger or smaller dopants to free surfaces or interfaces.[21,22,44,45] On the other hand, oxygen vacancies and the net positive charge at the surface of perovskite oxides[46] drive the negatively charged dopants in the form of point defects, such as Sr'_{La} or Ba'_{La}, to the surface. Based on our understanding of these two key mechanisms, we propose that the secondary phase segregation can be hindered by choosing suitable dopants. The total elastic energy driver can be minimized by a dopant that is similar in size to the host; for example, La on the La site was shown to minimize the surface segregation on $LaMnO_3$ cathode films.[32]

To minimize the electrostatic driver to the surface, the excess positive charge of the surface must be reduced. While the oxygen vacancies at the surface are important in determining the total reactivity to ORR,[47] they also contribute to attracting the aliovalent dopants, such as Sr'_{La} or Ba'_{La} defects, to the surface. This paper aims to test the hypothesis that decreasing the reducibility of the surface (thereby, reducing the concentrating of positively charged oxygen vacancies at the surface) can decrease the amount of Sr segregation and formation of blocking SrO-related phases. We modify the $La_{0.2}Sr_{0.8}CoO_3$ (LSC) film surfaces with a small amount of Ti deposited from a chemical bath at room temperature.[48–50] These films are denoted as LSC-T in this paper. We found that both the surface chemical stability and the electrochemical performance of the LSC-T films were superior compared to the unmodified LSC films. We propose that Ti-modification of the surface is one candidate for enabling better electrochemical performance and stability on cobaltite based ORR or OER electrodes. We believe that the enhanced stability of the LSC-T cathodes is due to the weakening of the electrostatic attraction of the to the surface. Indeed, the vacancy formation energy in Ti-oxides[51] is significantly higher than in LSC.[52] Other surface compositions to modulate the surface electrostatic driving force to segregation are currently being investigated systematically in our ongoing work.

Experimental

Dense LSC films were deposited onto substrates using a pulsed laser deposition (PLD) method with a KrF excimer laser of 248 nm wavelength. The $La_{0.8}Sr_{0.2}CoO_3$ target was purchased from MTI Corp., USA. The films were deposited at 650 °C under an oxygen pressure of 10 mTorr on the single crystalline $Yr_{0.08}Zr_{0.92}O_2$ (YSZ) substrates (MTI Corp., USA) with 20 nm $Gd_{0.2}Ce_{0.8}O_2$ interlayer grown with the same conditions as LSC films. The LSC film thickness was around 40 nm. After the growth process, the films were cooled down to room temperature in 2 Torr oxygen. To remove excess Sr and to deposit Ti at the surface of LSC thin films, the samples were dipped into the 2 mM $TiCl_4$ solution for 5–30 s at room temperature.[49,53] For samples which were not treated with Ti, the excess surface Sr-rich phase formed during the PLD process was removed by dipping of the films into a 0.1 M HCl aqueous solution for 10 s at room temperature.[54]

The Helios Nanolab 600 dual beam scanning electron microscope operated at a beam voltage of 15 keV and a current of 86 pA was used for imaging the surface morphology. A Veeco/Digital Instrument Nanoscope IV was used to perform tapping mode atomic force microscopy (AFM) for characterizing the surface morphology.

A Physical Electronics Model 700 scanning nanoprobe Auger electron spectroscopy (AES) instrument was used to identify lateral heterogeneities in cation compositions with high spatial resolution at the surface. Electron beam settings of 20 keV and 10 nA were used for both SEM imaging and the Auger electron excitation. The La MNN, Sr LMM, and Co LMM Auger emissions were measured for quantifying the surface cation composition of the LSC films. The sampling depths of these Auger electrons are ~8.0 nm for Sr LMM, ~4.0 nm for La MNN, and ~4.5 nm for Co LMM. The sampling depths for both the Auger and X-ray photoelectrons are estimated based on 3 times the inelastic mean free path (3× IMFP) of emitted electrons. However, we note that the Auger electron and photoelectron spatial profiles are exponentially dependent on the distance to the surface, so the data is always more sensitive to the near-surface region. It was not possible to quantify Ti due to low Ti content and overlapping of the most intense Ti Auger emissions with secondary Auger emissions of Sr and La. The smoothing and differentiation of the AES spectra collected were carried out using the Savitsky–Golay algorithm. Quantification of the AES differential spectra is performed using peak-to-peak intensities of the tight-scans of the noted emissions from the constituent cations. The AES sensitivity factors were obtained by using the fractured surface of the LSC target as an internal standard of stoichiometry.

X-ray photoelectron spectroscopy (XPS) measurements have been performed in order to estimate the surface cation composition using a Perkin-Elmer PHI-5500 ESCA Spectrometer with monochromated Al Kα (1486.65 eV) X-ray radiation under a base pressure of 10^{-9} Torr. Angle-resolved XPS measurements were performed by changing the emission angle from 0° to 70° relative to the surface normal. The probing depth (3× IMFP) at 0° was 6 nm, 5 nm, 4 nm and 4 nm for Sr 3d, Ti 2p, La 3d, and Co 2p, respectively. Tilting the sample to the higher emission angle, α, decreases the effective probing depth by $\sin(\alpha)$. For example, tilting the sample to 70° decreases the probing depth down to 2.0, 1.7, 1.4, and 1.4 nm for Sr 3d, Ti 2p, La 3d, and Co 2p, respectively. The quantitative analysis of the Co 2p, La 3d, Sr 3d, and Ti 2p XPS spectra was performed using Multipack 9.0 software.

Electrochemical impedance spectroscopy (EIS) measurements were performed on asymmetrical cells with the LSC thin film electrodes grown on YSZ single crystal substrates. Dense platinum current collectors in the form of a grid (which covers 75% of electrode surface with 25 μm × 25 μm openings) were deposited on the LSC thin films by means of photolithography and RF sputtering. A porous Ag layer served as the counter electrode. Due to high electrical conductivity of LSC,[55,56] we do not expect a significant potential gradient across the LSC film from one Pt grid to the next. Furthermore, the key point of our paper is the chemical segregation process and its effect on the electrochemical performance. The stage of the experiments that involve surface cation segregation is simply annealing at open circuit voltage, so no effect of electrochemical potential is involved on the segregation itself in this work. A Parstat 2273 potentiostat was used to perform the EIS measurements in the frequency range of 100 kHz to 1 mHz with an AC amplitude of 5 mV and 0 V DC bias from 420 °C to 470 °C and to 520 °C in air. After measuring the EIS response of each film at a given temperature in air for 32 hours, the dependence of the surface exchange resistance on oxygen partial pressure was measured in the range of 1 atm to 10^{-3} atm at the same temperature. Then, the temperature was increased up to the next level and the same measurement sequence was repeated. ZView software was used for the data fitting and analysis.

Results and discussion

Surface morphology on the LSC films with chemical bath deposition of TiO_2

We examined the effect of Ti treatment on the surface morphology of LSC films by imaging these surfaces using scanning electron microscopy (SEM) and AFM. The XPS measurements at a 45° emission angle on the LSC-T films treated with $TiCl_4$ for 5, 10, 20 and 30 s have a Ti content of 2%, 6%, 10% and 13%, respectively, quantified as Ti/(La + Sr + Co + Ti) in their as-prepared state. These samples are denoted as LSC-T2, LSC-T6, LSC-T10, and LSC-T13. Even though surface roughness is relatively small and consistent among the samples (Fig. 1), it is comparable to the IMPF of the Ti photo-electrons measured in this work. Therefore, one can expect an underestimation of the actual Ti content compared to what it should be at the very top surface, especially for low takeoff angle measurements.

Thus, for quantification of Ti, we take the XPS results as "trends" and "lower bounds" rather than actual quantities.

Surface morphology of LSC-T2, T6 and T10 films is similar to that on LSC films treated only with HCl (Fig. 1a–c and e). As seen in the AFM images (Fig. 1a–c), these surfaces do not show any detectable Ti-induced morphological features. Thus, we hypothesize that the Ti is deposited in the form of either a thin and smooth wetting layer of Ti-oxide, or very small nanoparticles with sizes below the instrumental resolution of ~1 nm, or is dissolved into the surface lattice layer of the LSC film. This observation is also in line with the measured root mean square (RMS) roughness values which are almost the same for the LSC and the LSC-T2, LSC-T6 and LSC-T10 films. Further increase in Ti content up to 13% leads to an increase in the film RMS roughness. The increase of the film's roughness is also accompanied by the visible formation of the nanoparticles with a diameter of 3–10 nm at the LSC-T13 surface. It is likely that the particles observed on LSC-T13 are TiO_2.

Surface oxygen exchange kinetics over time

We compared the surface oxygen exchange coefficients, k^q, on LSC and LSC-T thin films as a function of time and temperature. We found that the stabilized k^q values of LSC-T films are 2–8 times higher than that of LSC at 420–530 °C, respectively, with the best results for LSC-T10. A typical impedance response of the cells with the LSC and LSC-T10 thin film electrodes, and the equivalent circuit[57,58] used for modeling the EIS data is shown in Fig. 2a. We obtained the surface polarization resistance, R_S, values from the low frequency impedance

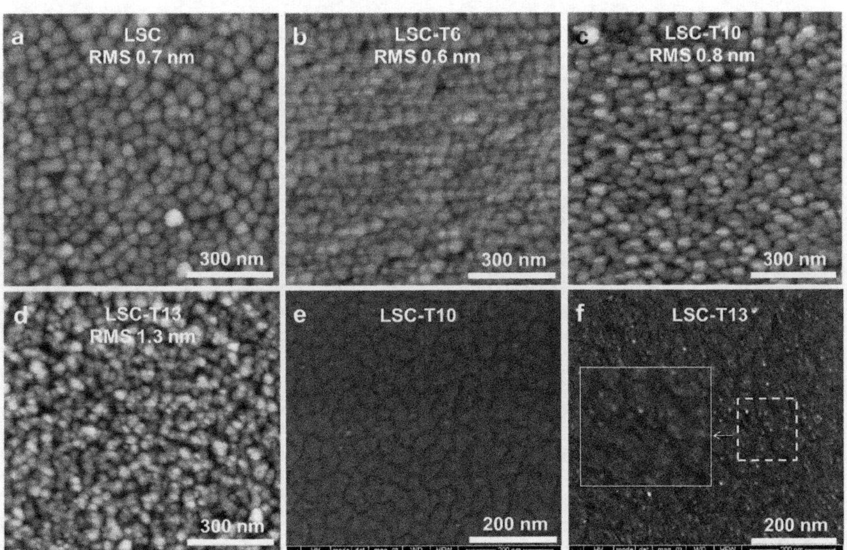

Fig. 1 Surface topography characterized by atomic force microscopy on (a) LSC, (b) LSC-T6, (c) LSC-T10, and (d) LSC-T13. Secondary electron microscopy images of (e) LSC-T10 and (f) LSC-T13 films. The inset in (f) marked with solid lines shows the magnified image of the area marked with the dashed lines.

response, and used those values to calculate the surface oxygen exchange coefficient, k^q, as:[59]

$$k^q = \frac{k_B T}{4e^2 c_O R_S},$$

where k_B is the Boltzmann constant, T is the temperature, e is the electronic charge, and c_O is the total concentration of lattice oxygen determined according to data from the study by Mizusaki *et al.*[60]

The measured k^q values as a function of time and annealing temperature are given in Fig. 2b. At the beginning, at 420 °C, all the samples have similar k^q values. Within the first few hours of annealing, the surface exchange kinetics degrades, and the k^q values decrease on all specimens. The LSC electrodes degrade most severely, with almost a one order of magnitude decrease of k^q within the first 30 hours at 420 °C. The stability of k^q improves by increasing the Ti content from 2% to 10% on LSC-T films. The LSC-T10 cathodes were the most stable and showed the highest k^q, which were 2 and 8 times higher than that on LSC at 420 and 530 °C, respectively. Increasing the Ti content from 10 to 13% degrades the stability of the LSC surface again. This loss in the improvement of stability can be related to the formation of small TiO_2 particles at the surface, as was shown in Fig. 1d and f. When the Ti content is in the range of 2–10%, the Ti-deposition uniformly wets and affects the LSC surface, preventing the degradation of the LSC-T. With the increase in Ti content, the TiO_2 at the surface is aggregated into nanoparticles (Fig. 1d and f), which may not provide complete surface coverage and cannot prevent surface degradation. We explain in the next section that the stability and improvement in the surface oxygen exchange kinetics is correlated to the relatively better stability of the surface cation chemistry.

Surface chemistry of LSC and LSC-T films at high temperature

In order to evaluate the changes in the surface cation chemistry and phase composition upon annealing at elevated temperatures, we performed combined

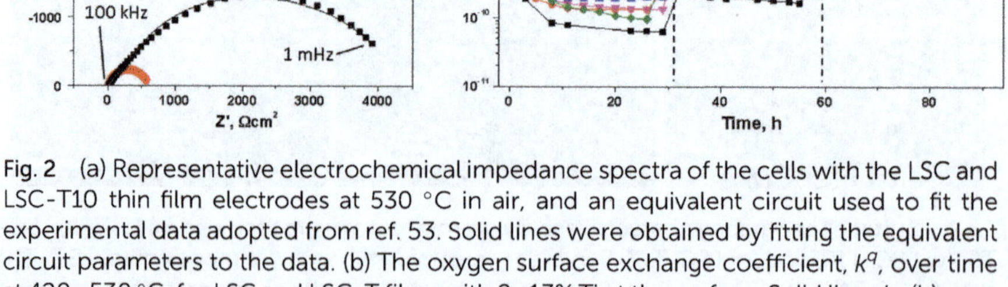

Fig. 2 (a) Representative electrochemical impedance spectra of the cells with the LSC and LSC-T10 thin film electrodes at 530 °C in air, and an equivalent circuit used to fit the experimental data adopted from ref. 53. Solid lines were obtained by fitting the equivalent circuit parameters to the data. (b) The oxygen surface exchange coefficient, k^q, over time at 420–530 °C, for LSC and LSC-T films with 2–13% Ti at the surface. Solid lines in (b) are a guide to the eye.

SEM and AES analysis on the LSC and LSC-T films. Sr-rich and Co-poor particles that were found at the surface of the films served as LSC cathodes. For the LSC, LSC-T2, and LSC-T13 films which demonstrated the lowest k^q, we found that the surfaces are largely covered with Sr-rich particles with sizes varying from several tens up to several hundreds of nanometers. The LSC-T10 surface which demonstrated the highest electrochemical performance and stability had significantly less particle coverage and higher Sr content within the particle-free (non-segregated) areas of the surface.

Fig. 3 shows the SEM images of the surface morphology on LSC and LSC-T films after annealing up to 530 °C in air. These are the same specimens whose surface exchange kinetics were measured by electrochemical impedance spectroscopy, as shown in Fig. 2. The average Sr/(Sr + La) and (Sr + La)/Co ratios deduced from the AES analysis on particles and on particle-free regions at the LSC and LSC-T surfaces are summarized in Table 1. While the as-deposited films (Fig. 1) had a homogeneous surface morphology and chemistry, the annealed LSC and LSC-T film surfaces show the presence of Sr-rich and Co-poor particles (Fig. 3). The segregated particle size varies from few tens up to several hundreds of nm. The Co content was lower and the Sr content was higher within particle-free zones compared to that on the as-fabricated LSC films or the particle-free zones of the annealed films, with the (Sr + La)/Co ratio in the particles measured as 2–3. Because of the difficulty of focusing the AES analysis volume due to sample drift and vibrations, the Sr/(La + Sr) and (Sr + La)/Co ratios on the particles reported in Table 1 should be taken as lower bounds to what the actual ratios should be on the particles. The LSC, LSC-T2, and LSC-T13 films which demonstrated the lowest k^q (Fig. 2) are largely covered with such Sr-rich particles (Fig. 3a, b and e). The LSC-T10 film, which has the fastest surface oxygen exchange kinetics (Fig. 2), had the most stable surface chemistry with the lowest coverage of the Sr-rich particles

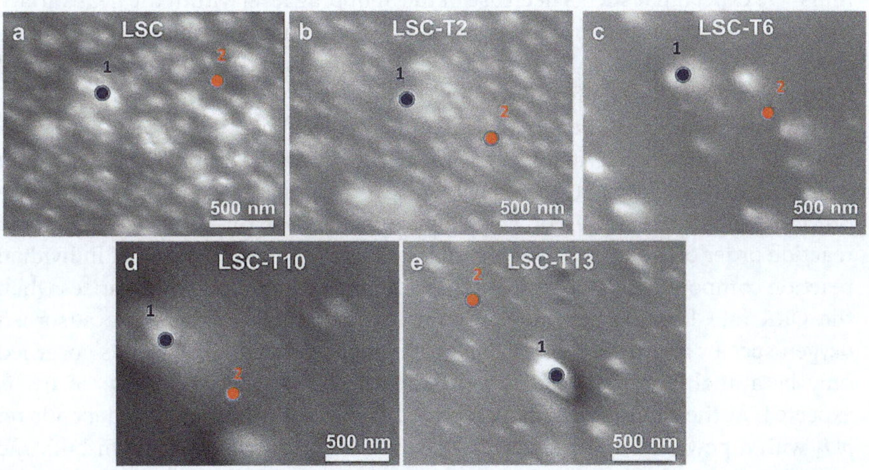

Fig. 3 Scanning electron microscopy images of the surface morphology of (a) LSC, (b) LSC-T2, (c) LSC-T6, (d) LSC-T10, and (e) LSC-T13 film cathodes after electrochemical testing up to 530 °C. Points 1 and 2 on each image indicate the particles and particle-free zones of the surface, respectively, where the chemical composition was probed by Auger electron spectroscopy as summarized in Table 1.

Table 1 Sr/(La + Sr) and (Sr + La)/Co ratios on particles and on particle-free regions on the LSC and LSC-T film cathodes after electrochemical testing at up to 530 °C. The values given in table are averages of several AES measurements at different points. The ± ranges represent the composition variations among different measurement points. The Sr/(La + Sr) and (Sr + La)/Co ratios are 0.20 ± 0.02 and 1.09 ± 0.09 respectively, on all the as-prepared samples before annealing and electrochemical testing up to 530 °C. Because of spatial resolution limits of the surface chemical analysis by the AES nanoprobe, the Sr/(La + Sr) and (Sr + La)/Co ratios on the particles should be taken as lower bounds

Sample	Sr/(Sr + La) (particle-free zone)	Sr/(Sr + La) (particle)	(Sr + La)/Co (particle-free zone)	(Sr + La)/Co (particle)
LSC	0.19 ± 0.02	0.51 ± 0.04	1.19 ± 0.10	2.1 ± 0.5
LSC-T2	0.16 ± 0.03	0.46 ± 0.06	1.30 ± 0.11	2.5 ± 0.9
LSC-T6	0.23 ± 0.02	0.66 ± 0.04	1.11 ± 0.07	2.6 ± 0.5
LSC-T10	0.25 ± 0.04	0.55 ± 0.04	1.04 ± 0.12	2.3 ± 0.5
LSC-T13	0.16 ± 0.04	0.43 ± 0.06	1.53 ± 0.11	2.8 ± 0.9

(Fig. 3d). This film also has a higher Sr fraction within the particle-free areas of the surface, indicating that a higher Sr fraction may remain within the LSC lattice. By combining the results shown in Fig. 2 and 3 and in Table 1, we conclude that the phase separated Sr-rich particles block the surface and increase its resistance toward ORR. This chemical and electrochemical stability at the surface of LSC is enhanced significantly by providing a small amount of Ti to the surface.

Mechanism behind the degradation of surface oxygen kinetics upon surface segregation

An evident effect of the segregated particles, as seen in Fig. 3, is the blockage of the LSC surface by an insulating Sr-rich phase that is inactive to ORR. In addition, formation of such Sr-rich particles decreases the Sr doping level within the LSC films. We expect that such a decrease of the Sr doping level within the near surface region of LSC reduces the availability of oxygen vacancies within LSC and slows down the ORR kinetics. To test this degradation mechanism, we performed measurements of k^q as a function of pO_2.

The double-logarithmic plots of the k^q *vs.* pO_2 for the LSC and LSC-T10 cathodes are given in Fig. 4. At 420 °C, the slope of the log–log plots for the LSC and the LSC-T10 cathodes are 0.38 and 0.43, respectively. These values are close to the previously reported slopes of k^q *vs.* pO_2 as 0.37–0.43 on LSC.[61,62] The overall reaction order of around 0.4 can be predicted from the contribution of individual reaction components.[61] According to mass action laws, a slope of 0.4 arises when the ORR rate limiting step depends on the availability of both the adsorbed oxygen species and of oxygen vacancies. If the rate determining step is governed only by availability of the adsorbed oxygen, the slope of k^q *vs.* pO_2 of 0.5 is expected. At the same time the oxygen vacancy concentration in LSC depends on pO_2 with a power of around −0.1.[60] By combining contributions from both the adsorbed oxygen species and the oxygen vacancy availability as reaction components, the overall reaction order of 0.4 can be deduced.

In the case of LSC in Fig. 4, we think that the change of the slope from 0.38 to 0.20 after annealing at higher temperatures is due to the increased effect of the oxygen vacancy availability as the rate limiting step. Indeed, previous studies on

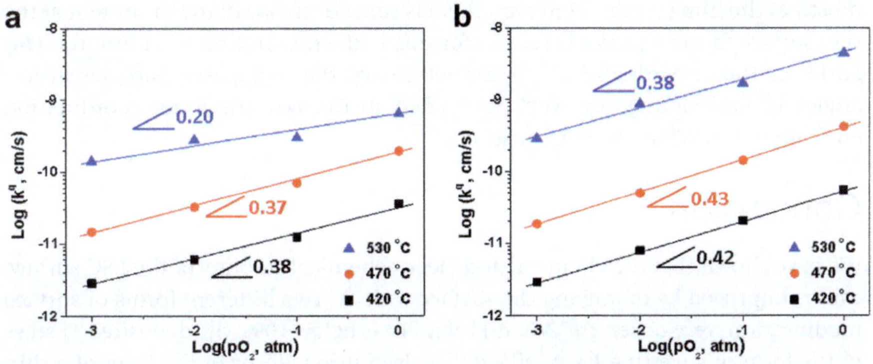

Fig. 4 The oxygen exchange coefficient, k^q, as a function of oxygen partial pressure, pO_2, for (a) LSC and (b) LSC-T10 cathodes at temperatures increasing from 420 °C to 530 °C. Before each pO_2 dependent measurement, the samples were kept at the corresponding temperatures for 32 h in air as shown in Fig. 2.

LSC showed that the power in the dependence of oxygen vacancy concentration on pO_2 becomes stronger with decreasing Sr fraction in LSC: from −0.06 for $La_{0.3}Sr_{0.7}CoO_3$ to −0.50 for $LaCoO_3$.[60] Thus, with the lowering of the Sr content within the LSC lattice upon Sr-rich particle separation (Fig. 3 and Table 1), the ORR rate limiting step, and consequently, k^q, depends more significantly on the oxygen vacancy availability. This effectively decreases the power in the dependence of k^q on pO_2 while slowing down the ORR kinetics at high oxygen partial pressure.

Ti distribution in LSC films

Fig. 5 shows the Ti content as a function of probing depth (photoelectron emission angle) for LSC-T10 in its as-fabricated conditions (described in Experimental section). It can be seen that even after annealing after 32 h, the Ti tends to stay

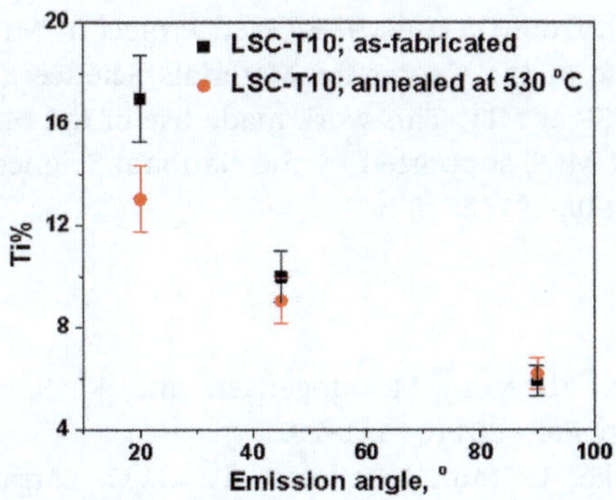

Fig. 5 Ti content quantified from the angle-resolved X-ray photoelectron spectroscopy measurements for the as-prepared LSC-T10 and after annealing at 530 °C in air for 32 h. Increasing the emission angle represents deeper probing depth from the surface. Error bars reflect 10% error generally associated with XPS measurements.

mostly at the film surface. However, there is some decrease in the Ti content at the very surface layer. A possible reason for this is the dissolution of Ti into the LSC lattice or the screening of Ti photoelectrons by the segregated particles at low angles of emission. More work is needed to uncover the exact coordination environment of Ti at the LSC surface.

Conclusion

We have shown that the chemical and electrochemical stability of the LSC surface can be improved by modifying the surface with Ti. Two different forms of surface modification were observed. When Ti content is below 10%, the deposited Ti stays in the form of a wetting layer, either dissolved into LSC, or in the form of a thin conformal TiO_2 layer. Higher Ti concentration leads to formation of dispersed TiO_2 nanoparticles at the surface. We found that the Ti surface content up to 10% (quantified by XPS) on LSC improved the surface oxygen exchange kinetics up to 8-fold by hindering the detrimental segregation of Sr-rich phases at 530 °C in air. The Sr-rich phase segregation slows the oxygen exchange kinetics, both by blocking the LSC surface with insulating particles and by decreasing the Sr-doping level and the oxygen vacancy availability within the LSC lattice. We believe that the presence of a small amount of Ti that dissolves into the LSC lattice at the surface decreases the oxygen vacancy concentration, and this effect decreases the electrostatic attraction of the negatively charged Sr'_{La} from bulk towards the LSC surface. As a result, the cation chemistry on LSC-T10 is more stable and gives rise to faster oxygen exchange kinetics. Ti-modification of the surface is proposed as one candidate for enabling better chemical stability and reaction kinetics on cobaltite based catalyst and electrocatalyst materials. In our ongoing work, we are investigating other surface compositions to systematically modulate the surface electrostatic driving force to segregation of the dopant cations.

Acknowledgements

The authors are grateful for the funding support from the NSF CAREER Award of the National Science Foundation, Division of Materials Research, Ceramics Program, Grant no. 1055583, from the US-DOE Basic Energy Sciences, Grant no. DE-SC00026333, and from the BP-MITEI Seed Project at MIT. The authors also acknowledge the use of the Center for Materials Science and Engineering, an MRSEC facility of NSF at MIT. This work made use of the MRSEC Shared Experimental Facilities at MIT, supported by the National Science Foundation under award number DMR-08-19762.

References

1 C. Graves, S. D. Ebbesen, M. Mogensen and K. S. Lackner, *Renewable Sustainable Energy Rev.*, 2011, **15**, 1–23.
2 E. V. Kondratenko, G. Mul, J. Baltrusaitis, G. O. Larrazabal and J. Perez-Ramirez, *Energy Environ. Sci.*, 2013, **6**, 3112–3135.
3 Z. Cai, M. Kubicek, J. Fleig and B. Yildiz, *Chem. Mater.*, 2012, **24**, 1116–1127.
4 Z. Cai, Y. Kuru, J. W. Han, Y. Chen and B. Yildiz, *J. Am. Chem. Soc.*, 2011, **133**, 17696–17704.

5 Y. Chen, W. Jung, Z. Cai, J. J. Kim, H. L. Tuller and B. Yildiz, *Energy Environ. Sci.*, 2012, **5**, 7979–7988.
6 E. J. Crumlin, E. Mutoro, S.-J. Ahn, G. J. la O', D. N. Leonard, A. Borisevich, M. D. Biegalski, H. M. Christen and Y. Shao-Horn, *J. Phys. Chem. Lett.*, 2010, **1**, 3149–3155.
7 S. Jiang, *J. Solid State Electrochem.*, 2007, **11**, 93–102.
8 W. Jung and H. L. Tuller, *Energy Environ. Sci.*, 2012, **5**, 5370–5378.
9 M. Kubicek, A. Limbeck, T. Frömling, H. Hutter and J. Fleig, *J. Electrochem. Soc.*, 2011, **158**, B727–B734.
10 E. Mutoro, E. J. Crumlin, M. D. Biegalski, H. M. Christen and Y. Shao-Horn, *Energy Environ. Sci.*, 2011, **4**, 3689–3696.
11 T. H. Shin, S. Ida and T. Ishihara, *J. Am. Chem. Soc.*, 2011, **133**, 19399–19407.
12 A. H. McDaniel, E. C. Miller, D. Arifin, A. Ambrosini, E. N. Coker, R. O'Hayre, W. C. Chueh and J. Tong, *Energy Environ. Sci.*, 2013, **6**, 2424–2428.
13 C.-Y. Tsai, A. G. Dixon, Y. H. Ma, W. R. Moser and M. R. Pascucci, *J. Am. Ceram. Soc.*, 1998, **81**, 1437–1444.
14 L. Yang, L. Tan, X. Gu, W. Jin, L. Zhang and N. Xu, *Ind. Eng. Chem. Res.*, 2003, **42**, 2299–2305.
15 E. Levi, A. Mitelman, D. Aurbach and M. Brunelli, *Chem. Mater.*, 2007, **19**, 5131–5142.
16 N. Yabuuchi, Y.-C. Lu, A. N. Mansour, S. Chen and Y. Shao-Horn, *J. Electrochem. Soc.*, 2011, **158**, A192–A200.
17 W.-S. Yoon, Y. Paik, X.-Q. Yang, M. Balasubramanian, J. McBreen and C. P. Grey, *Electrochem. Solid-State Lett.*, 2002, **5**, A263–A266.
18 S. W. Lee, C. Carlton, M. Risch, Y. Surendranath, S. Chen, S. Furutsuki, A. Yamada, D. G. Nocera and Y. Shao-Horn, *J. Am. Chem. Soc.*, 2012, **134**, 16959–16962.
19 D. Serrate, J. M. De Teresa, J. Blasco, M. R. Ibarra, L. Morellon and C. Ritter, *Appl. Phys. Lett.*, 2002, **80**, 4573–4575.
20 C. N. R. Rao and P. V. Vanitha, *Curr. Opin. Solid State Mater. Sci.*, 2002, **6**, 97–106.
21 S. Estrade, J. Arbiol, F. Peiro, I. C. Infante, F. Sanchez, J. Fontcuberta, F. de la Pena, M. Walls and C. Colliex, *Appl. Phys. Lett.*, 2008, **93**, 112505.
22 S. Estrade, J. M. Rebled, J. Arbiol, F. Peiro, I. C. Infante, G. Herranz, F. Sanchez, J. Fontcuberta, R. Cordoba, B. G. Mendis and A. L. Bleloch, *Appl. Phys. Lett.*, 2009, **95**, 072507.
23 I. Hamada, A. Uozumi, Y. Morikawa, A. Yanase and H. Katayama-Yoshida, *J. Am. Chem. Soc.*, 2011, **133**, 18506–18509.
24 F. Stavale, X. Shao, N. Nilius, H.-J. Freund, S. Prada, L. Giordano and G. Pacchioni, *J. Am. Chem. Soc.*, 2012, **134**, 11380–11383.
25 S. P. Jiang, *J. Solid State Electrochem.*, 2005, **11**, 93–102.
26 G. J. la O, S.-J. Ahn, E. Crumlin, Y. Orikasa, M. D. Biegalski, H. M. Christen and Y. Shao-Horn, *Angew. Chem., Int. Ed.*, 2010, **49**, 5344–5347.
27 J. Hayd, H. Yokokawa and E. Ivers-Tiffée, *J. Electrochem. Soc.*, 2013, **160**, F351–F359.
28 S. P. Jiang, *Solid State Ionics*, 2002, **146**, 1–22.
29 Z. Shao and S. M. Haile, *Nature*, 2004, **431**, 170–173.
30 Z. Cai, M. Kubicek, J. Fleig and B. Yildiz, *Chem. Mater.*, 2012, **24**, 1116–1127.

31 M. Kubicek, Z. Cai, W. Ma, B. Yildiz, H. Hutter and J. Fleig, *ACS Nano*, 2013, **7**, 3276–3286.

32 W. Lee, J. W. Han, Y. Chen, Z. Cai and B. Yildiz, *J. Am. Chem. Soc.*, 2013, **135**, 7909–7925.

33 H. Dulli, E. W. Plummer, P. A. Dowben, J. Choi and S. H. Liou, *Appl. Phys. Lett.*, 2000, **77**, 570–572.

34 H. Dulli, P. A. Dowben, S. H. Liou and E. W. Plummer, *Phys. Rev. B: Condens. Matter Mater. Phys.*, 2000, **62**, R14629–R14632.

35 J. Druce, H. Tellez, M. Burriel, M. D. Sharp, L. J. Fawcett, S. N. Cook, D. S. McPhail, T. Ishihara, H. H. Brongersma and J. A. Kilner, *Energy Environ. Sci.*, 2014, **7**, 3593–3599.

36 H.-S. Lee, C.-S. Park and H.-H. Park, *Appl. Phys. Lett.*, 2014, **104**, 191604.

37 R. Waser and M. Aono, *Nat. Mater.*, 2007, **6**, 833–840.

38 H. Jalili, J. W. Han, Y. Kuru, Z. Cai and B. Yildiz, *J. Phys. Chem. Lett.*, 2011, **2**, 801–807.

39 T. T. Fister, D. D. Fong, J. A. Eastman, P. M. Baldo, M. J. Highland, P. H. Fuoss, K. R. Balasubramaniam, J. C. Meador and P. A. Salvador, *Appl. Phys. Lett.*, 2008, **93**, 151904.

40 K. Katsiev, B. Yildiz, K. Balasubramaniam and P. A. Salvador, *Appl. Phys. Lett.*, 2009, **95**, 092106.

41 Y. Gong, D. Palacio, X. Song, R. L. Patel, X. Liang, X. Zhao, J. B. Goodenough and K. Huang, *Nano Lett.*, 2013, **13**, 4340–4345.

42 D. Lee, Y.-L. Lee, A. Grimaud, W. T. Hong, M. D. Biegalski, D. Morgan and Y. Shao-Horn, *J. Phys. Chem. C*, 2014, **118**, 14326–14334.

43 X. Zhu, D. Ding, Y. Li, Z. Lü, W. Su and L. Zhen, *Int. J. Hydrogen Energy*, 2013, **38**, 5375–5382.

44 A. Lussier, J. Dvorak, S. Stadler, J. Holroyd, M. Liberati, E. Arenholz, S. B. Ogale, T. Wu, T. Venkatesan and Y. U. Idzerda, *Thin Solid Films*, 2008, **516**, 880–884.

45 S. Estrade, J. Arbiol, F. Peiro, L. Abad, V. Laukhin, L. Balcells and B. Martinez, *Appl. Phys. Lett.*, 2007, **91**, 252503.

46 W. C. Chueh and S. M. Haile, *Annu. Rev. Chem. Biomol. Eng.*, 2012, **3**, 313–341.

47 M. M. Kuklja, E. A. Kotomin, R. Merkle, Y. A. Mastrikov and J. Maier, *Phys. Chem. Chem. Phys.*, 2013, **15**, 5443–5471.

48 N. Tsvetkov, L. Larina, O. Shevaleevskiy and B. T. Ahn, *Energy Environ. Sci.*, 2011, **4**, 1480–1486.

49 N. A. Tsvetkov, L. L. Larina, O. Shevaleevskiy, E. A. Al-Ammar and B. T. Ahn, *Prog. Photovoltaics*, 2012, **20**, 904–911.

50 B. Yildiz, N. Tsvetkov, and Q. Lu, U.S. Patent Application no.: 62/140009, 2015.

51 M. M. Islam, T. Bredow and A. Gerson, *Phys. Rev. B: Condens. Matter Mater. Phys.*, 2007, **76**, 045217.

52 Y.-L. Lee, J. Kleis, J. Rossmeisl, Y. Shao-Horn and D. Morgan, *Energy Environ. Sci.*, 2011, **4**, 3966–3970.

53 P. M. Sommeling, B. C. O'Regan, R. R. Haswell, H. J. P. Smit, N. J. Bakker, J. J. T. Smits, J. M. Kroon and J. A. M. van Roosmalen, *J. Phys. Chem. B*, 2006, **110**, 19191–19197.

54 M. Kubicek, A. Limbeck, T. Frömling, H. Hutter and J. Fleig, *J. Electrochem. Soc.*, 2011, **158**, B727–B734.

55 J. X. Wang, J. Shao, Y. K. Tao and W. G. Wang, *ECS Trans.*, 2009, **25**, 595–600.

56 E. Koep, D. S. Mebane, R. Das, C. Compson and M. Liu, *Electrochem. Solid-State Lett.*, 2005, **8**, A592–A595.
57 S. B. Adler, *Chem. Rev.*, 2004, **104**, 4791–4844.
58 J. Jamnik and J. Maier, *Phys. Chem. Chem. Phys.*, 2001, **3**, 1668–1678.
59 F. S. Baumann, J. Fleig, H.-U. Habermeier and J. Maier, *Solid State Ionics*, 2006, **177**, 1071–1081.
60 J. Mizusaki, Y. Mima, S. Yamauchi, K. Fueki and H. Tagawa, *J. Solid State Chem.*, 1989, **80**, 102–111.
61 R. E. van Doorn, I. C. Fullarton, R. A. de Souza, J. A. Kilner, H. J. M. Bouwmeester and A. J. Burggraaf, *Solid State Ionics*, 1997, **96**, 1–7.
62 P. Hjalmarsson, M. Søgaard and M. Mogensen, *Solid State Ionics*, 2008, **179**, 1422–1426.

Faraday Discussions

PAPER

Oxygen exchange and transport in dual phase ceramic composite electrodes

John Druce,*[a] Helena Téllez,[a] Tatsumi Ishihara[a] and John A. Kilner[ab]

Received 24th February 2015, Accepted 26th February 2015
DOI: 10.1039/c5fd00028a

Composites consisting of a perovskite-based electronic or mixed conductor with a fluorite-structured ionic conductor are often used as electrodes in solid oxide electrochemical energy conversion devices. After sintering the materials, there is often evidence for inter-reaction between the two phases, or inter-diffusion of cations or impurities between the two phases. We studied the ^{18}O exchange properties of a composite consisting of CGO and LSCF in a 50 : 50 ratio. High resolution ToF-SIMS mapping reveals that the ^{18}O fraction at the very outer surface of grains of the CGO phase is much higher than expected from D^* and k^* values for the single-phase parent material. Surface compositional analysis by ToF-SIMS and low energy ion scattering (LEIS) spectroscopy suggests that the surfaces of the CGO grains in the composite do not show the impurities which typically segregate to the surface in single-phase CGO. Thus, the "cleaning" of impurities from the CGO surface by dissolution into the perovskite phase may be one explanation for the apparent enhanced surface exchange for CGO in these composites.

Introduction

Oxygen electrodes for solid oxide electrolyser cells (SOECs) should possess mixed ionic-electronic conductivity (MIEC) under a range of operating conditions. In the case of reversible cells, this entails both oxidising and reducing conditions. Whilst single-phase materials such as $La_{1-x}Sr_xMnO_{3+\delta}$ (LSM) and $La_{1-x}Sr_xCo_{1-y}Fe_yO_{3-\delta}$ (LSCF) show MIEC behaviour under solid oxide fuel cell (SOFC) conditions, their ionic conductivities often decrease in the high oxygen activities encountered during SOEC operation due to a drop off in the concentration of charge-carrying oxygen vacancies. This is one of the typical modes of failure of electrodes by delamination.[1]

[a]International Institute for Carbon Neutral Energy Research (wpi-I2CNER), Kyushu University, Nishi-ku, Fukuoka, 819-0395, Japan. E-mail: john.druce@i2cner.kyushu-u.ac.jp; Fax: +81-92-802-6738; Tel: +81-92-802-6738

[b]Department of Materials, Imperial College London, London, UK, SW7 2BP. Fax: +44-(0)207-594-6745; Tel: +44-(0)207-594-6745

Although single-phase MIEC materials are attractive for electrodes, another way to achieve mixed conduction in both SOEC and SOFC modes is to use macroscopically mixed conducting dual phase ceramic composites. These offer the ability to combine high oxygen diffusivity, normally contributed by an ionically conducting fluorite phase such as $Zr_{1-x}Y_xO_{2-x/2}$ (YSZ) or $Ce_{1-x}Gd_xO_{2-x/2}$ (CGO), with electronic conduction, typically from a perovskite-structured phase such as LSM or LSCF. In this case, the ionic conductivity is dominated by the fluorite phase and expected to be high in both SOFC and SOEC modes (*i.e.* the oxidising conditions experienced as anodes under electrolysis mode as well as the reducing conditions for cathodes under SOFC mode).

However, the high temperatures required to sinter the materials can lead to inter-reaction. In the case of LSM with YSZ, this can result in the formation of secondary phases such as $La_2Zr_2O_7$ which block oxygen transport.[2,3] Although CGO and LSCF are not believed to react to form secondary phases,[4,5] there is evidence that cation inter-diffusion does occur between the two phases.[6–9] Particular attention has been paid to diffusion of both transition metals and the larger La and Sr cations from the perovskite into the CGO,[7] although the solubility of Ce and Gd in the perovskite is likely to be low.[10,11] However, the effect of such inter-reaction on the overall material properties, such as oxygen surface exchange, are not yet known. It has even been proposed that the diffusion of transition metal cations into the fluorite phase could enhance its electro-catalytic activity towards oxygen incorporation.[12,13]

The combination of ^{18}O isotopic labelling experiments with secondary ion mass spectrometry (SIMS) is an attractive strategy to study the mechanisms of oxygen exchange. Methods such as electrochemical impedance studies and electrical conductivity relaxation provide information averaged over an entire macroscopic sample, and as such it is difficult to extract any microstructurally-resolved information on the roles of the two phases. On the other hand, SIMS instruments can provide a probe on the order of 100 nm, allowing for localised information to be extracted, highlighting the active sites for oxygen incorporation.[14,15]

Although previous reports of IEDP-SIMS studies on composites used quadrupole SIMS instruments,[12,16] this type of SIMS suffers from the drawback that mass detection is sequential. That is, the masses to be studied must be chosen beforehand, and only a single signal can be monitored at any one time. Therefore, it is impossible to obtain, say, both the ^{16}O and ^{18}O intensities from the same point – one must be recorded first, which entails the removal of some material (on the order of 50 nm in depth) by sputtering. The second channel is then recorded by rastering the ion beam a second time across the area of interest. The information in the second scan hence comes from a slightly greater depth. Therefore, obtaining information about the distribution of cations as well as the oxygen isotopes would consume a large depth of the sample. On the other hand, in contrast to quadrupole SIMS, time of flight SIMS (ToF-SIMS) offers parallel ion detection, where the entire mass spectrum is recorded for every pixel of the image at each depth.[17,18] Therefore, it is possible to extract far more information about the distribution of the chemical and isotopic composition in the sample by the use of ToF-SIMS. In this work, we exploit this combination of isotope exchange and surface analysis by two complementary ion beam techniques, namely ToF-SIMS and low energy ion scattering spectroscopy (LEIS), to study oxygen exchange in these materials. The former technique offers high chemical sensitivity and lateral resolution, whilst the latter quantitatively probes the elemental composition of the very outer atomic surface of a material.[19–21]

Experimental

Commercially available powders of $Ce_{0.9}Gd_{0.1}O_{1.95}$ (CGO, Nextech, USA) and $La_{0.6}Sr_{0.4}Co_{0.2}Fe_{0.8}O_{3-\delta}$ (LSCF, Praxair, USA) were mixed in a 50 : 50 (by weight) ratio by ball milling overnight (longer than 12 hours) in ethanol. After drying the powder blends, they were pressed into pellets and sintered at 1250 °C for 4 hours.

The sintered pellets, with densities (as estimated by the Archimedes method) greater than 95% of the theoretical value, were ground flat with SiC and then polished with diamond suspension. One set of these samples was analysed in the "as-polished" state (see below for experimental details of the characterisation techniques), whilst the other samples were used for the ^{18}O exchange experiments.[19,22] This methodology involves a pre-anneal in research grade oxygen (of nominally natural isotopic abundance) for at least 10 times the duration of the subsequent exchange anneal, before the introduction of 200 mbar of oxygen gas enriched in ^{18}O and the (shorter) exchange anneal. The ^{18}O diffusion profiles were then measured by ToF-SIMS, and the surface compositions of the samples analysed by LEIS spectroscopy.

The durations of the isotope exchanges (900 s at 500 °C and 1200 s at 700 °C) were chosen based on data in the literature for the two constituent phases.[23,24] Key parameters for the exchange are summarised in Table 1. The diffusivity of LSCF is lower than that of CGO by one to two orders of magnitude at the temperatures of interest. Therefore, the normalized surface isotopic fractions will essentially be governed by the surface exchange coefficient, k^*. Note that under the exchange conditions chosen, the isotopic fraction in the CGO is expected to be virtually zero, whilst that in the LSCF should be 30–40%. In other words, we expect significant contrast in the ^{18}O isotopic fractions between the two phases.

Measurement of oxygen isotope distributions

Time of flight secondary ion mass spectrometry (ToF-SIMS V, Ion-ToF GmbH., Germany) was used to analyse the lateral distribution of various elements of interest (*e.g.* cations characteristic of the two phases) at the sample surface, as well as the distribution of the ^{18}O isotopic tracer, as it provides isotopically-resolved information of the elemental composition. The depth profiles were performed in dual beam mode, with sputtering by a 2 keV Ar^+ beam interlaced

Table 1 Oxygen-18 exchange conditions used in this work

Temp.	Phase	D^* ($cm^2\ s^{-1}$)	k^* ($cm\ s^{-1}$)	Anneal time (s)	Ld (μm)	h'	$C's$ [Cs] (%)	Reference
500 °C	CGO	5×10^{-10}	2×10^{-9}	1058	14.5	0.0029	0.33 [0.51]	24
	LSCF	6×10^{-12}	3×10^{-8}		1.6	0.3984	32.8 [31.3]	23
700 °C	CGO	2×10^{-8}	4×10^{-8}	1590	112.8	0.0113	1.3 [1.4]	24
	LSCF	3×10^{-9}	1×10^{-6}		43.7	0.7280	48.4 [46.2]	23

with the analysis cycles. After profiling, the sputter rates were calibrated by measuring the final crater depth by 3D microscopy (LEXT OLS 4000, Olympus, Japan). To avoid inaccuracies in the determination of the oxygen isotope fractions due to detector saturation, selective attenuation of secondary ions (SASI)[25] was applied to the $^{16}O^-$ and $^{18}O^-$ signals, with the attenuation level determined by the instrument software for each species, adjusted automatically throughout the profile.

After depth calibration, the intensities for the $^{16}O^-$ and $^{18}O^-$ ions were extracted from the data, and values for the kinetic parameters D^* and k^* (oxygen diffusion and surface exchange coefficients, respectively) were determined by fitting Crank's solution to the diffusion equation for a semi-infinite medium including surface exchange limitation (eqn (1)[26]) in MATLAB (The Mathworks, USA). In this equation, $C_{(x)} = \frac{I(^{18}O^-)}{I(^{16}O^-) - I(^{18}O^-)}$, which is calculated from the secondary ion intensities $I(^{16}O^-)$ and $I(^{18}O^-)$ and is the observed isotope fraction at depth x, C_{bg} is the fraction of the ^{18}O isotope in the solid immediately prior to the exchange (taken as the natural isotopic abundance of ^{18}O), C_g is the fraction of the ^{18}O tracer in the gas phase (95%), and t is the duration of the isotope exchange anneal.

$$C'_{(x)} = \frac{C_{(x)} - C_{bg}}{C_g - C_{bg}} = \operatorname{erfc}\left(\frac{x}{2\sqrt{D^*t}}\right) - \exp\left(x\frac{k}{D^*} + \frac{k^2t}{D^*}\right)\operatorname{erfc}\left(\frac{x}{2\sqrt{D^*t}} + \sqrt{\frac{k^2t}{D^*}}\right) \quad (1)$$

For the analysis of the lateral distribution of the ^{18}O tracer at the sample surfaces, ToF-SIMS imaging analysis was conducted with the instrument operating in the "fast imaging" mode, offering high lateral resolution (approximately 100 nm spot size) and unit mass resolution. Rather than bursting the beam to reduce the primary ion current, and hence the secondary ion currents,[27] saturation of the detector by the intense oxygen secondary ion currents was again avoided by attenuation of the signals at m/z 16 u (attributed to $^{16}O^-$) and 18 u (attributed to $^{18}O^-$) by SASI.[25] For the imaging analysis, the attenuation factors were fixed as "high" (attenuation by a factor of 87) for m/z 16 u and "low" (attenuation by a factor of 7.8) for m/z 18 u.

Surface analysis by low energy ion scattering (LEIS) spectroscopy

Low energy ion scattering (LEIS) spectroscopy measurements of the compositions of the very outer monoatomic layers of the samples were performed using a Qtac^{100} instrument (Ion-ToF GmbH., Germany). Primary ion beams of $^4He^+$ (accelerated to 3 keV) and $^{20}Ne^+$ (accelerated to 6 keV) were directed to the sample surface at normal incidence, and the energy distribution of ions backscattered through 145° was measured by a double toroidal electrostatic analyser. Further details of the LEIS technique in general are given in ref. 28 and 29, and more details on the specific experimental setup are given in ref. 30. Prior to analysis, adventitious atmospheric contamination (*e.g.* adsorbed moisture and hydrocarbons), which would otherwise obscure information on the composition of the very outer monolayer, was cleaned from the sample surfaces by exposure to reactive atomic oxygen.[31]

Results

Microstructure and assignment of phases

Fig. 1 shows ToF-SIMS negative ion maps representing the microstructures of the composites. The image on the left shows the intensity at *m*/*z* 88 u. Although the mass of the most abundant isotope of Sr is 87.91 u (85.58% abundance), the signal at *m*/*z* 88 u is most likely due to $^{56}Fe^{16}O_2^-$ fragment ions rather than negatively charged $^{88}Sr^-$ ions. However, since both Sr and Fe arise from the LSCF phase, the resolution of this possible interference is not necessary for the purposes of distinguishing between the two phases in this image, and the map at *m*/*z* 88 u is deemed characteristic of the LSCF. On the other hand, the signal at *m*/*z* 188 u is attributed predominantly to GdO_2^-, and so taken as characteristic of the CGO phase. The two negative secondary ion maps presented are clearly the inverse of one another, and the overlay of the two signals shows that the distribution of these two elements is complementary, as expected for the two phases.

The grain sizes of the two phases measured from SEM images were 620 nm for the LSCF and 750 nm for the CGO. In comparison, the grain sizes in the single-phase CGO and LSCF samples, sintered under the same conditions, were around 2 μm and 300 nm, respectively. Grain growth of the LSCF grains during sintering of the composite appear to be suppressed by the presence of neighbouring CGO grains. On the other hand, the size of the CGO grains in the composite sample are more than twice that in the single-phase parent material. Cobalt is often added to CGO as a sintering aid,[32,33] and it is possible that Co diffusing from the LSCF may act in a similar way.

^{18}O isotope exchange experiments

"Macroscopic" isotope diffusion profiles at 500 °C. Isotope exchange profiles in the pure LSCF and the CGO–LSCF composite exchanged at 500 °C and obtained by dual beam ToF-SIMS sputter depth profiling are shown in Fig. 2. The fit of the solution to the diffusion equation (eqn (1)) to the experimentally determined profile for the LSCF agrees reasonably well, and returns values of 9.3×10^{-12} cm^2 s^{-1} and 3.0×10^{-8} cm s^{-1} for the diffusivity, D^*, and surface exchange, k^*, coefficients, respectively, which are very close to those previously obtained.[23] As is clear from the inset in Fig. 2, at greater depths, the experimental profile deviates from the analytical solution. This is probably indicative of a fast grain boundary

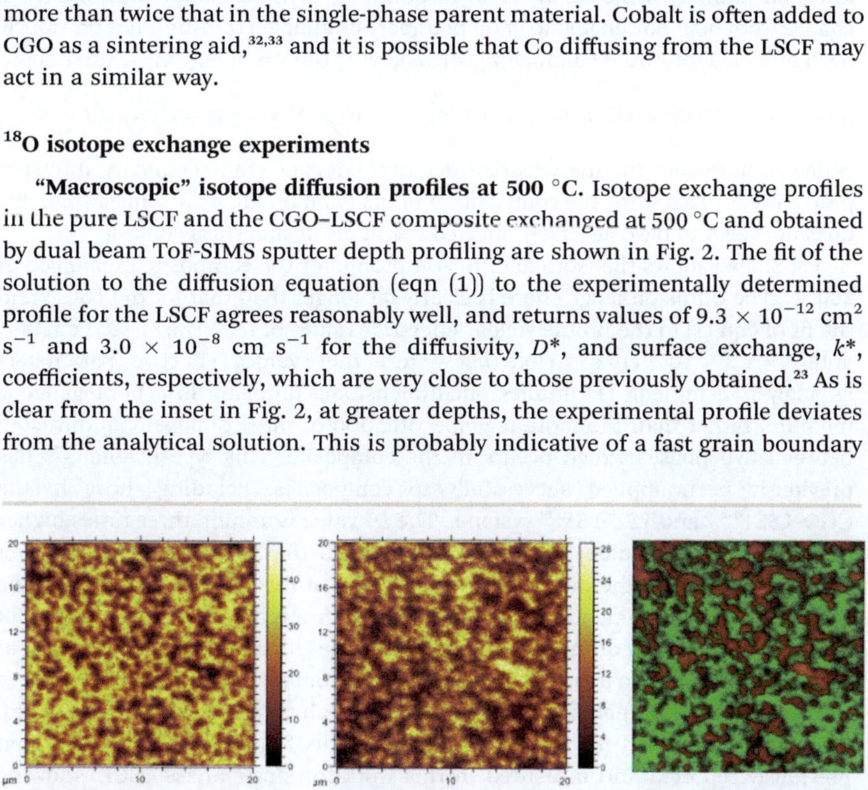

(a) m/z = 88 u (b) m/z = 188 u (c) Overlay of (a) and (b)

Fig. 1 ToF-SIMS negative secondary ion images for (a) *m*/*z* 88 u (FeO_2^-) and (b) *m*/*z* 188 u (GdO_2^-), and (c) two-channel overlay of the two in green and red, respectively.

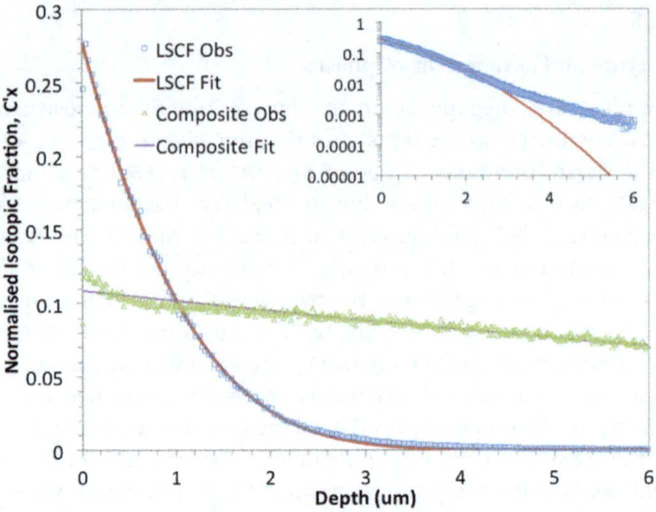

Fig. 2 Oxygen-18 diffusion profiles and associated fits for pure LSCF and a 50 : 50 CGO–LSCF composite subjected to isotope exchange at 500 °C. Inset shows data for LSCF plotted on a logarithmic scale to highlight the deviation at greater depths.

diffusion pathway, which may begin to dominate at lower temperatures. Such a change from bulk-dominated to grain boundary-dominated transport has previously been observed for mixed conducting perovskites.[23] Unfortunately, the sputter depth profile was terminated before reaching the region $6 \leq \left(\eta = x \Big/ \sqrt{D_b^* t}\right) \leq 10$, which is necessary for the accurate determination of grain boundary diffusion product.[25,34,35] Therefore, the contribution of the grain boundary was neglected; this does not seem to have adversely affected the fit of the analytical solution.

Fig. 2 also shows the isotope diffusion profile for the composite exchanged at 500 °C. The diffusion length here is clearly far longer than that for the LSCF, and the fit of eqn (1) to the profile yields "effective" values of $D^* = 1.6 \times 10^{-9}$ cm^2 s^{-1} and $k^* = 1.3 \times 10^{-7}$ cm s^{-1}. Note that we term these values "effective" parameters as analysis using eqn (1) implies one-dimensional diffusion in a homogeneous medium, rather than the more realistic situation of three dimensional diffusion between two phases which occurs in the composite. This type of analysis has previously been applied successfully to composites including those in the CGO–LSCF[16,36] and YSZ–LSM[13] systems. The D^* value is around three times higher than the literature value for CGO.[24] Although in the present work it was not possible to measure the diffusion profile in CGO at this temperature, due to the low k^* value, measurements on samples prepared from this CGO powder as part of another study showed diffusivities of around 5 times higher than those reported in ref. 24, possibly due to its relatively low impurity content.[37] Thus, the diffusivity of the composite is comparable to that of the single-phase CGO. On the other hand, the k^* value obtained for the present composite is four times higher than that previously reported and measured in this work for single-phase LSCF, showing some enhancement in the effective surface exchange rate in this composite.

This type of analysis has previously been applied to fit diffusion profiles and extract "effective" D^* and k^* values for composites,[13,16,36] and the fit of eqn (1) to the present data agrees well at greater depths into the sample. However, a feature is evident in the profile for the composite at shallow depths, showing an increased isotopic fraction not described by the analytical solution of eqn (1). The artefact is not seen in the diffusion profile for the single-phase LSCF (also in Fig. 2), and is likely to be related to the mechanism of oxygen exchange in the composite. Since the length of the feature is comparable to the diffusion length in LSCF, it may indicate that oxygen is incorporated into the LSCF and diffuses a short distance before being transferred to the CGO. Although this feature was not reported in the previous work, this may be because linescan analyses of cross sections were performed,[38] and the lateral resolution in these measurements (around 10 μm) is insufficient to resolve features on this length scale.

^{18}O isotopic fraction mapping at low temperatures (500 °C). Fig. 3 shows results of the ToF-SIMS isotopic mapping of the composite exchanged at 500 °C. Parts (a) and (b) are the secondary ion maps for the $^{16}O^-$ and $^{18}O^-$ oxygen isotopes, respectively. Part (c) shows the intensity map of the signal at *m/z* 88 u (as discussed above, this is attributed to FeO_2^- ions, taken to be characteristic of the LSCF phase). Although some contrast between the two phases is evident in the oxygen images, *i.e.* the CGO regions are brighter, this appears in both the $^{16}O^-$ and $^{18}O^-$ maps and is therefore probably related to differences in the secondary ion yields of oxygen between the two phases,[39] and not the exchange properties.

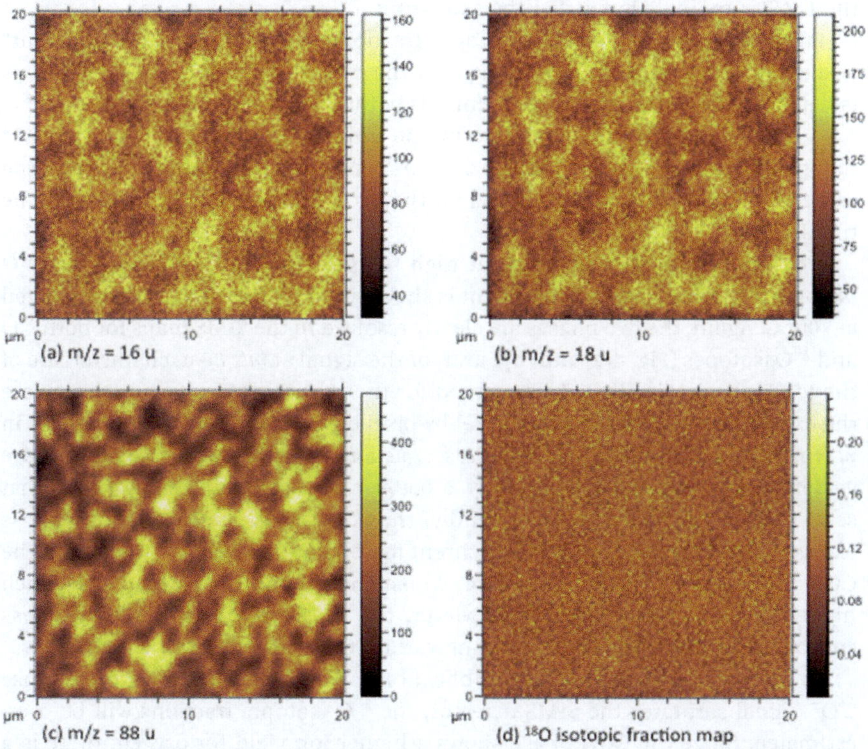

Fig. 3 ToF-SIMS negative secondary ion images of a 50 : 50 CGO–LSCF composite exchanged at 500 °C. (a) *m/z* 16 u ($^{16}O^-$), (b) *m/z* 18 u ($^{18}O^-$), (c) *m/z* 88 u (FeO_2^-, indicative of LSCF phase) and (d) oxygen isotopic fraction map.

On the other hand, we can account for this by calculating the ^{18}O isotopic fraction on a pixel-by-pixel basis $\left(C_{x,y} = \frac{^{18}\text{O}_{x,y}}{^{16}\text{O}_{x,y}+^{18}\text{O}_{x,y}}\right)$, which normalises out the effect of different secondary ion yields for the two phases. The resulting image is shown in Fig. 3(d). Note that the values shown in this figure have not been normalised to the isotopic fraction in the exchange gas, nor to the natural isotopic abundance of ^{18}O. However, since the former is rather high (95 at.% ^{18}O), and the latter low (0.204% ^{18}O), this normalisation would not qualitatively affect the image.

Considering the rather large difference in the expected isotopic fractions for the two phases under the present annealing conditions (0.005 for CGO and 0.313 for LSCF, from Table 1), relatively little contrast is evident in Fig. 3(d). Close scrutiny of the image suggests some marginally brighter patches corresponding to grains of LSCF indicated in Fig. 3(c), although this is not as significant as might be expected from the literature values.

The surface ^{18}O isotopic fraction may also be determined for each phase by extracting data from regions of interest corresponding to the two phases. This was performed in the SurfaceLab "Measurement Explorer" Software, provided by IonToF GmbH., Germany. This procedure gives average isotopic fraction values of 0.099 for the CGO and 0.101 for the LSCF regions. The ^{18}O surface isotopic fraction for the CGO is around 20 times higher than that expected for the single-phase material under the same exchange conditions, whilst that for the LSCF is about a third of that expected for the LSCF, and indeed of that seen in Fig. 2 for the LSCF sample exchanged at the same time.

From the extracted surface isotopic fractions, we can calculate "apparent" surface isotopic fractions, assuming that the two constituent phases behave as isolated phases in the exchange. From this point of view, we obtain values of $k^* = 6.8 \times 10^{-8}$ cm s^{-1} for the CGO (around 30 times the value of 2×10^{-9} cm s^{-1} reported by Manning *et al.*[24]) and 9.5×10^{-9} cm s^{-1} for the LSCF (a third of the value of 3×10^{-8} cm s^{-1} determined for the LSCF sample exchanged at the same time).

^{18}O isotopic fraction mapping at high temperatures (700 °C). A similar ^{18}O isotopic fraction mapping experiment is shown in Fig. 4 for the sample exchanged at 700 °C. Again, the two phases are clearly resolved in the SIMS maps for both ^{16}O and ^{18}O isotopes (Fig. 4(a) and (b)), and for the signal at m/z 88 u, characteristic of the LSCF phase (Fig. 4(c)). As discussed for the previous case, we must calculate the oxygen isotopic fraction on a pixel-by-pixel basis to account for differences in useful ion yield between the two phases, as shown in Fig. 4(d). Two things are evident from this latter image; firstly, a better microstructural contrast than that seen in Fig. 3(d) is achieved. Secondly, the contrast is the inverse of what is expected from the properties of the parent materials – the isotopic fraction in the CGO phase should be around 0.014, whilst that in the LSCF should be much higher, around 0.462 (Table 1). However, the ratios in Fig. 4(d) show far less contrast than expected, and surprisingly, the CGO is brighter in the ^{18}O tracer.

As discussed in several recent publications,[25,27,40] if the typically more intense ^{16}O$^-$ signal saturates the SIMS detector, the ^{18}O isotopic fractions will be over-estimated. Since the CGO phase shows a higher ion yield for oxygen, there is a danger that this may induce some detector saturation and lead to an overestimate of the ^{18}O isotopic fraction for the CGO phase. Care has been taken to exclude the

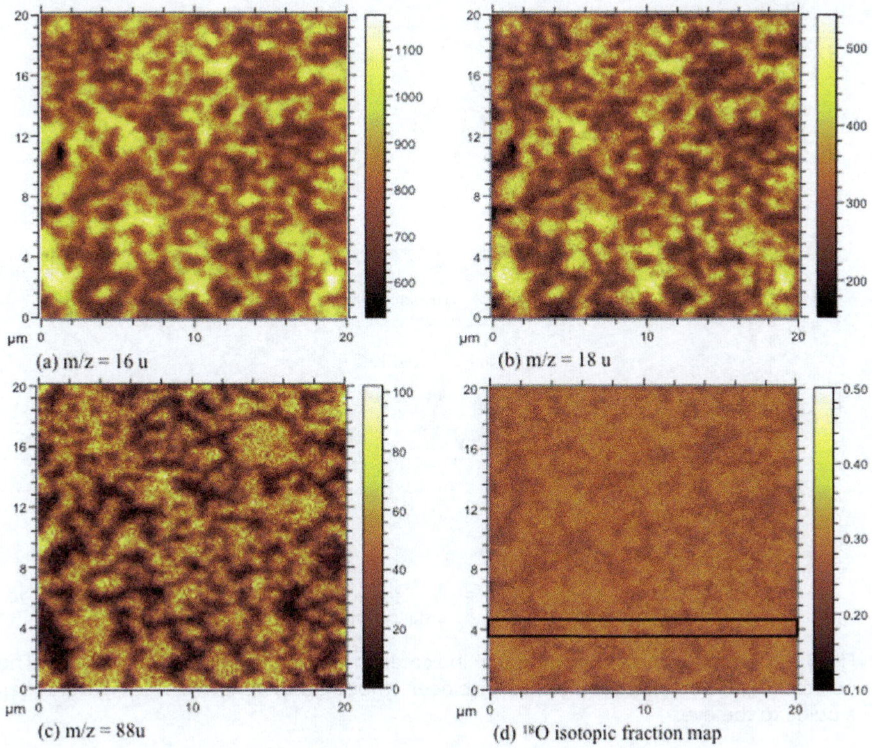

Fig. 4 ToF-SIMS negative secondary ion images of the CGO–LSCF composite exchanged at 700 °C. (a) m/z 16 u ($^{16}O^-$), (b) m/z 18 u ($^{18}O^-$), (c) m/z 88 u (LSCF: FeO_2^-) and (d) oxygen isotopic fraction map. The box in (d) indicates the region from which the linescans shown in Fig. 5 were extracted.

possibility of saturation, by the use of SASI[41] during data acquisition, and by cross checking that the same distribution is obtained when calculating the isotopic fraction from the (unsaturated) peaks at $m/z = 32$ u ($^{16}O^{16}O$), 34 u ($^{16}O^{18}O$) and 36 u ($^{18}O^{18}O$). Therefore, the isotope fraction map in Fig. 4(d) does indeed reflect the true distribution in the composite.

Fig. 4(d) indicates a region from which the linescans shown in Fig. 5 were extracted. The top plot in Fig. 5 shows the intensity profile across Fig. 4(c), indicating the intensity of the signal characteristic of the LSCF phase (m/z 88 u, attributed to Fe_2O^-). The bottom plot shows the profile of the ^{18}O isotopic fraction (C_x') in Fig. 4(d) in the same region. The variation in the oxygen isotope fractions is not as large as the variation of the signal at m/z 88 u, because ^{18}O has been incorporated in both phases. Despite this, the anti-correlation between the intensity of the m/z 88 u signal (*i.e.* the LSCF) and the ^{18}O isotopic fraction is clear from these two plots; valleys in the plot of the oxygen isotope fraction match peaks in the plot of the signal at m/z 88 u, and *vice versa*. Hence, it becomes clear from this analysis that the isotopic fraction in grains of the CGO phase is in fact surprisingly higher than that in the LSCF phase. Since the diffusivity of CGO is still higher than that of LSCF at this temperature, the value of k^*, the surface exchange coefficient, controls the surface isotopic fraction. Therefore, the higher

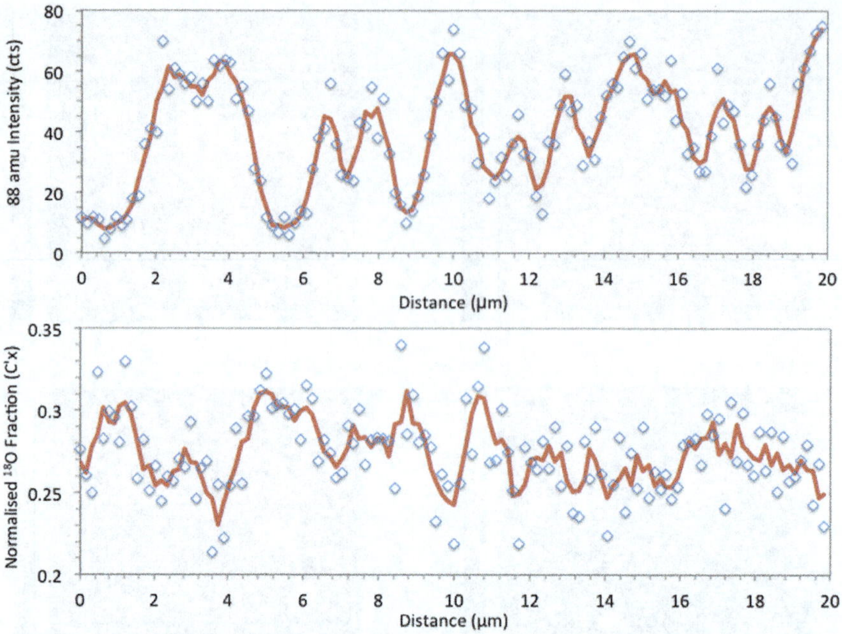

Fig. 5 Linescans extracted from region indicated in isotopic mapping data of Fig. 4. The points show the data, whilst the lines have been smoothed by a 3-point moving average as a guide to the eye.

^{18}O isotopic fraction for the CGO grains implies that k^* is higher for the CGO grains in the composite than for the LSCF grains.

To obtain more precise representative values of the surface isotopic fractions in each phase, the same procedure as for the sample exchanged at 500 °C was followed, extracting from the data only those regions corresponding to CGO or LSCF, and taking the intensities of $^{16}O^-$ and $^{18}O^-$ in those subsets of the data. This gave surface isotopic fraction values of 0.291 for the CGO phase and 0.267 for the LSCF phase. These values are averaged over all of the grains corresponding to each phase (*i.e.* not the maximum values), and agree well with the linescans plotted in Fig. 5. Using these surface isotopic fractions along with literature values of D^* for the two phases,[23,24] in order to estimate their apparent surface exchange coefficients, gives values of 4.2×10^{-7} cm s^{-1} for LSCF, about 0.42 of the 1×10^{-6} cm s^{-1} reported for single-phase LSCF,[23] and 1.2×10^{-6} cm s^{-1} for CGO, which is a factor of 30 times higher than the value of 4×10^{-8} cm s^{-1} reported for the single-phase parent material,[24] and indeed even around 3 times higher than the apparent value for the LSCF phase in this composite.

Surface analysis of composites and pure end members

Before discussing possible reasons for this apparent enhancement, we must also consider the surface composition of the composites. It is known that cation interdiffusion can occur between the two phases at sintering and operation temperatures,[7–9] and this may explain some of the observed changes in surface exchange behaviour.

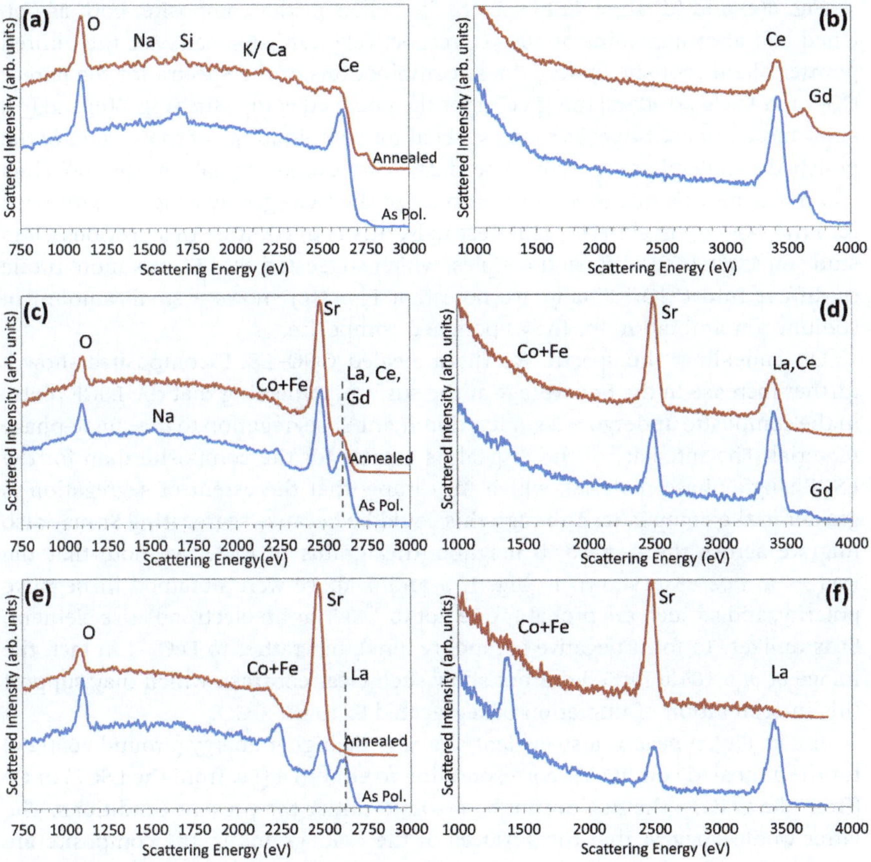

Fig. 6 LEIS analysis of samples as-polished and after annealing at 700 °C for (a and b) CGO, (c and d) the 50 : 50 composite and (e and f) LSCF. Spectra were taken using $^{4}He^{+}$ at 3 keV (a, c and e) and $^{20}Ne^{+}$ at 6 keV (b, d and f).

Low energy ion scattering (LEIS) spectroscopy. The $^{4}He^{+}$ (@3 keV) LEIS spectrum for the polished CGO pellet (Fig. 6(a)) shows two peaks, one for oxygen (~1100 eV) and a second at higher energy representing both Ce and Gd (~2600 eV), which are too close in mass to be resolved under the present scattering conditions. A small amount of Si is also evident as a surface contaminant. On annealing (Fig. 6(b)), the coverage of the rare earth elements decreases, whilst several contamination peaks (Na, Si and K or Ca) appear. These are believed to be impurities segregating from the bulk, which is consistent with previous reports both on CGO[42] and on the closely related fluorite yttria stabilised zirconia.[43]

The evolution of the surface composition of the LSCF, shown in Fig. 6(e) and (f), is similar to that already reported in ref. 44. The as-polished spectra show peaks for all elements expected at the surface, although the Fe and Co components are not resolved under the present scattering conditions. Following annealing, Sr segregates towards the surface, at the expense of the coverage of the transition metals and lanthanum, which are virtually undetectable at the surface after annealing at 700 °C, and the outer layer is comprised of Sr and O, as generally observed in most commonly used solid oxide electrode materials.[41]

Fig. 6(c) and (d) show LEIS spectra for pellets of the composite, both as-polished and after annealing at 700 °C, respectively. Whilst spectra for the unfired powder blend (not show) were linear combinations of the spectra for the parent CGO and LSCF powders, the spectra for the polished composite (Fig. 6(c) and (d)) show more intense Sr peaks than expected for combinations of the spectra of the polished single-phase parent materials. The excess signal in the polished composite may be due to Sr at the surface of the CGO grains which has diffused into the CGO upon sintering. Interestingly, this is in contrast to a previous SIMS study on CGO–LSCF diffusion couples, which suggested that La was more prone to diffuse into CGO.[8] Finally, we note that Fig. 6(c) shows a small amount of sodium contamination for the as-polished composite.

On annealing, the spectra for the annealed CGO–LSCF composites show a further increase in the Sr coverage at the surface, indicating that the LSCF phase in the composite undergoes a similar dominant Sr segregation to the single-phase material. The intensity of the Sr peak is greater for the composite than for the LSCF single-phase material, which may imply that the extent of segregation is greater in the composite. Although this could be because segregating Sr may also migrate across the surface to neighbouring grains of CGO, we note that the images at *m*/*z* 88 u shown in Fig. 1, 3 and 4 above were obtained in negative polarity, and as such are probably due not to $^{88}Sr^-$ (as an electropositive element, Sr is unlikely to form negative secondary ions), but rather to FeO_2^-. In fact, the image at *m*/*z* 104 u (SrO^-) did not show such clear contrast, which may support this interpretation of migration of segregated Sr to the CGO.

In Fig. 6(c), a peak is also evident at a slightly higher energy (around 2600 eV) for the annealed composite, corresponding to either La (*i.e.* from the LSCF) or Ce (from the CGO) – the two cannot be resolved under the present conditions. The latter would suggest that the surfaces of the CGO grains in the composite are cleaner than those in the single-phase parent CGO material.

This is also consistent with the absence of impurities such as Na, Si or Ca, which are seen for the CGO pellet, but not the composite. Fig. 7 is an enlarged version of the energy range corresponding to Na, Si and Ca impurities for the CGO and composite pellets, both before (*i.e.* as polished) and after annealing. The as-polished CGO already shows a small Si impurity, which increases upon annealing, with the appearance of peaks for Na and Ca. These contaminants are typical for fluorite electrolyte materials after annealing.[42,43,45] In contrast, although the as-polished pellet of the composite did show a small Na contamination peak, none of the impurity peaks (Na, Si and Ca) seen for the single-phase CGO are evident after annealing. This also suggests that the surfaces of the CGO in the composites are indeed cleaner. It is possible that grains of the neighbouring perovskite phase may incorporate these impurities in solid solution, effectively "soaking up" the impurities which tend to segregate to the surface of the fluorite grains and block surface exchange.

The transition metals might also be expected to play a significant role in the surface exchange properties. However, these are barely detected at the surface of the composites (Fig. 6(d)), and seem to be in proportion to their concentration in the single-phase material. This suggests that the signal is arising only from transition metal cations in the outer surface of the LSCF grains, which disappears with increasing annealing temperature. Secondary ion maps of signals characteristic of the two transition metals, Co and Fe (Fig. 8), show strong

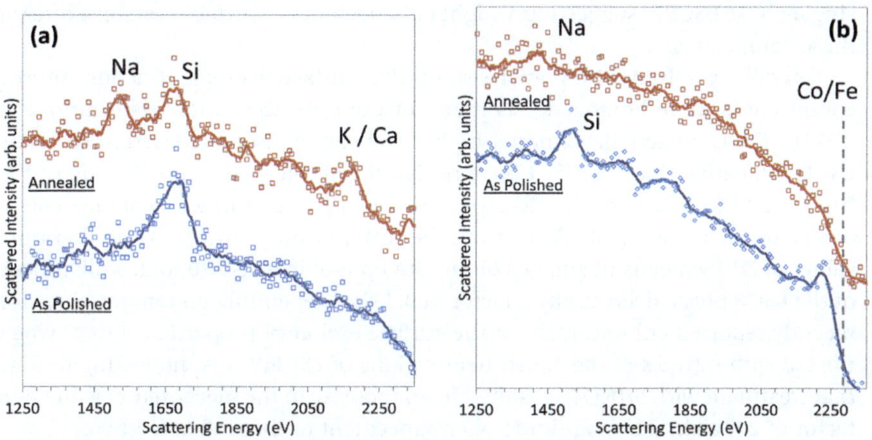

Fig. 7 LEIS spectra ($^4He^+$ @3 keV) of light contaminants on (a) CGO and (b) the CGO–LSCF composite, as-polished and after annealing at 700 °C, and detail showing the effect of annealing on the composite and CGO surfaces.

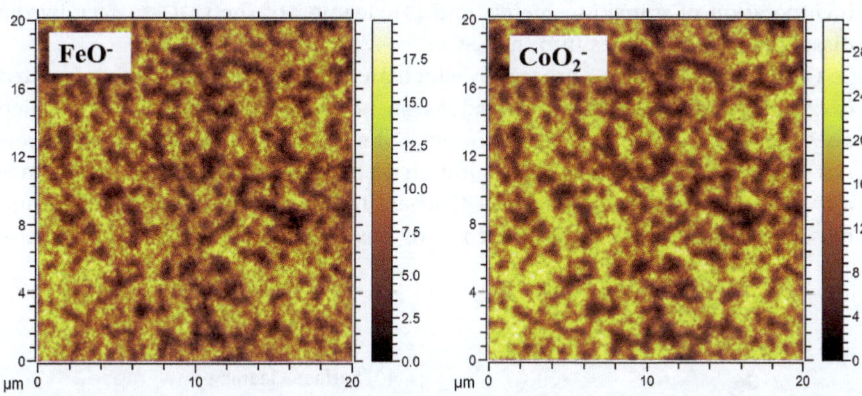

Fig. 8 Secondary ion maps at $m/z = 72$ u ($^{56}Fe^{16}O^-$) and $m/z = 91$ u ($^{59}Co^{16}O_2^-$).

contrast between the LSCF and CGO phases, with virtually no signal detected in the CGO grains. The edges of the grains are sharp, implying that there has been no significant diffusion of transition metals from the LSCF phase to the CGO phase.

Discussion

The combination of ^{18}O isotopic exchange with high lateral resolution SIMS measurements shows some differences in oxygen exchange between the composites at 500 °C and at 700 °C. For both temperatures, "effective" values of D^* and k^* were obtained by fitting the diffusion profiles for the composite as if they came from a single-phase material. The effective diffusivities were higher than the values reported for single-phase CGO,[24] which was attributed to a lower impurity content in the present material.[37] The k^* values were also higher than for single-phase LSCF,[23] suggesting a slight enhancement in surface exchange rate for these composites.

Laterally resolved SIMS analyses of the surface isotopic fraction of the composite is more homogeneous than, for example, the cation distribution. At 700 °C, the CGO actually shows a higher ^{18}O surface isotopic fraction than the LSCF, indicating that the effect is more significant than a simple diffusion of ^{18}O from the LSCF phase to the CGO phase. The apparent surface exchange coefficients for each phase (calculated from the surface isotopic fraction and literature values of D^* by means of eqn (1)) of the CGO phase increase 30 fold, whilst those of the LSCF phase decrease by a factor of 0.3–0.4, depending on temperature. We recently reported enhancement of the surface exchange properties of CGO when treated with nitrates of the constituent cations of LSCF;[46] it is interesting to note that treatment with $Sr(NO_3)_2$ resulted in an increase in the measured k^* value by a factor of 24, a similar magnitude of enhancement to that indicated here.

In order to aid interpretation of these results, we characterised the surface composition of the composites by both ToF-SIMS and low energy ion scattering (LEIS) spectroscopy. In light of the changes in surface composition, we consider three mechanisms which may explain this, summarised in Fig. 9. Briefly, these mechanisms are; (1) catalytic spill-over, (2) activation of the CGO surface by incorporation of transition metals and (3) cleaning of the CGO surface by the dissolution of impurities into the LSCF.

Catalytic spill-over is a process by which an active species, such as dissociated oxygen, is formed on one surface and is transported to an adjacent surface which does not form that species.[47] For example, if the dissociation of oxygen is relatively fast on the LSCF surface, but incorporation there is slow, the activated oxygen may diffuse across the surface to the CGO where incorporation is relatively fast and dissociation is the rate limiting step. This has also been proposed as the active

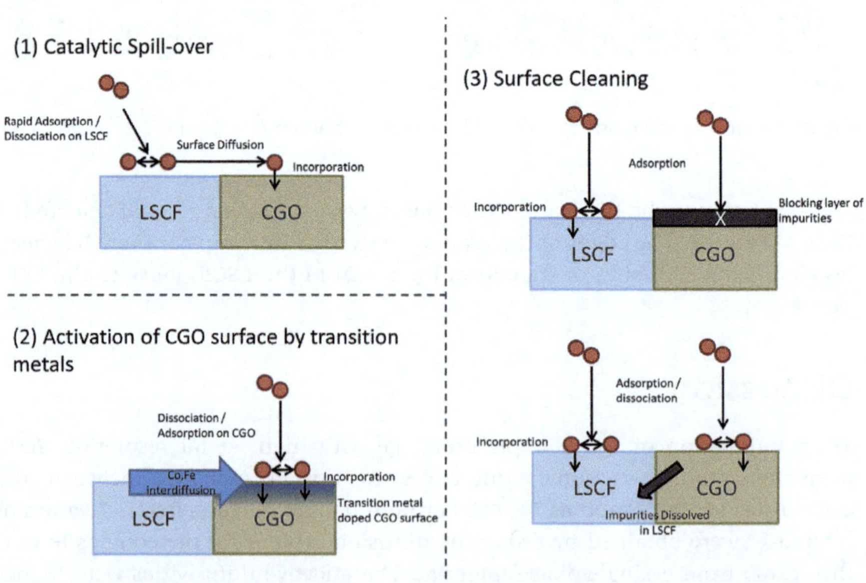

Fig. 9 Schematic of possible mechanisms for surface exchange enhancement for CGO grains in composite.

mechanism for surface exchange in YSZ–LSM composites.[12] We may also extend the spill-over mechanism to include sub-surface transfer of ^{18}O which has been incorporated from the LSCF into the CGO. Such a mechanism may be consistent with the more homogeneous isotopic fraction seen for the composite exchanged at 500 °C, but may not explain the increased surface isotopic fraction in the CGO grains of the composite exchanged at 700 °C.

The second possibility to consider is that the transition metal cations could diffuse to the surfaces of the CGO, which may make the surface of the CGO catalytically active towards oxygen incorporation. In the present composites, SIMS mapping of transition metal related species (Fig. 8) did not suggest any inter-diffusion of these cations into the CGO, and the concentration of transition metals at the surface measured by LEIS (Fig. 6) seemed to follow the same trend as in the single-phase LSCF. It is also worth noting that although in ref. 46treatment of CGO surfaces with transition metal nitrates, which were subsequently decomposed to oxides, gave a factor of 15–20 enhancement, the greatest enhancement was actually seen for the lanthanum and strontium nitrate treatments. Therefore, there does not seem to be much evidence to support this explanation for the observed enhancement.

Previous studies of the fluorite structured oxides by LEIS and ToF-SIMS indicated that there is a strong tendency for impurities such as sodium, silicon and calcium to segregate from the oxide bulk to the surface.[42,43,45,48] Furthermore, these impurities form a layer which blocks surface exchange,[43] and removal of this layer leads to an increase in the surface exchange coefficient. Given the solubilities of the typically observed impurities (Na, Si and Ca) in the perovskite structure, and the fact that these weren't observed in the outer surface of the composite (Fig. 7), we believe that the LSCF has dissolved these impurities from the CGO surface, leaving a bare "intrinsic" CGO surface, with a high surface exchange coefficient.

We note that mechanisms (1) and (3) are by no means mutually exclusive. Since the impurities can block oxygen incorporation, it is likely that a "clean" CGO surface would also be necessary for a catalytic spill-over to occur. Although further studies are warranted, *e.g.* using different grain sizes to change the surface diffusion lengths, which may control these processes, we believe that mechanisms (1) and (3), or a combination of the two, are most supported by the present data. We do not find any evidence to support mechanism (2) — activation of the CGO surface by incorporation of transition metal cations.

Conclusions

We present results of an ^{18}O isotope exchange-SIMS study combined with surface compositional analysis by the emerging technique of low energy ion scattering (LEIS) spectroscopy. Isotope exchanges were conducted at two temperatures, 500 °C and 700 °C, with subsequent analysis by ToF-SIMS, and the oxygen exchange behaviour was interpreted with consideration of differences in surface composition between the single-phase end members and the composites.

It was found that for the exchange at 500 °C, the ^{18}O isotope fractions at the surface of the composite were homogeneous across both phases, even though the surface isotopic fractions of single-phase materials exchanged at the same time showed very different values. A feature in the ^{18}O isotopic fraction was evident near the surface, which may be due to predominant ^{18}O incorporation into the LSCF, some of which is subsequently transferred into the CGO phase.

At 700 °C, the ^{18}O isotopic fraction, and apparent surface exchange coefficient, was higher for the CGO phase in the composite than for the LSCF. Several mechanisms were considered for this, in light of surface compositional analysis by LEIS and ToF-SIMS, and we believe that a combination of a catalytic spillover and surface cleaning mechanism may be responsible for such enhancement.

Acknowledgements

The authors acknowledge support from the International Institute for Carbon Neutral Energy Research (wpi-I2CNER), funded by the World Premier Research Center Initiative of the Ministry of Education, Culture, Sports, Science and Technology. HT also acknowledges the Japanese Society for Promotion of Science for its financial support through the JSPS postdoctoral fellowship and Kakenhi Grant-in-Aid project 25-03770.

References

1 A. V. Virkar, *Int. J. Hydrogen Energy*, 2010, **35**, 9527–9543.
2 A. Mitterdorfer and L. J. Gauckler, *Solid State Ionics*, 1998, **111**, 185–218.
3 M. Mori, T. Abe, H. Itoh, O. Yamamoto, G. Q. Shen, Y. Takeda and N. Imanishi, *Solid State Ionics*, 1999, **123**, 113–119.
4 J. M. Ralph, C. Rossignol and R. Kumar, *J. Electrochem. Soc.*, 2003, **150**, A1518–A1522.
5 E. Perry Murray, M. J. Sever and S. A. Barnett, *Solid State Ionics*, 2002, **148**, 27–34.
6 N. Sakai, H. Kishimoto, K. Yamaji, T. Horita, M. E. Brito and H. Yokokawa, *J. Electrochem. Soc.*, 2007, **154**, B1331–B1337.
7 J. D. Sirman, D. Waller and J. A. Kilner, *Cation diffusion into a CGO electrolyte*, Electrochemical Society Inc, Pennington, 1997.
8 M. Izuki, M. E. Brito, K. Yamaji, H. Kishimoto, D. H. Cho, T. Shimonosono, T. Horita and H. Yokokawa, *J. Power Sources*, 2011, **196**, 7232–7236.
9 V. V. Kharton, A. V. Kovalevsky, A. P. Viskup, A. L. Shaula, F. M. Figueiredo, E. N. Naumovich and F. M. B. Marques, *Solid State Ionics*, 2003, **160**, 247–258.
10 V. V. Kharton, A. A. Yaremchenko and E. N. Naumovich, *J. Solid State Electrochem.*, 1999, **3**, 303–326.
11 H. Yokokawa, N. Sakai, T. Kawada and M. Dokiya, *Solid State Ionics*, 1992, **52**, 43–56.
12 M. Dhallu, PhD, University of London, 2006.
13 M. Dhallu and J. A. Kilner, *J. Fuel Cell Sci. Technol.*, 2005, **2**, 29–33.
14 T. Horita, K. Yamaji, N. Sakai, H. Yokokawa, T. Kato and T. Kawada, *Solid Oxide Fuel Cells Vii (Sofc Vii)*, 2001, vol. 2001, pp. 539–546.
15 A. K. Opitz, A. Schintlmeister, H. Hutter and J. Fleig, *Phys. Chem. Chem. Phys.*, 2010, **12**, 12734–12745.
16 J. Druce and J. A. Kilner, 2015, in preparation.
17 B. Hagenhoff, *Mikrochim. Acta*, 2000, **132**, 259–271.

18 R. N. S. Sodhi, *Analyst*, 2004, **129**, 483–487.
19 J. A. Kilner, S. J. Skinner and H. H. Brongersma, *J. Solid State Electrochem.*, 2011, **15**, 861–876.
20 H. Tellez, A. Aguadero, J. Druce, M. Burriel, S. Fearn, T. Ishihara, D. S. McPhail and J. A. Kilner, *J. Anal. At. Spectrom.*, 2014, **29**, 1361–1370.
21 J. A. Kilner, H. Tellez, M. Burriel, S. Cook and J. Druce, *Solid Oxide Fuel Cells 13 (Sofc-Xiii)*, 2013, vol. 57, pp. 1701–1708.
22 J. A. Kilner, B. C. H. Steele and L. Ilkov, *Solid State Ionics*, 1984, **12**, 89–97.
23 S. J. Benson, PhD, University of London, 1999.
24 P. S. Manning, J. D. Sirman and J. A. Kilner, *Solid State Ionics*, 1996, **93**, 125–132.
25 H. Tellez, J. Druce, J. E. Hong, T. Ishihara and J. A. Kilner, 2015.
26 J. Crank, *The Mathematics of Diffusion*, Clarendon, Oxford University Press, Oxford, UK, 1979.
27 R. A. De Souza, J. Zehnpfenning, M. Martin and J. Maier, *Solid State Ionics*, 2005, **176**, 1465–1471.
28 H. H. Brongersma, M. Draxler, M. de Ridder and P. Bauer, *Surf. Sci. Rep.*, 2007, **62**, 63–109.
29 H. H. Brongersma, in *Characterization of Materials*, John Wiley & Sons, Inc., 2012.
30 H. H. Brongersma, T. Grehl, P. A. van Hal, N. C. W. Kuijpers, S. G. J. Mathijssen, E. R. Schofield, R. A. P. Smith and H. R. J. ter Veen, *Vacuum*, 2010, **84**, 1005–1007.
31 M. de Ridder, R. G. van Welzenis and H. H. Brongersma, *Surf. Interface Anal.*, 2002, **33**, 309–317.
32 E. Jud, Z. Zhang, W. Sigle and L. Gauckler, *J. Electroceram.*, 2006, **16**, 191–197.
33 S. Taub, R. E. A. Williams, X. Wang, D. W. McComb, J. A. Kilner and A. Atkinson, *Acta Mater.*, 2014, **81**, 128–140.
34 R. A. De Souza, J. A. Kilner and J. F. Walker, *Mater. Lett.*, 2000, **43**, 43–52.
35 Y. C. Chung and B. J. Wuensch, *J. Appl. Phys.*, 1996, **79**, 8323–8329.
36 A. Esquirol, J. Kilner and N. Brandon, *Solid State Ionics*, 2004, **175**, 63–67.
37 J. Druce, PhD, University of London, 2010.
38 R. J. Chater, S. Carter, J. A. Kilner and B. C. H. Steele, *Solid State Ionics*, 1992, **53**, 859–867.
39 D. McPhail and M. Dowsett, in *Surface Analysis – the Principal Techniques*, John Wiley & Sons, Ltd, 2009, pp. 207–268.
40 G. Holzlechner, M. Kubicek, H. Hutter and J. Fleig, *J. Anal. At. Spectrom.*, 2013, **28**, 1080–1089.
41 J. Druce, H. Tellez, M. Burriel, M. D. Sharp, L. J. Fawcett, S. N. Cook, D. S. McPhail, T. Ishihara, H. H. Brongersma and J. A. Kilner, *Energy Environ. Sci.*, 2014, **7**, 3593–3599.
42 P. J. Scanlon, R. A. M. Bink, F. P. F. van Berkel, G. M. Christie, L. J. van Ijzendoorn, H. H. Brongersma and R. G. Van Welzenis, *Solid State Ionics*, 1998, **112**, 123–130.
43 M. de Ridder, A. G. J. Vervoort, R. G. van Welzenis and H. H. Brongersma, *Solid State Ionics*, 2003, **156**, 255–262.
44 J. Druce, T. Ishihara and J. Kilner, *Solid State Ionics*, 2014, **262**, 893–896.
45 M. de Ridder, R. G. van Welzenis, H. H. Brongersma, S. Wulff, W. F. Chu and W. Weppner, *Nucl. Instrum. Methods Phys. Res., Sect. B*, 2002, **190**, 732–735.

46 J. Druce and J. A. Kilner, *J. Electrochem. Soc.*, 2014, **161**, F99–F104.
47 W. C. Conner and J. L. Falconer, *Chem. Rev.*, 1995, **95**, 759–788.
48 K. V. Hansen, K. Norrman and M. Mogensen, *Surf. Interface Anal.*, 2006, **38**, 911–916.

Faraday Discussions

PAPER

Calcium manganite as oxygen electrode materials for reversible solid oxide fuel cell

Chengsheng Ni* and John T. S. Irvine

Received 18th February 2015, Accepted 7th April 2015
DOI: 10.1039/c5fd00026b

For an efficient high-temperature reversible solid oxide fuel cell (RSOFC), the oxygen electrode should be highly active for the conversion between oxygen anions and oxygen gas. $CaMnO_{3-\delta}$ (CM) is a perovskite that can be readily reduced with the formation of Mn^{3+} giving rise to oxygen defective phases. CM is examined here as the oxygen electrode for a RSOFC. $CaMn_{0.9}Nb_{0.1}O_{3-\delta}$ (CMN) with Nb doping shows superior electric conductivity (125 S cm^{-1} at 700 °C) compared with CM (1–5 S cm^{-1} at 700 °C) in air which is also examined for comparison. X-ray diffraction (XRD) data show that CM and CMN are compatible with the widely used yttria-stabilized zirconia (YSZ) electrolyte up to 950 °C. Both materials show a thermal expansion coefficient (TEC) close to 10.8–10.9 ppm K^{-1} in the temperature range between 100–750 °C, compatible with that of YSZ. Polarization curves and electrochemical impedance spectra for both fuel cell and steam electrolysis modes were investigated at 700 °C, showing that CM presented a polarization resistance of 0.059 Ω cm^2 under a cathodic bias of −0.4 V while CMN gave a polarization resistance of 0.081 Ω cm^2 under an anodic bias of 0.4 V. The phase stability up to 900 °C of these materials was investigated with thermogravimetric analysis (TGA) and variable temperature XRD.

Introduction

Renewable energy sources have attracted much attention in recent years to reduce the dependency on fossil fuels that cause pollutant gas emission and climate change.[1] The lack of synchronization between electricity production and consumption makes energy storage crucial for the grid to function effectively. A reversible solid oxide fuel cell (RSOFC) that incorporates the production of H_2 fuel from electricity *via* a solid oxide electrolysis cell (SOEC) and electricity generation *via* a solid oxide fuel cell (SOFC) could be an efficient and low-cost solution for energy conversion and storage. Operation at high temperatures (500–1000 °C) reduces the electrical energy requirement for the electrolysis and also increases the thermal efficiency of the power-generating cycle.[2–4] It is expected that through the combination of a high-temperature reactor and a high temperature electrolyser, the process will achieve an overall thermal conversion efficiency of 48–59%

School of Chemistry, University of St Andrews, Fife KY16 9ST, Scotland, UK. E-mail: cn24@st-andrews.ac.uk while avoiding the challenging chemistry and corrosion issues associated with the thermochemical processes.[4]

Yttria-stabilized zirconia (YSZ) is widely used as the solid oxide electrolyte due to its high stability at low oxygen partial pressure at elevated temperature. Sr-doped $LaMnO_3$ (LSM) typically fulfils the requirement for use as a cathode in solid oxide fuel cells (SOFCs) operating at high temperature ($T > 850$ °C) due to its high thermal and chemical stability and catalytic activity for oxygen reduction in composites with YSZ, but it exhibits much higher polarization losses as a cathode than its Fe and Co analogues, Sr-doped $LaFeO_3$ (LSF) and Sr-doped $LaCoO_3$ (LSCo)[5–7] at intermediate temperatures (500–800 °C). For an SOEC with a configuration of Ni(O)–YSZ|YSZ|LSM–YSZ at 700 °C, the loss from the oxygen electrode accounts for most of the total polarization loss,[5,6,8] so it would be useful to minimize the energy loss from the oxygen electrode for a better performance. Although high performance has been demonstrated, neither LSF nor LSCo are compatible with the YSZ electrolyte due to the high thermal expansion coefficient and reactivity during the high-temperature firing of the electrode on electrolyte during processing.[9]

Novel materials compatible with YSZ need to be explored for better performance as the oxygen electrode of RSOFCs at temperatures below 800 °C for the purpose of using metallic interconnects and also reducing degradation and maintenance costs. $AMnO_{3-\delta}$ (A = Ba, Sr or Ca) is a perovskite without expensive rare-earth element where most Mn shows a valence of IV and gives rise to the formation of an oxygen vacancy when Mn^{IV} is partially reduced to Mn^{III}. Both $BaMnO_{3-\delta}$ and $SrMnO_{3-\delta}$ were reported to have a hexagonal structure[10] and show very low electric conductivity,[11] but $CaMnO_{3-\delta}$ (CM) has an orthorhombic $GdFeO_3$ structure and shows a conductivity of higher than 1 S cm^{-1} at 700 °C.[12] The addition of oxyanion dopants (such as Si, B and P) in to a $Ca_{1-y}Sr_yMnO_{3-\delta}$ system was reported to increase the conductivity by increasing the ratio of Mn^{III} and inhibit the formation of hexagonal $SrMnO_{3-\delta}$.[11,13,14] However, these nonmetallic-oxide-doped perovskites showed a reactivity even with CGO at 1000 °C,[14] a common buffer layer to prevent the reaction between YSZ and the oxygen electrode materials. Oxygen-deficient $CaMn_{1-x}Nb_xO_{3-\delta}$ ($x < 0.6$) has been used as catalyst thermoelectric materials and sensors[12,15] thanks to the change in valence of Mn, crystal structure and oxygen content. Since $CaMn_{0.9}Nb_{0.1}O_{3-\delta}$ (CMN) showed the optimum Nb doping content in respect to the high electric conductivity (125 S cm^{-1} at 700 °C),[12,16] it is also studied as a novel oxygen electrode for more practical benefits in the current collection. CM and CMN are for the first time electrochemically characterized as oxygen electrodes for an RSOFC: they both show high catalytic activity for the reduction and oxidation of oxygen at 700 °C. A clear relation between the physical/chemical property variation, *e.g.* TEC, microstructure and oxygen content thanks to the addition of Nb on the B site of a perovskite, and the difference in electrochemical performance of the oxygen electrode from both materials under an anodic and cathodic bias were established in this study.

Experimental

Stoichiometric amounts of $CaCO_3$ (99%, Alfa Aesar), MnO_2 (98%, Alfa Aesar, metal basis) and Nb_2O_5 (99.9985%, Alfa Aesar, metal basis) were mixed in an agate tray with a pestle and pre-calcined in a alumina crucible at 1000 °C for 12 hours. Then the 20 gram admixture was ground in a planetary ball-mill at 300 rpm for 30 min for better mixing and then pelletized and fired at 1300 °C for 20 hours for the correct perovskite phase. The heating and cooling ramp rates were 10 °C min^{-1} for both the calcination and final firing state. The pellets were then crushed and ground using an agate mortar for characterization with thermogravimetric analysis (TGA) and room temperature X-ray diffraction (XRD). Dense pellets of dilatometry measurement were obtained by calcining the ball-milled powder at 1300 °C for 20 hours and then cooling to room temperature at 1 °C min^{-1} to avoid any cracking during the cooling process. The dimensions for the pellet were approximately 10 mm in diameter and 2 mm in thickness. The thermal expansion of the dense pellets was analyzed with a high-temperature dilatometer (Netzsch, DIL 402C) in a flowing air atmosphere. The samples were heated and cooled to 900 °C at a rate of 3 °C min^{-1} and dwelled at 900 °C for 1 hour. For the compatibility test between the materials and YSZ, the 1.0 g powder of CM(N) and 1.0 g YSZ was agate-mixed, pressed under 150 MPa and then calcined at 950 °C for 5 hours at a ramp rate of 10 °C min^{-1}. XRD was performed at room temperature on a PANalytical Empyrean diffractometer operated in reflection mode using Cu-$K_{\alpha 1}$ (wavelength: 1.5406 Å) radiation for the analysis of unit cell and structure. The XRD data were simulated with GSAS simulation software, in order to determine the unit cell and arrangement of atoms in the materials. TGA was performed on a NETZSCH 49C thermo-gravimeter up to 900 °C at a ramp rate of 10 °C min^{-1}.

Pellets of 8% Y_2O_3-stabilized zirconia (8YSZ) powder were pressed uni-axially at a pressure of 80 MPa and calcined at 1500 °C for 10 hours to give a >95% dense electrolyte. The electrolyte diameter was controlled to be 20 mm. In order to improve the adhesion between the electrode and electrolyte, the hand-ground powder was milled with zirconia balls (~1 mm in diameter) again at 650 rpm in isopropanol for 30 minutes in a planetary ball-mill to get the fine source powders for the fabrication of the electrode. High temperature XRD measurements were performed on a Panalytical diffractometer operated in reflection mode using a Mo target (wavelength of $K_{\alpha 1} = 0.7093$ Å and $K_{\alpha 2} = 0.71359$ Å). The samples were stabilized at each temperature for 15 minutes before each scanning in the 2θ range from 10 to 50°. Three grams of the mixed powder was mixed with a vehicle containing polyvinyl butyral (PVB) and fish oil to make the slurry that would be applied onto a dense 8YSZ electrolyte through screen printing. Three screen-printed layers with intermediate dryings at 80 °C were utilized to give an electrode of around 20 μm thickness after firing. A good bonding between the electrode and electrolyte was obtained after calcination at 950 °C for 2 hours at a ramp rate of 3 °C min^{-1}.

Two configurations of cells were used to test the performance of the air electrode in this paper. A symmetrical cell contains two identical oxygen electrodes on both sides of the 8YSZ electrolyte of 1.8 mm in thickness and 2 cm in diameter. The symmetrical cells were employed to measure the polarization resistance of the electrode in static air in the temperature range between 550 °C and 700 °C. The other configuration, called a three-electrode system, was composed of an oxygen electrode, a Ni(O)–YSZ fuel electrode and a reference electrode on the edge of the electrolyte (~1.0 mm thick). The Ni(O)–YSZ fuel electrode (13 mm in diameter) was prepared by screen printing of the slurry containing 55 wt% NiO and 45 wt% YSZ onto the electrolyte plus a firing at 1350 °C for 3 hours. A

$(La_{0.8}Sr_{0.2})_{0.95}MnO_3$ slurry was painted onto the edge of the electrolyte and dried at 80 °C for 30 minutes and then fired at 950 °C for two hours along with the firing of the oxygen electrode. A silver wire (0.5 mm thick) was bundled the LSM coating and then fired at 700 °C again to provide the reference. The humidified gas (N_2–50% H_2O–3% H_2) in the fuel electrode was produced by injecting deionised water into an evaporator to mix with the nitrogen carrier gas containing 5% H_2, as described previously.[2]

Silver paste was used for the current collection for both the symmetrical-cell and three-electrode testing. The impedance was measured with a Solartron electrochemical workstation equipped with a lock-in frequency analyzer (Solartron 1260) with a sine wave of 10 mV. The three-electrode testing procedure commenced with sealing of the fuel electrode side of cells to the alumina tube of the testing jig, which was then placed in a vertical furnace. Scanning electron microscopy (SEM) and transmission electron microscopy (TEM) of the powders and tested samples are taken on a Jeol 6700F microscope and Jeol 2011, respectively.

Result and discussion

Crystal structure and compatibility with YSZ

The crystal structure of CM and CMN was regarded as an orthorhombic structure,[12,15,17] with the cell parameters of $\sqrt{2}a_{\text{cubic}} \times 2a_{\text{cubic}} \times \sqrt{2}a_{\text{cubic}}$ (a_{cubic} refers to the cubic structure), involving successive cooperative rotations of MnO_6 octahedra around the $\langle 111 \rangle$ crystallographic axis. The cell parameters of calcium manganite depend on the oxygen content that can be affected by the synthesis route in terms of firing temperature and cooling rate.[18,19] The cell parameters determined from powder XRD are shown in Table 1, showing that the addition of Nb on the Mn site enlarges the cell parameters.

The reaction between zirconia and the cathode perovskite materials, especially those containing cobalt on the B-site, would cause the formation of insulating products, such as $La_2Zr_2O_7$ or $SrZrO_3$ that would block the transport of electrons/ions between the electrode and electrolyte. The XRD patterns (Fig. 1) show no observable reaction after 5 hour annealing in air in contact with YSZ at 950 °C. No obvious variation of lattice parameter (Table 1) was observed. Thermodynamic calculation has predicted the formation of $CaMn_2O_4$ for the CM–YSZ system at

Table 1 Cell parameters (in Å) and volume (Å^3) of CM and CMN with symmetry of *Pnma*, oxygen nonstoichiometry, (δ) from TGA measurement and the Goldschmidt tolerance factors, τ^a

	a	b	c	V	δ	$\frac{Mn^{IV}}{Mn^{IV}+Mn^{III}}$	τ
CM	5.2808(3)	7.45525(1)	5.2668(2)	207.258(4)	0.028	0.944	0.878
CMN	5.3316(2)	7.5119(2)	5.3074(2)	212.504(4)	0.017	0.851	0.871
CM^a	5.2825(1)	7.4563(2)	5.2706(4)	207.603(4)	—	—	—
CMN^a	5.3326(2)	7.5112(2)	5.3058(3)	212.532(9)	—	—	—

^{a} Cell parameters after calcined with YSZ at 950 °C for 5 hours.

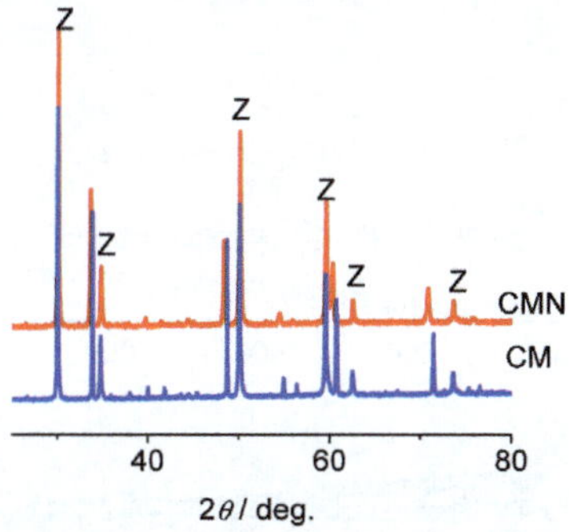

Fig. 1 XRD patterns of CM and CMN with YSZ after calcination at 950 °C for 5 hours. The peaks indexed with "Z" belong to YSZ and the rest to perovskites.

1400 °C,[20] which was not observed in our case due to the relatively low temperature. The compatibility of LSM is limited to 1200 °C and can be improved by A-site deficiency. Attempts have been made to prepare A-site deficient $Ca_{0.95}MnO_{3-\delta}$ and $Ca_{0.95}Mn_{0.9}Nb_{0.1}O_{3-\delta}$ using the same method as for their stoichiometric counterparts, but $CaMn_2O_4$ persisted as the impurity in both materials, even though the phase diagram[21] of CaO–Mn_2O_3 indicated that 10% A-site deficient was achievable. The A-site deficiency of a perovskite was reported to be controlled by the bond energy of ⟨B–O⟩[22] and 5% A-site deficiency for LSM was reported.

Chemical and structural transformations with temperature variation

The oxygen non-stoichiometries, δ, were obtained by H_2-reduction TGA. In order to analyze the TGA data, the initial assumption was made that the final products of the reduction contained Mn in the +2 state and Nb in the +5 state.[15] The oxygen non-stoichiometry, δ, is listed in Table 1, and the ratio of Mn^{IV} to Mn^{IV} + Mn^{III} is also calculated, assuming the charge neutrality of the perovskite. It can be seen that the oxygen content in CMN is higher than that of CM. A higher oxygen content in niobium doped CM was also reported by Kruth *et al.*[15] for the samples synthesized in air, due to the higher thermal stability with Nb doping during the high temperature synthesis. TGA under air and oxygen (Fig. 2) can provide information on structural and chemical changes occurring in the volume of CM and CMN with temperature variation. It can be seen that in oxygen atmosphere, CM also picks more weight than CMN to make the δ value near zero after calcination at 900 °C for 1 hour. CMN is more stable against thermal treatment in air up to 900 °C. The δ values of CM and CMN change to 0.099 and 0.025 at 900 °C in air, respectively. The changes of oxygen content caused by the valence variation of the transition metal were anticipated to be beneficial to the electrochemical performance of the oxygen electrode.[23]

The Goldschmidt tolerant factor for perovskite structure, τ, of the materials was calculated according to the equation:[24]

$$\tau = \frac{R_A + R_O}{\sqrt{2}(R_B + R_O)} \tag{1}$$

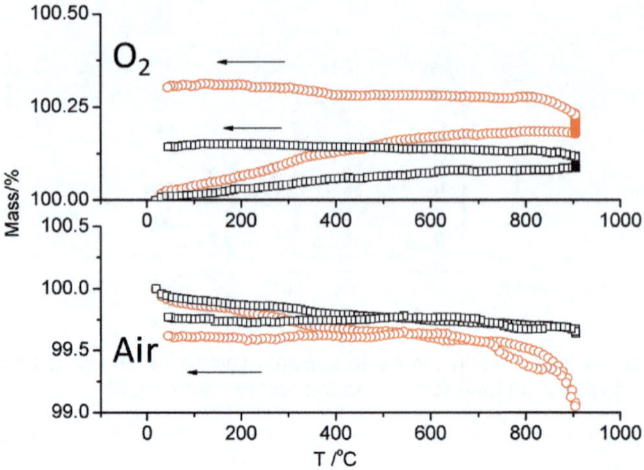

Fig. 2 TGA curves CM (open circles) and CMN (open squares) in air and O_2. The heating or cooling direction is indicated with the arrow. The ramp rate is 10 °C min^{-1}.

where, R_A = ionic radius of larger cation, R_B = ionic radius of smaller cation, R_O = ionic radius of oxygen (=1.40 Å).

The R_B for CM and CMN was calculated *via* the composition weighted average ionic radii of Mn^{IV}, Mn^{III} and Nb^{V}. It can be seen that the addition of Nb into the lattice of CM increases the percentage of Mn^{III} (Table 1) and decreases the tolerance factor thanks to the larger radii of Mn^{III} (72 pm for low-spin) and Nb^{V} (78 pm) than Mn^{IV} (67 pm for low spin) under the similar oxygen content.[15] The tolerance factors of both materials are *ca.* 0.87, less than 1 for a cubic perovskite. On the one hand, Dabrowski *et al.*[25] found that the thermal stability of $Sr_{1-x-y}Ca_xBa_yMnO_3$ perovskites was governed by the magnitude of tolerance factor ($\tau < 1$) which can be adjusted by the oxygen content. Similarly, the stabilizing energy of a perovskite structure was found to be in a linear relationship to the tolerance factor (when $\tau < 1$).[26] On the other hand, the substitution of a more electronegative Nb (electronegativity, $\chi_{Nb} = 1.60$) for Mn ($\chi_{Mn} = 1.55$) reduces the tendency to form oxygen vacancies.[27] For example, Nb and Mo are reported to efficiently stabilize the cubic perovskite crystal structure of $SrCoO_{3-\delta}$ oxide, which is noted for both high oxide ionic and electronic conductivity at elevated temperature.[28,29] A balance of the above two factors should strike a balance to control the thermal stability of CM and CMN.

The values of TEC in flowing air are presented in Fig. 3. One can distinguish two segments with different TECs for both CM and CMN. The thermal expansion of CMN is more linear than CM in the whole temperature range. Both materials show quite similar TECs in the temperature range 25–750 °C, but CM shows a much higher TEC between 750–900 °C. A hysteresis of the shrinkage in the heating and cooling period was observed, and can be explained by the non-equilibrated release and recovery of oxygen of the bulk materials in air. Two TECs delimited at temperature *ca.* 750 °C were assumed for better fitting of the expansion data. At temperatures below 750 °C, both materials show similar TECs, *ca.* 13 ppm K^{-1} for heating and *ca.* 11 ppm K^{-1} for cooling. On the contrary, CM shows much larger

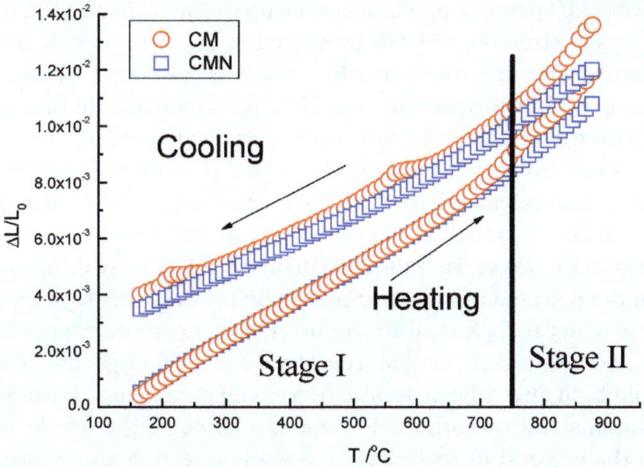

Fig. 3 Dilatometry of the CM and CMN samples in the temperature range between 150 and 900 °C. The ramp rate for the heating and cooling was 3 °C min^{-1} and the samples dwelled in flowing air at 900 °C for 1 hour before cooling down.

thermal expansion than CMN at temperatures above 750 °C, reaching 21 ppm K^{-1} for heating and 25 ppm K^{-1} for cooling. The heavy doping of Nb on the B-site was reported to decrease the thermal expansion of $LaCoO_3$,[30] but this is not obvious in our study unless at temperatures higher than 750 °C. The additional expansion at higher temperatures can be a result of chemical expansion due to the change of Mn^{IV} to the larger Mn^{III}, similar to the phenomenon observed in Sr-doped lanthanum cobalt oxide.[31] It should be noted that the thermal history of the sample is also very important to the TECs of the samples, as can be seen from the discrepancy between the heating-up and cooling-down process. More importantly, the thermal expansion in the cooling stage is closer to that of YSZ and more meaningful since the stress could arise from the strain in the cooling process after the calcination of powder on the electrolyte.

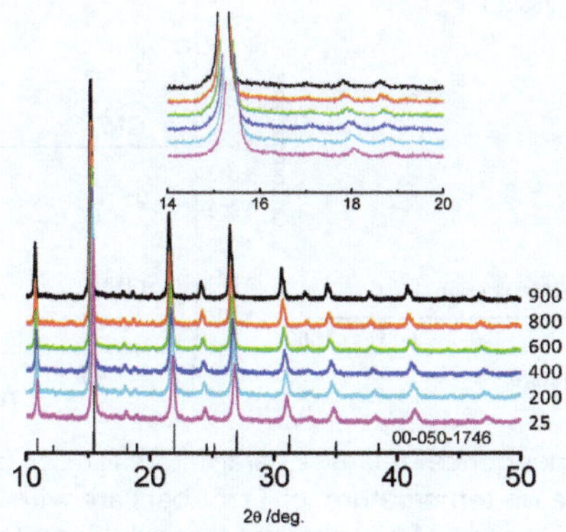

Fig. 4 High temperature XRD spectra of CMN from 25 to 900 °C in static air. The insert is the magnified region in between 14–20°. The temperature (°C) is indicated beside the graph. A standard powder diffraction file (pdf) of CM is provided to show the rough position of peaks.

The *in situ* XRD patterns at temperatures up to 900 °C in air were measured to study the crystal structure of CMN presented in Fig. 4 to provide a better equilibrium than dilatometry since very fine powders (as shown afterwards in this paper) were used for the determination of lattice parameters in this study. The *in situ* XRD patterns of CM and CMN were performed in static air with an environmental X-ray diffractometer with Mo target. The cell parameters of CMN at room temperature calculated from XRD with Mo target are $a = 5.3226(2)$ Å, $b = 7.5103(2)$ Å and $c = 5.3110(3)$ Å, and the volume of unit cell is 212.301(5) $Å^3$, which are quite similar to those from the XRD using a Cu $K_{\alpha1}$ X-ray diffractometer. The unit cell of the powder at room temperature before and after high-energy milling was checked with Cu $K_{\alpha1}$ X-ray diffractometer, and no variation was observed. The fluorescence of Mn-ion under characteristic X-ray of Cu target and $K\alpha_2$ of the Mo target would both affect the determination of cell parameters. It can be seen that CMN maintains the orthorhombic structure through the whole temperature range and the peaks shift toward the low angle direction along with increasing temperatures, indicating an expansion of the unit cells. The unit cell parameters (Fig. 5) of both materials during the heating stage were calculated by fitting the peaks into an orthorhombic structure. The linear fitting formula of unit cell parameters and parabolic fitting parameters of cell volumes are inserted in the

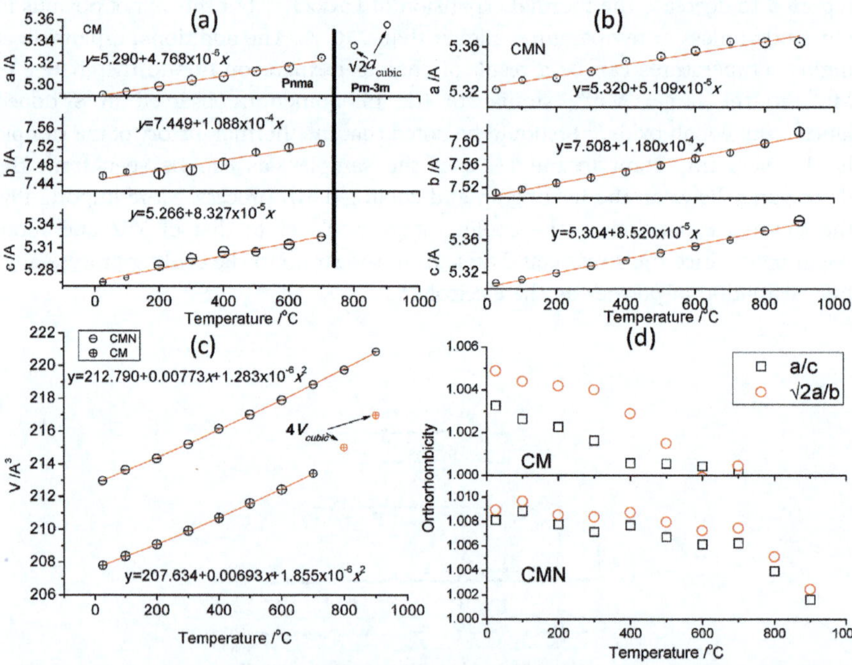

Fig. 5 Temperature dependence of lattice parameters for CM (a) and CMN (b) and their cell volume dependence on temperature (c). Error bars are within the symbols. (d) Shows the orthorhombicity of CM and CMN in terms of *a*/*c* and $\sqrt{2}a/c$. The linear fitting of the unit cell in selected temperature ranges is indicated in the respective graphs of (a) and (b) and the parabolic simulation of cell volume with temperature is indicated in (c). The unit cell parameter (a_{cubic}) and volume (V_{cubic}) of the cubic phase ($Pm\bar{3}m$) for CM at 800 and 900 °C is scaled by $\sqrt{2}$ and 4, respectively, for better comparison.

diagram and the thermal expansion from *a*-, *b*- and *c*-directions are listed in Table 2. The TECs calculated from $V^{1/3}$ (V = volume of unit cells) are in correspondence to those from dilatometry in the heating up stage. Previous studies[17] on CM reported the existence of successive structural transitions from orthorhombic (*Pnma*) to tetragonal (*I4/mcm*) at 896 °C and further to cubic ($Pm\bar{3}m$) at 913 °C, but we observed the change from orthorhombic to cubic structure at temperatures higher than 800 °C by analyzing the orthorhombicity of the structure, possibly due to the fast loss of oxygen due to the high surface area. It is obvious from the orthorhombicity that the addition of Nb clearly stabilizes the orthorhombic structure. A recent study has shown that chemical expansion is the result of two competing processes:[32] the lattice contraction primarily due to electrostatic interaction and the lattice expansion due to the cation radius change. The exceptionally large unit cell parameter for CM at 800 and 900 °C could be the lattice expansion thanks to the higher concentration of Mn^{3+} cation in the lattice due to the loss of oxygen. The thermal expansion is anisotropic and the thermal expansion in *a*-direction is much smaller than those in the *b*- and *c*-direction. Basically, the addition of 10% Nb in CM increases slightly the thermal expansion in the *a*-, *b*- and *c*-direction from the high temperature XRD, but wider temperature range for CMN could bring in more contribution from chemical expansion. The variation in temperature on CM between 25 °C and 800 °C showed a volumetric increase of 2.45%,[33] slightly smaller than the value of this study (3.26%). The high temperature XRD analysis of $CaMn_{0.98}Nb_{0.02}O_3$ in the temperature between 25 °C and 900 °C showed a volumetric increase of 4.40% (ref. 27) and turned out to be a cubic structure at 900 °C. The discrepancy in thermal expansion between different studies could be explained by a difference in the oxygen content in CM due to thermal history that can contribute to the chemical expansion at elevated temperatures.[31]

Microstructure of the powders and oxygen electrode

In order to improve the sinterability of the powder at 950 °C for better bonding between the oxygen electrode and electrolyte, the hand-ground powder was milled in a high-energy planetary milling machine for finer particles. The powder and milling media was mixed in isopropanol milled at a high speed of 650 rpm for 30 min. Transmission electron microscopy (TEM) and scanning electron microscopy (SEM) were used to analyze the fine powder after high energy milling and the results are shown in Fig. 6 and 7, respectively. It can be seen that the particle size of both materials is in the range between 10 nm to 2 μm and the CMN particles are slightly finer than those of CM. Amorphous phases are found in both CM and CMN. In particular, the former consists of small polycrystalline particles and

Table 2 Thermal expansion values of CM and CMN from high temperature XRD in selected temperature ranges. The unit of thermal expansion is ppm K^{-1}

	A	*b*	*c*	$V^{1/3}$
CM 25–700 °C	9.0	14.6	15.8	13.6
CMN 25–900 °C	9.5	15.7	16.1	13.8

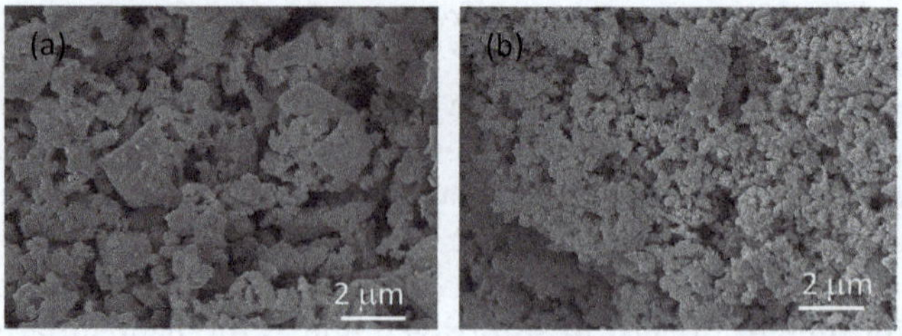

Fig. 6 SEM images of CM (a) and CMN (b) after milling with high-energy planetary milling.

covered with an amorphous layer while the latter composes of mostly a single crystal. The calculated *d*-spacings of the selected areas of CM and CMN are 0.257 nm and 0.271 nm respectively, as indicated in the Fig. 7, which could correspond to the (121) crystal planes.

The screen-printed layers (20–30 μm in thickness) composed of CM and CMN after calcination at 950 °C for two hours show good bonding to the YSZ electrolyte and similar grain size compared to the raw powder, as shown in Fig. 8. The grains lost their sharp edges in the sintering process and connected well to each other, which is beneficial to the transport of electrons.

Electrochemical characterization

Electrochemical impedance spectroscopy was used to evaluate a symmetrical-electrode system (CM(N)|YSZ|CM(N)) to study the polarization resistance of the electrode at OCV in static air, as shown in Fig. 9. The Nyquist and Bode plots of the symmetrical electrode at 700 °C are shown in Fig. 9(a). One arc can be distinguished for the CMN electrode, while two arcs for CM after analyzing the Bode plots. The Arrhenius plots of the series resistance and polarization resistance are shown in Fig. 9(b) and (c). R_s is obtained from the high-frequency intersection between the Nyquist plots and *x*-axis, and R_p is calculated by the reduction of R_s from the low frequency intersection between the Nyquist plots and *x*-axis. The $\log(R_s)$ shows a linear plot against the reciprocal of temperature ($1000/T$) and activation energy of 0.8 eV, which is slightly lower than the activation energy (>1.0 eV below 650 °C)[34] of YSZ in the temperature below 700 °C. The R_s may compose of two parts: the ohmic resistance of YSZ electrolyte and CM or CMN

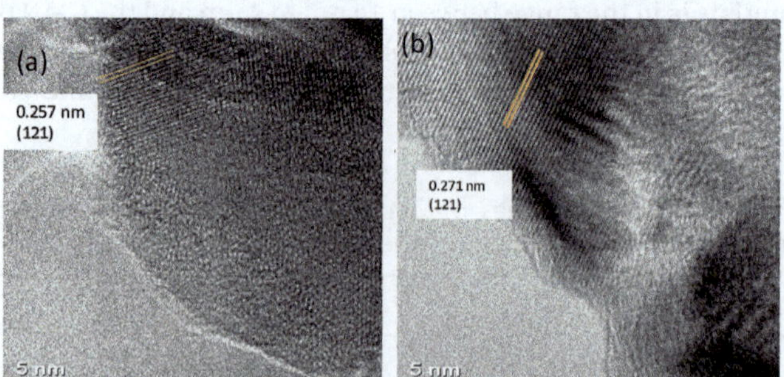

Fig. 7 Transmission electron microscopy images of CM (a) and CMN (b).

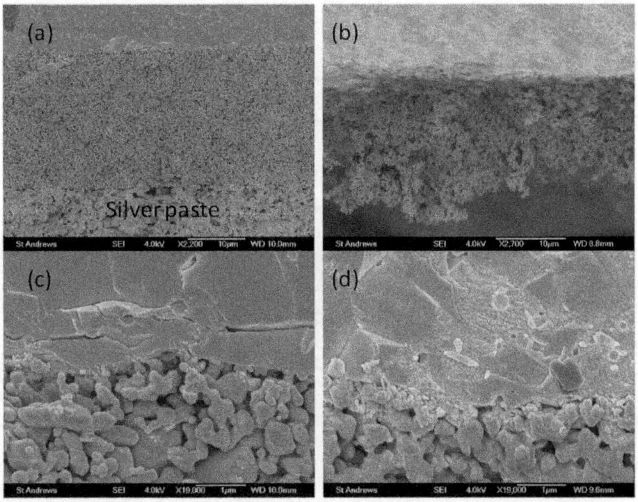

Fig. 8 SEM images of post-test CM (a and c) and CMN (b and d) after calcination at 950 °C for 2 hours in air and testing at 700 °C for symmetrical cells.

electrode. Even though the conductivity of CMN (125 S cm^{-1} at 700 °C) is higher than that of CM (1–5 S cm^{-1} at 700 °C),[12,19] the R_s for both electrodes are similar, indicating that the ohmic loss from the electrode is negligible. The R_p values for the CM and CMN electrodes are 3.40 and 5.53 Ω cm^2, respectively and are quite close to that of LSM/YSZ oxygen electrode on YSZ electrolyte (~4 Ω cm^2) at 700 °C.[5,35] The polarization resistance of CM or CMN on YSZ at 700 °C is much smaller than that of $SrMnO_3$ or CM/$Gd_{0.1}Ce_{0.9}O_{2-\delta}$ (CGO) composite on CGO electrolyte at 700 °C (~8 Ω cm^2)[11,14] and LSM electrode on YSZ (~10 Ω cm^2 at 800 °C).[6] Porras-Vazquez *et al.*[11] found that the composite electrode $CaMn_{0.95}Si_{0.05}O_{3-\delta}$/CGO showed a lower area-specific resistance (ASR) (0.35 Ωcm^2 at 800 °C) than CM/GDC (1.5 Ωcm^2 at 800 °C) and the increase of electric conductivity due to Si doping was thought to be the reason. With the benefit of higher ionic conductivity, the addition of CGO can decrease the electronic conductivity of the electrode. It was reported that the doping of 10% Ta, 10% Nb or 5% Mo on B site stabilized the low-temperature cubic structure of (La)SrCo(Fe)$O_{3-\delta}$ to avoid the negative effect of intermediate hexagonal structure for better performance,[28,29,36–38] while in our case the addition of Nb stabilized the low temperature-orthorhombic structure that showed inferior performance than the undoped CM. The addition of YSZ into LSM as oxygen electrode was reported to enhance the performance of electrode by a factor of two at 800 °C (ref. 39) due to the enlargement of triple phase boundary (TPB), since LSM showed an ionic conductivity as low as 10^{-7} S cm^{-1}.[9] Two linear regions are implied in the Arrhenius plots and the activation energies in the temperature range 650–700 °C are twice as large as those in 540–620 °C. The changes in activation energy are ascribed to the shifting of rate-limiting step from bulk diffusion limiting to surface diffusion limiting as the temperature increases since the activation energy of the bulk diffusivity is normally half of the surface diffusivity.[40,41]

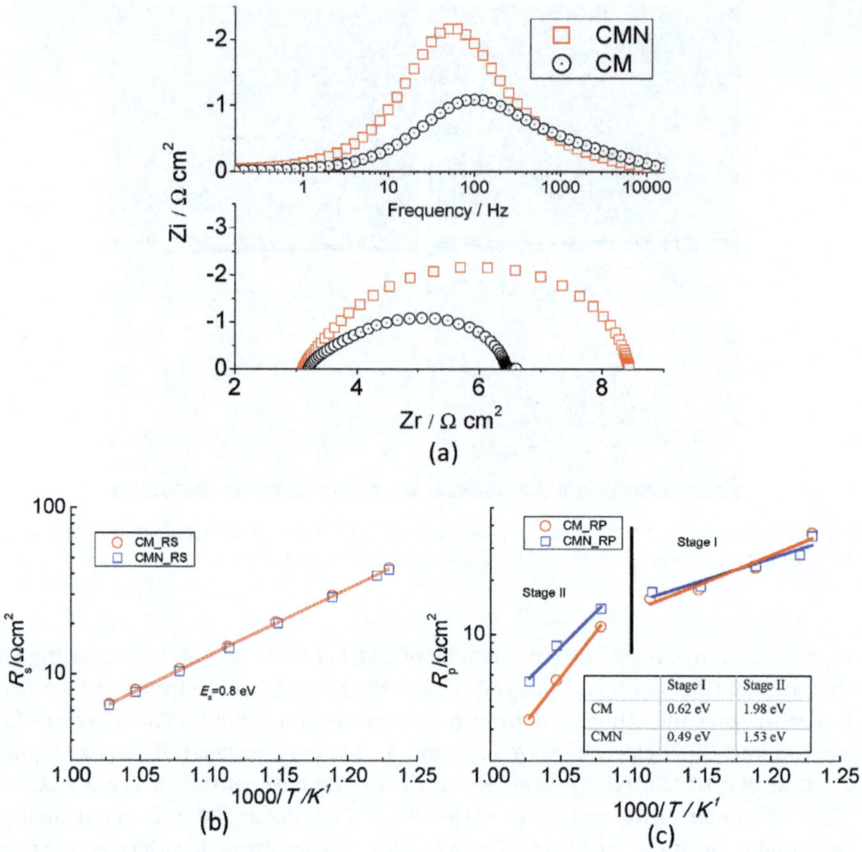

Fig. 9 (a) The impedance spectra (Nyquist plots and bode plots) of CM and CMN at 700 °C in static air. Arrhenius plots of the R_s (b) and R_p (c) at OCV before activation for the symmetrical cells in the temperature range between 550 to 700 °C. The activation energies of R_s and R_p are indicated in the respective diagrams.

In order to study the properties of the electrodes under current load, a three-electrode system with a reference electrode was used for the electrolysis measurement. The fuel electrode was Ni(O)–YSZ under N_2–50% H_2O–3% H_2 at a flow rate of 500 mL min^{-1}. Fig. 10 shows the typical impedance spectrum of the electrolysis cell with CMN as the oxygen electrode at OCV. The impedance spectrum between the reference and either anode or cathode allows one to determine the electrode loss from the nonohmic part of the spectrum. In a three-electrode system, the improper placement of the reference and working electrodes will give erroneous results. It is noteworthy that the ohmic resistances in both spectra are quite close, and almost one-half of the total ohmic resistance of the cell as indicated from the two-electrode testing, showing that the reference electrode is sampling the potential near the centre of the electrolyte. Furthermore, the characteristic frequencies, ν^*, for the fuel electrode spectra, 200 and 2 kHz, are different from that for the oxygen electrode, 80 Hz. For this cell, the fuel electrode and oxygen electrode impedances are 0.5 and 1.5 Ω cm^2, respectively, showing the fuel electrode contributes a minor amount to the polarization resistance of the whole cell. The difference in R_p (1.5 *vs.* 5.53 Ω cm^2)

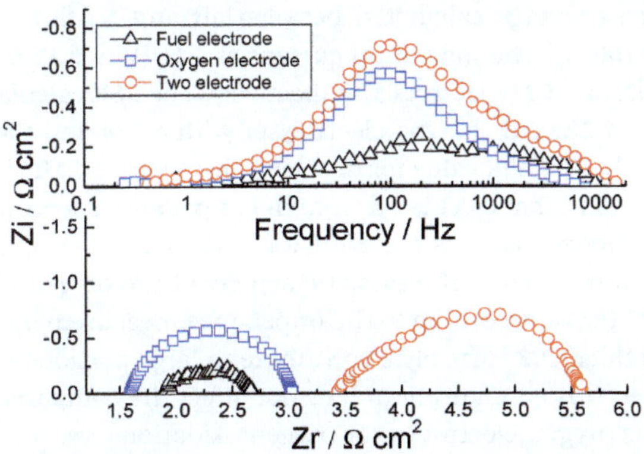

Fig. 10 Impedance with three-electrode system for the CMN oxygen electrode, Ni/YSZ hydrogen electrode and the whole electrolyser.

between the three electrode measurement and the previous symmetrical-cell measurement is caused by the additional activation of the oxygen electrode under 100 mA cm^{-2} cathodic current for 24 hours in the former case. It should be noted that the characteristic frequency of the CMN electrode is the same, 80 Hz, in both two electrode and three electrode systems. The electrodes can also be activated with anodic current since the same R_p of 1.5 Ω cm^2 was observed if a 100 mA cm^{-2} current was passed through the symmetrical cell where a cathodic current for one electrode is anodic for the other. Similar polarization hysteresis has been observed in the LSM/YSZ electrode,[5,42] in which polarization resistance decreased after both cathodic and anodic current passage.

The *IV* curves of the electrolysis in the voltage range between −1.7 V to −0.82 V are shown in Fig. 11, where both *IV* curves show good linearity. The open circuit voltages (OCV) of both electrodes are 0.82 V, which is consistent with the

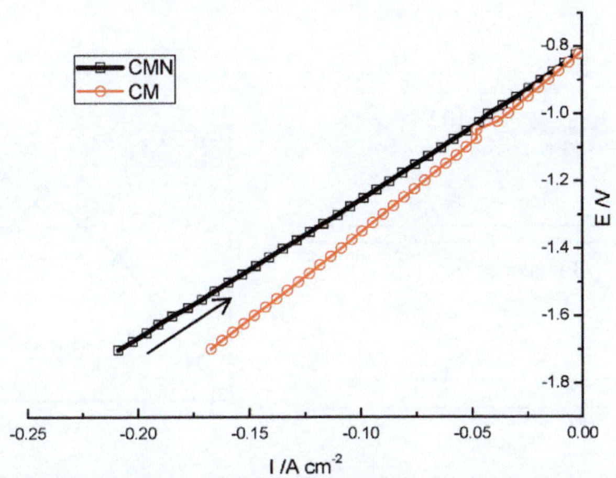

Fig. 11 *IV* curves of the electrolyser with CM and CMN oxygen electrode after activation for 24 hours under cathodic current 100 mA cm^{-2}. The steam electrode is Ni/YSZ in N_2–50% H_2O–3% H_2 at a flow rate of 500 mL min^{-1}.

theoretical Nernst voltage calculated between air and N_2 flow containing 50% H_2O–3% H_2 at 700 °C. The maximum current at −1.7 V is 0.21 A cm^2 and 0.17 A cm^2 for the CMN and CM electrodes, respectively. The ASR calculated from the *IV* curve is 5.3 and 4.2 Ω cm^2 for the electrolyser with a CM and CMN oxygen electrode, respectively. The ASR value for the electrolyser with a CMN oxygen electrode are 20% smaller than that calculated from the impedance spectrum at OCV, 5.5 Ω cm^2. Since the anode is kept as the same for both electrolysers, the difference in performance is attributed to the catalytic activity of the oxygen electrode for the oxidation of O^{2-} to O_2. According to the impedance measurement, the ohmic loss from the 1 mm thick YSZ electrolyte contributes a large portion of the total loss at 700 °C. However, this configuration does not affect the comparison between the two materials as oxygen electrodes for oxygen oxidation.

The Nyquist and Bode plots of the CM electrode are shown in Fig. 12(a). The characteristic frequencies for CM increase to a higher frequency after activation under 100 mA cm^{-2} cathodic current, which could be attributed to the change of surface and capacitance. The polarization resistance of the oxygen electrode under bias was measured using impedance and presented in Fig. 12(b). However, CM presents a polarization resistance of 0.059 Ω cm^2 under −0.4 V bias as the cathode for fuel cell operation while CMN gives a polarization resistance of 0.081 Ω cm^2 under 0.4 V bias as the anode for the electrolysis process. Considering the R_s is higher than R_p for both electrodes at all biases, the maximum actual voltage loss on the electrode under a bias of 0.4 V is lower than 0.2 V, which is quite a reasonable loss for the oxygen electrode in either fuel cell or electrolyser modes. The reversible decrease in impedance upon cathodic polarization of LSM-based electrodes has been observed by several groups, and the mechanism was not definitely clear,[43,44] but the reduction of LSM surface could be involved in the activation process. Similarly, the loss of oxygen on the surface of CM during the reduction process could result in the formation of vacancies through eqn (2),[45] which are beneficial to fast transport of the oxygen ion in the electrode.

$$\frac{1}{2}O_2(g) + V_O^{\cdot\cdot} + 2Mn'_{Mn} = O_O^{\times} + 2Mn_{Mn}^{\times} \quad (2)$$

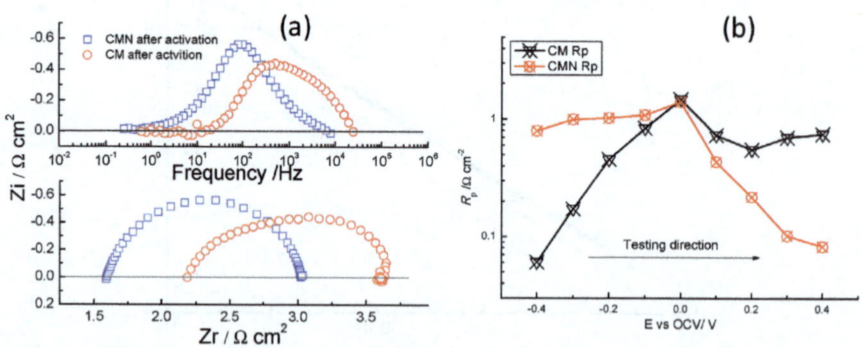

Fig. 12 Nyquist and Bode plots (a) of the CM and CMN cathodes at 700 °C in static air at OCV immediately after the activation at 100 mA cm^{-2} cathodic current for 24 hours. (b) is the R_p measured under bias for both electrodes with three-electrode system. The counter electrode is a Ni/YSZ electrode in N_2–50% H_2O–3% H_2 at a flow rate of 500 mL min^{-1}.

Here, the Kroger–Vink notations are used.[46] The oxygen vacancies could also be ordered to be immobile and the highest conductivity for CM was achieved at $\delta = 0.08$ where no oxygen vacancy was observed.[47,48] Under anodic current, the decrease in the non-stoichiometry due to the absorption of O_2 means the loss of oxygen vacancies, and explains the resultant decrease of performance in the bias range of 0.2–0.4 V. Aguadero *et al.*[29] stipulated that the performance increase of $SrCo_{1-x}Mo_xO_3$ oxygen electrode under anodic current was due to the increase of p-type carriers that improves the electrocatalytic properties. This may help to explain the decrease of performance of CM a typical n-type oxide[48] under anodic current. On the other hand, because CMN is more difficult to be reduced as demonstrated in the TGA curves, there is only a slight improvement in the performance under cathodic current due to the negligible increase of oxygen vacancies. CMN can be more stable under anodic current that preserves the oxygen vacancy for the transport of oxygen ions. Moreover, an oxygen-excess CMN ($\delta = -0.1$) was synthesized with high oxygen pressure,[15] and it was not very clear if the interstitial oxygen could also be a route of oxygen transport.[47,49]

Conclusion

CM and CMN were examined as oxygen electrodes for SOEs and SOFCs. The addition of 10 mol% Nb on the B site of perovskite improved the thermal stability of CM in air and oxygen up to 900 °C. Both materials showed a thermal expansion of 10.8–10.9 ppm K^{-1}, close to that of YSZ (~11 ppm K^{-1}). At 700 °C, CM as a cathode for SOFCs could achieve a polarization resistance less than 0.1 Ω cm^2, but CMN showed superior performance ($R_p < 0.1$ Ω cm^2 at 0.4 V bias) as an anode for SOECs. The performance of the electrode can be further improved by mixing with ionic conductors to make a composite electrode. A possible exploration of the materials with a Nb content less than 10 mol% doping on Mn site would lead to a better oxygen electrode performance in both SOFC and SOE mode.

Acknowledgements

We thank EPSRC for funding under the contract of H2FC hub EPSRC: EP/J016454/1 and Platform EPSRC: EP/K006800/1 and JTSI thanks Wolfson Merit of Royal Society of Chemistry: WRMA 2012/R2.

References

1 Q. Cai, C. S. Adjiman and N. P. Brandon, *J. Power Sources*, 2014, **268**, 212–224.
2 G. Tsekouras and J. T. S. Irvine, *J. Mater. Chem.*, 2011, **21**, 9367–9376.
3 S. H. Jensen, P. H. Larsen and M. Mogensen, *Int. J. Hydrogen Energy*, 2007, **32**, 3253–3257.
4 Y. Shin, W. Park, J. Chang and J. Park, *Int. J. Hydrogen Energy*, 2007, **32**, 1486–1491.
5 W. Wang, Y. Huang, S. Jung, J. M. Vohs and R. J. Gorte, *J. Electrochem. Soc.*, 2006, **153**, A2066–A2070.
6 Y. Bo, Z. Wenqiang, X. Jingming and C. Jing, *Int. J. Hydrogen Energy*, 2008, **33**, 6873–6877.

7 W. Wang, M. D. Gross, J. M. Vohs and R. J. Gorte, *J. Electrochem. Soc.*, 2007, **154**, B439–B445.

8 H. Uchida, N. Osada and M. Watanabe, *Electrochem. Solid-State Lett.*, 2004, 7, A500–A502.

9 J. M. Vohs and R. J. Gorte, *Adv. Mater.*, 2009, **21**, 943–956.

10 K. Kuroda, N. Ishizawa, N. Mizutani and M. Kato, *J. Solid State Chem.*, 1981, **38**, 297–299.

11 J. M. Porras-Vazquez, E. R. Losilla, P. J. Keenan, C. A. Hancock, T. F. Kemp, J. V. Hanna and P. R. Slater, *Dalton Trans.*, 2013, **42**, 5421–5429.

12 G. Xu, R. Funahashi, Q. Pu, B. Liu, R. Tao, G. Wang and Z. Ding, *Solid State Ionics*, 2004, **171**, 147–151.

13 C. A. Hancock and P. R. Slater, *Dalton Trans.*, 2011, **40**, 5599–5603.

14 J. M. Porras-Vazquez, T. F. Kemp, J. V. Hanna and P. R. Slater, *J. Mater. Chem.*, 2012, **22**, 8287–8293.

15 A. Kruth, U. Guth and A. R. West, *J. Mater. Chem.*, 1999, **9**, 1579–1583.

16 C. A. Hancock, A. L. Ong, P. R. Slater and J. R. Varcoe, *J. Mater. Chem. A*, 2014, **2**, 3047–3056.

17 H. Taguchi, M. Nagao, T. Sato and M. Shimada, *J. Solid State Chem.*, 1989, **78**, 312–315.

18 T. E. F. Lichtenthaler, *Master*, Universitetsbiblioteket i Oslo, 2006.

19 M. E. Melo Jorge, A. Correia dos Santos and M. R. Nunes, *Int. J. Inorg. Mater.*, 2001, **3**, 915–921.

20 H. Yokokawa, N. Sakai, T. Kawada and M. Dokiya, *Solid State Ionics*, 1990, **40–41**(Part 1), 398–401.

21 P. V. Riboud and A. Muan, *J. Am. Ceram. Soc.*, 1963, **46**, 33–36.

22 E. Y. Konysheva, X. Xu and J. T. S. Irvine, *Adv. Mater.*, 2012, **24**, 528–532.

23 E. Konysheva and J. T. S. Irvine, *J. Mater. Chem.*, 2008, **18**, 5147–5154.

24 R. S. Roth, *J. Res. Natl. Bur. Stand.*, 1957, **58**, 75–88.

25 B. Dabrowski, O. Chmaissem, J. Mais, S. Kolesnik, J. D. Jorgensen and S. Short, *J. Solid State Chem.*, 2003, **170**, 154–164.

26 H. Yokokawa, N. Sakai, T. Kawada and M. Dokiya, *Solid State Ionics*, 1992, **52**, 43–56.

27 L. Bocher, M. H. Aguirre, R. Robert, D. Logvinovich, S. Bakardjieva, J. Hejtmanek and A. Weidenkaff, *Acta Mater.*, 2009, **57**, 5667–5680.

28 W. Zhou, Z. Shao, R. Ran, W. Jin and N. Xu, *Chem. Commun.*, 2008, 5791–5793.

29 A. Aguadero, D. Pérez-Coll, J. A. Alonso, S. J. Skinner and J. Kilner, *Chem. Mater.*, 2012, **24**, 2655–2663.

30 V. Oygarden and T. Grande, *Dalton Trans.*, 2013, **42**, 2704–2715.

31 X. Chen, J. Yu and S. B. Adler, *Chem. Mater.*, 2005, **17**, 4537–4546.

32 D. Marrocchelli, S. R. Bishop, H. L. Tuller and B. Yildiz, *Adv. Funct. Mater.*, 2012, **22**, 1958–1965.

33 Q. Zhou and B. J. Kennedy, *J. Phys. Chem. Solids*, 2006, **67**, 1595–1598.

34 J. T. S. Irvine, I. R. Gibson and D. P. Fagg, *Ionics*, 1995, **1**, 279–285.

35 J. Kim, H.-I. Ji, H. P. Dasari, D. Shin, H. Song, J.-H. Lee, B.-K. Kim, H.-J. Je, H.-W. Lee and K. J. Yoon, *Int. J. Hydrogen Energy*, 2013, **38**, 1225–1235.

36 V. Cascos, R. Martínez-Coronado and J. A. Alonso, *Int. J. Hydrogen Energy*, 2014, **39**, 14349–14354.

37 B. Qu, W. Long, F. Jin, S. Wang and T. He, *Int. J. Hydrogen Energy*, 2014, **39**, 12074–12082.

38 S. Yoo, J. Kim, S. Y. Song, D. W. Lee, J. Shin, K. M. Ok and G. Kim, *RSC Adv.*, 2014, **4**, 18710–18717.

39 S. Wang, Y. Jiang, Y. Zhang, J. Yan and W. Li, *J. Electrochem. Soc.*, 1998, **145**, 1932–1939.

40 H. K. B. W. David Kingery and D. R. Uhlmann, *Introduction to Ceramics*, John Wiley & Sons, Inc, 2nd edn, 1976.

41 Z. Shao and S. M. Haile, *Nature*, 2004, **431**, 170–173.

42 M. Liang, B. Yu, M. Wen, J. Chen, J. Xu and Y. Zhai, *J. Power Sources*, 2009, **190**, 341–345.

43 S. P. Jiang and J. G. Love, *Solid State Ionics*, 2001, **138**, 183–190.

44 M. J. Jørgensen, S. Primdahl and M. Mogensen, *Electrochim. Acta*, 1999, **44**, 4195–4201.

45 E. Goldyreva, I. Leonidov, M. Patrakeev and V. Kozhevnikov, *J. Solid State Electrochem.*, 2012, **16**, 1187–1191.

46 F. A. Kröger and H. J. Vink, in *Solid State Physics*, ed. S. Frederick and T. David, Academic Press, 1956, vol. 3, pp. 307–435.

47 A. Reller, J. M. Thomas, D. A. Jefferson and M. K. Uppal, *Proc. R. Soc. London, Ser. A*, 1984, **394**, 223–241.

48 M. Schrade, R. Kabir, S. Li, T. Norby and T. G. Finstad, *J. Appl. Phys.*, 2014, **115**, 103705.

49 J. B. Goodenough, *Annu. Rev. Mater. Res.*, 2003, **33**, 91–128.

Faraday Discussions

DISCUSSIONS

Materials development: general discussion

Raymond Gorte, John Vohs, Theis L. Skafte, Robert Kee, John Varcoe, Ian Metcalfe, Sune Dalgaard Ebbesen, Guntae Kim, Dehua Dong, San Ping Jiang, Ming Li, Tatsumi Ishihara, John Bøgild Hansen, Beatriz Molero-Sanchez, Steven McIntosh, Helena Téllez, Alex Morata, Viola Birss, Xiangling Yue, John Druce, Joseph Hartvigsen, Ann Call, John Irvine, Tae Ho Shin, Mogens Mogensen, Chengsheng Ni, Marc Torrell, Christopher Graves and Bilge Yildiz

DOI: 10.1039/C5FD90074C

Robert Kee opened a general discussion of the paper by John Vohs: You are developing dual-phase electrolytes, with one or more of the phases being mixed conductors. Can you offer any insight concerning the possibility of different internal electrostatic potentials within the participating phases, and thus the possibility of electrical double layers at the interfaces between the phases. Any charge-transfer chemistry across the phase boundaries could affect the defect concentrations and electrostatic potential profiles within the phases, and thus influence local charge neutrality and migration contributions for ion transport within each of the phases.

John Vohs answered: You raise an interesting question. It is not something that we have thought much about, but I suspect that for the relatively large grain sizes that are present in our dual-phase electrolyte membranes, the presence of electrical double layers at the interfaces between the two phases would have little effect on the transport of electrons and oxygen ions through the bulk of the membrane. The effects might be more important on the surfaces of the membranes, especially for the ones that we have infiltrated LSCr where the grain sizes are much smaller, but we are not able to detect the presence or absence of double layers so it is hard to say for sure. This might be something that a modeling approach could provide insight into.

Tae Ho Shin asked: In your manuscript you mentioned that theses materials show high ionic conductivity, thus, I am wondering what portion you have in your system with ionic conductivity, and how can you distinguish between ionic and electronic values from your measuring? In addition, during high temperature processing or long term operation over 800 °C, is there a chemical reaction?

John Vohs replied: The estimates for ambipolar conductivity, which is essentially equal to the ionic conductivity for materials that have high electronic conductivity, were obtained from the resistance *versus* thickness data presented in Fig. 5. For the conditions we used to synthesize the membranes we did not observe any reaction between the LSCr and YSZ phases by XRD.

Steven McIntosh questioned: Platinum addition does not seem to have a dramatic impact on the observed oxygen flux, is this due to the membrane now being limited by the air side or by bulk transport?

Raymond Gorte responded: While the effect of platinum does not appear to be that large, it is actually significant. As we point out in the manuscript, at 1073 K, the resistance of the membrane decreased from 3.77 to 2.85 Ω cm^2 upon the addition of platinum. In previous work on SOFCs,[1] the effect of adding platinum was actually quite similar to the difference shown here.

1 S. McIntosh, J. M. Vohs and R. J. Gorte, *Electrochem. Solid-State Lett.*, 2003, **6**, A240–A243.

John Vohs replied: While not dramatic, platinum addition did cause modest improvement. We did not test a cell in which we only added platinum rather than platinum and ceria. As the data in the paper shows, ceria addition did cause a dramatic increase in performance and as you suggest it may be that the added activity provided by the ceria was such that the air side became limiting. It should also be noted that on the fuel side we were doing CO oxidation which is not a very catalytically demanding reaction. For less facile reactions, such as methane oxidation, I suspect that platinum addition would have a much bigger effect.

Ming Li said: Is the infiltration process reproducible and controllable? Is the infiltrated electrode stable?

John Vohs answered: The infiltration process is very reproducible. While we pioneered this method of electrode fabrication it is now used by many groups and has proven to be an effective way to synthesize high performance electrodes. Electrodes produced by this method have good stability, but some sintering of the high surface area infiltrated film can occur over time and this can cause some loss in activity.

Viola Birss remarked: It would be very helpful to know/determine the real (active) surface area of SOFC/SOEC electrodes, as some of the differences seen in the literature may be due to surface/active area differences more than other factors. Normally, the surface areas of electrodes synthesized using ceramic methods are just not high enough to determine by gas sorption methods, and so other techniques are needed. Do you have any suggestions regarding how to determine the real surface area of MIEC electrodes, after sintering, perhaps even after operation, *etc.*? It would have been nice to be able to use the chemical capacitance for this purpose, but this will likely not work as it is not known to what depth into MIECs the capacitance arises from.

John Vohs responded: The fact that surface area plays an important role in the overall performance of an electrode was one of the primary points I was trying to get across in my talk. I agree with your point that this is something that is often not considered when one is comparing results in the literature. As you note, the surface areas for electrodes are quite small which makes them challenging to measure using the standard BET method. We have, however, had some success using this technique for low surface area electrodes. It requires a well-designed BET apparatus with a low volume and the use of krypton as the adsorbent. We use a home built apparatus for this purpose and can measure total surface areas as low as 0.1 m^2.

Viola Birss asked: A previous question was asked about whether there could be a potentially interfering double layer issue in the mixed phase structures that you showed. My comment/question was whether this could be explored by examining the impedance response at somewhat lower temperatures, where the grain boundary impedance (resistance and interfacial capacitance, with the capacitance being a reflection of double layer properties) should appear.

John Vohs replied: I suspect that it could but this is not something that we have attempted to do.

Tatsumi Ishihara enquired: You used sintering agents for preparing composites, however, is this an issue at the interface? Why choose bromide? Is this because bromide easily deactivates the catalytic property?

John Vohs answered: We did not use a sintering agent.

Tatsumi Ishihara then said: You used La(Sr)CrO_3–YSZ for the electrode, however, it is well-known that $LaCrO_3$ reacts with ZrO_2 to form $La_2Zr_2O_3$ or $SrZrO_3$ which have low activity and low conductivity. Why wasn't a more active and stable catalyst chosen for the modified surface CO oxidation catalyst?

John Vohs replied: We agree that reactions between La(Sr)CrO_3 and the YSZ are potentially problematic. We chose LSCr, however, because we know that it is the electronically conductive material that is being used in a mixed conducting membrane development program in industry. It should also be noted that the high oxygen flux rates that we were able to obtain with our membranes suggest that this issue is manageable.

Raymond Gorte commented: With regards to using a Sr-doped $LaCrO_3$ (LSCr)–YSZ composite, this is very similar to the system being developed for MIEC membranes at Praxair. The system is very stable, which Praxair has stated is the main reason why they are looking at it. I do not believe that interactions between the two phases affect the conductivity of either phase significantly. Firstly, the crystallite sizes are relatively large (on the order of 500 nm) compared to the size one would expect for a boundary effect. Secondly, we were able to calculate the "overall" ionic conductivity of the membrane in the paper, based on the oxygen fluxes and the P_{O_2} gradient. That conductivity was close to what we would expect for the YSZ phase, assuming a reasonable tortuosity factor (the LSCr provides sufficient conductivity so that electronic conductivity is not limiting).

Robert Kee replied: My comment was not to suggest that the interactions between phases should affect the intrinsic conductivity of the constituent phases or that the interactions should degrade stability. It is clear that the ionic and electronic transport within the individual phases is qualitatively different, which should lead to different electrostatic potential fields within the two constituent phases. Thus, there should be electrical double layers throughout the composite at all of the phase interfaces. One at least wonders how any charge transfer across these double layers might affect the electrostatic potential fields, and thus how net ion migration may be affected. The intent of the query was not to suggest that dramatic results should be anticipated, but rather to raise a potentially interesting aspect of electrochemical transport in dual-phase composites.

John Irvine addressed Raymond Gorte and John Vohs: Have you considered using Fe-doped $La_{0.8}Sr_{0.2}CrO_3$, as this has better gas exchange function than the pure chromite?

Raymond Gorte responded: In our more recent follow-up work, we have been using the Mn-doped material, $La_{0.8}Sr_{0.2}Cr_{0.75}Mn_{0.25}O_3$ (LSCM). LSCM-YSZ composites are significantly easier to sinter, but they do not seem to have improved oxygen-exchange properties. The catalytic properties are still critical. It is interesting to notice that work from Praxair has previously shown that the addition of a catalyst to the syngas side (ceria in this case) significantly improved the performance of membranes made with the Fe-doped material, demonstrating that Fe doping is not sufficient.[1]

1 G. M. Christie, J. R. Wilson and B. A. van Hassel, *US Patent*, 8323463, 2012.

John Vohs answered: Yes, we have considered using Fe-doped LSCr for the reason you indicate and we may do this in the future. For our initial work we decided to use LSCr because we know that this material is being used in at least one commercial development program.

Mogens Mogensen queried: Could segregation of thin layers of chromium oxide to grain boundaries be the reason for the rather low conductivity with the strange temperature dependency, or is it only due to an insufficient contact between the LSCr particles?

John Vohs responded: We don't have any evidence for chromium oxide deposits at the grain boundaries, but due to the high temperatures used during fabrication, this possibility can not be ruled out.

John Irvine commented: I would concur that the chromites have little reactivity with zirconia, we typically mix YSZ with them to anchor them when using as SOFC electrodes as they tend to delaminate otherwise.

Mogens Mogensen opened the discussion of the paper by Tae Ho Shin: What is the reduction capability of the 5% hydrogen? Is it "dry" hydrogen? If the hydrogen is dry enough no oxide will be thermodynamically stable. In general, there must be a well-known ratio between steam and hydrogen in order to have defined thermodynamic conditions.

Tae Ho Shin answered: I used dry 5% hydrogen and 95% argon gas for reduction, without a water bubbler. Unfortunately, I didn't measure the steam and hydrogen contents in that gas and I agree that some oxide would be unstable under extremely reducing conditions. However, this material was obviously stable and worked under SOFC/SOEC modes with dry H_2, hydrocarbon fuel or CO/CO_2 conditions in our study. With respect to the stability of this oxide, we should study more material properties under various defined thermodynamic conditions in our future work. Thank you for your worthy comments.

John Irvine commented: 5% hydrogen from a cylinder is not a well-defined system as humidity is uncertain. Bubbling through water gives a well defined oxygen particle pressure.

Beatriz Molero-Sanchez asked: Regarding the CV in Fig. 5b of your paper, what is the scan rate? Is that increase in OCV reproducible in many experiments and in different cells? If you were to perform 6–10 cycles would you still see that increase in OCV? And would it be the same in each cycle?

Tae Ho Shin replied: I gave the scan rate as 10 mV s^{-1}, so very slow. I didn't see exactly the same value during several cycles nor 10 cycles but it showed the same tendency. Obviously, if I gave it enough time, the OCV would recover to nearly the initial value within 2 h. It might be explained by temporary carbon adsorption from the Boudouard reaction at high temperature. However, temporary carbon adsorption would be oxidized on the surface to CO_2 or CO by oxygen from the PBMO lattice. Thus, it shows a slow OCV recovery time. It would be worthy future work to find a detailed surface reaction mechanism, if possible I will do this future work.

Steven McIntosh enquired: Do you understand why the material performs relatively poorly in SOFC mode when compared with SOEC mode? Is it becoming oxidized under anodic polarization?

Tae Ho Shin responded: First of all thank you for your interest in my results. You said it performs poorly in SOFC mode. However, when it comes to a CO/CO_2 30 : 70 atmosphere in SOFC mode, the *I–V* curve would not be poor. In addition, when I measured *I–V* in SOFC mode with a H_2 atmosphere, very high current density was recorded. Anyway, referring to Fig. 5b in our paper, the resistance from the *I–V* curve in SOFC mode looks to be a higher value compared with the one in SOEC mode. I agree with you. It might be explained by different conductivities of PBMO5 and PBMO6 under different oxygen partial pressures. As we reported on the electrical conductivity in Nature materials,[1] the O5 phase has a higher conductivity value than the O6 phase, and therefore, in SOEC mode the PBMO electrode would be under reducing conditions, while in SOFC mode the PBMO electrode would be under oxidizing conditions, and nearly the O6 phase. Thus, they show different resistance values from the *I–V* curves for SOEC and SOFC modes. I hope this answer will be helpful for you to understand my results. Thanks again.

1 S. Sengodan, S. Choi, A. Jun, T. H. Shin, Y.-W. Ju, H. Y. Jeong, J. Shin, J. T. S. Irvine and G. Kim, *Nat. Mater.*, 2015, **14**, 205.

Theis L. Skafte remarked: In Fig. 5b the conditions are very far from the thermodynamic Boudouard threshold, so I doubt that the hysteresis observed is due to carbon deposition. The threshold at 850 °C is $pCO = 0.943$, while in the presented case, taking the gas flows and the small cell into account, only 3% CO_2 is converted at the shown maximum current density, reaching a maximum pCO of 0.33. If it was indeed carbon deposition, you would have a huge stagnant diffusion layer above the cell or very low electrode porosity, thereby causing large gas concentration gradients. My question is, do you know the porosity of the electrode?

Tae Ho Shin responded: Unfortunately, I don't have a porosity value for the electrode. However, I fully understand your comments. In my previous experience, the screen printed electrode would have about over 40% porous microstructure under my experimental setup. When I measured the *I–V* curves, OCV recovery time was very slow. This might be explained by temporary carbon adsorption. Anyway, I will do more to consider the porosity as in your comment in my future work. Thank you for your useful comments.

Mogens Mogensen said: The change in OCV in Fig. 5 is equal to *ca* 315 mV, more than a 30% increase in OCV. This is equivalent to a change in CO/CO_2 of a factor of *ca* 1000. You would be able to calculate whether this might cause carbon formation, *i.e.* move the gas composition inside the carbon precipitation area of the thermodynamic diagram. Furthermore, why not calculate (by integration of the current–time curve) from the measurements how many coulombs have been transferred during the different stages of the reducing part of the sweep? This might also give some hint as to what could have happened. Was the outlet gas analysed during the hold at OCV?

Tae Ho Shin replied: That would be a great point for future work. Unfortunately, I didn't analyse the exhaust gas because of equipment set-up limitations, even though I had tried to do so in this study. However, I hope to try again for a more detailed analysis using *in situ* exhaust gas analysis, and it should be helpful to shed further light on the issue in future work. Thank you again for your worthy comments.

Xiangling Yue queried: In the *I–V* curve in Fig. 5b, the very high OCV value was interpreted by the temporary carbon deposition or CO adsorption; did you try different compositions of CO/CO_2 mixtures, especially CO/CO_2 mixtures with higher CO fractions? It would be helpful to support your theory if the high OCV occurred at higher CO concentrations.

Tae Ho Shin replied: We didn't use CO/CO_2 mixtures with various compositions, we just used optimized the CO/CO_2 gas mixture, as for yours and John's previous experimental conditions for CO_2 electrolysis.[1,2] A CO/CO_2 mixture of 70% CO_2 and 30% CO vol% was employed for this experiment. However, I agree with you that it would be helpful to gain further insight if we used various conditions in our future work. Thank you for your comments.

1 X. Yue and J. T. S. Irvine, *Electrochem. Solid-State Lett.*, 2012, **15**, B31.
2 X. Yue and J. T. S. Irvine, *Solid State Ionics*, 2012, **225**, 131.

Sune Dalgaard Ebbesen continued: You claim that carbon formation during the *I–V* measurement is causing the relaxation at OCV. If carbon were formed the structure would break up and permanent degradation would be observed. Thus, the *I–V* curves cannot be reproduced if carbon is formed. Can the *I–V* curves be reproduced?

Tae Ho Shin responded: I fully agree with your comments. If I performed more long term or more *I–V* cycles, serious degradation would be found in my experiment. However, during my measurements over about 2 days, there was no breakdown. Obviously, when I measured at least 5 cycles, the curve looked roughly reproduced without serious degradation within 2 days of measuring. However, long term stability will be studied in my future work as per your useful comments.

Ann Call communicated: Referring to Fig. 5b, can you comment on how long it took for the voltage to drop from 1.141 V to the OCV value of 0.825 V? The paper just says "slow and strongly time dependent". Knowing how long it took to return to OCV may give you a clearer idea of what sort of process is occurring in the system.

Tae Ho Shin communicated in reply: When I measured the *I–V* curve, the OCV recovery time was very slow. It roughly took over 2 h to drop from 1.141 V to the initial OCV value of 0.8. As discussed, it might be explained by temporary carbon adsorption from the Boudouard reaction at high temperature. It would be worthy future work to find a detailed surface reaction mechanism.

Mogens Mogensen asked: Comparing the O5 and O6 figures in Table 1, from these numbers I get a linear relative dimension change of about 1.6%. Is this ceramic material mechanically stable when redoxed? How can it survive such a change in dimension?

Tae Ho Shin answered: That's a good point, I understand your concern. When I used fully dense specimens of PBMO, I could obviously find the huge volume change under redox, and then it looked mechanically poor under redox cycles. However, there were no serious mechanical cracks in the porous bodies such as the SOFC/SOEC electrodes after redox cycles.

John Irvine commented: I disagree, the transition from O6 to O5 is not so easy. The layer structure once formed is very stable and will not go back to the cubic structure easily, so the volume change is not so critical under operation conditions.

Mogens Mogensen said: Can the very slow change of OCV after cathodic polarization (Fig. 5) really be due to carbon formed? I would expect this to be a fast process at 850 °C.

Tae Ho Shin responded: The slow chemical kinetics are not fully understood in this work, but there would be complex reasons for that; absorption of carbon or CO/CO_2 gas molecules on the surface of the oxide might be just one of the reasons for the slow chemical kinetics because that would also be related with oxygen mobility in the PBMO, which has oxygen vacancy ordering in the lattice. Real time impedance analysis shouldn't have to be measured for that. Unfortunately, in this work, we didn't measure *in situ* impedance curves under loading current conditions. If I were to do that in future work, it would be helpful in order to understand the slow change of OCV. Many thanks for your useful comments.

Guntae Kim commented: The PBM fuel electrode for SOE experiences a phase transition from simple perovskite to double perovskite under reduction conditions at around 400 °C. This double perovskite doesn't return back to simple perovskite until temperatures of higher than 900 °C in air. It shows higher conductivity under reduced and air conditions and shows higher sulphur and carbon coking resistance. It also has the highest oxygen storage.

Mogens Mogensen asked: Fig. 6 shows big low frequency arcs. Don't you think that this points to chemical capacitance redoxing from O6 to O5 and back rather than to any carbon formation that is oxidized at more anodic potentials? Did you try to calculate the equivalent capacitance from the low frequency arc?

Tae Ho Shin replied: I didn't do that because of the ceria buffer layer between LSGM and PBMO. It would be complex to analyse only the chemical capacitance of PBMO from that impedance curve. However, if I get more reliable impedance data, it would be worthy work. Anyway, I thought that oxygen mobility in the lattice vacancy ordering under redox would be variable under different cathodic potentials and then it might cause the slow change and this would also be one of the reasons aside from carbon formation. If I use other electrolytes and PBMO electrodes with simple features, it would be useful to answer that in future work. Many thanks for your useful comments and worthy suggestion.

John Vohs opened discussion of Albert Tarancón's paper†: What is providing the catalytic activity for the reverse water gas shift reaction? Could the current collector play a role here? What were you using for the current collector, platinum, gold?

Marc Torrell replied: We were using gold paint as the current collector of the cell, but the station possesses a platinum mesh to collect the current. We'll switch to gold mesh to look at the effect.

Raymond Gorte remarked: Generally speaking, LSCM is not very catalytic. We showed this some time ago in a joint paper with John Irvine.[1] You also need to be very careful about using platinum as the current collector for oxide-based anodes. This was shown by a number of people but perhaps most clearly in a paper by Bi and Zhu.[2]

1 G. Kim, S. Lee, J. Y. Shin, G. Corre, J. T. S. Irvine, J. M. Vohs and R. J. Gorte, *Electrochem. Solid-State Lett.*, 2009, **12**, B48–52.
2 Z. H. Bi and J. H. Zhu, *J. Electrochem. Soc.*, 2011, **158**, B605–613.

Marc Torrell answered: Thanks for the comments. We used to collect the current with platinum mesh on the ProboStat stations, however we try to use gold paint to avoid the platinum paint. I will check the recommended papers and change the platinum mesh for gold mesh.

† Albert Tarancón's paper was presented by Marc Torrell, *Catalonia Institute for Energy Research, Sant Adrià de Besòs, Spain.*

John Bøgild Hansen asked: On page 9 of your paper you claim that you can not calculate the reverse water gas shift equilibrium "because of the complex balances among parallel catalytic reactions." If you know the outlet gas composition why can you not calculate the equilibrium constant?

Marc Torrell responded: We'll find references about this and calculate it in future tests. Thanks.

John Bøgild Hansen said: You state that the decrease in hydrogen to carbon monoxide ratio seen as the cell degrades is due to the higher operating voltage. Could it not be due to a higher cell temperature as you progressively operate at more exothermal conditions when departing from the thermoneutral voltage? Have you measured or estimated the real cell temperatures?

Marc Torrell replied: We monitored the cell temperature with a thermocouple close to the cell surface. We didn't observe any increase in temperature but we'll try to consider this for future measurements. Thanks.

John Irvine commented: For LSCM, I disagree with Ray that this is an inactive material, indeed we have published a paper with Ray that shows reasonable performance as well as one where it was less good. Performance is much better at lower temperatures with appropriate catalysts. To get good performance at lower temperature, high surface area is needed. Ceramic LSCM needs heating to get good performance but is not a totally dead material.

Marc Torrell responded: I totally agree.

San Ping Jiang asked: LSCM is known to have very low electrical conductivity and in the fuel environment its conductivity can be as low as 3–5 S cm^{-1}. Have you considered this factor in the development of practical LSCM based electrodes for solid oxide electrolysers?

Marc Torrell answered: LSCM has very low conductivity under low P_{O_2}, which was not the case in the presented results where a maximum of 6% H_2 was used. It is clear, and it is explained in the paper, that the presence of H_2 promotes the drop in the cell performance, but that the LSCM remains active.

Dehua Dong remarked: On page 7, the cell performance decreased after the introduction of hydrogen into the feed gas. You have attributed this to the reduction of LSCM. We noticed that the introduction of hydrogen increased the cell OCV, which might be the main contribution to the decrease in cell performance since the electrode is really inactive.

John Irvine addressed John Bøgild Hansen: It was previously suggested that adding hydrogen to the steam would minimise the corrosion of steel in the system. I am not sure that it will have that great a benefit. There are many applications using process steam which must be similar to the initial heating anyhow. Smaller scale units would be simpler without feeding hydrogen into steam feed.

John Bøgild Hansen replied: Pure steam without hydrogen is quite aggressive at elevated temperatures. In fact this is the limiting factor for the steam data in advanced power plants, so I believe that it is prudent to operate with certain hydrogen recycling also for that reason.

Xiangling Yue questioned: As you used pure LSCM as the electrode and LSCM was fired at 1200 °C, did you get a good interfacial binding with the electrolyte, *i.e.*, a good attachment of electrode to electrolyte? Is there any evidence of delamination? How do you manage this with pure LSCM? We used the same material, but found that it was very difficult to get good attachment of pure LSCM to YSZ electrolyte even at much higher temperatures.

Marc Torrell responded: This is one of the weakest points of the work. This attachment is not enough but it seems to improve with a diluted ceramic layer sprayed above the electrolyte to give some roughness to the tape casted electrolyte.

John Irvine asked: Do you find any problems with electrode delamination as you only use a pure perovskite rather than using something like ceria as a composite?

Marc Torrell replied: Yes, the anode side tends to delaminate. We want to improve this with the CGO–LSCM composite approach.

Mogens Mogensen said: In Fig. 2 of your paper, looking at conductivity compared to YSZ, what happened to the conductivity of YbScSZ? The Arrhenius curve looks wavy?

Marc Torrell responded: The slight fluctuation of the values can be ascribed to fitting mismatches of the EIS spectra, but it shows a clear linear trend.

Mogens Mogensen enquired: What is the reason for the disagreement between current densities at given voltages between the tables and curves? Are the data in Fig. 4 and 5 from the same cell or from different cells? The numbers are not the same when you go back to CO_2 (700 mA cm^{-2} *vs.* 600 mA cm^{-2}). Does the cell degrade relatively quickly with time?

Marc Torrell replied: There is no table in the paper with the current densities, and all the tables represent the same values for the same samples at given voltages, as is also described in the text. We do not ascribe the drop in current between the two different measurements of the sample due to the different conditions of operation. The degradation of the cell is given in Fig. 9 of the paper and it is relative fast, but not during the first 3 hours.

Tatsumi Ishihara asked: At 0.2 V, some current was observed in the *I–V* curves (as in Fig. 4); what kind of species are reduced under these conditions? What kind of products are observed?

Marc Torrell answered: At this voltage we are not expecting the reduction of the fuel species. That's probably the reason why the resistance in the first voltage steps is so high. Maybe the electrolyzer materials suffer some reduction until the minimum voltage to reduce the fuel is reached. The species were not characterized at these voltages.

Tatsumi Ishihara remarked: In the *I–V* curves, current was observed at very low applied potentials and I think that when the electrolysis starts and H_2 or CO is formed, P_{O_2} in the fuel site is immediately decreased and the potential should be immediately increased to the equilibrium potential. I wonder why the potential immediately increased when the current started to flow.

Marc Torrell responded: At the first steps of current, the voltage is too low to reduce steam or CO_2, but there is no chemical potential to have a voltage. So the voltage has to be increased due to some reduction of the materials involved in the cell. Then when the H_2 starts to evolve the potential increases and the performance of the cell is similar to the one operated with H_2.

Joseph Hartvigsen asked: Looking at Fig. 4 and 5, what is the explanation for the non-linearity? The shape of the curve is not consistent with diffusion overpotential and such strong activation overpotentials are not normally seen at this high temperature. Dr Christopher Graves pointed out that because the polarization sweep started with only steam and CO_2, the Nernst potential changed radically with the initial introduction of current. Once sufficient H_2/CO was produced to define and stabilize the Nernst potential in the vicinity of the cathode, the polarization curve was reasonably linear.

Christopher Graves replied: I think there are at least two effects leading to the activation in electrolysis mode. Firstly there is the activation of mixed conducting oxide electrodes by cathodic polarization, most likely due to increasing surface oxygen vacancy concentration, see my Faraday Discussions paper (DOI: 10.1039/c5fd00048c) and also another one presented earlier today (DOI: 10.1039/c5fd00029g). Secondly, at higher voltage, there may be electronic leak in the electrolyte, as the authors mention in this paper, resulting in increasingly higher current with higher voltage, but that current is not performing electrochemical reactions.

Marc Torrell answered: As we have already discussed, there is no OCV at the first steps of the *I–V* curves without H_2. That's why the curve presents a huge resistance at the beginning and then reaches the normal shape of the other *I–V* curves. Thanks.

John Vohs opened discussion of the paper by Bilge Yildiz: What is the origin of the decrease in the surface exchange coefficient with time at each temperature and how does the addition of titanium affect the stability of the electrode?

Bilge Yildiz replied: The origin of the decrease in the surface exchange coefficient is the segregation and phase precipitation of SrO-related phases at the surface, as we have shown by AES and AFM/SEM. The phase precipitation of SrO particles at the surface, in part, blocks the surface, and in part leaves behind a Sr-deficient lattice. Both of these detrimentally affect the surface exchange kinetics.

Addition of titanium makes the surface more oxidized, and kills excess oxygen vacancies. This reduces the positiveness of the surface and the electrostatic attraction of the strontium dopant (a negative defect) towards the surface. Keeping strontium in the bulk LSC ultimately avoids phase precipitation and reduces the instability of the LSC surface.

John Irvine asked: Could the titanium addition simply be inducing an A-site deficient stoichiometry. It is well established that A-site deficient manganite perovskites are less reactive than stoichiometric perovskites. Cobaltites are generally more reactive than manganites and inducing A-site deficiency should stabilise the surface.

Bilge Yildiz responded: It is possible that an A-site deficient surface could make the surface less prone to strontium segregation and phase precipitation, however we believe that this is not the dominant effect here. To test that we are not simply dealing with the effect of an A-site deficient surface, we have also added excess cobalt cations to the surface of LSC (data shown during my presentation but not in the paper). However, this did not improve the surface chemical and electrochemical stability. We found improvement only with oxidizing cations that are added to the surface. So the positive effect we found cannot be predominantly due to the formation of an A-site deficient perovskite at the surface. Furthermore, in the past we have examined the effect of A-site deficiency on the surface chemical stability of LSM,[1] and the improvement we obtained in that case is small compared to the large improvement we found here on LSC with oxidizing cations at the surface.

1 W. Lee and B. Yildiz, *ECS Trans.*, 2013, **57**, 2115–2123.

Ian Metcalfe enquired: If you modify the surface in terms of its tendency to segregate, surely you also modify its catalytic activity?

Bilge Yildiz replied: Yes, these additive elements at the surface, titanium, hafnium, zirconium *etc.*, should also modify the surface electronic and defect structure of LSC and affect the catalytic activity. However, their effect on pristine LSC would not actually be to increase the rate of oxygen exchange at the surface because they would tend to preserve a very oxidized surface. Contrary to that, we find a higher rate of oxygen exchange when we deposit these cations at the surface of LSC that simultaneously has a more oxidized surface and a higher stoichiometric strontium concentration. Therefore, we conclude that the dominant effect of these cations (titanium, hafnium, zirconium) is their role in stabilizing the cation chemistry and phase stability of LSC at the surface, and we ultimately benefit from a more LSC-like surface rather than one that is blocked by SrO-related insulating species.

Helena Téllez questioned: In your paper, you mentioned that it is not clear whether the titanium ions are incorporated into the perovskite structure or deposited as a smooth titanium oxide layer on top of the perovskite film. However, if the titanium ions were forming a thin layer on top of the perovskite, the surface exchange coefficient could then be attributed to the thin TiO_2 layer and not to the perovskite pristine surface. Could you comment on this?

Bilge Yildiz said: Since writing the paper we have performed more experiments, based on *in situ* synchrotron X-ray spectroscopy, that I presented during my talk. Based on X-ray absorption spectroscopy of titanium at the surface, we find a perovksite-like coordination above 500 °C, and not a TiO_2 structure. We are not entirely sure if titanium is doping LSC or if it is forming a thin $SrTiO_3$ layer. However, I believe that the former is more likely, *i.e.* titanium is doping LSC. This is because LSC and STO are completely solid-soluble and it would be hard to imagine that a sub-monolayer of STO would stay at the surface without mixing into the LSC surface.

Helena Téllez asked: Have you applied the same strategy to highly Sr-doped cobaltites (*e.g.*: $La_{0.6}Sr_{0.4}CoO_3$)? Is it still effective to suppress strontium segregation?

Bilge Yildiz answered: Yes, we have also implemented this strategy for highly doped cobaltites and ferrites as well as for BSCF, and found improvements in surface stability and electrochemical performance. However, the results aren't optimized yet, so I can't quote relative improvement values here.

Mogens Mogensen said: You say that titanium is stabilising the surface chemistry. I would expect that this isn't stabilisation of the original chemistry of the perovskite that exists without adding titanium, but changing the cation chemistry to another more stable chemistry. Strontium and titanium oxides may form the very stable $SrTiO_3$. Were any new crystalline phases observed?

Bilge Yildiz responded: I agree, and that's our point, that the surface cation chemistry changes. Based on our recent x-ray absorption spectroscopy data, in the case of Ti, the coordination is perovskite. So it either forms a $(La,Sr)TiO_3$, or Ti dopes into $(La,Sr)CoO_3$ B-site. It is unlikely that $SrTiO_3$ just stays as a sub-monolayer phase at the surface, because both Ti and Co are solid soluble into LSC or STO base structures. Therefore, we think that a more likely surface cation chemistry is $(La,Sr)(Co,Ti)O_3$-d.

Steven McIntosh queried: Do you have an estimate of what fraction of the strontium is coming out of these thin films and to what depth the Sr-depleted region reaches? How does this compare to the amount of titanium deposited onto the surface?

Bilge Yildiz replied: Based on our more recent work in collaboration with TEM and LEIS experts, this sub-surface affected zone appears to be on the order of 4 nm.[1] The amount of strontium that comes out on the unmodified LSC surface is more than the amount of oxidizing cations (*e.g.* titanium) that we put at the surface. For example, LEIS in much of the work of Téllez, Druce, Ishihara and Kilner *et al.* indicates that the surface is entirely covered by strontium. However, including all cations we put up to 16% titanium at the surface (measured by grazing emission XPS at 70 degrees off normal).

1 Y. Chen, H. Téllez, M. Burriel, F. Yang, N. Tsvetkov, Z. Cai, D. W. McComb, J. A Kilner and B. Yildiz, *Chem. Mater.*, 2015, **27**, 5436–5450.

John Druce enquired: Referring to Table 1 in your paper, which shows the quantified AES data, it seems that the surface concentration of strontium in the "particle-free" regions of the best performing composition (LSC-Ti10) is actually higher than for the other samples. Although the errors are quite large, this seems to suggest that the most electrochemically stable material does in fact show more segregation of the strontium, although not as many particles seem to form. Could the treatment be suppressing the precipitation of secondary phases, rather than the segregation of strontium itself? Also, if the surface is more heavily doped by the strontium segregation, one might expect a higher vacancy concentration, resulting in a more active surface.

I think as a field, we should be more careful when using the term "segregation". The phenomenon we often see is not just an enrichment of the element to the surface, but also precipitation of a secondary phase, or even decomposition of the material in the near surface. Your previous work shows the precipitation is detrimental for materials performance, but perhaps the effect of the surface enrichment alone is not quite as clear.

Bilge Yildiz answered: Yes, the LSC-Ti10 which has the least amount of particles appearing in AES has a larger fraction of strontium in the particle-free zones. We believe that this means that the remaining strontium in the perovskite lattice is actually higher in concentration at the surface of LSC-Ti10 and this contributes to its better performance. We have made this clear in the paper already, and also the reducing amount of strontium in the lattice in LSC (unmodified) is indicated by the reduction in the slope of the k^q *vs.* pO_2 plots in Fig. 4. But in effect, the LSC (unmodified) has a larger "total" amount of strontium that is segregated towards the surface and precipitated out. So the total strontium segregation must be greater with the LSC than with the LSC-Ti10.

We think that the surface treatment by titanium (or by hafnium, zirconium or niobium, which I showed in my presentation) reduces the "total" strontium that tends to enrich towards the surface, and keeping the near-surface concentration lower does not allow for as much phase precipitation as on LSC. It is also possible that the surface additives increase the surface phase stability of perovskite, for example by reducing the polarity of charged surfaces. But again, oxygen vacancies themselves, in theory, also serve for that purpose (*i.e.* reducing the polarity of CoO_2 surfaces and making them more stable), and even more so than the additive cations (oxygen vacancies are 2+ in charge while the titanium, hafnium, zirconium or cobalt sites would be 1+). On the other hand, we see a more oxidized surface from XAS measurements. As a result, we still believe that the positive effect of these surface additives is through reduction of the oxygen vacancy concentration at the surface and suppression of the electrostatic attraction, rather than phase stabilization of the surface.

I agree with your comment on the definition of segregation. We do distinguish the enrichment of the dopant within the lattice from its phase separation to dopant-oxides at the surface. I believe that if the dopant is only enriched within the perovskite lattice near the surface (but not phase separated out), the effect should be positive, not detrimental. The point is that a certain amount of oxygen vacancies (induced by the dopant, strontium here) is good for reactivity. However, having too many oxygen vacancies that charge the surface positively is not good because they pull the strontium out too much and lead to phase separation. So there must be an optimal surface concentration of oxygen vacancies, and the question is what is that optimal level.

Alex Morata remarked: When depositing titanium with $TiCl_4$, the layer thickness depends on the dipping time inside the solution. Nevertheless, this dipping is also producing an etching of the surface that is proportional to time. Do you think that this etching can affect the surface performance? How can you decouple the effect of the etching from the effect of the titanium deposition on the surface? Why do you use a chemical procedure to deposit titanium instead of using, for example, PLD?

Bilge Yildiz responded: Yes, the metal chloride solutions, like $TiCl_4$, etch the surface, too. Based on our chemical analysis, they preferentially etch strontium from the surface. To separate the effect of etching, we also treat LSC in a dilute HCl solution, to remove the excess strontium from the surface as well. The starting surface of LSC and LSC-Ti samples are then the same, except from the added titanium cations. And clearly, the etched LSC does segregate more strontium at its surface with time, but the LSC-Ti surface stays stable. So the effect of titanium is not because of etching but because of titanium itself playing a role at the surface.

We have used a solution route because it is more practical to implement and control than PLD, and it is also scalable to porous electrodes because it is just a water-based solution that could be infiltrated into the pore surfaces. If you started from an etched LSC surface and then deposited a small amount of titanium from (low-temperature) PLD or by sputtering, I would expect similar results.

Tatsumi Ishihara asked: How uniform are the additives distributed on the surface? Do these additives form a perovskite-like surface layer? If so, how dense are these films? Strontium continuously comes from the bulk to the surface and so removal of surface strontium by reaction is limited to an initial short period and after long term operation, I think the surface becomes the same. Do you think that the positive effects of titanium or hafnium can be observed after long term or further higher temperature operation?

Bilge Yildiz replied: The surface additive cations, when deposited as a small fraction (less than ~10%), appear to distribute uniformly at the surface, and do not form clusters or particles (within the detection limit of AFM and SEM). When they are uniformly deposited, we see the beneficial effect. When we deposit larger amounts (more than 10%), they cluster to form particles and they do not improve the surface chemical stability in that case.

Based on X-ray absorption spectroscopy of titanium at the surface, we find a perovksite-like coordination above 500 °C. We are not entirely sure if titanium is doping LSC or if it is forming a thin $SrTiO_3$ layer. However, as previously discussed, I believe the former is more likely. Based on the surface doping model, yes I do think that after longer term operation, or at higher temperature, we should still see a beneficial effect of these oxidizing surface-dopants. Especially larger dopants, like hafnium, would prefer to stay at the surface even at higher temperatures. In addition, their presence in essence alters the thermodynamics of defect formation at the surface, and that effect would hold also as a function of temperature.

Steven McIntosh questioned: If the Sr-depleted region reaches 4 nm deep from the surface of a 10 nm film, do you see significant changes in the impedance response during testing? Perhaps more importantly, do you think the assumption that the surface kinetics are the rate limiting step in oxygen incorporation holds when there are such dramatic changes in the film structure and composition? Can anything be learned from the changing shape and frequency of the impedance arcs as they change during strontium segregation?

Bilge Yildiz answered: The LSC film thickness in our paper is 40 nm, as reported in the experimental section. Affecting 4 nm out of 40 nm may not affect the bulk behavior significantly. Given that the bulk oxygen diffusion coefficient of LSC is quite high, depleting 4 nm from the surface should not change the rate limiting step from being that at the surface to being that in the bulk. In fact phase separation of strontium out of this surface zone makes the perovskite lattice at/near the surface Sr-deficient (as also indicated from the reduction in the slope of k^q *vs.* pO_2 in Fig. 4), thereby making the surface even less reactive to oxygen exchange. So overall, I do think that the surface limitation of the kinetics holds in the presence of strontium phase separation at the surface. Lastly, yes, perhaps the frequency of the impedance arcs can be useful during segregation, because the capacitance (specifically the chemical capacitance) is related to the oxygen non-stoichiometry, and the segregation of strontium out of the perovskite lattice should affect the oxygen non-stoichiometry.

John Vohs opened discussion of John Druce's paper: Does the mixing that occurs during SIMS analysis cause any uncertainty in the measurement of the relative amounts of oxygen exchange in the two different phases?

John Druce responded: Although ion beam mixing is a significant concern with sputter depth profiling measurements, these SIMS analyses are "plan-view" images of the very outer surface. They consume a negligible amount of material (*i.e.* we are analyzing a few nm), so mixing in depth is not an issue. The mixing laterally will be of the order of the lateral straggle, which is also a few nm, which is much lower than the beam spot size for these measurements (a minimum of 100 nm from the manufacturers specifications). Therefore the lateral resolution is not limited by any lateral mixing.

Mogens Mogensen remarked: Fig. 7 shows analyses of the impurities on as polished and as annealed sample surfaces for a) CGO alone and b) the CGO–LSCF composite. When you look at the data it is hardly in agreement with the cleaning hypothesis. The as polished samples should be clean but they have the most silica in both cases. Why is this? Is this silica in the grain boundaries in the as polished samples? And what removes the silica from the CGO in case a), where there is no "scavenger" to "clean" the silica away?

John Druce answered: We are not sure why the as-polished samples have the most silica. It may be due to silica in the grain boundaries, as the grain size of the single phase CGO is quite small (a few hundred nm), so the area density of grain boundaries is quite high. As for case (a), you are correct that the coverage of silica after annealing is lower. However, we must bear in mind that LEIS is measuring the area coverage of the elements exposed at the very outer surface; the silicon is not necessarily removed, but it may be covered by another oxide with a lower surface energy. Indeed, this is what the "annealed" spectrum appears to show; although the coverage of Si does appear to decrease, those of Na and an unresolved contribution for K and/or Ca do appear. Note that the sensitivity for lighter elements such as Na is lower, so the peak areas are not directly indicative of coverage.

In fact, previous studies of $Ce_{0.8}Gd_{0.2}O_{1.9}$ (**ref. 1** below) showed that the annealed surface was covered by CaO, whilst Si was not detected. Work by de Ridder, Brongersma *et al.*[2,3] on YSZ also showed that Ca was the most dominant surface impurity in some cases, whilst is was Na or Si in others. Thus we should not focus solely on silica impurities, but consider all these light element impurities. In other words, we do not believe that the silica has been scavenged from the surface of the pure CGO, but has been replaced by these other impurity elements. In contrast, even leaving aside the question of the source of the impurities, the surface of the composite is much cleaner, with only negligible amounts of impurities detected. Therefore, we believe that considering all of the impurities, rather than focusing just on silica, the data are consistent with our cleaning hypothesis, although we do not discount the contribution of another mechanism (e.g. spillover from the LSCF, which would also require a clean CGO surface for incorporation).

1 P. J. Scanlon, R. A. M. Bink, F. P. F. van Berkel, G. M. Christie, L. J. van IJzendoorn, H. H. Brongersma and R. G. van Welzenis, *Solid State Ionics*, 1998, **112**, 123.
2 M. de Ridder, A. G. J. Vervoort, R. G. van Welzenis and H. H. Brongersma, *Solid State Ionics*, 2003, **156**, 255.
3 M. de Ridder, R. G. van Welzenis, A. W. Denier van der Gon, H. H. Brongersma, S. Wulff, W.-F. Chu and W. Weppner, *J. Appl. Phys.*, 2002, **92**, 3056.

Raymond Gorte commented: LSCF is similar to the types of materials being developed for MIEC membrane reactors, implying that the ionic conductivities are very high. Our own measurements of Ca-doped $LaFeO_3$ imply that the ionic conductivity of these materials is higher than that of YSZ.[1] Do you have any indication of how the ionic conductivities measured by SIMS compare to numbers measured in flux measurements?

1 A. S. Yu, J. Kim, G. Kim, T.-S. Oh, R. J. Gorte and J. M. Vohs, *Appl. Catal., A*, 2014, **486**, 259–265.

John Druce responded: Indeed, LSCF is itself of interest as an MIEC material for membrane reactors.[1] Of course, diffusivities measured by SIMS and the oxygen permeation flux are linked through the ionic conductivity and the Nernst-Einstein relation. We do not have permeation flux data for our samples, but oxygen tracer diffusion coefficients determined from these two experimental approaches have previously been shown to agree well.[2] Therefore, we take the oxygen diffusivity measured by SIMS to be an accurate measure of the ionic conductivity (we note that this is specifically the oxygen ionic conductivity of the MIEC material — there is no issue in separating electronic contributions, nor does the oxygen tracer diffusion depend on the ambipolar conductivity).

With regard to your point about the ionic conductivity of $La_{0.9}Ca_{0.1}FeO_3$ being higher than that of YSZ, we would comment that for this LSCF–CGO system, the tracer diffusivity (*i.e.* oxygen conductivity) of CGO is two orders of magnitude higher than LSCF at 600 °C and around one order of magnitude higher at 800 °C, but the values for the two materials become comparable at 900–1000 °C, due to different activation energies. We are currently preparing these data for publication.

1 J. E. ten Elshof, H. J. M. Bouwmeester and H. Verweij, *Appl. Catal., A*, 1995, **130**, 195–212.
2 J. A. Lane, S. J. Benson, D. Waller and J. A. Kilner, *Solid State Ionics*, 1999, **121**, 201–208.

John Irvine enquired: It is a well established practice to clean zirconia grain boundaries by adding alumina to form alumina silicates in interstices. Could you do this with ceria? On adding alumina, the properties would be improved. Cobalt might be acting as a sintering aid, would a small amount of cobalt in ceria affect the exchange properties as it would change the surface a bit?

John Druce answered: Yes. In a similar way, there have also been reports of scavenging grain boundary impurities in cerias by additives such as La_2O_3,[1] SrO,[2] Fe_2O_3,[3–5] and even CaO[6] and MgO.[7] We have demonstrated that the addition of La_2O_3 to the surface of $Ce_{0.9}Gd_{0.1}O_{1.95}$ resulted in an enhancement of k^* by a factor of 50, which we attributed to the scavenging of surface impurities.[8]

Of course, cobalt has been reported as a sintering aid for CGO (*e.g.* **ref. 9** below), and our composite samples in the present work seem to show enhanced sintering behavior, which we attribute to the inter-diffusion of one of the LSCF cations, possibly cobalt, which would be located at the grain boundaries. However, from the SIMS and LEIS data presented in our paper, we do not detect cobalt at the surface of the CGO. Therefore, if it is present, it is only in very small amounts. We do note that addition of Co_3O_4 to $Ce_{0.8}Gd_{0.2}O_{1.9}$ did not show significant impurity scavenging behaviour,[5] and that in our previous work,[8] the addition of Co_3O_4 to $Ce_{0.9}Gd_{0.1}O_{1.95}$ was seen to enhance k^* by a factor of 14, which was less than the effect of La_2O_3. Therefore, we would tend to favour the arguments for surface cleaning by La_2O_3 or LSCF.

1 D. Ivanova, A. Kovalevsky, V. V. Kharton and F. M. B. Marques, *Bol. Soc. Esp. Ceram. Vidrio*, 2008, **47**, 201.
2 P. S. Cho, Y. H. Cho, S. Y. Park, S. B. Lee, D. Y. Kim, H. M. Park, G. Auchterlonie, J. Drennan and J. H. Lee, *J. Electrochem. Soc.*, 2009, **156**, B339–B344.
3 Q. Dong, Z. H. Du, T. S. Zhang, J. Lu, X. C. Song and J. Ma, *Int. J. Hydrogen Energy*, 2009, **34**, 7903.
4 T. S. Zhang, J. Ma, L. B. Kong, S. H. Chan, P. Hing and J. A. Kilner, *Solid State Ionics*, 2004, **167**, 203.
5 T. S. Zhang, J. Ma, Y. J. Leng, S. H. Chan, P. Hing and J. A. Kilner, *Solid State Ionics*, 2004, **168**, 187.
6 J. A. Lane, J. L. Neff and G. M. Christie, *Solid State Ionics*, 2006, **177**, 1911.
7 Y. H. Cho, P. S. Cho, G. Auchterlonie, D. K. Kim, J. H. Lee, D. Y. Kim, H. M. Park and J. Drennan, *Acta Mater.*, 2007, **55**, 4807.
8 J. Druce and J. A. Kilner, *J. Electrochem. Soc.*, 2014, **161**, F99.
9 D. P. Fagg, J. C. C. Abrantes, D. Pérez-Coll, P. Núñez, V. V. Kharton and J. R. Frade, Electrochim. Acta, 2003, **48**, 1023.

John Varcoe asked: Have you got any direct evidence of the impurities going into the LSCF phase?

John Druce replied: No, we do not have any direct experimental evidence that the impurities are dissolved in the LSCF in the composites, only that they seem to be absent from the CGO. We do have evidence from other studies of LSCF micropatterned electrode structures on YSZ substrates that there is significant mobility of impurities between the two phases.[1] In that case, we do detect the impurities from the YSZ going into the LSCF. We consider these LSCF micro-electrode structures to be analogous to the situation in the composites.

1 H. Téllez, J. Druce, Y. Shi, M. Kubicek, N. J. Simrick, J. L. M. Rupp, T. Ishihara and J. A. Kilner, *ECS Trans.* 2015, **66**, 69–77.

John Varcoe opened discussion of the paper by Chengsheng Ni: How big is your reference electrode (in the 3 electrode cell) relative to the 1 mm thickness of your electrolyte? How confident are you that this will give you the true anode and cathode impedance response?

Chengsheng Ni answered: The diameter of the 1 mm thick YSZ electrolyte is 2 cm and the working electrode (CM or CMN) or counter electrode (Ni-YSZ) is 1.3 cm. The reference electrode covered the rim of the electrolyte with LSM paste, so there was sufficient distance between the reference electrode and working electrode. This geometry is designed to make sure that there is no short circuit of the reference to either electrode. The validity of the reference is justified by the close ohmic resistance values from the impedance spectra for the working electrode and counter electrode.

Steven McIntosh queried: Can you comment on the effectiveness of the reference electrode placement and how you ensure that the anode and cathode impedance spectra are truly separated. Summation of both spectra to reconstitute the full spectrum should occur even if the reference is misplaced. Accurate reference electrode placement is extremely difficult for these systems and it can be hard to fully deconvolute the electrode responses. I don't think this will change the observed trends in your paper but it may influence the reported values.

John Irvine said: This is an intrinsic issue to three electrode measurements, all one can do is minimise possible errors by using thick pellets, careful alignments *etc.* In a series of experiments with the same geometry there is useful comparative information to be obtained.

Chengsheng Ni responded: I agree with you on the deviation of the measured polarization resistances from the actual values. To be cautious with the data, we used a symmetrical cell to measure the polarization resistances at OCV with and without activation under current; these values agree with those of the three-electrode measurements. When the oxygen electrode is under positive bias, the variation in polarization resistance can also be seen from the *I–V* curve of the electrolysis cell, though ohmic resistance contributed to a large portion of the total area-specific resistance (ASR).

John Irvine remarked: No three electrode measurements are an exact science, splitting ohmic resistance in half as a quality insurance is surprising, unless you drill a hole in the middle of the electrolyte, which would not be done perfectly, why should it be exactly half? Getting a robust and reliable reference electrode for high temperature electrochemistry would be a wonderful advance.

Chengsheng Ni replied: The purpose of the configuration of putting the reference on the edge is to measure the electrolyte potential along an equipotential surface somewhere between the two active electrodes. Ideally, the potential drop from the ohmic resistance could be divided into two halves since it is a symmetrical design, *i.e.* the potential of the reference is equal to the potential in the middle of the electrolyte. However, the splitting of resistance in half is not the ultimate goal of the three electrode measurement in practice, but an indicator that the reference is reliably separated from both working electrode and counter electrode. In the manuscript, the ohmic resistance values for both electrodes are similar, but not the same. In order to get a good separation between the reference electrode and the working electrode, the distance of the reference from the electrode is 3 mm, significantly larger than the thickness of the electrolyte, 1mm. The thick electrolyte is also beneficial to attenuate the effect of misalignment between the working electrode and the counter electrode.

Steven McIntosh commented: I completely agree — reference electrodes in these systems are incredibly problematic and very hard to get exactly correct. An advance in this area would be a true breakthrough. With regard to the ohmic resistance split, I was referring to the work in **ref. 1 and 2** below where, under the assumption of negligible ohmic losses in the electrodes, accurate separation of the electrode impedance response occurs when the reference samples the potential at the centre of the electrolyte. This leads to splitting of the ohmic resistance between the two electrode responses. Of course, this idealization is from a simulation. The issues surrounding the problems with experimental systems are discussed in **ref. 2** below. Again, I agree that these experiments are extremely difficult to perform, and we typically have to interpret imperfect data for these systems.

1 S. B. Adler, *J. Electrochem. Soc.*, 2002, **149**, E166–E172.
S. McIntosh, J. M. Vohs and R. J. Gorte, *J. Electrochem. Soc.*, 2003, **150**, A1305–A1312.

John Vohs asked: Have you measured the electrode impedance during polarization and does the split of the ohmic contribution (when using the reference electrode) remain the same as a function of current density?

Chengsheng Ni replied: We do see some change in the ohmic resistance when we measure the electrode impedance during polarization, but the change is minor (less than 0.05 Ω cm^2 from -0.4 to 0.4 V bias). In addition, the change of inductance would also affect the apparent ohmic resistance.

Mogens Mogensen commented: With regard to the reference electrode, there is no reason to believe that you have to split the series resistance in half. It depends on the electrodes. If they are exactly the same then you can just divide the two-electrode measured impedance by two. No reference electrode is needed. In the case of a 1 mm thick electrolyte there is a good chance that a relatively (compared to the electrolyte thickness) good symmetry of working and counter electrodes can be obtained, i.e. a small misalignment of these two, which gives a small error on the measured overpotential obtained in the three-electrode set-up, see **ref. 1** below.

1 J. Winkler, P. V. Hendriksen, N. Bonanos and M. Mogensen, *J. Electrochem. Soc.*, 1998, **145**, 1184.

Steven McIntosh responded: I agree that it does depend on the geometry and the electrodes themselves. Splitting the ohmic resistance is a good guide for a geometry where the reference electrode is on the surface of the electrolyte and under the assumption that the electrodes have negligible ohimc resistance.[1,2]

1 S. McIntosh, J. M. Vohs and R. J. Gorte, *J. Electrochem. Soc.*, 2003, **150**, A1305–A1312.
2 S. B. Adler, *J. Electrochem. Soc.*, 2002, **149**, E166–E172.

Joseph Hartvigsen remarked: This paper has generated considerable discussion on the use and interpretation of reference electrodes. While there are potentially some very complex configurations where reference electrodes have been used to separate anode and cathode overpotential, in general this is practically impossible as a simple add-on to a conventional cell. We have done analysis showing this. As noted in the discussion so far, Stu Adler, a former Ceramatec colleague, has publications in this area. While he was at Ceramatec I showed Stu, using finite element analysis to compute iso-potential lines, that the potential lines cross through the electrolyte biasing the opposite reference electrode in such a way that all of the electrolyte and electrode losses on one electrode become erroneously attributed to the opposite electrode. Analysis showed that bending of potential lines occurred even with mathematically exact alignment of anode and cathode on electrolyte with perfect uniformity of thickness and parallelism. Modeling a slight misalignment of the electrodes shows that the phenomenon is even more pronounced. In practice, perfect registration of printing anode and cathode on electrolyte supported cells is not possible, and electrode supported cells naturally eliminate the possibility of using two reference electrodes and bias a single reference non-supporting electrode to the potential of the supporting electrode. With well aligned electrodes on an electrolyte supported cell, solid state diffusion will effectively make one electrode larger than the other by spreading around its edge. In the Ni–YSZ/YSZ/LSM material set, cations from the LSM electrode move enough to make it behave as if it were larger than the Ni–YSZ electrode. Dr Adler was able to apply this observed phenomenon to eliminate seal failure due to potential across the seal by biasing both sides of the seal to the same potential using size mis-matched electrodes.[1,2] The cross over of potential lines can be used for benefit if understood, or can be misleading if ignored.

1 S. B. Adler *et al.*, *US Patent*, 6042703, 1998.
2 S. B. Adler *et al.*, *US Patent*, 5868918, 1996.

Chengsheng Ni responded: I agree that the cross over of potential lines can cause problems, but there are some general indicators that we can use to minimize this problem, as shown in the literature by Adler, by using a thick electrolyte. The 1 mm thick electrolyte, in my opinion, could avoid the cross over of potential lines. Another indicator is the distance of the reference from the electrode in a macroscopic design, which can avoid the detection of potential lines tilting to one side. The distance of the reference in my case is 2 times longer than the thickness of the electrolyte, and this could be beneficial in avoiding errors. Thank you for your comments and suggestions.

Raymond Gorte said: These materials are interesting, with conductivities approaching that of LSM. The performances of the pure compounds are not that great but are likely comparable to what you would get with LSM. Have you tried to make composite electrodes from these materials? How well do they perform? Could you make electrodes by infiltration?

Chengsheng Ni answered: I have tried to make a composite electrode with YSZ using this material, but I do not get a better performance if I use 50% CM(N) and 50% YSZ. This is probably because the YSZ content is too high. We might need to do some optimization. We have not tried any impregnation on this material, but it could be done since it is compatible with YSZ up to 950 °C, and can be synthesized by solution methods at temperatures lower than 900 °C.

San Ping Jiang commented: The effect of the reference electrode position or cell configuration on the reliability of the anodic and cathodic overpotential separations of these film cells is an important issue and is often debated in the SOFC field. We investigated this issue in great detail on 50 by 50 mm planar cells, using different cell configurations with different reference electrodes and on electrolytes with different thicknesses.[1] The results showed that the reliable separation of anodic and cathodic overpotential depends on the position of the reference electrodes as well as the electrolyte thickness. In the case of conventional Ni–YSZ anodes and LSM cathodes, the anodic and cathodic overpotential can be accurately separated using a cell configuration with the reference electrode positioned along the edge of the working electrode and with YSZ electrolyte thicker than ~200 μm. However, the symmetry contact of the anode and cathode also has a significant effect on the accuracy of the separation of the overpotential.

1 S. P. Jiang, *J. Appl. Electrochem.*, 2004, **34**, 1045–1055.

Chengsheng Ni responded: We agree that the reference electrode on a thin electrolyte (*e.g.* thinner than 200 μm for a 5 cm by 5 cm cell) will cause problems due to the misalignment and current constriction. Our reference electrode is on the edge of a 1 mm thick electrolyte while the area of the electrode is as small as ~1 cm^2, so the reference electrode should be reliable. Thank you for your information and backup.

Faraday Discussions

PAPER

A techno-economic model of a solid oxide electrolysis system

Daniel G. Milobar,[a] Joseph J. Hartvigsen*[b] and S. Elangovan[c]

Received 6th February 2015, Accepted 4th March 2015
DOI: 10.1039/c5fd00015g

Solid oxide cells can play a vital role in addressing energy and environmental issues. In fuel cell mode they are capable of producing electric energy at high efficiency using hydrocarbon fuels and in the electrolysis mode can produce hydrogen from steam or synthesis gas from a mixture of steam and carbon dioxide. The solid oxide electrolysis cells (SOECs) can operate at a wide range of conditions. A capable means by which to select operating conditions in the application of solid oxide electrolyzers is a necessity for successful commercial operation. Power and efficiency can be determined over a wide range of operating conditions by applying fundamental electrochemical principles to a SOEC system. Operating conditions may be selected based on power requirements or with efficiency as a priority. Operating cost for electricity which is a function of both power and efficiency can also be used to determine optimal operating conditions. Performance maps based on closed form isothermal parametric models for both hydrogen and natural gas fueled SOFC stacks have been demonstrated previously. This approach applied to a SOEC stack is shown. This model was applied to generate performance maps for a solid oxide cell stack operated in the electrolysis mode. The functional form of the model and the boundaries of the operating envelope provide useful insight into the SOEC operating characteristics and a simple means of selecting conditions for electrolysis operation.

1 Introduction

Solid oxide cells can play a vital role in addressing energy and environmental issues. As the demand for energy continues to increase, the need for technological solutions to provide energy at low environmental impact as well as mitigation and reuse of emissions becomes critical. Recent efforts to increase the use of renewable energy sources to provide green energy are hampered by the difficulty in efficiently storing energy to accommodate their intermittency. In that regard, SOEC technology offers a method to convert and store electrical energy in chemical form at an exceptionally high efficiency. The renewable energy can be

[a]*Tucson, AZ, USA. E-mail: dmilobar@lasertel.com*
[b]*Ceramatec, Inc., 2425 S. 900 W., Salt Lake City, UT, USA. E-mail: jjh@ceramatec.com*
[c]*Ceramatec, Inc., 2425 S. 900 W., Salt Lake City, UT, USA. E-mail: elango@ceramatec.com*

stored in the form of hydrogen *via* the steam electrolysis process or synthesis gas *via* co-electrolysis of steam and carbon dioxide.[1] The synthesis gas from the co-electrolysis process can be further converted to common fuels such as methane and liquid fuels or chemicals. One such scheme of storing renewable energy in the form of liquid fuel using co-electrolysis technology is shown in Fig. 1.

Not only does the high temperature electrolysis process convert steam and carbon dioxide at high efficiency, it can also operate over a wide range of conditions in endothermic, thermoneutral, or exothermic mode. The selection of operating condition largely depends upon availability of external heat, for example from a solar concentrator[2] *vs.* wind energy, and power available depending on other load requirements. When external heat is available, the SOEC can be operated in endothermic mode at electrical efficiencies exceeding 100% and when external heat is not available it can be operated in exothermic mode to provide the heat of reaction in which case the electrical efficiency will be lower than 100% but the hydrogen or syngas production rate is higher. Thus, having the SOEC as the load allows tremendous flexibility in managing the distribution and storage of intermittency of renewable energy sources. Most previous economic analyses of high temperature electrolysis were life cycle cost analysis of full systems including the power generation cycle.[3-6] The cost of produced fuel, hydrogen or syngas, also depends upon the operating condition.[3] In this paper a methodology to derive a performance map of an SOEC as a function of the two main variables is presented. While what is shown is for a specific stack performance (*i.e.* area specific resistance) for the production of hydrogen, the map can easily be translated to production of syngas using stacks of any defined performance behaviour. The performance map forms the basis for estimating the cost of fuel generated given the cost of electricity and capital cost of the stack and the balance of plant.

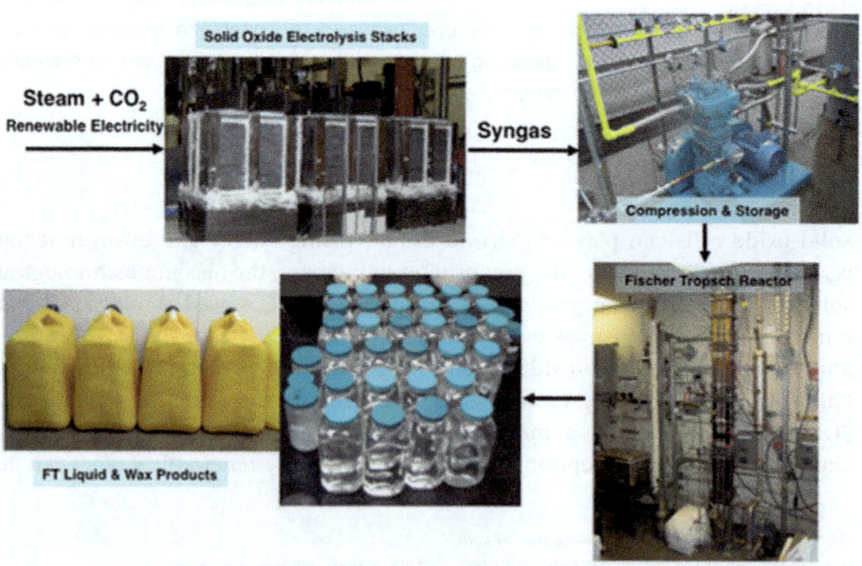

Fig. 1 Fischer–Tropsch process utilizing SOEC technology.

The electrochemical production rate, electrolysis efficiency and electrical power consumed by a solid oxide electrolysis system are not solely determined by stack characteristics but by operating conditions as well. These quantities can be determined by applying various operating conditions to a SOEC system experimentally. However, to gain a meaningful and comprehensive representation of the performance of a SOEC system one must formulate relationships among, electrochemical production, power consumption and efficiency and present them clearly in terms of controllable operating conditions. The costs related to SOEC operation can be as important as the performance attributes of an electrolysis system. This work looks at the financial costs associated with up front capital as well as operational expense related to hydrogen production. However the context of this work could easily be reframed for the production of syngas. In certain situations it is feasible that the importance of resource consumption could be of critical importance, *i.e.* oxygen production in human space flight applications. It would be an easy task to modify the model at hand to account for resource cost rather than financial cost. This work provides a map of the performance of an SOEC system as well as a cost map associated with controllable operating conditions. Having a clear view of performance attributes and cost of operation would provide a greater opportunity for system optimization prior to capital investment. Such relationships have been developed previously for SOFC stacks.[7–9] These relationships for solid oxide systems operating in the electrolysis mode are addressed in the present work.

2 Stack performance model

The efficiency of a fuel cell stack is the ratio of the electrical power produced to the thermochemical power of the fuel consumed. In this instance fuel utilization is important and justifiably so due to the premium placed on fuel. Reactant utilization is however not a contributing factor in the electrolysis efficiency equation. In the electrolysis case efficiency is the ratio of the thermochemical power produced in the way of hydrogen production to the electrical power consumed. It has been shown that the efficiency (η_{stack}) of an electrochemical cell stack can be represented in terms of the operating voltage (V_{op}) and thermoneutral voltage (E_{tn}).[10]

$$\eta_{stack} = \frac{E_{tn}}{V_{op}} \tag{1}$$

In eqn (1) V_{op} is the only independent variable. Thermoneutral voltage, defined in eqn (2) below, is the heating value (ΔH) of the reactant divided by the number of Faradays of charge required to completely consume this reactant. In the electrolysis mode the reactant is steam.

$$E_{tn} = \frac{\Delta H}{nF} \tag{2}$$

Steam utilization is defined below as the ratio of total current applied to the cell to stoichiometric current. It is shown in eqn (8) that steam utilization (U) is a function of steam flow rate, operating voltage and stack resistance.

$$U = \frac{I}{\dot{n}nF} \tag{3}$$

Given that inlet gas composition and stack resistance are known, the only independent variables that will affect power consumption and efficiency are operating voltage and steam flow rate. Stack resistance being a function of temperature is not considered a variable due to the assumption that the stack will operate close to isothermal conditions at the highest operating temperature possible with respect to stack life and balance of plant limitations. In the same approach covered previously for a SOFC system[7–9] relationships are developed between inlet reactant flow, steam utilization and operating voltage. With these relationships a closed form parametric model is created and used to generate performance and cost maps. These maps can then be used to optimize operating conditions.

The reactant flow rate can be expressed as an electrochemical term which represents the average current density (j_f) at which steam utilization is 100%.

$$j_f = \frac{\dot{n}nF}{A} \tag{4}$$

The driving potential V_f, which is the difference between the operating voltage V_{op} and the Nernst potential E_N, required to sustain full steam utilization, current density can then be calculated using the stack area specific resistance.

$$V_f = j_f R'' \tag{5}$$

Here the term R'' represents the cell area specific resistance (Ω cm^2), the steam flow rate (mol s^{-1}) is represented by $\dot{n}$. The term nF represents the number of Faradays (C mol^{-1}), and the cell area (cm^2) is represented by A. By using the term V_f as the driving potential to sustain full steam utilization several simple expressions can be parameterized with respect to steam utilization (U) and operating voltage (V_{op}).

2.1 Model parameterization

When steam utilization is described in terms of fuel side electrode species concentrations[7] the Nernst potential can be rewritten as a function of utilization (U).

$$E_N(U) = E^o + \frac{RT}{nF} \ln \frac{U{O_2}^{1/2}}{1-U} \tag{6}$$

The average Nernst potential can then be calculated as follows.

$$\overline{E_N}(U) = \frac{\int_{U_o}^{U} E_N(U)\mathrm{d}U}{U} \tag{7}$$

Integrating eqn (6) yields a closed form solution for $\overline{E_N}$. While evaluating eqn (7) is easily accomplished, obtaining the inverse function to solve explicitly for U is not possible. The need to solve the exit utilization (U) can be eliminated by developing a parametric representation in U and V_f rather than working with the true physical variables V_{op} and steam flow (V_f). Combining eqn (4), (5) and (7) steam utilization can be expressed with the following equation.

$$U = \frac{V_{op} - \overline{E_N}(U)}{V_f} \tag{8}$$

Which can be solved to obtain V_{op} as a function of U and V_f.

$$V_{op}(U,V_f) = E_N(U) + UV_f \tag{9}$$

The average current density follows from Ohm's law.

$$j_i(U, V_f) = \frac{V_{op} - \overline{E_N}(U)}{R''} \tag{10}$$

While power density is the average current density (j_i) multiplied by operating voltage (V_{op}).

$$P_i(U, V_f) = V_{op}\frac{V_{op} - \overline{E_N}(U)}{R''} \tag{11}$$

Eqn (1) and (8) require no parameterization. With these expressions a comprehensive set of performance maps can be generated which show efficiency, current density, power density and utilization in the independent variable space

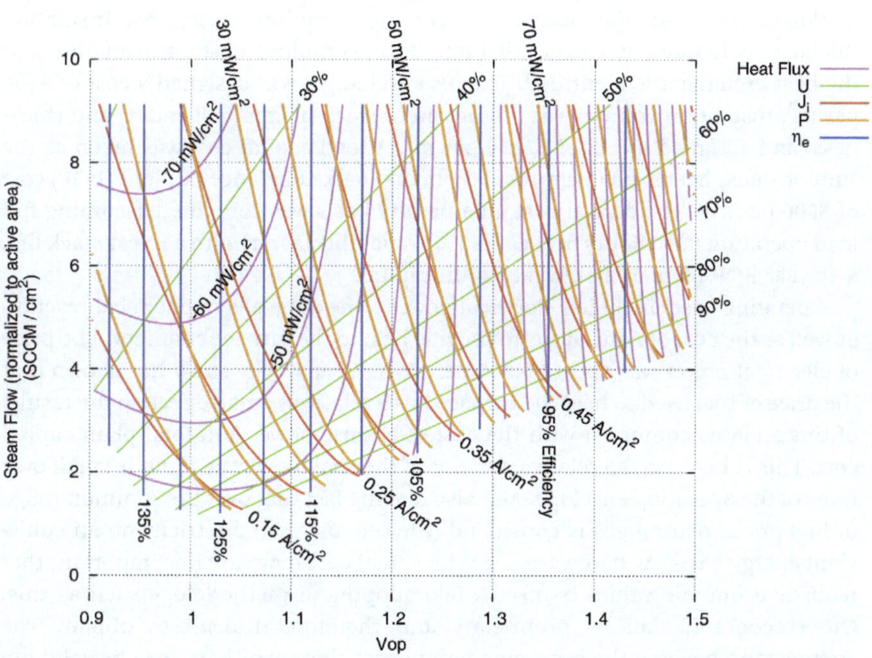

Fig. 2 SOEC performance map.

(V_{op}, V_f) representing operating voltage, area specific resistance, steam flow and composition. Parameterization in (V_{op},V_f) is done simply for mathematical convenience. An example performance map of a SOEC stack operating at 800 °C, with an ASR of 1.0 Ω-cm^2, is shown in Fig. 2.

3 Cost model

The performance model along with mass and energy balances were used to determine the cost of the major system components for a SOEC system with 25 kW electrolysis capacity. This system was comprised of four stacks of solid oxide cells each having 225 cm^2 of active area. The stack height at a given point in the (V_{op},V_f) plane was determined by the total number of cells that would be required to consume 25 kW of electrical power at the power density obtainable at that given operating point. Stack cost was assumed to be $35 per cell-interconnect repeat unit within each stack. Air flow was calculated to maintain a 100 °C temperature difference between the 900° C furnace chamber outlet and a constant 800 °C stack air inlet. Energy required to heat the inlet streams from 25 °C to 800 °C was added to the energy needed to maintain a stack outlet temperature of 800 °C. The energy required to maintain stack temperature will depend on how far above or below the operating voltage (V_{op}) is from thermoneutral voltage (V_{tn}). Steam flow rates are proportional to V_f as shown in eqn (5).

Once the air flow is known, the heat exchanger and insulation were sized on the basis of their combined cost. Starting at an insulation thickness of 5 cm, heat transferred through the insulation was used to determine the temperature of the furnace chamber outlet. The heat exchanger area required is then calculated based on this new furnace chamber outlet and the air inlet flow rate. As insulation thickness increases the heat exchanger area required decreases. Insulation thickness is iterated by 1 cm until a minimum combined cost for insulation and the heat exchanger is identified. The heat exchanger was assigned a cost of $400 per m^2. Insulation volume was a function of stack height (cell count) and thickness, and assigned a cost of $2180 per m^3. A total capital cost was taken as the sum of stack, heat exchanger, insulation, and a fixed balance of plant (BOP) cost of $500 per kW. The capital cost on a per kW basis was computed assuming full load operation 8600 hours per year (98% availability factor) with a 5 year stack life, a 10 year BOP life and a 10% return on capital.

Operating costs included electrical power to feed the electrochemical reaction as well as the cost of natural gas to provide heat to the furnace chamber. The price of electrical power was set at $0.045 per kWh. A sensitivity study has shown that the price of fuel used to heat the furnace has a relatively small effect on the results of this analysis compared with the cost of electric power and total plant capital cost. This is because the relative amount of thermal to electric input is small over most of the operating envelope, and also due the fact that over the common range of fuel prices natural gas is considerably lower cost than electricity on an equivalent energy basis. As hardware, electricity, fuel cost may vary it is important that realistic economic values be used when adopting a methodology such as this. Often economic data is proprietary and therefore difficult to obtain. The circumstances where the economic parameters presented here may be valid are rather specific, and particular values may not be widely applicable. The

Table 1 SOEC cost model computation sequence

1. Set Parameters U, V_f
2. Compute V_{op}, j_i, power density, and efficiency
3. Size/cost stack for 25 kW target
4. Compute air flow for temperature requirements
5. Compute inlet steam flow from V_f
6. Size/cost heat exchanger and insulation
• Cost insulation
• Compute heat loss
• Update chamber temperature
• Size/cost heat exchanger
• Increment insulation thickness until heat exchanger + insulation cost minimum is found
7. Calculate capital and component cost contribution \$ per kgH_2 (stack, heat exchanger, insulation)
8. Calculate auxiliary heat cost contribution \$ per kgH_2
9. Calculate electric power cost contribution \$ per kgH_2

Table 2 SOEC cost model assumptions

1. Stacks: 15 cm, 0.5 ohm-cm^2, \$35 per repeat unit
2. Insulation: \$2180 per m^3
3. Heat exchanger: \$400 per m^2
4. Other BOP: \$500 per kW
5. Fuel: \$4 per MMBTU (LNG)
6. Depreciation: stack 5 years, BOP 10 years, 10% continuous ROI
7. Duty factor: full load 8600 hours per year (98%)

computation sequence is outlined in Table 1. Assumed cost factors are contained in Table 2.

4 Performance and cost maps

As can be seen in Fig. 2 the domain is bounded to the right by high operating voltage and to the left by operating voltage below the steam inlet Nernst potential (E_N). Most of the lower boundary is due to the max steam utilization approaching 100%. It can be seen in Fig. 2 that efficiency as a function of operating voltage decreases as operating voltage increases. Conversely the electrical basis electrolysis efficiency increases higher than 100% when the stack is operated lower than the thermoneutral voltage. Power density on the other hand increases with operating voltage. As can be seen in Fig. 3 when the stack operating voltage is high, the electrical power required to drive the electrochemical reaction is the dominant cost factor. When stack operating voltage is low, capital costs shown in Fig. 4 relative to production are high. Taking these competing factors into account one can determine a minimum cost of operation. The total cost to produce hydrogen in the given system is given in Fig. 5. If cost is a primary concern then operating the stack at 1.1 V and at a steam utilization of approximately 20% would yield the lowest cost per unit produced for the given system.

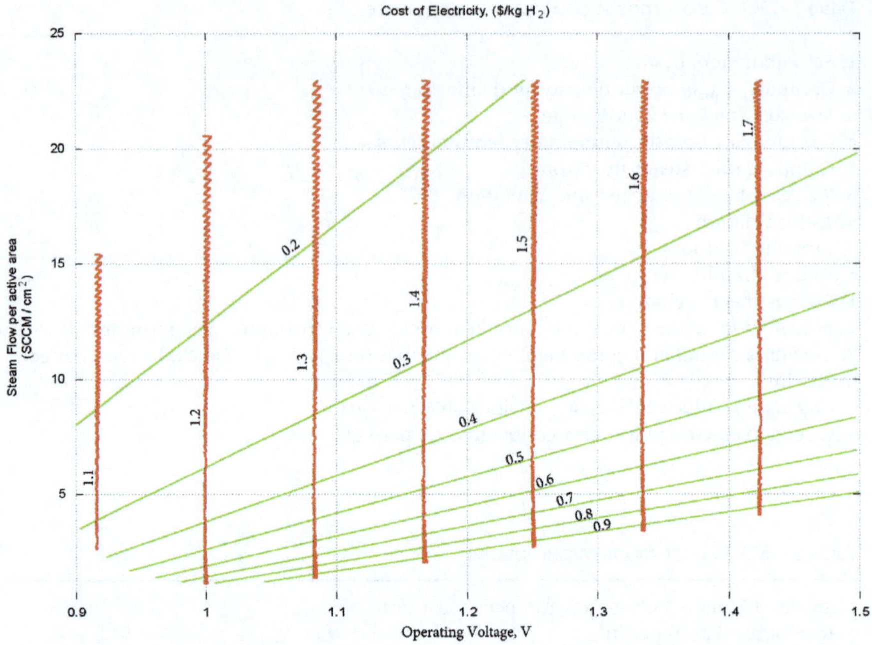

Fig. 3 Cost of electricity, $ per kg of H_2 produced.

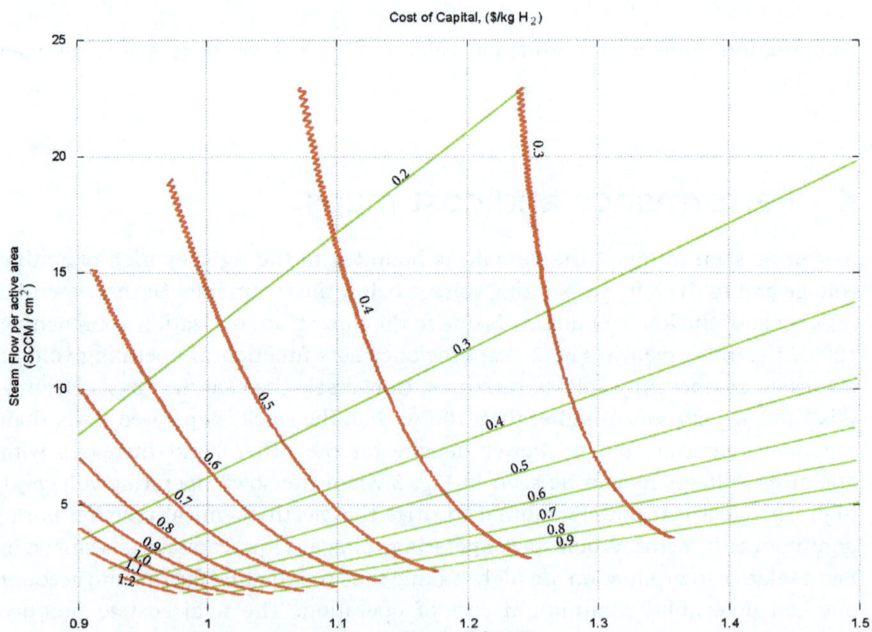

Fig. 4 Cost of capital, $ per kg of H_2 produced.

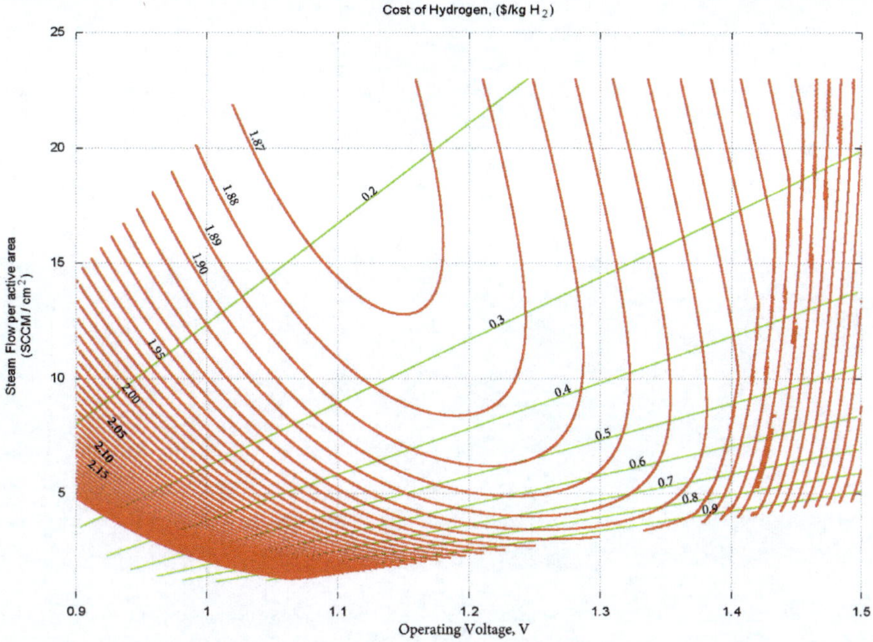

Fig. 5 Cost to produce hydrogen, \$ per kgH_2.

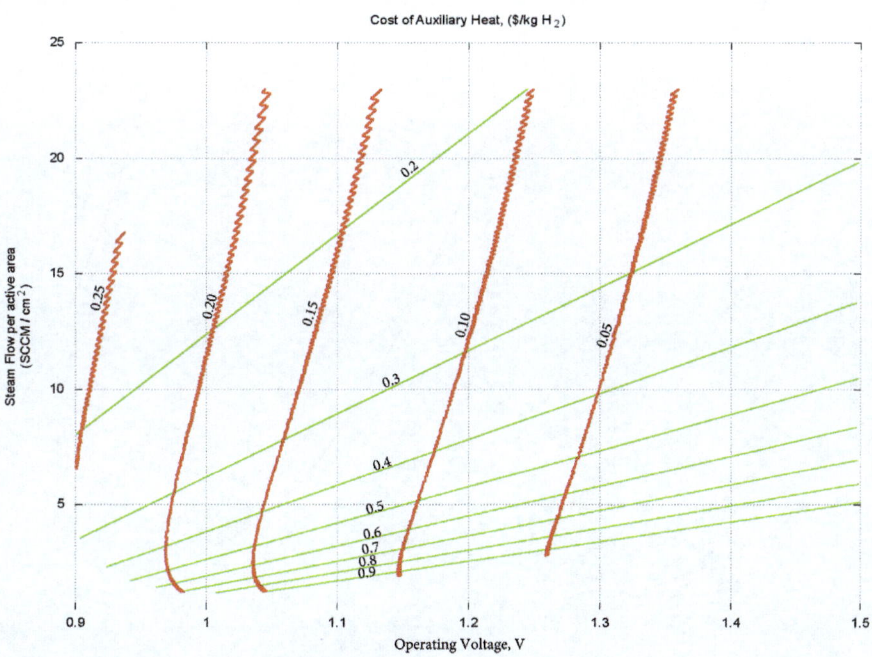

Fig. 6 Cost of fuel, \$ per kg of H_2 produced.

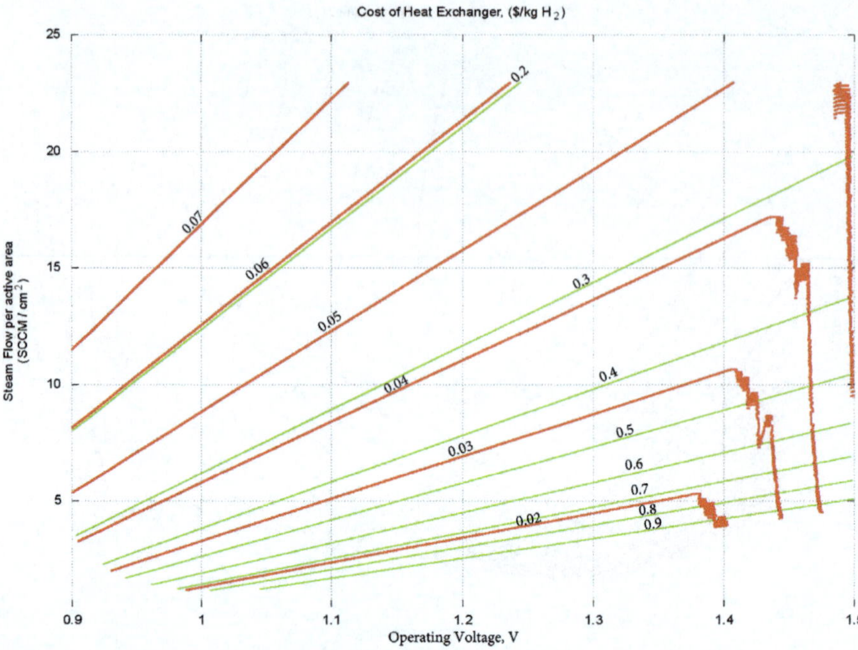

Fig. 7 Cost of heat exchanger, $ per kg of H_2 produced.

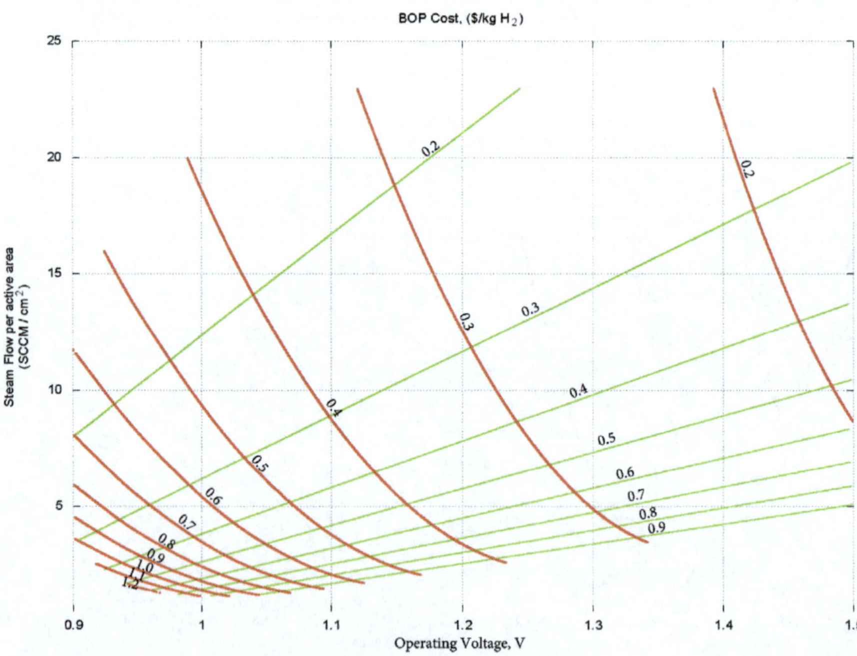

Fig. 8 Cost of BOP, $ per kg of H_2 produced.

Additional factors included in the cost buildup include the cost of fuel, Fig. 6 the heat exchanger cost, Fig. 7 and additional BOP cost, Fig. 8. It turns out in this analysis that the cost of insulation accounts for at most $0.007 per kg of H_2 produced so that map is not shown.

5 Conclusions

A set of relationships combining specified operating and performance parameters for an SOEC system, such as steam flow rates, operating voltage and power density, was developed for a steam fed SOEC hydrogen production system. These were then combined with estimated costs of principal components and sized for specific system design criteria to obtain a cost/performance model. The model was used to map the performance of a stack and the cost elements for a given set of performance parameters across the operating space in voltage and specific steam flow rate (V_{op}, SCCM per cm^3). Component sizing and cost estimates were used to obtain a prediction of total system capital cost. The cost per unit hydrogen produced for this set of assumptions was then calculated. A cost analysis framework is a useful tool to reveal opportunities for system optimization and cost minimization.

References

1 C. Stoots, J. O'Brien and J. Hartvigsen, *Int. J. Hydrogen Energy*, 2009, **34**, 4208–4215.

2 A. Steinfeld, *Sol. Energy*, 2005, **78**, 603–615.

3 E. Harvego, M. McKellar, M. Sohal, J. O'Brien and J. Herring, *ASME 2008 2nd International Conference on Energy Sustainability Collocated with the Heat Transfer, Fluids Engineering, and 3rd Energy Nanotechnology Conferences*, 2008, pp. 549–558.

4 M. G. McKellar, E. A. Harvego and A. M. Gandrik, *Proceedings of HTR*, 2010, p. 093.

5 E. A. Harvego, J. E. O'Brien and M. G. McKellar, *ASME 2012 International Mechanical Engineering Congress and Exposition*, 2012, pp. 875–884.

6 P. Krull, J. Roll and R. Varrin Jr, HTSE Plant Cost Model for the INL HTSE Optimization Study, R-6828-00-01, DEI Record Locator (DEI-1156), Dominion engineering inc. technical report, 2013.

7 J. Hartvigsen, A. Khandkar and S. Elangovan, *Solid Oxide Fuel Cells*, Pennington, NJ, 1999, p. 1135.

8 J. Hartvigsen, S. Elangovan and A. Khandkar, *Third European Solid Oxide Fuel Cell Forum Proceedings*, Nantes, France, 1998, p. 517.

9 A. Khandkar, J. Hartvigsen and S. Elangovan, *Solid State Ionics*, 2000, **135**, 325–330.

10 J. Hartvigsen, S. Elangovan and A. Khandkar, *Solid Oxide Fuel Cells*, Pennington, NJ, 1993, p. 878.

Conclusions

References

Faraday Discussions

PAPER

Syngas production by high temperature steam/CO_2 coelectrolysis using solid oxide electrolysis cells

Xinbing Chen, Chengzhi Guan, Guoping Xiao, Xianlong Du and Jian-Qiang Wang*

Received 9th February 2015, Accepted 23rd March 2015
DOI: 10.1039/c5fd00017c

High temperature (HT) steam/CO_2 coelectrolysis with solid oxide electrolysis cells (SOECs) using the electricity and heat generated from clean energies is an important alternative for syngas production without fossil fuel consumption and greenhouse gas emissions. Herein, reaction characteristics and the outlet syngas composition of HT steam/CO_2 coelectrolysis under different operating conditions, including distinct inlet gas compositions and electrolysis current densities, are systematically studied at 800 °C using commercially available SOECs. The HT coelectrolysis process, which has comparable performance to HT steam electrolysis, is more active than the HT CO_2 electrolysis process, indicating the important contribution of the reverse water-gas shift reaction in the formation of CO. The outlet syngas composition from HT steam/CO_2 coelectrolysis is very sensitive to the operating conditions, indicating the feasibility of controlling the syngas composition by varying these conditions. Maximum steam and CO_2 utilizations of 77% and 76% are achieved at 1.0 A cm^{-2} with an inlet gas composition of 20% H_2/40% steam/40% CO_2.

1. Introduction

Recently, HT steam/CO_2 coelectrolysis using SOECs, which electrochemically electrolyze CO_2 and steam into syngas (CO + H_2), has provided an alternative to traditional syngas production technologies such as coal gasification and natural gas reforming, which release a lot of greenhouse gases and cause serious environmental concerns.[1–4] It is also highly valued as a crucial and limiting step in the CO_2 recycling process, including CO_2 capture, dissociation and reuse, due to its improved stability owing to the reduced possibility of carbon deposition associated with HT CO_2 electrolysis.[5,6] When coupled with clean energies, such as solar, wind and nuclear energies, HT steam/CO_2 coelectrolysis can store the generated heat and power in the form of chemical energy in syngas, which can be further

Shanghai Institute of Applied Physics, Chinese Academy of Sciences, Shanghai, 201800, China. E-mail: wangjianqiang@sinap.ac.cn; Fax: +86 21 3919 4051; Tel: +86 21 3919 4051

processed into liquid fuels such as synthetic diesel or dimethyl ether (DME) *via* catalytic reactions, thereby resolving the related energy storage issues as well as the supply-and-demand conflictions.[7,8] Therefore, the development of HT coelectrolysis technologies bears great significance for the large-scale application of clean energies and the reduction of CO_2 emissions.

Christopher Graves[9] *et al.* reviewed the different routes for steam/CO_2 dissociation, including thermolysis, thermochemical cycles, electrolysis and photoelectrolysis, and they concluded that HT steam/CO_2 coelectrolysis with SOECs is the most efficient path with a potential electricity-to-syngas efficiency of $\sim$100%. The process model established by Carl M. Stoots[10] at Idaho National Lab (INL) predicts an overall thermal-to-syngas efficiency of 48.3 $\pm$ 5.0% for large-scale HT steam/CO_2 coelectrolysis. Qingxi Fu[5] *et al.* economically analysed synthetic fuel production by using the syngas produced from HT steam/CO_2 coelectrolysis, and the diesel production cost was estimated to be 0.86€ per L, comparable to that of the biomass-to-liquid process. Therefore, HT steam/CO_2 coelectrolysis is a promising technology for syngas production with high efficiency and feasibility.

Compared with HT steam or CO_2 electrolysis, the study of HT steam/CO_2 coelectrolysis is much more complicated, and the reasons for this include the following: (1) apart from electrochemical reactions (eqn (1) and (2)), the chemical reverse water-gas shift reaction (RWGS, eqn (3)) also plays an important part in the formation of CO, (2) the equilibrium of the RWGS is highly sensitive to the outlet gas temperature from the electrolysis cell, and (3) the performance and syngas composition change significantly with the microstructure of the SOECs and the operating conditions.[11–13] This complexity also explains why tremendously different phenomena have been observed by different groups.

$$H_2O(g) \xrightarrow{\text{electricity}} H_2(g) + 1/2O_2(g) \tag{1}$$

$$CO_2(g) \xrightarrow{\text{electricity}} CO(g) + 1/2O_2(g) \tag{2}$$

$$CO_2(g) + H_2(g) \rightarrow CO(g) + H_2O(g) \tag{3}$$

Until now, the information available on HT steam/CO_2 coelectrolysis has been very limited and previous studies primarily focus on the development of novel materials, performance characterization, and system efficiency analysis.[14–18] A number of fundamental questions still need to be answered. For example, the contribution of the RWGS reaction in the HT coelectrolysis process is still unclear. Carl M. Stoots[19] *et al.* reported an area specific resistance (ASR) of 1.38 Ω cm^2 for HT coelectrolysis at 800 °C, close to that of 1.36 Ω cm^2 for HT steam electrolysis and much lower than that of 3.84 Ω cm^2 for HT CO_2 electrolysis, concluding that during HT steam/CO_2 coelectrolysis, CO is primarily formed *via* the RWGS reaction. Nevertheless, M. Mogensen[1] and Zhongliang Zhan[20] *et al.* report similar ASR values with only small differences under the three different modes for SOECs, inconsistent with the results of Carl M. Stoots *et al.* Another issue regarding the relationship between the outlet syngas composition and the inlet gas composition, which can provide useful information on the control of the syngas composition, is still scarcely studied. Herein, a systematic study is conducted on HT steam/CO_2 coelectrolysis using commercially available SOECs from Ningbo

Institute of Industrial Technology to evaluate the performance, reaction characteristics, outlet gas composition and steam/CO_2 utilization of HT steam/CO_2 coelectrolysis under different operating conditions, including different inlet gas compositions and electrolysis current densities.

2. Experimental

All the button cells and single cells used in this study were purchased from Ningbo Institute of Industrial Technology. The SOECs are cathode-supported cells composed of Ni–YSZ (Y_2O_3-doped ZrO_2) cathode/YSZ electrolyte/GDC (Gd_2O_3-doped CeO_2) buffer layer/LSCF ($(La_{0.6}Sr_{0.4})(Co_{0.2}Fe_{0.8})O_3$) anode. The button cells are ~2.4 cm in diameter with an active area of 1 cm^2 at the anode side. A single cell (10 × 10 cm^2) with an active area of 64 cm^2 after the application of sealant was used during the study on the effect of the inlet gas composition on the performance and outlet gas composition under a constant electrolysis current density of 0.15 A cm^{-2}. A single cell (5 × 5 cm^2) with an active area of 16 cm^2 was used during the study on the influence of the electrolysis current density on the HT steam/CO_2 coelectrolysis between 0–1.0 A cm^{-2}, considering the upper current limit of 22 A of the electrochemical workstation. A cross-sectional SEM image and photograph of the single cell are shown in Fig. 1.

A schematic of the apparatus used for HT steam/CO_2 coelectrolysis analysis is shown in Fig. 2. The system is primarily composed of the corresponding gas bottles, mass flow controllers (MFC), a bubbler as the humidifier, humidity indicators (H), pressure gauges (P), thermometers (T), a high temperature furnace, the solid oxide electrolysis cell, a dehumidifier to remove the steam in the outlet gas by cooling, a volumetric flow meter and an Agilent 7820A gas chromatograph (GC). H_2 is supplied to the single cell at a flow rate of 300 sccm for the reduction of the NiO particles in the cathode prior to the HT electrolysis measurements. In order to prevent the partial oxidation of the Ni particles in the cathode, 20 mol% of H_2 is used as the protection gas. In the anode, air is used as a sweeping gas at a flow rate of 300 sccm. CO_2 and H_2 at selected flow rates are controlled by the mass flow controller and directed into the humidifier to be mixed with the steam, whose content is changed and controlled by the water temperature in the humidifier. The accurate steam content is calculated and calibrated by the total pressure and humidity indicator placed at the exit of the humidifier *via* eqn (4) below:

$$C_{s,i} = p_s H / p_t \tag{4}$$

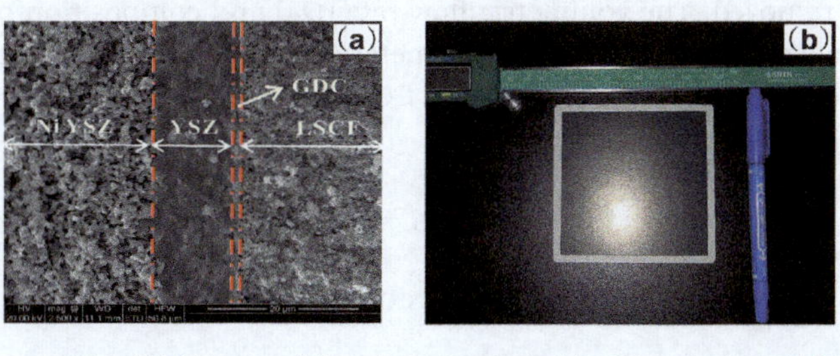

Fig. 1 Cross-sectional SEM image (a) and photograph (b) of the single cell (10 × 10 cm^2).

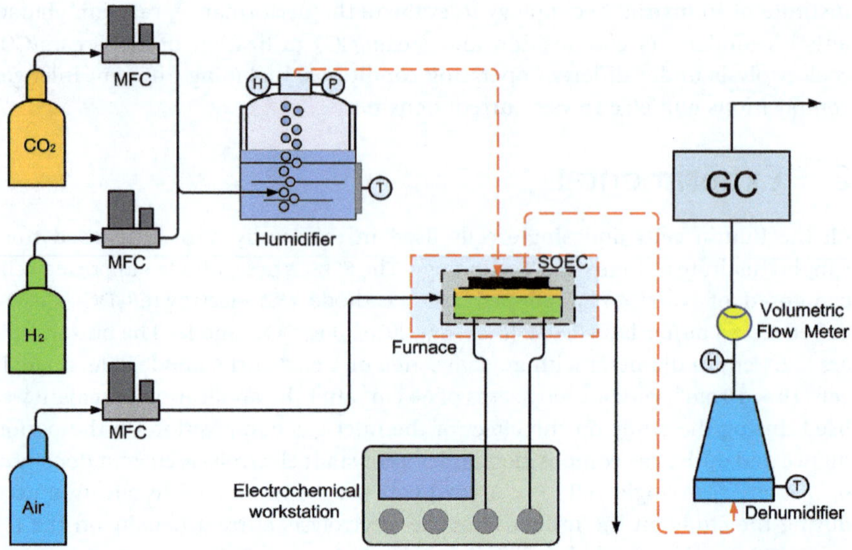

Fig. 2 Schematic illustration of the apparatus for HT steam/CO_2 coelectrolysis analysis.

where $C_{s,i}$ is the steam content in the inlet gas, H is the relative humidity measured by the humidity indicator, p_s is the corresponding saturated steam pressure at the measured temperature and pressure, and p_t is the total pressure measured by the pressure gauge. The flow rate of steam ($Q_{s,i}$) under the normal conditions is determined by eqn (5):

$$Q_{s,i} = C_{s,i}\,(Q_{CO_2,i} + Q_{H_2,i})/(1 - C_{s,i}) \tag{5}$$

where $Q_{CO_2,i}$ and $Q_{H_2,i}$ are the corresponding flow rates of CO_2 and H_2 controlled by the mass flow controller. The total flow rate (Q_t) of the inlet gas is fixed at 100 sccm for the button cell test and 200 sccm for the single cell test. The temperature of the mixed gas from the humidifier is further elevated to ~150 °C and a layer of heat insulating material is applied to the outside of the gas channel in order to avoid steam condensation, as indicated by the dashed line in Fig. 2. The inlet gas is then directed into the SOEC for HT coelectrolysis. The input and output electrical signals are monitored by a Solatron 1260 & 1287 electrochemical workstation coupled with a 12 V/20 A power booster. The SOEC is heated and kept at a constant temperature of ~800 °C by a furnace with integrated thermocouples. The outlet gas from the SOEC is directed into a dehumidifier, where the steam in the gas is removed. The volumetric flow rate (Q_O) and composition of the outlet gas are analysed by a volumetric flow meter and an Agilent 7820 gas chromatograph. The utilization of steam (U_s) and CO_2 (U_{CO_2}) is calculated according to eqn (6) and (7) below:

$$U_s = (Q_O C_{H_2,O} - Q_{H_2,i})/Q_{s,i} \tag{6}$$

$$U_{CO_2} = Q_O C_{CO,O}/Q_{CO_2,i} \tag{7}$$

where $C_{H_2,O}$ is the H_2 content in the outlet gas, while $C_{CO,O}$ is the CO content in the outlet gas.

The impedance curves of the button cell at different atmospheres are measured with the Solatron 1260 impedance analyser in the frequency range of 100 kHz to 0.1 Hz, with an amplitude of 10 mV at open circuit. The ohmic resistance (R_{Ω}) is measured from the high frequency intercept and the polarization resistance (R_p) is obtained from the differences between the high- and low-frequency intercepts on the impedance spectra. The *I–V* curve is measured in the voltage range of 0–0.6 V at a scan rate of 50 mV s^{-1}.

3. Results and discussion

One significant advantage of HT steam/CO_2 coelectrolysis lies in the flexibility of adjusting the outlet syngas composition by varying the operating conditions, such as the inlet gas composition and the electrolysis current density.[11,17] Therefore, the effect of the inlet gas composition on the electrochemical performance, outlet syngas composition, and the steam/CO_2 utilization was investigated. Fig. 3 shows the electrochemical performance of HT steam/CO_2 coelectrolysis with different inlet gas compositions at 800 °C using button cells. In the figure, the inlet gas with a CO_2 concentration of '0' corresponds to the atmosphere for HT steam electrolysis while the inlet gas with a CO_2 concentration of '80%' corresponds to the atmosphere for HT CO_2 electrolysis. In different atmospheres with varying steam/

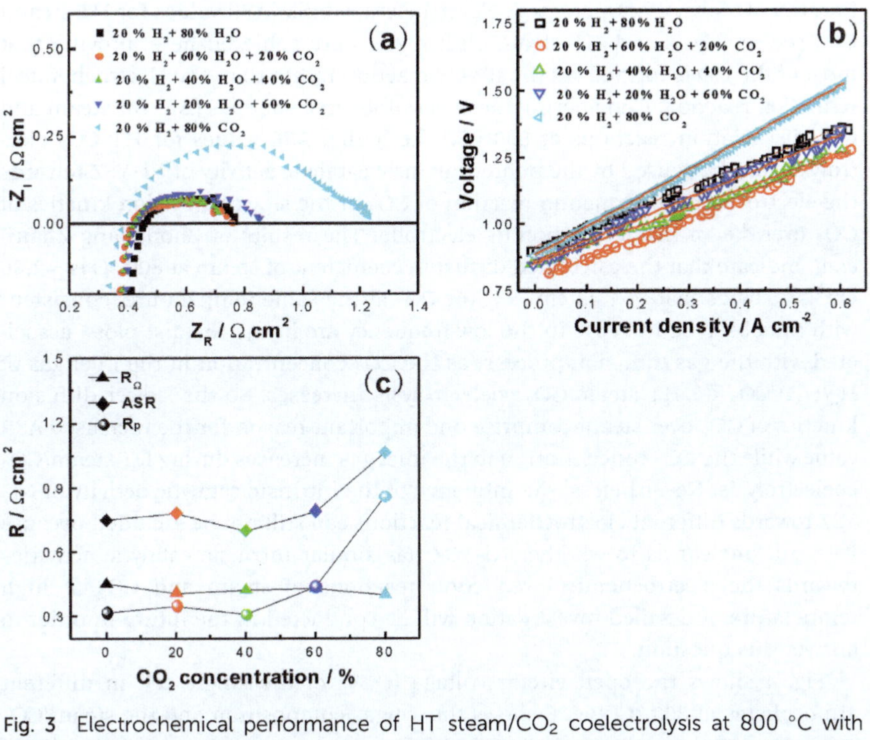

Fig. 3 Electrochemical performance of HT steam/CO_2 coelectrolysis at 800 °C with different inlet gas compositions: (a) Nyquist plots; (b) *I–V* curves; (c) plots of R_{Ω}, ASR and R_p as functions of the CO_2 concentration in the inlet gas.

CO_2 concentration ratios, the cell maintains nearly constant R_Ω values with only minor differences at 800 °C, illustrating the stable electrode/electrolyte contact condition and ionic conductivity of the electrolyte.[21] The polarization resistance (R_p) for the cell is 0.32, 0.35, 0.31, 0.44 and 0.86 Ω cm^2 when the CO_2 content in the inlet gas is 0%, 20%, 40%, 60% and 80%, respectively. This indicates that the inlet gas composition only has a significant influence on the related reaction kinetics when the CO_2 content is above 60%. This is also supported by the area specific resistance (ASR) derived from the slopes of the *I–V* curves in the electrolysis current density range of 0–0.6 A cm^{-2} (Fig. 3b). The ASR for the cell in different atmospheres is 0.75, 0.78, 0.70, 0.79 and 1.06 Ω cm^2 when the CO_2 content in the inlet gas is 0%, 20%, 40%, 60% and 80%, respectively. The results confirm the superior performances of HT steam electrolysis and steam/CO_2 coelectrolysis in comparison to HT CO_2 electrolysis, in agreement with the report by Carl. M. Stoots[19] *et al.*

During the HT steam/CO_2 coelectrolysis process, two possible reactions, the electrochemical reduction of CO_2 (eqn (2)) and the RWGS chemical reaction (eqn (3)), may be responsible for the formation of CO. The ASR value, which has no direct relationship with the kinetics of the RWGS reaction, has a strong dependency on the performance of the electrochemical reduction of CO_2 and steam. As a result, it can be used to evaluate the contribution of the electrochemical reduction reaction in the formation of CO during HT steam/CO_2 coelectrolysis, considering the much slower kinetics of HT CO_2 electrolysis over HT steam electrolysis. An increased ASR value should have been observed for HT steam/CO_2 coelectrolysis as the contribution of the electrochemical reduction reaction towards CO formation increased. Nevertheless, similar ASR values for HT steam/CO_2 coelectrolysis and HT steam electrolysis were achieved here, proving that most of the CO is formed *via* the RWGS reaction rather than the electrochemical reduction reaction. Considering the close Gibbs free energies (ΔG) for steam and CO_2 dissociation reactions at 800 °C,[9] the higher ASR values for HT CO_2 electrolysis may be caused by the poorer intrinsic catalytic activity of Ni–YSZ towards the electrochemical reduction reaction of CO_2 or the slower diffusion kinetics of CO_2 over steam inside the porous electrode. The results of Zhongliang Zhan[20] *et al.* indicate that the calculated diffusion coefficient of steam at 800 °C is ~2.86 cm^2 s^{-1}, twice that of 1.43 cm^2 s^{-1} for CO_2 at the same temperature, consistent with the observed increase in the low-frequency arc in the Nyquist plots, associated with the gas diffusion process, as the CO_2 concentration in the inlet gas of $H_2/H_2O/CO_2$ for HT steam/CO_2 coelectrolysis increases. So the slower diffusion kinetics of CO_2 over steam comprise one important reason for the increased ASR value while the CO_2 concentration in the inlet gas increases during HT steam/CO_2 coelectrolysis. Nevertheless, the influence of the intrinsic catalytic activity of Ni–YSZ towards different electrochemical reactions can still not be excluded because it is still unclear as to whether Ni–YSZ has similar intrinsic catalytic activities towards the electrochemical reduction reactions of steam and CO_2 at high temperature. A detailed investigation will be conducted in the future in order to answer this question.

Fig. 4 shows the open circuit voltage (OCV) of the single cell in different atmospheres at 800 °C. Fig. 5 shows the syngas composition and the steam/CO_2 utilization of the cell operated at 800 °C under a constant current density of 0.15 A cm^{-2} after the removal of steam in the outlet gas. The single cell in pure

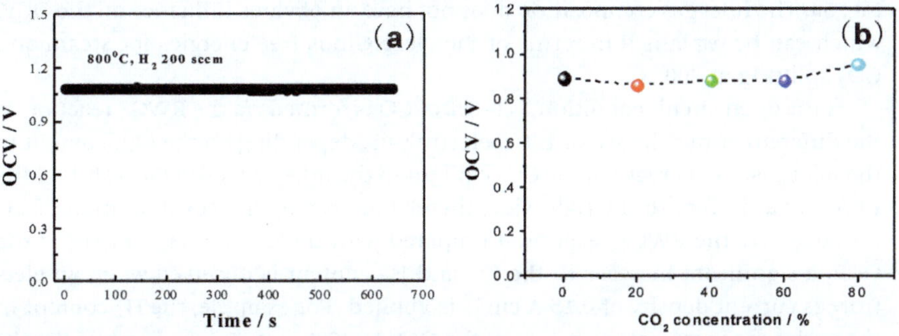

Fig. 4 The OCV of the single cell at 800 °C: (a) in pure hydrogen; (b) in different atmospheres for HT steam/CO_2 coelectrolysis.

hydrogen delivers a stable OCV value of ~1.10 V during the test, indicating a good sealing effect. In the atmospheres for HT electrolysis, the OCV is 0.89, 0.86, 0.88, 0.88 and 0.95 V when the CO_2 content in the inlet gas is 0%, 20%, 40%, 60% and 80%, respectively. The slightly higher OCV value of 0.95 V for HT CO_2 electrolysis at a CO_2 content of 80% is probably caused by the presence of ~8% of CO formed *via* the RWGS reaction, as illustrated by the outlet syngas composition analysis in

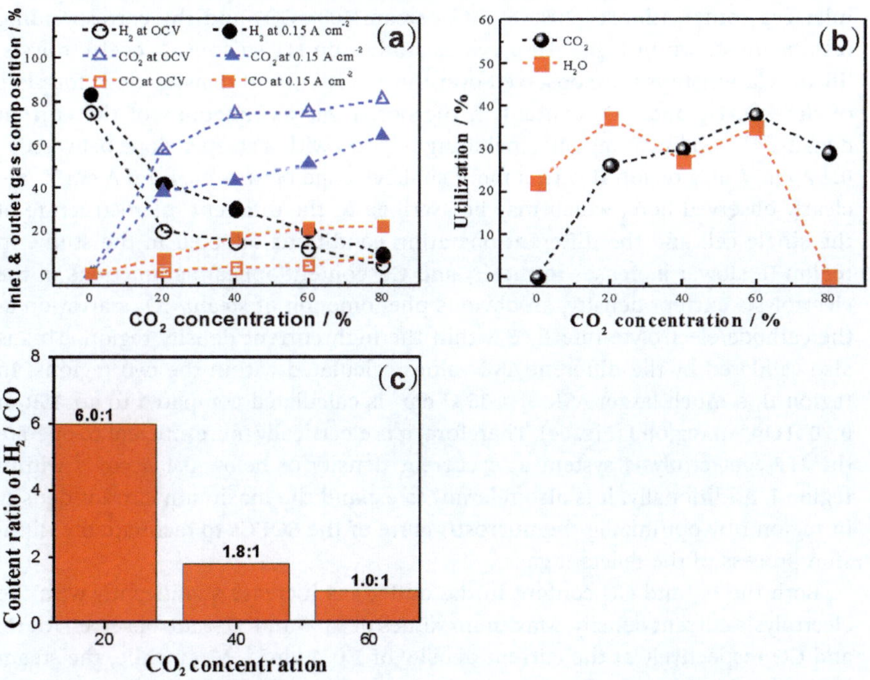

Fig. 5 Outlet gas composition analysis of HT steam/CO_2 coelectrolysis at 800 °C with different inlet gas compositions: (a) plots of the H_2, CO_2 and CO concentrations as functions of CO_2 concentration in the inlet gas, (b) plots of steam/CO_2 utilization as functions of CO_2 concentration in the inlet gas, and (c) content ratio of H_2/CO in the outlet gas for different inlet gas compositions.

Fig. 5a. The inlet gas composition does not have an obvious influence on the OCV, which can be explained in terms of the close Gibbs free energies for steam and CO_2 splitting at 800 °C.[15]

At the open circuit condition, ~1–8% of CO is formed *via* the RWGS reaction in the different atmospheres for HT coelectrolysis, depending on the CO_2 content in the inlet gas. An increase in the CO_2 content in the inlet gas results in a shift of the RWGS equilibrium to the right side, thereby increasing the concentration of CO produced *via* the RWGS reaction. Compared with the CO and H_2 content at the OCV, a significant increase in the CO and H_2 content is observed when an electrolysis current density of 0.15 A cm^{-2} is applied. For example, the H_2 content in the outlet gas increases from 15% at the OCV to 30% at 0.15 A cm^{-2} while the CO content increases from 3% to 17% correspondingly with an inlet gas composition of 20% H_2/40% steam/40% CO_2. This clearly demonstrates the effect of the input electric power on the outlet gas composition of the HT steam/CO_2 coelectrolysis system. However, a low steam/CO_2 utilization of ~20–40% is obtained due to the low electrolysis current density of 0.15 A cm^{-2} (Fig. 5b). The H_2/CO content ratio in the outlet gas during HT steam/CO_2 coelectrolysis ranges from 6 : 1 to 1 : 1 with the inlet gas composition, confirming the feasibility of adjusting the syngas composition by controlling the inlet gas composition (Fig. 5c). It should be noted that the 20% H_2 as the protection gas is also included in the outlet H_2 during the calculation process of the H_2/CO content ratio.

The outlet syngas composition and steam/CO_2 utilization were also measured at 800 °C at different electrolysis current densities of 0–1.0 A cm^{-2} with a fixed inlet gas composition of 20% H_2/40% steam/40% CO_2 and the corresponding results are shown in Fig. 6. In previous studies on HT steam/CO_2 coelectrolysis, linear relationships were observed over the whole current density range for plots of the CO, H_2 and CO_2 content in the outlet gas as functions of the current density.[8,19] Two linear regions, including region I with a steeper slope between 0–0.4 A cm^{-2} and region II with a more shallow slope between 0.4–1.0 A cm^{-2}, are clearly observed here, which may be ascribed to the different microstructure of the single cell and the different operating conditions adopted in the study. In region II, slower increases in the H_2 and CO content appear as functions of the electrolysis current density, an obvious phenomenon of steam/CO_2 starvation at the cathode/electrolyte interface within the high current density region. This is also validated by the different ASR values calculated within the two regions. In region II, a much larger ASR of 1.35 Ω cm^2 is calculated compared to an ASR of 0.70 Ω cm^2 in region I (Fig. 6c). Therefore, it is electrically more efficient to operate the HT coelectrolysis system at a current density of below 0.4 A cm^{-2} within region I. Additionally, it is also relevant to expand the maximum current density in region I by optimizing the microstructure of the SOECs to facilitate the diffusion process of the reactant gas.

Both the H_2 and CO content in the outlet gas increase significantly with the electrolysis current density. Maximum values of 41% and 37% are observed for H_2 and CO respectively at the current density of 1.0 A cm^{-2}. Meanwhile, the steam utilization increases from −8.0% at the OCV to 76% at 1.0 A cm^{-2} and the CO_2 utilization increases from 13% to 77% accordingly. The negative steam utilization is attributed to the formation of CO through the RWGS reaction at the open circuit condition, which consumes part of the 20% H_2 in the inlet gas as the protection gas of the Ni–YSZ composite cathode. Similarly, the H_2/CO content

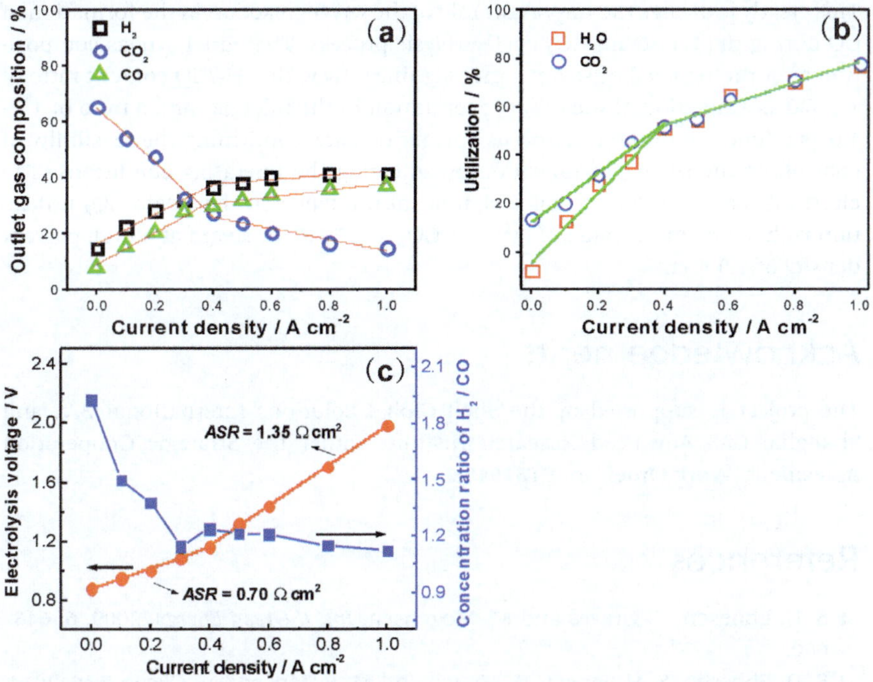

Fig. 6 Outlet gas composition, electrolysis voltage and H_2/CO content ratio of HT steam/CO_2 coelectrolysis at 800 °C under different electrolysis current densities: (a) plots of the H_2, CO_2, CO content in the outlet gas as functions of current density; (b) plots of steam/CO_2 utilization as functions of current density; (c) plots of electrolysis voltage and H_2/CO content ratio as functions of current density.

ratio also clearly varies with the current density, from 1.9 at the OCV, to 1.5 at 0.1 A cm^{-2}, 1.2 at 0.4 A cm^{-2} and 1.1 at 1.0 A cm^{-2}, indicating the high sensitivity of the outlet syngas composition to the electrolysis current density applied. During the synthesis of hydrocarbons or methanol from syngas *via* catalytic processes such as the Fischer–Tropsch process, the desired H_2/CO content ratio varies from ~1.4 to 2.1, which can be flexibly achieved by adjusting the operating conditions such as the inlet gas composition or the electrolysis current density.[22,23] This also comprises one important advantage of HT steam/CO_2 coelectrolysis. Therefore, the inlet gas composition and electrolysis current density are two important parameters which should be fully considered in the operation.

4. Conclusions

The reaction characteristics of syngas production by HT steam/CO_2 coelectrolysis with commercially available solid oxide electrolysis cells are systematically studied. For the inlet gas composition range studied (20% H_2/20–80% CO_2/remaining steam), the electrochemical performance of HT steam/CO_2 coelectrolysis is close to that of HT steam electrolysis and significantly superior to that of HT CO_2 electrolysis, supported by ASRs of 0.75 Ω cm^2 for HT steam electrolysis, 0.70–0.79 Ω cm^2 for HT coelectrolysis, and 1.06 Ω cm^2 for HT CO_2 electrolysis.

This clearly indicates the important role of the RWGS reaction in the formation of CO during the HT steam/CO_2 coelectrolysis process. The outlet syngas composition after the removal of steam varies significantly with a H_2/CO content ratio of 1.0–6.0 as a function of the CO_2 concentration in the inlet gas and a ratio of 1.9–1.0 as a function of the electrolysis current density, confirming the feasibility of controlling the syngas composition by adjusting the operating conditions. The electrolysis current density has a significant influence on the steam/CO_2 utilization with a maximum value of 76% for CO_2 and 77% for steam at a high current density of 1.0 A cm^{-2}.

Acknowledgements

The project is supported by the Shell Global Solutions International B.V. and Shanghai CAS Advanced Research Institute under the Strategic Cooperation Agreement (Work Order no. PT31046).

References

1 S. D. Ebbesen, C. Graves and M. Mogensen, *Int. J. Green Energy*, 2009, **6**, 646–660.
2 S. D. Ebbesen, S. H. Jensen, A. Hauch and M. B. Mogensen, *Chem. Rev.*, 2014, **114**, 10697–10734.
3 G. Hawkes, J. O'Brien, C. Stoots and B. Hawkes, *Int. J. Hydrogen Energy*, 2009, **34**, 4189–4197.
4 S. Li, Y. Li, Y. Gan, K. Xie and G. Meng, *J. Power Sources*, 2012, **218**, 244–249.
5 Q. Fu, C. Mabilat, M. Zahid, A. Brisse and L. Gautier, *Energy Environ. Sci.*, 2010, **3**, 1382–1397.
6 S.-E. Yoon, S.-H. Song, J. Choi, J.-Y. Ahn, B.-K. Kim and J.-S. Park, *Int. J. Hydrogen Energy*, 2014, **39**, 5497–5504.
7 W. L. Becker, R. J. Braun, M. Penev and M. Melaina, *Energy*, 2012, **47**, 99–115.
8 Q. Fu, J. Dailly, A. Brisse and M. Zahid, *ECS Trans.*, 2011, **35**, 2949–2956.
9 C. Graves, S. D. Ebbesen, M. Mogensen and K. S. Lackner, *Renewable Sustainable Energy Rev.*, 2011, **15**, 1–23.
10 J. E. O'Brien, M. G. McKellar, E. A. Harvego and C. M. Stoots, *Int. J. Hydrogen Energy*, 2010, **35**, 4808–4819.
11 C. M. Stoots, *High-temperature co-electrolysis of H_2O and CO_2 for syngas production*, 2006.
12 V. Menon, Q. Fu, V. M. Janardhanan and O. Deutschmann, *J. Power Sources*, 2015, **274**, 768–781.
13 Y. Xie and X. Xue, *Solid State Ionics*, 2012, **224**, 64–73.
14 K. Xie, Y. Zhang, G. Meng and J. T. S. Irvine, *Energy Environ. Sci.*, 2011, **4**, 2218–2222.
15 X. Sun, M. Chen, S. H. Jensen, S. D. Ebbesen, C. Graves and M. Mogensen, *Int. J. Hydrogen Energy*, 2012, **37**, 17101–17110.
16 C. Graves, S. D. Ebbesen and M. Mogensen, *Solid State Ionics*, 2011, **192**, 398–403.
17 M. Ni, *J. Power Sources*, 2012, **202**, 209–216.
18 P. Kim-Lohsoontorn and J. Bae, *J. Power Sources*, 2011, **196**, 7161–7168.

19 C. Stoots, J. O'Brien and J. Hartvigsen, *Int. J. Hydrogen Energy*, 2009, **34**, 4208–4215.

20 Z. Zhan, W. Kobsiriphat, J. R. Wilson, M. Pillai, I. Kim and S. A. Barnett, *Energy Fuels*, 2009, **23**, 3089–3096.

21 X. Chen, C. Jin, L. Zhao, L. Zhang, C. Guan, L. Wang, Y.-F. Song, C.-C. Wang, J.-Q. Wang and S. P. Jiang, *Int. J. Hydrogen Energy*, 2014, **39**, 15728–15734.

22 G. A. Mills, *Fuel*, 1994, **73**, 1243–1279.

23 E. Iglesia, *Appl. Catal., A*, 1997, **161**, 59–78.

Faraday Discussions

PAPER

Synthesis of ammonia directly from wet air using $Sm_{0.6}Ba_{0.4}Fe_{0.8}Cu_{0.2}O_{3-\delta}$ as the catalyst

Rong Lan,[a] Khaled A. Alkhazmi,[a] Ibrahim A. Amar[a] and Shanwen Tao*[ab]

Received 20th March 2015, Accepted 7th April 2015

DOI: 10.1039/c5fd00033e

Ammonia was directly synthesised from wet air at 400 °C at atmospheric pressure. A new perovskite $Sm_{0.6}Ba_{0.4}Fe_{0.8}Cu_{0.2}O_{3-\delta}$ was used as the electrocatalyst for electrochemical synthesis of ammonia. Ammonia formation rates of 9.19×10^{-7} mol s^{-1} m^{-2} and 1.53×10^{-6} mol s^{-1} m^{-2} were obtained at 400 °C when wet air and wet N_2 were introduced into a simple single chamber reactor, respectively. The perovskite catalyst is low cost compared to the previously reported Ru/MgO and Pt/C catalysts. This experiment indicates that ammonia can be directly synthesised from wet air, a very promising simple technology for sustainable synthesis of ammonia in the future.

1 Introduction

Nitrogen fertiliser has supported approximately 27% of the world's population over the last century, equivalent to around 4 billion people born (or 42% of the estimated total births) since 1908.[1] Although over 78% of the atmosphere is composed of nitrogen, it is difficult to use N_2 directly from plants as plants can only use fixed nitrogen. The breakthrough in nitrogen fixation took place a century ago, the well-known Haber–Bosch process, operating at high temperatures (~500 °C) and high pressures (150–300 bar).[2] Fritz Haber and Carl Bosch won the Nobel Prize in Chemistry in 1918 and 1931, respectively, for their contributions in industrial production of ammonia. This innovation became a key driver in the development of the industrialized society and is still securing the nutrition of billions of people today. In 2007, the Nobel Prize in Chemistry was awarded to Gerhard Ertl for his studies of chemical processes on solid surfaces. A major contribution to this prize was Gerhard Ertl's understanding of the reaction mechanism of H_2 and N_2 on solid ammonia synthesis catalysts.[3]

In the Haber–Bosch process, the required hydrogen is produced through the steam reforming of natural gases or coal of which consumed more than 1% of

[a]*Department of Chemical & Process Engineering, University of Strathclyde, Glasgow G1 1XJ, UK*

[b]*School of Engineering, University of Warwick, Coventry, CV4 7AL, UK. E-mail: S.Tao.1@warwick.ac.uk; Fax: +44 (0) 24 764 18922; Tel: +44 (0) 24 761 51680*

world power generation.[4] In addition, CO_2 emitted from the ammonia industry is equivalent to 0.77% of the world total CO_2 emission. In order to reduce the current dependency on fossil fuels and to reduce carbon emissions associated with their use, it is essential to introduce new ammonia synthesis processes and break the link between fossil fuels and the ammonia industry.[5]

To reduce global CO_2 emission, renewable electricity generated from solar, wind, marine and other sources are becoming more and more important. Due to the intermittent nature of those renewable resources, energy storage, particularly renewable electricity storage has become a big challenge. One of the possible solutions for renewable electricity storage is to produce basic chemical feedstocks such as ammonia. The electrochemical synthesis processes are among the promising alternatives in particular if the hydrogen required for ammonia synthesis is produced from non-fossil fuel resources.[6–8] Recently, it has been reported that ammonia can be directly synthesised from air and water at ambient conditions.[2] It has been widely reported that ammonia can be synthesised from N_2 and H_2 (or steam) at a temperature around 500 °C.[9–15] The possible reason is that, the decomposition of ammonia at high temperature is kinetically slow thus ammonia can still be collected if it is removed from the reactor in time. Ammonia will be able to be synthesised from air and water at elevated temperatures if the oxidation of ammonia is also kinetically slow.

In our previous papers, we reported the electrochemical synthesis of ammonia based on electrochemical cells using a $Ce_{0.8}Sm_{0.2}O_{2-\delta}$–$(Li,Na,K)_2CO_3$ composite electrolyte.[12] The eutectic point of $(Li,Na,K)_2CO_3$ ternary molten salts is 396 °C.[16,17] The low melting point of mixed $(Li,Na,K)_2CO_3$ salts causes the composite to exhibit high ionic conductivity at relatively low temperatures which can minimise the operating temperature of the cell, reducing the thermal decomposition and oxidation of ammonia. It has been reported that the ionic conductivity of the $Ce_{0.8}Sm_{0.2}O_{2-\delta}$–$(Li,Na,K)_2CO_3$ composite reached 0.1 S cm^{-1} at a temperature around 400 °C.[18] The total ionic conductivity includes those from Li^+, Na^+, K^+, H^+, HCO_3^-, CO_3^{2-} and O^{2-} ions. Good fuel cell performance based on a $Ce_{0.8}Sm_{0.2}O_{2-\delta}$–$(Li,Na,K)_2CO_3$ composite electrolyte indicates higher H^+/O^{2-} ionic conductivity in this type of composite material.[18–20] Either H^+ or O^{2-} ionic conduction can be used for electrochemical synthesis of ammonia.[6] Therefore a similar composite electrolyte $Ce_{0.8}Gd_{0.2}O_{2-\delta}$–$(Li,Na,K)_2CO_3$ was used as the electrolyte in this study for ammonia synthesis. Recently, it has been reported that ammonia was successfully synthesised directly from wet air using $La_{0.8}Cs_{0.2}Fe_{0.8}Ni_{0.2}O_{3-\delta}$ [21] or $Pr_{0.6}Ba_{0.4}Fe_{0.8}Cu_{0.2}O_{3-\delta}$ [22] as the catalyst. It has been reported that Cu and Ni co-doped $SmFeO_3$, $SmFe_{0.7}Cu_{0.1}Ni_{0.2}O_{3-\delta}$ is a very good ammonia synthesis catalyst and an ammonia formation rate of 1.13×10^{-8} mol cm^{-2} s^{-1} was observed at 80 °C when N_2 and H_2 were used as the reactants.[23] This inspired us to use doped $SmFeO_3$ catalysts for the electrochemical synthesis of ammonia. On the other hand, BaO is a typical promoter of ammonia synthesis catalysts.[24,25] However, BaO itself is unstable in wet air because of the reaction between BaO and H_2O to form $Ba(OH)_2$. Therefore, in this paper, we introduce some Ba at the A-site of perovskite oxide $SmFeO_3$. A new perovskite, $Sm_{0.6}Ba_{0.4}Fe_{0.8}Cu_{0.2}O_{3-\delta}$ (SBFCu), was synthesised as an electro-catalyst for the synthesis of ammonia from air/N_2 and water. $Sm_{0.6}Ba_{0.4}Fe_{0.8}Cu_{0.2}O_{3-\delta}$ was used as both the cathode and anode and a single chamber reactor was used for the synthesis.

2 Experimental

Materials synthesis

Calculated amounts of Gd_2O_3 were dissolved in hot dilute nitric acid to form a gadolinium nitrate aqueous solution. Then an appropriate amount of $Ce(NO_3)_3 \cdot 6H_2O$ was added to the solution. A 1 M $(NH_4)_2CO_3$ solution was added dropwise with vigorous stirring until the pH value reached 7–8, after that vigorous stirring was continued for another 30 min. The formed precipitate was then collected by vacuum filtration and washed several times with deionised water then dried at 100 °C overnight. The obtained powder was calcined in air at 650 °C for 2 hours with a heating/cooling rate of 2 °C min^{-1}. Finally, ultrafine powder of GDC was obtained.

$Ce_{0.8}Gd_{0.2}O_{2-\delta}$ and ternary carbonate mixture $(Li,Na,K)_2CO_3$ (32.1 wt% Li_2CO_3; 34.5 wt% K_2CO_3; 33.4 wt% Na_2CO_3) powders were mixed together with a weight ratio of oxide to carbonate of 7 : 3. The mixture was mixed and ground with the use of 25 ml isopropanol. Then the mixture was put in a ball miller (Planetary Mono Mill Pulverisette 6) for mixing with a speed of 400 rpm for 4 hours. The weight ratio of balls to powders was roughly 1 : 1. Then the mixture was heated by a hot plate magnetically stirred at 150 °C to release and remove the isopropanol. The powder was heated at 600 °C for one hour then quenched in air to room temperature. The as-prepared oxide–carbonate composite electrolyte will be used for ammonia synthesis.

Samarium(III) oxide was dissolved in diluted nitric acid to form samarium nitrate under heating and stirring. Calculated amounts of nitrates $Ba(NO_3)_2$, $Fe(NO_3)_3 \cdot 9H_2O$ and Cu $(NO_3)_3 \cdot 2.5H_2O$ were dissolved in deionised water and were added to the above prepared solution. Appropriate amounts of citric acid and EDTA (ethylenediaminetetraacetic acid) were then added as complexing agents with a molar ratio of citric acid : EDTA : metal cations of 1.5 : 1 : 1. Dilute aqueous ammonia solution was then added to the mixed solution to adjust the pH value to around 6 and a dark green clear solution was obtained. The mixed solution was evaporated on a hot-plate at 200–250 °C under stirring which gradually changed into a dark sticky gel before complete drying. By further heating the gel converted to a black colour. The brown ash was grounded and subsequently calcined in air at 900 °C for 2 hours with heating/cooling rates of 5 °C min^{-1} to obtain a single phase of $Sm_{0.6}Ba_{0.4}Fe_{0.8}Cu_{0.2}O_{3-\delta}$.

Materials characterisation

X-ray data were collected on a PANanalytical X' Pert Pro in the Bragg–Brentano reflection geometry with a Ni-filtered Cu Kα source (1.5405 Å), fitted with a X'Celerator detector and an Empyrean CuLFF XRD tube. Absolute scans in the 2θ range of 5–100° with step sizes of 0.0167° were used during data collection. The surface area of $Sm_{0.6}Ba_{0.4}Fe_{0.8}Cu_{0.2}O_{3-\delta}$ powder was measured by a N_2 adsorption method using Micromeritics ASAP 2420.

TG/DSC analyses were performed using a Stanton Redcroft STA/TGH series STA 1500 operating through a Rheometric Scientific system interface controlled by the software RSI Orchestrator in flowing air or N_2 at a flow rate of 50 ml min^{-1}.

Total conductivity measurements were carried out by a pseudo-four-terminal method using a computer-controlled Solartron Analytical® SI 1470A electrochemical interface by applying a constant current. The $Sm_{0.6}Ba_{0.4}Fe_{0.8}Cu_{0.2}O_{3-\delta}$ powder fired at 900 °C was pressed into pellets with diameters of 13 mm and thicknesses of around 2 mm which were then fired at 1100 °C for 5 hours. A silver coated pellet was fitted into the measuring apparatus and the conductivity measurement was carried out in air between room temperature and 750 °C.

Fabrication of the single cell and ammonia synthesis measurements

The cell was placed in a self-designed single chamber reactor as described elsewhere.[22] The electrolytic cell for ammonia synthesis was a tri-layer single cell which was fabricated by a cost-effective one-step dry-pressing and co-firing process. The anode and cathode material was 0.8 g $Sm_{0.6}Ba_{0.4}Fe_{0.8}Cu_{0.2}O_{3-\delta}$ plus 0.2 g composite electrolyte to form composite electrode materials. The composite anode, composite electrolyte and composite cathode (SBFCu/CGO–(Li,Na,K)$_2$CO$_3$/SBFCu) were fed into the die, layer by layer, with the aid of a sieve to ensure uniform powder distribution, and then unixally pressed at a pressure of 300 MPa by cold pressing into 13 mm pellets. The pellets were sintered at 600 °C for 2 h. The thickness of the anode, electrolyte and cathode was approximately 0.6, 1.0 and 0.6 mm, respectively. The catalyst surface area of the cathode and anode was 1.281 cm^2. Silver paste was painted on each electrode surface of the cell as a current collector. Ag wires were used as output terminals for both electrodes. It has been reported that silver itself has negligible catalytic effects on ammonia synthesis.[9]

Compressed air or N_2 was passed through room temperature water before inputting into the single chamber reactor as described in previous reports.[21,22] The flow rate of air (or N_2) was 50 ml min^{-1}. The steam concentration in air (or N_2) was 3 mol%. A DC voltage was applied to the cell using a Solartron 1287A electrochemical interface. The synthesised ammonia was absorbed by 25 ml of diluted sulphuric acid (0.001 M) for 30 min as described before.[2,12] The concentration of NH_4^+ in the absorbed solution was analysed using the Orion Application solution for low adjusting ISA. The produced ammonia was detected using an ammonia meter (ISE Thermo Scientific Orion Star A214) and the rate of ammonia formation was calculated using the following equation.[2]

$$r_{NH_3} = \frac{[NH_4^+] \times V}{t \times A} \tag{1}$$

where: $[NH_4^+]$ = is the measured NH_4^+ ion concentration, V = is the volume of the dilute H_2SO_4 for ammonia collection, t = is the adsorption time and A = is the effective area of the cell.

3 Results and discussion

Thermal analysis of the brown ash was carried out in air (Fig. 1A). It was found that a weight loss of ~52% occurred at a temperature between 300 and 430 °C. This is due to the decomposition of the ash and loss of organic components. The sample weight became stable at a temperature above 800 °C indicating the minimum formation temperature for $Sm_{0.6}Ba_{0.4}Fe_{0.8}Cu_{0.2}O_{3-\delta}$ is 800 °C. To guarantee the formation of a single phase, we choose 900 °C as the firing

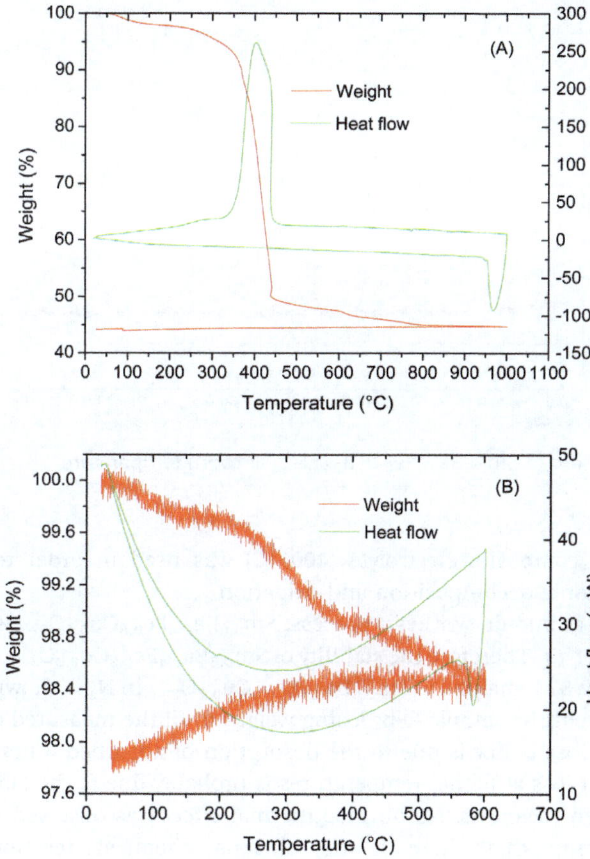

Fig. 1 The STA analysis of the $Sm_{0.6}Ba_{0.4}Fe_{0.8}Cu_{0.2}O_{3-\delta}$ ash (A) and the $Sm_{0.6}Ba_{0.4}Fe_{0.8}Cu_{0.2}O_{3-\delta}$ powder in N_2 (B).

temperature for $Sm_{0.6}Ba_{0.4}Fe_{0.8}Cu_{0.2}O_{3-\delta}$. The ash was fired at 900 °C for 2 hours to form a perovskite phase. XRD analysis of the powder after firing at 900 °C indicates it is a single phase (Fig. 2). It exhibits an cubic structure with a space group of $Pm\bar{3}m$(221); $a = 3.9148(1)$ Å, $V = 60.00(1)$ Å^3 (Table 1). In our previous report, perovskite oxide $La_{0.6}Sr_{0.4}Fe_{0.8}Cu_{0.2}O_{3-\delta}$ exhibits a cubic structure with a space group of *Pm*-3*m*(221).[12] SEM pictures indicate that the prepared $Sm_{0.6}Ba_{0.4}Fe_{0.8}Cu_{0.2}O_{3-\delta}$ powders agglomerated together with a secondary particle size of ~0.1 μm (Fig. 3). The surface area of the $La_{0.6}Sr_{0.4}Fe_{0.8}Cu_{0.2}O_{3-\delta}$ prepared at 900 °C was 7.94(1) m^2 g^{-1}.

The electronic conductivity of $Sm_{0.6}Ba_{0.4}Fe_{0.8}Cu_{0.2}O_{3-\delta}$ is very important in order to be used as both the cathode and anode for the electrochemical cells for ammonia synthesis. The DC conductivity measurement indicated that the conductivity increased before 550 °C, reaching a value of 2.29 S cm^{-1} and then started to decrease (Fig. 4). This is probably due to the semi-conductor to metal transition which is a common phenomenon in perovskite oxides.[26] The conductivity of $Sm_{0.6}Ba_{0.4}Fe_{0.8}Cu_{0.2}O_{3-\delta}$ was 1.88 S cm^{-1} at 400 °C. Regarding the operating temperature of the cell, the lowest working temperature of the CGO–

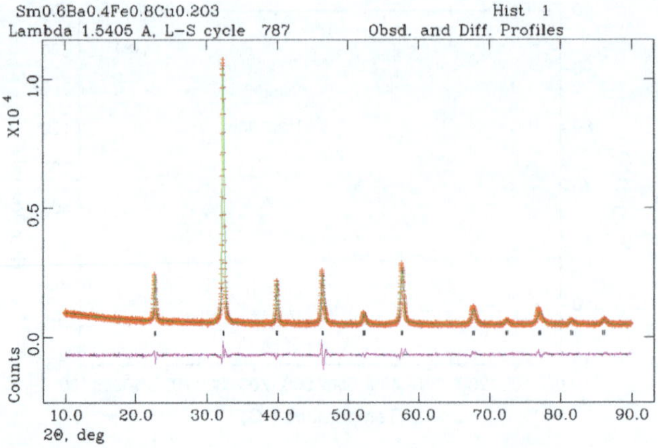

Fig. 2 XRD pattern of $Sm_{0.6}Ba_{0.4}Fe_{0.8}Cu_{0.2}O_{3-\delta}$ at room temperature.

$(Li,Na,K)_2CO_3$ composite electrolyte (400 °C) was used in order to minimise possible ammonia decomposition and oxidation.

During the ammonia synthesis process, $Sm_{0.6}Ba_{0.4}Fe_{0.8}Cu_{0.2}O_{3-\delta}$ was exposed to wet N_2 or wet air. Therefore the stability of $Sm_{0.6}Ba_{0.4}Fe_{0.8}Cu_{0.2}O_{3-\delta}$ in N_2 is very important. The STA analysis of $Sm_{0.6}Ba_{0.4}Fe_{0.8}Cu_{0.2}O_{3-\delta}$ in N_2 is shown in Fig. 1B. It was found that the sample kept losing weight until the measured 600 °C. The initial weight loss below is due to the desorption of adsorbed water and gases. Further weight loss at higher temperatures is probably due to the lattice oxygen, forming oxygen vacancies. No obvious thermal effect was observed on the DSC curve indicating that there is no obvious chemical reaction between $Sm_{0.6}Ba_{0.4}Fe_{0.8}Cu_{0.2}O_{3-\delta}$ and N_2 at a temperature below 600 °C.

For ammonia synthesis, wet air was fed into the single chamber reactor as described elsewhere.[22] The gas composition was approximately 3 mol% H_2O, 97 mol% air. The effect of the applied potential on ammonia formation rate was investigated at 400 °C and the order for applied voltage was from low to high, between 1.2 and 1.7 V with and interval of 0.1 V. Fig. 5A shows the recorded current change against time for the electrolytic cell during the ammonia synthesis process at different potentials at 400 °C over a period of 3600 seconds (was 3300 seconds due to an interruption when applying 1.6 V). As can be seen, the current density decreased at the beginning but tended to become stable. The ammonia

Table 1 Structure parameters of $Sm_{0.6}Sr_{0.4}Fe_{0.8}Cu_{0.2}O_{3-\delta}$[a]

Atom	Site	Occupancy	x	y	z	U_{iso}/Å^2
Sm	1a	0.6	0	0	0	0.0181(5)
Ba	1a	0.4	0	0	0	0.0181(5)
Fe	1b	0.8	0.5	0.5	0.5	0.0171(9)
Cu	1b	0.2	0.5	0.5	0.5	0.0171(9)
O	3c	1	0	0.5	0.5	0.796(27)

[a] Space group $Pm\bar{3}m$(221); $a = 3.9148(1)$ Å, $V = 60.00(1)$ Å^3. $R_{wp} = 6.11\%$, $R_p = 4.48\%$, $\chi_{red}{}^2 =$ 2.52.

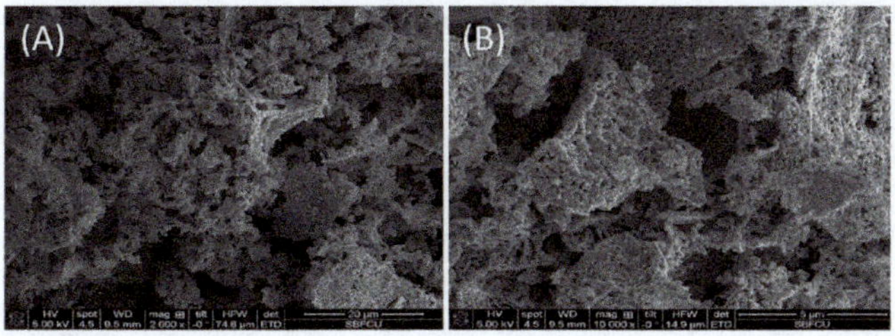

Fig. 3 SEM images of $Sm_{0.6}Ba_{0.4}Fe_{0.8}Cu_{0.2}O_{3-\delta}$ powder after calcining at 900 °C for 2 hours. (A) at ×2000 magnification, (B) at 10 000 magnification.

produced in the single chamber reactor was absorbed by 20 ml of dilute H_2SO_4 (0.001 M) for 60 minutes. It is surprising that ammonia was successfully synthesised at atmospheric pressure from wet air (~3% H_2O). As can be seen in Fig. 6A the ammonia formation rate increased significantly with increasing the applied potential and reached a maximum value at 1.4 V (9.19×10^{-7} mol s^{-1} m^{-2} at 400 °C), showing that 1.4 V was the optimum potential for ammonia synthesis in this study. In terms of catalyst weight, the ammonia formation rate was 1.48×10^{-10} mol s^{-1} $g_{cat}{}^{-1}$. This is two orders of magnitude higher than the reported ammonia formation rates (around 1.0×10^{-12} mol s^{-1} $g_{cat}{}^{-1}$ at ~600 °C) when either a H^+ ($SrCe_{0.95}Yb_{0.05}O_{3-\delta}$) or O^{2-} (8 mol% yttria-stabilized zirconia) conductor was used as the electrolyte with an industrial Ru/MgO catalyst used at the cathode.[9] It was 400 °C in our experiment which is roughly 200 °C lower than that reported by Skodra.[9] Higher working temperatures may cause the thermal decomposition or oxidation of the produced ammonia, particularly at a temperature above 500 °C.[2] The $Sm_{0.6}Ba_{0.4}Fe_{0.8}Cu_{0.2}O_{3-\delta}$ catalyst used in this study is much cheaper than the Ru/MgO catalyst reported by Skodra and Stoukides[9] and the Pt/C catalyst used for low temperature synthesis of ammonia from air and water.[2,4]

For comparison, ammonia was synthesised from wet nitrogen using the same cell. Fig. 5B shows the recorded current change against time for the electrolytic

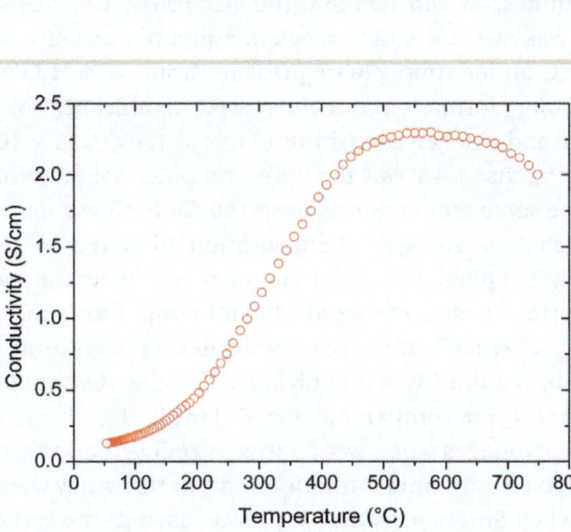

Fig. 4 The DC conductivity of $Sm_{0.6}Ba_{0.4}Fe_{0.8}Cu_{0.2}O_{3-\delta}$ in air.

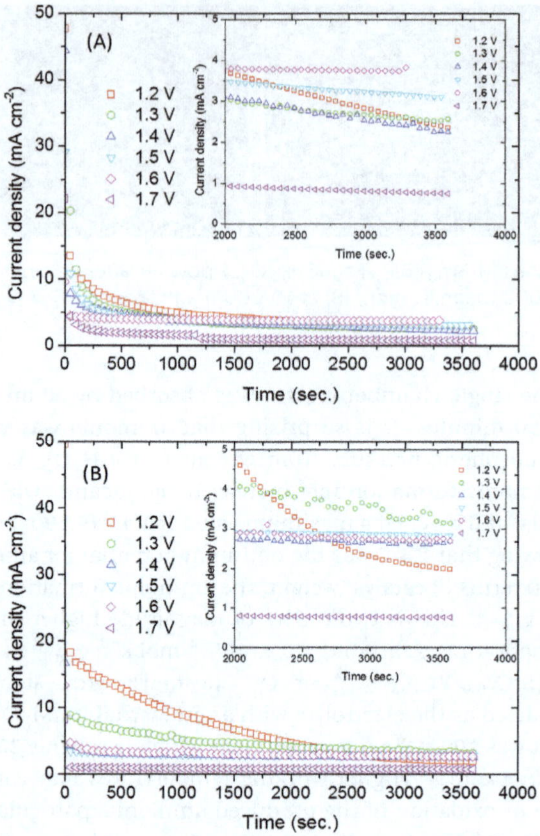

Fig. 5 The current density across the cell in wet air (A) and wet N_2 (B) when applied with different voltages.

cell during the ammonia synthesis process at different potentials at 400 °C over a period of 60 minutes. As can be seen, the electrolytic cell was stable when the applied voltage was over 1.4 V. As expected, ammonia was also successfully synthesised at 400 °C under atmospheric pressure from ~3% H_2O and N_2 (Fig. 6A). Again the ammonia formation rate increased significantly by increasing the applied potential and reached a maximum value at 1.4 V (1.53×10^{-6} mol s^{-1} m^{-2} at 400 °C), showing that 1.4 V was the optimum potential for ammonia synthesis in this study. The same tendency was observed for both wet air and wet N_2. This could be attributed to nitrogen chemisorption hindered by the high rate of electrochemically supplied H^+, which in turn poisoned the catalyst (cathode surface).[6,27] In terms of catalyst weight, the ammonia formation rate was 2.45×10^{-10} mol s^{-1} $g_{cat}{}^{-1}$ which is more than two orders of magnitude higher than the values reported by Skodra.[9] When applying 1.7 V, the ammonia formation rates for wet air and wet N_2 are comparable (Fig. 6A) while slightly more ammonia was produced at lower voltages when wet N_2 was introduced into the single chamber reactor. The observed ammonia formation rates in this study were lower than the reported value when $SmFe_{0.7}Cu_{0.1}Ni_{0.2}O_{3-\delta}$ was used as the cathode catalyst but the experiments were operated under different conditions.[23] H_2 and N_2 were used

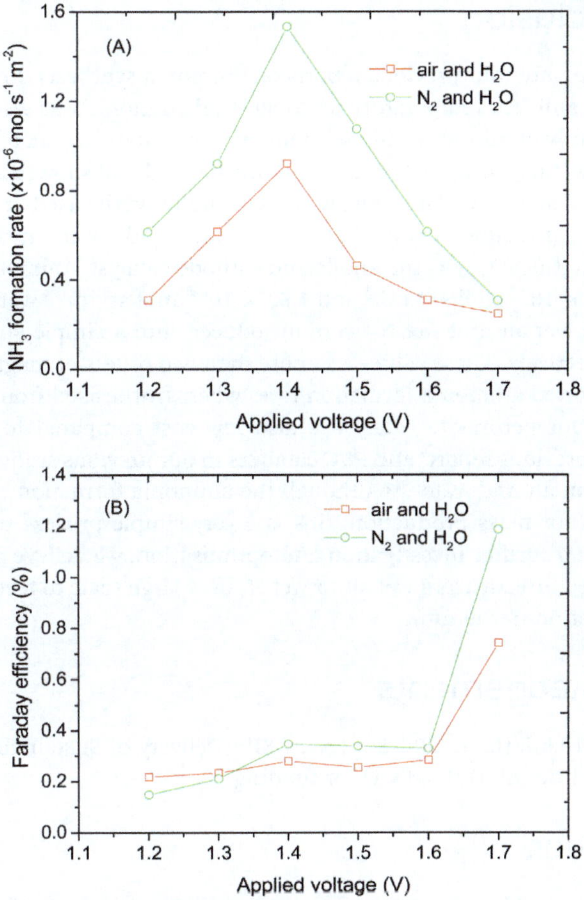

Fig. 6 The ammonia formation rate (A) and Faraday efficiency (B) at 400 °C in wet air and wet N_2 when applied with different voltages.

as the reactant while we used wet N_2/air as the precursor. The cell operating temperature was 400 °C while it was near room temperature in a previous report.[23]

The Faraday efficiencies of the electrochemical synthesis process under different conditions are also calculated. As shown in Fig. 6B, when wet air was used, the highest Faraday efficiency was 0.74% when a voltage of 1.7 V was applied. In the case of wet N_2, the highest Faraday efficiency was 1.19% at a voltage of 1.7 V. At lower applied voltages, the Faraday efficiencies were comparable for wet air and wet N_2, indicating the possibility to use wet air instead of wet N_2 for direct synthesis of ammonia at temperature or below 400 °C. Although the Faraday efficiencies in our experiments are lower than the high efficiency of 78% when H_2 and N_2 was used as the precursor,[13] steam instead of H_2, and air instead of N_2 were used as the precursors in this study. It is believed that both the formation rate and Faraday efficiency can be improved when better electro-catalysts are identified.

4 Conclusion

In conclusion, after our previous reports on ammonia synthesis directly from air and water at ambient temperature up to 80 °C,[2,4] ammonia was also synthesised directly from wet air at a higher temperature (400 °C). Perovskite oxide, $Sm_{0.6}Ba_{0.4}Fe_{0.8}Cu_{0.2}O_{3-\delta}$, was successfully synthesised and showed a conductivity of 1.88 S cm^{-1} at 400 °C. Ammonia was successfully synthesised at 400 °C under atmospheric pressure from both wet air and wet nitrogen using $Sm_{0.6}Ba_{0.4}Fe_{0.8}Cu_{0.2}O_{3-\delta}$ as the anode and cathode catalyst. Ammonia formation rates of 9.19×10^{-7} mol s^{-1} m^{-2} and 1.53×10^{-6} mol s^{-1} m^{-2} were obtained at 400 °C when wet air and wet N_2 were introduced into a simple single chamber reactor, respectively. These values are more than two orders of magnitude higher than the reported ammonia formation rates when synthesised from N_2 and H_2O at 600 °C.[9] Our perovskite catalyst is also low cost compared to the Ru/MgO catalyst in a previous report[9] and Pt/C catalysts in our previous study on ammonia synthesis from air and water.[2,4] Although the ammonia formation rates were not high enough for mass production, this is a very simple process using low-cost materials. After further investigation and optimisation, we believe ammonia can be synthesised directly from wet air or wet N_2 on a large scale to feed the growing world population in the future.

Acknowledgements

The authors thank the EPSRC SuperGen XIV 'Delivery of Sustainable Hydrogen' project (Grant no. EP/G01244X/1) for funding.

References

1 J. W. Erisman, M. A. Sutton, J. Galloway, Z. Klimont and W. Winiwarter, *Nat. Geosci.*, 2008, **1**, 636–639.
2 R. Lan, J. T. S. Irvine and S. W. Tao, *Sci. Rep.*, 2013, **3**, 1145.
3 http://www.nobelprize.org/nobel_prizes/chemistry/laureates/2007/.
4 R. Lan and S. W. Tao, *RSC Adv.*, 2013, **3**, 18016–18021.
5 R. Lan, J. T. S. Irvine and S. W. Tao, *Int. J. Hydrogen Energy*, 2012, **37**, 1482–1494.
6 I. A. Amar, R. Lan, C. T. Petit and S. W. Tao, *J. Solid State Electrochem.*, 2011, **15**, 1845–1860.
7 S. Giddey, S. P. S. Badwal and A. Kulkarni, *Int. J. Hydrogen Energy*, 2013, **38**, 14576–14594.
8 I. Garagounis, V. Kyriakou, A. Skodra, E. Vasileiou and M. Stoukides, *Front. Energy Res.*, 2014, **2**, 1–10.
9 A. Skodra and M. Stoukides, *Solid State Ionics*, 2009, **180**, 1332–1336.
10 C. Chen and G. L. Ma, *J. Mater. Sci.*, 2008, **43**, 5109–5114.
11 I. A. Amar, R. Lan, C. T. Petit, V. Arrighi and S. W. Tao, *Solid State Ionics*, 2011, **182**, 133–138.
12 I. A. Amar, C. T. Petit, L. Zhang, R. Lan, P. J. Skabara and S. W. Tao, *Solid State Ionics*, 2011, **201**, 94–100.
13 G. Marnellos and M. Stoukides, *Science*, 1998, **282**, 98–100.
14 G. Marnellos, S. Zisekas and M. Stoukides, *J. Catal.*, 2000, **193**, 80–87.

15 I. A. Amar, C. T. G. Petit, G. Mann, R. Lan, P. J. Skabara and S. W. Tao, *Int. J. Hydrogen Energy*, 2014, **39**, 4322–4330.
16 G. J. Janz and M. R. Lorenz, *J. Chem. Eng. Data*, 1961, **6**, 321–323.
17 R. I. Olivares, C. Chen and S. Wright, *J. Sol. Energy Eng.*, 2012, 134.
18 C. Xia, L. Li, Y. Tian, Q. Liu, Y. Zhao, L. Jia and Y. Li, *J. Power Sources*, 2009, **188**, 156–162.
19 L. Fan, C. Wang, M. Chen and B. Zhu, *J. Power Sources*, 2013, **234**, 154–174.
20 L. Zhang, R. Lan, A. Kraft and S. W. Tao, *Electrochem. Commun.*, 2011, **13**, 582–585.
21 R. Lan, K. A. Alkhazmi, I. A. Amar and S. W. Tao, *Electrochim. Acta*, 2014, **123**, 582–587.
22 R. Lan, K. A. Alkhazmi, I. A. Amar and S. W. Tao, *Appl. Catal., B*, 2014, **152–153**, 212–217.
23 G. Xu, R. Liu and J. Wang, *Sci. China, Ser. B: Chem.*, 2009, **52**, 1171–1175.
24 C. H. Liang, Z. B. Wei, Q. Xin and C. Li, *Appl. Catal., A*, 2001, **208**, 193–201.
25 W. Rarog-Pilecka, E. Miskiewicz, L. Kepinski, Z. Kaszkur, K. Kielar and Z. Kowalczyk, *J. Catal.*, 2007, **249**, 24–33.
26 T. Sugahara and M. Ohtaki, *Appl. Phys. Lett.*, 2011, **99**, 062107.
27 W. Wang, X. Cao, W. Gao, F. Zhang, H. Wang and G. Ma, *J. Membr. Sci.*, 2010, **360**, 397–403.

Faraday Discussions

PAPER

Degradation of $(La_{0.8}Sr_{0.2})_{0.98}MnO_{3-\delta}$–$Zr_{0.84}Y_{0.16}O_{2-\gamma}$ composite electrodes during reversing current operation

Gareth A. Hughes, Justin G. Railsback, Kyle J. Yakal-Kremski, Danielle M. Butts and Scott A. Barnett*

Received 9th February 2015, Accepted 6th May 2015

DOI: 10.1039/c5fd00020c

Reversing-current operation of solid oxide cell $(La_{0.8}Sr_{0.2})_{0.98}MnO_{3-\delta}$–$Zr_{0.84}Y_{0.16}O_{2-\gamma}$ (LSM–YSZ) oxygen electrodes is described. Degradation was characterized by impedance spectroscopy in symmetric cells tested at 800 °C in air with a symmetric current cycle with a period of 12 hours. No change in cell resistance could be detected, in 1000 h tests with a sensitivity of ~1% per kh, at a current density of 0.5 A cm^{-2} corresponding to an overpotential of 0.18 V. At a current density to 0.6 A cm^{-2} (0.33 V overpotential) measurable resistance degradation at a rate of 3% per kh was observed, while higher current/overpotential values led to faster degradation. Degradation was observed mainly in the ohmic resistance for current densities of 0.6, 0.8 and 0.9 A cm^{-2}, with little change in the polarization resistance. Polarization degradation, mainly observed at higher current density, was present as an increase in an impedance response at ~30 kHz, apparently associated with the resistance of YSZ grain boundaries within the electrode. Microstructural and chemical analysis showed significant changes in electrode structure after the current cycling, including an increase in LSM particle size and a reduction in the amount of YSZ and LSM at the electrode/electrolyte interface – the latter presumably a precursor to delamination.

1 Introduction

The increasing utilization of intermittent renewable electricity sources is focusing attention on electricity storage technologies that can help match the fluctuating supply with varying demand.[1] The energy storage capacity requirements of this application are quite demanding because of the large grid power levels and the need to store for periods of several hours or more. Pumped hydroelectric is the only storage technology currently being extensively utilized, but geographic requirements will likely limit its more widespread use. A number of different storage methods, including batteries and reversible fuel cells, are also being

Department of Materials Science and Engineering, Northwestern University, 2220 Campus Dr., Evanston, IL 60208, USA. E-mail: s-barnett@northwestern.edu; Tel: +1 8474912447

investigated. An advantage of reversible fuel cells is that the energy is stored using low-cost feedstocks (*e.g.* water and CO_2), and energy capacity can be increased by increasing storage tank size, while the stack size determines the power capacity. In contrast, in batteries the energy is stored within the electrodes, limiting the extent to which energy capacity can be increased relative to power capacity. Furthermore, the energy is typically stored in relatively expensive electrode materials (*e.g.* oxides such as $LiCoO_2$). However, reversible fuel cells usually have relatively low round-trip efficiency, too low to compete with batteries and other storage technologies.[1a,d]

Recently, a higher-efficiency storage method utilizing a reversible solid oxide cell (ReSOC) has been proposed.[2] One key question regarding ReSOCs is the long-term durability. Degradation in ReSOCs is similar to that observed during solid oxide electrolysis, which has been studied in some detail.[3] Delamination of the oxygen electrode (electrolysis anode), typically $(La_{0.8}Sr_{0.2})_{0.98}MnO_{3-\delta}$–$Zr_{0.84}Y_{0.16}O_{2-\gamma}$ (LSM–YSZ), is the dominant degradation mechanism.[4] Furthermore, reversing current operation does not introduce new degradation mechanisms, and in fact reduces the delamination degradation when compared with DC electrolysis operation.[4,5] It was possible to essentially eliminate degradation at 1.0 A cm^{-2} by using a cycle with very short periods of electrolysis separated by fuel cell operation,[5] but such a cycle would not be practical for most storage applications. Current density J (or the electrode overpotential η) has a large impact on the degradation rate, with rapid degradation at 1.5 A cm^{-2} ($\eta = 0.37$ V) but no detectable degradation at 0.5 A cm^{-2} ($\eta = 0.18$ V).[6] Based on these ReSOC reports, and also prior reports on solid oxide electrolysis cells,[3g,k] it appears that there is a threshold current density of ~0.5–1.0 A cm^{-2} above which oxygen electrode degradation is fast.

In order to design a ReSOC system and assess its performance and economic viability,[7] it is important to know more accurately the maximum J and η values that are consistent with long-term stable operation. In order to be an economically viable storage method, cell lifetimes should exceed 40 000 h and 2000 cycles (assuming a ~24 h storage cycle). Thus, degradation rates should be <0.5% per kh and <0.0125% per cycle, assuming that a cell resistance increase of <25% over the stack life is acceptable. The prior reports discussed above do not provide such detailed information on degradation rates; in particular, there is no information in the critical range from 0.5 to 1.0 A cm^{-2}.

In the present report, LSM–YSZ composite oxygen electrodes were investigated in order to more accurately determine the effect of J and electrode η on degradation rate in this critical range. (Note that the relative importance of J as opposed to η in causing degradation has not yet been elucidated, so both quantities are included in the present report.) In order to focus on the oxygen electrode, which appears to be the cell component most prone to degradation, symmetrical LSM–YSZ/YSZ/LSM–YSZ cells were tested. This has the advantage of minimizing extraneous degradation effects in full cells that may obfuscate the LSM–YSZ electrode degradation of interest here. Furthermore, symmetric applied current cycles, with equal current densities and times in each direction, were utilized in order to maintain symmetrical electrodes throughout the test despite the tendency for more rapid degradation in electrolysis mode. This enables more straightforward electrochemical impedance analysis.

2 Experimental

LSM–YSZ/YSZ/LSM–YSZ symmetrical cells were fabricated by first die pressing YSZ electrolyte pellets from YSZ powder mixed with 1% PVB binder, and then sintering at 1400 °C for 4 hours. The resulting electrolyte pellets were ~0.6 mm thick and 19 mm in diameter. LSM–YSZ electrode functional layers were then screen printed on both sides of the pellets using an ink that was produced as follows. A 1 : 1 mixture (by weight) of $Zr_{0.84}Y_{0.16}O_{2-\gamma}$ (Tosoh) with A-site deficient $(La_{0.8}Sr_{0.2})_{0.98}MnO_{3-\delta}$ (Praxair) was ball milled in ethanol for 24 h. The solution was then dried, ground, and passed through a 125 μm mesh sieve. This powder was then added to a Heraeus organic vehicle 1 : 1.18 by weight and fully mixed with a three roll mill. After applying the ink by screen printing, the electrodes were fired at 1175 °C for 1 hour. A current collector layer of LSM ink, made similarly to the LSM–YSZ ink, was then screen printed on top of the LSM–YSZ layer and fired at 1025 °C. The final as-fired thicknesses of the LSM–YSZ functional layer and LSM current collector were both ~20 μm thick and had a circular area of 0.5 cm^2. Note that this electrode fabrication procedure is essentially identical to that used in making full cells.

All cells were tested in an open-air furnace at 800 °C. Electrical contacts were made by pressure contacting circular porous LSM current collector pellets, with 0.5 cm^2 area, to the electrodes using a weight of 14 g. LSM was utilized to avoid artifacts that may arise from the transport of noble-metal current collector materials into the electrode active region.[4,8] Current densities between 0.5 A cm^{-2} and 1.5 A cm^{-2} were applied. Current was applied galvanostatically using a Keithley source meter controlled by a LabView program. The current cycle period was 12 h, with equal times and equal currents in each current direction. The program was paused roughly every 24 hours to run an EIS measurement at open circuit voltage. All cells were tested for ~1000 hours.

EIS analysis is simplified in two ways through the use of symmetric cells and current cycles. First, only the oxygen electrode is measured, with no interference from fuel electrode EIS responses as encountered in full cells. Second, by cycling the current with equal periods and amplitudes, each electrode experiences identical conditions, keeping them essentially symmetrical. Thus, electrode impedances are obtained by simply dividing the measured impedances by two. Impedance values are then normalized to the active electrode area, to yield the area specific polarization resistance (R_P) of one electrode and the area specific ohmic resistance (R_Ω) of half the electrolyte.

Although the symmetrical cell test is quite convenient, it is important to consider how well this represents the behaviour of an electrode in a full cell. In a symmetrical cell test, the current/overpotential arises solely from an applied electrical potential, different than a full cell where it arises from a combination of an electrical potential and the chemical potential difference between fuel and oxidant. Nonetheless, a given J should yield the same η for symmetrical and full cells, assuming the same electrode, electrolyte, and oxidant composition. Furthermore, the electrochemical potential gradient across the electrolyte will be the same if both cells have the same current density. Thus, the symmetrical cell tests should provide a very good approximation of the electrode conditions during full cell operation.

Post-test microstructural evaluation was performed on cross sections prepared by first fracturing the cells and then polishing mechanically. Fractured cells were vacuum infiltrated with epoxy (EpoThin, Buehler), cured at 42 °C overnight and the epoxy puck diced with an Accutom 5 saw for scanning electron microscopy (SEM) mounting. The epoxy helps to give sharp pore contrast and allows for more stable polishing. A cross-sectional surface of the epoxy imbedded cell was then polished to 1 micron with a diamond slurry. The cell was then mounted on an SEM stub for imaging. To prevent charging in SEM, samples were sputter coated with 15 nm of osmium and carbon tape placed close to the cross-sectional surface. Measurements were made using a Hitachi SU8030 Scanning Electron Microscope and EDS collected using an Oxford X-max 80 SDD EDS detector. Images and EDS measurements were collected at 20 and 30 keV with a 20 μA probe at a 15 mm working distance in order to gain adequate contrast through the osmium and collect EDS signal from the underlying material.

3 Results

An initial exploratory test was done wherein the current density J was increased incrementally in steps of ~150 h, starting at 0.5 A cm^{-2} and finishing at 1.5 A cm^{-2}; this provided an approximate overview of the degradation rate *versus* J, and the results suggest that the degradation rate becomes significant just below ~1 A cm^{-2}. Based on this result, 1000 h life tests were done at current densities of 0.6, 0.8 and 0.9 A cm^{-2}. Prior data taken with identical cells and an identical test setup, but different current densities of 0.5 and 1.5 A cm^{-2}, are also shown for comparison.[6] Section 3.1 describes the increasing current test, Section 3.2 the constant current test, and Section 3.3 the SEM-EDS analysis.

3.1 Incrementally stepped cycled current density

The incremental test was started at 0.5 A cm^{-2} and the current was then increased to 0.7, 0.9, 1.0, 1.1, 1.2, 1.3, 1.4, and 1.5 A cm^{-2}. The cell was maintained at each J for at least 150 hours. The results from selected EIS measurements are shown in Fig. 1. The data were fit using the equivalent circuit shown in Fig. 1, and an example fit is shown for 1.4 A cm^{-2}. The fit using two R–Q elements is reasonably good; better fits can be achieved by adding more elements to the model, but the intent here was to quantify the main electrode responses with a reasonably simple model. The two responses have peak frequencies at ~100 Hz and 30 kHz. As seen in the Nyquist plot, the ohmic resistance (approximately the high frequency intercept) first decreased during operation at up to ~1.2 A cm^{-2}, before increasing at higher J. As seen in the Bode plot, the main change with increasing time and J was an increase in the high-frequency response, while the lower frequency response remained approximately constant.

Fig. 2 shows a plot of measured total, ohmic, and polarization resistances *versus* time. The values were obtained from the EIS fitting, rather than by observing the real-axis intercepts, in order to minimize errors in determining the intercepts. The same trends that were apparent in Fig. 1 are also seen here: the total resistance changes relatively little at the lower J values, but then the resistance increase accelerates with increasing J. The polarization resistance increase begins early on, near 0.5 A cm^{-2}. However, there was considerable variability of

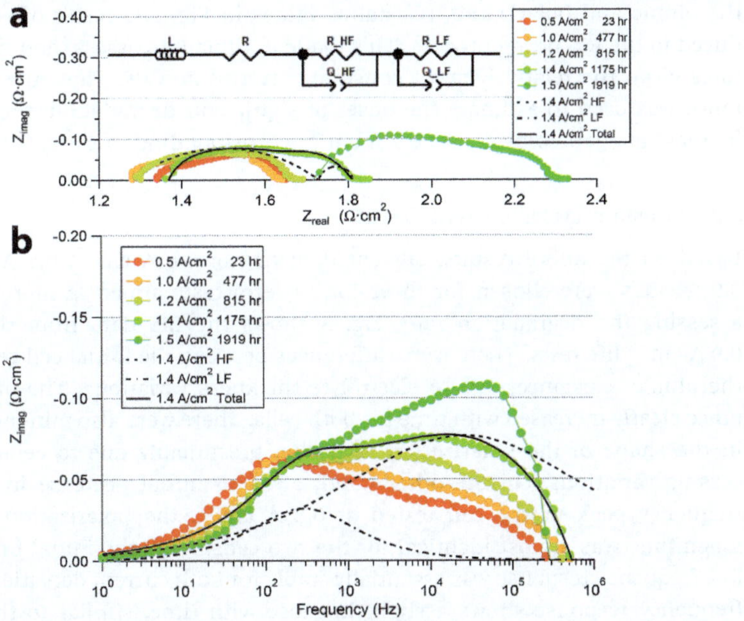

Fig. 1 (a) Nyquist and (b) Bode plots of selected EIS measurements, taken at zero current during brief interruptions in a reversing-current life test in which J was incrementally increased to higher values.

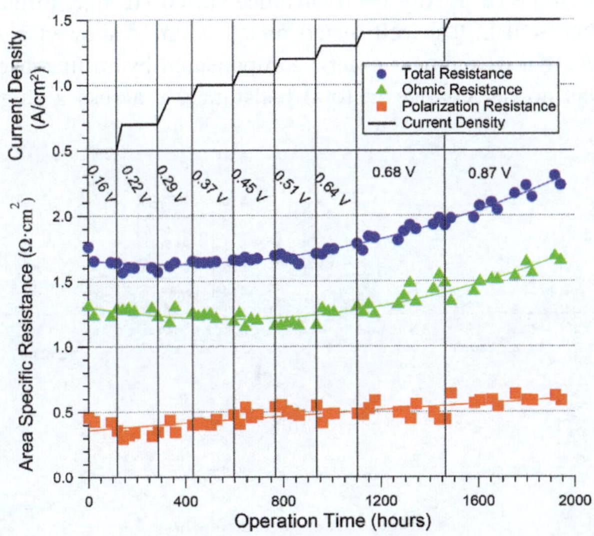

Fig. 2 Total, ohmic, and polarization resistance values *versus* time during a reversing-current life test in which the current density was incrementally stepped up to higher values. The values are derived from EIS data as shown in Fig. 1 from circuit models. The vertical lines separate regions with different constant current densities, with values indicated. Voltage labels are electrode overpotentials at the start of each current step.

the ohmic and polarization resistance values in Fig. 2, a result of errors introduced in the EIS fitting process. This made it difficult to assess their degradation rates given the relatively short constant-current intervals. Thus, the total resistance was used to estimate the onset of significant degradation: the resistance increase appears to begin at >0.7 A cm^{-2}, corresponding to η > 0.22 V.

3.2 Constant cycled current density

Based on the above results, current densities of 0.6 A cm^{-2}, 0.8 A cm^{-2} and 0.9 A cm^{-2} were chosen for three longer-term tests aimed at more accurately assessing the degradation rate. Fig. 3 shows the EIS data from the 0.8 and 0.9 A cm^{-2} life tests. There were differences between the initial cell responses in the ohmic resistance, due to electrolyte thickness variations. The ohmic resistance clearly increased with time for both cells. There were also minor differences in the shape of the polarization response, presumably due to cell-to-cell processing variations. On the other hand, after an initial decrease in the lower-frequency peak of the cell tested at 0.8 A cm^{-2}, the polarization resistance magnitude was almost identical for the two cells. After the initial break-in, the low-frequency response was essentially stable for both current densities. The high frequency response shows a slight increase with time, similar to the changes observed in the cell tested at multiple current densities (Fig. 1), but the increases are much less, as expected given the lesser overall degradation observed.

The total, ohmic, and polarization resistance values, obtained from fits to the data in Fig. 3, are plotted *versus* time in Fig. 4. Prior data[6] also shown in Fig. 4 are from life tests at 0.5 A cm^{-2}, showing no detectable change in resistance, and 1.5 A cm^{-2}, showing substantial increases in both ohmic and polarization resistance. In some cases, the cell resistance varied strongly initially, but then stabilized after ~70 h. The cell tested at 0.5 A cm^{-2} showed a slight steady decrease in R_P that was almost exactly compensated by an increase in R_Ω, such that there was no net change in total resistance. η across a single electrode,

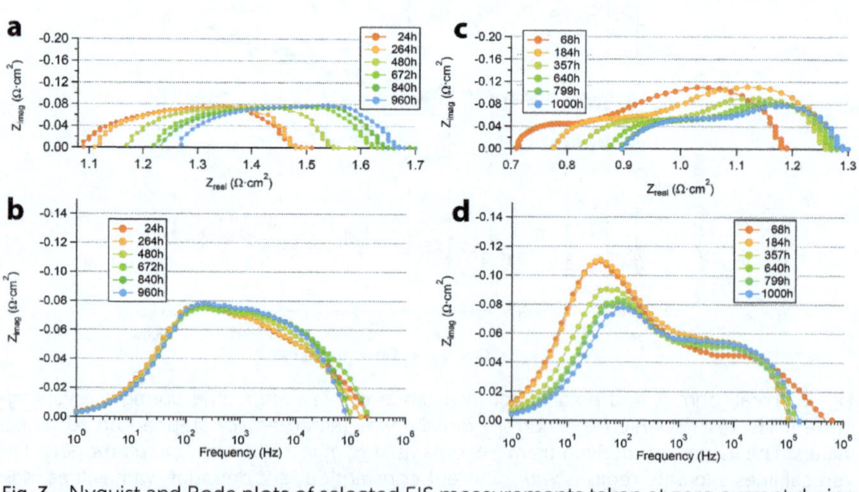

Fig. 3 Nyquist and Bode plots of selected EIS measurements taken at zero current during brief interruptions in reversing-current life tests in which the current density was maintained at (a and b) 0.9 or (c and d) 0.8 A cm^{-2}.

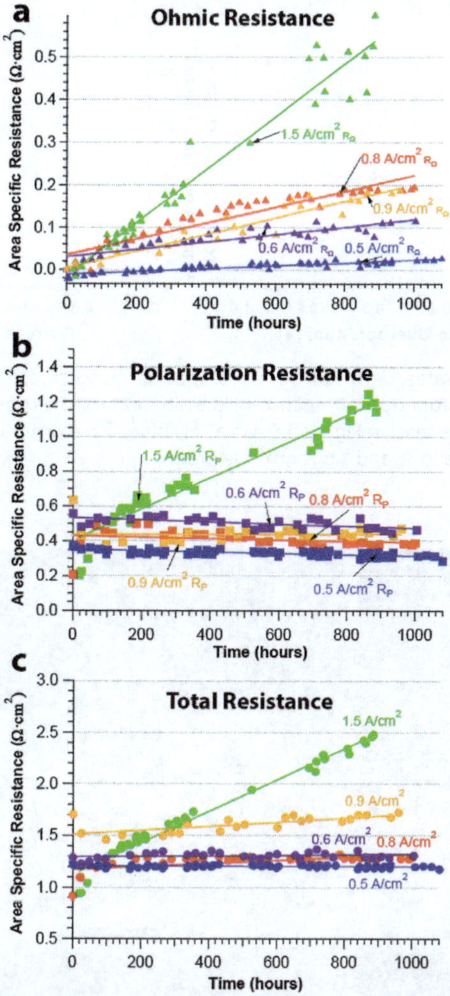

Fig. 4 (a) Ohmic (b) polarization and (c) total resistance *versus* time during reversing-current life tests at current density values of 0.6, 0.8 and 0.9 A cm^{-2}. The values are derived from fits to the EIS data shown in Fig. 3. Also shown are data from a previous report in which identical cells were tested in the same way in the same test setup at 0.5 A cm^{-2} and 1.5 A cm^{-2}.[6] The initial Ohmic resistance, which varied with electrolyte thickness, has been subtracted in part (a) to allow for easier comparison of the resistance changes.

calculated from J and polarization resistance, was 0.17 V. For J increased from 0.6 to 0.9A cm^{-2} (initial η value from 0.33 to 0.38 V), the rate of decrease of R_P lessened, while the rate of increase of R_Ω increases, resulting in a faster total resistance increase. At $J = 1.5$ A cm^{-2} (initial $\eta = 0.93$ V), both R_P and R_Ω increased rapidly with time.

Fig. 5 plots the measured total resistance degradation rates, obtained from best fits to the cell resistance *versus* time data in Fig. 4 neglecting any initial break-in effects, *versus* η (a) and J (b). The constant-current life tests have relatively small error bars – they appear able to resolve ~1% per kh degradation. However, the life

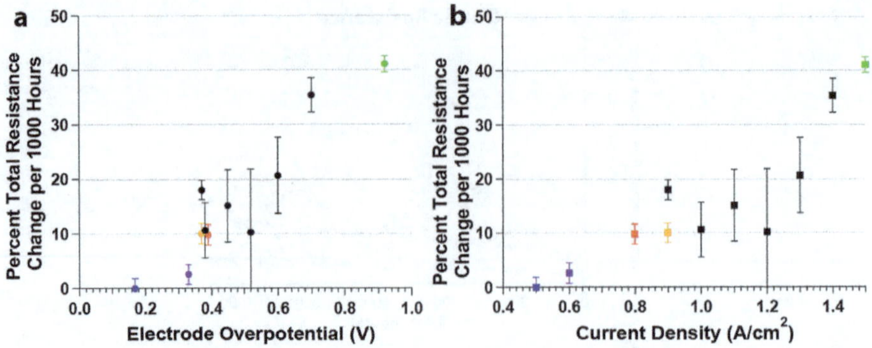

Fig. 5 Degradation rates, derived from the slopes of the total resistance *versus* time data in Fig. 2 and 4, as a function of (a) current density and (b) overpotential. Colored points correspond to the life tests in Fig. 4: 0.5 A cm^{-2} is blue, 0.6 A cm^{-2} is purple, 0.8 A cm^{-2} is red, 0.9 A cm^{-2} is yellow, and 1.5 A cm^{-2} is green.

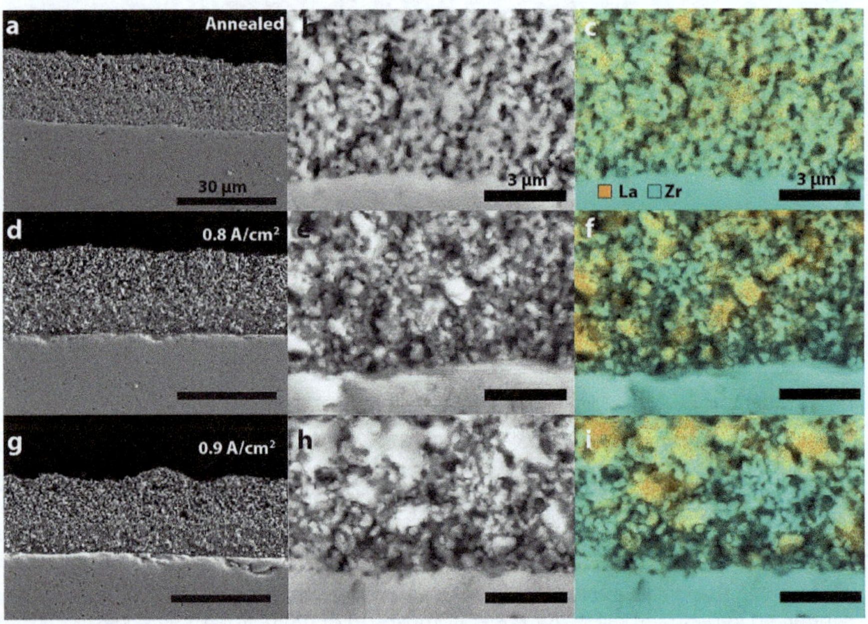

Fig. 6 A summary of: low magnification SEM images (left column a, d, g), higher magnification SEM images (middle column b, e, h), and EDS compositional maps (right column c, f, i) from the electrode/electrolyte regions of cells with no current (top row a, b, c), 0.8 A cm^{-2} (middle row d, e, f), and 0.9 A cm^{-2} (bottom row g, h, i). The color scheme of the composition maps is shown in the upper right frame.

test with incrementally increased current had only ~150 h hold times at each J, leading to much larger errors. Despite the relatively large errors in some of the data points, Fig. 5 gives a reasonable idea of how the degradation rate increases with current. Furthermore, the constant-current 0.5 A cm^{-2} and 0.6 A cm^{-2} tests provide a fairly clear indication that measurable degradation starts at a current between these values.

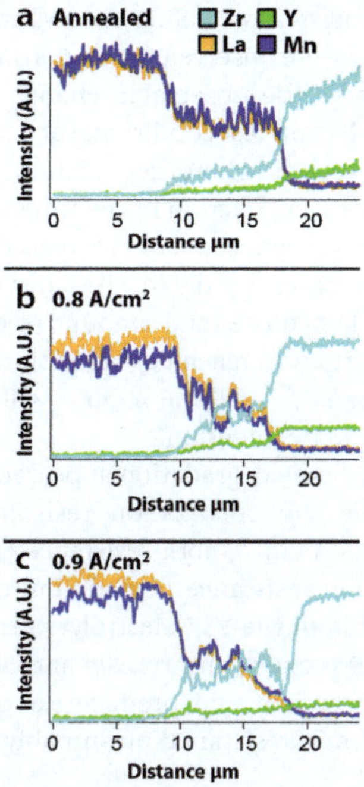

Fig. 7 EDS linescans for (a) annealed (b) 0.8 A cm^{-2} and (c) 0.9 A cm^{-2} cells.

3.3 SEM-EDS analysis

SEM-EDS analysis was completed on a cell that was annealed with zero current along with the 0.8 and 0.9 A cm^{-2} cells. Example images from these are shown in Fig. 6. The cell annealed without current shows uniform electrode structure right up to the interface, whereas in the 0.8 A cm^{-2} and 0.9 A cm^{-2} cases there are microstructural changes near the electrode/electrolyte interface and throughout the active layer. The electrode structure near the electrode–electrolyte interface appears to be depleted of material, containing fewer electrode particles and larger pores. There is significant coarsening of the LSM particles in the active layer for the cells run at 0.8 A cm^{-2} and 0.9 A cm^{-2}, compared to the no-current cell, but the YSZ particles remain un-coarsened. Also, the electrolyte shows pits and pores near the electrode/electrolyte interface, especially in the 0.9 A cm^{-2} cell.

The line scans in Fig. 7 quantitatively show that La and Mn (*i.e.*, LSM) are depleted near the electrode–electrolyte interface in the 0.8 and 0.9 A cm^{-2} samples, in agreement with the images in Fig. 6. The La and Mn signals increase with increasing distance from the electrolyte. On the other hand, the main effect of current on the Zr composition profile is a stronger local variation in intensity, presumably related to the presence of enlarged LSM particles.

4 Discussion

The present results show that the overall resistance degradation rate increases with increasing current density J (overpotential η), increasing above the present detection limit of 1% per kh for J somewhere between 0.5 and 0.6 A cm^{-2} ($\eta = 0.17$

to 0.33 V). This result is consistent with SEM observations: substantial changes in the near-interface structure were observed for $J = 0.8$ and 0.9 A cm^{-2} ($\eta = 0.39$ and 0.37 V), whereas there was little observable change in the structure reported previously (for the same cells and test conditions) at 0.5 A cm^{-2}.[6] These results are consistent with a recent report[5] where degradation was observed for $J = 1.0$ A cm^{-2} ($\eta = \sim 0.25$ V), unless the fraction of the time in electrolysis mode during the reversing-current cycle was reduced. For J increased above 0.8 A cm^{-2} (0.39 V), the rate of degradation increases rapidly to >10% per kh, fast enough to severely limit device lifetime and hence the total amount of energy that can be stored/converted. Since it is important to reach degradation rates well below 1% per kh, longer-term measurements at $J \sim 0.5$–0.6 A cm^{-2} will be useful in the future to provide greater sensitivity to degradation.

For all J below 1.5 A cm^{-2}, the degradation appeared to be mainly in the ohmic resistance (Fig. 4a), while the polarization resistance showed little change (Fig. 4b). In contrast, at 1.5 A cm^{-2}, both resistance components contributed to the degradation. The ohmic resistance degradation can be understood *via* the observed formation of voids in the YSZ electrolyte near the electrode/electrolyte interface (see Fig. 6 for 0.9 A cm^{-2}, and ref. 3*m* and 5). At higher J (1.5 A cm^{-2}), where grain boundary pores join and produce extended regions of electrode/electrolyte separation,[6] ohmic resistance presumably also increases by current constriction.

In a previous report on similar LSM–YSZ symmetric cells, the response peaking at ~30 kHz was attributed to the grain boundary transport losses in the electrode YSZ phase.[9] EIS measurements of the present LSM–YSZ cells with varied oxygen partial pressure (not shown) indicated no change in this high-frequency response, providing further evidence that this response arises from an interfacial (*e.g.*, grain boundary transport) resistance. Given that pore formation on YSZ grain boundaries within LSM–YSZ electrodes has been observed after reversing-current operation,[5] it is reasonable to attribute the polarization resistance increase to increasingly resistive YSZ pathways because of degraded electrode grain boundaries. The SEM images in Fig. 6 and 7 indicate depletion of YSZ from the electrode near the electrolyte, consistent with this interpretation. That is, formation of voids in the YSZ corresponds to a net local depletion of YSZ.

The SEM-EDS results for 0.8 and 0.9 A cm^{-2} suggest that the amount of LSM in the interface region also decreases after cell operation under current. There appears to be a gradual increase in the amount of LSM with increasing distance from the interface, perhaps indicating that material has been transported away from the interface during cell operation. Finally, the LSM particle size within the functional layers has increased after degradation. There does not appear to be a prior report of such changes in the LSM phase in LSM–YSZ functional layers. While there is no obvious explanation, cation transport has been observed in similar perovskite oxygen electrode materials under an oxygen gradient, as was present in the present experiments during cell operation.[10] Given the extensive structural/chemical changes within the electrode functional layer for 0.8 and 0.9 A cm^{-2}, it is perhaps surprising that the increases in polarization resistance are not larger. Indeed, SEM-EDS analysis of the electrode/electrolyte interface appears to be a very sensitive means for detecting low levels of degradation, perhaps more so than the EIS measurements.

The apparent removal of material from near the electrode/electrolyte interface presumably represents an early stage of the oxygen-electrode delamination that was reported for LSM–YSZ electrodes after ~1000 h at a higher J of 1.5 A cm^{-2}, both in current switching and electrolysis modes.[6] That is, continuing removal of material ultimately results in a complete interfacial opening or delamination. Such delamination is likely even at the lower J of 0.8 A cm^{-2} for sufficiently long-term cell operation.

As discussed previously,[6] the degradation during reversing current operation appears to be similar to that seen in dc electrolysis operation, although the periods of fuel cell operation can partially mitigate degradation.[4,5] The degradation presumably occurs during the electrolysis periods of reversible operation. Thus, the present results should be qualitatively similar to dc electrolysis. However, there is relatively little data that is directly comparable. Knibbe *et al.*[11] showed degradation of full Ni–YSZ/YSZ/LSM–YSZ cells at $J = 1$ A cm^{-2}, but the initial η was only 0.19 V. While this is consistent with the J values causing degradation in the present results, no degradation would be expected at this η based on the present data. Note that most of the degradation was observed in the ohmic resistance,[11] similar to the present results. Reports on degradation of LSM electrodes are difficult to compare directly with the present LSM–YSZ results. In one case, delamination was observed after operation at 0.5 A cm^{-2},[3l] but the polarization resistance was quite high such that $\eta \sim 3$ V, well above the value where degradation is expected. Tests on symmetric LSM electrode cells (on YSZ) showed rapid resistance increases and delamination after 100 h at ~2.0 A cm^{-2} and $\eta \sim 0.76$ V.[3i]

Although there is no definitive experimental evidence showing whether J or η controls oxygen electrode degradation, models of the degradation mechanism suggest that η is the controlling parameter. Models are generally based on the large electrochemical oxygen potential developed near the electrode/electrolyte interface.[3i,l,m] This high effective oxygen pressure can drive the observed nucleation of oxygen bubbles in the region near the electrode–electrolyte interface.[3m] In one case, a critical η required to nucleate oxygen bubbles in YSZ was predicted to be 0.28 V.[12] The present results, where the η value sufficient to cause degradation was between 0.18 V and 0.33 V, are consistent with this predicted value. Other models are based on $La_2Zr_2O_7$ formation[3i] and LSM lattice contraction[3l] due to the high oxygen potential, but these do not predict potential values where degradation is expected.

5 Conclusions

The present results represent an attempt to determine the J and η values below which LSM–YSZ electrode degradation becomes negligible during reversing current operation. The following conclusions can be made:

(1) EIS tests over ~1000 h on symmetrical cells appear able to detect a degradation rate of 1% per kh.

(2) SEM observations appear to be quite sensitive to the early stages of degradation, as seen by a decrease in the apparent amount of LSM and YSZ at the interface, and an increase in the size of LSM particles in the functional layer.

(3) The present results show that degradation caused by reversing-current operation can be reduced below ~1% per kh, the present detection limit, by

operating at a current density between 0.5 A cm^{-2} and 0.6 A cm^{-2} with an oxygen electrode overpotential between 0.18 V and 0.33.

(4) At higher current densities and overpotentials, the degradation rate increases rapidly, leading to device lifetimes that would be too short for practical application.

(5) EIS measurements show a unique signature associated with the degradation – an increase in a high-frequency ($\sim 10^4$ Hz) response – which appears to be associated with YSZ grain boundary transport losses. This could be a sign of oxygen bubble formation on grain boundaries.

Acknowledgements

The authors gratefully acknowledge financial support from the Global Climate and Energy Project at Stanford University under award 51922. The development of the cell test setup and the image interpretation were also supported by the Department of Energy Basic Energy Science program (Grant # DE-FG02-05ER46255). SEM analysis made use of the EPIC facility (NUANCE Center-Northwestern University), which has received support from the MRSEC program (NSF DMR-1121262) at the Materials Research Center, and the Nanoscale Science and Engineering Center (EEC-0118025/003), both programs of the National Science Foundation; the State of Illinois; and Northwestern University.

References

1 (*a*) Z. G. Yang, J. L. Zhang, M. C. W. Kintner-Meyer, X. C. Lu, D. W. Choi, J. P. Lemmon and J. Liu, *Chem. Rev.*, 2011, **111**, 3577; (*b*) D. O. E., ARPA-E, *Grid Scale Energy Storage Workshop*, Seattle, 2009; (*c*) R. M. Dell and D. A. J. Rand, *J. Power Sources*, 2001, **100**, 2; (*d*) B. Dunn, H. Kamath and J. M. Tarascon, *Science*, 2011, **334**, 928; (*e*) J. Liu, J. G. Zhang, Z. G. Yang, J. P. Lemmon, C. Imhoff, G. L. Graff, L. Y. Li, J. Z. Hu, C. M. Wang, J. Xiao, G. Xia, V. V. Viswanathan, S. Baskaran, V. Sprenkle, X. L. Li, Y. Y. Shao and B. Schwenzer, *Adv. Funct. Mater.*, 2013, **23**, 929.

2 (*a*) D. M. Bierschenk, J. R. Wilson and S. A. Barnett, *Energy Environ. Sci.*, 2011, **4**, 944; (*b*) P. Kazempoor and R. J. Braun, *Int. J. Hydrogen Energy*, 2014, **39**, 5955.

3 (*a*) J. Malzbender, P. Batfalsky, R. Vaßen, V. Shemet and F. Tietz, *J. Power Sources*, 2012, **201**, 196; (*b*) Y. L. Liu, K. Thydén, M. Chen and A. Hagen, *Solid State Ionics*, 2012, **206**, 97; (*c*) H. Yokokawa, H. Y. Tu, B. Iwanschitz and A. Mai, *J. Power Sources*, 2008, **182**, 400; (*d*) M. J. Jørgensen, P. Holtappels and C. C. Appel, *J. Appl. Electrochem.*, 2000, **30**, 411; (*e*) J. R. Wilson, J. S. Cronin and S. A. Barnett, *Scr. Mater.*, 2011, **65**, 67; (*f*) Z. Zhan, W. Kobsiriphat, J. R. Wilson, M. Pillai, I. Kim and S. A. Barnett, *Energy Fuels*, 2009, **23**, 3089; (*g*) A. Hauch, S. H. Jensen, S. Ramousse and M. Mogensen, *J. Electrochem. Soc.*, 2006, **153**, A1741; (*h*) A. Hauch, S. D. Ebbesen, S. H. Jensen and M. Mogensen, *J. Electrochem. Soc.*, 2008, **155**, B1184; (*i*) M. Keane, M. K. Mahapatra, A. Verma and P. Singh, *Int. J. Hydrogen Energy*, 2012, **37**, 16776; (*j*) J. R. Mawdsley, J. D. Carter, A. J. Kropf, B. Yildiz and V. A. Maroni, *Int. J. Hydrogen Energy*, 2009, **34**, 4198; (*k*) C. Graves, S. D. Ebbesen and M. Mogensen, *Solid State Ionics*, 2011, **192**, 398;

(*l*) K. Chen and S. P. Jiang, *Int. J. Hydrogen Energy*, 2011, **36**, 10541; (*m*) A. V. Virkar, *Int. J. Hydrogen Energy*, 2010, **35**, 9527.

4 G. A. Hughes, K. Yakal-Kremski, A. V. Call and S. A. Barnett, *J. Electrochem. Soc.*, 2012, **159**, F858.

5 C. Graves, *Solid Oxide Fuel Cells*, 2013, **57**, 3127.

6 G. A. Hughes, K. Yakal-Kremski and S. A. Barnett, *Phys. Chem. Chem. Phys.*, 2013, **15**, 17257.

7 (*a*) P. Kazempoor, C. H. Wendel and R. J. Braun, *ECS Transactions*, 2014, **58**, 45; (*b*) S. D. Ebbesen, S. H. Jensen, A. Hauch and M. B. Mogensen, *Chem. Rev.*, 2014, **114**, 10697; (*c*) U. S. Department of Energy, *Grid Energy Storage*, 2013.

8 S. P. Simner, M. D. Anderson, L. R. Pederson and J. W. Stevenson, *J. Electrochem. Soc.*, 2005, **152**, A1851.

9 J. S. Cronin, K. Muangnapoh, Z. Patterson, K. J. Yakal-Kremski, V. P. Dravid and S. A. Barnett, *J. Electrochem. Soc.*, 2012, **159**, B385.

10 M. Oh, A. Unemoto, K. Amezawa and T. Kawada, *Solid Oxide Fuel Cells*, 2011, **35**, 2249.

11 R. Knibbe, M. L. Traulsen, A. Hauch, S. D. Ebbesen and M. Mogensen, *J. Electrochem. Soc.*, 2010, **157**, B1209.

12 O. Comets and P. W. Voorhees, *Electrochemical Synthesis of Fuels 1*, 2012, **41**, 123.

Faraday Discussions

PAPER

Evaluation of oxide ceramics as anodes for SOECs

A. Egger,* N. Schrödl and W. Sitte

Received 10th February 2015, Accepted 26th March 2015
DOI: 10.1039/c5fd00021a

$La_2NiO_{4+\delta}$ is characterised as an example for a potential anode material for high-temperature solid oxide electrolyser cells (SOECs). Short-term characterisation is performed from 700 °C to 850 °C between 0.01 and 1 bar oxygen partial pressure (pO_2) on asymmetrical cells using $Ce_{0.9}Gd_{0.1}O_{2-\delta}$ as the electrolyte. Long-term degradation studies over more than 3000 hours are conducted at 800 °C and 0.2 bar pO_2 in dry and humid atmospheres with and without a Cr-source placed in close vicinity to the cell. The SOEC anode performance is investigated by current–voltage curves combined with impedance spectroscopy. Current densities of up to −410 mA cm^{-2} are applied in current–voltage measurements and during long-term degradation studies. A total increase in anode resistance by 350% is observed over the course of the degradation measurements in an increasingly harsh environment. Post-test analyses by SEM/EDX on a polished cross section of the cell show the presence of several contaminants in the electrode structure. However, chromium has not been identified by EDX even after prolonged exposure to Cr-sources in humid atmospheres, which is attributed to the anodic polarisation of the electrode. Electrode delamination appears to be the main factor for the strong loss in performance.

Introduction

Increasing production of electrical energy from renewables such as wind power or photovoltaics requires large-scale storage capacities to compensate for production fluctuations and to bridge the gap between energy supply and demand. Currently, intense research is devoted to high-temperature (HT) electrolysers, which operate at temperatures above 600 °C. They feature high conversion efficiencies, especially when thermally coupled to existing infrastructure such as power plants or industrial facilities. HT-electrolysers can be used to generate hydrogen by water splitting as well as for the production of syngas when operated with H_2O/CO_2-mixtures. SOEC research and development can draw on the extensive body of knowledge acquired in the field of solid oxide fuel cells (SOFCs).

Montanuniversitaet Leoben, Chair of Physical Chemistry, Franz-Josef-Strasse 18, 8700 Leoben, Austria. E-mail: andreas.egger@unileoben.ac.at

Similar to SOFC technology, long-term stability of SOE cells is a major issue. A prevalent degradation mode for SOECs is delamination of the anode (oxygen electrode) observed during steam electrolysis.[1] Another issue – of high importance in SOFC electrodes – is poisoning of the air electrode by impurities such as Cr released from metallic interconnects or balance-of-plant components.[2] The transport of chromium into the porous electrode proceeds through gas phase transport of volatile Cr species. Chromium in its highest oxidation state has relevant partial pressures with CrO_3 being the dominant gas species under dry and oxidising conditions. Moreover, Cr transport is strongly enhanced in the presence of water due to the higher volatility of the corresponding mono-hydrate H_2CrO_4 in humid atmospheres.[3] For SOFC cathodes the deposition of chromium at the surface or triple phase boundaries of the electrode has often been attributed to the cathodic polarisation mode,[3–7] where electrochemical reduction or low oxygen partial pressures under operating conditions appear to enhance deposition of Cr(III) species. It is thus interesting to see if Cr-poisoning poses less problems to SOEC oxygen electrodes under anodic polarisation.

In this work, $La_2NiO_{4+\delta}$ (LNO) is investigated with respect to its suitability and long-term stability as a SOEC anode. The investigated compound is free of alkaline earth elements which is considered to be beneficial for long-term stability since many degradation modes in SOFCs are known to originate from – or to be aggravated by – the presence of alkaline earth metals.[8–13] Long-term durability is tested at 800 °C for 140 days (~3300 hours) applying current densities up to −410 mA cm^{-2} under pure Ar/O_2 atmospheres as well as in the presence of chromium and humidity.

Experimental

Powders of $La_2NiO_{4+\delta}$ (LNO) and $Ce_{0.9}Gd_{0.1}O_{2-\delta}$ (CGO) were obtained from Treibacher Industrie AG. Screen printing pastes of LNO were produced externally (Fraunhofer IKTS, Germany) as well as in-house by dispersing the powder in a commercial ink vehicle and using a three roll mill (Exakt 50I) for homogenisation.

Dense electrolyte substrates were obtained by uniaxially pressing CGO powder in a stainless steel die (2 cm Ø) at 1 ton and sintering for 10 hours at 1450 °C in air. Typical dimensions of the sintered electrolyte pellets were 16 mm diameter and 1.6 mm thickness with a relative density of above 95%. A thin groove was cut along the circumference of the CGO tablets and a Pt-wire was fixed with Pt-paste into the notch serving as the reference electrode. On one side of the CGO substrates a Pt-grid was attached with Pt-paste as the counter electrode. On the other side of the substrates an LNO layer was applied *via* screen printing and sintered for 2 h between 1200 °C and 1250 °C in air. The thickness of the LNO electrodes ranged between 17 and 30 μm as determined by SEM from cell cross sections. The LNO layers were contacted by means of a gold grid attached with gold paste, since gold is known to be catalytically inactive for the oxygen redox reaction.

Each cell was mounted in an alumina tube located inside a horizontal tubular furnace. Short-term measurements were conducted between 700 and 850 °C in an atmosphere of 20% O_2 in argon at a flow rate of 2 L h^{-1}.

Electrochemical characterisation was performed by means of current–voltage curves and impedance analysis in open circuit conditions (OCV) and under load. Current bias was applied such that oxygen was pumped through the CGO electrolyte from the Pt-side to the LNO side, thus operating the LNO electrode as an SOEC anode under current bias. For impedance analysis the frequency response analyser (Alpha-A with POT/GAL interface, Novocontrol) was set to three-electrode, 4-wire galvanostatic mode with perturbation amplitudes between 1 and 10 mA (rms) and a frequency range of 100 mHz to 1 MHz.

Long-term investigations of Cr-degradation at the anode side were performed at 800 °C in dry and humid atmospheres. Cr-poisoning was achieved by placing small chromium chips (99.996% purity) in the reactor upstream of the sample cell. Humidification of the gas stream was accomplished by passing the gas stream through a water bubbler thermostated at 6 °C which corresponds to an equilibrium partial pressure of water of 9.4 mbar or 30% relative humidity (r.h.) at 25 °C.

Cell cross sections were obtained by cutting the sample with a diamond wire saw, embedding in cold-curing resin, grinding and polishing with diamond suspensions with a final particle size of 1 μm. Post-test microstructural analysis was performed using scanning electron microscopy (SEM), Zeiss EVO 50, Germany, with LaB_6-cathode and EDX-detector (Oxford Instruments).

Methodology

A schematic of the cell design employed in this work is shown in Fig. 1. In an early stage of the degradation study the cell design was changed from symmetric to asymmetric due to delamination issues of the LNO counter electrode after applying large bias currents. Since a reference electrode was used for all electrochemical measurements, the cell's counter electrode was replaced by a porous Pt-layer which is expected to have little effect on results obtained for the SOEC anode side.

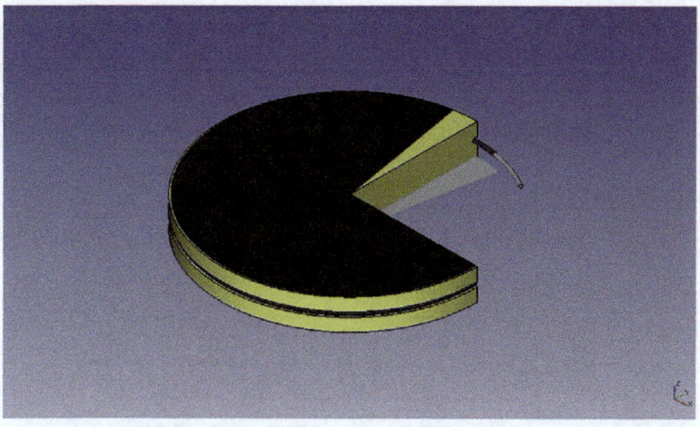

Fig. 1 Schematic of the asymmetric cell design used to evaluate SOEC anode performance. Layer structures from bottom to top are: Pt-counter electrode, CGO electrolyte with Pt-reference along the circumference, LNO anode as working electrode.

CGO was used as electrolyte material due to its high ionic conductivity and low reactivity with LNO at higher temperatures.[14] Electrolytes of real SOEC cells are not expected to be made of ceria due to electronic contributions to the total conductivity under reducing conditions, resulting in internal short-circuiting of the cell. However, a protective barrier layer of CGO at the oxygen electrode side is very common and thus the SOEC anode side is well represented by the cell design used in this work.

All electrochemical measurements were conducted against a Pt-reference electrode mounted at the circumference of the electrolyte substrate. The equivalent circuit used to model the impedance response is shown in Fig. 2, where L_0 is the inductance of the setup, R_0 the bulk resistance of the electrolyte, R_1 the grain boundary resistance of the electrolyte, and R_2, R_3 resistances of processes occurring at the SOEC anode.

The validation of the above assignment was based mainly on the order-of-magnitude values obtained for capacitances C. The capacitances were calculated from the fitting values of the constant phase elements (CPEs) used to model the capacitive contributions of the processes (Fig. 2) according to eqn (1) (ref. 15)

$$C_i = R_i^{\frac{1}{n_i}-1} Q_i^{\frac{1}{n_i}} \tag{1}$$

where n_i and Q_i are parameters defining the CPE admittance[16] *via* $Y_{CPE} = Q(jw)^n$. Values of C of grain boundaries could be obtained from measurements below 800 °C and were in the range of several tens of nF, while capacitances of processes in the low-frequency part of the spectra were above 100 μF and thus attributed to anode processes. No attempts were made for a more detailed assignment of anode processes to impedance elements in the equivalent circuit. The main point of the impedance measurements was to separate the lead inductance and electrolyte contributions from the SOEC electrode part.

Anode performance was investigated by means of current–voltage (I–U) curves up to current densities of -410 mA cm^{-2}. I–U measurements were complemented by impedance analysis under load over the same current range to correct for the electrolyte contribution.

Fig. 3 shows examples of impedance spectra under increasing current bias. Blue lines correspond to fitting curves based on the equivalent circuit given in Fig. 2. Shifts of the high frequency intercepts on the real axis are caused by Joule heating of the sample during current load. The temperature increase of the sample at different current bias was measured separately with a thermocouple pressed onto the anode layer. A maximum temperature change of +15 °C was

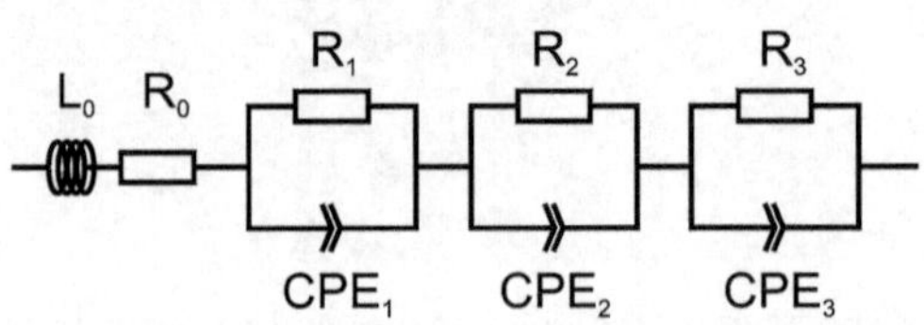

Fig. 2 Equivalent circuit used to model impedance spectra with inductance L_0, resistances R_i and constant phase elements CPE_i.

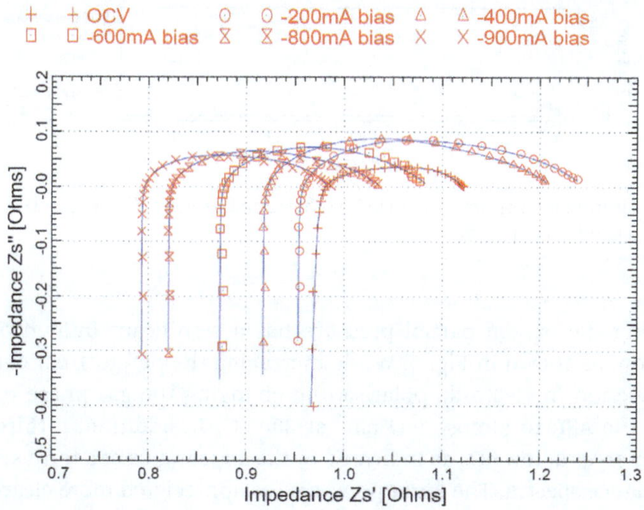

Fig. 3 Impedance spectra of an LNO electrode as a SOEC anode at 800 °C under current load. Data points are plotted in red colour, blue lines are fitting curves obtained from non-linear complex regression analysis.

observed at 800 °C for current densities of around −450 mA cm^{-2}, which is consistent with the observed decrease in the electrolyte contribution in Fig. 3.

Results and discussion

Short-term characterisation

SEM images of sintered anode layers (Fig. 4) confirm good connectivity of the LNO grains while retaining sufficient porosity for gas diffusion of oxygen through the electrode to the surroundings.

Impedance spectra under OCV conditions at 0.2 bar pO_2 are shown in Fig. 5 where contributions from the electrolyte and measurement setup have been subtracted. ASR values of the LNO electrode are 2.16, 0.78, 0.30 and 0.19 Ω cm^2 at 700, 750, 800 and 850 °C, respectively. Impedance data indicate at least two sub-processes on the anode side with an overall activation energy of 240 kJ mol^{-1}. *C*-values of the low and high-frequency arcs are around 1×10^{-2} and 5×10^{-4} F cm^{-2}, respectively, and do not change significantly within the temperature range investigated.

Fig. 4 Top view of screen printed LNO anode layers used for short-term characterisation (left) and Cr-poisoning studies (right) sintered at 1200 °C and 1250 °C, respectively, for 2 h in air.

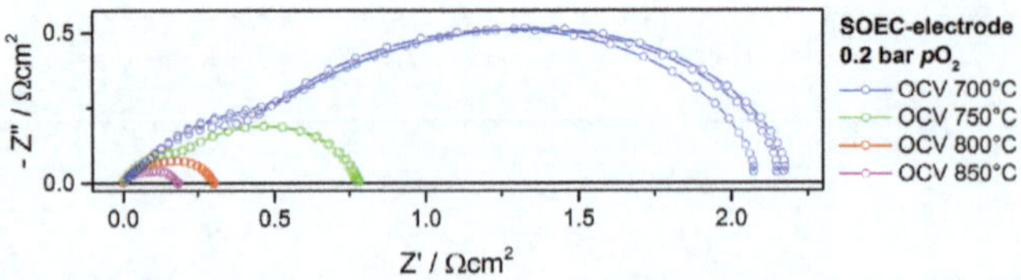

Fig. 5 OCV impedance spectra of an LNO electrode between 700 and 850 °C in 20% O_2/Ar. Lines are guides to the eye.

Reducing the oxygen partial pressure has a significant influence on anode polarisation, as shown in Fig. 6. While increasing the pO_2 to 1 bar causes only a slight reduction in electrode polarisation, changing the gas phase to 1% O_2/Ar increases the ASR to almost 1 Ω cm^2 at 800 °C. In addition, a third electrode process emerges at low pO_2 as indicated by the appearance of a third semicircle in the impedance spectra. The latter point can be appreciated more clearly in Fig. 7, where the imaginary part Z'' is plotted as a function of frequency.

Peak frequencies of the sub-processes can be estimated from Fig. 7. Electrolyte and setup contributions have been removed in Fig. 6 and 7. For fitting the impedance data at 0.01 bar pO_2 an additional R||CPE-element has been incorporated in the equivalent circuit shown in Fig. 2.

Long-term investigation

The long-term stability of LNO as a SOEC anode was investigated at 800 °C over a period of 140 days (3300 h) under continuous current load of -410 mA cm^{-2}. After 46 days (1100 h) in pure 20% O_2/Ar atmosphere Cr was added under dry conditions and after another 41 days (1000 h) 30% relative humidity was introduced into the gas stream. Fig. 8 and 9 show OCV impedance data and *I–U* curves in dry and humid atmospheres with a Cr-source placed close to the sample. Electrolyte contributions are separated and subtracted from *I–U* data by means of impedance measurements. Current–voltage curves show significant non-linearity at higher current densities which is probably caused by Joule heating of the cell.

The results in Fig. 8 show that electrode deterioration proceeds rather slowly in the first 30 days after Cr-insertion but accelerates in the subsequent 10 days. The total increase in anode ASR is around 50% in this period.

A similar trend is observed after humidification of the gas stream (Fig. 9). Degradation rates are comparatively low at the beginning but increase after prolonged exposure to 30% relative humidity. Based on OCV impedance data the

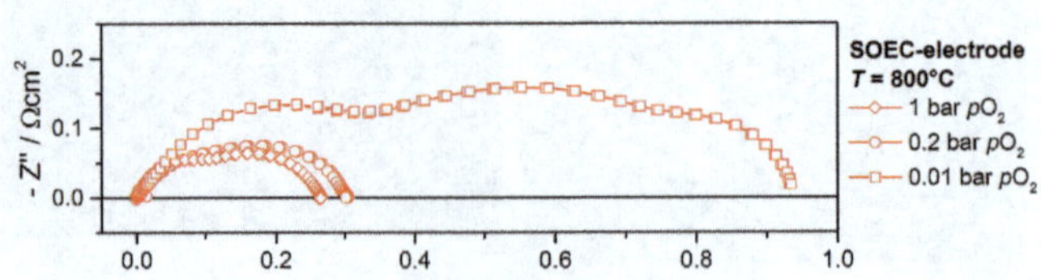

Fig. 6 OCV impedance spectra of an LNO-electrode at 800 °C and oxygen partial pressures between 0.01 and 1 bar. Lines are guides to the eye.

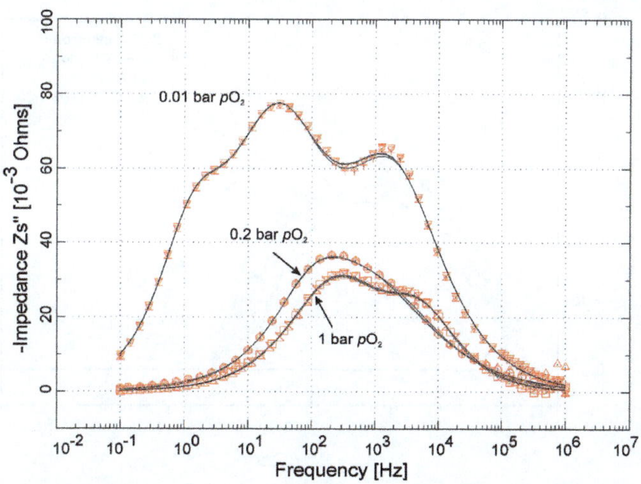

Fig. 7 Reactance of impedance data plotted in Fig. 6 *vs.* frequency at 800 °C and different pO_2. Red symbols are data points, fitting curves are shown as black lines.

increase in anode ASR after 41 days in Cr/H_2O-containing atmospheres amounts to ~35%.

ASR values of the SOEC anode (R_{pol}) and the CGO electrolyte (R_s) were determined from OCV impedance spectra over the whole degradation period (Fig. 10). R_{pol} of the fresh LNO anode is around 0.2 Ω cm^2 and was found to decrease in the first 10 days, after which a continuous increase in polarisation resistance is observed. After three weeks, distortions appearing in the impedance spectra indicated cell failure and the cell was cooled to room temperature and visually inspected. Complete delamination of the LNO counter electrode was found and cell tests were continued after applying Pt-paste/Pt-grid as a new counter electrode. The reason for the delamination of the SOFC electrode is unclear but has been reproduced in a separate experiment with another symmetrical LNO cell under current load. Reducing conditions on the cathode side might have caused slow decomposition of the LNO counter electrode. However, based on results from Nakamura *et al.*,[17] such decomposition would require effective oxygen partial pressures of below 10^{-15} bar at 800 °C. A discontinuity in R_{pol} and – less pronounced – in R_s is observed after replacement of the counter electrode (Fig. 10), which suggests that data before the delamination issue are of limited accuracy and may have been shifted to too low ASR values. Furthermore, the sintering step necessary to attach the freshly applied Pt-counter electrode to the electrolyte might also have affected the microstructure of the porous SOEC anode.

General trends regarding the effect of Cr and humidity on anode deterioration are difficult to establish based on impedance and *I–U* data obtained in this work. Degradation rates for R_{pol} are similar between all three time segments shown in Fig. 10 and are thus likely to be caused by the current load rather than by poisoning processes. A strong increase in anode ASR toward the end of the second stage (*i.e.* Cr in dry atmospheres) might indicate a Cr-poisoning effect setting in after some induction period. Shortly after humidification of the gas stream the anode ASR was found to recover, but after several days degradation resumed at

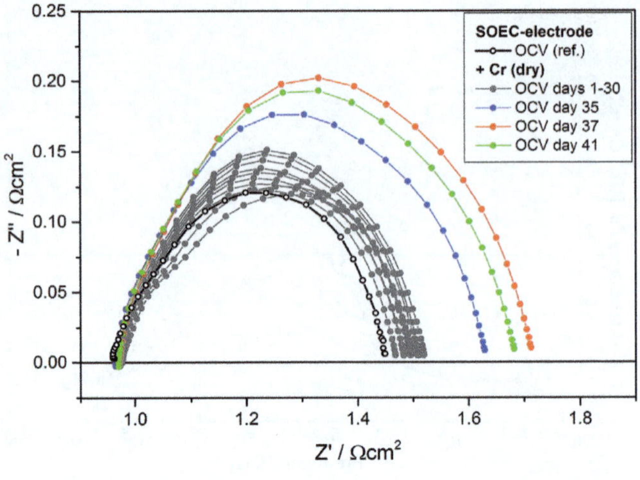

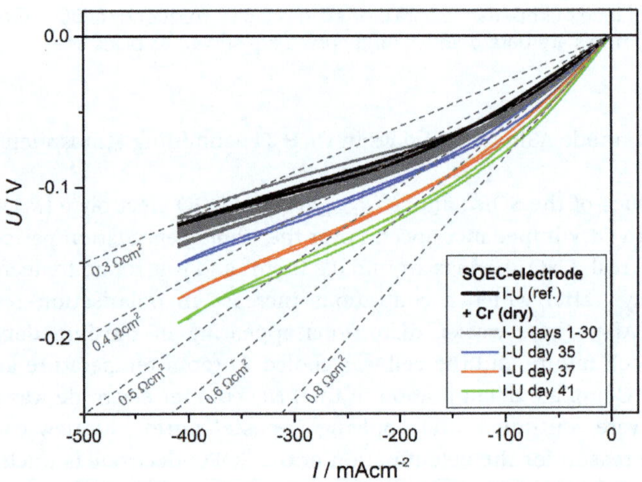

Fig. 8 OCV impedance spectra (top) and *I*–*U* curves (bottom) of an LNO anode at 800 °C in dry 20% O_2/Ar in the presence of a Cr-source. Electrolyte resistance has been subtracted in *I*–*U* curves. Results from the last measurement before introducing the Cr-source are shown for comparison ("ref.").

rates similar to the previous ones. During the whole degradation study R_{pol} increased by 350% from 0.22 to 1.0 Ω cm², whereas the electrolyte contribution R_s remained practically constant. More insight regarding the contribution of chromium and humidity to the anode degradation was obtained from analytical examination of cross sections of the tested cell.

Post-test analysis

Visual inspection of the cell after the measurements showed no signs of delamination of the anode layer from the electrolyte. An SEM picture of a polished cross section of the cell after finishing long-term investigations are given in Fig. 11. The image shows the screen printed anode layer of uniform thickness (29 μm)

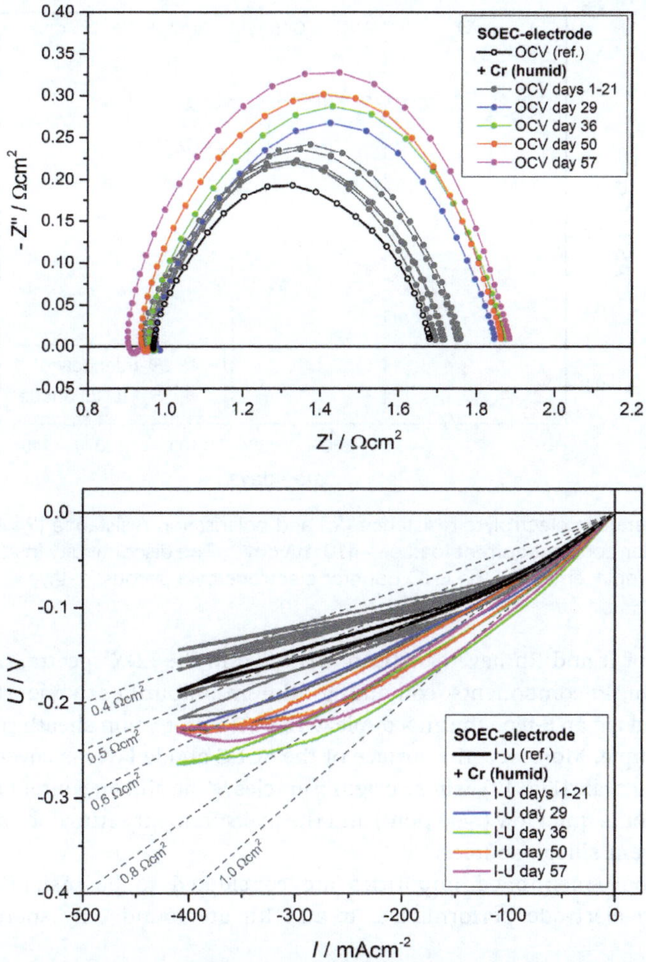

Fig. 9 OCV impedance spectra (top) and *I*–*U* curves (bottom) of an LNO anode exposed to a Cr-source in 20% O_2/Ar with 30% relative humidity at 800 °C. Electrolyte contributions have been removed from *I*–*U* curves. Results from the last measurement before starting gas humidification are given for comparison ("ref.").

deposited on the CGO electrolyte containing some residual porosity. Foreign phases can be discerned in the anode layer by material contrast.

Several impurity phases have been identified by EDX analysis in the upper half of the anode layer (Fig. 12). Bright patches are due to gold, probably caused by the polishing procedure since the Au-current collector has been cut and polished together with the sample. In the upper half of the anode layer light-grey textures are visible in the surface region of the LNO particles containing significant amounts of bismuth. Bi is used as additive in the gold paste to improve sinterability but is found to diffuse into the porous electrode structure and further into the LNO bulk. From the rather finely dispersed distribution of the Bi-containing phase it appears that diffusion proceeds mainly along grain boundaries.

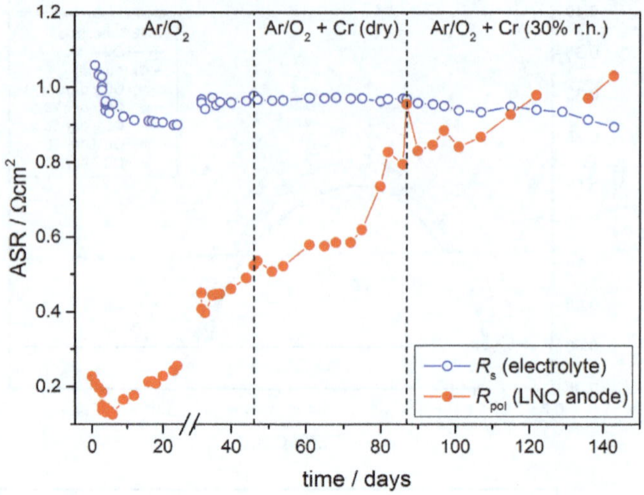

Fig. 10 Trends in electrolyte resistance (R_s) and polarisation resistance (R_{pol}) of an LNO anode under constant current load of −410 mA cm^{-2}. The discontinuity in ASR at day 24 marks the replacement of the LNO counter electrode by a porous Pt-layer.

Peaks of Ir and Rh have also been identified in the EDX spectra. Ir may originate from Pt-components containing iridium impurities while rhodium is introduced *via* an S-type thermocouple with Pt/Rh-protection sheath placed close to the sample. Moreover, the surface of the SOEC anode layer is covered by a Si-containing scale (Fig. 12), whose origin is unclear. The silicon content of the LNO raw powder is quite low (~90 ppm) and the measurement setup does not contain any apparent silicon sources.

The above-mentioned impurities are considered to be insignificant with respect to electrode performance. Ir and Rh are found only sporadically as

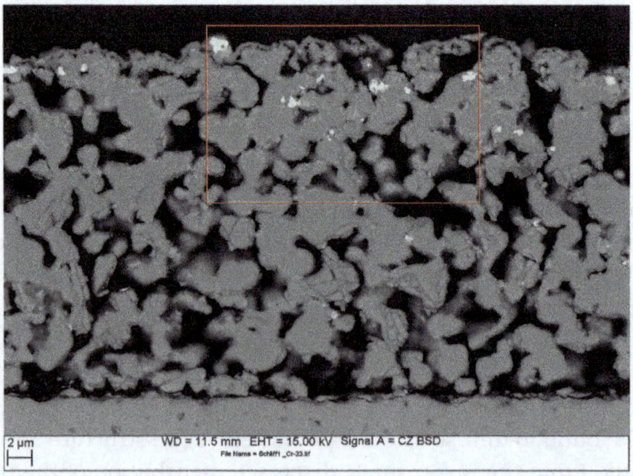

Fig. 11 Backscatter SEM image of a cross section of the LNO anode after long-term measurements. The picture shows a close-up view of the SOEC anode layer on top of the CGO electrolyte, the red outline marks a region analysed by EDX (see Fig. 12).

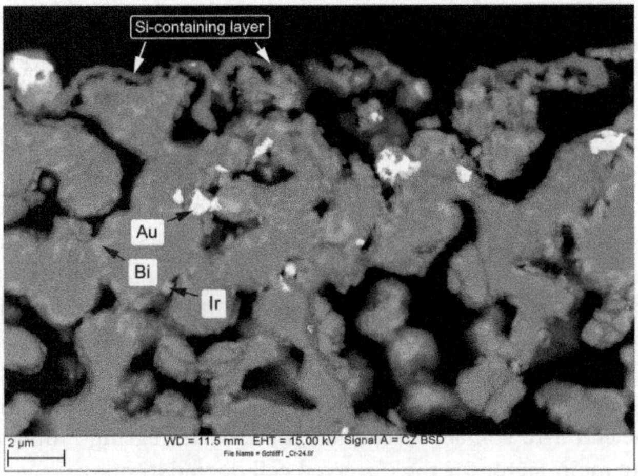

Fig. 12 SEM-EDX analysis of foreign phases identified in the SOEC anode layer after testing.

isolated inclusions within the SOEC anode. The presence of Si is restricted to the gas–anode interface which seems to be less critical with respect to the drastic loss in anode performance. Bi-contaminations are spread throughout the porous electrode structure but appear to be accumulated in the outer part of the anode layer and less concentrated in the electroactive region close to the electrolyte. Moreover, superficial Bi-contamination of the related nickelate compound $Nd_2NiO_{4+\delta}$ has been shown previously to be uncritical with respect to electrical conductivity and oxygen surface exchange activity.[18]

Surprisingly, chromium could not be identified by EDXS even after long time exposure to Cr/H_2O-containing atmospheres. Detection of low concentrations of Cr next to a La/O-rich bulk phase by EDXS is difficult due to strong overlap of the Cr X-ray K_α and L_α emission lines with those of La and O.[19] However, larger amounts of Cr can be detected by a shift in the La L_α/L_β peak intensity ratios.[20,21]

EDX point spectra of the degraded anode yield intensity ratios of La-L signals which appear to be as expected for Cr-free material and thus chromium impurities – if present – may occur only in low concentrations in the anode structure. This finding is rather unexpected, since long-term investigations on dense LNO samples at 700 °C (ref. 22) and 800 °C (ref. 23) in Cr-containing atmospheres have shown significant impact on oxygen surface exchange kinetics and the deposition of Cr-species on the material surface has been clearly established.[22]

The observed absence of chromium might be a consequence of the anodic polarisation of the SOEC electrode. This mode of operation could prevent degradation processes observed in Cr-poisoning of SOFC cathodes, where the conversion of gaseous Cr(VI)-species[24] to less volatile Cr(III)-compounds (*e.g.* Cr_2O_3) has been ascribed to electrochemical reduction[3–5] or low effective pO_2 at the electrode–electrolyte interface under cathodic current load.[6,7]

Cracks in the electrolyte, either intragranular or along grain boundaries, do not occur near the electrolyte–anode–interface, as has been reported in the literature for SOEC operation at high current densities.[25,26] This is consistent with the fairly constant electrolyte resistance obtained from impedance measurements (see Fig. 10 and EIS spectra in Fig. 8 and 9).

EIS results clearly indicate the anode contribution being the main factor in the increasing performance loss. Close inspection of the anode micrographs in Fig. 11 and 12 shows several LNO grains in the SOEC layer to have been fractured but the main part of the porous anode framework appears to be undamaged. The electrode–electrolyte contact, however, appears to be rather poor which is a well-known degradation mode caused by long-term SOEC operation. Although mechanical delamination has not been observed upon removing the cell after testing, loss of electrode–electrolyte contact area may well account for the ongoing degradation under anodic load. This degradation mode may be even more dominant in the current type of single cell testing since the cell was not kept under compressive load during operation. It is thus concluded that an increase in the anode–electrolyte contact resistance due to a decreasing number of contact points is the main factor for the observed cell degradation.

Conclusions

$La_2NiO_{4+\delta}$ was characterised as a SOEC anode material on a $Ce_{0.9}Gd_{0.1}O_{2-\delta}$ electrolyte between 700 and 850 °C and 0.2 bar oxygen partial pressure. ASR-values of the oxygen electrode at 800 °C were determined as 0.22 and 0.30 Ω cm^2 for two different cells. Long-term investigations under constant current load of −410 mA cm^{-2} were performed over a period of 140 days at 800 °C under dry conditions as well as in Cr/H_2O-containing atmospheres. A total increase in anode ASR by 350% from 0.22 to 1.0 Ω cm^2 was observed. A clear effect of Cr in dry/humid conditions on the degradation behaviour could not be determined from cell measurements. Post-test analysis on a cell cross section by EDXS revealed the presence of several impurities in the anode layer which are considered to be unlikely the reason for the marked degradation phenomena. Cr has not been detected in the SOEC anode after long-term operation which indicates that Cr-poisoning may be less severe for LNO electrodes under anodic polarisation. Poor electrode–electrolyte contact or partial delamination – if caused during cell operation and not being a result of post-test sample preparation – is probably the largest contribution to the observed performance degradation of the LNO anode.

Acknowledgements

This work has been financially supported by the Austrian Klima- und Energiefonds, AVL List GmbH and Plansee SE within the project HydroCell (project no. 838770, Neue Energien 2020). Treibacher Industrie AG is acknowledged for supplying powders of La_2NiO_4 and $Ce_{0.9}Gd_{0.1}O_2$. Preparation of screen printing pastes by Fraunhofer IKTS is greatly appreciated. The authors thank the Department of Physical Metallurgy and Materials Testing (Montanuniversitaet Leoben) for providing SEM facilities as well as Gerhard Hawranek and Johannes Hofer for SEM operation.

Notes and references

1 M. A. Laguna-Bercero, *J. Power Sources*, 2012, **203**, 4–16.

2 J. R. Mawdsley, J. D. Carter, A. J. Kropf, B. Yildiz and V. A. Maroni, *Int. J. Hydrogen Energy*, 2009, **34**, 4198–4207.

3 K. Hilpert, D. Das, M. Miller, D. H. Peck and R. R. Weiss, *J. Electrochem. Soc.*, 1996, **143**, 3642–3647.

4 Y. Matsuzaki and I. Yasuda, *J. Electrochem. Soc.*, 2001, **148**, A126–A131.

5 S. P. S. Badwal, R. Deller, K. Foger, Y. Ramprakash and J. P. Zhang, *Solid State Ionics*, 1997, **99**, 297–310.

6 S. Taniguchi, M. Kadowaki, H. Kawamura, T. Yasuo, Y. Akiyama, Y. Miyake and T. Saitoh, *J. Power Sources*, 1995, **55**, 73–79.

7 M. Krumpelt, T. A. Cruse, B. J. Ingram, J. L. Routbort, S. Wang, P. A. Salvador and G. Chen, *J. Electrochem. Soc.*, 2010, **157**, B228–B233.

8 E. Bucher, A. Egger, G. B. Caraman and W. Sitte, *J. Electrochem. Soc.*, 2008, **155**, B1218–B1224.

9 E. Bucher, W. Sitte, F. Klauser and E. Bertel, *Solid State Ionics*, 2012, **208**, 43–51.

10 Y. Xiong, K. Yamaji, H. Teruhisa, H. Yokokawa, J. Akikusa, H. Eto and T. Inagaki, *J. Electrochem. Soc.*, 2009, **156**, B588–B592.

11 Z. Wuillemin, A. Nakajo, A. Müller, A. J. Schuler, S. Diethelm, J. Van Herle and D. Favrat, in *11th Intern. Symp. Solid Oxide Fuel Cells (SOFC-XI)*, ed. S. C. Singhal, The Electrochemical Society, Vienna, Austria, 2009, vol. 25(2), pp. 457–466.

12 M. Kubicek, A. Limbeck, T. Frömling, H. Hutter and J. Fleig, *J. Electrochem. Soc.*, 2011, **158**, B727–B734.

13 E. Bucher, C. Gspan, F. Hofer and W. Sitte, *Solid State Ionics*, 2013, **230**, 7–11.

14 A. Montenegro-Hernández, J. Vega-Castillo, L. Mogni and A. Caneiro, *Int. J. Hydrogen Energy*, 2011, **36**, 15704–15714.

15 J. Fleig, *Solid State Ionics*, 2002, **150**, 181–193.

16 B. A. Boukamp, *Solid State Ionics*, 1986, **20**, 31–44.

17 T. Nakamura, K. Yashiro, K. Sato and J. Mizusaki, *Solid State Ionics*, 2009, **180**, 368–376.

18 A. Egger, W. Sitte, F. Klauser and E. Bertel, *J. Electrochem. Soc.*, 2010, **157**, B1537–B1541.

19 J. Andreas Schuler, P. Tanasini, A. Hessler-Wyser and J. V. herle, *Scr. Mater.*, 2010, **63**, 895–898.

20 M. C. Tucker, H. Kurokawa, C. P. Jacobson, L. C. De Jonghe and S. J. Visco, *J. Power Sources*, 2006, **160**, 130–138.

21 X. Chen, L. Zhang and S. P. Jiang, *J. Electrochem. Soc.*, 2008, **155**, B1093–B1101.

22 N. Schrödl, E. Bucher, A. Egger, P. Kreiml, C. Teichert, T. Höschen and W. Sitte, *Solid State Ionics*, 2015, **276**, 62–71.

23 N. Schrödl, E. Bucher and A. Egger, in preparation.

24 B. B. Ebbinghaus, *Combust. Flame*, 1993, **93**, 119–137.

25 F. Tietz, D. Sebold, A. Brisse and J. Schefold, *J. Power Sources*, 2013, **223**, 129–135.

26 R. Knibbe, M. L. Traulsen, A. Hauch, S. D. Ebbesen and M. Mogensen, *J. Electrochem. Soc.*, 2010, **157**, B1209–B1217.

Faraday Discussions

PAPER

Understanding the processes governing performance and durability of solid oxide electrolysis cells

Sune Dalgaard Ebbesen,* Xiufu Sun and Mogens Bjerg Mogensen

Received 17th March 2015, Accepted 8th May 2015
DOI: 10.1039/c5fd00032g

Operation of a Ni–YSZ electrode supported Solid Oxide Cell (SOC) was studied in both fuel cell mode (FC-mode) and electrolysis cell mode (EC-mode) in mixtures of H_2O/H_2, CO_2/CO, $H_2O/H_2O/CO_2/CO$ at 750 °C, 800 °C and 850 °C. Although the SOCs are reversible, the polarisation characterisation shows that the kinetics for the reduction of H_2O and CO_2 is slower compared to oxidation of H_2 and CO, and that oxidation/reduction in CO_2/CO mixtures is slower than in H_2O/H_2 mixtures. The kinetic differences are partly related to the polarisation heating and the entropy change. Also the diffusion resistance is larger in EC-mode as compared to FC-mode and the low frequency concentration resistance (which is affected by diffusion), is asymmetric around the open circuit voltage (OCV), and is significantly higher in the EC-mode. Both the increased diffusion resistance and the asymmetric low frequency concentration resistance result in a decreased activity in the EC-mode. Changing the porosity of the support structure shows a significant change in both the diffusion resistance and low frequency concentration resistance when applying current, showing that diffusion limitations cannot be neglected for SOCs operated in the EC-mode. Also the Ni–YSZ TPB resistance is affected by changing the support porosity, indicating that kinetic investigations under current and even at OCV, and the chase for a general expression for "all" Ni–YSZ electrodes may be pointless. The diffusion limitations through the support and active electrode structure create an increased reducing atmosphere at the interface which may be related to the degradation of the cells.

Introduction

The reversibility of solid oxide cells(SOCs), *i.e.* that they can be operated as both a solid oxide fuel cell (SOFC) and a solid oxide electrolysis cell (SOEC), was demonstrated for both H_2 [1–9] and CO_2 [1,9–13] in the early 1980s. Although the cells are reversible a slightly lower performance and an increased degradation is normally observed in the electrolysis cell mode (EC-mode),[14–17] and reactant

Department of Energy Conversion and Storage, Technical University of Denmark, Roskilde, Denmark. E-mail: sune.dalgaard.ebbesen@innofond.dk

starvation is generally observed at lower current densities and degrees of conversion in EC-mode as compared to FC-mode. For both H_2O/H_2 and CO_2/CO mixtures a marginally higher activity towards H_2 or CO oxidation *i.e.* in fuel cell mode (FC-mode) is found in comparison to H_2O or CO_2 reduction (operation in EC-mode).[14,18–23] Based on a theoretical study, it was suggested that this difference between oxidation and reduction lies in the slower diffusion of H_2O compared to H_2.[24] A recent study has shown that the lower performance in EC-mode may be caused by an increased low frequency concentration resistance in EC-mode as well as a local temperature decrease in EC-mode as compared to FC-mode due to the exothermic nature of the oxidation and endothermic nature of the reduction.[25] It was further speculated that the increased low frequency concentration resistance was caused by diffusion to the active Ni–YSZ electrode, depleting the reactants at the active Ni–YSZ electrode.[25] Beside the increased diffusion, the lower performance may be related to the increased Ni–YSZ polarisation resistance in dry conditions.[26,27]

The performance during CO_2 electrolysis is found to be significantly lower during H_2O electrolysis under identical conditions.[1,14,18,19,27] Beside the significantly different reaction mechanisms for the two electrolysis reactions,[22,27–29] the performance during CO_2 electrolysis may be due to the slower diffusion of CO_2/CO compared to H_2O/H_2,[24,30–34] leading to an increased diffusion and low frequency concentration resistance, and a larger Ni–YSZ electrode TPB resistance due to the increased reducing atmosphere caused by diffusion limitations.

A number of possible rate determining steps have been identified for the Ni–YSZ electrode when operating in both FC-mode and EC-mode in mixtures of hydrogen, carbon monoxide, steam, carbon dioxide and methane[22,27–29,35–40] which are very dependent on electrode structure and composition. In general there is a large disagreement regarding understanding of the processes occurring in the SOCs[14] and the apparent activation energy, current–voltage relationship (*i–V* curve), measured impedance, dependence of the *i–V* curve on partial pressure of reactants and products, vary hugely.[26,41–51] The partial pressure dependency seems to be related to both the performance of the cell and level of degradation[25] as well as on the electrode microstructure.[52–54]

Although no direct evidence has been presented, it is generally believed that SOECs show larger degradation during CO_2 electrolysis (and co-electrolysis of steam and CO_2) as compared to steam electrolysis.[14] Further it is believed that increasing the steam/CO_2 concentration may increase degradation,[55,56] and that an increased conversion degree may increase in the degradation rate.[57,58] The observed degradation may further be dependent on both the initial cell performance and microstructure.[14,59–63]

In this paper we try to correlate some of the differences observed when operating the SOC in either fuel cell or electrolysis cell mode as well as the differences between the reaction rates observed during steam electrolysis and carbon dioxide electrolysis with variations in the cell structure and the degradation of these cells.

Experimental

Solid oxide cells

Planar Ni–YSZ supported SOCs of 5×5 cm^2 with an active electrode area of 4×4 cm^2 were used for the experiments. Cells with an LSM–YSZ oxygen electrode (denoted: Ni–YSZ|YSZ|LSM–YSZ cells) and cells with an LSCF–CGO (denoted: Ni–YSZ|YSZ|CGO|LSCF–CGO cells) oxygen electrode were applied for the study.[15,64] The Ni–YSZ support/electrode was nominally the same for all the cells. The cells were supported by a ~300 μm thick porous Ni–YSZ layer,[64,65] and have a 10–15 μm thick Ni–YSZ cermet electrode; a 10–15 μm thick YSZ electrolyte and a 15–20 μm thick either strontium-doped lanthanum manganate (LSM–YSZ) or strontium-doped lanthanum ferrite partially substituted with cobalt (LSCF–CGO) composite oxygen electrode. For the cell with the LSCF–CGO oxygen electrode, a CGO barrier layer (5–10 μm thick) between the electrolyte and the LSCF–CGO oxygen electrode was applied. Although the Ni–YSZ support/electrode were nominally the same for all the cells, different processing parameters resulted in a support structure porosity of 34% for the Ni–YSZ|YSZ|LSM–YSZ cells whereas a lower porosity of 28% for the Ni–YSZ|YSZ|CGO|LSCF–CGO cells.

For the characterisation of the cell as well as durability testing, a test house as shown in Fig. 1 and described in detail elsewhere[66] was used. A total of 15 cells were tested for the experiments shown in the present study. Reduction of nickel oxide in the Ni–YSZ electrode was performed in hydrogen at 1000 °C for 13 of the cells, whereas the remaining two cells (used to compare the degradation rate for CO_2 and steam electrolysis) were reduced at 850 °C. The experiments shown in this study have been reproduced on several Ni–YSZ based SOCs and all initial *i*–*V* curves and impedance spectra were reproducible. On the other hand, when performing durability studies at high current (at or above -1 A cm^{-2}) structural degradation occurs, which makes the degradation rate difficult to reproduce.[67]

The DC characterisation of the cells was performed by recording polarisation curves (*i*–*V* curves) in both electrolysis and fuel cell mode by varying the current. AC characterisation was performed by Electrochemical Impedance Spectroscopy

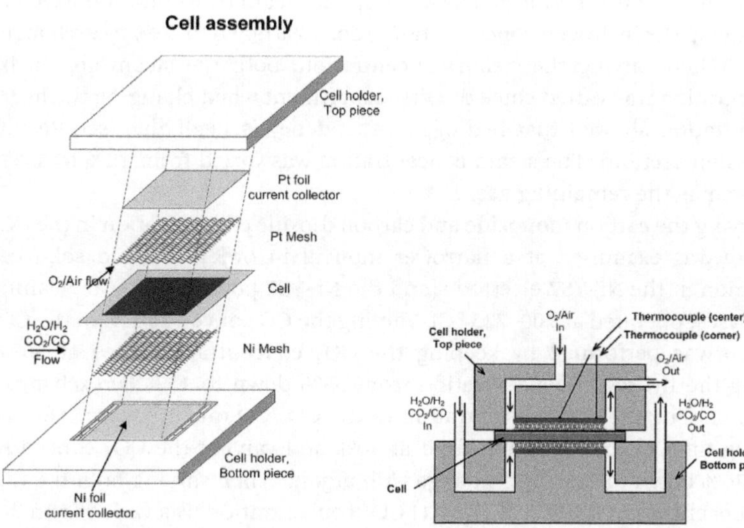

Fig. 1 Cell assemblies in a cross-flow pattern. (A) Exploded view of the cell assembly, and (B) side view of the cell assembly showing the gas flow directions and the placement of thermocouples, it has to be noticed that the flow directions shown in (B) is in co-flow pattern, whereas the actuarial flow during the experiments were cross-flow.

(EIS) using an external shunt and a Solartron 1255B or 1260 frequency analyser at frequencies from 82 kHz to 0.08 Hz (when applying 6 points pr. decade) or 96 kHz to 0.096 Hz (when applying 12 points pr. decade) with an amplitude of 60 mA. The impedance data were corrected using the short-circuit impedance response of the test set-up.

We have previously shown that the gas purity has a significant effect on the degradation of the SOCs operated in both EC and FC-mode.[66,68,69] In order to eliminate impurity related degradation, the inlet gases to the Ni–YSZ electrode compartment were cleaned by flowing the inlet gasses (except oxygen) over porous Ni–YSZ particles at 700–725 °C in order to catch/adsorb all impurities that would otherwise be adsorbed on the active Ni–YSZ TPBs in the SOC.[70,71]

Effect of porosity in the Ni–YSZ support and electrode

Ni–YSZ|YSZ|LSM–YSZ cells with different Ni–YSZ cermet porosity were produced by sintering identical half cells (Ni–YSZ support, Ni–YSZ active electrode and electrolyte) at temperatures ranging from 1255 to 1335 °C. Several cells are shown in the present study, with porosity varying from 23% to 31% in the support structure and 13–18% in the active Ni–YSZ electrode.[72,73] The structural parameters for each cell are given in the text. The stated porosity is based on analysis of SEM images on reference cells and should be taken as indications only. These experiments are not used to give any specific values, but are used to show specific trends related to the porosity.

Characterisation

Performance as a function of gas composition (pH_2, pH_2O, pCO, and pCO_2). After an initial AC and DC characterisation, additional EIS were recorded while varying pH_2, pH_2O, pCO, and pCO_2. Varying the hydrogen concentration in the H_2O/H_2 mixture was performed by keeping the steam concentration fixed at 15% and varying the hydrogen concentration from 85% down to 35% by exchange with argon. When varying the steam concentration, both the steam and hydrogen concentration was varied since the first experiment when changing the hydrogen concentration showed that hydrogen dependency is negligible (see Result and Discussion section). The steam concentration was varied from 15% to 85% with hydrogen as the remaining gas.

Varying the carbon monoxide and carbon dioxide concentration in the CO_2/CO mixture was examined at a narrower interval in order to avoid solid carbon formation in the Ni–YSZ electrode (and the Ni–YSZ powder used for cleaning the inlet gases, operated at 700–725 °C). Varying the CO concentration in the CO_2/CO mixture was performed by keeping the CO_2 concentration fixed at 30% and varying the hydrogen concentration from 65% down to 15% by exchange with argon. Varying the CO_2 concentration in the CO_2/CO mixture was performed by keeping the CO concentration fixed at 30% and varying the CO_2 concentration from 70% down to 20% by exchange with argon. When varying both the CO and CO_2 concentration ($pCO + pCO_2 = 1$) CO_2 concentration was varied from 30% to 80% with CO as the remaining gas.

Performance as a function of current density. The performance as a function of applied current was examined in both EC-mode and FC-mode. Electrochemical Impedance Spectra (EIS) were recorded in 50% H_2O/50% H_2 or 50% CO_2/50% CO at current densities varying from either +0.5 or +0.7 A cm^{-2} (FC-mode) to either −0.5 or −0.7 A cm^{-2} (EC-mode), with steps of 0.1 A cm^{-2}. After setting the current, the cell was left at the applied current density for 15 minutes before the impedance measurement was initiated.

Flow variations under current. The performance under current at varying flowrates was examined by performing EIS measurements under current while keeping the average pH_2O constant at 0.50, while varying the current density and flowrate according to Table 1.

Performance as a function of temperature. The performance as a function of temperature was investigated in steam/hydrogen mixtures by recording impedance spectra with a constant pH_2O of 0.50 at temperatures ranging from 850 to 900 °C.

Durability testing

The durability of the Ni–YSZ based SOCs was examined for H_2O and CO_2 electrolysis and co-electrolysis of H_2O and CO_2. All cells were operated with oxygen flown to the oxygen electrode, and a current density of −1.0 A cm^{-2}. Both the flowrate and gas composition to the Ni–YSZ electrode was varied.

One series of experiments to investigate the temperature dependency on the degradation was carried out for six cells (three Ni–YSZ|YSZ|LSM–YSZ cells, and three Ni–YSZ|YSZ|CGO|LSCF–CGO cells), operated at 750 °C, 800 °C or at 850 °C. Three experiments were performed on the Ni–YSZ|YSZ|LSM–YSZ cells, and Ni–YSZ|YSZ|CGO|LSCF–CGO cells. For these six experiments the durability was examined during co-electrolysis, with 45% H_2O/45% CO_2/10% H_2 flown to the Ni–YSZ electrode. The tests were carried out with a H_2O + CO_2 conversion at 60%. Additionally, one experiment was carried out at 800 °C with a H_2O + CO_2 conversion of 30% by increasing the flowrate to the Ni–YSZ electrode (keeping the current density of −1.0 A cm^{-2}). The effect of reactant conversion was further examined during steam electrolysis (to avoid carbon deposition at high conversion) in 90% H_2O/10% H_2 at 800 °C by changing the flowrate to the Ni–YSZ electrode (keeping the current density of −1.0 A cm^{-2}) with an initial H_2O conversion degree of 30% and subsequently increased by steps of 10% conversion. Finally two tests were conducted in order to examine the degradation rate of steam electrolysis (90% H_2O/10% H_2) and CO_2 electrolysis (90% CO_2/10% CO) at 800 °C.

Table 1 Conditions for flow variations under current

Flow rate		Fuel cell mode			Electrolysis cell mode		
Flow rate (L h^{-1})	1/flow rate (L^{-1} h)	Current density (A cm^{-2})	Inlet gas composition (pH_2O)	Average gas composition (pH_2O)	Current density (A cm^{-2})	Inlet gas composition (pH_2O)	Average gas composition (pH_2O)
15.0	0.067	0.30	0.43	0.5	−0.30	0.57	0.5
17.5	0.057	0.35	0.43	0.5	−0.35	0.57	0.5
20.0	0.050	0.40	0.43	0.5	−0.40	0.57	0.5
22.5	0.044	0.45	0.43	0.5	−0.45	0.57	0.5
25.0	0.040	0.50	0.43	0.5	−0.50	0.57	0.5

Modelling the current and overpotential along the cell

The distribution of the current and overvoltage along the cell in the flow direction was modelled in order to estimate the effect of both support structure and conversion degree on the overvoltage at the Ni–YSZ electrode TPB. The modelling was performed for steam electrolysis only in order to simplify the calculations. For the modelling, the cell was divided into 16 slabs of 1 cm^2 (length of 0.25 cm) along the steam flow direction. Negligible in-plane voltage along the cell was assumed, thus a constant cell voltage along the cell applies. The overpotential cell voltage in each slab along the cell was calculated as following:

$$U_{\text{slab1}} = \text{OCV}_{\text{slab1}} + \eta_{\text{slab1}} \tag{1}$$

with the overpotential calculated as:

$$\eta_{\text{slab1}} = R_{\text{slab1}} i_{\text{slab1}} \tag{2}$$

The resistance of each slab was calculated by adding all resistances for the individual processes in the cell:

$$R_{\text{slab1}} = R_{\text{serial,slab1}} + R_{\text{Ox,slab1}} + R_{\text{Ni–YSZ,slab1}} + R_{\text{diffusion,slab1}} + R_{\text{conversion,slab1}} \tag{3}$$

A breakdown of the cell contributions (see Table 3) are used for the serial resistance and the oxygen electrode resistance. The conversion and diffusion resistance was calculated according to the equations described below.

Diffusion resistance. The diffusion resistance ($R_{\text{diffusion}}$) is related to the change in Nernst potential due to gas diffusion limitations within the porous Ni–YSZ support/electrode. In H_2O/H_2 mixtures $R_{\text{diffusion}}$ can be calculated as:[74]

$$R_{\text{diffusion}} = \frac{\eta_{\text{D}}}{i} = \left(\frac{RT}{2F}\right)^2 \frac{l}{PD_{\text{eff}}} \left(\frac{1}{x_{\text{i,H}_2\text{O}}} + \frac{1}{x_{\text{i,H}_2}}\right) \tag{4}$$

where η_D is the diffusion overpotential, P is the pressure, D_{eff} the effective diffusion coefficient, l is the diffusion length (support thickness), i is the current, R is the gas constant, F is Faraday's constant, T is the temperature in Kelvin, $x_{\text{i,H}_2\text{O}}$ and $x_{\text{i,H}_2}$ are the inlet mole fractions of steam and hydrogen. The diffusion resistance in CO_2/CO mixtures is calculated in a similar manner. For diffusion in porous structures, a correlation for the average percolating porosity in the diffusion direction must be made, and in pores smaller than a few microns (Kn > 10, Kn = diffusion length/pore diameter) Knudsen diffusion dominates. The Knudsen diffusion coefficient (D_{Kn}) can be calculated as:

$$D_{\text{Kn}} = \frac{d_{\text{p}}}{3}\sqrt{\frac{8RT}{\pi M}} \tag{5}$$

where d_p is the pore diameter, M is molar weight of the gases. T is the temperature in Kelvin. Based on the Knudsen diffusion, the effective diffusion can be calculated as:

$$D_{\text{Kn, eff}} = \frac{\varepsilon}{\tau} D_{\text{Kn}} \tag{6}$$

where ε is the porosity and τ is the tortuosity. The actual diffusion length through the porous structure is extremely difficult to estimate, and the tortuosity factor describes that the diffusion path through the porous structure is longer than a direct path (electrode/support thickness). Microstructural studies (using *e.g.* Focused Ion Beam Scanning Electron Microscopy) of fuel electrode supported cells provide evidence that the tortuosity for typical fuel electrode supported SOCs with a porosity around 30% is 1–3.[75–79] In the modelling a tortuosity factor of 1 is used and changes are made in porosity only (Fig. 14). Later, an analysis of the actual/experimental determined that a relation between porosity and tortuosity exists and is discussed.

Conversion resistance. The conversion resistance ($R_{\text{conversion}}$) is based on the change in Nernst potential due to conversion and is therefore related to the gas composition. Assuming a continuously stirred tank reactor (CSTR) the impedance caused by gas conversion in H_2O–H_2 mixtures at the Ni/YSZ electrode can be expressed as:[80]

$$R_{\text{conversion}} = \frac{\eta_C}{i} = \frac{RT}{4F^2 J_i}\left(\frac{1}{x_{i,H_2O}} + \frac{1}{x_{i,H_2}}\right) \quad (7)$$

where η_C is the conversion overpotential, J_i is the inlet area specific flow rate (mol m^{-2} s^{-1}). i, R, F, T, x_{i,H_2O} and x_{i,H_2} have the same meaning as described above. The conversion resistance in CO_2/CO and $H_2O/CO_2/H_2$/CO mixtures may be calculated in a similar manner. Assuming a plug-flow reactor (PFR), the impedance caused by gas conversion is half of the impedance when assuming a continuously stirred tank reactor (CSTR).[81] It has been well proven that the conversion resistance can be calculated as stated in eqn (7) at OCV,[80] which will also be shown in the present article, whereas under current, this has not been fully investigated,[25] and is a part of the present article. Thus at OCV, the conversion resistance is denoted conversion resistance, whereas under current, the changes in Nernst potential due to the changes in concentrations during the AC perturbations caused by the impedance measurements will be denoted "low frequency concentration resistance".

Ni–YSZ TPB resistance. The Ni–YSZ resistance in H_2O/H_2 mixtures can be calculated based on the pH_2O dependency as shown in Fig. 2. The $R_{\text{Ni-TPB}}$ follows a power law dependent on the steam and hydrogen concentration ($pH_2{}^a \cdot pH_2O^b$) where the dependence on H_2 is negligible[25] and the exponent for pH_2O is between −0.30 and −0.32.[25]

The current distribution along the cell was calculated by minimizing the error between the calculated cell voltage (according to eqn (1)) and the actual constant cell voltage by non-linear least-square. The output of the modelling gives the

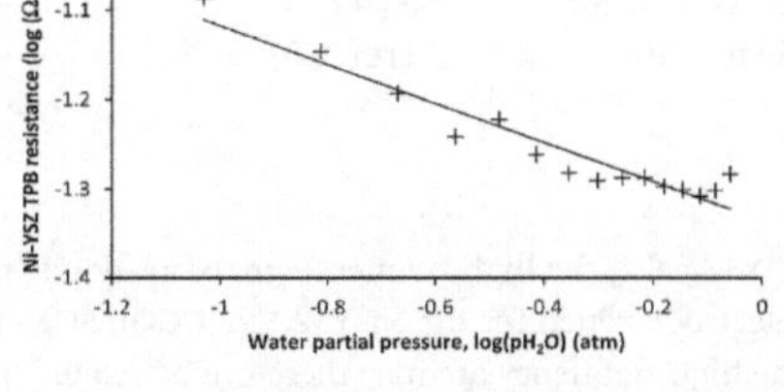

Fig. 2 Dependency of Ni–YSZ resistance on the steam concentration.

distribution of current along the cell as well as the overpotential of each of the processes occurring in the cell.

Break down of the impedance by fitting to an equivalent circuit model

Cells with an LSM–YSZ oxygen electrode. To break down the impedance contributions from each of the two electrodes, the impedance spectra were fitted to an equivalent circuit model consisting of an inductance, a ohmic resistance (serial resistance R_S), and five RQ-equivalent circuits in parallel – as previously described for this type of Ni–YSZ based SOCs.[50] When characterising the cells with 25% H_2O–75% H_2 supplied to the Ni–YSZ electrode and air to the LSM–YSZ electrode the five circuits represent a high frequency LSM–YSZ electrode arc† ($R_{LSM\text{-}high}$), a contribution from the TPB reaction in the Ni–YSZ electrode (R_{TPB}), a low frequency LSM–YSZ electrode TPB arc ($R_{LSM\text{-}low}$), a diffusion arc ($R_{diffusion}$) and a gas conversion arc ($R_{conversion}$). $R_{diffusion}$ and $R_{conversion}$ originate from the Ni–YSZ electrode. The resistances and frequencies for each of the five circuits when characterising the SOCs produced at DTU Energy with 25% H_2O–75% H_2 supplied to the Ni–YSZ and air supplied to the LSM–YSZ electrode at 850 °C are listed in Table 2.[50]

Cells with an LSCF–CGO oxygen electrode. Also for the cells with an LSCF–CGO oxygen electrode, the impedance contributions were broken down to the contributions from each of the two electrodes, by fitting the impedance spectra to an equivalent circuit model. For the cells with an LSCF–CGO oxygen electrode, the equivalent circuit model consisted of an inductance, a ohmic resistance (serial resistance R_S), and four constant phase elements (CPE). One for the oxygen ion transfer (R_{O_2trans}, comparable to the $R_{LSM\text{-}high}$), one for the TPB reaction in the Ni–YSZ electrode (R_{TPB}), a diffusion arc ($R_{diffusion}$) and a gas conversion arc ($R_{conversion}$) similar to the cells with an LSM–YSZ oxygen electrode. In contrast to

Table 2 Break down of the polarisation resistances for the SOCs produced at DTU Energy (former Risø DTU) SOCs when characterised in 25% H_2O–75% H_2 at the Ni–YSZ electrode and air at the LSM–YSZ electrode at 850 °C. Activation energies are summarised based on literature

Process	Resistance (Ω cm^2)	Characteristic frequency (Hz)	Activation energy (eV)
R_S	0.03–0.13 (ref. 50)	—	0.7–0.9 (ref. 50 and 84–86)
LSM_{Hi}	0.05–0.07 (ref. 50)	50000 (ref. 50)	1.0 (ref. 50, 86 and 87)
Ni–YSZ TPB	0.05–0.08 (ref. 50)	8000 (ref. 50)	0.8–1.1 (ref. 46 and 50)
LSCF–CGO		100–200 (ref. 88)	1.4 (ref. 88)
LSM_{low}	0.03–0.06 (ref. 50)	1100 (ref. 50)	1.4–2.0 (ref. 50, 53, 86, 87 and 89)
Diffusion	0.02–0.03 (ref. 50)	20 (ref. 50)	—
Conversion	~0.06 (ref. 50)	3 (ref. 50)	—

† For the Ni–YSZ|YSZ|LSM–YSZ cells, the high frequency arc corresponds to the O^{2-} transport through the structure of LSM–YSZ electrode which for the Ni–YSZ|YSZ|CGO|LSCF–CGO cells is normally related to the Ni–YSZ electrode. The high frequency arc may therefore be related to the entire O^{2-} transport in the composite electrodes and thus not only in a single electrode.

the cells with an LSM–YSZ oxygen electrode, the contribution for the TPB reaction in the LSCF–CGO oxygen electrode was fitted with a Gerischer element (G).

The resistances and characteristic frequencies listed in Table 2 can only be used as a guideline because variation in performance between the produced cells occur, and since other resistances and frequencies will certainly be observed when changing the gas compositions (see Table 3), *i.e.* steam partial pressure to the Ni–YSZ electrode and oxygen partial pressure to the LSM–YSZ electrode.[50,82] An initial fit of the spectrum recorded at OCV was obtained by keeping the exponent of the constant phase elements constant[50] and applying the published values for the resistance of each circuit as a starting point. For the subsequent fitting routines a free fit of the resistances and characteristic frequency of the respective processes were performed. Subsequently, the spectra when changing the operation conditions (current and gas composition) were fitted. We are aware that the description of the break-down has been much improved recently,[83] but the available break down data to be used here were obtained as described.

Results

Performance as a function of current density

Fig. 3 shows sets of two *i–V* curves measured in FC-mode and EC-mode at 850 °C with 50% H_2O/50% H_2 or 50% CO_2/50% CO mixtures supplied to the Ni–YSZ electrode while pure oxygen was supplied to the LSM–YSZ electrode. From the *i–V* characteristic shown in Fig. 3 it is observed that no discontinuity occurs in the shift from FC-mode to EC-mode, although the slopes (Area Specific cell Resistance, ASR) were higher in EC-mode compared to FC-mode. In FC-mode, the ASR was 0.20 Ω cm^2 in agreement with previous measurements for identical SOCs.[90] The ASR in EC-mode was 0.21 Ω cm^2, showing a marginal higher activity towards oxidation of H_2 than reduction of H_2O for the Ni–YSZ electrode as reported previously.[20,21] The ASR for oxidation of CO, when supplying 50% CO_2/50% CO to the Ni–YSZ electrode was 0.24 and an ASR of 0.25 Ω cm^2 were found for reduction

Table 3 Break down of the polarisation resistances for the SOCs produced at DTU Energy (former Risø DTU) when characterised in 25% H_2O–75% H_2 at the Ni–YSZ electrode and air at the LSM–YSZ electrode at 850 °C (ref. 50)

Circuit		H_2O/H_2		CO_2/CO	
		Resistance (Ω cm^2)	Characteristic frequency (Hz)	Resistance (Ω cm^2)	Characteristic frequency (Hz)
Ohmic resistance		0.068	—	0.068	—
1	LSM_{high}	0.039	53 000	0.040	53 000
2	Ni–YSZ TPB	0.037	7500	0.051	7000
3	LSM_{low}	0.012	1900	0.012	1900
4	Diffusion	0.009	110	0.013	60
5	Conversion	0.037	$3.9^{20\%}$–$4.0^{50\%}$ ($3.4^{70\%}$)	0.056	$1.5^{50\%}$–$2.6^{70\%}$
Total		0.202		0.240	

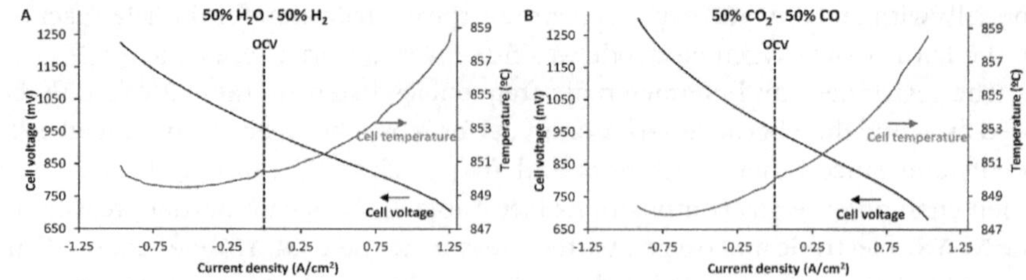

Fig. 3 DC characterisation at 850 °C, (A) in 50% H_2O/50% H_2, (B) in 50% CO_2/50% CO. The measured cell temperature while performing the DC characterisation is indicated.

of CO_2. The cells thereby showed lower activity for CO_2 reduction than for CO oxidation as previously observed, and lower activity for CO oxidation compared to H_2 oxidation in agreement with literature.[91] Finally the cells also showed slower kinetics for CO_2 reduction than for H_2O reduction as have also previously been reported.

Fig. 3 also shows the temperature evolution during the measurement of the i–V curves. Also the production/consumption of heat showed differences when comparing operation in FC-mode and EC-mode, where heat is produced in FC-mode, and consumed in EC-mode, and that production/consumption is larger in the CO_2/CO mixture compared to the H_2O/H_2 mixture.

Electrochemical impedance spectra measured at current densities varying from −0.5 A cm^{-2} (operation in EC-mode) to +0.5 A cm^{-2} (operation in FC-mode) while supplying 50% H_2O/50% H_2 to the Ni–YSZ electrode are shown in Fig. 4C and D. The measured cell voltage and cell temperature while recording the impedance spectra are shown in Fig. 4A. Similar sets of impedance spectra, cell voltage and cell temperature while supplying 50% CO_2/50% CO to the Ni–YSZ electrode are shown in Fig. 4B, D and F.

The cell temperature and cell voltage follows the same pattern as in Fig. 3. The impedance spectra in Fig. 4C–F shows that during reduction (EC-mode) both the ohmic and polarisation resistance increases with increasing electrolysis current density, whereas during oxidation (FC-mode) both the ohmic and polarisation resistance decreases with increasing positive current density. The evolution in impedance is further analysed by a breakdown of the resistance of the individual electrode processes (Fig. 5).

To establish the characteristic frequencies for each of the processes contributing to the total polarisation of the cell, DRT analysis for the initial impedance spectra while varying pH_2O, pH_2, pCO_2, and pCO at the Ni–YSZ and pO_2 at the oxygen electrode was performed at 850 °C. As anticipated a change in resistance was observed at three distinctive frequency regions (~5 Hz, 100–110 Hz, and 7000–7500 Hz) for the Ni–YSZ electrode when changing the H_2O/H_2 ratio. These frequencies and are assigned to the resistance caused by gas conversion (~5 Hz), gas diffusion (100–110 Hz) and resistance at the TPB (7000–7500 Hz). Similar characteristic frequencies are observed when performing the CO_2/CO gas-shift (~2 Hz, 55–60 Hz, and 6500–7000 Hz) and are also assigned to the resistance caused by gas conversion (~2 Hz), gas diffusion (55–60 Hz) and resistance at the TPB (6500–7000 Hz). For the pO_2 shift, also two characteristic frequencies are observed (~5 Hz, and 1900–2000 Hz). The change observed at ~5 Hz has been

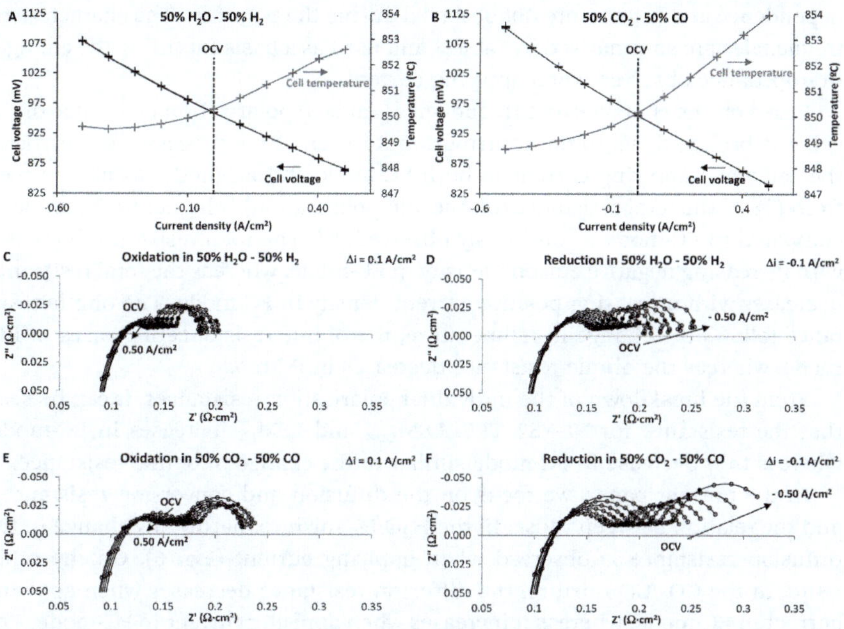

Fig. 4 DC (A and B) and AC (C–F) characterisation at 850 °C in mixtures of 50% H_2O/50% H_2 (A, C and E) and 50% CO_2/50% CO (B, D and F). The points on the curves for DC characterisation (A and B) indicate the measuring points for the AC characterisation. AC characterisation was performed at OCV and during oxidation at current densities up to +0.5 A cm^{-2} (C and E) and during reduction at current densities to −0.5 A cm^{-2} (D and F).

assigned to gas conversion at the Ni–YSZ electrode and may be caused by a gas-leak, whereas the change observed at 1900–2000 Hz is characteristic for the LSM–YSZ electrode. Further, one characteristic frequency (50 000 Hz (ref. 50)) was reported for the LSM–YSZ electrode, this response shows only a small or no pO_2

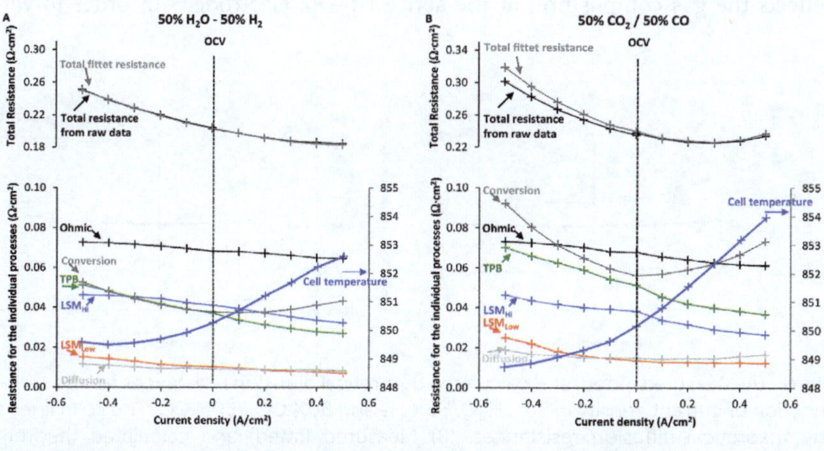

Fig. 5 Measured ohmic resistance and break down of the polarisation resistances for the electrochemical impedance spectra shown in Fig. 4 during oxidation and reduction at 850 °C in (A) 50% H_2O/50% H_2 and (B) 50% CO_2/50% CO.

dependence and is therefore not observed during the pO_2 shift. The characteristic frequencies are summarised in Table 3 and used as a basis for fitting the changes in impedance observed when applying current.

Based on the characteristic frequencies (Table 3) polarisation resistance of the cell was broken down to the contribution for each of the processes occurring in the cell while applying current in both EC-mode and FC-mode. As can be seen from Fig. 5, the total resistance (ohmic and polarisation) is higher in EC-mode as compared to FC-mode as previously observed.[18–22] The total resistance increases with increasing negative current density in EC-mode, whereas the total resistance decreases with increasing positive current density in FC-mode. The ohmic resistance follows the change in temperature; the ohmic resistance increases in EC-mode, whereas the ohmic resistance decreases in FC-mode.

From the breakdown of the individual polarisation resistances, it can be seen that the resistance for Ni–YSZ TPB, LSM_{Low} and LSM_{Hi} increases in EC-mode, whereas they decrease in FC-mode similar to the change in ohmic resistance.

In the present paper, we focus on the diffusion and conversion resistances, and the relation between those. In the H_2O/H_2 mixture, hardly any change in the diffusion resistance is observed when applying current (Fig. 6). On the other hand, in the CO_2/CO mixture, the diffusion resistance decreases when applying current in EC-mode, whereas it increases when applying current in FC-mode. The bold lines in Fig. 6A represents the theoretical diffusion resistances. Finally, the low frequency concentration resistance increases with both increasing positive and negative current densities according to the changes in gas composition. The bold lines in Fig. 6B represent the theoretical conversion resistances based on the CSTR or PFR model.

The main difference in H_2O/H_2 mixtures and CO_2/CO mixtures lies in the higher diffusion, and the higher low frequency concentration resistance. The higher diffusion resistance in CO_2/CO mixtures implies that the actual gas composition at the active electrode is different from the bulk gas composition. We have previously speculated the increased low frequency concentration resistance was caused by diffusion to the active Ni–YSZ electrode, depleting the reactants at the active Ni–YSZ electrode, and that the low frequency concentration resistance reflects the gas composition at the active Ni–YSZ electrode.[25] In order to verify

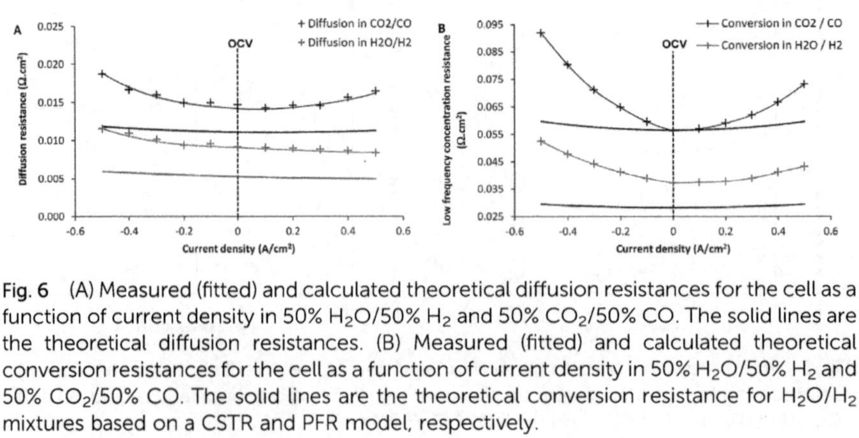

Fig. 6 (A) Measured (fitted) and calculated theoretical diffusion resistances for the cell as a function of current density in 50% H_2O/50% H_2 and 50% CO_2/50% CO. The solid lines are the theoretical diffusion resistances. (B) Measured (fitted) and calculated theoretical conversion resistances for the cell as a function of current density in 50% H_2O/50% H_2 and 50% CO_2/50% CO. The solid lines are the theoretical conversion resistance for H_2O/H_2 mixtures based on a CSTR and PFR model, respectively.

these hypotheses, the performance as a function of current density was examined for cells with different porosity.

Dependency on Ni–YSZ support structure porosity

Electrochemical impedance spectra measured at OCV for three different cells with a support structure porosity of either 25%, 27% or 31% while supplying 50% H_2O/ 50% H_2 to the Ni–YSZ electrode and pure oxygen to the oxygen electrode are shown in Fig. 7A.

The most distinct difference between the three spectra is in the Ni–YSZ TPB resistance which increases with decreasing porosity. Also, a trained eye can see that significant diffusion resistances are visible, and that the diffusion resistance increase with decreasing porosity. Further, the conversion resistance also increases with decreasing porosity. A visualisation of the Ni–YSZ TPB resistance and diffusion resistance at varying porosity is shown in Fig. 7B. The conversion resistance was measured on a large number of cells; including those shown in Fig. 7A and the measured conversion resistance at varying porosity is shown in Fig. 7C. All measured conversion resistances shown in Fig. 7C fall within the values predicted by the CSTR (0.057 Ω cm^2) or PFR (0.028 Ω cm^2) models.

The resistance caused by diffusion through the porous support structure depend on the support porosity as shown in Fig. 7. The impedance breakdown of the diffusion (Fig. 8A) and low frequency concentration resistance (Fig. 8B) under current as a function of support structure is shown in Fig. 8. Clearly the diffusion resistance increases when applying current in EC-mode. On the other hand, in FC-mode, the diffusion resistance remain close to unchanged. The resistance for the low frequency concentration arc show a similar asymmetric behaviour (similar to the results shown in Fig. 5 and 6B) with a large increase in EC-mode and much smaller increase in FC-mode. Further, the resistance for the low frequency concentration arc shows a clear dependency on the support structure, with an

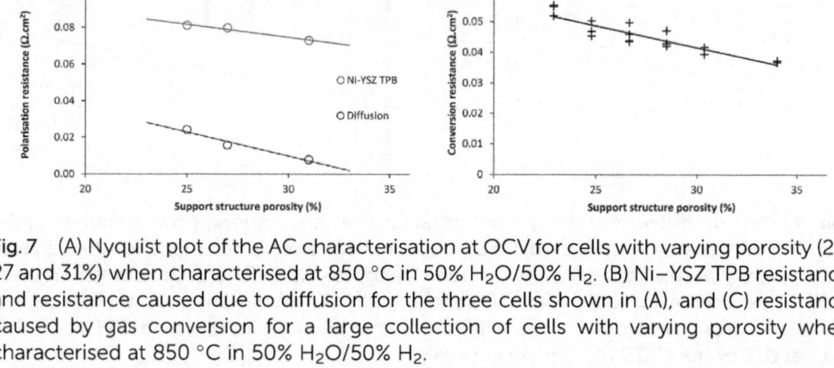

Fig. 7 (A) Nyquist plot of the AC characterisation at OCV for cells with varying porosity (25, 27 and 31%) when characterised at 850 °C in 50% H_2O/50% H_2. (B) Ni–YSZ TPB resistance and resistance caused due to diffusion for the three cells shown in (A), and (C) resistance caused by gas conversion for a large collection of cells with varying porosity when characterised at 850 °C in 50% H_2O/50% H_2.

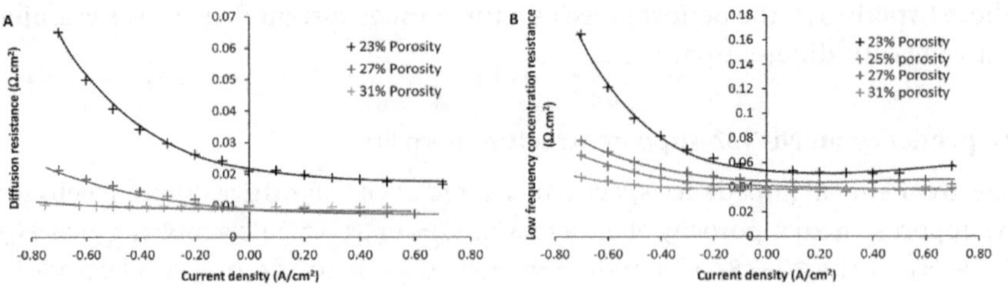

Fig. 8 (A) Measured (fitted) diffusion and (B) low frequency concentration resistance at 850 °C under current (in 50% H_2O/50% H_2).

increased resistance at low porosity and an increased asymmetry with lowering the support structure porosity. This is a clear indication that a significant part of this low frequency resistance originates from diffusion limitations, and may be related to the gas composition at the active Ni–YSZ electrode.[25]

The resistance for the low frequency concentration arc under current in both FC-mode and electrolysis cell mode at varying flowrates are shown in Fig. 9A for the cell with 25% porosity in the support structure. The resistance for the low frequency concentration arc under current in EC-mode for cells with varying porosity is shown in Fig. 9B. The experiments were conducted such that the average steam composition was kept constant by also changing, beside the flowrate, the current density, and steam/hydrogen ratio (when shifting between FC-mode end EC-mode) according to the conditions stated in Table 1. The bold lines in Fig. 9A represent the theoretical conversion resistances based on the CSTR or PFR model.

For the cell with 25% porosity in the support structure (Fig. 9A) a higher low frequency concentration resistance is measured in EC-mode as compared to FC-mode at the same current density and flowrate. Further, a linear dependency between the low frequency concentration resistance and the inverse flowrate is observed in FC-mode, whereas in EC-mode, this relation is only observed at low flowrates and low current density (a similar trend is observed for the remaining cells, not shown). For the cell with a high porosity in the support structure, an almost linear dependency is also observed in EC-mode (Fig. 9B), while decreasing the porosity, and the deviation at high current increases.

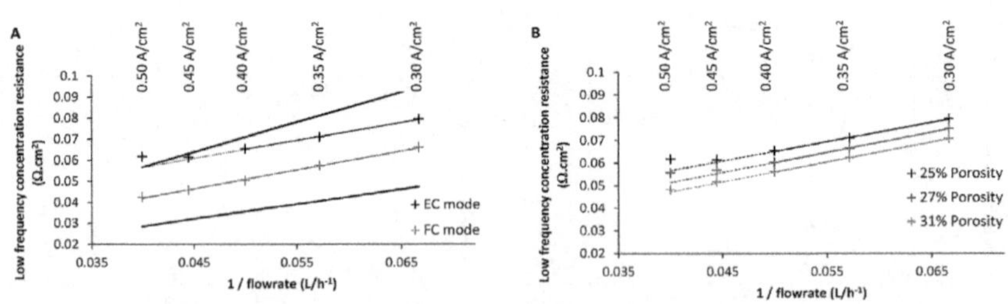

Fig. 9 Measured low frequency concentration resistance during flow variations under current. (A) Resistance measured in FC-mode and electrolysis cell mode for the cell with 25% porosity and (B) resistance measured during flow variations under current in electrolysis cell mode for cells with different porosity. The bold lines are the theoretical conversion resistance due to the different Nernst potential based on a CSTR and PFR model (1/2 of the CSTR model), respectively.

The conversion resistance at temperatures varying from 850 °C to 900 °C is shown in Fig. 10. Again the bold lines represent the theoretical conversion resistances based on the CSTR or PFR model. As can be seen, an increase in the resistance for the low frequency concentration arc increases with temperature.

Degradation

After testing the initial characterisation (iV and EIS at OCV) of the cells, durability in EC-mode was tested at 750, 800, or 850 °C with 45% H_2O/45% CO_2/10% H_2 supplied to the Ni–YSZ electrode, oxygen supplied to the oxygen electrode, and a current density of -1.0 A cm^{-2}. The cell voltage evolution with time for Ni–YSZ|YSZ|LSM–YSZ (Fig. 11A) and Ni–YSZ|YSZ|CGO|LSCF–CGO cells (Fig. 11B) operate at a reactant conversion of 60% is shown in Fig. 11. From the experiments shown in Fig. 11, degradation during electrolysis was observed for all tested cells, and a clear increase in degradation rate is observed when decreasing the operation temperature for the Ni–YSZ|YSZ|LSM–YSZ cells. Similar, decreasing the temperature to 750 °C for the Ni–YSZ|YSZ|LSCF–CGO cells show an increase in degradation rate. Surprisingly, similar degradation rates were found at 800 °C and 850 °C for the Ni–YSZ|YSZ|CGO|LSCF–CGO cells. The Ni–YSZ|YSZ|CGO|LSCF–CGO shows an increased performance as compared to the Ni–YSZ|YSZ|LSM–YSZ (lower initial cell voltage), whereas the degradation rates were comparable. When operating these SOECs at high current density (at or above -1 A cm^{-2}) structural degradation occurs (evident by a change in the ohmic resistance, see Fig. 12), which makes the degradation rates difficult to reproduce,[67] thus, when looking at the cell voltage only, Fig. 11 does not show any evidence supporting an increased durability (as reported in literature[59,60,62]) of the cell with an increased performance (Ni–YSZ|YSZ|CGO|LSCF–CGO).

The contribution to the degradation in overpotential from the different cell components was analysed by breaking down the impedance contribution measured during the durability test, Fig. 12. From the data in Fig. 12 it can be seen that the ohmic contribution to the degradation is higher for the Ni–YSZ|YSZ|LSM–YSZ cells. A clear relation between the ohmic contribution to the degradation and the oxygen electrode with temperature occur for both cell types (increased degradation at lower temperature). On the other hand, the fuel electrode degradation seems independent of the operating temperature (beside the fuel electrode degradation observed for the Ni–YSZ|YSZ|CGO|LSCF–CGO cell

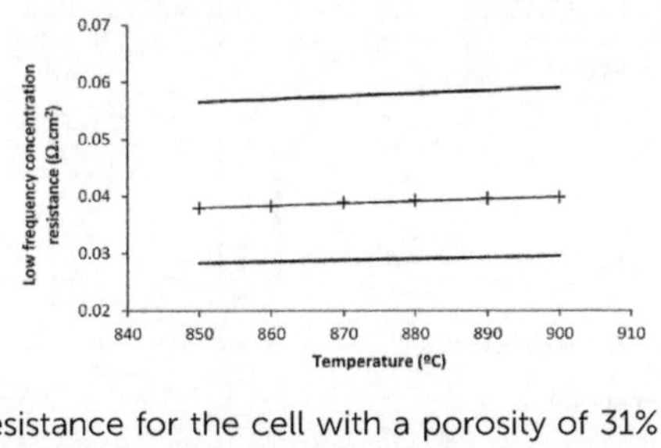

Fig. 10 Conversion resistance for the cell with a porosity of 31% when characterised in 50% H_2O/50% H_2 at temperatures ranging from 850 °C to 900 °C. The bold lines are the theoretical conversion resistance due to the different Nernst potential based on a CSTR and PFR model, respectively.

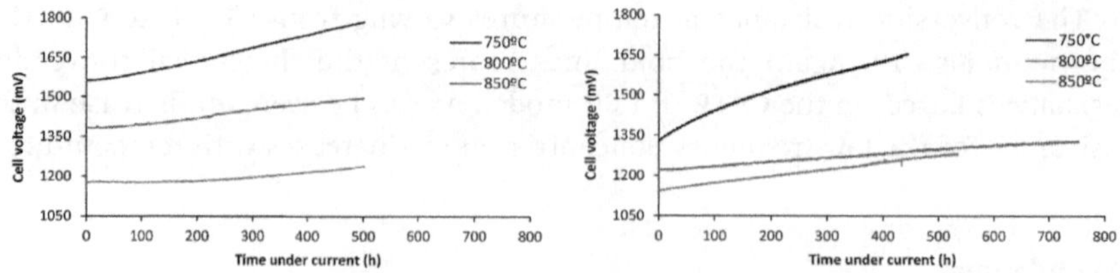

Fig. 11 Cell voltage measured during co-electrolysis at 750 °C, 800 °C, and 850 °C at −1.00 A cm^{-2} and with 45% H_2O/45% CO_2/10% H_2 flown to the Ni–YSZ electrode. The reactant (H_2O + CO_2) conversion was 60%.

operated at 850 °C, which is surprisingly high, but reproducible) and is similar for all temperatures (for either Ni–YSZ|YSZ|LSM–YSZ cells or Ni–YSZ|YSZ|CGO|LSCF–CGO cells). The fuel electrode degradation for the Ni–YSZ|YSZ|CGO|LSCF–CGO cells are around 4 times higher than the fuel electrode degradation for the Ni–YSZ|YSZ|LSM–YSZ cells.

The effect of the conversion degree on the degradation rate was investigated by performing two additional tests with YSZ|YSZ|CGO|LSCF–CGO cells; one durability test was conducted at 800 °C with 45% H_2O/45% CO_2/10% H_2 supplied to the Ni–YSZ electrode, and a current density of −1.0 A cm^{-2} (identical to the test at 800 °C, shown in Fig. 11B), but with an increased flowrate as compared to the tests shown in Fig. 11B, thus the reactant conversion was 30% (Fig. 13A). Another test

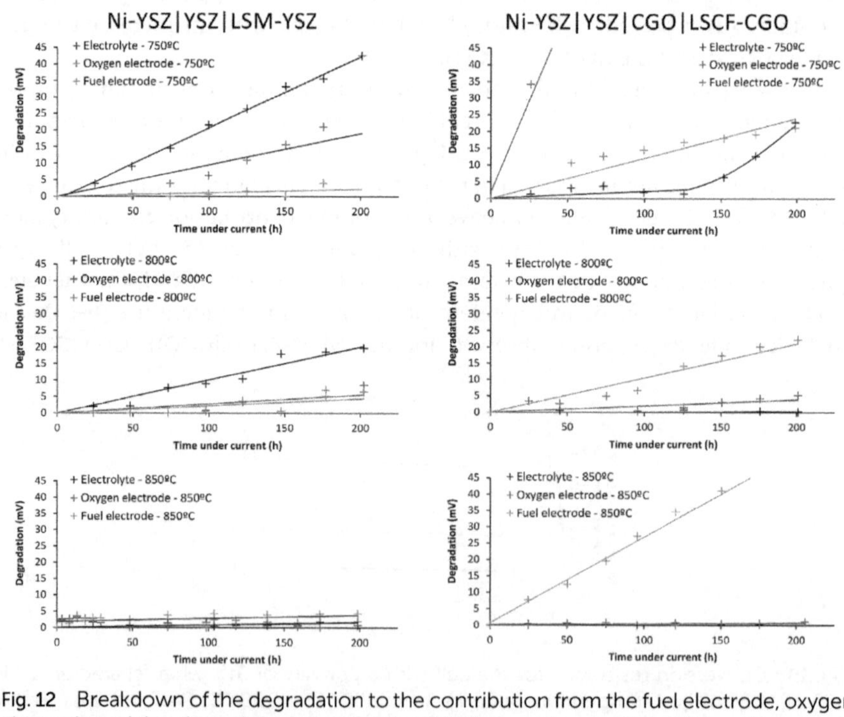

Fig. 12 Breakdown of the degradation to the contribution from the fuel electrode, oxygen electrode and the electrolyte during the first 200 hours of operation of the cells shown in Fig. 11.

was carried out with 90% H_2O/10% H_2 supplied to the Ni–YSZ electrode, and a current density of -1.0 A cm^{-2}, but with an initial reactant conversion of 30% and subsequently increased by steps of 10% conversion to reach a conversion degree of 90% (not shown). At both 30% and 60% reactant conversion (Fig. 13A), the cell voltage during co-electrolysis shows initially an increased degradation whereafter a linear long-term degradation was observed. From the experiments shown in Fig. 13A, it can be seen that increasing the reactant conversion from 30% to 60% does not influence the degradation rate. The test with increasing conversion degree showed that a reactant conversion degree of around 80–90% is necessary in order to observe an increase in degradation rate (not shown here).

Two additional tests were conducted to investigate the different degradation rates during steam and carbon monoxide electrolysis. When comparing the degradation rate during steam and carbon dioxide electrolysis (Fig. 13B) it can be seen that a similar degradation rate is observed for the two tests, especially taking the reproducibility into account (if any difference is present a slightly higher degradation is measured during electrolysis of carbon dioxide).

Modelling the current and overpotential along the cell

A simple model to evaluate the effect of both porosity and reactant conversion on the Ni–YSZ TPB resistance was set up. Assuming a negligible in-plane voltage (which is normally below 1 mV), the cell voltage is identical along the cell, whereas the resistance and overpotential (due to the Nernst potential as well as the relation between the steam partial pressure and Ni–YSZ TPB resistance) varies along the cell in the flow direction. Changing the reactant conversion will, although keeping the total current constant, change the current distribution along the cell (Fig. 14A), which in turn alters the overpotential at the Ni–YSZ electrode. According to the relation between the resistances of the individual processes occurring in the SOC and the current distribution, the overpotential at the Ni-TPB can be found (Fig. 14B) when neglecting diffusion limitations. Similarly, the overpotential at the Ni–YSZ TPB with varying support structure can be found; here the difference is not due to the different conversion degree, but due to an altered current distribution and steam/hydrogen ration at the active nickel electrode due to diffusion (Fig. 14C). Depending on the degradation mechanism, the degradation rate may be related to the overpotential. This dependency will be the case for

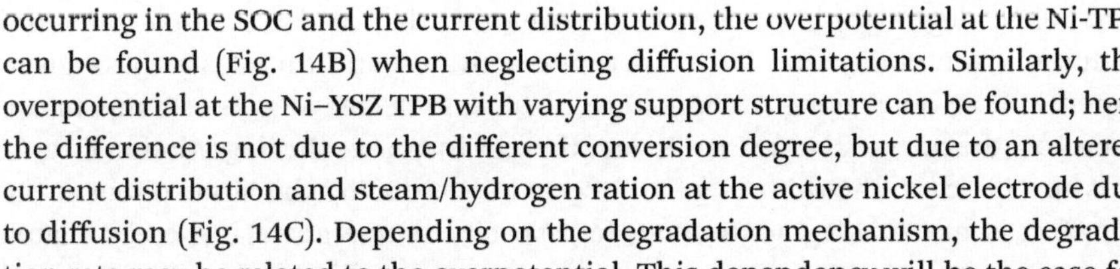

Fig. 13 (A) Cell voltage measured during co-electrolysis at 30% and 60% reactant (H_2O + CO_2) conversion with 45% H_2O/45% CO_2/10% H_2 flown to the Ni–YSZ electrode (the cell with 60% conversion is the same cell as shown in Fig. 11B). (B) Cell voltage measured during either steam or carbon dioxide electrolysis at 60% reactant conversion with 90% H_2O/10% H_2 or 90% CO_2/10% CO flown to the Ni–YSZ electrode. All cells were operated at 800 °C, and a current density of -1.0A cm^{-2}.

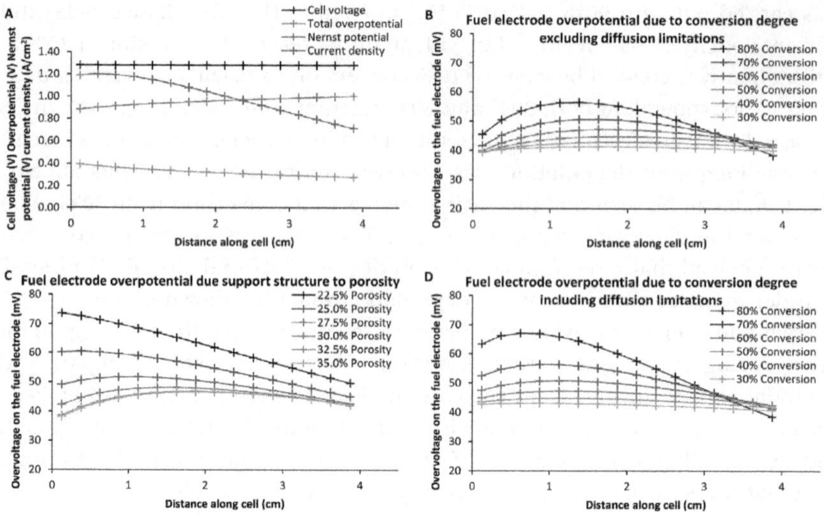

Fig. 14 Calculated theoretical overpotential at the Ni–YSZ TPB along the cell. (A) Variation of cell voltage, overpotential, Nernst potential and current density along the cell. (B) Fuel electrode overpotential at various reactant conversion degrees neglecting diffusion limitations. (C) Fuel electrode overpotential at various support structure porosities. (D) Fuel electrode overpotential at various reactant conversion degrees taking diffusion through the support into account, based on a cell with a support structure porosity of 28%.

electrochemical degradation phenomena (Potential Induced Degradation). On the other hand, degradation mechanisms such as sintering, nickel coarsening and corrosion are thermally activated, and thereby increase with increasing temperature (decreasing potential). It is very difficult to separate the phenomena, and relate degradation rates with the overpotential, as for example potential induced ion mobility may be temperature activated and thus faster (at a given potential) at increasing temperature, whereas decreasing the temperature increases the potential (due to ASR) and thereby increases the degradation rate. In this specific case, the overpotential reflects how reducing the atmosphere at the interface is, and the degradation rate (at a specific temperature) is thereby related to the overpotential due to gas transport and current distribution (Fig. 14), and not the temperature (Fig. 12). The overpotential (overvoltage) at the Ni–YSZ TPB increases when increasing the conversion degree due to the altered current distribution as well as the more reducing gas at the interface when increasing the reactant conversion. Thus, changing the reactant conversion may affect the degradation rate if, and only if degradation is observed in all cases. Similar increase, although much larger, in the Ni–YSZ TPB overpotential, and thus increased degradation may be observed when decreasing the porosity of the support layer because the gas composition at the active electrode becomes increasingly reducing. Further, it is seen that decreasing the porosity the maximal Ni–YSZ TPB overpotential moves towards the inlet of the cell, thus if the degradation of the cell is correlated with the diffusion, an increased degradation at the inlet should be observed. Since both the reactant conversion and support structure affects the Ni–YSZ TPB resistance by changing the pO_2 at the active electrode, an interplay between the conversion degree and structural parameters does apply, and the diffusion limitations enhance the observed changes at the Ni–YSZ TPBs (Fig. 14D). The results shown in Fig. 14D are based on a cell with a porosity of 28% (Ni–YSZ|YSZ|CGO|LSCF–CGO) in the support structure.

Discussion

Cell performance

Continuity of the *i–V* curves across OCV was observed for both H_2O/H_2 and CO_2/CO mixtures, showing that these Ni–YSZ SOCs can work as reversible SOCs in H_2O/H_2 and CO_2/CO mixtures as previously shown.[18–20,68,92] Each *i–V* curve yields two cell resistances, Area Specific cell Resistance (ASR) – one during oxidation of the reactants when applying positive current densities (operation in FC-mode, $ASR_{fuel\ cell}$) and the other during reduction of the reactants when applying negative current densities (operation in EC-mode, $ASR_{electrolysis}$). The observed ASR values in both H_2O/H_2 and CO_2/CO mixtures are lower for oxidation than for reduction, showing a marginally higher activity towards oxidation in comparison to reduction as has been previously shown by several research groups.[18–22] Further, the ASR (both $ASR_{fuel\ cell}$ and $ASR_{electrolysis}$) were higher for CO_2/CO mixtures compared to the ASR values for H_2O/H_2 mixtures showing lower activity for CO oxidation/CO_2 reduction compared to H_2 oxidation/H_2O reduction in agreement with literature.[18,19,68,91,92] Fig. 3–5 shows that the electrode temperature drops when applying a negative current density (H_2O reduction, operated in EC-mode) and increases when applying a positive current density (H_2 oxidation, operated in FC-mode). During both oxidation and reduction, a temperature increase caused by polarisation heating of the cell will occur, and the polarisation heating will be identical during oxidation and reduction. As can be seen from Fig. 3, the temperature increase during oxidation and decrease during reduction, and the change in temperature can therefore not solely be explained by polarisation heating. Besides the polarisation heating, the exothermic oxidation of H_2 and CO oxidation and the endothermic reduction of H_2O and CO_2 will inevitably result in a temperature increase during oxidation and a temperature decrease during reduction. Based on the reaction enthalpy more heat will be produced/consumed when applying current in CO_2/CO mixtures which is in good agreement with Fig. 3–5. Further, at increasing cell voltage closer to or above the thermoneutral voltage, the temperature starts to increase. At 850 °C, the thermoneutral voltage for H_2O electrolysis is 1.29 V. The current density where the temperature starts to increase was -0.55 A cm^{-2} in the H_2O/H_2 mixture and -0.73A cm^{-2} in the CO_2/CO mixture. Assuming an ASR of 0.21 Ω cm^2 for reduction of H_2O, the current density where the cell voltage is equal to the thermoneutral voltage (1.29 V) is -1.57 A cm^{-2} and the current density where the minimum temperature would be measured would thus be -0.78 A cm^{-2}. The temperature profile shows the expected trend, although the minimum temperature is observed at current densities slightly lower than expected, which may be because the ASR increases at high current densities due to conversion (Fig. 3 and 5), and because of the added polarisation heating. This suggests that the temperature change is caused both by polarisation heating and the exothermic/endothermic nature of the reactions.

The decreasing temperature will inevitably lead to an increased resistance of the thermal activated electrochemical reactions (Table 2) when operated in EC-mode as compared to when operated in FC-mode where heating will result in a lower ASR. The temperature profile measured during the DC characterisation may at least partly explain the higher ASRs calculated for electrolysis *versus* the ASRs for fuel cell mode. The total resistances measured by the impedance measurements are in good agreement with the ARSs measured by DC characterisation (Fig. 3, 4 and Table 3).

The impedance spectra was analysed by breaking down the losses into an ohmic loss and five polarisation losses ($R_{LSM\text{-}high}$, R_{TPB}, $R_{LSM\text{-}low}$, $R_{diffusion}$, and $R_{conversion/low\ frequency\ concentration}$). Although the resistances for the individual arcs contributing to the Ni–YSZ electrode are larger in CO_2/CO mixtures, comparison of the impedance spectra and especially a DRT analysis for the two mixtures (Fig. 4, Table 3), show that the spectra contain the same arcs. This suggests that the process in principle consists of the same reaction type at the Ni–YSZ electrode irrespective of the gas mixture, and that the model may be applicable for both H_2O/H_2 and CO_2/CO mixtures. The ohmic resistance is comparable with the values obtained when establishing the model, whereas the resistances obtained for the electrochemical arcs ($R_{LSM\text{-}high}$, R_{TPB} and $R_{LSM\text{-}low}$) are typically 10% lower.[50] The lower resistances in the present may be a consequence of oxygen flow to the LSM–YSZ electrode (instead of air), and the higher steam concentration at the Ni–YSZ electrode, which increase the activity of these SOCs[50,82] (Fig. 2).

The ohmic resistance shows notable differences when applying current in both H_2O/H_2 and CO_2/CO mixtures (Fig. 5). The most distinct temperature profile can be seen during reduction of H_2O in H_2O/H_2 mixture, where the temperature reaches a minimum, whereafter it starts to increase again. The ohmic loss follows the exact same trend, clearly showing that the cell (and active electrodes) experiences a change in temperature. During oxidation in H_2O/H_2 the ohmic resistance decreased from 0.067 Ω cm^2 to 0.064 Ω cm^2, and increased to 0.072 Ω cm^2 during reduction. During oxidation in CO_2/CO the ohmic resistance decreased from 0.067 Ω cm^2 to 0.061 Ω cm^2, and increased to 0.073 Ω cm^2 during reduction. A slightly larger temperature difference was measured during oxidation/reduction in the CO_2/CO mixture as compared to the H_2O/H_2 mixture, consistent with the larger change in ohmic resistance. The larger temperature change in the CO_2/CO mixture is consistent with the higher reaction enthalpy for CO oxidation compared to H_2 oxidation (and for CO_2 reduction compared to H_2O reduction), and may observe some of the differences observed when comparing the reaction kinetics in H_2O/H_2 and CO_2/CO mixtures.

Based on both the AC and DC characterisation, the heat of reaction and polarisation heating during application of current definitely has an effect on the reaction kinetics, and may account for some of the differences, during operation in EC-mode as compared to FC-mode. The ohmic resistance may be used as a measure of the actual electrode temperature, and show that the electrodes experience a significantly higher temperature change as measured by the thermocouple.[25]

Due to the temperature changes, no reliable correlation/discussion of the electrode kinetics can be made under current. On the other hand, diffusion and conversion which is close to independent of temperature changes (eqn (4) and (7), and Fig. 10) can be analysed as a function of applied current.

The calculated theoretical diffusion resistances compared to the experimentally determined diffusion resistances are shown in Fig. 6A. Because of the higher molar weight of CO_2/CO compared to H_2O/H_2 a higher $R_{diffusion}$ is observed in

CO_2/CO mixtures compared to H_2O/H_2 mixtures (Fig. 4–6). In general, the measured diffusion resistances are close to the theoretical values (Fig. 6), although the experimental diffusion resistances are higher than the theoretical ones. Theoretically, the diffusion resistance is expected to be slightly higher in EC-mode compared to FC-mode for both H_2O/H_2 and CO_2/CO mixtures, in accordance with the experiments (Fig. 6). For H_2O/H_2 mixtures, a gradual decrease occurs when decreasing the current in EC-mode and increasing current in FC mode, as predicted theoretically. For the CO_2/CO mixture, an increase at high FC-current is observed, also theoretically, this is predicted. A similar trend is found at OCV when varying either pH_2O or pCO_2 as shown in Fig. 15.

Because the conversion resistance is only associated with the gas composition in the gas phase and not the molar masses the conversion resistance is expected to be identical for mixtures of H_2O/H_2 and CO_2/CO at OCV, with a minimum in a 50% H_2O/50% H_2 mixture, and is expected to increase symmetrically around this minimum with increasing/decreasing H_2O/H_2 ratio (eqn (7)). Further, a slight increase with temperature as well as a linear relation with the inverse flow rate is expected. At OCV (Fig. 9 and 16), as well as in FC-mode (Fig. 9), the measured conversion/low frequency concentration resistance follows the theoretically predicted trends. As can be seen from Fig. 5 and 6B, the observed low frequency concentration resistance is certainly not identical in H_2O/H_2 and CO_2/CO mixtures, and a significantly higher low frequency concentration resistance is observed in CO_2/CO mixtures as compared to H_2O/H_2 mixture (Fig. 6B). Fig. 16 shows the measured conversion resistance at OCV for the cell with a porosity of 25% compared to the measured low frequency concentration resistance under current as a function of the average pH_2O along the cell. Further, the theoretical conversion resistances based on the CSTR or PFR model are shown by the bold lines in Fig. 16. It is seen that the measured conversion resistance at OCV follows the trend predicted by the CSTR/PFR model, where the minimum conversion resistance is observed in a 50% H_2O/50% H_2 mixture, and increases symmetrically around this minimum with increasing/decreasing H_2O/H_2 ratio. In Fig. 16, the observed conversion resistances at OCV lie between the values predicted by the two models, which makes sense when recalling the flow pattern for the set-up (Fig. 1). Under current, the measured low frequency concentration resistance is highly asymmetric (Fig. 5, 6B, and 16) with an increased low frequency concentration resistance in EC-mode. This asymmetry is not due to the steam concentration, varying the inlet steam concentration from 45% to 55% shows the exact same trend. Further, the asymmetry is higher in CO_2/CO mixtures as compared to

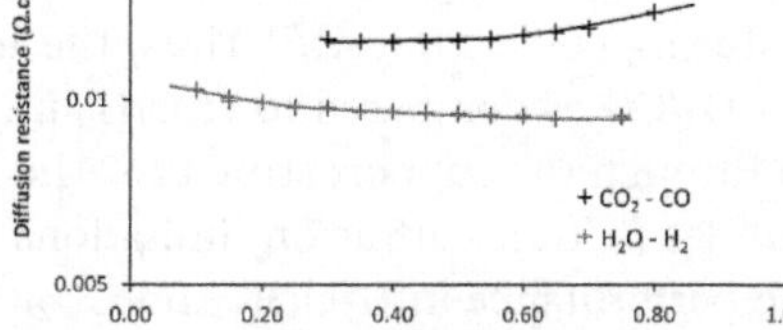

Fig. 15 Measured (fitted) diffusion resistances in H_2O/H_2 and CO_2/CO mixtures as a function of pH_2O or pCO_2.

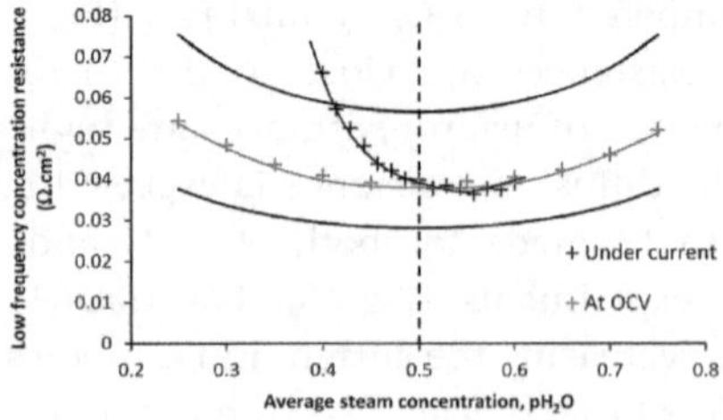

Fig. 16 Comparison of the low frequency concentration resistance at 850 °C at OCV with varying the steam concentration with the low frequency concentration resistance measured under current (in 50% H_2O/50% H_2) as a function of the average steam concentration.

H_2O/H_2 mixtures (Fig. 5 and 6B), and the asymmetry becomes more pronounced in CO_2/CO mixtures as compared to H_2O/H_2 mixtures and increases with decreasing porosity (Fig. 8). This shows that the low frequency concentration resistance is affected by diffusion and may reflect the gas composition at the active Ni–YSZ electrode, and that a more precise description of the detailed nature of this arc remains to be modelled.

Even at OCV, the diffusion and the low frequency concentration resistance changes with support structure porosity, this means that the actual concentration at the active fuel electrode varies with the structural parameters (Fig. 7). Although the cells presented in this study were manufactured by sintering the half-cell at different temperatures, and thereby creating very different structural parameters for the active Ni–YSZ electrode, an increasing Ni–YSZ TPB resistance was observed with decreasing porosity (Fig. 7B). This increased resistance may very well be due to diffusion, changing the gas composition at the active Ni–YSZ TPB sites during the measurements, even at OCV. In general it is found that there is no pH_2 dependency for Ni–YSZ supported cells, whereas a significant dependency on pH_2O, pCO, and pCO_2 is found. Surprisingly, the Ni–YSZ TPB resistance dependency on apparent partial pressure of reactants and products, seem to correlate with the molar mass of the species in question, and the observed dependency may thus be related to the diffusion, and thereby also the electrode structure. Therefore it may not be possible to investigate electrode kinetics and make a general expression for "all" Ni–YSZ electrodes as this dependency will be dependent on the structural parameters of the specific cell.

The difference in performance for CO_2 electrolysis and H_2O electrolysis lies in both the Ni–YSZ TPB resistance, but also the slower diffusion, and thus higher low frequency concentration resistance.[27] When comparing H_2O electrolysis and co-electrolysis of H_2O and CO_2, almost identical Ni–YSZ TPB resistance and diffusion resistance is observed, whereas a slightly higher low frequency concentration resistance is observed during co-electrolysis.[27] Thus, the increased Ni–YSZ TPB resistance observed in CO_2/CO as compared to H_2O/H_2 mixtures may be associated with the diffusion through the support structure. This may be supported by studies on model electrodes (with no diffusion limitations) that show much less differences between the performance in H_2O/H_2 and CO_2/CO mixtures.[35,49,93,94]

When comparing the co-electrolysis performance on fuel electrode supported cells and electrolyte supported cells[27,95–97] significant differences are found. For fuel electrode supported cells (for which diffusion through the support structure plays a role), a large difference between co-electrolysis and steam electrolysis performance is observed.[27] On the other hand, similar performance for co-electrolysis and steam electrolysis performance is observed for electrolyte supported cells[95–97] (where no diffusion restriction occurs). These differences may be explained by diffusion, affecting both the Ni–YSZ TPB and the low frequency concentration resistance in a Ni–YSZ supported cell. Probably, a diffusion limiting current density is reached in the active Ni–YSZ layer next to the electrolyte, *i.e.* pH_2O and pCO_2, respectively, approaches zero and then the Ni–YSZ support starts to become electrochemically active, but it has a much higher polarisation resistance due to its coarse structure, and this high resistance results in a very long time constant (a large low frequency arc). A modelling of this will require a very detailed 3D, which is not available to us.

The large effect of the structural parameters indicates that electrolyte supported or oxygen electrode supported cells may be preferable as electrolysis cells as these will minimise the diffusion limitations and decrease the Ni–YSZ TPB resistance (caused by diffusion limitations). Similarly, pressurised operation may significantly increase the performance of these cells, again by decreasing the diffusion limitations and Ni–YSZ TPB resistance.

Cell degradation

Long-term degradation of the SOCs during electrolysis operation was observed for all tests through the course of the cell voltage (Fig. 11) and the increase in impedance (Fig. 12). These cells show degradation in both the serial resistance as well as the polarisation resistances. From the cell voltages shown in Fig. 11 it can be seen that lowering the operation temperature increases the degradation rate of the cells. There is a clear correlation between the operation temperature and the degradation rate for the oxygen electrode (and electrolyte) whereas the degradation for the fuel electrode does not seem to depend on the operation temperature. The degradation at the oxygen electrode (and electrolyte) is therefore most likely related to the overpotential as previously discussed.[14,61,63] On the other hand, the degradation at the fuel electrode cannot be correlated to the temperature. It has previously been suggested that the observed degradation is caused by metallic inter-diffusion of Zr formed by reduction of YSZ at low oxygen partial pressures (a critical pO_2 of $3.2 \times 10^{-29\pm3}$ bar at 850 °C (ref. 98)). Such low oxygen partial pressures imply an overpotential at the Ni–YSZ TPB of several hundred mV.[98] It is difficult to imagine such low oxygen partial pressures/high overpotential at the Ni–YSZ TPB (Fig. 14) unless significant diffusion limitations occur. The Ni–YSZ degradation is higher for the Ni–YSZ|YSZ|CGO|LSCF–CGO cells as compared to the Ni–YSZ|YSZ|LSM–YSZ cells. Although an increased polarisation at the oxygen electrode does smoothen the current distribution along the cell for the Ni–YSZ|YSZ|LSM–YSZ cells, this is far from enough to explain the difference observed. On the other hand, the support structure porosity for the Ni–YSZ|YSZ|LSM–YSZ cells (34%) is higher than for the Ni–YSZ|YSZ|CGO|LSCF–CGO (28%). Based on the discussion above, such a decrease in porosity will inevitably create a decreased pO_2 at the interface due to diffusion limitations (Fig. 14), and the porosity may be critical for the observed degradation. This is consistent with the carbon formation observed in the Ni–YSZ electrode close to the YSZ electrolyte in cells operated far outside the carbon deposition regime,[99,100] indicating a very reducing atmosphere.

Careful analysis of the degradation of these cells has shown that degradation in the Ni–YSZ electrode is found close to the inlet of the cell.[101] The increased diffusion limitations, will, due to the lower steam concentration at the interface, move the maximal overpotential at the Ni–YSZ electrode closer to the inlet (Fig. 14), thus supporting that diffusion limitation may play a significant role for the degradation of these cells. Due to changes in diffusion, similar increased degradation may be found by increasing the reactant conversion (Fig. 14) as well an increased degradation may be observed during CO_2 electrolysis (as compared to steam electrolysis). Within the experimental uncertainties, no evidence of an increased degradation was observed when increasing the conversion degree from 30% to 60% (Fig. 13A), nor an influence of CO_2 electrolysis as compared to steam electrolysis (Fig. 13B) was found. A reactant conversion of at least 80% was necessary in order to observe an increase in degradation rates. The effect of support structure and the effect of reactant conversion are related, and without significant diffusion limitations, the conversion degree has only a minor effect on the degradation rate. In this study we did not find any clear correlation between the long term degradation and the conversion degree, whereas in literature it has been speculated that the conversion degree influences the degradation rate of the cells.[14] Besides the obvious effect of impurities in the gas phase (and in raw materials) which will be affected by the steam partial pressure (the decreasing steam partial pressure due to the conversion will cause an increased deposition of the steam borne impurities in the Ni–YSZ cathode), and that the impurity level may be higher in CO_2/CO mixtures as compared to H_2O/H_2, the presence of impurities may explain the increased degradation with increased reactant conversion as well as in carbon dioxide, sometimes observed in literature, but for CO_2 the degradation rate should not be dependent on the degree of conversion as long as there is no significant diffusion limitation. The similar degradation rates were observed in steam and carbon dioxide mixtures and at 30% or 60% reactant conversions in the present study may be due to the fact that the inlet gases were cleaned to eliminate impurity related degradation.

Adsorption of impurities[18,66,102] as well as structural changes[101,103,104] during durability tests have shown to alter the measured low frequency concentration resistance. Assuming that the low frequency concentration resistance is in fact dependant on the electrode structure, such changes may be explained since the active Ni–YSZ TPB alters in these cases. For example if degradation occurs at the electrode|electrolyte interface, the active Ni–YSZ TPBs move away from the interface, changing the diffusion distance.

When comparing the tests with different reduction temperature (Fig. 13A compared to Fig. 13B), it can be seen that a slightly lower degradation rate is found for the cells reduced at 1000 °C (850 °C for the two cells shown in Fig. 13A, as compared to 1000 °C for the two cells shown in Fig. 13B). The decreased degradation rate for the cells reduced at a higher temperature may be due to the increased mechanical strength induced by the higher reduction temperature.[105]

Conclusion

A Ni–YSZ electrode supported Solid Oxide Cell (SOC) was operated in both fuel cell mode (FC-mode) and electrolysis cell mode (EC-mode) in mixtures of H_2O/H_2, CO_2/CO, $H_2O/H_2O/CO_2$/CO at 750 °C, 800 °C and 850 °C. Polarisation (*i*–*V*) characterisation show that the kinetics for the reduction of H_2O and CO_2 is slightly slower compared to oxidation of H_2 and CO. The slower kinetics for reduction was found to be partly related to the polarisation heating and the entropy change. Due to the exothermic reaction in FC-mode and endothermic reaction in EC-mode, both the ohmic resistance as well as the electrochemical electrode resistances are higher in EC-mode as compared to FC-mode. The measured diffusion resistance is slightly larger in EC-mode as compared to FC-mode whereas the low frequency concentration resistance is significant higher in EC-mode at high current densities, probably related to diffusion limiting current densities. Both the increased diffusion resistance and the asymmetric low frequency concentration resistance result in an increased area specific resistance (decreased activity) in EC-mode. The temperature profile shows a decrease in temperature when going from FC-mode to EC-mode due to the exothermic nature of the oxidation and endothermic nature of the reduction. The lower performance in EC-mode compared to FC-mode close to the open circuit voltage (OCV) may therefore be explained by the temperature changes caused by the reactions. At higher current densities the increase caused by diffusion limitations become more pronounced and may explain the apparent fuel starvation normally observed at lower conversion degrees in EC-mode as compared to FC-mode. Changing the porosity of the support structure shows a change the Ni–YSZ TPB resistance and a significant change in the low frequency concentration (partial pressure) related resistance when applying high current densities. The observed increased polarisation in EC-mode at decreasing porosity shows that diffusion limitations cannot be neglected for support structures with porosity below 30% (with a support thickness of 300 μm). These diffusion limitations through the support and active electrode structure create an increased reducing atmosphere at the interface which may be related to the degradation of the cells.

Acknowledgements

The authors would like to thank Karen Brodersen for producing the Ni–YSZ|YSZ|LSM–YSZ cells with different Ni–YSZ cermet porosity and Anne Hauch for valuable discussions. This work was financially supported by Energy Technology Development and Demonstration Programme (EUDP) under the Danish Energy Agency *via* the project "Green Natural Gas", the Danish Council for Strategic Research, *via* the Strategic Electrochemistry Research Center (SERC, http://www.serc.dk), and by Energienet.dk through the ForskEL project "Solid Oxide Electrolysis for Grid Balancing"

References

1 A. O. Isenberg, *Solid State Ionics*, 1981, **3–4**, 431.
2 W. Donitz and E. Erdle, *Int. J. Hydrogen Energy*, 1985, **10**, 291.
3 W. Dönitz, E. Erdle, R. Schaumm and R. Streicher, *Adv. Hydrogen Energy*, 1988, **6**, 65.
4 W. Dönitz, E. Erdle and R. Streicher, High Temperature Electrochemical Technology For Hydrogen Production and Power Generation, in *Electrochemical hydrogen technologies: Electrochemical production and*

combustion of hydrogen, ed. H. Wendt, Elsevier, Amsterdam, The Netherlands, 1990.

5 W. Dönitz, G. Dietrich, E. Erdle and R. Streicher, *Int. J. Hydrogen Energy*, 1988, **13**, 283.

6 W. Doenitz and R. Schmidberger, *Int. J. Hydrogen Energy*, 1982, **7**, 321.

7 K. H. Quandt and R. Streicher, *Int. J. Hydrogen Energy*, 1986, **11**, 309.

8 H. S. Spacil and C. S. Tedmon, *J. Electrochem. Soc.*, 1969, **116**, 1618.

9 E. Erdle, W. Donitz, R. Schamm and A. Koch, *Int. J. Hydrogen Energy*, 1992, **17**, 817.

10 L. Elikan, J. P. Morris and C. K. Wu, Development of a Solid Electrolyte Carbon Dioxide and Water Reduction System for Oxygen recovery, *NASA Report CR-2014; Report for NASA Research Center*, 1972.

11 A. O. Isenberg and R. J. Cusick, Carbon Dioxide Electrolysis with Solid Oxide Electrolyte Cells for Oxygen Recovery in Life Support System, *18th Intersociety Conference on Environmental Systems*, San Franscisco, California, USA, 1989, p. 881040.

12 A. O. Isenberg, Three-man Solid Electrolyte Carbon Dioxide Electrolysis Breadboard, *NASA Report CR-185612*; Report for NASA Research Center, 1989.

13 K. R. Sridhar, *J. Br. Interplanet. Soc.*, 1996, **49**, 435.

14 S. D. Ebbesen, S. H. Jensen, A. Hauch and M. Mogensen, *Chem. Rev.*, 2014, **114**, 10697.

15 A. Hagen, R. Barfod, P. V. Hendriksen, Y. L. Liu and S. Ramousse, *J. Electrochem. Soc.*, 2006, **153**, A1165.

16 A. Hauch, Solid Oxide Electrolysis Cells – Performance and Durability, PhD thesis, Risø National Laboratory, Risø, Roskilde, Denmark, 2007.

17 S. H. Jensen, A. Hauch, P. V. Hendriksen, M. Mogensen, N. Bonanos and T. Jacobsen, *J. Electrochem. Soc.*, 2007, **154**, B1325.

18 S. D. Ebbesen and M. Mogensen, *J. Power Sources*, 2009, **193**, 349.

19 S. D. Ebbesen, C. Graves and M. Mogensen, *Int. J. Green Energy*, 2009, **6**, 646.

20 A. Hauch, S. H. Jensen, S. Ramousse and M. Mogensen, *J. Electrochem. Soc.*, 2006, **153**, A1741.

21 O. A. Marina, L. R. Pederson, M. C. Williams, G. W. Coffey, K. D. Meinhardt, C. D. Nguyen and E. C. Thomsen, *J. Electrochem. Soc.*, 2007, **154**, B452.

22 M. Mogensen, Electrode kinetics of SOFC anodes and cathodes, *Proceedings of the 14th Risø international symposium on materials science, High temperature electrochemical behaviour of fast ion and mixed conductors*, ed. F. W. Poulsen, J. J. Bentzen, T. Jacobsen, E. Skou and M. J. L. Østergård, Risø National Laboratory, Risø, Roskilde, Denmark, September 6–9, 1993, p. 117.

23 G. B. Jung, J. Y. Chen, C. Y. Lin and S. Y. Sun, *Int. J. Hydrogen Energy*, 2012, **37**, 15801.

24 M. Ni, M. K. H. Leung and D. C. Leung, *J. Power Sources*, 2006, **163**, 460.

25 S. D. Ebbesen and M. Mogensen, *ECS Trans.*, 2013, **50**, 167.

26 J. C. Njodzefon, D. Klotz, A. Kromp, A. Weber and E. Ivers-Tiffée, *J. Electrochem. Soc.*, 2013, **160**, F313.

27 S. D. Ebbesen, R. Knibbe and M. Mogensen, *J. Electrochem. Soc.*, 2012, **159**, F482.

28 T. H. Etsell and S. N. Flengas, *J. Electrochem. Soc.*, 1971, **118**, 1890.

29 C. J. Wen and D. M. Mason, *J. Appl. Electrochem.*, 1978, **8**, 81.

30 R. Suwanwarangkul, E. Croiset, M. W. Fowler, P. L. Douglas, E. Entchev and M. A. Douglas, *J. Power Sources*, 2003, **122**, 9.
31 M. Ni, *Chem. Eng. J.*, 2010, **164**, 246.
32 M. Ni, *Int. J. Hydrogen Energy*, 2009, **34**, 7795.
33 M. Ni, *Int. J. Hydrogen Energy*, 2012, **37**, 6389.
34 Y. Matsuzaki and I. Yasuda, *J. Electrochem. Soc.*, 2000, **147**, 1630.
35 A. Utz, K. V. Hansen, K. Norrman, E. Ivers-Tiffée and M. Mogensen, *Solid State Ionics*, 2011, **183**, 60.
36 M. S. Schmidt, K. V. Hansen, K. Norrman and M. Mogensen, *Solid State Ionics*, 2009, **180**, 431.
37 A. Leonide, S. Hansmann, A. Weber and E. Ivers-Tiffée, *J. Power Sources*, 2011, **196**, 7343.
38 A. Bieberle, L. P. Meier and L. J. Gauckler, *J. Electrochem. Soc.*, 2001, **148**, A646.
39 M. Mogensen, S. Sunde and S. Primdahl, SOFC anode kinetics, *Proceedings of the 17th Risø International Symposium on Materials Science: High Temperature Electrochemistry: Ceramisc and Metals*, ed. F. W. Poulsen, N. Bonanos and S. Linderoth, Risø National Laboratory, Risø, Roskilde, Denmark, September 2–9, 1996, p. 77.
40 J. Mizusaki, T. Yamamura, K. Mori, H. Tagawa, K. Hirano, S. Ehara, T. Takagi, M. Hishinuma, H. Sasaki, T. Sogi, Y. Nakamura and K. Hashimoto, Kinetic Studies on Ni/YSZ Anodes Reaction of SOFC in H_2-H_2O Atmospheres By the use of Nickel Pattern Electrodes, *Proceedings of the 17th Risø International Symposium on Materials Science: High Temperature Electrochemistry: Ceramisc and Metals*, ed. F. W. Poulsen, N. Bonanos and S. Linderoth, Risø National Laboratory, Risø, Roskilde, Denmark, September 2–9, 1996; p. 363.
41 M. Mogensen, K. V. Jensen, M. J. Jørgensen and S. Primdahl, *Solid State Ionics*, 2002, **150**, 123.
42 M. Mogensen, J. Høgh, K. V. Hansen and T. Jacobsen, *ECS Trans.*, 2007, **7**, 1329.
43 S. P. Jiang and S. P. S. Badwal, *Solid State Ionics*, 1999, **123**, 209.
44 J. Mizusaki, H. Tagawa, T. Saito, K. Kamitani, T. Yamamura, K. Hirano, S. Ehara, T. Takagi, T. Hikita, M. Ippommatsu, S. Nakagawa and K. Hashimoto, *J. Electrochem. Soc.*, 1994, **141**, 2129.
45 M. Ni, M. K. H. Leung and D. Y. C. Leung, *Int. J. Hydrogen Energy*, 2007, **32**, 2305.
46 S. Primdahl and M. Mogensen, *J. Electrochem. Soc.*, 1997, **144**, 3409.
47 S. P. Jiang and S. P. S. Badwal, *J. Electrochem. Soc.*, 1997, **144**, 3777.
48 M. Brown, S. Primdahl and M. Mogensen, *J. Electrochem. Soc.*, 2000, **147**, 475.
49 A. Utz, A. Leonide, A. Weber and E. Ivers-Tiffée, *J. Power Sources*, 2011, **196**, 7217.
50 R. Barfod, M. Mogensen, T. Klemenso, A. Hagen, Y. L. Liu and P. V. Hendriksen, *J. Electrochem. Soc.*, 2007, **154**, B371.
51 V. Sonn, A. Leonide and E. Ivers-Tiffée, *J. Electrochem. Soc.*, 2008, **155**, B675.
52 M. Juhl, S. Primdahl and M. Mogensen, Characterisation of Composite SOFC Cathodes by Impedance Spectroscopy, *Proceedings of the 17th Risø International Symposium on Materials Science: High Temperature Electrochemistry: Ceramisc and Metals*, ed. F. W. Poulsen, N. Bonanos and S. Linderoth, Risø National Laboratory, Risø, Roskilde, Denmark, September 2–9, 1996, p. 295.

53 M. J. Jørgensen and M. Mogensen, *J. Electrochem. Soc.*, 2001, **148**, A433.
54 C. Peters, A. Weber and E. Ivers-Tiffée, *J. Electrochem. Soc.*, 2008, **155**, B730.
55 J. E. O'Brien, G. K. Housley, D. G. Milobar and N. Petigny, Performance of Single Electrode-Supported Cells Operating in the Electrolysis Mode, *Proceedings of the ASME 2009 International Mechanical Engineering Congress and Exposition, IMECE2009*, ASME, Lake Buena Vista, Florida, USA, November 13–19, 2009, p. 617.
56 T. Matsui, R. Kishida, J. Y. Kim, H. Muroyama and K. Eguchi, *J. Electrochem. Soc.*, 2010, **157**, B776.
57 A. Mansuy, J. Mougin, M. Petitjean and F. Mauvy, Durability studies of Solid Oxide Electrolysis Cells (SOEC), *Proceedings of the 10th European SOFC forum*, European Fuel Cell Forum AG, Luzern, Switzerland, July 26–29, 2012, p. B0704.
58 E. Tang, T. Wood, S. Benhaddad, C. Brown, H. He, J. Nelson, O. Grande, B. Nuttal, M. Richards and R. Petri, Advanced Materials for RSOFC Dual Operation with Low Degradation, Report for United States Department of Energy, 2012.
59 J. Guan, N. Minh, B. Ramamurthi, J. Ruud, J.-K. Hong, P. Riley and D. Weng, High Performance Flexible Reversible Solid Oxide Fuel Cell, *Final Technical Report*, Report for United States Department of Energy, 2006.
60 N. Q. Minh, *ECS Trans.*, 2011, **35**, 2897.
61 K. Chen, N. Ai and S. P. Jiang, *J. Electrochem. Soc.*, 2010, **157**, P89.
62 P. Hjalmarsson, X. Sun, Y.-L. Liu and M. Chen, *J. Power Sources*, 2013, **223**, 349.
63 A. Momma, T. Kato, Y. Kaga and S. Nagata, *J. Ceram. Soc. Jpn.*, 1997, **105**, 369.
64 P. H. Larsen, C. Bagger, S. Linderoth, M. Mogensen, S. Primdahl, M. J. Jørgensen, P. V. Hendriksen, B. Kindl, N. Bonanos, F. W. Poulsen and K. A. Maegaard, *Proceedings of Solid Oxide Fuel Cell VII (SOFC VII)*, ed. H. Yokokawa and S. C. Singhal, Electrochem. Soc., Pennington, NJ, 2001, p. 28.
65 M. J. Jørgensen and M. Mogensen, *J. Electrochem. Soc.*, 2001, **148**, A433.
66 S. D. Ebbesen, C. Graves, A. Hauch, S. H. Jensen and M. Mogensen, *J. Electrochem. Soc.*, 2010, **157**, B1419.
67 S. D. Ebbesen, Durability of Solid Oxide Electrolysis Cells, in *Green Natural Gas, Energy Technology Development and Demonstration Programme (EUDP) under the Danish Energy Agency*, 2014.
68 S. D. Ebbesen and M. Mogensen, *Electrochem. Solid-State Lett.*, 2010, **13**, B106.
69 A. Hauch and M. Mogensen, *Solid State Ionics*, 2010, **181**, 745.
70 S. D. Ebbesen and M. Mogensen, Method and System for Purification of Gas Streams for Solid Oxide Cells, EP 2362475 A1, 2010.
71 S. D. Ebbesen and M. Mogensen, Method and System for Purification of Gas Streams for Solid Oxide Cells, EP 2537203 A1, 2011.
72 C. Birkl, Eects of anode sintering temperature on microstructure and performance of multilayer tape cast SOFC, MSc thesis, DTU Energy Conversion, Risø Campus, Technical University of Denmark, Roskilde, Denmark, 2012.
73 A. Hauch, C. Birkl, K. Brodersen and P. S. Jørgensen, Multilayer tape cast SOFC – effect of anode sintering temperature, *Proceedings of the 10th European SOFC forum*, European Fuel Cell Forum AG, Luzern, Switzerland, July 26–29, 2012, p. A1007.

74 S. Primdahl and M. Mogensen, *J. Electrochem. Soc.*, 1999, **146**, 2827.
75 J. R. Izzo, A. S. Joshi, K. N. Grew, W. K. S. Chiu, A. Tkachuk, S. H. Wang and W. Yun, *J. Electrochem. Soc.*, 2008, **155**, B504.
76 R. E. Williford, L. A. Chick, G. D. Maupin, S. P. Simner and J. W. Stevenson, *J. Electrochem. Soc.*, 2003, **150**, A1067.
77 H. Iwai, N. Shikazono, T. Matsui, H. Teshima, M. Kishimoto, R. Kishida, D. Hayashi, K. Matsuzaki, D. Kanno, M. Saito, H. Muroyama, K. Eguchi, N. Kasagi and H. Yoshida, *J. Power Sources*, 2010, **195**, 955.
78 J. R. Wilson, J. S. Cronin and S. A. Barnett, *Scr. Mater.*, 2011, **65**, 67.
79 P. S. Jørgensen, Quantitative Data Analysis Methods for 3D Microstructure Characterization of Solid Oxide Cells, PhD thesis, Technical University of Denmark, Kongens Lyngby, Denmark, 2010.
80 S. Primdahl and M. Mogensen, *J. Electrochem. Soc.*, 1998, **145**, 2431.
81 S. H. Jensen, A. Hauch, P. V. Hendriksen and M. Mogensen, *J. Electrochem. Soc.*, 2009, **156**, B757.
82 A. Hagen, M. Menon, R. Barfod, P. V. Hendriksen, S. Ramousse and P. H. Larsen, *Fuel Cells*, 2006, **6**, 146.
83 J. Nielsen and J. Hjelm, *Electrochim. Acta*, 2014, **115**, 31.
84 S. Primdahl, P. V. Hendriksen, P. H. Larsen, B. Kindl and M. Mogensen, Evaluation of SOFC test data, *Proceedings of Solid Oxide Fuel Cell VII (SOFC VII)*, ed. S. C. Singhal and H. Yokokawa, Electrochemical Society, Inc., Pennington, New Jersey, USA, Tsukuba, Japan, June 5–8, 2001, p. 932.
85 S. P. S. Badwal, *Solid State Ionics*, 1992, **52**, 23.
86 E. P. Murray, T. Tsai and S. A. Barnett, *Solid State Ionics*, 1998, **110**, 235.
87 M. Juhl, S. Primdahl and M. Mogensen, Characterisation of Composite SOFC Cathodes by Impedance Spectroscopy, *Proceedings of the 17th Risø International Symposium on Materials Science: High Temperature Electrochemistry: Ceramisc and Metals*, ed. F. W. Poulsen, N. Bonanos, S. Linderoth , Roskilde, Denmark, 1996, p. 295.
88 A. Leonide, V. Sonn, A. Weber and E. Ivers-Tiffée, *J. Electrochem. Soc.*, 2008, **155**, B36.
89 E. Siebert, A. Hammouche and M. Kleitz, *Electrochim. Acta*, 1995, **40**, 1741.
90 N. Christiansen, J. B. Hansen, H. Holm-Larsen, S. Linderoth, P. H. Larsen, P. V. Hendriksen and A. Hagen, Solid Oxide Fuel Cell Development at Topsoe Fuel Cell A/S and Risø National Laboratory, *Proceedings of Risø International Energy Conference, Energy Solutions for Sustainable Development*, ed. S. L. Petersen and H. Larsen, Risø National Laboratory, Risø, Roskilde, Denmark, May 22–24, 2007, p. 357.
91 P. Holtappels, L. G. J. de Haart, U. Stimming, I. C. Vinke and M. Mogensen, *J. Appl. Electrochem.*, 1999, **29**, 561.
92 S. H. Jensen, P. H. Larsen and M. Mogensen, *Int. J. Hydrogen Energy*, 2007, **32**, 3253.
93 A. Utz, H. Störmer, A. Leonide, A. Weber and E. Ivers-Tiffée, *J. Electrochem. Soc.*, 2010, **157**, B920.
94 A. Ehn, J. Høgh, M. Graczyk, K. Norrman, L. Montelius, M. Linne and M. Mogensen, *J. Electrochem. Soc.*, 2010, **157**, B1588.
95 C. M. Stoots, J. E. O'Brien, J. S. Herring and J. J. Hartvigsen, *J. Fuel Cell Sci. Technol.*, 2009, **6**, 011014.
96 C. Stoots, J. O'Brien and J. Hartvigsen, *Int. J. Hydrogen Energy*, 2009, **34**, 4208.

97 C. M. Stoots, J. E. O'Brien, K. G. Condie and J. J. Hartvigsen, *Int. J. Hydrogen Energy*, 2010, **35**, 4861.
98 M. Chen, Y. L. Liu, J. J. Bentzen, W. Zhang, X. Sun, A. Hauch, Y. Tao, J. R. Bowen and P. V. Hendriksen, *J. Electrochem. Soc.*, 2013, **160**, F883.
99 Y. Tao, S. D. Ebbesen and M. Mogensen, *J. Electrochem. Soc.*, 2014, **161**, F337.
100 Y. Tao, S. D. Ebbesen, W. Zhang and M. B. Mogensen, *ChemCatChem*, 2014, **6**, 1220.
101 X. Sun, M. Chen, Y. L. Liu, P. V. Hendriksen and M. Mogensen, *ECS Trans.*, 2013, **57**, 3229.
102 A. Hauch, A. Hagen, J. Hjelm and T. Ramos, *J. Electrochem. Soc.*, 2014, **161**, F734.
103 Y. Tao, Durability of the Solid Oxide Cells for Co-Electrolysis of Steam and Carbon Dioxide under High Current Densities, PhD thesis, DTU Energy Conversion, Risø Campus, Technical University of Denmark, Roskilde, Denmark, 2014.
104 P. Hjalmarsson, X. Sun, Y. L. Liu and M. Chen, *J. Power Sources*, 2014, **262**, 316.
105 A. Faes, H. L. Frandsen, M. Pihlatie and D. R. Goldstein, *J. Fuel Cell Sci. Technol.*, 2010, **7**, 051011.

Faraday Discussions

PAPER

Synthesis of mesoporous nanocomposites for their application in solid oxide electrolysers cells: microstructural and electrochemical characterization

M. Torrell, L. Almar, A. Morata and A. Tarancón*

Received 29th March 2015, Accepted 17th April 2015
DOI: 10.1039/c5fd00035a

Fabrication routes for new SOEC mesoporous nanocomposite materials as electrodes are presented in this paper. NiO–CGO and SDC–SSC are described as possible new materials and their synthesis and the cell fabrication are discussed. The obtained materials are microstructurally characterised by SEM, TEM and XRD, where the mesoporous structures are observed and analysed. The obtained samples are electrochemically analysed by I–V polarisation curves and EIS analysis, showing a maximum current density of 330 mA cm^{-2} at 1.7 V. A degradation of the cell performance is evidenced after a potentiostatic test at 1.4 V after more than 40 hours. Oxygen electrode delaminating is detected, which is most probably the main cause of ageing. Even taking into account the lack of durability of the fabricated cells, the mesoporous electrodes do not seem to be altered, opening the possibility for further studies devoted to this high stability material for SOEC electrodes.

Introduction

Nowadays renewable energy sources are an important aspect of the present energy scenario. In 2012, the world relied on renewable sources for around 13% of its total primary energy supply, and in 2013 renewables accounted for almost 22% of global electricity generation, a 5% increase from 2012.[1,2] These clean energy sources (*i.e.* solar, wind, geothermal, hydro) present an intrinsic problem of fluctuant production. These fluctuations do not match with the demand of the final consumers and industry. Considering this scenario, energy storage technology becomes a key factor in new energy routes to achieve a free or low carbon energy cycle. Unlike battery storage, chemical energy storage represents a relatively low cost alternative with high capacity and long term storage. Water electrolysis technology presents several advantages of hydrogen production as an energy vector and has shown as the preferred alternative, the so called power-to-

Catalonia Institute for Energy Research (IREC), Jardins de les Dones de Negre, 1, 08930-Sant Adrià de Besòs, Barcelona, Spain. E-mail: atarancon@irec.cat fuel technology. The reasons for this are the high energy density of hydrogen and the high efficiencies that electrolysis can achieve. This efficiency is especially high when high temperature electrolysers are used to reduce steam to hydrogen, *i.e.* solid oxide electrolyser cells (SOECs). Low temperature water electrolysers such as alkaline electrolysers or proton exchange membranes are already available in the market. However, these systems present lower efficiencies than SOECs mainly because their higher electrical energy demand and internal resistances. The amount of fuel produced per unit of injected electricity and the hydrogen production rates are remarkably higher in SOECs resulting in more cost-effective hydrogen generation. Moreover, the increase in the operation temperature allows avoiding the use of noble metals as catalysts.[3,4] Nevertheless, the exploitation of the favourable thermodynamics and small resistances of SOECs is up to now hindered by the lack of maturity of this technology. One of the major technical issues is the degradation that the functional materials used in the cell electrodes suffer.[5,6] Thermal mismatch,[7,8] redox cycling,[9] nickel agglomeration,[10,11] electrode delamination[12] are some of the main problems of durability on solid oxide electrolyser cells. A decrease in the operation temperature has been one of the main objectives of the scientific community to reduce the degradation processes previously, however temperature reduction in SOECs would promote an undesired effect on the final efficiency due to the increase in the electrical demand of the electrolyser.

It is well-known that some losses associated with the electrode processes can be minimised by engineering the triple phase boundary (TPB). The maximization of the TPB density can promote not only an improvement in the performance but also an improvement in the cell durability. One of the most interesting approaches involves the preparation of interpenetrated nanocomposites showing a high TPB density that avoids the accumulation of the current density in conduction paths.[13] The use of nanostructured composites improves different critical aspects of the cell performance. The nanostructured composite electrode ensures a more homogeneous distribution of both phases multiplying the active TPB with two main effects on the electrochemical behaviour of the electrolyser. There are more catalytic active points where the steam reaction can take place and the current density in each point is drastically reduced. In this way the performance and durability of the cell can be improved. Various strategies, such as electrodes impregnation, infiltration and gradual composition of composite electrodes, have been developed to improve the catalytic activity and stability of the SOEC electrodes.[8,14–16] Recent published papers by the authors have been devoted to explore a new family of electrodes based on the infiltration or impregnation of mesoporous scaffolds.[17–20] Good performance, thermal stability and, more interestingly, excellent durability under real conditions have been recently published in SOFC operation mode, opening the door for their use in more demanding applications such as high temperature electrolysis.

In this work, the use of nanostructured composites based on mesoporous materials is presented for fuel and oxygen electrodes for solid oxide electrolysers cells (SOECs). Ceramic-metal (cermet) mesoporous composites of Ni–CGO obtained after the reduction of the mesoporous NiO impregnated with CGO were employed as a fuel electrode while Samarium Strontium Cobaltite ($Sm_{0.5}Sr_{0.5}CoO_{3-\delta}$/SSC) impregnated scaffolds of mesoporous samarium doped ceria ($Ce_{0.8}Sm_{0.2}O_{1.9}$/SDC) were used as the oxygen electrode. These electrodes were implemented in electrolyte supported cells of $(Sc_2O_3)_{0.1}(CeO_2)_{0.01}(ZrO_2)_{0.89}$ (ScCeSZ). The mesoporous materials have been characterised and their synthesis route described. SOECs based on mesoporous electrodes were characterized by impedance spectroscopy and in operation conditions under different fuel compositions in order to characterise their performance.

Experimental method

Fabrication of electrolysis cells based on mesoporous electrodes

Solid oxide electrolyser cells based on mesoporous electrodes were fabricated using $(Sc_2O_3)_{0.1}(CeO_2)_{0.01}(ZrO_2)_{0.89}$ 150 μm commercial tapes (Kerafold) as an electrolyte. Scandia and ceria-doped zirconia electrolytes were used to achieve better performances of the electrolyser at temperatures below 900 °C. The use of multi-valence cations, *i.e.* cerium, as the dopant of the zirconia electrolyte has been proposed as a solution to avoid oxygen electrode interphase degradation.[12] A diffusion barrier of 2 μm of $Ce_{0.8}Gd_{0.2}O_{1.9}$ (CGO) was deposited between the oxygen electrode and the electrolyte by Pulsed Laser Deposition (PLD) to prevent the formation of insulating phases observed when zirconia is in contact with La- and/or Sr-based cathodes.[21–24]

Mesoporous powders of a cermet of nickel and gadolinium doped ceria were synthesized to fabricate the fuel electrode of the SOEC electrolyser. Impregnation of powder of silica KIT-6 with *Ia3d* symmetry was carried out using a Ni precursor of $(Ni(NO_3)_2\cdot 6H_2O)$ in an ethanol solution and annealed at 600 °C for 5 h to obtain the NiO crystalline desired mesoporous phase. Afterwards, the silica template was removed by employing a solution of NaOH 2 M at 70 °C. A detailed description of the whole process can be found elsewhere.[17] The obtained NiO replicas were impregnated with cerium nitrate and gadolinium nitrate salts with the stoichiometry necessary to obtain the $Ce_{0.9}Gd_{0.2}O_2$ (CGO) and crystallize inside the NiO mesoporous structure at $T = 1100$ °C, 3 h. The final obtained powder corresponds to a mesoporous interpenetrated structure of NiO : CGO (50 : 50 wt%). An ethanol based ink for airbrushing was prepared to deposit the fuel electrode. 30 μm thick NiO–CGO layers were airbrushed onto the ScCeSZ surface and attached by an annealing treatment at 1200 °C for 3 h. A commercial nickel ink (fuel-cell materials) layer was painted and attached at 900 °C for 1 h above the electrode to improve the current collection of the electrode.

Mesoporous powder of samarium doped cerium (SDC) were synthesised and attached to the CGO barrier layer. Mesoporous SDC was fabricated from KIT-6 hard template which was impregnated in an ethanol solution of the SDC precursors and annealed at 600 °C for 5 h. The silica template was removed using a NaOH 2 M solution cleaning process. The so-obtained powder was used to prepare an ethanol-based airbrushing ink to deposit 15 μm thick layers of porous SDC scaffold onto the CGO barrier deposited by PLD. The attachment was carried out at 1000 °C for 3 h in air. After the attachment step, the SDC was functionalized as a catalytic cathode by infiltration of samarium strontium cobaltite (SSC). SDC was impregnated with stoichiometric amounts of $Sm(NO_3)_3$, $Sr(NO_3)_2$, $Co(NO_3)_2\cdot 6H_2O$ previously mixed in a water and ethanol solution of 0.1 M. Three impregnation steps of 20 μL of solution were carried out in order to ensure good infiltration followed by a final annealing treatment at 800 °C for 2 h. SSC commercial ink layers were finally painted and attached at 900 °C for 1 h.

Microstructural and electrochemical characterisation

The electrodes were micro structurally characterised using a Zeiss Auriga scanning electron microscope. X-Ray diffraction studies were performed in a Bruker D8 Advance diffractometer in 2θ reflection mode from 20 to 80° using a copper $K_{\alpha 1}$ radiation. The electrochemical and fuel cell tests were carried out in a commercial ProboStatTM (NorEcs AS) sample holder placed inside a high temperature tubular furnace. A galvanostat/potentiostat with EIS capabilities (Parstat 2273, PAR) was employed for the electrochemical measurements. A frequency range from mHz to MHz was covered by applying ac voltages of 50 mV over OCV or operation voltages (1.4 V). The test station allows a compositional control of the water content in the fuel gas inlet mixture. Ultrapure water is evaporated by a controlled evaporator and mixing liquid delivery system (Bronkhorst-CEM system) that regulates and monitors the steam flow by a liquid flow controller.

Results

Structural characterisation of SOEC electrodes

Mesoporous NiO was microstructurally characterised by scanning electron microscopy (SEM) after the nanocasting synthesis process. Fig. 1a shows the powder microstructure, which is a direct replica of the mesoporous structure of the original hard template of KIT-6. The NiO powder presents nanoscaled porosity with periodicity and homogeneity throughout the observed powder. Low angle transmission XRD (not shown here) was also carried out to ensure the mesoporosity of the prepared powder. The XRD pattern presented in Fig. 1b corroborates the crystallization of CGO after impregnation of the NiO scaffold. The intimate interpenetration of the two phases was analyzed by the authors by TEM-EELS in a previous publication.[18] The SEM cross section image presented in Fig. 1c shows a 12 μm thick homogeneous and uniform porous electrode after the attachment of the interpenetrated NiO–CGO to the electrolyte at 1200 °C for 3 h.

Mesoporous SDC employed as the scaffold for the oxygen electrode was also characterised by SEM after the nanocasting synthesis based on the KIT-6 hard template and the subsequent KIT-6 removal step. The SEM micrograph presented in Fig. 2a shows the mesoporous structure of the as-synthesized SDC powder. The mesoporous structure of the grains and the periodicity of the pores can be clearly appreciated. The XRD pattern presented in Fig. 2b was acquired after the crystallization step of the SSC–SDC nanocomposite and confirms the formation of both compounds. A diffraction peak corresponding to a parasitic phase can be observed at $2\theta = 37°$. This was attributed to Co_3O_4 and it is not expected to significantly affect the final performance of the electrode. Fig. 2c shows the SEM cross section of the SDC–SSC layer after being attached to the CGO diffusion barrier. A uniform and porous layer of 18 μm in thickness is observed. The 2 μm thick PLD diffusion layer can also be clearly appreciated. The use of a dense and continuous layer of CGO avoids the formation of insulating layers at the operation temperatures.

Electrochemical testing

Mesoporous based electrolyser cells were electrochemically characterised using EIS and their performance evaluated by acquisition of *I–V* polarization curves

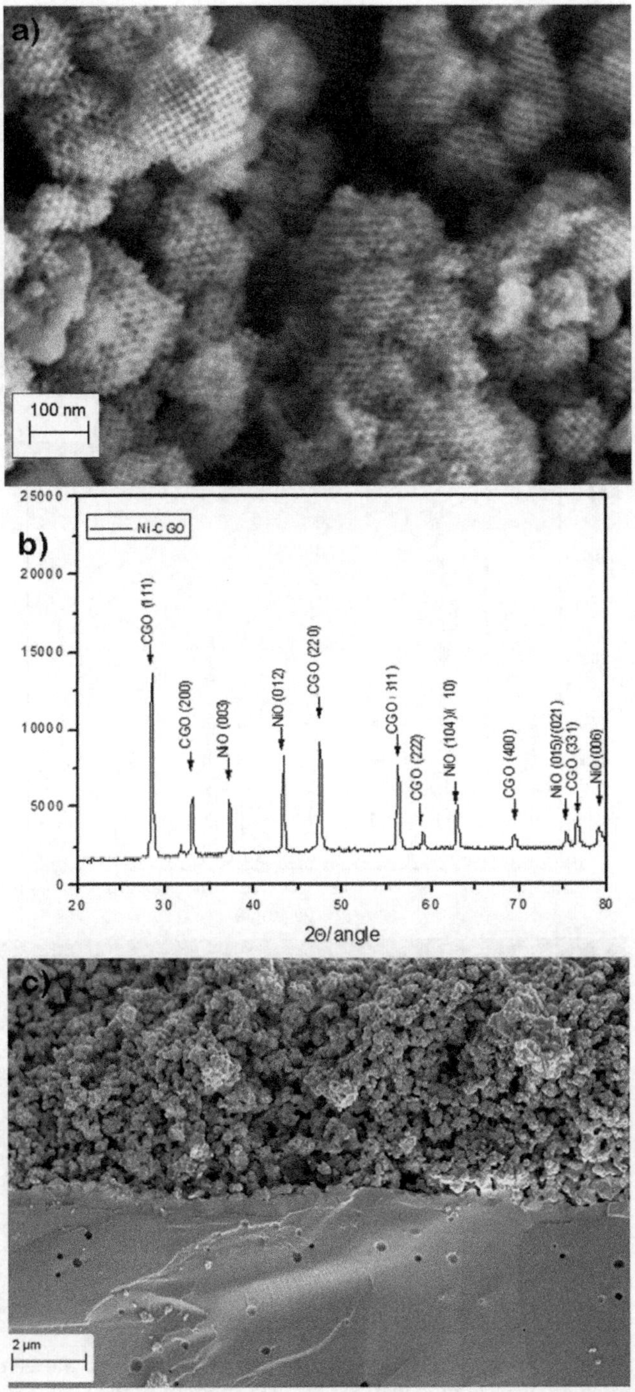

Fig. 1 (a) SEM images of the mesoporous as synthesised NiO powder. (b) XRD pattern of the NiO after CGO impregnation and calcination. (c) NiO–CGO mesoporous nanocomposite attached to the ScCeSZ electrolyte at 1200 °C.

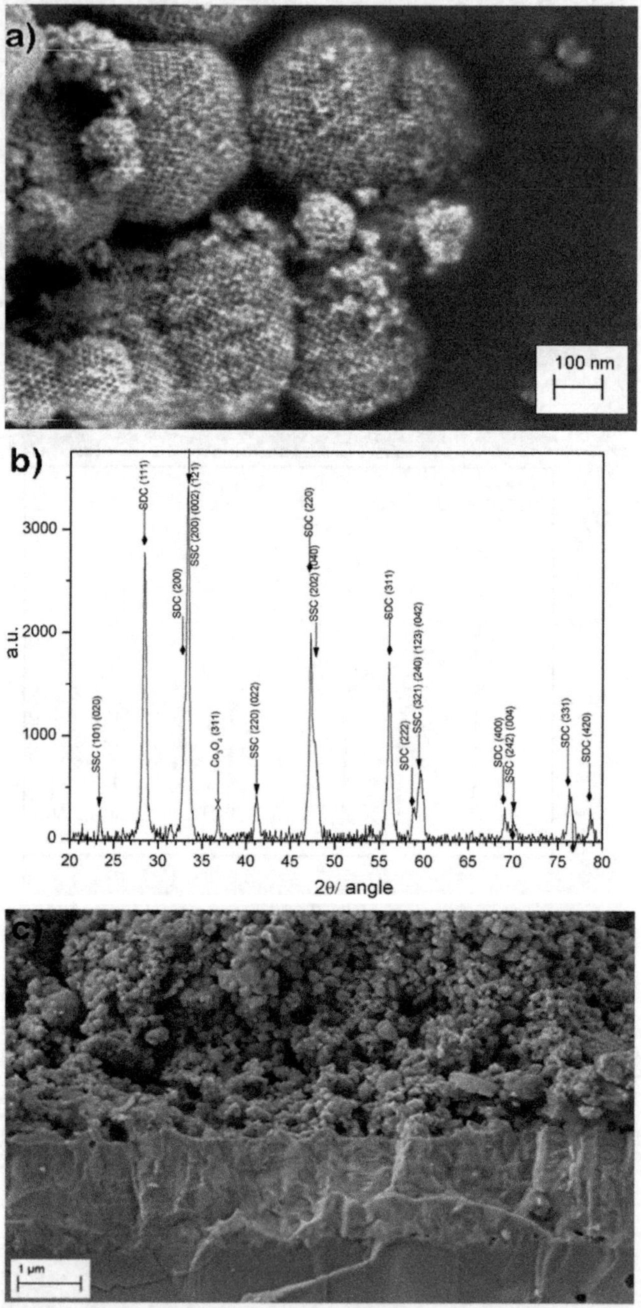

Fig. 2 (a) SEM images of the SDC powder before attachment to the electrolyte as a ceramic scaffold for its infiltration with SSC. (b) XRD pattern of the SDC–SSC composite. (c) SDC–SSC infiltrated mesoporous layer attached to the CGO diffusion barrier growth over the ScCeSZ electrolyte.

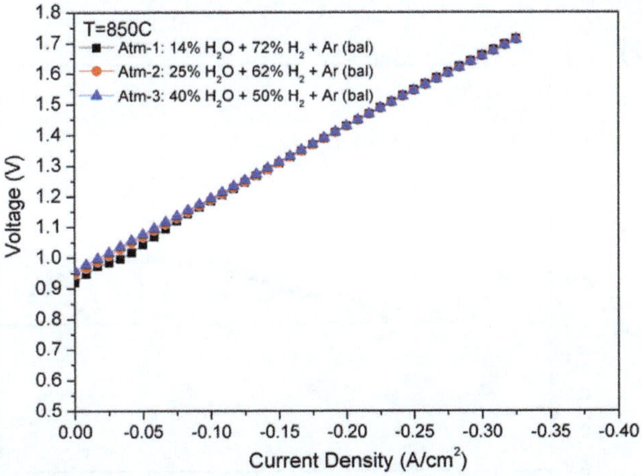

Fig. 3 *I–V* polarization curves at different fuel compositions obtained at 850 °C.

when operating under three different fuel compositions at 850 °C. The fuel compositions used were balanced with Argon, obtaining the following compositions: (i) 14%H_2O–72%H_2–14%Ar, (ii) 25%H_2O–62.5%H_2–12.5%Ar and (iii) 40% H_2O–50%H_2–10%Ar. Fig. 3 shows the injected current density as a function of the voltage and for the different fuels under study. No significant differences were observed on the solid oxide cell electrolyser performance with the different fuel compositions indicating a certain fuel flexibility of the SOEC. An open circuit voltage (OCV) of 0.94 V demonstrates gas tight chambers for the oxygen and the fuel. At 1.7 V, a maximum value of injected current of 330 mA cm^{-2} was reached. This injected current density compares well with previously reported values for SOECs.[25–29] M. Laguna-Bercero *et al.* reported about 400 mA cm^{-2} at 1.7 V with 80%H_2O using Ni-YSZ/ScCeSZ/Pt cells at 800 °C with ASR of 1.22 Ω cm^2.[26] Basile *et al.* present for 50%H_2O at 800 °C current densities below 300 mA cm^{-2} at 1.5 V.[27] At temperatures higher than 900 °C, X. Yue and J. T. S. Irvine injected more than 900 mA cm^{-2} at voltages below 1 V in a LSCM-CGO/YSZ/LSM with only 3% H_2O.[25] The authors recently published a symmetric cell where 650 mA cm^{-2} were injected at 850 °C and 1.6 V.[30] The linear shape of the *I–V* curve does not evidence a gas diffusion problem or an electrode high activation barrier.

A more detailed study of the different contributions to the total resistance of the cells was carried out by Electrochemical Impedance Spectroscopy (EIS). Fig. 4 shows the Nyquist plots corresponding to the EIS measurements carried out at OCV and 1.4 V. The spectra can be adjusted by employing a simple $R_s(R_1Q_1)(R_2Q_2)$ equivalent circuit. More than half of the total resistance is associated to in series contributions (Fig. 4a). This high resistance was ascribed to an insufficient adhesion between the electrolyte and the electrodes (the electrolyte itself showed a low resistance for independent measures). The high thermal stability of the mesoporous SDC previously demonstrated by the authors[18] could be in the origin of this phenomenon. In order to focus the analysis on the resistance strictly associated to the electrode performance, deconvoluted versions of the spectra were included in Fig. 4b and c for OCV and 1.4 V. Similar polarization resistances with an area specific resistance (ASR) below 1 Ω cm^2 were observed for the

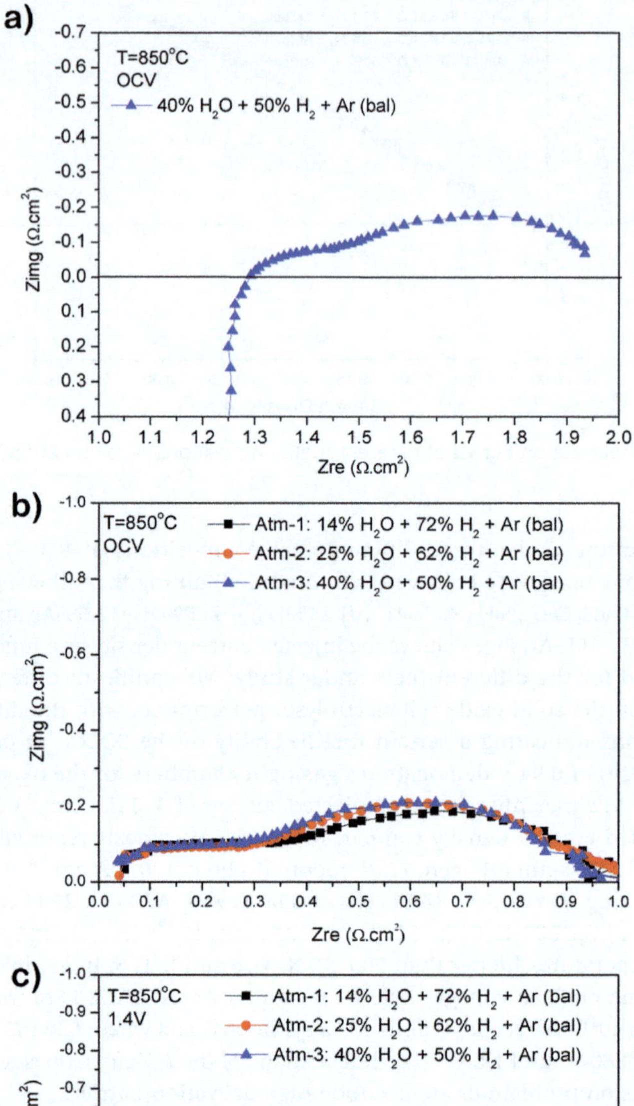

Fig. 4 (a) Complete Nyquist diagram of the EIS obtained at 40% of H_2O and 50% of H_2 (Ar bal.) operating at 850 °C (b and c) OCV and 1.4 V at different fuel compositions where inductance and ohmic resistance have been subtracted.

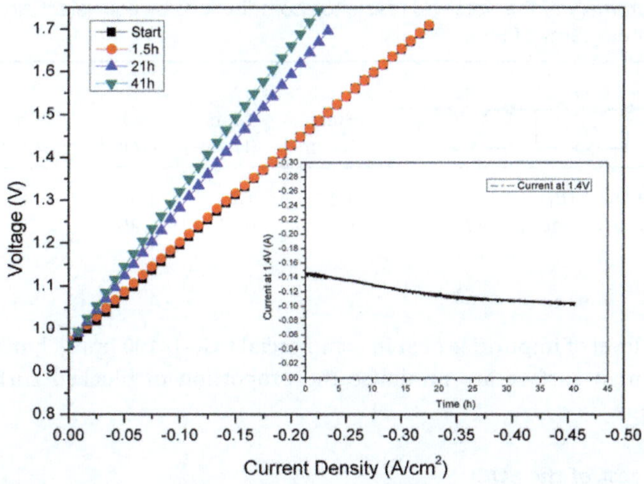

Fig. 5 Evolution of the mesoporous based electrolyser during the first 40 hours and studied through the evolution of the *I–V* curves and potentiostatic tests (inset of the figure).

different fuel compositions under study.[25–30] Although these are remarkable values, a better performance is expected after overcoming the adhesion issues previously mentioned. Further work is ongoing to reach a reasonable trade off in terms of attachment temperature to ensure a proper adhesion while maintaining the mesoporous nature of the electrodes. Another possible cause of the low performance could be the SiO_2 content remaining from the KIT-6 template after the alkaline washing step, and its negative effect on the electrode performance. Si content was evaluated by EDX after 4 alkaline washing steps at 70 °C with NaOH 2 M, followed by 3 washing steps with deionised pure water and was presented elsewhere.[19] The percentage of remaining Si is reduced to 1.16 wt%. This value is

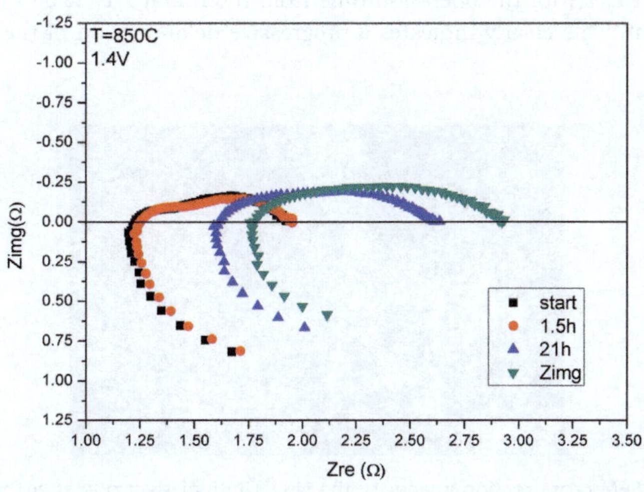

Fig. 6 Nyquist plots at 1.4 V measured at different operation times of the aging tests.

Table 1 Summary of the obtained resistance contributions by equivalent circuit fitting of the Nyquist diagrams of EIS

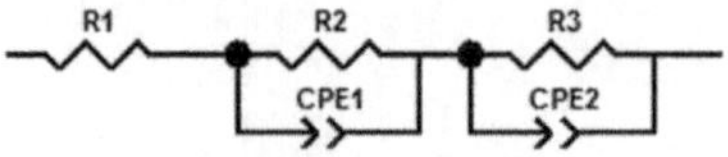	Start (Ω cm^2)	1.5 h (Ω cm^2)	20 h (Ω cm^2)	41 h (Ω cm^2)	ΔR (Ω cm^2)
Ohmic resistance (R_1)	1.17	1.21	1.56	1.71	0.54
Polarisation resistance ($R_2 + R_3$)	0.74	0.72	1.30	1.35	0.61

above the level of impurities of Si in commercial CGO (<100 ppm), but at the same time the high surface are minimise the proportion of blocked surface in the mesoporous.

Mid-term test of the SOEC

The SOEC cell was also tested during 40 h at 1.4 V in potentiostatic mode when subjected to an atmosphere of 25%H_2O–62.5%H_2–12.5%Ar at 850 °C. The evolution of the *I–V* polarisation curves with the operation time can be observed in Fig. 5. A significant drop in the electrolyser performance takes place during the first 25 hours of operation. The increase of the resistance shown in the gradient change in the *I–V* curves is corroborated with the continuous evolution of the current during the potentiostatic test (insert of Fig. 5). The current density during the potentiostatic test decreases from the initial 150 mA cm^{-2} to the final 110 mA cm^{-2} after 40 hours of testing.

Impedance spectra were acquired at the operation voltage of 1.4 V at different times during the experiment. Corresponding Nyquist plots are presented in Fig. 6. Similar spectra shapes are observed with time. Accordingly, the equivalent circuit employed before was used for deconvolution of the whole range of data. The values obtained by fitting with the equivalent circuit elements have been summarised in Table 1. As previously mentioned, ohmic contribution represents a significant amount of the total resistance. Moreover, this contribution continuously increases with the operation time from the initial 1.17 Ω cm^2 to the final 1.71 Ω cm^2. This clearly indicates a progressive delamination of the electrodes

Fig. 7 SE-SEM cross section images of the Ni–CGO fuel electrode after the aging test during 41 hours.

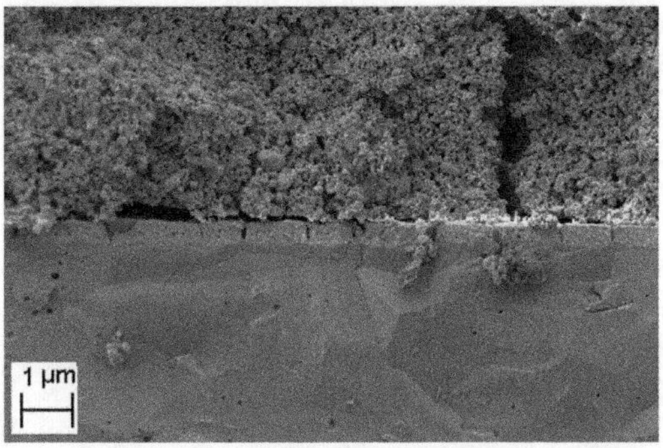

Fig. 8 SE-SEM cross section images of the SDC–SSC oxygen electrode after the aging test during 41 hours.

(probably due to the weakness suggested in the preceding section). The ASR associated to the electrode is correspondingly increasing from 0.74 Ω cm^2 to 1.35 after 41 hours of test.

In order to confirm the source of this degradation, a post mortem structural characterization was carried out. Microstructures of the fuel and oxygen electrodes after the mid-term test are shown on Fig. 7 and 8, respectively. While the internal microstructure of the electrodes was preserved, an evident degradation of the oxygen electrode/CGO-layer interface is present when compared with Fig. 2c. Therefore, the origin of this delamination will be likely associated to previously reported issues observed in the oxygen electrode. This behaviour was described by Virkar[12] as a consequence of the increase of the internal oxygen pressure within the diffusion barrier and oxygen electrode. These degradation has been already described on solid oxide electrolysers by different authors.[31,32] Once the contact points between electrode and electrolyte are affected, the electrode performance decreased as well as the actual active area decreases. Graves *et al.* recently proposed the reversible alternative use in SOFC–SOEC mode of the cell showing important improvements on the cell degradation resistance.[31]

Conclusions

Mesoporous nanocomposite electrodes have been presented as a novel material for electrodes in solid oxide electrolyser cells. Mesoporous nanocomposite electrodes of NiO–CGO mesoporous cermet have been synthesised by a wet impregnation route and deposited ion a ScCeSZ electrolyte as a fuel electrode and their fabrication is described. The SDC–SSC mesoporous nanocomposite has been obtained by infiltration of the scaffold attached to the electrolyte and used as oxygen electrode. The maximum value of current density injected at 1.7 V was 330 mA cm^{-2} at a temperature of 850 °C. A significant ohmic resistance was observed for all the fabricated cells. This is probably due an insufficient adhesion of the electrode to the electrolyte caused by the high thermal stability of the mesoporous materials previously shown by the authors. The optimisation of the electrode attachment temperatures seems to be a crucial factor especially when high thermal stability materials like this are used. Therefore, although the high stability of the mesoporous materials has great potential to reduce degradation of SOEC electrodes, further work is required to ensure the good adhesion that is required to take advantage of this.

Acknowledgements

The authors want to acknowledge MINECO for the CHISTE project (ENE2013-47826-C4-3-R) that funded the acquisition of presented results.

References

1 M. Ni, M. Leung and D. Leung, Technological development of hydrogen production by solid oxide electrolyzer cell (SOEC), *Int. J. Hydrogen Energy*, 2008, **33**, 2337–2354, 131–135(2012).

2 http://www.iea.org/aboutus/faqs/renewableenergy/.

3 A. Nechache, M. Cassir and A. Ringuedé, Solid oxide electrolysis cell analysis by means of electrochemical impedance spectroscopy: a review, *J. Power Sources*, 2014, **258**, 164–181.

4 A. Tarancón, C. Fábrega, A. Morata, M. Torrell and T. Andreu, *Materials for Energy, Chapter: power to fuel and artificial photosynthesis for chemical energy storage*, ed. D. Muñoz and X. Moya, Pan Stanford Publishing, 2015.

5 K. Wang, *et al.* A Review on solid oxide fuel cell models, *Int. J. Hydrogen Energy*, 2011, **36**, 7212–7228.

6 H. Yokokawa, H. Tu, B. Iwanschitz and A. Mai, Fundamental mechanisms limiting solid oxide fuel cell durability, *J. Power Sources*, 2008, **182**, 400–412.

7 K. S. Howe, *et al.* Performance of microtubular SOFCs with infiltrated electrodes under thermal cycling, *Int. J. Hydrogen Energy*, 2013, **38**, 1058–1067.

8 S. P. Jiang, Nanoscale and nano-structured electrodes of solid oxide fuel cells by infiltration: advances and challenges, *Int. J. Hydrogen Energy*, 2012, **37**, 449–470.

9 A. Faes, A. Hessler-Wyser, A. Zryd and J. Van herle, A Review of Redox Cycling of Solid Oxide Fuel Cells Anode, *Membranes*, 2012, **2**, 585–664.

10 B. Iwanschitz, L. Holzer, A. Mai and M. Schütze, Nickel agglomeration in solid oxide fuel cells: the influence of temperature, *Solid State Ionics*, 2012, **211**, 69–73.

11 D. Simwonis, Nickel coarsening in annealed Ni/8YSZ anode substrates for solid oxide fuel cells, *Solid State Ionics*, 2000, **132**, 241–251.

12 A. V. Virkar, Mechanism of oxygen electrode delamination in solid oxide electrolyzer cells, *Int. J. Hydrogen Energy*, 2010, **35**, 9527–9543.

13 S. P. Jiang, Nanoscale and nano-structured electrodes of solid oxide fuel cells by infiltration: advances and challenges, *Int. J. Hydrogen Energy*, 2012, **37**, 449–470.

14 E. P. Murray, T. Tsai and S. A. Barnett, Oxygen transfer processes in (La,Sr)MnO_3/Y_2O_3-stabilized ZrO_2 cathodes: an impedance spectroscopy study, *Solid State Ionics*, 1998, **110**, 235–243.

15 Y. M. Park, J. H. Kim and H. Kim, High-performance composite cathodes for solid oxide fuel cells, *Int. J. Hydrogen Energy*, 2011, **36**, 9169–9179.

16 Y. Matsuzaki, Electrochemical properties of a SOFC cathode in contact with a chromium-containing alloy separator, *Solid State Ionics*, 2000, **132**, 271–278.

17 L. Almar, *et al.* High-surface-area ordered mesoporous oxides for continuous operation in high temperature energy applications, *J. Mater. Chem. A*, 2014, **2**, 3134.

18 L. Almar, *et al.* High-Temperature Long-Term Stable Ordered Mesoporous Ni–CGO as an Anode for Solid Oxide Fuel Cells, *J. Mater. Chem. A*, 2013, **1**, 4531.

19 L. Almar, T. Andreu, A. Morata and A. Tarancón, Mesoporous NiO–CGO Obtained by Hard Template as High Surface Area Anode for IT-SOFC, *ECS Trans.*, 2011, **35**, 1647–1654.

20 L. Almar, A. Morata, M. Torrell, M. Gong, M. Liu, T. Andreu and A. Tarancon, Synthesis and characterisation of robust, mesoporous electrodes for solid oxide fuel cell, *Nano Energy*, 2015, Submitted.

21 R. Knibbe, J. Hjelm, M. Menon, N. Pryds, M. Søgaard, H. J. Wang and K. Neufeld, *J. Am. Ceram. Soc.*, 2010, **93**, 2877–2883.

22 M.-B. Choi, B. Singh, E. D. Wachsman and S.-J. Song, Performance of $La_{0.1}Sr_{0.9}Co_{0.8}Fe_{0.2}O_{3-\delta}$ and $La_{0.1}Sr_{0.9}Co_{0.8}Fe_{0.2}O_{3-\delta}$–$Ce_{0.9}Gd_{0.1}O_2$ oxygenelectrodes with $Ce_{0.9}Gd_{0.1}O_2$ barrier layer in reversible solid oxide fuel cells, *J. Power Sources*, 2013, **239**, 361–373.

23 P. Hjalmarsson, X. Sun, Y.-L. Liu and M. Chen, Influence of the oxygen electrode and inter-diffusion barrier on the degradation of solid oxide electrolysis cells, *J. Power Sources*, 2013, **223**, 349–357.

24 T. Klemensø, J. Nielsen, P. Blennow, A. H. Persson, T. Stegk, B. H. Christensen and S. Sønderby, *J. Power Sources*, 2011, **196**, 9459–9466.

25 X. Yue and J. T. S. Irvine, (La,Sr)(Cr,Mn)O_3/GDC cathode for high temperature steam electrolysis and steam-carbon dioxide co-electrolysis, *Solid State Ionics*, 2012, **225**, 131–135.

26 M. A. Laguna-Bercero, S. J. Skinner and J. A. Kilner, Performance of solid oxide electrolysis cells based on scandia stabilised zirconia, *CONAPPICE 2008*, Zaragoza, Spain, 24–26 September 2008, 2009, vol. 192, pp. 126–131.

27 A. Basile, *et al.* Gadolinium doped ceria-impregnated nickel–yttria stabilised zirconia cathode for solid oxide electrolysis cell, *Int. J. Hydrogen Energy*, 2011, **36**, 9420–9427.

28 Q. Liu, C. Yang, X. Dong and F. Chen, Perovskite $Sr_2Fe_{1.5}Mo_{0.5}O_{6-\delta}$ as electrode materials for symmetrical solid oxide electrolysis cells, *Int. J. Hydrogen Energy*, 2010, **35**, 10039–10044.

29 P. Hjalmarsson, X. Sun, Y.-L. Liu and M. Chen, Influence of the oxygen electrode and inter-diffusion barrier on the degradation of solid oxide electrolysis cells, *J. Power Sources*, 2013, **223**, 349–357.

30 M. Torrell, S. García-Rodríguez, A. Morata, G. Penelas and A. Tarancon, Co-electrolysis of steam and CO_2 in full-ceramic symmetrical SOECs: a strategy for avoiding the use of hydrogen as a safe gas, *Faraday Discuss.*, 2015, DOI: 10.1039/c5fd00018a.

31 S. Li, Y. Li, Y. Gan, K. Xie and G. Meng, Electrolysis of H_2O and CO_2 in an oxygen-ion conducting solid oxide electrolyzer with a $La_{0.2}Sr_{0.8}TiO_{3+\delta}$ composite cathode, *J. Power Sources*, 2012, **218**, 244–249.

32 C. Graves, S. D. Ebbesen, S. H. Jensen, S. B. Simonsen and M. B. Mogensen, Eliminating degradation in solid oxide electrochemical cells by reversible operation, *Nat. Mater.*, 2015, **14**, 239–244.

Faraday Discussions

System studies and understanding durability: general discussion

Raymond Gorte, Ian Metcalfe, John Vohs, Sune Dalgaard Ebbesen, Christopher Graves, San Ping Jiang, Dehua Dong, Tatsumi Ishihara, Rachael Elder, Beatriz Molero-Sánchez, Denis Cumming, Lisa Kleiminger, Steven McIntosh, Göran Lindbergh, John Bøgild Hansen, Viola Birss, Robert Kee, John Irvine, Bilge Yildiz, Joseph Hartvigsen, Mogens Mogensen, Marc Torrell, Shanwen Tao, Andreas Egger, Scott Barnett and Chen Xinbing

DOI: 10.1039/C5FD90075A

John Bøgild Hansen opened discussion of the paper by Joseph Hartvigsen: In your paper you state that you use an electricity price of 9.6 cents per kW h. Yet on your performance map (Fig. 3 in the manuscript), a production cost of approx. 1.55 per kg hydrogen is indicated at the thermoneutral voltage (100% LHV efficiency). How can that be reconciled when the energy requirement at 100% efficiency is 33.3 kW h kg^{-1} hydrogen?

Joseph Hartvigsen replied: As noted (after a brief period of thought) in the discussion, this was an editing error in the paper, which has since been corrected. The graphs were produced assuming 4.5 cents per kW h. This has been confirmed by referring back to the source files used to generate the graphs. However, much of the performance map domain is to the left (lower operating voltage) of the thermoneutral voltage, with supplemental heat being provided by fuel gas (e.g. natural gas, FT tail gas, etc.). Therefore it is possible to operate somewhat below thermal neutral voltage, supplementing lower cost heat for higher cost electricity and produce hydrogen at a cost below the electric price multiplied by 33.3 kW h kg^{-1} hydrogen. As shown in Fig. 5 in the manuscript, the full rolled-up cost of hydrogen (CAPEX, steam, supplemental heat, electric) is 1.87 kg^{-1} at the operating voltage in the 1.03 V to 1.16 V range, where the cost of electricity contribution is about 1.3 kg^{-1} and the cost of heat is 0.15 kg^{-1}.

Göran Lindbergh commented: A practical point of view of operating the electrolyser at an optimal point is that factors such as cell voltage and temperature will have a large impact on the lifetime of the system and consequently the capital cost. Is this, or could this aspect in some way be integrated in the model?

Joseph Hartvigsen responded: Yes, it has been widely shown that operating at high voltage, or more specifically high over potential increases the degradation rate relative to operation at a reduced voltage. However that effect is strongly dependent on the materials set and was therefore not considered in this analysis. If one had a correlation of lifetime *vs.* operating voltage it could very simply be factored into the CAPEX map.

This is an isothermal analysis, but sets of maps can be generated for a variety of operating temperatures, and again lifetime dependence of the operating temperature can be folded into the CAPEX. A graph of minimum hydrogen cost *vs.* temperature could be extracted from such a set of performance maps to select the optimal operating temperature.

It would be important to include these effects when considering the optimal operating point of a large scale electrolysis system.

Rachael Elder remarked: Have you also tried modelling co-electrolysis in the same way you modelled steam electrolysis?

Joseph Hartvigsen replied: We have not generated performance/economics maps for co-electrolysis. We published SOFC performance maps first for hydrogen, then natural gas reformate, followed by the techno-economics maps. Since the methodology has previously been extended to include CO/CO_2 with H_2/H_2O in fuel cell mode, it could easily be applied in co-electrolysis mode. The underlying Nernst *vs.* $H_2O–CO_2$ conversion curve is identical to steam electrolysis for co-electrolysis, with the assumption of the water–gas shift reaction equilibrium. Even without the shift equilibrium assumption, if the operating temperature is near ~818 °C, where the shift equilibrium constant crosses 1.0, the standard state free energies of formation for $H_2–H_2O$ and $CO–CO_2$ are identical and therefore the Nernst curve is identical. Co-electrolysis does shift the thermal neutral voltage from about 1.29 V for $H_2–H_2O$ to a weighted average of that and 1.46 V for $CO–CO_2$. Using Fischer–Tropsch for syngas conversion at a 2.1 : 1 H_2/CO ratio the mixture thermal neutral voltage is about 1.35 V. The stack ASR in steam electrolysis and co-electrolysis modes are virtually identical, so the main difference in the maps would be the need for more heat or higher voltage due to the higher thermal neutral voltage.

John Vohs opened discussion of the paper by Chen Xinbing: Ni is a good methanation catalyst. Did you observe any methane as a product for the conditions used in your experiments?

Chen Xinbing responded: Thanks for the question. In our experiments, no methane was observed despite the fact that Ni is an active catalyst for methanation. It can be explained in terms of the low total pressure of ~1 bar and the high temperature up to 800 °C during our study. Firstly, the low total pressure of 1 bar is unfavorable for the formation of methane. Secondly, methane is not very stable and usually decomposes into solid carbon and hydrogen at the high operation temperature of the solid oxide cells.

Sune Dalgaard Ebbesen remarked: When we (at DTU, Risø campus) do this kind of experiment (on Ni-based SOCs), the outlet gas composition is always in WGS/RWGS equilibrium. In Fig. 3 in your manuscript (impedance spectra), the performance is much lower for 80% CO_2 electrolysis than the other compositions. Please explain why this is bad compared to the other one? The performance decrease should be a gradual change, with a decreased activity with increasing CO_2 concentration. When supplying 80% CO_2 and 20% H_2, you will have a significant amount of steam in the system due to the WGS/RWGS equilibrium. Thinking about the steam content, please comment on your justification for the high resistance.

Finally, since the WGS/RWGS reaction is fast, there will be a significant amount of steam at the interface, and you cannot claim that the performance in 80% CO_2 and 20% H_2 is for pure CO_2 electrolysis, it is the performance for co-electrolysis.

Chen Xinbing answered: Thanks for the constructive suggestion. The outlet gas composition here does not represent the equilibrium composition at 800 °C. In the experiments, the composition of the outlet gas exiting the solid oxide cells was analyzed after the removal of steam by cooling. The temperature of the outlet gas decreased significantly in the gas pipe and during the dehumidification process. Indeed, it is not accurate to define it as pure CO_2 electrolysis because a certain amount of steam would probably be produced *via* the reverse water gas shift reaction when we directed 20% H_2 and 80% CO_2 into the solid oxide cells operated at 800 °C. This was also proven by the 8% of CO detected under open circuit conditions, as shown in Fig. 5. The increase in the resistance at high CO_2 concentration was caused by the slower diffusion kinetics of CO_2 over steam according to the study of C. Graves *et al.*[1] and Zhongliang Zhan *et al.*[2] I am still not very sure about the high resistance observed at high CO_2 concentration while we still have a certain amount of steam at the interface. But the possible reason may be that the diffusion barrier for the reactant gas with high CO_2 concentration may be sharply increased when the CO_2 concentration exceeds a certain value.

1 C. Graves, S. D. Ebbesen and M. Mogensen, *Solid State Ionics*, 2011, **192**, 398–403.
2 Z. Zhan, W. Kobsiriphat, J. R. Wilson, M. Pillai, I. Kim and S. A. Barnett, *Energy Fuels*, 2009, **23**, 3089–3096.

Sune Dalgaard Ebbesen also commented: In Fig. 4b in the manuscript, showing the CO_2/CO mixture *vs.* the steam/H_2 mixture, the thermodynamic parameters are close. The OCV should not change significantly with the increase of CO_2, it should be a gradual change of only a few mV (2–3 mV). The increase in OCV at high CO_2 (80% CO_2) cannot be due to WGS (as you have WGS/RWGS in all cases), and is most likely due to incorrect gas composition/flow (calibration of the MFC).

Chen Xinbing responded: Thanks for the comments. Indeed, the OCV value should not fluctuate with the CO_2 concentration. In the study, the gas composition was accurately controlled by the mass flow controller with high accuracy. So I am still not very sure of the reason and I will repeat the experiment to find the reason in the future.

Ian Metcalfe asked: If you feed 80% carbon dioxide and 20% hydrogen and only the RWGS occurs then 8% water production will also result in 8% carbon monoxide production. Is this an equilibrium mixture at 800 °C? I do not believe it is as the equilibrium constant for the WGS is only slightly above unity at this temperature.

Chen Xinbing responded: Thanks for the question. The outlet gas composition provided in the paper is not an equilibrium mixture at 800 °C. In the experiments, the composition of the outlet gas exiting the solid oxide cells, which is very sensitive to pressure and temperature, was analyzed after the removal of steam by cooling, as described in the experimental section. The temperature of the outlet gas varied significantly in the gas pipe and during the dehumidification process. Therefore, the outlet gas composition analyzed at room temperature does not represent the equilibrium mixture at 800 °C. At this stage, we cannot analyze the outlet gas composition at the operation temperature of 800 °C and we are planning to carry out this work by improving our test equipment.

Joseph Hartvigsen asked: If the shift reaction ($CO + H_2O \rightleftarrows CO_2 + H_2$) is in equilibrium, the Nernst potentials for CO/CO_2 and H_2/H_2O are identical. Substitution of the equilibrium expression for the shift reaction into the Nernst expression for either CO/CO_2 or H_2/H_2O degenerates the system to a single expression. There is no difference in the Nernst potential for either couple, so long as the shift reaction is in equilibrium. Another way to see this is to consider that $\Delta G_{CO-CO_2} + \Delta G_{shift} = \Delta G_{H_2-H_2O}$, and note that with the shift reaction at equilibrium $\Delta G_{shift} = 0$ and $\Delta G_{CO-CO_2} = \Delta G_{H_2-H_2O}$ and therefore the Nernst potential of both $CO-CO_2$ and H_2-H_2O reactions are identical. The typical Ni cermet electrode material is a very active shift catalyst at a typical 800 °C SOEC operating temperature. Even if the shift reaction were not in equilibrium, or if the system only contained CO/CO_2 and no H_2 or H_2O, thereby precluding the shift reaction pathway, the shift equilibrium constant is 1 at about 818 °C, so at temperatures close to 800 °C the standard state ΔG of the CO/CO_2 and H_2/H_2O reactions is still virtually identical, as is the resulting Nernst potential.

Chen Xinbing responded: Thanks very much for the detailed explanation. Indeed, the OCV values should not vary significantly with the CO_2 concentration. Now I am still not very sure why an increase in OCV was observed at high CO_2 concentration. I will repeat the experiment in the future and try to find where the problem lies.

Denis Cumming asked: As discussed during the session, open circuit voltages seem unusual. My question is regarding the high steam content used in the experiments. How sure are you that there was no condensation of any steam in the pipework downstream of the humidifier but before the cell?

Chen Xinbing responded: Thanks for the question. In the experiments, three measures were adopted to avoid steam condensation. First, the gas exiting the humidifier is further heated to ~150 °C. Second, a heating element, whose temperature was maintained at 150 °C, was coated outside the gas pipe to avoid steam condensation. Third, a layer of heat insulating materials was applied outside to avoid heat loss. During the experiments, the temperature of the mixed gas at the entrance of the solid oxide cells was measured to be ~115 °C. As a result, I do not think that serious steam condensation occurred in our study.

John Bøgild Hansen opened discussion of Shanwen Tao's paper: Have you calculated the ammonia equilibrium exiting your cells?

Shanwen Tao answered: Yes, we have calculated this. It is three orders of magnitude lower than the measured value. Please see the details on the calculation process below in our response to the question by Ian Metcalfe.

John Bøgild Hansen asked: Have you observed any NO_x in the exit gas in your experiments using wet air?

Shanwen Tao replied: It is certainly possible to form NO_x when wet air is used for electrochemical synthesis of ammonia. Unfortunately we did not measure NO_x in our experiments.

Raymond Gorte asked: In your measurements, you detected ammonia through neutralisation of sulfuric acid. Are you confident that you have actually made ammonia and that the neutralization is not associated with some other base? Can you be confident that you are actually making ammonia?

Shanwen Tao replied: We used a standard commercial ammonia detector based on ion selective electrode (ISE) technology. The meter was calibrated by a standard solution before the measurements. This is also a standard method used by other groups as well. Due to the use of ISE, interruption by other ions is possible but very unlikely.

Mogens Mogensen queried: 1. Did you do a blind experiment passing the input compressed air through the single cell chamber and the sulphuric acid without having a cell in it? 2. In the case of oxide ions taking part in the anodic reaction, which kind of reaction do you expect to see? Can you write the chemical formula?

Shanwen Tao responded: Unfortunately we did not carry out blind experiments. Thermodynamically, water and air will not react to form ammonia under conventional conditions. It is expected that no ammonia would be produced in blind experiments. However, it would be worthy to have a blind test in the future. In the case of O^{2-} ions taking part in the anode reaction, O^{2-} ions will lose electrons to form O_2 at the anode: $O^{2-} \rightarrow 1/2O_2 + 2e^-$. Under the circumstance, O_2 will be pumped from the cathode to the anode, reducing the overall efficiency, as mentioned in the paper. From this point of view, using a pure protonic conducting electrolyte is a better choice in order to synthesise ammonia directly from wet air due to the presence of oxygen.

Ian Metcalfe said: Have you compared your outlet ammonia mole fraction with the value that could be achieved at equilibrium?

Shanwen Tao replied: To make things simple, let us estimate the situation of wet N_2 only, according to our previous paper.[1]

For the reaction:

$N_2(g) + 3H_2(g) = 2NH_3(g)$.

At 400 °C, $\Delta G = 47.7786$ kJ mol^{-1}.

$\Delta G = -RT\ln K_p$,

$\ln K_p = -\Delta G/RT = -47.778.6 \times 1000/8.314/673 = -8.539$,

$K_p = 1.96 \times 10^{-4}$.

Let us presume the current density is 3 mA cm^{-2} (the current was fairly close to 3 mA cm^{-2} in wet N_2). The cell area was 1.281 cm^2. The current passing through the cell was:

$3 \times 1.281 = 3.843$ mA.

The electricity passing through the cell per second is:

$3.843/1000 \times 1 = 3.843 \times 10^{-3}$ As $= 3.843 \times 10^{-3}$ C.

The produced H_2 is:

$3.843 \times 10^{-3}/2/96485 = 1.99 \times 10^{-8}$ mol s^{-1}.

The volume of 1.99×10^{-8} mol H_2 is:

$V = nRT/p = 1.99 \times 10^{-8} \times 0.082 \times 673/1 = 1.1 \times 10^{-6}$ L

The diameter of the quartz is 30 mm. Length 300 mm. The volume of the quartz reactor is estimated as:

$V = 3.14 \times (30/2)^2 \times 300 = 211950$ mm^3 $= 212$ ml.

The gas flow rate was 50 ml min^{-1}. For 212 ml, it takes

$212/50 = 4.24$ min $= 4.24 \times 60 = 254.4$ s.

In 254.4 seconds, the produced H_2 will fill the whole reactor.

Therefore the H_2 concentration is:

1.99×10^{-8} mol s^{-1} $\times$ 254.4 s / 0.212 L $= 2.39 \times 10^{-5}$ mol L^{-1}.

The pressure of produced H_2 is:

$P = nRT/V = 1.99 \times 10^{-8}$ mol s^{-1} $\times$ 254.4 s $\times 0.082 \times 673/0.212$ L$= 1.32 \times 10^{-5}$ atm.

This is very low compared to the N_2 concentration of 97%. In other words, the N_2 partial pressure is fairly close to 0.97 atm.

Put the partial pressure of N_2 and H_2 in the equation

$K_p = [NH_3]_2/[N_2]/[H_2]^3$,

$1.96 \times 10^{-4} = [NH_3]^2/0.97/(1.32 \times 10^{-5})^3$,

$[NH_3] = 6.61 \times 10^{-9}$ atm.

At 400 °C, applied 1.4 V, wet N_2, the NH_3 formation rate was:

1.53×10^{-6} mol s^{-1} m^{-2} $= 1.53\times 10^{-10}$ mol s^{-1} cm^{-2}.

At 254.4 s, cell area of 1.281, the total produced NH_3 was:

$1.53 \times 10^{-10} \times 254.4 \times 1.281 = 4.99 \times 10^{-8}$ mol.

The corresponding pressure of NH_3 was:

$P = nRT/V = 4.99 \times 10^{-8} \times 0.082\times 673/0.212 = 1.3 \times 10^{-5}$ atm.

The corresponding concentration of ammonia is: $1.3 \times 10^{-5} \times 100\% = 0.0013\% = 13$ ppm.

Therefore the partial pressure of the produced ammonia (1.3×10^{-5} atm) is more than three orders of magnitude of the partial pressure of the equilibrium (6.61×10^{-9} atm).

This estimation also demonstrated that ammonia is likely to be formed through the one-step reaction between protons and nitrogen, rather than the two-step process, forming molecular hydrogen first, then reacts with N_2 to form NH_3 according to the Haber–Bosch process.

1 R. Lan, J. T. S. Irvine and S. W. Tao, *Sci. Rep.*, 2013, 3, 1145.

San Ping Jiang asked: What is the main reason for adding carbonates to CGO to form composite electrolytes?

Shanwen Tao answered: To avoid the thermal decomposition of the produced ammonia, we want to reduce the reaction temperature. Adding the carbonate is done to minimise the operating temperature of the cell.

Joseph Hartvigsen said: In the figure showing a schematic of cells, comparing two cells side by side, you stated that the oxygen is not a good conductor, as oxygen combines with hydrogen (single chamber configuration). What if you used a cell configuration where the oxygen was separated from the hydrogen? Have you looked at a two chamber cell configuration? If so are you liberating free hydrogen molecules in the product stream or does all produced hydrogen react with nitrogen to form ammonia? Or, in the more likely scenario where both molecular hydrogen and ammonia product results, how does the ratio of H_2/NH_3 compare to the case where you just pass hydrogen and nitrogen over your cathode as a catalyst? If the cathode material has any catalytic activity, when you flow hydrogen and nitrogen over your cathode material at appropriate flow rates it should approach equilibrium with respect to ammonia formation, even in the absence of electrochemical activity.

Because, in a single chamber cell configuration, oxygen evolved at the anode will react with hydrogen or ammonia formed at the cathode, wouldn't a separated chamber cell give better results?

Shanwen Tao replied: To use an oxygen ion conductor as the electrolyte for electrochemical synthesis of ammonia when wet air is used is not ideal, as some of the charges passing the cell will be used for pumping oxygen. We carried out experiments using both two chamber and single chamber cells. The two chamber cell gave better results. In the single chamber cell, some of the produced ammonia may be oxidised by oxygen. From this point of view, two chambers is a better choice.

The produced ammonia is very little. The majority of the charge is used for splitting water into H_2 and O_2 (electrolysis). As for the question about equilibrium, I have addressed this in detail below, in answer to the question by Raymond Gorte. The conclusion is that ammonia produced from direct chemical reaction between H_2 and N_2 is very limited.

Göran Lindbergh asked: Is the electrolyte carbonated in a molten form? In that case it is not an oxide ion conductor, but in principle a carbonate ion conductor. This has been proven by many years operation of molten carbonate fuel cells where the majority of the charge is carried by carbonate ions, not oxide ions. We used high temperature NMR to look for protons in molten carbonates and these composite electrolytes, but we have not measured any hydrogen at all, so it seems very unlikely that the electrolyte is a proton conductor at these temperatures.

Shanwen Tao answered: The electrolyte we used is a CGO–carbonate composite. The oxygen ion conduction is mainly from CGO and the CGO–carbonate interface. This has been mentioned in many previous reports in other groups.

It is believed that the proton conduction of the composite is mainly associated with the HCO_3 ions when hydrogen, water or a hydrogen-containing chemical is present, along with an applied current. I am a little surprised that you did not observed protons in the oxide–carbonate composite through high temperature NMR.

Raymond Gorte commented: The primary argument for why one can make NH_3 electrochemically is that electrochemical pumping can create very high effective partial pressures at the electrode surface. That does not seem to be the case in your system. What do you envision to be the driving force for the reaction in your system?

Shanwen Tao answered: To the best of my knowledge, the first report on solid state electrochemical synthesis of ammonia at atmospheric pressure was made by George Marnellos and Michael Stoukides.[1]

In this paper, the authors claim that 'greater than 78% of the electrochemically supplied hydrogen was converted into ammonia'. This is well above the equilibrium number for synthesis of ammonia from N_2 and H_2. We think the possible reason is that charged protons generated in the electrochemical cell are much more reactive than the neutral hydrogen molecules. This may allow the protons to break the strong nitrogen–nitrogen bonds, forming ammonia. In other words, charged protons are different from neutral hydrogen molecules. This may also cause another effect, called the non-faradaic electrochemical modification of catalytic activity (NEMCA effect), also known as electrochemical promotion of catalysis (EPOC).

Of course, direct experimental evidence is required to demonstrate this assumption.

1 G. Marnellos and M. Stoukides, *Science*, 1998, **282**, 98–100.

John Bøgild Hansen enquired: Do you have any hypothesis about which entity in your electrodes is responsible for the alleged ammonia synthesis activity and can you speculate anything about the reaction mechanism? The dinitrogen bond is very hard to activate.

Shanwen Tao replied: We believe that the perovskite oxide catalyst in the composite cathode may have some catalytic activity to ammonia synthesis although the activity is not very high. That is why the observed ammonia formation rate is low. The nitrogen–nitrogen bond is the 2nd strongest bond in chemicals and we have noticed it is not an easy job to break it.

John Bøgild Hansen communicated: You prepare your electrodes from nitrates and, although you calcine them at elevated temperature, could it be imagined that the ammonia stems from the nitrates as your cells start producing hydrogen?

Shanwen Tao communicated in reply: The precursors were nitrates. However, the oxide catalyst was formed after firing at 900 °C for 2 hours. Unreacted nitrates are not impossible but very unlikely. However, it is worth double checking in the future.

John Bøgild Hansen opened discussion of the paper by Scott Barnett: You use 6 hours in electrolysis mode followed by 6 hours in fuel cell mode. Graves *et al.* use

1 hr electrolysis and 4 to 5 hours in fuel cell mode and they observe no degradation.[1,2] Could you speculate on the maximum time in electrolysis mode you could operate without degradation in your cells without reversing the current, and what are the possible degradation mechanism(s) you could alleviate?

1 C. Graves, *ECS Trans.*, 2013, **57**, 3127–3136.
2 C. Graves, S. D. Ebbesen, S. H. Jensen, S. B. Simonsen and M. B. Mogensen, *Nat. Mater.*, 2015, **14**, 239–244.

Scott Barnett answered: We have carried out a 0.5 h electrolysis/0.5 h fuel cell test at 1.5 A cm^2 and still observed degradation. Presumably both the cycle time and the overpotential/current density will determine the degradation rate. For example, the critical time at which degradation starts will likely be shorter for higher overpotentials.

It appears there are at least two separate mechanisms: (1) oxygen bubble formation in YSZ and (2) rupture in the perovskite electrode due to chemical expansion/contraction with varying oxygen partial pressure.

John Vohs asked: Does deactivation still occur but is reversible on the time scale that you are cycling, or is it being prevented from occurring by the cycling?

Scott Barnett answered: I'm not sure whether there is really a difference, other than semantic, between these two.

Christopher Graves addressed the delegates: I agree there are open questions about the degradation mechanism and how much time can be spent in the high-overpotential electrolysis mode before it becomes impossible to recover and maintain stability. In my paper[1] we also reported the effect of different cycling periods, such as 5 hours in each mode, which showed faster degradation than the cycle with 1 h electrolysis mode and 5 h fuel cell mode. And at one point after significant degradation by the longer cycling, we ran the fuel-cell mode for 2 days, which mostly recovered the ohmic resistance. But the limits are an open question that should be further studied.

1 C. Graves, S. D. Ebbesen, S. H. Jensen, S. B. Simonsen and M. B. Mogensen, *Nat. Mater.*, 2015, **14**, 239–244.

Christopher Graves addressed Scott Barnett: Regarding the degradation mechanism and the evolution of the ohmic resistance (R_s) and polarization resistance (R_p) with time, we should keep in mind that damage/delamination that is causing R_s increase can also be causing R_p increase by the loss of the active area effect. Checking the ratio of R_p/R_s over time would be interesting for the high current (1.5 A cm^{-2}) test – if it remains the same then it is likely dominated by an active area effect. From the Fig. 4 data it is not possible to see the ratio only, because you have set the R_s to zero at $t = 0$.

Scott Barnett answered: This mechanism is likely occurring to some extent, and it will be worthwhile to plot the resistance ratio. However, it cannot be the dominant mechanism in all cases, for example at 0.9 A cm^{-2} where the ohmic and polarization resistances change in opposite directions.

Christopher Graves asked: How did you determine the electrode overpotential in the symmetric cell?

Scott Barnett responded: Since the cells remain essentially symmetric during the symmetric current cycles, the overpotential on each electrode is just the current density times the measured polarization resistance, divided by two.

Steven McIntosh asked: Is the additional ohmic resistance observed after degradation due to the extension of the 'electrolyte' into the LSM-free YSZ region of the porous electrode? Can you make a rough calculation of the ohmic loss in this region?

With regard to the reversibility, is it feasible that the LSM could migrate back into the porous region over this distance?

Scott Barnett answered: Increases in effective electrolyte thickness due to the interfacial loss of LSM should not impact the ohmic resistance, because the electrolyte is so thick (>0.5 mm) compared to the denuded zone (a few microns).

The LSM migration may indeed be reversible, and experiments observing interfacial composition with and without cycling would be interesting to explore this.

Joseph Hartvigsen said: Have you thought about whether the disappearance of the LSM in the composite electrode near the electrolyte and the incipient delamination shown by voids developing in the electrolyte are related, and are they particular to one material or even one cation in the electrode composition? Can you avoid that by changing the material system, or is it a mechanism that can be ascribed solely to over potential? Following your response that you have looked at other materials but do not have the micrographs, do they have a similar appearance (voids developing suggesting incipient delamination) or do you not know? We have seen similar problems resolved by changing the electrode composition. I do not fully understand the underlying mechanisms, but have empirically shown it can be countered by changing the materials.

Scott Barnett answered: It appears that these two different features appear under similar conditions, but they are not necessarily directly related. It appears that similar degradation can occur in other materials systems; as reported by a number of groups, electrolysis cells with LSCF electrode also degrade at sufficiently high current density/overpotential. My group also has preliminary results on LSCF–GDC electrodes, showing degradation versus current results somewhat similar to those for LSM–YSZ, but we do not yet have micrographs for comparison with the LSM–YSZ electrodes. These degradation mechanisms are probably general, and will occur in most materials under sufficiently extreme conditions (high current density/overpotential). Of course, some materials may be more resistant, providing reduced or no degradation for reasonable operating conditions.

John Irvine said: Comment on cycling: how much current comes from the fuel cell compared to how much comes from electrolysis? A net charge of zero would be expected for a reversible system. If one is removing chemicals the efficiency would need to be re-evaluated. Is the optimisation of a reversible fuel cell much different than that for electrolysis?

Scott Barnett answered: It is certainly reasonable to envision hybrid systems that combine electricity storage and fuel production, and experiments exploring this (such as those reported recently by Graves *et al.*[1]) are very valuable. In the experiments reported here, however, it was expedient to use symmetrical current cycles because it maintained the cells in symmetric conditions for simplified analysis.

1 C. Graves, S. D. Ebbesen, S. H. Jensen, S. B. Simonsen and M. B. Mogensen, *Nat. Mater.*, 2015, **14**, 239–244.

Christopher Graves commented: Regarding maintaining equal charge and discharge energy for reversible operation, it does not need to be equal unless you want to operate it exactly like a conventional rechargeable battery. In the long-term 2050 scenario in Denmark, for example, there will need to be an excess of electrolysis mode to produce fuel for the transportation sector, in addition to electrical grid load-balancing.

Mogens Mogensen said: How do you calculate your overpotential from the current density? Fig. 3 in the paper shows 2 clear arcs, *i.e.* at least 2 processes contributing to the electrode resistance. The low frequency one has such a high capacitance that it must be some sort of chemical capacitance, *i.e.* at a significant current density the electrodes probably change compositions, but differently. How do you then find the overvoltage?

Scott Barnett responded: The overpotential was calculated by multiplying the current density times the area-specific electrode polarization resistance, which was obtained from the impedance data and included both arcs. Does a response at ~100 Hz necessarily indicate chemical capacitance? I'm not convinced of that. In any case, the overpotential values quoted do not depend on the response frequency values, but rather are those expected during DC fuel cell operation.

San Ping Jiang commented: The significant change in the composition and microstructure of LSM oxygen electrodes after the reversed operation as shown in Fig. 6 could be related to the dependence of surface segregation, in particular Sr on the polarization. There are independent reports of the accelerated Sr segregation under anodic polarization and Sr depletion or incorporation into the lattice under cathodic polarizations. Our results on Cr deposition on LSM under anodic polarization conditions also indicate the accelerated Sr segregation under anodic polarization conditions (see this issue, DOI: 10.1039/c5fd00010f). The repeated Sr segregation/depletion can have a significant detrimental effect on the LSM composition and microstructure, particularly at the LSM/YSZ interface.

Beatriz Molero-Sanchez enquired: Regarding the high frequency (104 Hz) response, it was stated in the conclusions that it is associated with the degradation of the oxygen electrode, in particular that it appears to be associated with YSZ grain boundary transport losses. Is this a general statement seen in many samples? Also, have you seen this response with different materials or is this specific to this composite?

Also, since this could be a sign of oxygen bubble formation on grain boundaries, have you also seen delamination of the oxygen electrode in all the post-mortem cells that showed this particular response?

Scott Barnett replied: This response was associated with YSZ grain boundaries in a previous study of LSM-YSZ composite electrodes prepared under a range of firing temperatures. It is not clear whether this would transfer to other electrode materials. Extensive interfacial delamination was observed in cases where the increases in this high frequency response were large; in cases where the increases were small, some reduction in the EDS signal from the interface – probably a precursor of delimitation – was observed.

John Vohs opened discussion of Andreas Egger's paper: Please comment on the mechanism of chromium deposition and why it does not occur under SOEC anode conditions.

Andreas Egger replied: It has been stated in the literature that Cr-poisoning of SOFC cathodes under load is promoted by the reducing conditions on the SOFC side. This is to be expected, since Cr-poisoning is often associated with a reduction of volatile Cr^{+6}-species to less volatile Cr^{+3}-compounds (*e.g.* Cr_2O_3). Thus, Cr deposition mechanisms might be suppressed on anodically polarised La_2NiO_4.

Joseph Hartvigsen addressed Andreas Egger and John Vohs: Another explanation for observed differences in species migration through electrodes between SOFC and SOEC mode is vapor transport due to an oxygen potential gradient. The oxygen potential gradient is in the direction of the oxygen flux such that in SOFC mode there is a lower p_{O_2} at the electrode–electrolyte interface than in the flow field channel, while in SOEC mode there is a higher p_{O_2} at the electrode–electrolyte interface. In SOFC mode we have seen species transport and deposition due to the oxygen potential gradient. Consider this mechanism as a possible alternative explanation for the observed migration you have noted.

John Vohs asked: You indicated that the Ir impurity may have come from the thermocouple. If this were the case shouldn't you also have seen some Pt? Was Pt detected?

Andreas Egger answered: The S-thermocouple (Pt-Pt/Rh wires within Pt/Rh-mantle) does not contain any Ir, but is rather the source of small amounts of Rh found on the SOEC anode. Iridium may originate from other Pt-components in the setup (leads, grid), since Pt is sometimes alloyed with Ir to improve mechanical strength. Interestingly, no Pt was found on the anode side.

John Irvine said: What is in the gold paste as a glazing/binding agent? Could there be silicates or bismuth which could contaminate it?

Andreas Egger replied: We have performed XRF-analysis on the gold paste and identified Bi and Cd. It seems likely that the manufacturer added small amounts of a glass phase as a bonding aid, however, we could not detect any silicon by XRF and EDX.

San Ping Jiang queried: In your experimental setup, you placed Cr chips in the reactor upstream of the sample cell. Did you independently verify that such Cr placement will introduce Cr to the sample? Did you investigate the Cr deposition on LNO electrodes under cathodic polarization conditions? There are reports that LNO is a Cr-tolerant cathode under SOFC operation conditions. If this is the case, the reason for not being able to detect the Cr deposition on LNO would be due to the Cr tolerance properties of the LNO materials rather than the anodic polarization conditions.

Andreas Egger replied: We have tested dense samples of LNO under non-polarising conditions in a similar setup and found unambiguous evidence for Cr-deposition (see paper for references). Cr-poisoning of a porous LNO electrode under SOFC-polarisation is currently under investigation.

Tatsumi Ishihara enquired: During operation, did you check the change in valence number of the nickel? Ni is sensitively changed in valence number from +3 to +2, resulting in coarsening of the powder or a change in volume. Therefore, I recommend that not only CGO but also La_2NiO_4 should be analyzed carefully. Change in valence number could be easily done by XPS.

Andreas Egger replied: The valence charge of Ni in La_2NiO_4 is mainly +2, however, small amounts of Ni^{3+} (assuming small-polaron formalism) are present due to oxygen overstoichiometry in the compound. Oxidising conditions on the SOEC side might increase the fraction of Ni^{3+}. So far, we have not investigated the oxidation state of Ni in La_2NiO_4 after testing.

John Vohs asked: In Fig. 10 there is a sharp drop on the ohmic resistance during the first 10 hours of operation. What is the origin of this decrease?

Andreas Egger answered: The origin of this drop is unclear. It might be caused by a lowering of contact resistances in the cell when the current is drawn.

Raymond Gorte opened discussion of the paper by Sune Ebbesen: I was surprised that diffusion was so much more important in electrolysis than in fuel-cell mode. Based on the square-root dependence of diffusivities on molecular weight, one would expect that the diffusion gradients would be three times greater with electrolysis, but I was still surprised that the effect was so dramatic. Could you elaborate more on your results?

Sune Dalgaard Ebbesen responded: The effect of diffusion is reflected in, beside an increased diffusion resistance, an increased conversion resistance, as the conversion resistance reflects the change in Nernst potential due to conversion at the active electrode. The increase in total resistance is more affected by the change in conversion resistance. Consequently, the increase in total resistance is larger than the square-root dependency for diffusion.

John Irvine addressed Raymond Gorte: We see similar limitations on oxide electrodes, especially with steam. Water diffusion is rate limiting in electrolysis, with very clear starvation of steam on the oxide electrode. Oxides are definitely better performing with CO_2, where we see less sensitivity. This suggests some chemical interaction between water and metal oxide, but with CO_2 it will be a lot less likely to stick that way. It is interesting to see the limitation in steam, without seeing the same limitation for CO_2.

Raymond Gorte responded: Water could certainly be 'stickier' on the surface than CO_2 but I do not see how this would affect diffusive fluxes in a steady-state situation. Kinetic theory of gases in both molecular and Knudsen regimes shows that diffusivities are proportional to the inverse square-root of the molecular weight, implying that the diffusion of water should be faster than that of CO_2. Any surface diffusion process, if it occurs at all, will certainly be much slower than the gas-phase diffusion. Because diffusion fluxes are also proportional to the concentration gradient, it is important to compare water to CO_2 using the same feed concentrations. If rates are higher with CO_2 than with water for the same feed concentrations, I would have to conclude that diffusion is not the limiting process.

Joseph Hartvigsen asked: If the support structure is driving diffusive losses as shown in your presentation, and if a thinner, more open structure is mechanically problematic since the electrode also functions as the mechanical structure of the cell, might an anode (oxygen evolution electrode, *e.g.* SOFC cathode material) supported cell configuration be more appropriate for SOEC application?

In follow up to the comment that higher resistance is not all bad, since the resulting heat can supply the reaction endotherm, the counter argument is that any cell can be operated at thermal neutral voltage and match the endotherm with dissipative losses to achieve thermal balance in the cell. A lower resistance cell will produce more at that operating voltage and will require less interconnect area and occupy a smaller insulated volume. So there is no value to retaining resistance that could be avoided by an alternative cell architecture.

Is there anyone looking at oxygen electrode supported structures specifically for their mass transfer advantages in SOEC applications?

Sune Dalgaard Ebbesen answered: It is certainly true that supported oxygen electrodes would be preferable if only taking diffusion limitations into account. Furthermore, electrolyte supported cells would be preferable with respect to minimizing diffusion limitations. Oxygen electrode supported planar cells have been investigated in the literature, e.g. **ref. 1** below (and also by our group at DTU energy), although the mechanical stability and performance is not satisfying when compared with standard Ni–YSZ supported cells. This is partly because of lacking structural optimisation and the lower conductivity of the oxide material. Most of these investigations are performed at OCV or at very low current densities, which is normal for characterizing new materials. Characterising the cells at high current densities may show better performance for the oxygen electrode supported cells than for the Ni–YSZ supported cells. A better performing cell will of course always be advantageous to increase production, but for many cells, the performance is so good that operation at the thermal neutral voltage requires a high current density (in the order of 3 A cm^{-2}).[2] These high current densities may cause degradation of the cells, and an increased resistance at the electrolyte may not be disadvantageous when considering long term durability or performance of the cells. In fact many recent studies at high current density show better durability of electrolyte supported cells, than for fuel electrode supported cells, see for example publications by European Institute for Energy Research (EIFER) (I am sorry, I do not remember the exact references).

1 C. Bernuy-Lopez, R. Knibbe, Z. He, X. Mao, A. Hauch, K. A. Nielsen, *J. Power Sources*, 2011, **196**, 4396.
2 S. D. Ebbesen, S. H. Jensen, A. Hauch and M. B. Mogensen, *Chem. Rev.*, 2014, **114**, 10697.

Ian Metcalfe asked: Can you comment on the role that surface diffusional processes may have? I am not sure that such processes will always have a strong temperature dependence. As the temperature goes up the diffusion coefficient will increase but the coverage of the diffusing species will fall.

Sune Dalgaard Ebbesen replied: We have ourselves speculated on the role of surface diffusion,[1] but we are of the opinion that the surface diffusion can be neglected at the operation point of these cells.

1 S. Dalgaard Ebbesen, R. Knibbe and M. Mogensen, *J. Electrochem. Soc.*, 2012, **159**, F482.

Mogens Mogensen addressed Ian Metcalfe and Sune Dalgaard Ebbesen: We will have a clearer picture soon when we have finished our analysis. Until then, gas diffusion resistance has a temperature dependence very close to zero and surface diffusion resistance is clearly thermally activated. As the temperature dependence of the resistance of the low frequency arc here is close to zero or slightly negative, it is not originating from surface diffusion.

Robert Kee remarked: In addition to surface diffusion, another possibility is that finite-rate adsorption–desorption kinetics could be playing contributing roles. Albeit in a somewhat different context, we find that detailed heterogeneous kinetics in Ni-based SOFC anodes can have a significant effect on the measured low-frequency impedance spectra.[1] The ability of the surface to hold adsorbates produces capacitive behavior. The gas-transport and surface-kinetics EIS features were generally in the same frequency range, making the unique identification of impedance features difficult to unravel. If finite-rate surface kinetics were playing a role, then the surface area would be an important parameter. Thus, one would likely observe the derivative influences of porosity, pore dimensions, particle size, *etc.*

1 H. Zhu, A. Kromp, A. Leonide, E. Ivers-Tiffée, O. Deutschmann and R. J. Kee, *J. Electrochem. Soc.*, 2012, **159**, F255–F266.

John Irvine asked: With the laminar flow across your cell and about 15% difference in H_2 concentration, how does this influence the impedance response? Which parts of the impedance response are most sensitive?

Sune Dalgaard Ebbesen responded: It is correct that there is a significant difference in the H_2O/H_2 ratio (H_2 concentration) along the cell. The largest effect on the impedance spectra is seen for the contribution from conversion, and to a lesser effect, diffusion. Also the contribution from the Ni-TPB reaction is affected but only very little (see Fig. 2, 15 and 16 in the paper). The impedance contribution reflects an average of the entire cell. The average gas composition is therefore stated in the paper when compared to the impedance, to account for the concentration differences across the cell.

Lisa Kleiminger communicated: Have the authors investigated the effect of electrode porosity on the CO_2/CO electrolysis and co-electrolysis performance? The paper only presents data for H_2O/H_2 in Fig. 8. Based on diffusion coefficient calculations, CO_2/CO should be even more severely affected than H_2O/H_2. However, in co-electrolysis mode, would the authors expect behaviour similar to H_2O/H_2, CO_2/CO, or somewhere in between? *I.e.* if the CO_2/CO diffusion is slower than the H_2O/H_2 diffusion to reach the electrochemically active triple phase boundaries, would the electrochemical reaction mainly proceed *via* H_2O reduction followed by chemical reverse water gas shift reaction of the CO_2 with the electro-generated H_2? And how was this reflected in the diffusion resistances or, in the absence of experimental data, have the authors modelled the effect of porosity on gas diffusion for co-electrolysis gas compositions? At very low porosities, was the (reverse) water gas shift equilibrium still reached or did the low porosity restrict CO_2/CO penetration into the chemically active region to some degree during co-electrolysis? This is with reference to the statement that oxygen-electrode supported cells might be more suited. Would a thinner electrode still provide enough catalytic area for the (reverse) water gas shift reaction to reach equilibrium during high performing co-electrolysis? This is under the assumption that the kinetics of CO_2 reduction are reported to be slower than for H_2O reduction (for Ni-YSZ, see DOI: 10.1039/c5fd00048c); thus, some CO_2 reduction is expected to take place *via* electrogenerated H_2 at 850 °C, and not just direct CO_2 reduction. Currently, (R) WGS is assumed to be fast, but if the electrode thickness is drastically reduced, would this still be fast enough?

Sune Dalgaard Ebbesen communicated in reply: The effect of porosity was also investigated for CO_2/CO mixtures and for $CO_2/COH_2O/H_2$ mixtures, and the effect is more pronounced in CO_2/CO mixtures with $CO_2/COH_2O/H_2$ mixtures in between. The effect of gas composition, although not at varying porosity, was previously shown in *J. Electrochem. Soc.*[1] Based on the results presented in ref. 1 we concluded that electrolysis of both H_2O and CO_2 occur, and that the gas composition is always in WGS equilibrium. The WGS reaction is fast at the operating temperature, and the gas composition is expected to be in WGS equilibrium, although starvation may occur. Even though the WGS reaction was as slow as the electrolysis reaction, the amount of TPB-boundaries to drive the electrolysis reaction will inevitably be much lower that the active sites for the WGS reaction, and it should therefore be acceptable to expect that the WGS reaction is in equilibrium.

1 S. Dalgaard Ebbesen, R. Knibbe and M. Mogensen, *J. Electrochem. Soc.*, 2012, **159**, F482.

Mogens Mogensen opened discussion of Marc Torrell's paper: In your mesoporous powder, you have 5 nm pores in the powder. What is the particle size distribution of your powder? The size of the particles is very important, because, if the particle size is in the range of say 0.5 to 1 micron then it will be almost impossible to sinter the powder into a strong ceramic without closing the nano-pores by sintering. You still have nano-pores after sintering, so is your material badly sintered?

Marc Torrell replied: As can be seen in Fig. 1a and 2a in the paper, the particle size distribution is in the range of 1 micron. In Fig. 1c and 2c, it can be observed that the material is well sintered and we obtain good electrode layers. As stated, there is a loss in the nanoporosity of the material during the sintering process, but the most important thing is that the nanocomposite keeps the interpenetration between both materials (Ni-CGO or SCC-SDC), which ensures percolation, high TPB densities and thermal stability. The microstructural improvement of the interpenetration between the two phases compared with a standard electrode was shown in the presentation (Fig. a below). The microstructural improvements (Fig. 1 below) and thermal stability (Fig. 2 below) of the mesoporous ceramic powders have already been published in *J. Mater. Chem. A*,[1,2] and are referenced in the manuscript.

1 L. Almar, T. Andreu, A. Morata, M. Torrell, L. Yedra, S. Estradé, F. Peiró and A. Tarancón, *J. Mater. Chem. A*, 2014, **2**, 3134.
2 L. Almar, B. Colldeforns, L. Yedra, S. Estradé, F. Peiró, A. Morata, T. Andreu and A. Tarancón, *J. Mater. Chem. A*, 2013, **1**, 4531.

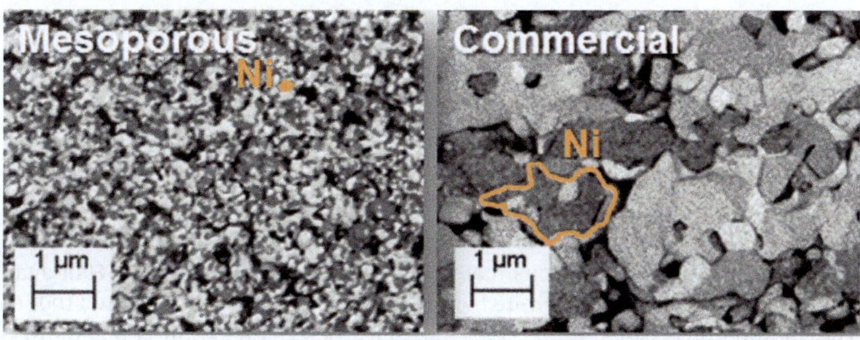

Fig. 1

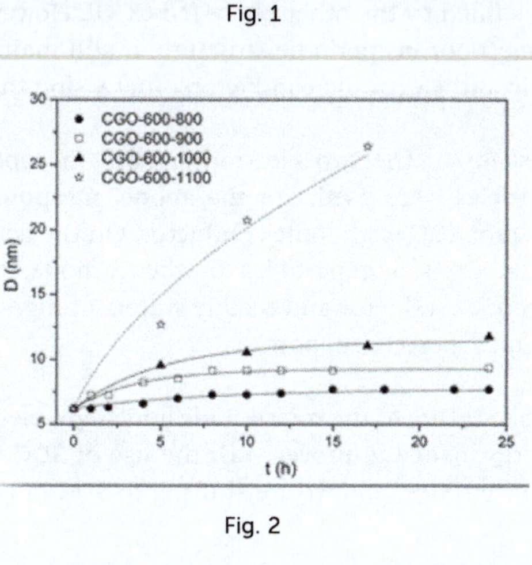

Fig. 2

John Irvine addressed Marc Torrell and Mogens Mogensen: I do not think there are issues with the processing. It looks like the authors have processed the materials well and they have a good microstructure; you would not get that with an agglomerated powder.

Mogens Mogensen replied: The microstructure seen on the micrographs does not look convincing to me, and I think that the cell performance indicates microstructure problems.

John Vohs remarked: You concluded that the high ohmic loss was due to poor attachment to the electrode. If this were the case, I would expect the electrode to also not perform very well. Are there any other potential explanations for the high ohmic loss?

Marc Torrell responded: Another possible explanation could be the electrolyte or the current collection. We have already demonstrated this and it is also stated in other publications that the electrolyte performs well. There is no evidence surrounding the current collection of the electrodes, but it seems that the most probable factor is the poor attachment. In fact the performance of the electrode is not outstanding in the reported results.

Raymond Gorte remarked: You have a remarkable material, but the real question is whether the high temperature application is the best place for it. How does the structure evolve with temperature? How do you stabilize the structure?

Marc Torrell replied: Thank you for your very constructive comments. We sincerely believe in the application of the mesoporous material for IT or HT SOFC and SOEC. In SOFC mode we achieve degradation below 1% in 1000 h. We characterised the evolution of symmetric cells with the anode and also with the cathode material and the ASR does not virtually increase or, in the case of the infiltrated electrode, improves with the time of operation during the first 1000h. Moreover, the CGO mesoporous material showed, under Williamson–Hall analysis of the XRD peaks, a self-limited grain growth at temperatures below 1100 °C. The mesoporous structure is stable at SOFC/SOEC operation temperatures, especially when it is filled by the other phase (Ni-CGO). However, even if we have a decrease in the mesoporous periodic structure, it still maintains the interpenetration of the ionic and electronic conductors, increasing the active sites.

Dehua Dong asked: In the two electrodes, both mesoporous support and infiltrated nanoparticles were used. For the anode, mesporous SDC supported SSC nanoparticles, and SDC is an ionic conductor. On the contrary, mesoporous NiO held infiltrated CGO nanoparticles in the cathode. Ni formed by NiO reduction is an electric conductor and readily sinters at high temperatures. Why not choose a CGO mesoporous support?

Marc Torrell replied: Due to the mixed ionic and slight electronic conductivity of the SDC, the performance improves with the use of SDC. Initially we started with CGO in both electrodes, and we are shifting to SDC in both electrodes. The presented work was performed when the SDC was used for the oxygen electrode (higher requirements) and now we are also using the SDC for the fuel electrode.

Joseph Hartvigsen queried: In the figure showing the processing steps, where do you etch out the silica? As shown they all look like deposition steps. Have you tried it the other way, that is to use CGO as the scaffold, and put nickel on the resulting CGO structure, or is nickel always the structure with CGO decorating the nickel backbone?

Marc Torrell responded: The silica is removed after the calcination of NiO, before impregnating with the CGO precursors. We follow this approach because we had better results than by doing it the other way round. It is not only a CGO decoration, it is an interpenetration of both structures.

Christopher Graves questioned: Could you clarify why it did not work to prepare the electrodes in the opposite order, with a ceria backbone and then infiltrating the nickel?

Marc Torrell replied: It seems that in the opposite order the mesoporous structure is more dense and it is difficult to infiltrate the NiO precursors on the CGO structure. The results were that the interpenetration between the phases was worse if we started by the CGO. But, of course, it is possible to start by the ceramic backbone as we have done with other materials.

Christopher Graves asked: Coming back to the high ohmic resistance that you report is possibly due to adhesion issues, you mentioned an active area shadowing effect due to poor adhesion that does not seem to affect the polarization resistance. But do you have a reference polarization resistance value without the poor adhesion problem? Isn't it possible that the area effect is affecting it and your true electrode performance is much better than the 1 Ω cm^2 measured at 850 °C (Fig. 4 in your paper)? 1 Ω cm^2 isn't very impressive for such mesoporous structured electrodes with a high concentration of reaction sites.

Marc Torrell answered: I agree, that is true. The only doubt is that I would expect a higher evolution of the polarization resistance for the depletion, and this is not observed. However I think that with a better adhesion the polarization resistance will be better than this 1 Ω cm^2.

Bilge Yildiz asked: It appears that adding the mesoporous frame to the electrolyte is challenging, and gives rise to issues related to adhesion. What are some new ways that you plan for making this interface adhesion better?

Marc Torrell replied: Maybe we can change the used powder, or add some adhesion layers below the mesoporous functional layer.

Viola Birss commented: Because you have a mesoporous structure, presumably also in your electrode layers, you do now have the chance to measure the real surface area of your electrodes using gas sorption methods, as the areas may be high enough to monitor. As discussed earlier in this meeting, differences in real surface areas may be part of the small differences seen in activities for many SOFC/SOEC electrodes. However, most of these ceramically processed electrodes do not have sufficiently high areas to measure by gas sorption methods. It is also critical that the mesoporosity is examined in the electrode structure and compared to the powders, to determine if the mesoporosity you see for your powders is retained after electrode preparation, sintering *etc.* To repeat, if your high mesoporous surface areas are retained in the electrodes, you would have a chance to try to correlate different activities seen at different electrodes with differences in real surface area.

San Ping Jiang asked: Have you tried to use mesoporous silica templates with large mesopores? Would the pore size of the silica templates affect the final distribution of the infiltrated Ni and YSZ after the high temperature sintering?

Marc Torrell replied: We have always used the same silica template, but we can tune the porous size through a hydrothermal treatment of the silica template synthesis. This of course can change the distribution of the Ni-CGO, but it is especially interesting for interpenetrating other materials such as LSM and others.

Faraday Discussions

PAPER

Chromium deposition and poisoning of $La_{0.8}Sr_{0.2}MnO_3$ oxygen electrodes of solid oxide electrolysis cells

Kongfa Chen,[a] Junji Hyodo,[b] Aaron Dodd,[c] Na Ai,[a] Tatsumi Ishihara,[bd] Li Jian[e] and San Ping Jiang*[a]

Received 25th January 2015, Accepted 2nd March 2015

DOI: 10.1039/c5fd00010f

The effect of the presence of an Fe–Cr alloy metallic interconnect on the performance and stability of $La_{0.8}Sr_{0.2}MnO_3$ (LSM) oxygen electrodes is studied for the first time under solid oxide electrolysis cell (SOEC) operating conditions at 800 °C. The presence of the Fe–Cr interconnect accelerates the degradation and delamination processes of the LSM oxygen electrodes. The disintegration of LSM particles and the formation of nanoparticles at the electrode/electrolyte interface are much faster as compared to that in the absence of the interconnect. Cr deposition occurs in the bulk of the LSM oxygen electrode with a high intensity on the YSZ electrolyte surface and on the LSM electrode inner surface close to the electrode/electrolyte interface. SIMS, GI-XRD, EDS and XPS analyses clearly identify the deposition and formation of chromium oxides and strontium chromate on both the electrolyte surface and electrode inner surface. The anodic polarization promotes the surface segregation of SrO and depresses the generation of manganese species such as Mn^{2+}. This is evidently supported by the observation of the deposition of $SrCrO_4$, rather than $(Cr,Mn)_3O_4$ spinels as in the case under the operating conditions of solid oxide fuel cells. The present results demonstrate that the Cr deposition is essentially a chemical process, initiated by the nucleation and grain growth reaction between the gaseous Cr species and segregated SrO on LSM oxygen electrodes under SOEC operating conditions.

[a]Fuels and Energy Technology Institute, Department of Chemical Engineering, Curtin University, Perth, WA 6102, Australia. E-mail: s.jiang@curtin.edu.au; Fax: +61 8 9266 1138; Tel: +61 8 9266 9804

[b]Department of Applied Chemistry, Faculty of Engineering, Kyushu University, Motooka 744, Nishi-ku, Fukuoka, 819-0395, Japan

[c]Centre for Microscopy, Characterisation and Analysis, The University of Western Australia, WA 6009, Australia

[d]International Institute for Carbon-Neutral Energy Research (WPI-I2CNER), Kyushu University, Motooka 744, Nishi-ku, Fukuoka, 819-0395, Japan

[e]Center for Fuel Cell Innovation, School of Materials Science and Engineering, State Key Laboratory of Material Processing and Die & Mould Technology, Huazhong University of Science and Technology, Wuhan 430074, China

1. Introduction

Solid oxide electrolysis cells or solid oxide electrolyzers (SOECs or SOEs) operating at high temperatures of 700–1000 °C are highly efficient to store the electrical energy generated by renewable sources such as solar and wind power into chemical energy of fuels such as hydrogen, syngas and methane. SOECs are reversible solid oxide fuel cells (SOFCs), and state-of-the-art SOFC materials such as nickel-based cermets, yttria-stabilized zirconia (YSZ) electrolyte and lanthanum strontium manganite (LSM) perovskites can be directly applied as electrode and electrolyte materials in SOECs.

A major issue for the development of SOECs is the significant performance degradation during long-term operation,[1,2] and the decay of oxygen electrodes has been regarded as the major cause, particularly at high currents.[3,4] The degradation of oxygen electrodes has been extensively studied and several degradation/failure modes of the oxygen electrodes have been proposed. The most common mode of failure is the occurrence of delamination at the oxygen electrode/electrolyte interface,[5–7] and it is generally believed that the main reason is the buildup of high internal oxygen pressure within the electrolyte close to the oxygen electrode/electrolyte interface.[8,9] On the other hand, Kim *et al.* suggested that the densification of porous LSM–YSZ oxygen electrodes caused by cation migration at a high anodic current is responsible for the complete delamination of oxygen electrodes.[10] Keane *et al.* proposed that the formation of highly resistive $La_2Zr_2O_7$ at the electrode/electrolyte interface is attributed to the LSM electrode delamination.[11] Our studies have shown that the delamination of LSM oxygen electrodes is mainly due to the disintegration or breaking of LSM particles and formation of nanoparticles at the electrode/electrolyte interface.[12,13] The formation of nanoparticles at the electrode/electrolyte interface is most likely caused by the LSM lattice shrinkage and thereby formation of local tensile strains due to the forced oxygen ion migration from the electrolyte into the LSM grains under the anodic polarization potential driving force.[13] For the (La,Sr) $(Co,Fe)O_3$ (LSCF) oxygen electrodes after the electrolysis test for 9000 h, the perovskite structure is partly demixed, accompanied by microstructure change and formation of Co_3O_4; and a dense $SrZrO_3$ layer is also observed at the Gd-doped ceria (GDC) interlayer/YSZ electrolyte interface.[14,15] Separation of GDC interlayer from the YSZ electrolyte has also been reported.[16,17]

A chromia-forming stainless steel alloy is the leading interconnect material for planar SOFCs due to its ease of fabrication, high thermal and electrical conductivity and low cost.[18] However, volatile Cr species such as CrO_3 and $CrO_2(OH)_2$ from the Cr_2O_3 protective scale on the metallic interconnect, deposit and poison the electrochemical activity and stability of LSM electrodes under SOFC operating conditions.[19] It is well known that the LSM electrodes are vulnerable to attack by chromium, forming products such as Cr_2O_3 and $(Cr,Mn)_3O_4$ spinel primarily at the electrode/electrolyte interface region and on the electrolyte surface under the cathodic polarization conditions.[20–22] Similar to that of SOFCs, Fe–Cr alloys are also commonly used as interconnect materials for SOECs.[23] Very different from the large amount of literature on Cr deposition and poisoning of SOFC cathodes,[19,24] studies on Cr poisoning and deposition on SOEC oxygen electrodes are very rare. Our early study on Cr poisoning of LSM electrodes showed that under anodic polarization at 200 mA cm^{-2} and 900 °C, fine grains of chromium oxide are deposited on the YSZ electrolyte surface.[21] Most recently, we studied Cr deposition and poisoning of LSCF oxygen electrodes of SOECs and showed that the deposition of chromium species such as $SrCrO_4$, Cr_2O_3 and Cr_2O_5 is preferential at the outermost surface of the LSCF electrodes, leading to a substantial increase of electrode polarization and ohmic resistances.[25] For a bi-layered oxygen electrode consisting of a (La,Sr)CoO_3 (LSC) contact layer, Cr deposition occurred throughout the whole LSC contact layer after polarization under SOEC conditions for 2000 h and transportation of Sr and Co cations to the electrode's outermost surface and formation of La–Cr–O phases were observed.[26] However, Cr deposition and interaction with oxygen electrodes such as LSM is largely unknown under SOEC conditions despite the significance of LSM in the development of SOEC technologies.

In this paper, the activity and stability of LSM oxygen electrodes is studied for the first time under SOEC operating conditions in the presence of an Fe–Cr interconnect. The results show the deposition of Cr species at the electrode/electrolyte interface under SOEC operating conditions and the presence of Cr species accelerates the degradation and delamination of LSM oxygen electrodes.

2. Experimental

Stoichiometric $La_{0.8}Sr_{0.2}MnO_3$ (LSM, Fuel Cell Materials, USA) oxygen electrodes were prepared on 0.18 mm thick dense YSZ wafers (Huazhong University of Science and Technology, China) by slurry coating, and sintered at 1150 °C for 3 h. The thickness of the LSM coating was 15–30 μm and the electrode surface area was 0.5 cm^2. Electrochemical measurements of LSM oxygen electrodes were carried out in a three-electrode configuration,[27] using a Gamry ref. 3000 Potentiostat. Chromia-forming alloy (RA446, 23–27% Cr, Rolled Alloy Co.) was used as the Fe–Cr interconnect in this study. Alloys were machined into coupons (12 mm × 12 mm × 5 mm thick). One side of the coupon was cut with channels (1.2 mm × 1.2 mm deep) with holes in the center. Air (industrial grade) was distributed to the electrode surface through channels by an alumina tube, simulating the air flow situation in an actual cell stack of planar-type SOEC. Two Pt wires were spot-welded to the coupon to serve as voltage and current probes, respectively. There was no additional Pt mesh placed between the chromia-forming alloy and the LSM electrode. In this arrangement, the chromia-forming alloy also acted as the current collector. The configuration for the arrangement of the electrode and interconnect is similar to that used in the SOFC studies.[21]

Electrochemical behavior of LSM oxygen electrodes was studied at 200 and 500 mA cm^{-2} and 800 °C with an air flow rate of 100 ml min^{-1}. Electrochemical impedance curves were measured under open circuit with frequency range from 0.1 Hz to 100 kHz and the signal amplitude of 10 mV. Electrode polarization resistance, R_E was measured by the differences between the high and low frequency intercepts, and electrode ohmic resistance, R_Ω was obtained from the high frequency intercept. For comparison, the electrochemical performance of LSM oxygen electrodes was also measured in the absence of Fe–Cr interconnect. In this case Pt mesh was used as the current collector. To confirm the reproducibility, at least two samples were tested under the same conditions.

The microstructure of the electrodes was examined using scanning electron microscopy (SEM, Zeiss Neon 40EsB) equipped with X-ray energy dispersion spectroscopy (EDS). In some cases, the electrodes after the electrochemical tests were removed by treatment in 3M HCl solution. Element imaging and line scan profiles on the polished cross sections were conducted using time of flight secondary ion mass spectroscopy (TOF.SIMS, ionToF, Germany) in the imaging mode with a 30 keV Bi^+ beam, a spot size of *ca.* 150 nm, and a raster size of 100 μm × 100 μm. Before the measurement, the cross-section was sputtered with 2 keV Ar^+ to remove polishing residues and adsorbed molecules on the sample surface. The phase of the YSZ electrolyte surface after the electrode delamination and/or after the acid treatment was characterized by a grazing incidence X-ray diffractometry (GI-XRD, Panalytical Empyrean XRD) with Cu K_α radiation at a step size of 0.005° and step time of 2.5 s. X-ray photoelectron spectroscopy (XPS) was carried out using a Kratos AXIS Ultra DLD system, with monochromated Al K_α X-rays (photon energy 1486.7 eV), a 700 μm × 300 μm aperture and a pass energy of 40 eV. The XPS spectra were calibrated with the C1s peak at 284.8 eV.

3. Results

3.1. Electrochemical performance

Fig. 1 shows the polarization curves of LSM oxygen electrodes under anodic currents of 200 and 500 mA cm^{-2} for 20 h in the absence and presence of an Fe–Cr interconnect. For the reaction under a low current of 200 mA cm^{-2}, R_E changes

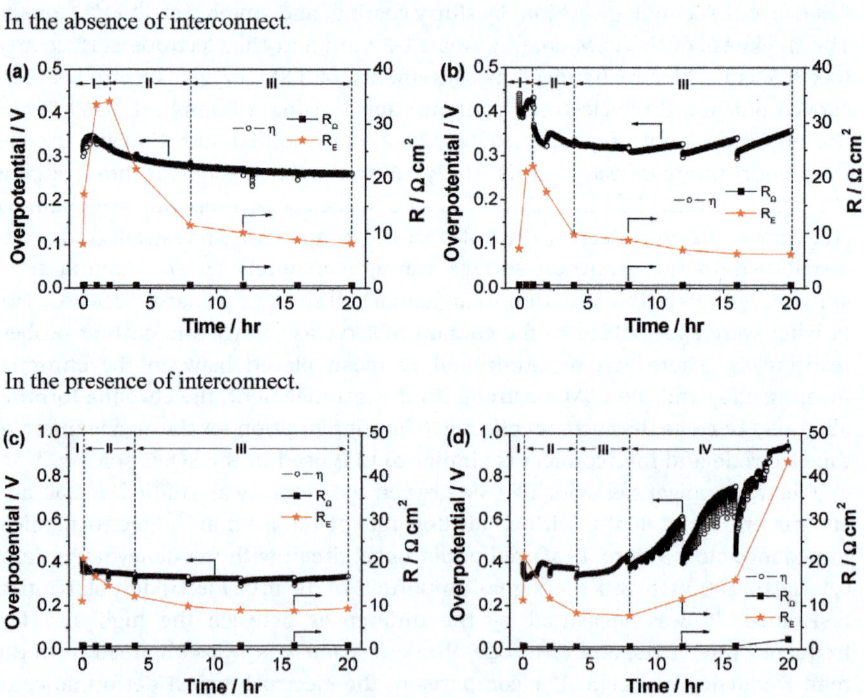

Fig. 1 Polarization curves of LSM oxygen electrodes as a function of anodic current passage time at (a,c) 200 mA cm^{-2} and (b,d) 500 mA cm^{-2}, 800 °C for 20 h in (a,b) the absence and (c,d) presence of Fe–Cr interconnect.

significantly with anodic polarization, showing three distinctive stages (Fig. 1a). R_E increases rapidly from 8.2 Ω cm^2 initially to 34.2 Ω cm^2 after anodic polarization for 2 h (stage I), then decreases rapidly to 11.5 Ω cm^2 after being polarized for 8 h (stage II). The decrease becomes much slower with further polarization (stage III). After anodic polarization for 20 h, R_E reaches 8.2 Ω cm^2, almost the same as the initial value. This indicates the initial activity decay of LSM electrode is recoverable by extended anodic polarization time. In the case of polarization at a high current of 500 mA cm^{-2} (Fig. 1b), R_E follows a similar trend of changes with three stages, but the change of R_E occurs faster under the higher current *e.g.* the duration of stages I and II is 1 and 3 h, respectively, significantly shorter than 2 and 6 h of the same stages in the case of the low currents. R_Ω is very stable during the test, and no delamination was observed after polarization at 200 and 500 mA cm^{-2}.

The polarization behavior of LSM oxygen electrodes in the presence of the Fe–Cr interconnect also shows three distinctive polarization regions (Fig. 1c). For the reaction in the presence of the interconnect, the initial R_E is ~11 Ω cm^2 and increases to 20.1 Ω cm^2 after being polarized for 5 min (stage I), followed by a rapid decrease to 12 Ω cm^2 after being polarized for 4 h (stage II), and stabilized at

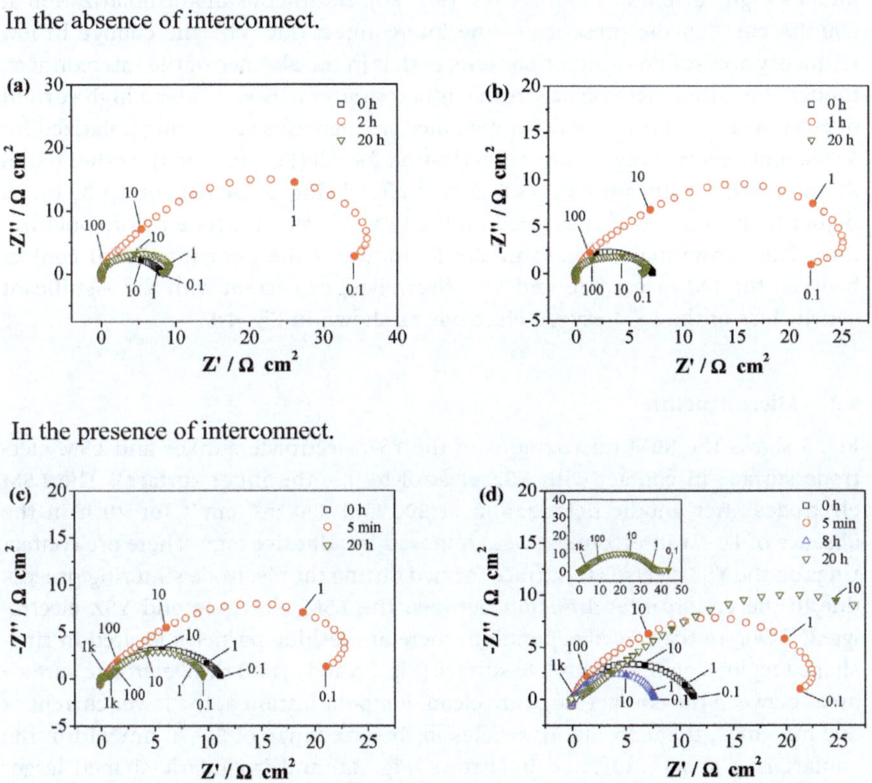

Fig. 2 Electrochemical impedance curves of LSM oxygen electrodes as a function of anodic current passage time at (a,c) 200 mA cm^{-2} and (b,d) 500 mA cm^{-2}, 800 °C for 20 h in (a,b) the absence and (c,d) presence of Fe–Cr interconnect. The high frequency intercepts of the EIS curves were adjusted to zero for the purpose of clarity.

9.5 Ω cm^2 after being polarized for 20 h (stage III). Compared to the reaction in the absence of the interconnect (Fig. 1a), the duration of stage I and II of the LSM oxygen electrodes in the presence of the interconnect is shortened. In the case of polarization under a high current of 500 mA cm^{-2} (Fig. 1d), the period for stage III is substantially reduced (*i.e.*, 4 h), followed by a dramatic increase of R_E from 7.7 Ω cm^2 to 43.5 Ω cm^2 after polarization treatment for 20 h. This is very different from that observed in the absence of the Fe–Cr interconnect, $R_Ω$ also increases substantially from 0.68 Ω cm^2 initially to 2.26 Ω cm^2 at the end of the experiment. Complete delamination of the LSM electrode was observed after polarization for 20 h, much shorter than 48 h observed in the absence of the interconnect under similar testing conditions.[13] The results clearly indicate that gaseous Cr species poisoned and degraded the electrocatalytic activity of LSM oxygen electrodes under SOEC operating conditions.

Fig. 2 shows the impedance responses of LSM oxygen electrodes during anodic polarization in the absence and presence of the Fe–Cr interconnect. The initial impedance responses of the LSM electrode in the absence of the interconnect are characterized by a large depressed arc (Fig. 2a). The impedance arcs (*i.e.*, R_E) increase significantly after anodic polarization at 200 mA cm^{-2} for 2 h, but decrease substantially after polarization for 20 h, and the change is mainly on the low frequency arc. Similar impedance behavior was also observed for the reaction under a high current of 500 mA cm^{-2} (Fig. 2b). During the anodic polarization at 200 mA cm^{-2} in the presence of the interconnect (Fig. 2b), the change in low frequency arcs follows similar patterns to that in the absence of the interconnect, though the initial increase occurs in a much shorter period. Under a high current of 500 mA cm^{-2}, the size of the impedance arc increases after being polarized for 5 min and decreases again after polarization for 8 h (Fig. 2d). On the other hand, R_E increased significantly to 43.5 Ω cm^2 after being polarized for 20 h, much higher than 11.3 Ω cm^2 measured on the LSM oxygen electrode before polarization. The substantial increase in the R_E indicates the poor interfacial contact between the LSM electrode and YSZ electrolyte, consistent with the significant activity loss of the LSM oxygen electrode as shown in Fig. 1d.

3.2. Microstructure

Fig. 3 shows the SEM micrographs of the YSZ electrolyte surface and LSM electrode surface in contact with YSZ electrolyte (*i.e.* the inner surface). The LSM electrodes after anodic polarization at 200 and 500 mA cm^{-2} for 20 h in the absence of Fe–Cr interconnect were removed by adhesive tape. There are contact rings on the YSZ electrolyte surface formed during the electrode sintering process due to the cation inter-diffusion between the LSM electrode and YSZ electrolyte.[28,29] Due to the forced separation, there are residue particles located in ring-shaped regions on the electrolyte surface (Fig. 3a and c) and the electrolyte surface areas between the contact rings are clean. For polarization at the lower current of 200 mA cm^{-2}, there are nanoparticles in the size range of 30–70 nm within the contact ring areas (indicated by arrows, Fig. 3a) and irregularly shaped larger residue particles. The existence of nanoparticles indicates the disintegration of LSM particles under SOEC anodic polarization conditions.[13] The irregularly shaped residue particles have an identical morphology observed on the inner surface of LSM electrode (Fig. 3b and d), indicating that LSM electrodes are still in

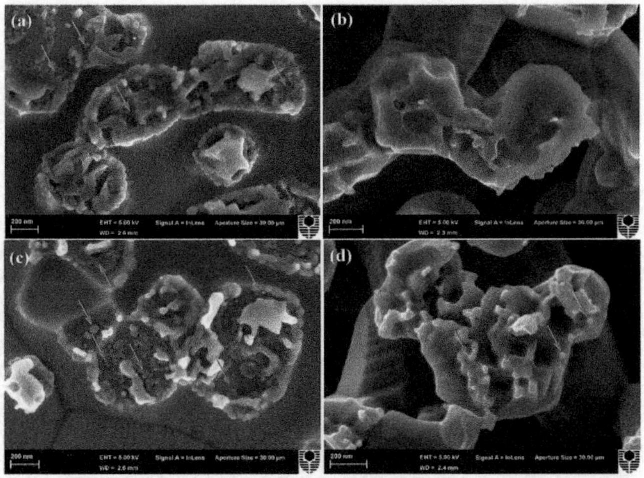

Fig. 3 SEM micrographs of the YSZ surface (left) and LSM inner surface (right) at the electrolyte/electrode interface region for an LSM oxygen electrode after anodic polarization at (a,b) 200 mA cm^{-2} and (c,d) 500 mA cm^{-2}, 800 °C for 20 h in the absence of the interconnect. The LSM electrode was removed by adhesive tape. Arrows indicate the nanoparticles formed.

intimate contact with the YSZ electrolyte. In the case of LSM electrode after being polarized at a high current of 500 mA cm^{-2}, the formation of nanoparticles is more pronounced and there are also fewer nanoparticles formed on the LSM electrode inner surface (Fig. 3b and d). This indicates that the disintegration of LSM particles and nanoparticle formation processes are much faster under a high current, which is consistent with electrochemical polarization results (Fig. 1).

Fig. 4 shows the SEM micrographs of the YSZ electrolyte surface and LSM electrode inner surface after anodic polarization at 200 mA cm^{-2} for 20 h in the presence of the Fe–Cr interconnect. The LSM oxygen electrode coating was removed by an adhesive tape. It was found that the electrode coatings in direct contact with the rib of the interconnect can be easily removed, while the coatings under the channel of the interconnect still adhered to the electrolyte (Fig. 4g). For the YSZ surface under the rib of the interconnect, the electrolyte surface between the contact rings is no longer clean, and is covered by deposits with a size range of 50–200 nm (Fig. 4a). The contact rings are occupied by a large number of nanoparticles in the size of 20–30 nm (indicated by arrows, Fig. 4c), and a few larger irregularly shaped residue particles. Such nanoparticles were also observed on the electrode inner surface (indicated by arrows, Fig. 4e). On the LSM inner surface there are also formations of particles with very different crystal facets and morphology to the LSM particles (circled in Fig. 4e). In the case of the electrolyte surface under the channel of the interconnect, the electrolyte surface between the rings is also covered by deposits (Fig. 4b). However, no nanoparticles were observed either in the contact rings or on the electrode inner surface (Fig. 4d and f). Instead there are large particles with a feature of forced fracture in the rings (Fig. 4b and d), indicating that LSM particles under the channels of the interconnect are intact with negligible disintegration. The significant nanoparticle formation under

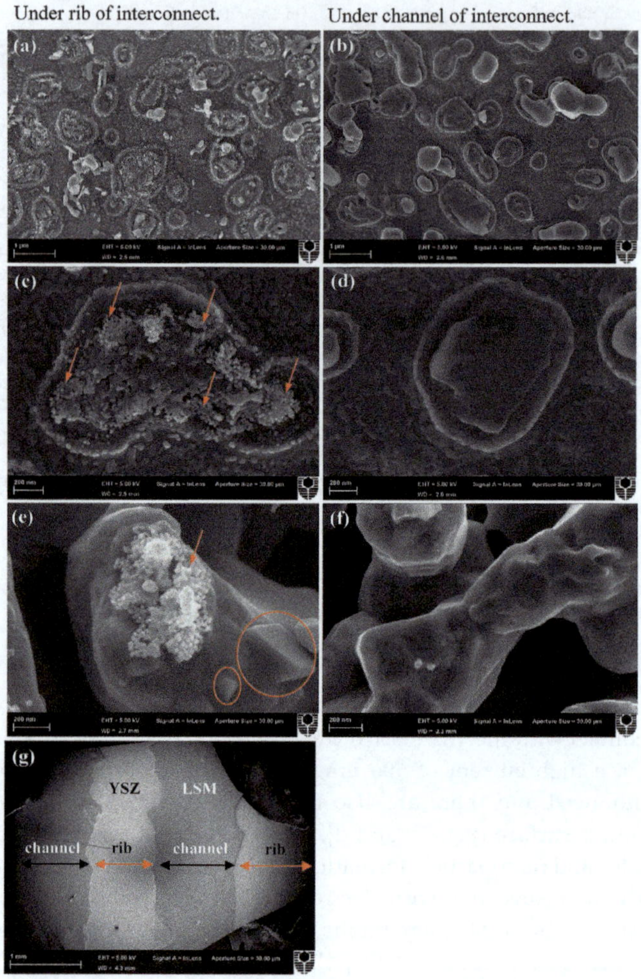

Fig. 4 SEM micrographs of the electrolyte/electrode interfaces under the rib (left) and channel (right) of the Fe–Cr interconnect, for a LSM oxygen electrode after anodic polarization at 200 mA cm^{-2}, 800 °C for 20 h: (a–d) the YSZ electrolyte surface, and (e, f) the LSM inner surface in contact with the YSZ electrolyte. Overview of the surface of the electrode partly removed by adhesive tape is shown in (g).

the rib of the interconnect as compared to that under the channel of the interconnect implies that the LSM electrode in direct contact with the interconnect is the major reaction site for the oxygen evolution reaction due to the much shorter current transfer pathway, thus giving rise to the preferential disintegration of LSM particles and occurrence of delamination of the portion of the LSM electrode under the rib of the interconnect (Fig. 4c and g).

Fig. 5 shows the SEM micrographs and EDS profiles of the YSZ electrolyte surface and LSM electrode surface under the rib of the interconnect, after anodic polarization at a high current of 500 mA cm^{-2} for 20 h. Similar to that observed on LSM electrodes polarized under a low current of 200 mA cm^{-2}, there are nanoparticles formed in the contact rings on the YSZ surface and on the LSM inner

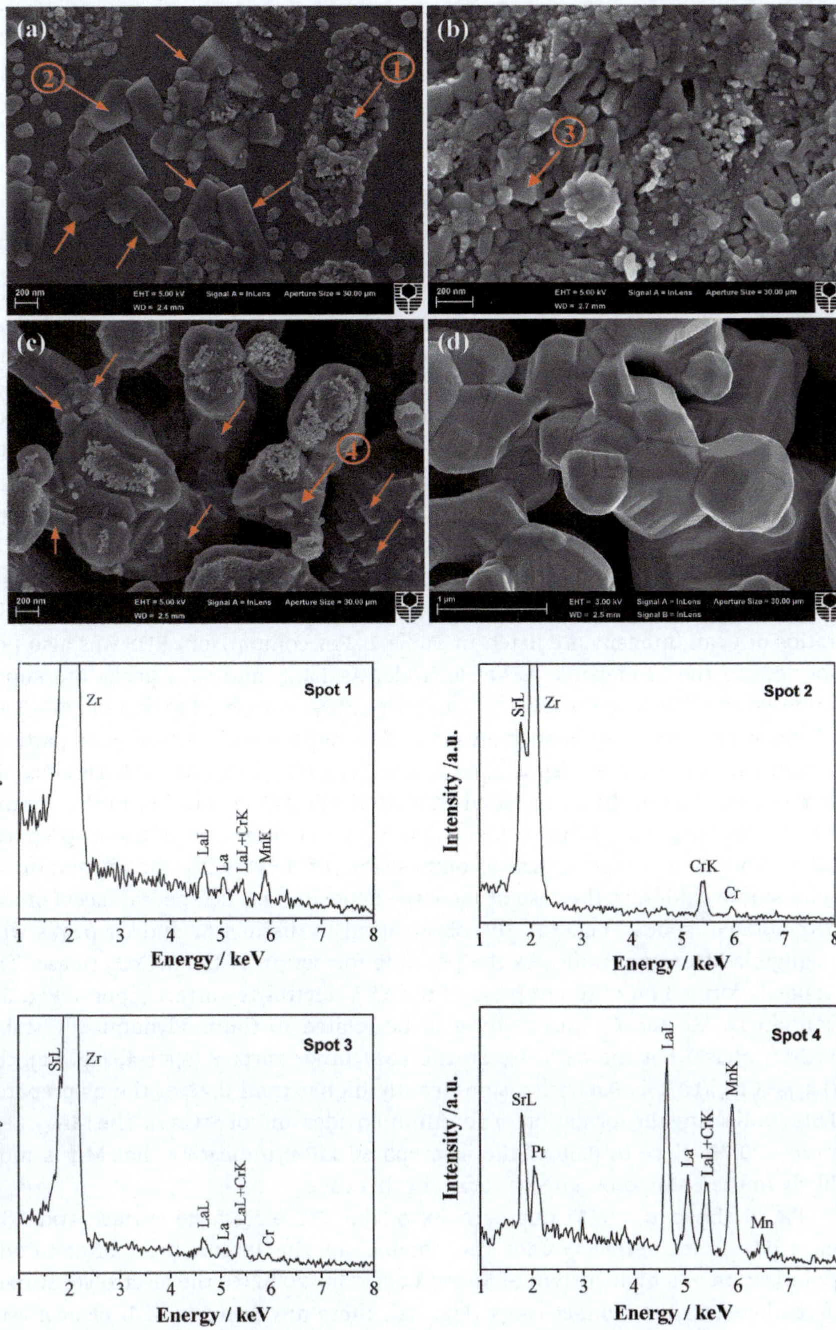

Fig. 5 SEM micrographs and spot EDS profiles of (a,b) the YSZ electrolyte surface and (c) the LSM inner surface and (d) the LSM outermost surface under the rib of the Fe–Cr interconnect after anodic polarization at 500 mA cm^{-2}, 800 °C for 20 h. The LSM electrode was delaminated after the test.

Table 1 Relative EDS peak intensity ratios of the selected areas as shown in Fig. 5

Locations	$Sr_{L\alpha}/La_{L\alpha}$	$(La_{L\beta2}+Cr_{L\alpha})/La_{L\alpha}$	$Mn_{K\alpha}/La_{L\alpha}$
As-prepared LSM surface	0.84	0.22	0.60
Spot 1, Fig. 5a	—	0.74	0.90
Spot 2, Fig. 5a	—	—	—
Spot 3, Fig. 5b	5.58	0.94	—
Spot 4, Fig. 5c	0.74	0.41	0.78

surface (Fig. 5a and c). Different from the numerous and fine nanoparticles formed on the electrolyte surface under a low current of 200 mA cm^{-2} (Fig. 4a), the deposits are large in size (200–800 nm) and have characteristics of octahedral crystal facets (indicated by allows, Fig. 5a). Such particles in the size range of 120–400 nm were also observed on the LSM inner surface (indicated by arrows, Fig. 5c). In some areas, the deposits were a mixture of nanoparticles and particles with clear crystal facets (Fig. 5b). On the other hand, the outermost surface of LSM oxygen electrode is clean and there is no deposition or formation of deposits after polarization at 500 mA cm^{-2} for 20 h (Fig. 5d).

EDS analysis was conducted on selected particles on the YSZ electrolyte surface and on the LSM inner surface (indicated by numbers 1 to 4, see Fig. 5), and the ratios of peak intensity are listed in Table 1. For comparison, EDS was also performed on the as-prepared LSM electrode. As $La_{L\beta2}$ and $Cr_{L\alpha}$ peaks are superimposed at ~5.88 keV, the ratio of $(La_{L\beta2}+Cr_{L\alpha})/La_{L\alpha}$ compared to that of $La_{L\beta2}/La_{L\alpha}$ of the as-prepared LSM is an indication of Cr deposition.[30] For the nanoparticles within the contact rings (spot 1, Fig. 5a), $(La_{L\beta2}+Cr_{L\alpha})/La_{L\alpha}$ is 0.74, significantly higher than 0.22 of the as-prepared LSM, indicating Cr deposition in the contact rings. The $Mn_{K\alpha}/La_{L\alpha}$ ratio is 0.90, slightly higher than 0.60 of the as-prepared LSM. This may suggest the decomposition of LSM with the formation of manganese oxides. In the case of the large deposits with octahedral facets on the YSZ surface (spot 2, Fig. 5a), the observation of distinct Sr and Cr peaks with negligible Mn signal indicates the possible formation of the $SrCrO_4$ phase. The deposits formed on different parts of the YSZ electrolyte surface (spot 3, Fig. 5b) contain La, Sr and Cr and could also be related to thermodynamically stable $SrCrO_4$ phase. For the deposits on the LSM inner surface (spot 4, Fig. 5c), the $(La_{L\beta2}+Cr_{L\alpha})/La_{L\alpha}$ is 0.41, also significantly higher than 0.22 of the as-prepared LSM, indicating the formation of chromium oxides and/or $SrCrO_4$. The $Mn_{K\alpha}/La_{L\alpha}$ ratio is 0.78, close to 0.60 of the as-prepared LSM, indicating that Mn is most likely in the LSM perovskite structure in this case.

Fig. 6 shows the SEM micrographs of the YSZ electrolyte surface and LSM electrode inner surface under the channel of the interconnect after anodic polarization at a high current of 500 mA cm^{-2} for 20 h. On the electrolyte surface areas between the contact rings (Fig. 6a), there are deposits with clear crystal facets but the amount is much lower compared to under the rib of the interconnect (Fig. 5). There are formations of relatively large nanoparticles in the range of 60–100 nm (Fig. 6b) and closely packed fine nanoparticles in the range of 30–60 nm (Fig. 6c). After acid treatment, the electrolyte surface becomes clean with distinctively visible contact rings (Fig. 6d). The electrode inner surface is

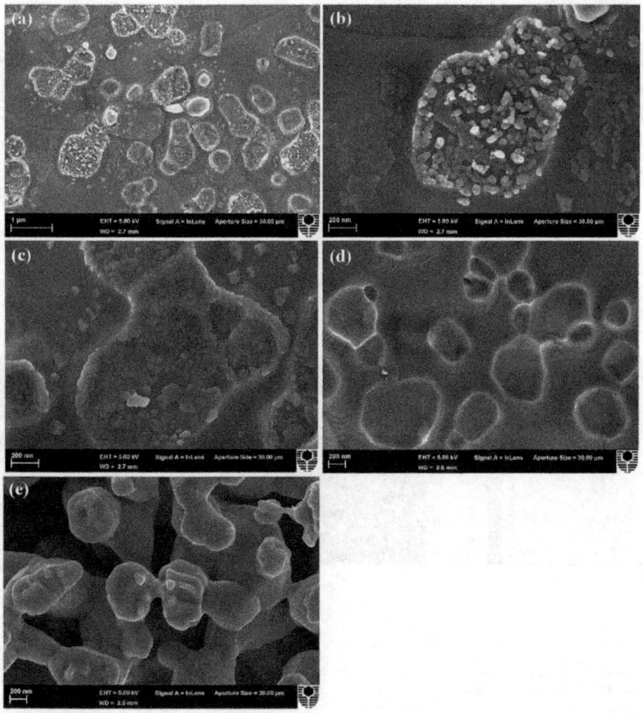

Fig. 6 SEM micrographs of the YSZ electrolyte surface and LSM inner surface under the channel of the Fe–Cr interconnect after anodic polarization at 500 mA cm^{-2}, 800 °C for 20 h: (a–c) YSZ surface before acid treatment, (d) YSZ surface after acid treatment, (e) LSM electrode inner surface in contact with the electrolyte. The LSM electrode was delaminated after the test.

generally clean with few fine particles (Fig. 6e). The deposition and formation of Cr particles on the YSZ electrolyte surface and LSM inner surface under the channel of the interconnect are much less, as compared to that under the rib of the interconnect, similar to that observed on the LSM electrode polarized under a low current of 200 mA cm^{-2} (Fig. 4).

3.3. Cr distribution and phase analysis

Fig. 7 shows the SIMS imaging and line scan profiles on the polished LSM electrode/YSZ electrolyte interfaces under the rib of the Fe–Cr interconnect after anodic polarization at 500 mA cm^{-2} for 1 and 8 h. In both cases, the SIMS image indicates chromium deposition along the entire electrode bulk (Fig. 7a and b). The line scan profiles show significantly higher deposition of chromium at the electrode/electrolyte interface region as compared to that in the bulk (Fig. 7c and d).

Fig. 8 shows the GI-XRD patterns of the YSZ electrolyte surface before and after anodic polarization at 800 °C. GI-XRD is a very sensitive technique used to probe the thin surface layer. For the as-prepared electrode, the electrode layer is removed by acid treatment, and there are typical YSZ diffraction peaks (PDF 04-001-9395) as well as minor $La_2Zr_2O_7$ peaks (PDF 00-017-0450) on the electrolyte

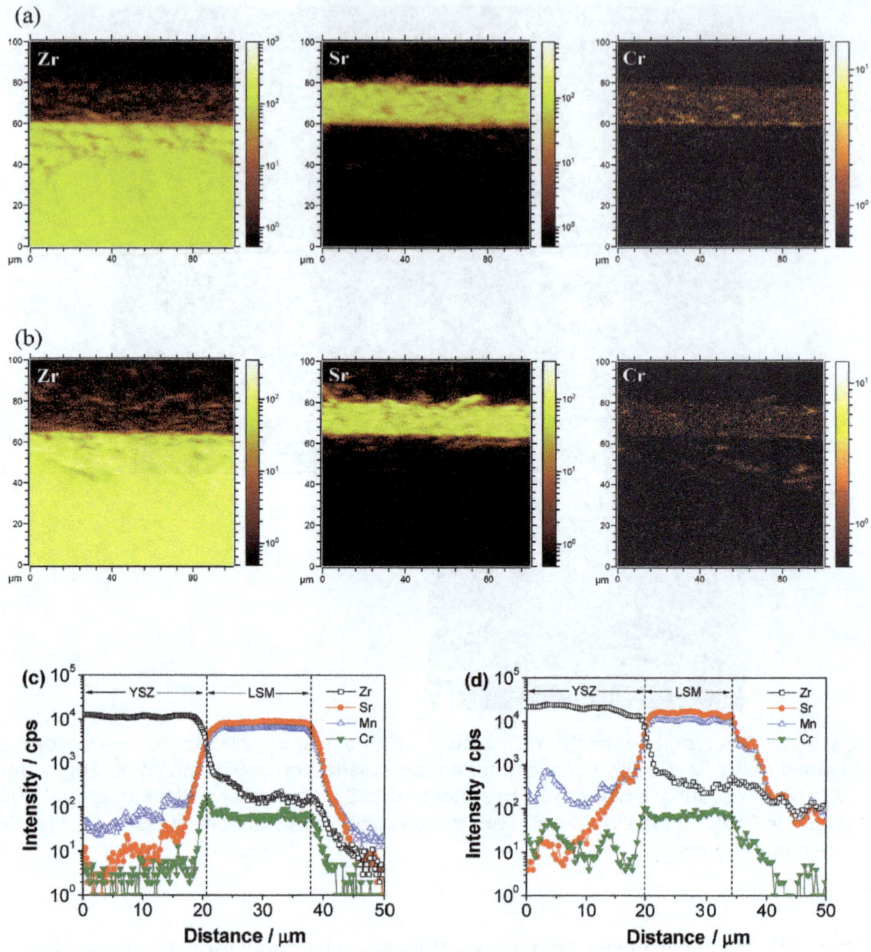

Fig. 7 SIMS imaging and line scan profiles on the cross section of the LSM oxygen electrodes under the rib of the Fe–Cr interconnect after anodic polarization at 500 mA cm^{-2}, 800 °C (a, c) for 1 h and (b, d) for 8 h.

surface (curve a, Fig. 8). The existence of minor $La_2Zr_2O_7$ phases indicates the chemical reaction between the LSM electrode and YSZ electrolyte during the electrode sintering process at 1150 °C.[31] After anodic polarization at 500 mA cm^{-2} for 48 h in the absence of the Fe–Cr interconnect, the entire electrode coating was peeled off of the electrolyte, and the LSM phase (PDF 04-015-1693) in addition to the $La_2Zr_2O_7$ phase (curve b, Fig. 8) was detected. The intensity of the LSM peak is substantially higher than that of the $La_2Zr_2O_7$ peak, indicating that the composition of the nanoparticles formed on the electrolyte surface after the electrode delamination (see Fig. 3c) is primarily LSM. This again confirms that nanoparticles formed within the contact rings originate from the disintegration or breaking of LSM particles with the LSM composition intact.[13] This is very different from the decomposition of LSM, as shown by Mizusaki *et al.*,[32] where LSM would decompose under low partial pressure of oxygen (*e.g.*, 10^{-18} atm at 800 °C),

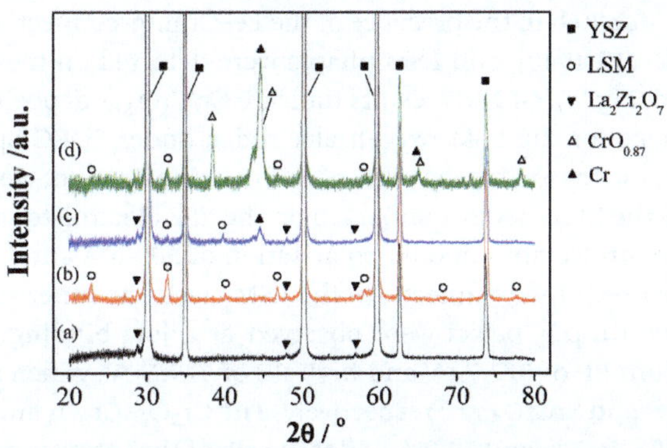

Fig. 8 GI-XRD patterns of the YSZ electrolyte surface in contact with the LSM oxygen electrodes (a) as-prepared electrode removed by acid treatment, and after the anodic current passage at 800 °C: (b) at 500 mA cm^{-2} for 48 h in the absence of the interconnect, (c) at 200 mA cm^{-2} for 20 h in the presence of Fe–Cr interconnect (prior to the measurement, the electrode was removed by acid treatment), and (d) at 500 mA cm^{-2} for 20 h in the presence of the Fe–Cr interconnect.

forming $La_{2-x}Sr_xMnO_4$, MnO and La_2O_3. After anodic polarization at 200 mA cm^{-2} for 20 h in the presence of the Fe–Cr interconnect and removal of the LSM electrode by acid treatment, there are Cr peaks (PDF 04-004-8467), $La_2Zr_2O_7$ peaks, and LSM peaks (curve c, Fig. 8). On the other hand, the LSM electrode was completely delaminated after anodic polarization at a high current

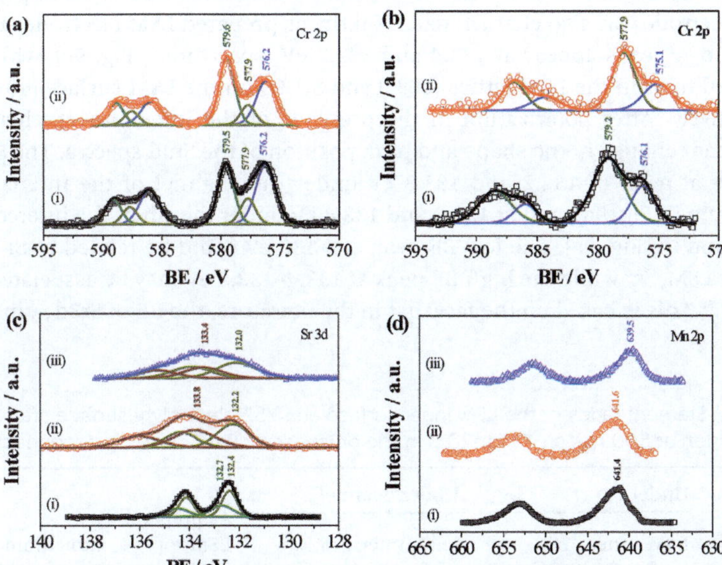

Fig. 9 XPS spectra of (a) Cr2p on the LSM electrode inner surface in contact with the YSZ electrolyte, (b) Cr2p, (c) Sr3d and (d) Mn2p on the YSZ electrolyte surface after anodic polarization at 500 mA cm^{-2} and 800 °C for 20 h in the presence of the Fe–Cr interconnect: (ii) under the channel and (iii) under the rib of the interconnect. The as-prepared LSM electrode surface is shown in (i).

of 500 mA cm^{-2} for 20 h in the presence of the Fe–Cr interconnect. In this case, Cr, $CrO_{0.87}$ (PDF 04-015-4940), and LSM phases were detected on the YSZ electrolyte surface (curve d, Fig. 8). GI-XRD results indicate $Cr/CrO_{0.87}$ deposition on the YSZ electrolyte surface for the LSM oxygen electrodes under SOEC operating conditions. However, the reason for the presence of metallic Cr is not clear at this stage.

Fig. 9 shows the XPS spectra conducted on the YSZ electrolyte surface and LSM electrode inner surface after anodic polarization at 500 mA cm^{-2} for 20 h in the presence of the Fe–Cr interconnect. On the LSM electrode inner surface (Fig. 9a), three distinctive $Cr2p_{3/2}$ peaks were observed at a low binding energy (BE) of 576.2 eV, medium BE of 577.7 eV and high BE of 579.5 eV, which are assigned to Cr_2O_3,[33] Cr_2O_5 [33] and $SrCrO_4$,[34,35] respectively. The Cr_2O_3, Cr_2O_5 and $SrCrO_4$ have a Cr ratio of 48.4%, 16.7% and 34.9% under the rib of the interconnect, and 40.9%, 15.3% and 43.8% under the channel of the interconnect, respectively. This indicates that the deposited Cr on the LSM inner surface is mainly in the form of Cr_2O_3 and $SrCrO_4$. On the electrolyte surface (Fig. 9b), there are only two $Cr2p_{3/2}$ peaks at a low BE of 576.1 eV related to Cr_2O_3 [33] and a high BE of 579.2 eV $SrCrO_4$ [34,35] under the channel of the interconnect, and at a low BE of 575.1 eV related to Cr_2O_3 [33] and at a medium BE of 577.9 eV related to Cr_2O_5 [33] under the rib of the interconnect. The Cr ratio is 32.5% for Cr_2O_3 and 67.5% for $SrCrO_4$ under the channel of the interconnect, and 35.4% for Cr_2O_3 and 64.6% for Cr_2O_5 under the rib of the interconnect, respectively. XPS results clearly show the formation of Cr_2O_3 and $SrCrO_4$ on both YSZ electrolyte surface and LSM electrode inner surface close to the interface.

The XPS spectra of Sr3d and Mn2p were also carried out on the electrolyte surface under the rib of the Fe–Cr interconnect after anodic polarization at 500 mA cm^{-2} for 20 h. For comparison, the outermost surface of an as-prepared LSM electrode was also characterized. On the as-prepared LSM electrode surface, two $Sr3d_{5/2}$ peaks appear at 132.4 and 132.7 eV (spectrum i, Fig. 9c), which are assigned to Sr in the LSM lattice (60%) and $SrCO_3$ on the LSM surface (40%),[36,37] respectively. After polarization in the presence of the interconnect, there is a significant change in the shape and peak position of the Sr3d spectra. The $Sr3d_{5/2}$ appears at a BE of 132.2 and 133.8 eV under the channel of the interconnect (spectrum ii, Fig. 9c), and at 132.0 and 133.4 eV under the rib of the interconnect (spectrum iii, Fig. 9c). The low BE peak at ~132 eV could be related to Sr in the bulk of LSM,[37,38] while the high BE peak at 133.4–133.8 eV may be associated with $SrCrO_4$.[38] This is based on the fact that in the literature, the BE of $Sr3d_{5/2}$ in SrO is

Table 2 Element ratios of the LSM inner surface and YSZ electrolyte surface after anodic polarization at 500 mA cm^{-2} for 20 h in the presence of Fe–Cr alloy interconnect

	Under rib		Under channel			
Elements	LSM inner surface/%	YSZ surface/%	LSM inner surface/%	YSZ surface/%	As-prepared LSM top surface/%	Stoichiometric $La_{0.8}Sr_{0.2}MnO_3$/%
La	46.4	44.6	39.6	47.8	47.8	40
Sr	6.8	8.5	13.2	8.5	8.1	10
Mn	33.6	36.5	33.1	31.0	44.1	50
Cr	13.2	10.4	14.1	12.7	0	0

135.0 eV,[39] significantly higher than a BE of 132.7 eV reported for $Sr3d_{5/2}$ in $SrCrO_4$.[38] In the case of the Mn2p XPS spectra, the $Mn2p_{3/2}$ peak occurs at 641.6 eV under the channel of the interconnect (spectrum ii, Fig. 9d), close to 641.4 eV of the as-prepared LSM electrode (spectrum i, Fig. 9d). This indicates the Mn valence state is between 3+ and 4+.[37,40] On the other hand, under the rib of the interconnect there is a shift of $Mn2p_{3/2}$ to a lower binding energy of 639.5 eV (spectrum iii, Fig. 9d), which suggests the presence of MnO.[41] This may indicate the partial decomposition of the LSM perovskite phase on the YSZ surface under the rib of the interconnect. Corresponding element ratios of LSM inner surface and YSZ electrolyte surface based on XPS analysis are given in Table 2.

4. Discussion

In the absence of the interconnect, the initial rapid increase in the electrode polarization resistance, R_E is most likely related to the deactivation effect of the anodic polarization on the electrocatalytic activity of LSM oxygen electrodes.[42,43] Our previous studies on the degradation and delamination of LSM oxygen electrodes have shown that the incorporation of oxygen ions from the electrolyte into the LSM grains under the anodic polarization potential driving force results in the shrinkage of the LSM lattice and thereby local tensile strains, leading to the microcracks and subsequent disintegration and breaking of LSM particles at the electrode/electrolyte interface.[12,13] The present study shows that the nanoparticles maintain the LSM phase and structure. The nanoparticle formation initially increases the three-phase boundaries (TPBs) and active surface areas for the oxygen oxidation reaction. This explains the rapid decrease in R_E in region II (Fig. 1a and b).

Similar polarization behavior was observed for the O_2 oxidation reaction in the presence of the Fe–Cr interconnect. However, the significantly reduced duration of the polarization changes in three regions and occurrence of the delamination at a high current of 500 mA cm^{-2} for the reaction in the presence of the Fe–Cr interconnect clearly demonstrate that the presence of the Fe–Cr interconnect poisons the electrocatalytic activity of the LSM oxygen electrode under SOEC operating conditions, similar to that reported for the LSM electrode under SOFC operating conditions.[44]

Very different from the preferential Cr deposition at the electrode/electrolyte interface region under SOFC operating conditions,[20–22,45] Cr deposition on the LSM oxygen electrode under SOEC operating conditions not only occurs on the YSZ electrolyte surface and at the electrode/electrolyte interface region but also in the LSM electrode bulk. The results indicate that the deposits with distinct crystal facets formed on the YSZ electrolyte surface between the contact rings and on the LSM electrode inner surface (see Fig. 5a and c) are most likely the $SrCrO_4$ phase and the small particles formed on the YSZ electrolyte surface within the contact rings and on LSM inner surface are a combination of Cr_2O_3 and Cr_2O_5. There is no evidence to indicate the formation of $(Cr,Mn)_3O_4$ spinels for the LSM oxygen electrodes under SOEC operating conditions. This is very different from the preferential Cr deposition at the electrode/electrolyte interface region and not in the electrode bulk, and the dominant formation of $(Cr,Mn)_3O_4$ spinels for the LSM electrodes under SOFC operating conditions.[20–22] However, such Cr deposition behavior of the LSM oxygen electrodes under SOEC operating conditions can be explained by the accelerated surface segregation of SrO species under the anodic polarization conditions.

SrO surface segregation is known to occur on the as-prepared LSM electrodes.[46,47] One of the main driving forces for the Sr segregation is probably due to the fact that the cation dopant, Sr^{2+}, has a larger ionic radius than the host cation La^{3+}. The rather large Sr^{2+} is under a compressive strain/stress state in the bulk, which can be relaxed when it moves to the surface, resulting in the migration and formation of SrO species on the surface of related perovskites such as LSM and LSCF.[48–51] Yildiz *et al.* reported that the elastic energy caused by the size mismatch between Sr dopant and La host of LSM is the main driving force for the SrO surface segregation.[52] They also found that the SrO surface segregation increases at a high oxygen partial pressure, which causes lattice shrinkage of LSM and thereby an increase in dopant strain energy.[52] However, as shown by Fister *et al.*[53] the strain state has only a small impact on the Sr segregation for LSM thin films. This appears to be supported by the fact that under open circuit conditions, no Cr deposition or formation of $SrCrO_4$ is observed on the surface of LSM electrodes.[21]

In addition to the relation between the intrinsic compressive strain in the bulk and the Sr segregation, polarization appears to have a significant effect on the surface segregation of perovskites. Backhaus-Ricoult *et al.*[54] studied the LSM electrode/YSZ electrolyte interface using *in situ* photoelectron microscopy and observed the increased manganese segregation/migration under cathodic polarization at the interface. This implies that under anodic polarization conditions, manganese segregation would be depressed. Huber *et al.* studied *in situ* the surface segregation of $La_{0.75}Sr_{0.25}Cr_{0.5}Mn_{0.5}O_3$ under cathodic and anodic dc bias by XPS, and observed that cathodic polarization leads to the depletion of Sr on the LSCM surface while Sr surface segregation is promoted by anodic polarization.[55] Fig. 10a shows the scheme of the nanoparticle formation at the electrode/electrolyte interface and accelerated Sr segregation from the bulk to the surface under anodic polarization. As LSM is a predominant electronic conductor with negligible oxygen ion conductivity,[56] the O_2 evolution reaction would also occur primarily at the electrode/electrolyte contact and/or interface areas.[13] High anodic polarization potential at the electrode/electrolyte interface region would lead to the higher Sr segregation on LSM inner surface in the vicinities of the interface as compared to that in the electrode bulk.

It is well known that segregated SrO plays a critical role in the deposition and formation of $SrCrO_4$ on the LSCF surface in contact with a chromia-forming alloy under SOFC operating conditions.[19] In the case of LSM oxygen electrodes under SOEC operating conditions, Cr deposition would be dominated by the interaction between the gaseous Cr species and segregated SrO, owing to the fact that generation of manganese species in particular, Mn^{2+} is depressed under anodic polarization conditions. This would lead to Cr deposition and $SrCrO_4$ formation in the bulk of the electrode. The high Sr segregation at the electrode/electrolyte interface region would lead to the substantially higher reaction activity between the segregated SrO and gaseous Cr species. This is supported by the observations that in the presence of the interconnect, a high polarization current of 500 mA cm^{-2} significantly accelerates Cr deposition particularly at the electrode/electrolyte interface regions under the rib of the interconnect. The degree of chromium deposition on the YSZ electrolyte surface is significantly higher than that under a low current of 200 mA cm^{-2}, indicated by the significant grain

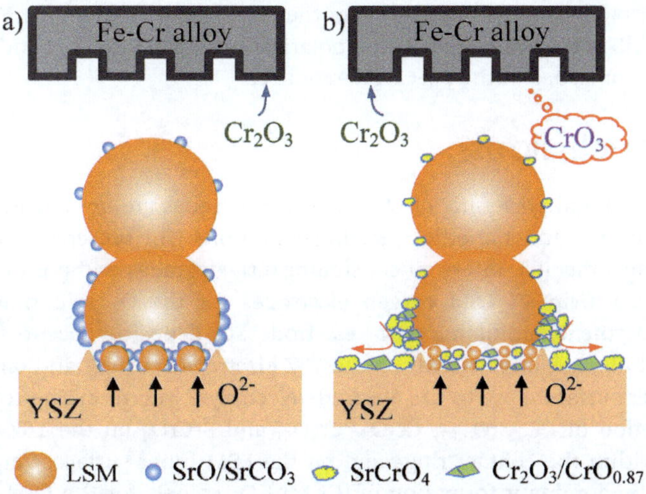

Fig. 10 Schemes of Cr deposition on LSM oxygen electrodes under SOEC operating conditions. (a) LSM nanoparticle formation at the electrode/electrolyte interface and accelerated SrO segregation from the bulk to the LSM surface under anodic polarization; (b) $Co_2O_3/CrO_{0.87}$ and $SrCrO_4$ formation on the LSM surface, at the electrode/electrolyte interface, and on the YSZ electrolyte surface. The red arrows indicate the migration of SrO/$SrCO_3$ from the LSM surface to the YSZ electrolyte surface.

growth of deposited particles on the YSZ electrolyte surface (Fig. 5a and b). The high activity between the Sr and Cr could also lead to the spread of Sr species from the LSM electrode to the YSZ surface. The high mobility of SrO in the presence of Cr has been reported in the case of LSCF and LSC oxygen electrode,[26,41,42] evident by the long-range transport of Sr from electrode bulk to the outermost surface. This is supported by high Cr deposition and significant $SrCrO_4$ formation on the YSZ electrolyte surface between the contact rings and on the LSM electrode inner surface (Fig. 4 and 7). The accelerated migration and segregation of SrO from the bulk to the free surface and subsequent reaction between SrO and gaseous Cr species based on the nucleation theory[19] are summarized below and schematically shown in Fig. 10b.

Sr segregation:

$$La_{0.8}Sr_{0.2}MnO_{3+\delta} \rightarrow La_{0.8}Sr_{0.2-x}MnO_{3+\delta} + x\mathrm{SrO} \quad (1)$$

Cr deposition *via* nucleation and grain growth route:

$$SrO + CrO_3 \rightarrow Cr\text{–}Sr\text{–}O_x \text{ (nuclei)} \quad (2)$$

$$Cr\text{–}Sr\text{–}O + CrO_3 \rightarrow CrO_y + Cr_2O_3 \quad (3)$$

$$Cr\text{–}Sr\text{–}O + SrO + CrO_3 \rightarrow SrCrO_4 \quad (4)$$

The segregated Sr species could also exist as $SrCO_3$ as indicated by the XPS results (Fig. 9c). The excess segregation of Sr could also result in the partial decomposition of LSM and this appears to be supported by the shift of $Mn2p_{3/2}$ to a lower BE of 639.5 eV, an indication of the presence of MnO on the YSZ surface under the rib of the interconnect after polarization at 500 mA cm^{-2} and 800 °C for 20 h in the presence of the Fe–Cr interconnect.

5. Conclusions

Cr deposition and poisoning of LSM oxygen electrodes are investigated in detail for the first time under anodic operating conditions. The presence of a chromia-forming alloy metallic interconnect significantly degrades and poisons the electrocatalytic activity of LSM oxygen electrodes for the O_2 oxidation reaction, accelerating the delamination of the electrodes. Cr deposition occurs in the LSM electrode bulk and in particular on the YSZ electrolyte surface and on LSM electrode inner surface close to the electrode/electrolyte interface. The results show the formation of Cr, $CrO_{0.87}$, Cr_2O_3, Cr_2O_5, and $SrCrO_4$ on the YSZ electrolyte surface, within the contact rings and on the LSM inner surface. Very different from the predominant formation of $(Cr,Mn)_3O_4$ spinels for the LSM electrodes under the SOFC operating conditions, the $(Cr,Mn)_3O_4$ spinel phase could not be found on the LSM oxygen electrodes under SOEC operating conditions. The fundamental reason for the observation of $SrCrO_4$ formation instead of $(CrMn)_3O_4$ on the LSM electrode under SOEC operating conditions is that anodic polarization promotes Sr segregation and depresses Mn segregation. Thus, under SOEC operating conditions, the nucleation agent for Cr deposition on LSM electrodes is the segregated SrO or $SrCO_3$ and not the manganese species, Mn^{2+}, generated under polarization conditions. Due to the high activity of gaseous Cr species with segregated Sr species, the segregated Sr species could be highly mobile and readily spread to the YSZ electrolyte surface. This may be the reason for significant Cr deposition and formation of $SrCrO_4$ on the YSZ electrolyte surface area between the contact rings.

Finally, the present study also provides clear evidence that the delamination or failure of LSM oxygen electrodes is primarily due to the disintegration or breaking of LSM particles attached to the YSZ electrolyte surface, forming nanoparticles at the interface under anodic polarization conditions as proposed early,[12,13] and not by the formation of highly resistive $La_2Zr_2O_7$ at the electrode/electrolyte interface.[11] This is supported by the correlation between the nanoparticle formation and easy delamination of the LSM electrodes under the rib of the interconnect as compared to that under the channel of the interconnect.

Acknowledgements

The project is supported by the Curtin University Research Fellowship, Australian Research Council (LP110200281), Australia, and National Natural Science Foundation of China (U1134001). The authors acknowledge the facilities, scientific and technical assistance of the Curtin University Electron Microscope Facility, the Australian Microscopy & Microanalysis Research Facility at the Centre for Microscopy, Characterisation & Analysis, The University of Western Australia, both of which are funded by the University, State and Commonwealth Governments, and the WA X-Ray Surface Analysis Facility, funded by the Australian Research Council LIEF grant (LE120100026). The authors also thank Dr Catherine Kealley, Curtin University, for the XRD data analysis.

References

1 J. Schefold, A. Brisse, M. Zahid, J. P. Ouweltjes and J. U. Nielsen, *ECS Trans.*, 2011, **35**, 2915–2927.

2 N. Q. Minh, *ECS Trans.*, 2011, **35**, 2897–2904.

3 X. Zhang, J. E. O'Brien, R. C. O'Brien and G. K. Housley, *J. Power Sources*, 2013, **242**, 566–574.

4 S. Elangovan, J. Hartvigsen, D. Larsen, I. Bay and F. Zhao, *ECS Trans.*, 2011, **35**, 2875–2882.

5 A. Momma, T. Kato, Y. Kaga and S. Nagata, *J. Ceram. Soc. Jpn.*, 1997, **105**, 369–373.

6 J. R. Mawdsley, J. D. Carter, A. J. Kropf, B. Yildiz and V. A. Maroni, *Int. J. Hydrog. Energy*, 2009, **34**, 4198–4207.

7 G. A. Hughes, K. Yakal-Kremski and S. A. Barnett, *Phys. Chem. Chem. Phys.*, 2013, **15**, 17257–17262.

8 A. V. Virkar, *Int. J. Hydrog. Energy*, 2010, **35**, 9527–9543.

9 R. Knibbe, M. L. Traulsen, A. Hauch, S. D. Ebbesen and M. Mogensen, *J. Electrochem. Soc.*, 2010, **157**, B1209–B1217.

10 J. Kim, H.-I. Ji, H. P. Dasari, D. Shin, H. Song, J.-H. Lee, B.-K. Kim, H.-J. Je, H.-W. Lee and K. J. Yoon, *Int. J. Hydrogen Energy*, 2013, **38**, 1225–1235.

11 M. Keane, M. K. Mahapatra, A. Verma and P. Singh, *Int. J. Hydrogen Energy*, 2012, **37**, 16776–16785.

12 K. Chen, N. Ai and S. P. Jiang, *Int. J. Hydrogen Energy*, 2012, **37**, 10517–10525.

13 K. F. Chen and S. P. Jiang, *Int. J. Hydrog. Energy*, 2011, **36**, 10541–10549.

14 F. Tietz, D. Sebold, A. Brisse and J. Schefold, *J. Power Sources*, 2013, **223**, 129–135.

15 D. The, S. Grieshammer, M. Schroeder, M. Martin, M. Al Daroukh, F. Tietz, J. Schefold and A. Brisse, *J. Power Sources*, 2015, **275**, 901–911.

16 H. Fan, M. Keane, P. Singh and M. Han, *J. Power Sources*, 2014, **268**, 634–639.

17 S. J. Kim and G. M. Choi, *Solid State Ionics*, 2014, **262**, 303–306.

18 W. Z. Zhu and S. C. Deevi, *Mater. Res. Bull.*, 2003, **38**, 957–972.

19 S. P. Jiang and X. B. Chen, *Int. J. Hydrog. Energy*, 2014, **39**, 505–531.

20 S. P. S. Badwal, R. Deller, K. Foger, Y. Ramprakash and J. P. Zhang, *Solid State Ionics*, 1997, **99**, 297–310.

21 S. P. Jiang, J. P. Zhang, L. Apateanu and K. Foger, *J. Electrochem. Soc.*, 2000, **147**, 4013–4022.

22 T. Horita, Y. P. Xiong, M. Yoshinaga, H. Kishimoto, K. Yamaji, M. E. Brito and H. Yokokawa, *Electrochem. Solid State Lett.*, 2009, **12**, B146–B149.

23 X. Zhang, J. E. O'Brien, R. C. O'Brien, J. J. Hartvigsen, G. Tao and G. K. Housley, *Int. J. Hydrogen Energy*, 2013, **38**, 20–28.

24 J. W. Fergus, *Int. J. Hydrog. Energy*, 2007, **32**, 3664–3671.

25 B. Wei, K. F. Chen, L. Zhao, Z. Lu and S. P. Jiang, *Phys. Chem. Chem. Phys.*, 2015, **17**, 1601–1609.

26 V. I. Sharma and B. Yildiz, *J. Electrochem. Soc.*, 2010, **157**, B441–B448.

27 L. Zhao, S. Amarasinghe and S. P. Jiang, *Electrochem. Commun.*, 2013, **37**, 84–87.

28 A. Mitterdorfer and L. J. Gauckler, *Solid State Ionics*, 1998, **111**, 185–218.

29 S. P. Jiang and W. Wang, *Electrochem. Solid State Lett.*, 2005, **8**, A115–A118.

30 X. B. Chen, L. Zhang, E. J. Liu and S. P. Jiang, *Int. J. Hydrog. Energy*, 2011, **36**, 805–821.

31 S. P. Jiang, J. P. Zhang, Y. Ramprakash, D. Milosevic and K. Wilshier, *J. Mater. Sci.*, 2000, **35**, 2735–2741.

32 J. Mizusaki, H. Tagawa, K. Naraya and T. Sasamoto, *Solid State Ionics*, 1991, **49**, 111–118.

33 U. Scharf, H. Schneider, A. Baiker and A. Wokaun, *J. Catal.*, 1994, **145**, 464–478.

34 P. Parhi and V. Manivannan, *J. Alloys Compd.*, 2009, **469**, 558–564.

35 T. L. Daulton, B. J. Little, K. Lowe and J. Jones-Meehan, *Journal of Microbiological Methods*, 2002, **50**, 39–54.

36 C. Knöfel, H.-J. Wang, K. T. S. Thydén and M. Mogensen, *Solid State Ionics*, 2011, **195**, 36–42.

37 Q.-H. Wu, M. Liu and W. Jaegermann, *Mater. Lett.*, 2005, **59**, 1980–1983.

38 F. Liu and J. Jia, *Tribol Lett*, 2013, **49**, 281–290.

39 E. Mutoro, E. J. Crumlin, M. D. Biegalski, H. M. Christen and Y. Shao-Horn, *Energy & Environmental Science*, 2011, **4**, 3689–3696.

40 E. Z. Kurmaev, M. A. Korotin, V. R. Galakhov, L. D. Finkelstein, E. I. Zabolotzky, N. N. Efremova, N. I. Lobachevskaya, S. Stadler, D. L. Ederer, T. A. Callcott, L. Zhou, A. Moewes, S. Bartkowski, M. Neumann, J. Matsuno, T. Mizokawa, A. Fujimori and J. Mitchell, *Physical Review B*, 1999, **59**, 12799–12806.

41 Y. Wang and Y. Ohtsuka, *Applied Catalysis A: General*, 2001, **219**, 183–193.

42 W. Wang and S. P. Jiang, *J. Solid State Electrochem.*, 2004, **8**, 914–922.

43 W. S. Wang, Y. Y. Huang, S. W. Jung, J. M. Vohs and R. J. Gorte, *J. Electrochem. Soc.*, 2006, **153**, A2066–A2070.

44 X. B. Chen, L. Zhang, E. Liu and S. P. Jiang, *Int. J. Hydrogen Energy*, 2011, **36**, 805–821.

45 T. Horita, Y. P. Xiong, H. Kishimoto, K. Yamaji, M. E. Brito and H. Yokokawa, *J. Electrochem. Soc.*, 2010, **157**, B614–B620.

46 N. Caillol, M. Pijolat and E. Siebert, *Appl. Surf. Sci.*, 2007, **253**, 4641–4648.

47 S. P. Jiang and J. G. Love, *Solid State Ionics*, 2001, **138**, 183–190.

48 S. P. Jiang, *J. Mater. Sci.*, 2008, **43**, 6799–6833.

49 H. P. Ding, A. V. Virkar, M. L. Liu and F. Liu, *Phys. Chem. Chem. Phys.*, 2013, **15**, 489–496.

50 L. Zhao, J. Drennan, C. Kong, S. Amarasinghe and S. P. Jiang, *Journal of Materials Chemistry A*, 2014, **2**, 11114–11123.

51 D. Oh, D. Gostovic and E. D. Wachsman, *J. Mater. Res.*, 2012, **27**, 1992–1999.

52 W. Lee, J. W. Han, Y. Chen, Z. Cai and B. Yildiz, *J. Am. Chem. Soc.*, 2013, **135**, 7909–7925.

53 M. Fister, *Appl. Phys. Lett.*, 2008, **93**, 151904.

54 M. Backhaus-Ricoult, K. Adib, T. S. Clair, B. Luerssen, L. Gregoratti and A. Barinov, *Solid State Ionics*, 2008, **179**, 891–895.

55 A.-K. Huber, M. Falk, M. Rohnke, B. Luerssen, L. Gregoratti, M. Amati and J. Janek, *Phys. Chem. Chem. Phys.*, 2012, **14**, 751–758.

56 S. Carter, A. Selcuk, R. J. Chater, J. Kajda, J. A. Kilner and B. C. H. Steele, *Solid State Ionics*, 1992, **53–6**, 597–605.

Faraday Discussions

ROYAL SOCIETY OF CHEMISTRY

PAPER

Electrochemical stability of $Sm_{0.5}Sr_{0.5}CoO_{3-\delta}$-infiltrated YSZ for solid oxide fuel cells/electrolysis cells

Hui Fan and Minfang Han*

Received 12th February 2015, Accepted 26th March 2015

DOI: 10.1039/c5fd00022j

Composite SSC ($Sm_{0.5}Sr_{0.5}CoO_{3-\delta}$)–YSZ (yttria stabilized zirconia) oxygen electrodes were prepared by an infiltration process. X-ray diffraction (XRD) analysis and scanning electron microscopy (SEM) of the composite electrodes showed the formation of SSC perovskite and a well-connected network of SSC particles in the porous YSZ backbone, respectively. The electrochemical performance of the cell was investigated under both fuel cell and steam electrolysis modes using polarization curves and electrochemical impedance spectroscopy (EIS). The cell experienced a large degradation rate at 700 °C with a constant voltage of 0.7 V for over 100 h under power generation operation. The subsequent post-cell SEM micrograph revealed that agglomeration of the infiltrated SSC particles was possibly the cause for the performance deterioration. Furthermore, the long-term stability of the cell was examined at 700 °C with a constant voltage of 1.3 V under steam electrolysis mode. SEM associated with energy dispersive X-ray spectroscopy (EDS) was employed to characterize the post-test cell after the long-term electrolysis operation and it indicated that besides the agglomeration of SSC particles, the delamination of the SSC–YSZ oxygen electrode from the YSZ electrolyte, as well as segregation of cobalt-enriched particles (particularly cobalt oxides) at the interface, was probably responsible for the cell degradation under the steam electrolysis mode.

1. Introduction

A solid oxide fuel cell (SOFC) is an electrochemical device that can continuously convert the chemical energy of a fuel directly into electrical energy.[1] SOFCs possess advantages of high efficiency and high fuel flexibility. At present, the most commonly used solid electrolyte in solid oxide fuel cells (SOFCs) is yttria-stabilized zirconia (YSZ),[2] which is predominantly an oxygen ion conductor with negligible electronic conductivity. A traditional Ni-based cermets anode has been extensively studied over the years in an effort to provide high performance for solid oxide fuel cells (SOFCs).[3,4] SOFCs show great promise as potential

State Key Laboratory of Power Systems, Department of Thermal Engineering, Tsinghua University, Beijing, 100084, P. R. China. E-mail: hanminfang@sina.com

alternatives in the near future for the efficient and benign conversion of chemical energy into electricity. Solid oxide electrolysis cells (SOECs) are the inverse of SOFCs and are a promising technology to achieve zero-emission and high-purity hydrogen production.[5] Compared with low temperature operation, high temperature steam electrolysis in SOECs has a reduced electrical energy demand and a lower internal resistance of the cell.[6]

LSM (Sr-doped $LaMnO_3$) has been widely used as an oxygen electrode for state-of-the-art SOFCs/SOECs.[7] However, it is well known that LSM exhibits poor performance at reduced temperature for SOFC operation and suffers delamination from the electrolyte for SOEC operation. Cobalt-containing perovskite oxides have been investigated as oxygen electrodes in SOFCs and have attracted much attention because of their mixed ionic/electronic conduction characteristics.[8] Strontium-doped cobaltites have been studied as cathode materials for SOFCs due to their high electrical conductivities and low overpotentials. Sr-doped $LaCoO_3$ (LSCo) was infiltrated into a porous YSZ backbone to prepare a composite LSCo–YSZ electrode.[9,10] The cell using the composite LSCo–YSZ electrode demonstrated high levels of performance and low cathode polarization. $Sm_{0.5}Sr_{0.5}CoO_{3-\delta}$ (SSC) perovskites have recently attracted attention as cathode materials for intermediate temperature SOFCs. In comparison with the LSCo electrode, SSC electrodes showed lower overpotentials.[8] Nevertheless, it has been reported that the Sr-doped cobaltite perovskites undergo a solid state reaction with YSZ at the sintering temperature required for standard electrode preparation (usually over 1000 °C).[9,11] Nano-sized composite electrodes were fabricated by infiltration technology and subsequently sintered at relatively low temperatures (700–900 °C),[12] which effectively hinder the formation of the insulating phases. A study by Nicholas *et al.*[13] reported that a SSC nitrate solution was infiltrated into porous GDC (gadolinia doped ceria) scaffolds to prepare a nano-composite SSC–GDC electrode, resulting in a low cathode polarization resistance. Furthermore, the influence of the infiltrate solution concentration, the infiltrate loading, and the number of infiltration steps was investigated with respect to the cathode polarization. Zhang *et al.*[14] reported a nano-sized SSC cathode which was fabricated by an ion-impregnation process and embedded into a porous SDC backbone for intermediate-temperature SOFCs with zirconia (ScSZ) electrolyte. The impregnated SSC–SDC composite electrode demonstrated an extended triple phase boundary (TPB) length, reduced interfacial resistance, and improved power density. In addition to their application as SOFC cathodes, Sr-doped cobaltites can be employed in SOECs for hydrogen production from steam electrolysis. A LSCo–YSZ composite was prepared and tested as a SOEC anode to show comparable impedance with that of LSF (Sr-doped $LaFeO_3$).[15] Furthermore, composite SSC–SDC ($Sm_{0.2}Ce_{0.8}O_{1.9}$) oxygen electrodes have been found to exhibit lower polarization resistance than pure SSC electrodes under steam electrolysis operating mode,[16] which indicated that the SSC composite was a promising oxygen electrode candidate for SOECs. Compared with the standard LSCF–GDC cell, the SSC-infiltrated cells exhibited improved performances and significantly reduced electrode polarization resistance under both SOFC and SOEC modes.

Perovskite SSC composites have been demonstrated to show enhanced electrochemical properties for both SOFC and SOEC operations. However, few investigations have focused on the electrochemical stability of the SSC-based oxygen electrode for both SOFCs and SOECs. In this study, nano-sized SSC–YSZ oxygen electrodes were prepared by an infiltration process for SOFCs/SOECs. The electrochemical performances of the cells were evaluated under both power generation and steam electrolysis modes. Long-term stability operation was investigated for both the oxygen reduction and oxidation reactions. Possible deterioration mechanisms of the cells under the two modes were proposed based on post-test scanning electron microscopy (SEM) analysis, associated with energy dispersive X-ray spectroscopy (EDS) characterization.

2. Experimental

2.1 Fabrication of porous cell substrates

As in our previous papers,[17] the porous cell substrates have a configuration of Ni–YSZ hydrogen electrode/YSZ electrolyte/porous YSZ layer. Tape casting, hot isostatic pressing and a one-step sintering process were employed to fabricate the substrates. A NiO–YSZ support was prepared by mixing nickel oxide powder, YSZ (TZ-8Y, Tosoh) and graphite pore former (Furunda Zirconium Material Co. Ltd., China) at a weight ratio of 50 : 50 : 10. Ethanol–butanone solvent, caster oil dispersant, dibutyl phthalate (DBP) plasticizer and polyvinyl butyral (PVB) binder were then added and ball-milled to form a NiO–YSZ slurry. The tape casting process for the NiO–YSZ support was reported in our previous work.[18] The YSZ electrolyte was prepared by the same tape casting process as used for the NiO–YSZ substrates, but without the addition of NiO powder or graphite powders. For the porous YSZ layer, 50 wt% graphite pore former was added to form the backbone. The slurries for the NiO–YSZ support, YSZ electrolyte and porous YSZ layer were cast separately *via* a tape casting machine (DR-150, made in Japan). These tapes were dried overnight in an air atmosphere. A single sheet of NiO–YSZ support (~300 μm), an YSZ electrolyte layer (20–30 μm), and a porous YSZ layer (~70 μm) were stacked together under vacuum conditions and laminated at 20 MPa for 10 min using a hot isostatic press (30Tm Shanxi, China) to form a NiO–YSZ supported tri-layer structure. The resulting tri-layer tape was punched into discs and then co-sintered at 1300 °C for 10 h in order to densify the electrolyte layer, during which graphite was burned out, leaving a well-formed porous YSZ network. The microstructural characteristics of the fired tri-layer tapes were analyzed using a scanning electron microscope, and the estimated porosity of the porous YSZ layer was 40–50%.

2.2 Single-cell preparation

A precursor solution of SSC ($Sm_{0.5}Sr_{0.5}CoO_{3-\delta}$), for the infiltration, was prepared by dissolving stoichiometric amounts of $Sm(NO_3)_3 \cdot 6H_2O$, $Sr(NO_3)_2$ and $Co(NO_3)_2 \cdot 6H_2O$ with a chelating agent in de-ionized water.[19] Ethanol, which has proven to be an effective additive to lower the surface tension of the infiltration solution, was added to the aqueous solution.[20] The solution was then infiltrated into the pre-treated porous YSZ backbone using a microsyringe. The infiltrated samples were dried at room temperature until the precursor solution was fully permeated, and they were then fired at 450 °C for 1 h in order to decompose the nitrate. The infiltration process was repeatedly carried out in order to achieve the desired SSC loading amount. The infiltrated samples were eventually sintered at 850 °C for 5 h to prepare nano-structured NiO–YSZ/YSZ/SSC–YSZ cells.

2.3 Cell characterization and electrochemical measurements

The hydrogen electrode-supported single cell was mounted onto a horizontal alumina tube. Ceramic paste (Aramco-552, USA) was adopted as the sealant. Silver paste (Ag ink, Beijing, China) was applied as a current collector onto the electrode surfaces. Silver wires were pressed against the oxygen electrode to obtain sufficient electronic contact with both electrodes in a four-wire set up. The cell was placed inside a temperature-controlled tube furnace and heated at a rate of 2 °C min^{-1}; when the temperature was increased to 700 °C, a low flow rate of hydrogen gas was introduced for the *in situ* reduction of the NiO–YSZ hydrogen electrode. The area of the electrolyte and hydrogen electrode was about 1.5 cm^2 and the geometrical area of the oxygen electrode was 0.2 cm^2. For SOFC operation, humidified hydrogen gas (3% H_2O, volume fraction, the same below) with a flow rate of 50 sccm (standard cubic centimeters per minute) was fed to the Ni-based hydrogen electrode, whilst the oxygen electrode was exposed to an ambient atmosphere. For SOEC operation, a 50% H_2O, 25% H_2 and 25% N_2 gas mixture was supplied to the hydrogen electrode at a total flow rate of 100 sccm. Similarly, the oxygen electrode was left open to the atmospheric environment. During the SOEC testing, a thermostatic water bath was used at a constant temperature to maintain the desired steam content. Current density *versus* voltage curves (polarization curves) and electrochemical impedance spectra (EIS) under open circuit were measured at 650–800 °C using electrochemical analyzers of Solartron 1287 and 1260, respectively. The EIS were recorded in the frequency range from 100 mHz to 100 kHz with an AC amplitude of 20 mV. Additionally, a high power X-ray diffractometer (XRD, PANalytical X'Pert PRO, Netherlands) loaded with CuKα radiation was employed to verify the crystallographic phase of the infiltrated SSC. Morphological and elemental characterizations of the pre- and post-test cells were carried out using a scanning electron microscope (SEM, FEI Quanta 250 FEG) attached with an energy dispersive X-ray spectroscope (EDS). Long-term stability experiments (*I–t* curves) were potentiostatically conducted under both fuel cell and steam electrolysis modes.

3. Results and discussion

3.1 Phase structure of the composite SSC–YSZ electrode

A $Sm_{0.5}Sr_{0.5}CoO_{3-\delta}$ (SSC) precursor solution was added into a YSZ matrix and sequentially sintered at 850 °C to prepare the composite oxygen electrode. Fig. 1 shows the X-ray diffraction pattern of the SSC-infiltrated YSZ electrode. It illustrates that the perovskite phase SSC was formed in the as-sintered YSZ layer. It was reported that SSC was susceptible to react with YSZ at high temperatures (>900 °C) and that there is a poor chemical compatibility between SSC and YSZ.[21] Here, no impurity phase is observable, which indicates that no chemical reaction occurs at this temperature between SSC and YSZ. In ref. 22, a porous $Sm_{0.7}Sr_{0.3}CoO_{3-\delta}$ cathode was deposited onto an YSZ electrolyte substrate by solution precursor plasma spraying. It demonstrated that a single perovskite phase was formed when the deposits were annealed at 900 °C for 10 h. Zhang *et al.*[14] fabricated a nano-structured SSC–SDC ($Sm_{0.2}Ce_{0.8}O_{1.9}$) electrode on a zirconia (ScSZ, $Sc_{0.1}Zr_{0.89}Ce_{0.01}O_{2-x}$) electrolyte by an ion-impregnation process.

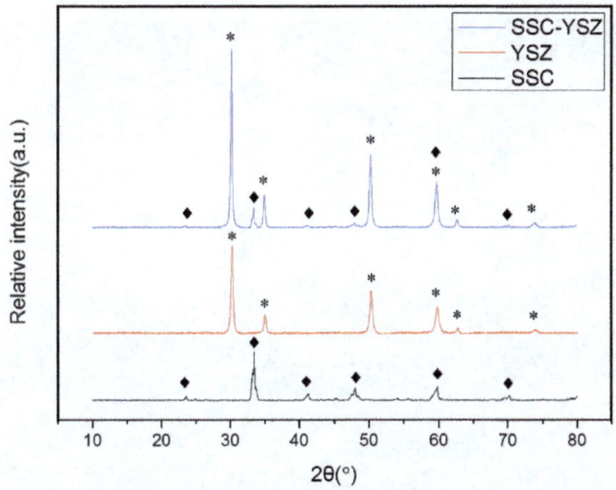

Fig. 1 XRD pattern of a SSC-infiltrated YSZ oxygen electrode after firing at 850 °C.

No chemical reaction was observed between SSC and SDC at relatively low sintering temperatures.

3.2 Microstructure characterization of the composite SSC–YSZ electrode

Fig. 2 illustrates the SEM cross-sectional morphologies of a freshly-prepared cell as well as its electrolyte and oxygen electrode. It can be seen from Fig. 2(a) that dense YSZ electrolyte and porous oxygen electrode were formed after co-sintering at 1300 °C for 10 h. The YSZ electrolyte with a thickness of ~20 μm maintains excellent adhesion with the SSC–YSZ oxygen electrode and no delamination/crack is presented at the electrolyte/oxygen electrode interface. We have reported that the porous YSZ layer prepared by the tape casting and co-sintering process has a porosity of 40–50% for the infiltration.[18] Fig. 2(b) shows a higher magnification SEM image of the composite SSC–YSZ electrode prepared by the infiltration method. It can be observed that nano-sized SSC particles uniformly cover the inner surface of the porous YSZ backbone. A well-connected network of SSC particles was formed through infiltration and *in situ* calcination. The introduction of SSC particles provides electron conductivity and an increased active area of TPBs, which consequently improves the electrochemical performance of the composite oxygen electrode. Surface modification through the solution-based infiltration process has been demonstrated to be an effective approach for the enhancement of surface electrocatalytic activity and stability.[23] The wetting property of the precursor solution plays a key role on the microstructure and morphology of the infiltrated catalyst coating. The addition of ethanol can dramatically improve the surface wetting property, resulting in the formation of uniform and continuous nano-catalyst particles.[20]

3.3 Electrochemical performance of the cell under SOFC mode

The Ni–YSZ/YSZ/SSC–YSZ cells prepared by the infiltration process were initially evaluated under SOFC mode. For the SOFC operation, 3% H_2O humidified hydrogen gas at a flow rate of 50 sccm was introduced into the Ni–YSZ hydrogen

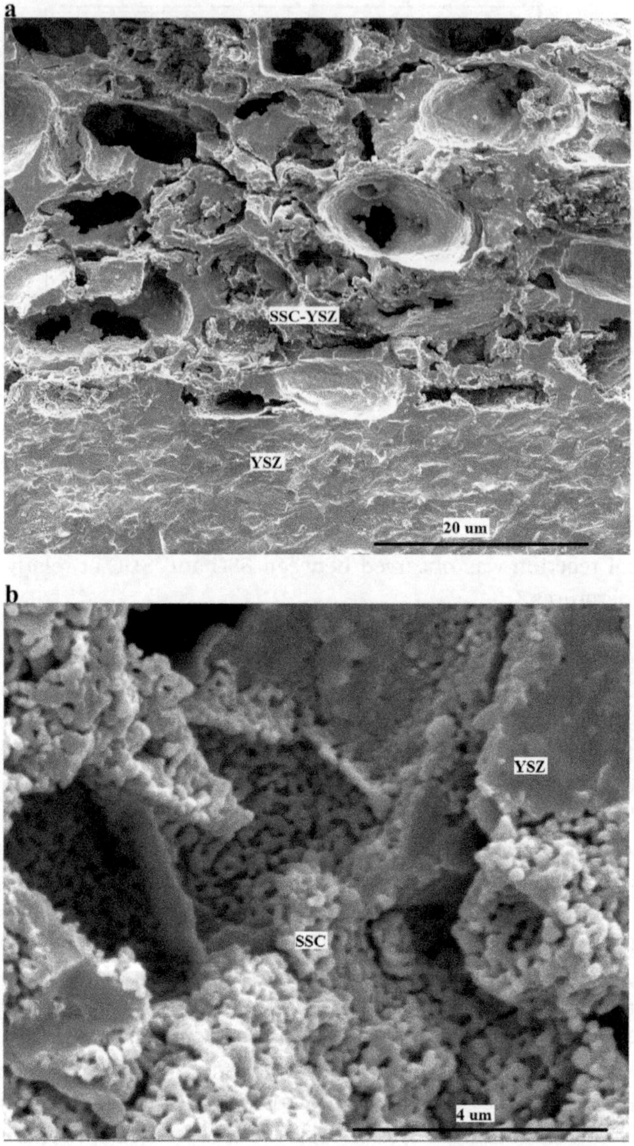

Fig. 2 a) Typical SEM micrograph of the cross section of the YSZ electrolyte and SSC–YSZ electrode and (b) a higher magnification image of the SSC-infiltrated composite electrode.

electrode after the NiO was reduced to Ni. Voltage and power density *vs.* current density curves of the cells in the operating temperature range 650–800 °C are shown in Fig. 3(a). The open circuit voltages (OCVs) of the cell are 1.047 V, 1.040 V, 1.035 V and 1.030 V at 650 °C, 700 °C, 750 °C and 800 °C, respectively, which are slightly lower than the theoretical OCV values predicted from the Nernst equation.[24] All of the OCV values are over 1 V, indicating that the YSZ electrolyte is dense and that good sealing was achieved for the SOFC operation.[25] The density of the electrolyte has also been demonstrated through the SEM image of the cell in

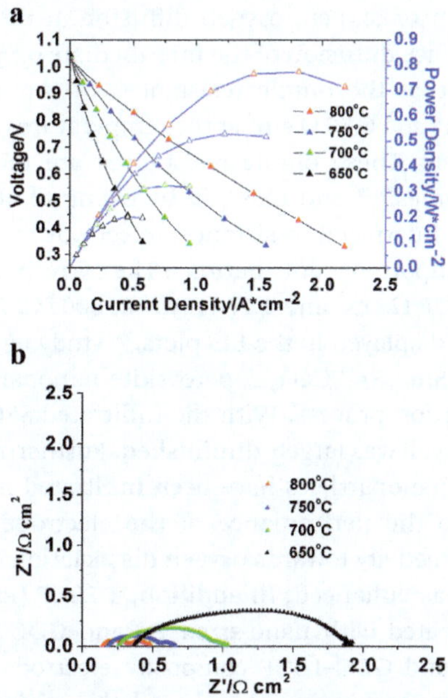

Fig. 3 (a) Voltage and power density *vs.* current density curves and (b) Nyquist electrochemical impedance spectra (EIS) under open circuit conditions at 650–800 °C for Ni–YSZ/YSZ/SSC–YSZ cell under SOFC mode.

Fig. 2(a). The cell shows maximum power densities of 776 mW cm^{-2}, 530 mW cm^{-2}, 327 mW cm^{-2} and 204 mW cm^{-2} at the temperatures measured from 800 °C to 650 °C, respectively. Huang *et al.*[26] have reported a composite LSF–YSZ cathode formed by an impregnation process. The interfacial reaction can be effectively precluded at low sintering temperatures. Compared with conventional electrode fabrication such as screen printing, the composite electrode generally showed improved performance over SOFCs.[27] In this work, the cell exhibited relatively low performance, which may be ascribable to a low SSC loading amount (~30 wt%). Fig. 3(b) shows the EIS plots of the cell under open circuit conditions at different temperatures. The EIS is composed of both ohmic and polarization resistances. The ohmic resistance (R_Ω) is usually obtained from the high-frequency intercept with the real axis, representing the sum of the electrolyte resistance of ionic transfer, lead resistance of electronic transfer, and electrode/electrolyte contact resistance.[28] During the measurements, silver wires were used as current collectors to minimize the contribution of the lead resistance. The polarization resistance (R_p), calculated from the difference between the high-frequency and low-frequency intercepts with the real axis, corresponds to catalytic oxygen reduction/oxidation, gas diffusion, ionic transport, and ion transfer between the electrodes and electrolyte.[29] In the Nyquist plots, two distinct depressed arcs were displayed as high-frequency resistance and low-frequency resistance, indicating at least two different electrode processes. The high-frequency resistance with a high temperature dependence is possibly associated with a charge transfer process, while the low temperature dependence of the low-frequency arc may be ascribed to diffusion processes composed of the adsorption–desorption of oxygen, oxygen diffusion at the gas-oxygen electrode interface, and the surface diffusion of the intermediate oxygen species.[30] It can be seen from Fig. 3(b) that the ohmic resistances of the cell were 0.13 Ω cm^2, 0.18 Ω cm^2, 0.25 Ω cm^2 and 0.36 Ω cm^2 at the temperatures measured from 800 °C to 650 °C, respectively. These ohmic resistances are much larger than those reported in other studies[31,32] and may not be owing to electrolyte contribution alone. Additionally, interfacial resistance, electrode resistance and contact resistance were also important contributions. The corresponding R_p values of 0.35 Ω cm^2, 0.48 Ω cm^2, 0.78 Ω cm^2 and 1.77 Ω cm^2 at 800 °C, 750 °C, 700 °C and 650 °C, respectively, were displayed in the EIS plots. A study by Lu *et al.*[33] investigated the incorporation of $Sm_{0.6}Sr_{0.4}CoO_{3-\delta}$ perovskite nanoparticles into a LSM–YSZ cathode by an infiltration process. With the infiltrated SSC nanoparticles in the cathode, the R_p of the cell was largely diminished. Furthermore, active-conductive $Ce_{0.8}Sm_{0.2}O_{1.9}$ (SDC) nanoparticles have been infiltrated into a porous SSC electrode.[34] Compared to the performance of the electrode fabricated by screen printing, the catalytic activity towards oxygen dissociation of SSC prepared by the infiltration process was enhanced. In addition, a LSCF ($La_{0.8}Sr_{0.2}Co_{0.5}Fe_{0.5}O_{3-\delta}$) cathode was impregnated with nano-sized Pd and GDC particles.[35] The nanostructured Pd–LSCF and GDC–LSCF composite electrodes demonstrated lower polarization resistances and higher electrocatalytic activities than the pure LSCF electrode.

3.4 Long-term durability of the cell under SOFC mode and corresponding microstructure of the post-test cell

The electrochemical stability of the Ni–YSZ/YSZ/SSC–YSZ cell was investigated under potentiostatic test conditions. Fig. 4 displays the long-term durability of the cell operated at 700 °C and 0.7 V under fuel cell mode with humidified H_2 as the fuel gas for over 100 h. The cell with the SSC-infiltrated YSZ oxygen electrode suffered a rapid and continuous degradation in the time range of 0 h to 50 h and then experienced a relatively stable current during the period from 50 h to 100 h. The operational stability of LSF-infiltrated YSZ electrodes was examined by Wang

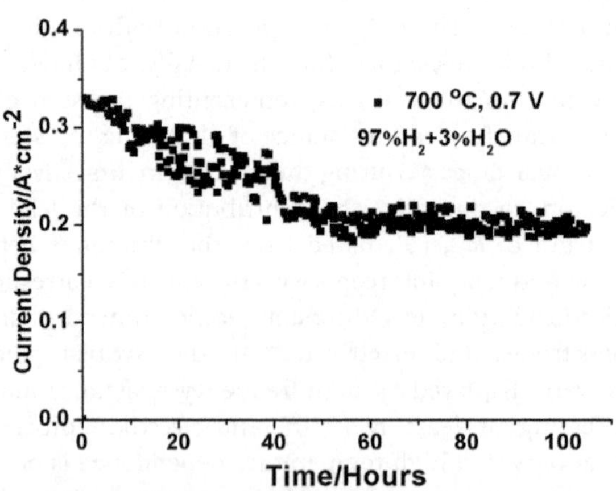

Fig. 4 Long-term durability operation of the Ni–YSZ/YSZ/SSC–YSZ cell at 700 °C and at a constant voltage of 0.7 V under the fuel cell mode.

et al.[31] They reported an appreciable increase in the electrode polarization loss for a period of more than 2500 h. Similarly, SSC-infiltrated YSZ cells were also investigated in ref. 36 with respect to evaluating their operational stability under SOFC mode. An electrode polarization resistance *vs.* time curve demonstrated that the SSC–YSZ composite electrode gradually deteriorated with increased polarization resistance. Microstructural changes play a vital role in the performance deterioration for the nano-sized electrode.[37] An SEM micrograph of the post-test SSC-infiltrated YSZ electrode under long-term power generation conditions is shown in Fig. 5. Compared with the microstructure of the freshly prepared SSC–YSZ in Fig. 2(b), the infiltrated SSC particles in the post-test cell are agglomerated and additionally partly separated from the YSZ scaffold. The performance degradation is likely associated with the growth and reduced interconnectivity of the nano-sized SSC particles.[38] Furthermore, the particle coarsening would lead to a decrease in the area of electrochemical reaction sites, *i.e.* the TPB, and an increase in charge transfer resistance.[37] In addition, the particle agglomeration also reduces the porosity of the infiltrated electrode and affects the mass transfer rate of the oxygen reduction reaction. These results reveal that the agglomeration of the infiltrated SSC nanoparticles is possibly responsible for the degradation of the cell's long-term performance.

3.5 Electrochemical performance of the cell under SOEC mode

The electrochemical reactions that take place in a SOEC are the inverse of those that take place in a solid oxide fuel cell (SOFC). Fig. 6 shows the electrochemical performance of the Ni–YSZ/YSZ/SSC–YSZ cell under steam electrolysis mode. For the SOEC operation, steam containing a mixture of 50% H_2O, 25% H_2 and 25% N_2 was supplied to the Ni–YSZ hydrogen electrode side, acting as a reactant, and the oxygen electrode was exposed to an air atmosphere. Hydrogen was then produced at the hydrogen electrode side, whilst oxygen ions were transported to the oxygen electrode through the dense YSZ electrolyte under the action of electrolysis

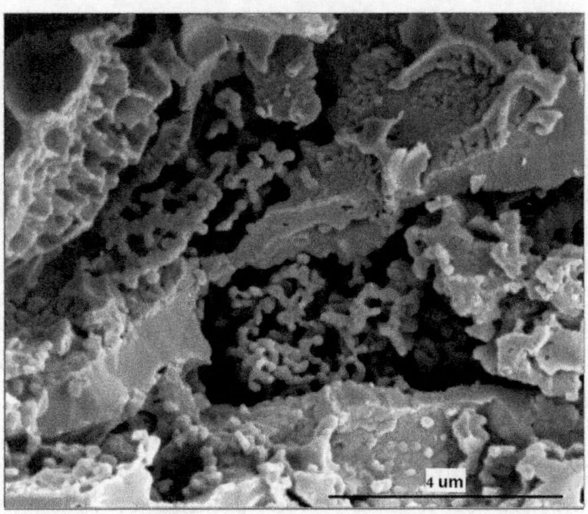

Fig. 5 SEM micrograph of post-test SSC–YSZ composite electrodes under long-term power generation mode.

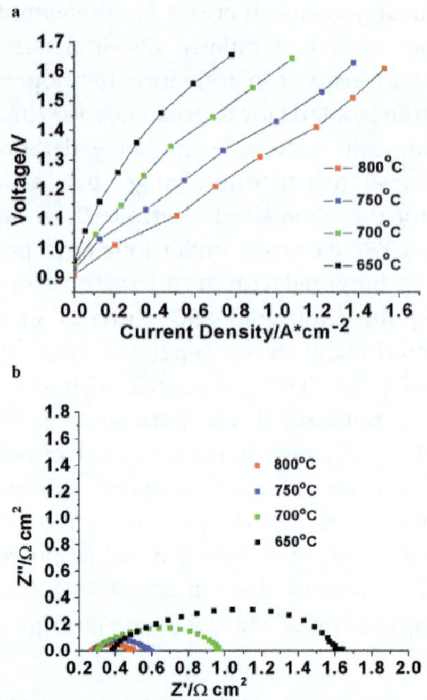

Fig. 6 (a) Voltage and current density curves, and (b) Nyquist EIS under OCV at 650–800 °C for the Ni–YSZ/YSZ/SSC–YSZ cell with 50% H_2O, 25% H_2 and 25% N_2 fed to the hydrogen electrode under SOEC mode.

current, and thus oxygen gas was produced at the SSC–YSZ oxygen electrode.[39] The current–voltage characteristics of the cell at temperatures of 800–650 °C were evaluated, as shown in Fig. 6(a). The measured OCVs under steam electrolysis conditions were 0.91 V, 0.93 V, 0.95 V and 0.96 V for the temperatures measured in the range of 800 °C to 650 °C, respectively. Both the steam/hydrogen gas ratio and the temperature affect the OCV values.[40] As expected, the open circuit voltage of the cell clearly decreases with increasing temperature. It can be noted that the current densities of the cell at an electrolysis voltage of 1.3 V are 0.91 A cm^{-2}, 0.70 A cm^{-2}, 0.45 A cm^{-2} and 0.27 A cm^{-2} at 800 °C, 750 °C, 700 °C and 650 °C, respectively. This indicates that a higher current density was achieved upon increasing the temperature at a given electrolysis voltage, contributing to improved hydrogen production.[41] Furthermore, an increased operation temperature can reduce the electrolyte electrical resistance and improve the electrode surface activity and electrolysis efficiency.[42] Typical Nyquist EIS at open circuit conditions for the cells under steam electrolysis mode are shown in Fig. 6(b). It can be seen that the R_Ω values of 0.27 Ω cm^2, 0.28 Ω cm^2, 0.30 Ω cm^2 and 0.40 cm^2 observed at 800 °C to 650 °C are similar to those obtained under fuel cell operation. The overall cell resistances were 0.50 Ω cm^2, 0.58 Ω cm^2, 0.96 Ω cm^2 and 1.64 Ω cm^2 at the temperatures measured in the range from 800 °C to 650 °C. The total cell resistances rapidly increased with decreasing temperature, mainly due to the significant increase in the electrode polarization. The R_p values changed significantly from 0.23 Ω cm^2 at 800 °C to 1.25 Ω cm^2 at 650 °C. It can be seen that the impedance spectra in Fig. 6(b) are nearly unchanged compared to those in

Fig. 3(b). In agreement with other studies,[43,44] the cell revealed no real change in electrode performance as its operation changed from anodic to cathodic behavior.

3.6 Long-term durability of the cell under SOEC mode and subsequent microstructure of the post-test cell by SEM coupled with EDS characterization

The long-term stability of the Ni–YSZ/YSZ/SSC–YSZ cell was potentiostatically tested under steam electrolysis mode. Fig. 7 depicts the long-term durability of the cell at 700 °C and at an electrolysis voltage of 1.3 V for more than 100 h. For the long-term durability operation under SOEC mode, a gas mixture of 50% H_2O, 25% H_2 and 25% N_2 was introduced to the hydrogen electrode as the reactant gas while the oxygen electrode was left open to the atmospheric environment. It can be seen that the cell under electrolysis conditions experienced a gradual degradation during the test period of over 100 h. The long-term durability of SOECs has been widely studied.[45] It has been reported that an overall voltage degradation rate of 3.8% kh^{-1} was achieved for a Ni–YSZ/YSZ/GDC/LSCF cell under SOEC operation.[32] It was found that SOEC degradation normally depends on the testing conditions, such as current density, steam content in fuel gas, temperature, *etc.* Moreover, an LSM–YSZ composite was investigated as the oxygen electrode for steam electrolysis over a period of 200 h and rapid deterioration was revealed to have occurred within 50 h.[6] The degradation of the cell performance was likely associated with its microstructural variation.[46] An SEM micrograph, as well as the EDS spectrum of a selected area of the cell after the long-term steam electrolysis operation, is shown in Fig. 8. As indicated by Fig. 8(a), a gap was clearly formed between the SSC–YSZ oxygen electrode and the YSZ electrolyte, indicating the delamination of the oxygen electrode and thus the cell failure. The oxygen electrode delamination, leading to cell deterioration under the SOEC mode, has been previously reported.[6,46–48] Chen *et al.*[46] reported that the delamination of a LSM oxygen electrode from YSZ electrolyte was due to the local tensile strains within the LSM particles and subsequent formation of microcracks and nanoparticles at the oxygen electrode/electrolyte interface. Additionally, high oxygen evolution rates, releasing gas

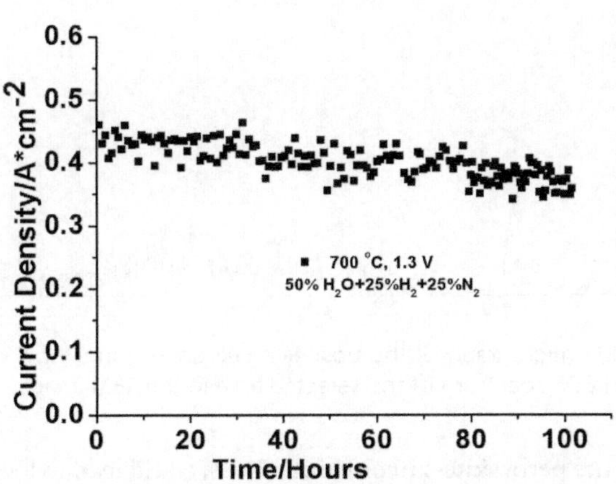

Fig. 7 Long-term durability operation of the Ni–YSZ/YSZ/SSC–YSZ cell at 700 °C and at a constant voltage of 1.3 V under steam electrolysis mode.

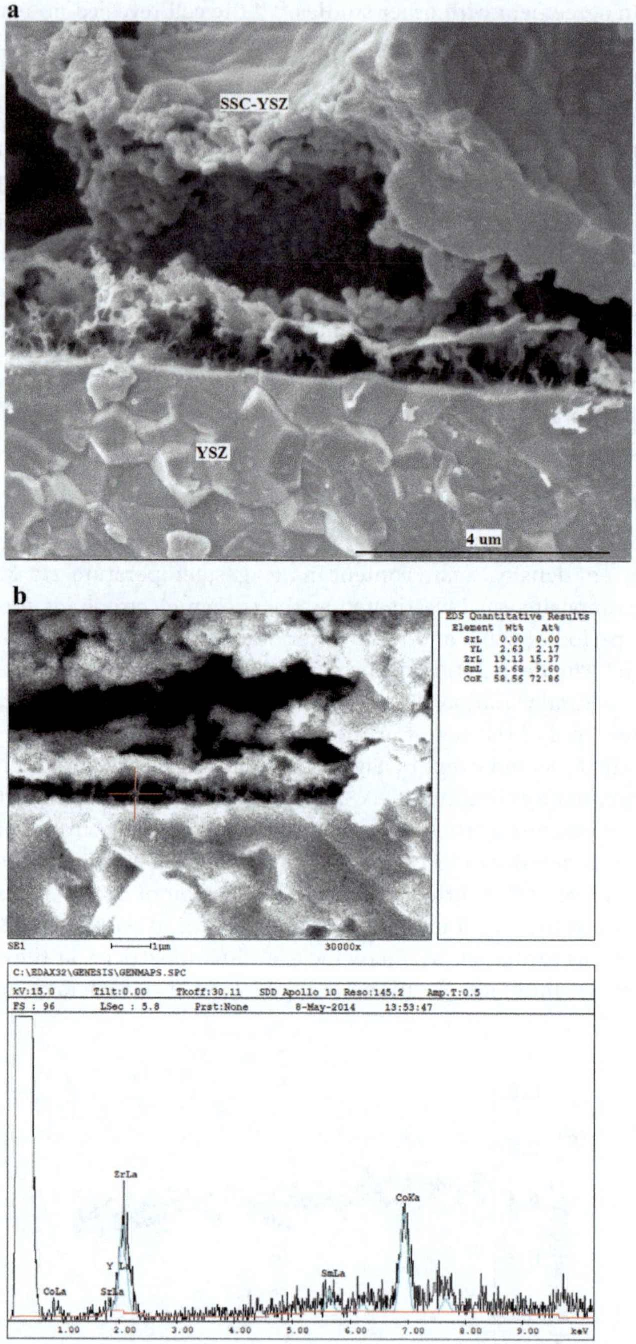

Fig. 8 (a) An SEM micrograph of the post-test cell under long-term steam electrolysis mode and (b) an EDS spectrum of the selected area in the SEM image.

into defects at the perovskite–zirconia interfaces, could lead to localized pressure-induced cracking at the interface.[11] Furthermore, agglomeration of the SSC particles occurred under steam electrolysis conditions, just as under fuel cell mode. Both the delamination of the oxygen electrode from the electrolyte and the agglomeration of SSC particles contributed to the cell degradation under steam electrode mode. An EDS spectrum of the selected area in the SEM image is exhibited in Fig. 8(b). It suggests that a high cobalt content is present at the oxygen electrode/electrolyte interface. Ref. 46 also revealed disintegration of the LSM particles at the oxygen electrode/electrolyte interface to form nanoparticles. The formation of lanthanum zirconate was evidenced at the interface of a YSZ electrolyte and a LSM oxygen electrode under anode polarization by Keane *et al.*[48] By applying a large cathode polarization to SOFCs, manganese oxides can be precipitated from the LSM electrode material.[49] Additionally, Co-rich particles were reported to be segregated on an LSCF electrode surface, as determined from an EDS spectrum.[50] Thus, according to the EDS spectrum, it is likely that under electrolysis, cobalt-enriched particles (particularly cobalt oxides) were segregated from the infiltrated SSC composite at the electrolyte/oxygen electrode interface; this requires further experimental verification.

4. Conclusion

An SSC–YSZ oxygen electrode was prepared by an infiltration process and used for both SOFCs and SOECs, with NiO–YSZ as the hydrogen electrode and YSZ as the electrolyte. The electrochemical performance and the long-term stability were investigated under power generation and steam electrolysis modes, using polarization curves and electrochemical impedance spectroscopy. SEM coupled with EDS was employed to examine the microstructure of the cells before and after long-term stability testing under both modes. Agglomeration of the infiltrated SSC particles resulted in degradation of the cell performance under long-term power generation mode. On the other hand, in addition to agglomeration of the SSC particles, delamination of the SSC–YSZ oxygen electrode from the YSZ electrolyte, along with deposition of cobalt-rich particles (most likely cobalt oxides) at the interface, was likely responsible for the cell deterioration under long-term steam electrolysis.

Acknowledgements

This work was supported by the National Basic Research Program of China (Grant No. 2012CB215404), State Key Laboratory of Power Systems in Tsinghua University (No. SKLD15Z02, Fuel Cell Distributed Power Generation System), Tsinghua University "Backbone Talent Support Plan" (No. 543404001), and One-hundred Leading Talents Development Project for Progress on Science and Technology of Beijing.

References

1 M. C. Williams, J. P. Strakey, W. A. Surdoval and L. C. Wilson, *Solid State Ionics*, 2006, **177**, 2039–2044.

2 P. Ried, C. Lorenz, A. Brönstrup, T. Graule, N. H. Menzler, W. Sitte and P. Holtappels, *J. Eur. Ceram. Soc.*, 2008, **28**, 1801–1808.

3 W. Z. Zhu and S. C. Deevi, *Mater. Sci. Eng., A*, 2003, **362**, 228–239.

4 M. Pihlatie, A. Kaiser and M. Mogensen, *J. Eur. Ceram. Soc.*, 2009, **29**, 1657–1664.
5 X. Wang, B. Yu, W. Q. Zhang, J. Chen, X. Luo and K. Stephan, *Int. J. Hydrogen Energy*, 2012, **37**, 12833–12838.
6 P. Kim-Lohsoontorn, Y.-M. Kim, N. Laosiripojana and J. Bae, *Int. J. Hydrogen Energy*, 2011, **36**, 9420–9427.
7 J. C. Ruiz-Morales, D. Marrero-López, J. Canales-Vázquez and J. T. S. Irvine, *RSC Adv.*, 2011, **1**, 1403–1414.
8 H. Fukunaga, M. Koyama, N. Takahashi, C. Wen and K. Yamada, *Solid State Ionics*, 2000, **132**, 279–285.
9 Y. Huang, K. Ahn, J. M. Vohs and R. J. Gorte, *J. Electrochem. Soc.*, 2004, **151**, A1592–A1597.
10 R. Küngas, F. Bidrawn, J. M. Vohs and R. J. Gorte, *Electrochem. Solid-State Lett.*, 2010, **13**, B87–B90.
11 J. R. Mawdsley, J. D. Carter, A. J. Kropf, B. Yildiz and V. A. Maroni, *Int. J. Hydrogen Energy*, 2009, **34**, 4198–4207.
12 J. M. Vohs and R. J. Gorte, *Adv. Mater.*, 2009, **21**, 943–956.
13 J. D. Nicholas and S. A. Barnett, *J. Electrochem. Soc.*, 2010, **157**, B536–B541.
14 H. Zhang, F. Zhao, F. L. Chen and C. R. Xia, *Solid State Ionics*, 2011, **192**, 591–594.
15 W. Wang, Y. Huang, S. Jung, J. M. Vohs and R. J. Gorte, *J. Electrochem. Soc.*, 2006, **153**, A2066–A2070.
16 W. Jiang, Z. Lü, B. Wei, Z. H. Wang, X. B. Zhu, Y. T. Tian, X. Q. Huang and W. H. Su, *Fuel Cells*, 2014, **14**, 76–82.
17 H. Fan, M. Keane, N. Li, D. Tang, P. Singh and M. F. Han, *Int. J. Hydrogen Energy*, 2014, **39**, 14071–14078.
18 Z. Liu, Z.-W. Zheng, M.-F. Han and M.-L. Liu, *J. Power Sources*, 2010, **195**, 7230–7233.
19 K. F. Chen, N. Ai and S. P. Jiang, *Electrochem. Commun.*, 2012, **19**, 119–122.
20 X. Y. Lou, Z. Liu, S. Z. Wang, Y. H. Xiu, C. P. Wong and M. L. Liu, *J. Power Sources*, 2010, **195**, 419–424.
21 H. Y. Tu, Y. Takeda, N. Imanishi and O. Yamamoto, *Solid State Ionics*, 1997, **100**, 283–288.
22 C.-X. Li, S. Liu, Y. Zhang and C.-J. Li, *Int. J. Hydrogen Energy*, 2012, **37**, 13097–13102.
23 D. Ding, X. X. Li, S. Y. Lai, K. Gerdes and M. L. Liu, *Energy Environ. Sci.*, 2014, **7**, 552–575.
24 M. A. Laguna-Bercero, J. A. Kilner and S. J. Skinner, *Solid State Ionics*, 2011, **192**, 501–504.
25 M. F. Liu, J. F. Gao, X. Q. Liu and G. Y. Meng, *Int. J. Hydrogen Energy*, 2011, **36**, 13741–13745.
26 Y. Huang, J. M. Vohs and R. J. Gorte, *J. Electrochem. Soc.*, 2004, **151**, A646–A651.
27 J. Chen, F. L. Liang, L. Liu, S. P. Jiang, B. Chi, J. Pu and J. Li, *J. Power Sources*, 2008, **183**, 586–589.
28 Y.-C. Chang, M.-C. Lee, W.-X. Kao, C.-H. Wang, T.-N. Lin and J.-C. Chang, *J. Taiwan Inst. Chem. Eng.*, 2011, **42**, 775–782.
29 T. Ishihara, M. Honda, T. Shibayama, H. Minami, H. Nishguchi and Y. Takita, *J. Electrochem. Soc.*, 1998, **145**, 3177–3183.

30 W. Zhou, Z. P. Shao, F. L. Liang, Z.-G. Chen, Z. H. Zhu, W. Q. Jin and N. P. Xu, *J. Mater. Chem.*, 2011, **21**, 15343–15351.
31 W. Wang, M. D. Gross, J. M. Vohs and R. J. Gorte, *J. Electrochem. Soc.*, 2007, **154**, B439–B445.
32 F. Tietz, D. Sebold, A. Brisse and J. Schefold, *J. Power Sources*, 2013, **223**, 129–135.
33 C. Lu, T. Z. Sholklapper, C. P. Jacobson, S. J. Visco and L. C. D. Jonghe, *J. Electrochem. Soc.*, 2006, **153**, A1115–A1119.
34 T. Z. Wu, Y. Q. Zhao, R. R. Peng and C. R. Xia, *Electrochim. Acta*, 2009, **54**, 4888–4892.
35 J. Chen, F. L. Liang, B. Chi, J. Pu, S. P. Jiang and L. Jian, *J. Power Sources*, 2009, **194**, 275–280.
36 F. C. Wang, D. J. Chen and Z. P. Shao, *J. Power Sources*, 2012, **216**, 208–215.
37 Y. H. Liu, F. Z. Wang, B. Chi, J. Pu, L. Jian and S. P. Jiang, *J. Alloys Compd.*, 2013, **578**, 37–43.
38 S. Lee, N. Miller and K. Gerdes, *J. Electrochem. Soc.*, 2012, **159**, F301–F308.
39 B. Zhu, I. Albinsson, C. Andersson, K. Borsand, M. Nilsson and B.-E. Mellander, *Electrochem. Commun.*, 2006, **8**, 495–498.
40 B. Annabelle, J. Schefold and M. Zahid, *Int. J. Hydrogen Energy*, 2008, **33**, 5375–5382.
41 C. H. Yang, C. Jin, A. Coffin and F. L. Chen, *Int. J. Hydrogen Energy*, 2010, **35**, 5187–5193.
42 S. K. Mazloomi and N. Sulaiman, *Renewable Sustainable Energy Rev.*, 2012, **16**, 4257–4263.
43 J. E. O'Brien, *J. Heat Transfer*, 2012, **134**, 031017.
44 M. A. Laguna-Bercero, N. Kinadjan, R. Sayers, H. El Shinawi, C. Greaves and S. J. Skinner, *Fuel Cells*, 2011, **11**, 102–107.
45 S. D. Ebbesen, S. H. Jensen, A. Hauch and M. B. Mongensen, *Chem. Rev.*, 2014, **114**, 10697–10734.
46 K. F. Chen and S. P. Jiang, *Int. J. Hydrogen Energy*, 2011, **36**, 10541–10549.
47 A. V. Virkar, *Int. J. Hydrogen Energy*, 2010, **235**, 9527–9543.
48 M. Keane, M. K. Mahapatra, A. Verma and P. Singh, *Int. J. Hydrogen Energy*, 2012, **37**, 16776–16785.
49 E. Park, S. Taniguchi, T. Daio, J.-T. Chou and K. Sasaki, *Int. J. Hydrogen Energy*, 2014, **19**, 1463–1475.
50 L. Zhao, J. Drennan, C. Kong, S. Amarasinghe and S. P. Jiang, *ECS Trans.*, 2013, **57**, 599–604.

Faraday Discussions

PAPER

Electrode kinetics of the NiO porous electrode for oxygen production in the molten carbonate electrolysis cell (MCEC)

Lan Hu,* Göran Lindbergh and Carina Lagergren

Received 30th January 2015, Accepted 5th February 2015
DOI: 10.1039/c5fd00011d

The performance of a molten carbonate electrolysis cell (MCEC) is to a great extent determined by the anode, *i.e.* the oxygen production reaction at the porous NiO electrode. In this study, stationary polarization curves for the NiO electrode were measured under varying gas compositions and temperatures. The exchange current densities were calculated numerically from the slopes at low overpotential. Positive dependency on the exchange current density was found for the partial pressure of oxygen. When the temperature was increased in the range 600–650 °C, the reaction order of oxygen decreased from 0.97 to 0.80. However, there are two different cases for the partial pressure dependency of carbon dioxide within this temperature range: positive values, 0.09–0.30, for the reaction order at lower CO_2 concentration, and negative values, −0.26–0.01, with increasing CO_2 content. A comparison of theoretically obtained data indicates that the oxygen-producing reaction in MCEC could be reasonably satisfied by the reverse of oxygen reduction by the oxygen mechanism I, an $n = 4$ electron reaction, assuming a low coverage of oxide ions at high CO_2 content and an intermediate coverage for a low CO_2 concentration.

Introduction

There is a growing interest in high-temperature electrolysis cell technology for producing hydrogen and/or syngas, since these fuels are considered important for future energy systems. Electrolysis in molten carbonate salts at high temperature (typically 650 °C) is a promising option, especially when in combination with renewable electricity resources such as wind power and/or solar energy. Some researchers have evidenced the possibility of electrolysis in molten carbonates, but mainly by converting CO_2 into CO.[1–4] However, most experiments were performed on flag electrodes, which differ from porous electrodes with regards to the electrode surface and mass-transport properties. Our previous study[5] showed that it is feasible to run the molten carbonate fuel cell reversibly with conventional Ni-

Applied Electrochemistry, Department of Chemical Engineering and Technology, KTH Royal Institute of Technology, SE-100 44 Stockholm, Sweden. E-mail: lanhu@kth.se; Tel: +46 8 7908244 based porous electrodes at 650 °C. A lower polarization loss has been found for the electrolysis cell compared to the fuel cell, mainly due to the NiO electrode performing much better as an anode in MCEC mode.[5] The activity of the NiO electrode provides a dominant contribution to the performance of the cell, regardless of fuel cell or electrolyzer operation. Thus, it is essential to elucidate the kinetics at the NiO porous electrode for oxygen reduction in the fuel cell, as well as for oxygen production in the electrolysis cell.

However, the kinetics and reaction mechanism at the NiO porous anode for oxygen production in MCEC mode are still unknown. A feasible path to overcome this is to understand whether the reversed process of oxygen reduction takes place at the NiO electrode or whether there are some other electrochemical/chemical reaction pathways in the electrolysis operation. Several reaction mechanisms for the oxygen reduction reaction in fuel cell mode have been discussed in the literature,[6–12] see Table 1. The corresponding reaction orders of oxygen and carbon dioxide for these five mechanisms have been reported by Makkus[13] for varying rate-determining steps (rds). According to Appleby and Nicholson,[6–8] the peroxide (O_2^{2-}) mechanism was suggested for oxygen reduction in pure lithium

Table 1 Reaction mechanisms reported in the literature for oxygen reduction[6–12]

The peroxide mechanism: $n = 2$

$O_2 + 2CO_3^{2-} \rightleftharpoons 2O_2^{2-} + 2CO_2$

$O_2^{2-} + e^- \rightleftharpoons O^{2-} + (O^-)$

$(O^-) + e^- \rightleftharpoons O^{2-}$

$O^{2-} + CO_2 \rightleftharpoons CO_3^{2-}$

The superoxide mechanism: $n = 3$

$O_2 + 2CO_3^{2-} \rightleftharpoons O_2^{2-} + 2CO_2$

$O_2 + O_2^{2-} \rightleftharpoons 2O_2^-$

$O_2^- + e^- \rightleftharpoons O_2^{2-}$

$O_2^{2-} + e^- \rightleftharpoons O^{2-} + (O^-)$

$(O^-) + e^- \rightleftharpoons O^{2-}$

$O^{2-} + CO_2 \rightleftharpoons CO_3^{2-}$

The peroxycarbonate mechanism: $n = 2$

$O_2 + 2CO_3^{2-} \rightleftharpoons 2CO_4^{2-}$

$CO_4^{2-} + e^- \rightleftharpoons CO_3^{2-} + (O^-)$

$(O^-) + e^- \rightleftharpoons O^{2-}$

$O^{2-} + CO_2 \rightleftharpoons CO_3^{2-}$

Oxygen mechanism I: $n = 4$

$O_2 + e^- \rightleftharpoons O_2^-$

$O_2^- + e^- \rightleftharpoons O_2^{2-}$

$O_2^{2-} + e^- \rightleftharpoons O^{2-} + (O^-)$

$(O^-) + e^- \rightleftharpoons O^{2-}$

$O^{2-} + CO_2 \rightleftharpoons CO_3^{2-}$

Oxygen mechanism II: $n = 2$

$O_2 \rightleftharpoons 2O$

$O + e^- \rightleftharpoons O^-$

$O^- + e^- \rightleftharpoons O^{2-}$

$O^{2-} + CO_2 \rightleftharpoons CO_3^{2-}$

and lithium rich melts, while the superoxide (O_2^-) path seemed to be involved in sodium and potassium-rich melts. Dunks and Stelman[9] and Takahashi[10] proposed a mechanism based on the peroxycarbonate (CO_4^{2-}) ion. Kunz *et al.*[11] suggested oxygen mechanism I to evaluate the concept of the homogeneous/agglomerate model. Makkus *et al.*[12] discussed oxygen mechanism II for the thin-film model. All proposed mechanisms consist of two or more single electron transfer reactions. All five of these reaction mechanisms show the same final recombination step:

$$O^{2-} + CO_2 \rightleftharpoons CO_3^{2-} \tag{1}$$

Kunz and Murphy,[14] Lu and Selman,[15] and Peelen *et al.*[16] suggested that this recombination reaction is possibly the rate-determining step at the porous electrodes under MCFC conditions, due to the slow reaction rate.

In MCEC mode, the main electrochemical reaction taking place at the NiO porous electrode for oxygen production is the electrolysis of the carbonate ion (CO_3^{2-}).

$$CO_3^{2-} \rightleftharpoons \frac{1}{2}O_2 + CO_2 + 2e^- \tag{2}$$

Sugiura *et al.*[17] suggested separation of CO_2 in a molten carbonate fuel cell (MCFC) device. One of the options suggested was to use reaction (2) as the anode reaction in the device.

This paper aims to study the kinetics and reaction mechanism at the NiO porous electrode for oxygen production in MCEC operation. For this purpose the general procedure is to determine the reaction orders of oxygen and carbon dioxide. Thus, steady-state polarization data are investigated for varying gas compositions and operating temperatures. The exchange current densities, which differ with the concentration of the electroactive species and the temperature, are calculated numerically from the slopes of the polarization curves at low overpotential. Based on this, the partial pressure dependencies of oxygen and carbon dioxide could be evaluated. Meanwhile, a possible reaction mechanism for oxygen production at the NiO porous electrode in MCEC mode is proposed.

Experimental

The cell set-up used in this study was identical to that described in a previous study.[5] Experimental data were obtained from the laboratory cell unit with a geometrical area of 3 cm^2, where the cell components were provided by Ansaldo Fuel Cells in Italy. The cell has a porous Ni anode, oxidized and lithiated *in situ*, and a porous Ni–Cr alloy as the cathode. The electrodes are separated by a porous $LiAlO_2$ matrix, which also supports the electrolyte, a eutectic mixture of 62/38 mol % Li_2CO_3/K_2CO_3. The current collectors used on the NiO and Ni electrode sides are made of stainless steel SS316 and nickel, respectively. Two current collector probes each separately measure the current and the potential. Two reference electrodes, consisting of Au wires in equilibrium with a gas mixture of 33/67% O_2/CO_2, are placed in separate chambers filled with the same electrolyte as that in the

Table 2 The inlet gas compositions, by percentage

Gas no.	O_2/%	CO_2/%	N_2/%
(1) Changing O_2			
1.1	1.5	30	68.5
1.2	2.5	30	67.5
1.3	5	30	65
1.4	10	30	60
1.5[a]	15	30	55
1.6	20	30	50
1.7	25	30	45
1.8	30	30	40
(2) Changing CO_2			
2.1	15	2.5	82.5
2.2	15	5	80
2.3	15	10	75
2.4	15	15	70
2.5	15	20	65
2.6	15	25	60
2.7[a]	15	30	55
2.8	15	35	50
2.9	15	40	45

[a] The standard inlet gas, 15/30/55% $O_2/CO_2/N_2$.

cell. These chambers are connected to the cell through a capillary with a gold plug.

The cell is operated at three different temperatures, 600, 625 and 650 °C. The inlet fuel gas mixture consisting of 25/25/25/25% $H_2/CO_2/H_2O/N_2$ is introduced to the Ni electrode. The concentration of H_2O in the fuel gas is controlled by the temperature of the water in the humidifier, 65 °C. The different gas mixtures of $O_2/CO_2/N_2$ for the NiO electrode are shown in Table 2. The concentrations of oxygen and carbon dioxide are varied separately in group 1 and group 2. The O_2 and CO_2 partial pressures increase in the ranges 1.5–30% and 2.5–40%, respectively. In order to reduce the effects of mass-transfer limitations in the gaseous phase, the flow rates of the inlet gases for the NiO porous electrode are held at as high as approximately 250 ml min^{-1}. The gas flow rate for the Ni electrode is also about 250 ml min^{-1}, while the reference gas has a flow rate of 20 ml min^{-1}. All of the gases used in the experiments are certified gas mixtures from AGA Gas AB, Sweden.

The polarization curves are measured using a Solartron Interface SI1287 supported by CorrWare software. The current-interrupt technique is used to separate the ohmic loss from other types of polarization. In this paper, the negative current density refers to the electrolysis mode, as shown in the figures below.

Theory

Elucidating the kinetics and mechanism of reaction (2) is generally done by determining the partial pressure dependencies of oxygen and carbon dioxide. The same method was used to understand the hydrogen production on the cathode side, the Ni electrode.[18] When analyzing kinetic data it is important to consider the current distribution along the depth of the NiO electrode due to its porous characteristics.[19] The current distribution is determined by a combination of the intrinsic electrode activity, its specific surface area, the effective diffusion path lengths of the electroactive species, and the effective conductivities of the electrolyte and electrode. In the present study the polarization data are analyzed only at a low overpotential range, where the electrode is assumed to be mainly under charge-transfer limitation and the influence of mass-transfer limitation is negligible. According to Lagergren and Simonsson,[20] in order to take the current distribution into account when analyzing data, the expression for the total polarization resistance is

$$\frac{d\eta}{di} = \frac{L}{k_1 + k_2} + \frac{L}{\sqrt{a}}\frac{1}{k_1 + k_2}\left[\frac{2}{\sinh\sqrt{a}} + \left(\frac{k_1}{k_2} + \frac{k_2}{k_1}\right)\coth\sqrt{a}\right] \tag{3}$$

where

$$a = \frac{nFL^2Si_0}{RT}\left(\frac{1}{k_1} + \frac{1}{k_2}\right) \tag{4}$$

L is the electrode thickness and k_1 and k_2 are the effective conductivities in the electrode and electrolyte phases, respectively. S is the exterior agglomerate surface area.

There are two limiting cases of eqn (3), in which a approaches either zero or infinity. These two cases refer to uniform and uneven current distributions, respectively.

According to eqn (3), the ohmic loss, $\frac{L}{k_1 + k_2}$, due to the finite conductivities of the electrolyte and electrodes, is included in the total polarization resistance. In this study the current-interrupt technique is used to correct the iR drop, and therefore the experimentally obtained polarization data does not contain the ohmic contribution.

The partial pressure dependencies of oxygen and carbon dioxide can be expressed as

$$i_0 = i_0^0(p_{O_2})^x(p_{CO_2})^y \tag{5}$$

where i_0^0 is the standard exchange current density, and x and y denote the reaction orders of oxygen and carbon dioxide, respectively.

Results and discussion

The iR-corrected polarization curves for different gas compositions at 650 °C are shown in Fig. 1. The blue curves in the figure correspond to the performance of the NiO electrode with the standard inlet gas 15/30/55% $O_2/CO_2/N_2$ in the electrolysis operation. The red symbols refer to gases with lower content than the standard level when changing O_2 and CO_2 partial pressures separately, while the green symbols refer to gases with higher concentrations. In Fig. 1(a) the polarization loss of the NiO electrode decreases gradually when increasing the

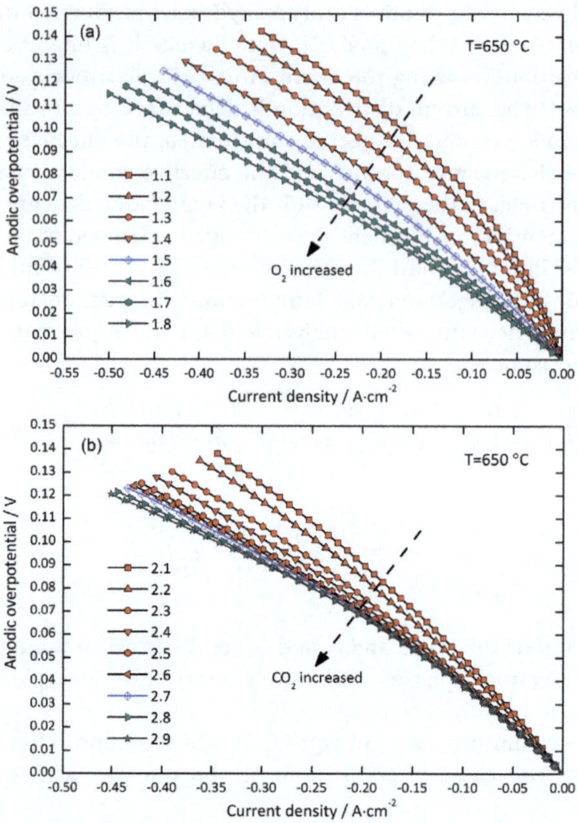

Fig. 1 iR-corrected polarization curves for the NiO electrode in electrolysis mode at 650 °C for different gas compositions, changing (a) O_2 and (b) CO_2 partial pressures.

concentration of O_2 from 1.5 to 30%. The polarization curves show a tendency to bend downward at high current densities, and under lower O_2 content that bending is observed even at low current density. It seems that mass-transfer limitation does not take place at high current density, as otherwise the curves would be bending upward. When increasing the CO_2 content in the range 2.5–20%, the activity of the NiO electrode shows a slow improvement in oxygen production at 650 °C, as shown in Fig. 1(b). However, with an increase in CO_2 content from 20 to 40%, only a very small impact on the polarization loss of the NiO electrode is observed also at high current density.

In order to more thoroughly analyze the polarization data, the electrochemical performance of the NiO electrode was investigated within a low overpotential range, <45 mV, at 600–650 °C. Varying the concentration of O_2 within the range 1.5–30% gives the quasi-linear characteristics of the iR-corrected curves at all measured temperatures, as visualized in Fig. 2. Increases in O_2 content improve the NiO electrode polarization performance for oxygen production in the electrolysis operation. However, the impact of CO_2 concentration on the polarization performance of the NiO electrode is strongly dependent on temperature, as shown in Fig. 3. At 600 °C, increasing the CO_2 content from 2.5 to 10% gives only a small improvement in the performance of the NiO electrode; in contrast, upon

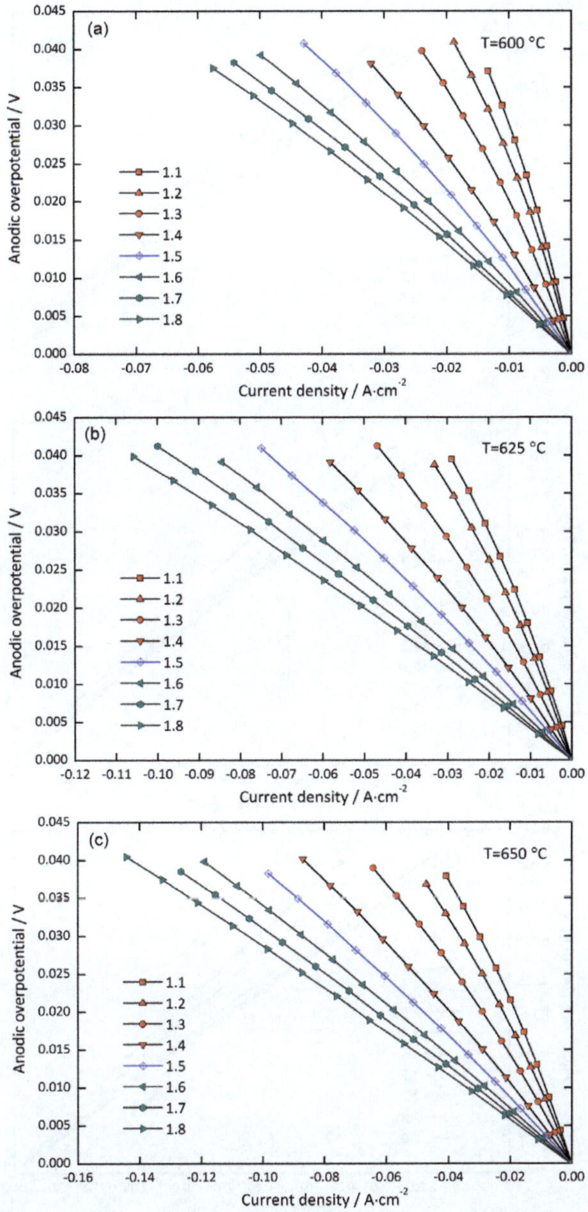

Fig. 2 iR-corrected polarization curves at low overpotential for the NiO electrode in electrolysis mode for different percentages of O_2: (a) 600 °C, (b) 625 °C and (c) 650 °C.

increasing the CO_2 content to 40%, increasingly poorer behavior is observed for the electrode. In Fig. 3(b) the polarization loss of the NiO electrode decreases gradually as the CO_2 concentration rises from 2.5 to 15%, and it then shows very small increases in the electrode polarization with increasing CO_2 content in the range 15–40% at 625 °C. When reaching 650 °C, the NiO electrode has an improved activity for oxygen production in the range 2.5–20% CO_2 content, and

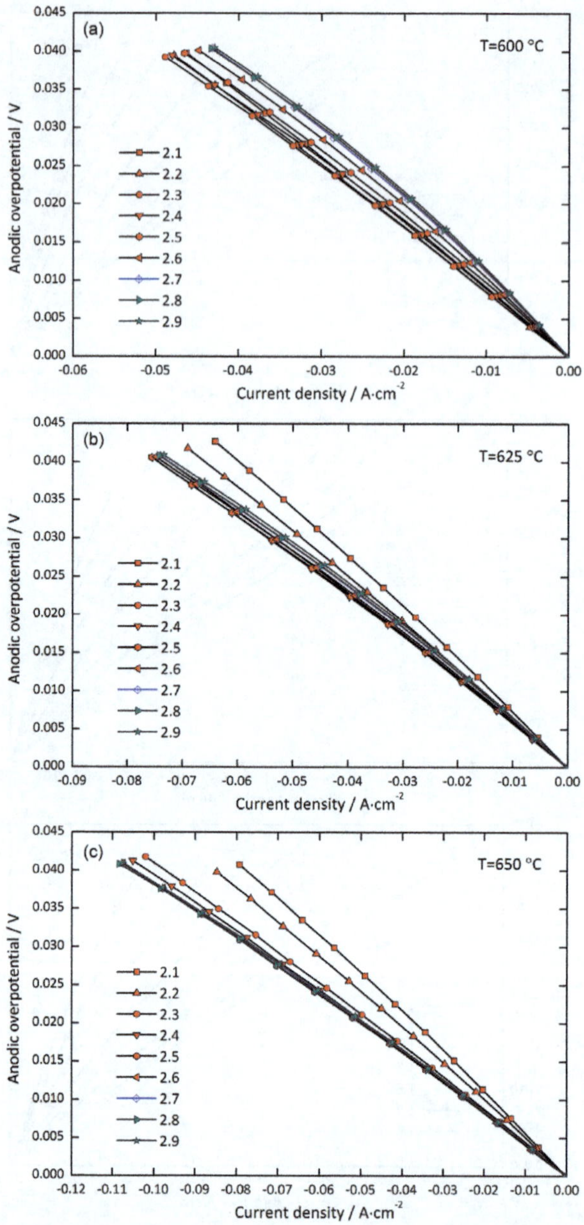

Fig. 3 iR-corrected polarization curves at low overpotential for the NiO electrode in electrolysis mode for different percentages of CO_2: (a) 600 °C, (b) 625 °C and (c) 650 °C.

the electrode performance remains almost stable within the partial pressure range 20–40%, as seen in Fig. 3(c). The carbon dioxide content affecting the performance of the NiO electrode for oxygen production varies with the temperature. It is assumed that in this low overpotential range the electrode is under kinetic control as it is not necessary to take mass-transfer limitations into consideration. All of the measured polarization curves are quasi-linear, and the polarization resistances, $d\eta/di$, could thus be evaluated from the slopes of these curves by linear regression analysis.

The exchange current densities for each gas composition and temperature could be obtained by solving eqn (3). Parameter values used for solving eqn (3) are listed in Table 3. It is assumed that the NiO electrode has a pseudo-homogeneous structure and that the electrode reaction takes place mainly at the exterior surface of the agglomerates.[19] Assuming that the agglomerates are spherical and have a radius of 5 μm, the external agglomerate area, S, is estimated to be 3×10^5 m^{-1}.[19] The effective conductivity of the porous NiO electrode is in the order of 1000 S m^{-1}.[20,21] The effective conductivity in the pore electrolyte was measured to be 0.9–2.2 S m^{-1} at 650 °C.[22] Thus, in this work the effective electrolyte conductivity at 650 °C was set to be 2.0 S m^{-1}, from which values of 1.7 S m^{-1} and 1.8 S m^{-1} were calculated for 600 and 625 °C, respectively, using the Arrhenius equation.[23] The exchange current density i_0 for the standard inlet gas, 15/30/55% $O_2/CO_2/N_2$ (1.5* and 2.7* in Table 4), shows an increase from 4.0 to 22.7 A m^{-2} when the temperature is raised from 600 to 650 °C. The value of i_0 for oxygen production in the electrolysis cell is in the same order of magnitude as that for oxygen reduction at the NiO porous electrode in fuel cell mode (22.4 A m^{-2} at 650 °C).[19]

A striking feature of Table 4 is that the exchange current density increases with the oxygen partial pressure when the NiO electrode is used as the anode in MCEC mode. The values are in the range 0.5–8.6, 1.9–25.0 and 3.8–41.4 A m^{-2} at 600, 625 and 650 °C, respectively. However, concerning the carbon dioxide partial pressure, the exchange current density should be analyzed at low content and at higher concentration. For example, the exchange current density increases slightly from 5.0 to 5.7 A m^{-2} in the range 2.5–10% CO_2 at 600 °C. The value then shows a slow decrease to 4.1 A m^{-2} as the CO_2 concentration reaches 40%. The content of carbon dioxide affects the exchange current density differently for different operating temperatures, corresponding with the results of the polarization data discussed above. At 625 °C, at CO_2 partial pressures of 2.5–15%, the exchange current density increases from 8.1 to 12.4 A m^{-2}; quite a small decrease in the value, to 12.4–11.6 A m^{-2}, is then observed for 15–40% CO_2 content. As the cell temperature goes to 650 °C, the values of i_0 increase from 12.5 to 22.8 A m^{-2} with increasing CO_2 concentration from 2.5 to 20%, but then the data remain almost stable within the CO_2 range 20–40%. The exchange current density is not as strongly affected by the partial pressure of carbon dioxide as by that of oxygen. This indicates that carbon dioxide may have a smaller impact on the charge-transfer polarization of the electrode, since the exchange current density reflects the rate of the charge transfer.

Table 3 Parameter values used when solving eqn (3)

		600 °C	625 °C	650 °C
L	0.6 mm			
r	5 μm			
S	3×10^5 m^{-1}			
n	4			
k_1	1000 S m^{-1}			
k_2		1.7 S m^{-1}	1.8 S m^{-1}	2.0 S m^{-1}

Table 4 Experimentally obtained exchange current densities at different gas compositions and temperatures

Gas no.	i_0/A m^{-2} 600 °C	i_0/A m^{-2} 625 °C	i_0/A m^{-2} 650 °C
(1) Changing O_2			
1.1	0.5	1.9	3.8
1.2	0.8	2.6	5.4
1.3	1.3	4.6	8.9
1.4	2.6	7.9	15.4
1.5*	4.0	12.0	21.6
1.6	5.9	16.5	29.2
1.7	7.3	20.8	35.2
1.8	8.6	25.0	41.4
(2) Changing CO_2			
2.1	5.0	8.1	12.5
2.2	5.4	9.7	15.0
2.3	5.7	11.8	19.7
2.4	5.5	12.4	21.4
2.5	5.1	12.4	22.8
2.6	4.6	12.2	22.6
2.7*	4.2	11.9	22.7
2.8	4.2	11.9	23.2
2.9	4.1	11.6	22.7

The reaction orders of oxygen and carbon dioxide at 600–650 °C are determined in Fig. 4 and 5. The linear regression results show that the oxygen partial pressure dependency decreases from 0.97 to 0.80 in the temperature range 600–650 °C. However, the dependency of the exchange current density on the carbon dioxide partial pressure shows two cases, as in Fig. 5. At lower CO_2 concentration, the reaction order of carbon dioxide gives positive values at all measured temperatures, and this then changes to negative values when increasing the CO_2

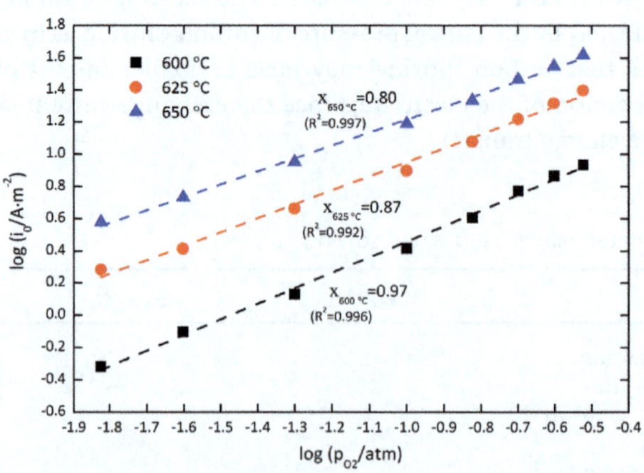

Fig. 4 Effect of oxygen partial pressure on the exchange current density, from 600 to 650 °C.

content. Specifically, at 600 °C the reaction order of carbon dioxide is 0.09 within the CO_2 content range 2.5–10%, and it becomes negative, −0.26, for 10–40% CO_2 content. In contrast to oxygen, the carbon dioxide partial pressure dependency increases with temperature. For 2.5–15% CO_2 content at 625 °C, the value of the reaction order is 0.25, and this becomes −0.07 within the concentration range 15–40% CO_2. When reaching 650 °C, the reaction order increases to 0.30 for 2.5–20% CO_2 content. The value then changes to 0.01 when increasing the CO_2 concentration to 40%. Thus, for all the measured temperatures from 600 to 650 °C, for a low concentration of CO_2, eqn (5) can be expressed as

$$i_0 = i_0^0 (p_{O_2})^{0.97-0.80} (p_{CO_2})^{0.09-0.30} \tag{6}$$

and for high CO_2 content it could be expressed as

$$i_0 = i_0^0 (p_{O_2})^{0.97-0.80} (p_{CO_2})^{-0.26-0.01} \tag{7}$$

As discussed above, the partial pressure dependencies obtained in this study are remarkably higher for oxygen than for carbon dioxide. The following reaction paths could probably be the mechanism for the oxygen production reaction in a molten carbonate electrolysis cell, on the basis of the reversed process of oxygen reduction in fuel cell operation (oxygen mechanism I) proposed by Kunz *et al.*[11]

$$CO_3^{2-} \rightleftharpoons O^{2-} + CO_2 \tag{8}$$

$$O^{2-} \rightleftharpoons (O^-) + e^- \tag{9}$$

$$O^{2-} + (O^-) \rightleftharpoons O_2^{2-} + e^- \tag{10}$$

$$O_2^{2-} \rightleftharpoons O_2^- + e^- \tag{11}$$

$$O_2^- \rightleftharpoons O_2 + e^-, \text{ rds} \tag{12}$$

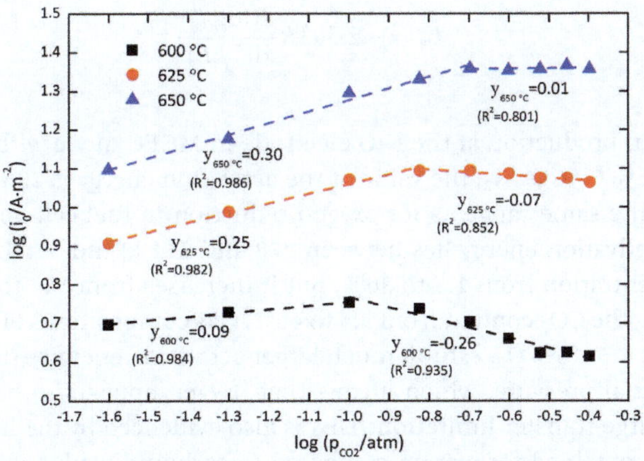

Fig. 5 Effect of carbon dioxide partial pressure on the exchange current density, from 600 to 650 °C.

It is assumed that (i) oxygen evolution reaction (12) is the rate-determining step and (ii) there is a low coverage of oxide ion. Assuming the symmetry factor $\beta = 0.5$ for the rate-determining step, the theoretical dependency of i_0 on the gas partial pressure is calculated in detail in the Appendix and is shown as

$$i_0 = i_0^0 (p_{O_2})^{0.875} (p_{CO_2})^{-0.25} \tag{13}$$

The theoretical reaction order is 0.875 for oxygen, satisfying the oxygen partial pressure dependency (0.97–0.80) determined from the experimental data. The theoretical value of −0.25 for carbon dioxide could reasonably fit the reaction order at high CO_2 concentration, as shown in eqn (7). However, it does not show agreement with the positive dependency at low CO_2 content. If assuming that the oxygen production is the reversed process of oxygen reduction in molten carbonate salts, the theoretical partial pressure dependencies for oxygen and carbon dioxide would be the same in both fuel cell and electrolyzer operations. Makkus[13] reported the values of the reaction orders for all of the reaction mechanisms for oxygen reduction mentioned in Table 1. According to Makkus, the exchange current density with a high coverage of O^{2-} can be expressed as

$$i_0 = i_0^0 (p_{O_2})^{0.875} (p_{CO_2})^{0.75} \tag{14}$$

The partial pressure dependency of carbon dioxide, y, could be affected by the chemical decomposition of $CO_3{}^{2-}$, producing O^{2-} and CO_2. For example, the reaction order is stable at 0.875 for oxygen and the value changes from negative (−0.25) to positive (0.75) for carbon dioxide when containing a high coverage of O^{2-}, see eqn (14). Upon lowering the carbon dioxide partial pressure, it makes sense from the perspective of chemical kinetics that the homogeneous reaction ($CO_3{}^{2-} \rightleftharpoons O^{2-} + CO_2$) will progress to produce the oxide ion. However, the theoretical value of 0.75 is still much higher than the experimental result (0.09–0.30), which indicates that reaction steps (8)–(12) with an intermediate coverage of O^{2-} could constitute a possible mechanism for oxygen production on the NiO porous electrode at a low CO_2 concentration.

The activation energy for the NiO electrode could be calculated by the Arrhenius equation, see Table 5.

$$E_a = -2.303R\frac{d(\log i_0)}{d(1/T)} \tag{15}$$

For oxygen production at the NiO electrode in MCEC mode with the standard gas, 15/30/55% $O_2/CO_2/N_2$, the value of the activation energy is 226 ± 1 kJ mol^{-1}, which is in the same range as for oxygen reduction in fuel cell operation.[24] The calculated activation energy lies between 278 and 211 kJ mol^{-1} when increasing the O_2 concentration from 1.5 to 30%, but it increases from 124 to 229 kJ mol^{-1} when varying the CO_2 content from 2.5 to 40%. Gas compositions of low O_2 partial pressure, *e.g.* 1.5–10% O_2, exhibit much larger activation energies than those with low CO_2 partial pressure, which means that oxygen appears to have a stronger effect on charge-transfer limitation. This is also evidenced by the higher reaction order values obtained for oxygen compared to carbon dioxide. As the activation energy is a clue in determining the rate-determining process,[25] it is probable that

Table 5 Calculation of the activation energy for the NiO electrode at different gas compositions

Gas no.	O_2/%	CO_2/%	N_2/%	E_a/kJ mol^{-1}
(1) Changing O_2				
1.1	1.5	30	68.5	278.3
1.2	2.5	30	67.5	258.5
1.3	5	30	65	254.3
1.4	10	30	60	238.7
1.5*	15	30	55	225.1
1.6	20	30	50	214.7
1.7	25	30	45	211.1
1.8	30	30	40	211.2
(2) Changing CO_2				
2.1	15	2.5	82.5	123.6
2.2	15	5	80	138.3
2.3	15	10	75	167.6
2.4	15	15	70	183.5
2.5	15	20	65	202.6
2.6	15	25	60	214.4
2.7*	15	30	55	226.5
2.8	15	35	50	229.0
2.9	15	40	45	229.3

the NiO electrode is under mixed control at quite low CO_2 concentrations, and is then limited by charge-transfer at high CO_2 concentrations. This confirms that at low CO_2 partial pressure, the reaction $CO_3^{2-} \rightleftharpoons O^{2-} + CO_2$ plays an important role in oxygen production in MCEC mode.

Conclusions

Stationary polarization data have been studied for the NiO electrode, as the anode in an electrolyzer, under varying gas compositions and temperatures. The slopes of the polarization curves at a low overpotential range were analyzed, while assuming the electrode to be under kinetic control.

The partial pressure dependencies on the exchange current density lie between 0.97 and 0.80 for oxygen in the temperature range 600–650 °C. There are two cases for the reaction order of carbon dioxide, positive values of 0.09–0.30 at low CO_2 concentration and then negative dependency (−0.26–0.01) with high CO_2 content.

The reverse process of oxygen reduction in MCFC operation, based on oxygen mechanism I, was found to suitably describe oxygen production in the electrolyzer. The theoretical reaction orders were 0.875 for oxygen and −0.25 for carbon dioxide under the assumption of a low coverage of the oxide ion. Due to the reaction $CO_3^{2-} \rightleftharpoons O^{2-} + CO_2$, the coverage of O^{2-} is dependent on the carbon dioxide concentration. For a high coverage of O^{2-}, the theoretical dependency changed to 0.75 for CO_2. Thus, according to this study, an intermediate coverage of the oxide ion might be achieved at low CO_2 content.

The activation energy of the NiO porous electrode for oxygen production lies between 278 and 211 kJ mol^{-1} for increasing O_2 content. In contrast, the values increased in the range 124–229 kJ mol^{-1} when increasing the CO_2 concentration. This indicates that the NiO electrode for oxygen production is generally under charge-transfer limitation, except at very low CO_2 partial pressure where it is under mixed control.

Appendix

Determination of reaction order

The reaction mechanism for oxygen production in a molten carbonate electrolysis cell follows

$$CO_3^{2-} \underset{k_{1c}}{\overset{k_{1a}}{\rightleftharpoons}} O^{2-} + CO_2 \quad \text{(A-1)}$$

$$O^{2-} \underset{k_{2c}}{\overset{k_{2a}}{\rightleftharpoons}} (O^-) + e^- \quad \text{(A-2)}$$

$$O^{2-} + (O^-) \underset{k_{3c}}{\overset{k_{3a}}{\rightleftharpoons}} O_2^{2-} + e^- \quad \text{(A-3)}$$

$$O_2^{2-} \underset{k_{4c}}{\overset{k_{4a}}{\rightleftharpoons}} O_2^- + e^- \quad \text{(A-4)}$$

$$O_2^- \underset{k_{5c}}{\overset{k_{5a}}{\rightleftharpoons}} O_2 + e^- \quad \text{rds} \quad \text{(A-5)}$$

For this mechanism it is assumed that (i) reaction (A-5) is the rate-determining step and (ii) there is a low coverage of the oxide ion.

At equilibrium, reaction (A-1) can be written

$$k_{1a}(1-\theta)C_{CO_3^{2-}} = k_{1c}\theta C_{CO_2} \quad \text{(A-6)}$$

where θ is the coverage of the oxide ion (O^{2-}).

Assuming a low coverage of O^{2-}, $(1-\theta) \sim 1$. From eqn (A-6) the fraction of θ is

$$\theta = \frac{k_{1a}}{k_{1c}}\frac{C_{CO_3^{2-}}}{C_{CO_2}} \quad \text{(A-7)}$$

The net current from reaction (A-2) is

$$i_2 = F\{k_{2a}\theta e^{\beta_2 FE/RT} - k_{2c}(1-\theta)C_{(O^-)}e^{-(1-\beta_2)FE/RT}\} \quad \text{(A-8)}$$

E is the potential difference between the electrode and the electrolyte. At equilibrium, when $i_0 = 0$

$$C_{(O^-)} = \frac{k_{1a}}{k_{1c}}\frac{k_{2a}}{k_{2c}}\frac{C_{CO_3^{2-}}}{C_{CO_2}}e^{FE/RT} \quad \text{(A-9)}$$

For reaction (A-3) the net current is

$$i_3 = F\{k_{3a}\theta C_{(O^-)}e^{\beta_3 FE/RT} - k_{3c}(1-\theta)C_{O_2^{2-}}e^{-(1-\beta_3)FE/RT}\} \quad \text{(A-10)}$$

When $i_3 = 0$ at equilibrium,

$$C_{O_2^{2-}} = \left(\frac{k_{1a}}{k_{1c}}\right)^2 \frac{k_{2a}}{k_{2c}} \frac{k_{3a}}{k_{3c}} \left(\frac{C_{CO_3^{2-}}}{C_{CO_2}}\right)^2 e^{2FE/RT} \tag{A-11}$$

The net current for reaction (A-4) is

$$i_4 = F\{k_{4a}C_{O_2^{2-}}e^{\beta_4 FE/RT} - k_{4c}C_{O_2^-}e^{-(1-\beta_4)FE/RT}\} \tag{A-12}$$

At equilibrium, when $i_4 = 0$

$$C_{O_2^-} = \left(\frac{k_{1a}}{k_{1c}}\right)^2 \frac{k_{2a}}{k_{2c}} \frac{k_{3a}}{k_{3c}} \frac{k_{4a}}{k_{4c}} \left(\frac{C_{CO_3^{2-}}}{C_{CO_2}}\right)^2 e^{3FE/RT} \tag{A-13}$$

For reaction (A-5) the net current is

$$i_5 = F\{k_{5a}C_{O_2^-}e^{\beta_5 FE/RT} - k_{5c}C_{O_2}e^{-(1-\beta_5)FE/RT}\} \tag{A-14}$$

Reaction (A-5) is assumed to be the rate-determining step, the equilibrium expression of eqn (A-14) for the equilibrium potential E^0 upon substitution for $C_{O_2^-}$ from eqn (A-13)

$$e^{FE^0/RT} = \left[\left(\frac{k_{1c}}{k_{1a}}\right)^2 \frac{k_{2c}}{k_{2a}} \frac{k_{3c}}{k_{3a}} \frac{k_{4c}}{k_{4a}} \frac{k_{5c}}{k_{5a}} \left(\frac{C_{CO_2}}{C_{CO_3^{2-}}}\right)^2 C_{O_2}\right]^{\frac{1}{4}} \tag{A-15}$$

Replacing $E = E^0 + \eta$, where η is the overpotential, in eqn (A-14), the following can be obtained

$$i_5 = F\left\{\begin{array}{l} k_{5a}\left[\left(\frac{k_{1a}}{k_{1c}}\right)^{\frac{1-\beta_5}{2}}\left(\frac{k_{2a}}{k_{2c}}\frac{k_{3a}}{k_{3c}}\frac{k_{4a}}{k_{4c}}\right)^{\frac{1-\beta_5}{4}}\left(\frac{k_{5a}}{k_{5c}}\right)^{\frac{3+\beta_5}{4}} C_{CO_3^{2-}}^{\frac{1-\beta_5}{2}}\right] C_{O_2}^{\frac{3+\beta_5}{4}} C_{CO_2}^{\frac{\beta_5-1}{2}} e^{(3+\beta_5)F\eta/RT} \\ -k_{5c}\left[\left(\frac{k_{1a}}{k_{1c}}\right)^{\frac{1-\beta_5}{2}}\left(\frac{k_{2a}}{k_{2c}}\frac{k_{3a}}{k_{3c}}\frac{k_{4a}}{k_{4c}}\right)^{\frac{1-\beta_5}{4}}\left(\frac{k_{5a}}{k_{5c}}\right)^{\frac{1-\beta_5}{4}} C_{CO_3^{2-}}^{\frac{1-\beta_5}{2}}\right] C_{O_2}^{\frac{3+\beta_5}{4}} C_{CO_2}^{\frac{\beta_5-1}{2}} e^{-(1-\beta_5)F\eta/RT} \end{array}\right\} \tag{A-16}$$

Eqn (A-16) may be compared with the general relationship for activation polarization according to the Butler–Volmer equation

$$i = i_0[e^{\alpha_A \eta F/RT} - e^{-\alpha_c \eta F/RT}] \tag{A-17}$$

The exchange current density may be written as

$$i_0 = i_0^0(C_{O_2})^{\frac{3+\beta_5}{4}}(C_{CO_2})^{\frac{\beta_5-1}{2}} \tag{A-18}$$

where

$$i_0^0 = \left(\frac{k_{1a}}{k_{1c}}\right)^{\frac{1-\beta_5}{2}} \left(\frac{k_{2a}}{k_{2c}}\frac{k_{3a}}{k_{3c}}\frac{k_{4a}}{k_{4c}}\right)^{\frac{1-\beta_5}{4}} \frac{(k_{5a})^{\frac{1-\beta_5}{4}}}{(k_{5c})^{\frac{3+\beta_5}{4}}} C_{CO_3^{2-}}^{\frac{1-\beta_5}{2}} \tag{A-19}$$

We obtain $\alpha_A = 3 + \beta_5$, assuming the symmetry factor $\beta_5 = 0.5$, so $\alpha_A = 3.5$ and $\alpha_c = 1 - \beta_5 = 0.5$.

The dependency of i_0 on the gas partial pressure (for $\beta_5 = 0.5$) is

$$i_0 = i_0^0 (p_{O_2})^{0.875} (p_{CO_2})^{-0.25} \tag{A-20}$$

List of symbols

F	Faraday's number, C mol^{-1}
i	Current density, A m^{-2}
i_0	Exchange current density, A m^{-2}
i_0^0	Standard exchange current density, A m^{-2}
L	Electrode thickness, m
n	Number of electrons
p	Partial pressure, atm
R	Gas constant, J mol^{-1} K^{-1}
S	Specific surface area, m^{-1}
T	Temperature, K
x	Partial pressure dependency of oxygen in MCEC mode
y	Partial pressure dependency of carbon dioxide in MCEC mode

Greek letters

α_A	Anodic transfer coefficient
α_c	Cathodic transfer coefficient
β	Symmetry factor
k_1	Effective conductivity of the electrode, S m^{-1}
k_2	Effective conductivity of the electrolyte, S m^{-1}
η	Overpotential, V

Acknowledgements

The financial support of the China Scholarship Council (CSC) is appreciated. The cell components were provided by Ansaldo Fuel Cells in Italy.

References

1 W. H. A. Peelen, K. Hemmes and J. H. W. De Wit, *Electrochim. Acta*, 1997, **43**, 763.

2 V. Kaplan, E. Wachtel, K. Gartsman, Y. Feldman and I. Lubomirsky, *J. Electrochem. Soc.*, 2010, **157**, B552.

3 V. Kaplan, E. Wachtel and I. Lubomirsky, *J. Electrochem. Soc.*, 2012, **159**, E159.

4 D. Chery, V. Albin, V. Lair and M. Cassir, *Int. J. Hydrogen Energy*, 2014, **39**, 12330.

5 L. Hu, I. Rexed, G. Lindbergh and C. Lagergren, *Int. J. Hydrogen Energy*, 2014, **39**, 12323.

6 A. J. Appleby and S. B. Nicholson, *J. Electroanal. Chem.*, 1974, **53**, 105.

7 A. J. Appleby and S. B. Nicholson, *J. Electroanal. Chem.*, 1977, **83**, 309.

8 A. J. Appleby and S. B. Nicholson, *J. Electroanal. Chem.*, 1980, **112**, 71.

9 G. B. Dunks and D. Stelman, *Inorg. Chem.*, 1983, **22**, 2168.

10 M. Takahashi, *Extended Abstracts of 21st Symposium on Molten Carbonate Salt Chemistry*, Yamanashi, Japan, 1989, p. 51.

11 H. R. Kunz, L. J. Bregoli and S. T. Szymanski, *J. Electrochem. Soc.*, 1984, **131**, 2815.

12 R. C. Makkus, H. Hemmes and J. H. W. De Wit, *J. Electrochem. Soc.*, 1994, **141**, 3429.

13 R. C. Makkus, Electrochemical studies on the oxygen reduction and NiO(Li) dissolution in molten carbonate fuel cells, Ph.D. thesis, Delft University of Technology, Netherlands, 1991.

14 H. R. Kunz and L. A. Murphy, *J. Electrochem. Soc.*, 1988, **135**, 1124.

15 S. H. Lu and J. R. Selman, *J. Electrochem. Soc.*, 1990, **137**, 1125.

16 W. H. A. Peelen, K. Hemmes and G. Lindbergh, *J. Electrochem. Soc.*, 2000, **147**, 2122.

17 K. Sugiura, K. Takei, K. Tanimoto and Y. Miyazaki, *J. Power Sources*, 2003, **118**, 218.

18 L. Hu, G. Lindbergh and C. Lagergren, *J. Electrochem. Soc.*, 2015, **162**, F1020.

19 C. Lagergren, Electrochemical performance of porous MCFC cathodes, Ph.D. thesis, KTH Royal Institute of Technology, Sweden, 1997.

20 C. Lagergren and D. Simonsson, *J. Electrochem. Soc.*, 1997, **144**, 3813.

21 L. Plomp and B. J. Veldhuis, *J. Power Sources*, 1992, **39**, 369.

22 C. Lagergren and G. Lindbergh, *Electrochim. Acta*, 1998, **44**, 503.

23 S. Tanase, Y. Miyazaki, M. Yanagida, K. Tanimoto and T. Kodama, *Prog. Batteries Sol. Cells*, 1987, **6**, 195.

24 C. Y. Yuh and J. R. Selman, *J. Electrochem. Soc.*, 1991, **138**, 3649.

25 C. Y. Yuh and J. R. Selman, *J. Electrochem. Soc.*, 1991, **138**, 3642.

Faraday Discussions

ROYAL SOCIETY OF CHEMISTRY

DISCUSSIONS

Oxygen electrodes: general discussion

John Irvine, Raymond Gorte, Scott Barnett, John Vohs, Tatsumi Ishihara, Viola Birss, Ann Call, Bilge Yildiz, Joseph Hartvigsen, Lan Hu, San Ping Jiang, Chengsheng Ni, Minfang Han and Mogens Mogensen

DOI: 10.1039/C5FD90076J

Scott Barnett opened the discussion of the paper by San Ping Jiang: How would the formation of the Cr compound be affected if A site deficient LSM was used instead of stoichiometric LSM? Would that tend to suppress the strontium chromite formation, due to the lower availability of Sr, and bias more towards Cr–Mn spinel?

San Ping Jiang responded: This is an important question. The stoichiometry of the composition of LSM may have an effect on the Cr deposition, but in this study we only used stoichiometric LSM. However, we already started studying Cr deposition on non-stoichiometric LSM under identical conditions to those used for the stoichiometric LSM. Once the study is completed, we will publish the results as soon as possible.

John Irvine remarked: I am concerned about the sintering of the ceramic interfaces, as the electrode is only fired at 1150 °C which is quite low. Have you looked at a composite? This would improve the electrochemistry and bonding to the electrolyte. With a weak interface there will be points of high current density.

San Ping Jiang responded: Thanks for the comments. Yes, you are right. The electrode–electrolyte interface depends on the sintering temperature. However, in the case of LSM electrode–YSZ electrolyte, the optimum sintering temperature is about 1100 to 1150 °C, as shown in the literature.[1] Sintering at temperatures of higher than 1150 °C would lead to an interfacial reaction between LSM and YSZ, forming an insulating lanthanum zirconate phase.

1 S. P. Jiang, *J. Mater. Sci.*, 2008, **43**, 6799–6833.

Bilge Yildiz asked: Based on the mechanism you proposed for chromium deposition (*i.e.* that it is correlated to excess Sr), it appears that avoiding Sr enrichment (and phase precipitation) can be helpful. Do you think that getting rid of Sr segregation or getting rid of Sr in the electrode composition could reduce or avoid chromium poisoning? And a comment; indeed both our work[1] and Juergen

Janek's work[2] have also shown less Sr segregation when the conditions are more reducing. In our analysis, we have shown this to be based on expansion of the lattice (under reducing conditions) which can accommodate the large Sr more easily in the bulk.

1 W. Lee, J. W. Han, Y. Chen, Z. Cai and B. Yildiz, *J. Am. Chem. Soc.*, 2013, **135**, 7909–7925.
2 A.-K. Huber, M. Falk, M. Rohnke, B. Luerßen, L. Gregoratti, M. Amati and J. Janek, *Phys. Chem. Chem. Phys.*, 2012, **14**, 751–758.

San Ping Jiang answered: Yes, you are right. Avoiding Sr enrichment or reducing Sr segregation is effective in reducing the Cr deposition. For example, adding Nb to the B-site and Ba to the A-site of LSCF can substantially decrease surface segregation of Sr, and thus increase the stability and tolerance of the LSCF-based electrodes towards Cr deposition.[1] In agreement with your and Juergen Janek's work, the current study shows that Sr segregation is in fact promoted when the conditions are more oxidative, *i.e.*, under anodic polarization conditions. This is evidently supported by the observation of the formation of $SrCrO_4$, rather than $(Cr,Mn)_3O_4$ as in the case of Cr deposition on LSM electrodes under cathodic polarization conditions.

1 X. Chen and S. P. Jiang, *J. Mater. Chem. A*, 2013, **1**, 4871–4878.

Viola Birss said: Yesterday, I discussed the promising characteristics of our MIEC (LSFCr-3), especially after replacing the Sr with Ca. What would be your prediction regarding the propensity for Cr poisoning in SOEC (oxygen evolution) mode for a Ca-substituted (no Sr content) material such as ours?

San Ping Jiang replied: So far we have not investigated chromium deposition and poisoning on Ca-substituted (no Sr content) electrodes under SOFC or SOEC operation conditions. As there is no Sr content in your material, I expect that Cr deposition would depend on the segregation of Ca and the activity between Cr and segregated Ca. I would be very interested in the Cr deposition and poisoning on your materials under both SOFC and SOEC conditions.

Viola Birss enquired: Have you observed any evidence that Cr deposition, resulting from Sr enrichment at positive overpotentials typical of SOEC anodes, can be reversed, perhaps by oscillating between SOEC and SOFC modes? In your EIS data in the paper, there may be some evidence for "activation" of the electrode during OCP EIS measurements (especially at low frequencies) after a period of time of oxygen evolution.

San Ping Jiang answered: Thanks for the comments. We did not study the Cr deposition under cyclic SOEC and SOFC operation conditions. However, as shown by the XRD and XPS data, the Cr deposition products on LSM under SOEC conditions are chromium oxides and strontium chromide; the reverse of Cr deposition by reversed operation under SOFC mode would thus be unlikely as both Cr deposits are thermodynamically stable. Also, Cr deposition also occurs on LSM electrodes under SOFC operation conditions.

Raymond Gorte opened discussion of Minfang Han's paper: I was very surprised to see that you had delamination problems with infiltrated electrodes. Please explain how you prepared the scaffold for your electrode.

Minfang Han answered: The porous YSZ scaffold was prepared by a tape casting process. YSZ and graphite pore former were mixed at a weight ratio of 50 : 50. Ethanol–butanone solvent, caster oil dispersant, dibutyl phthalate (DBP) plasticizer, and polyvinyl butyral (PVB) binder were then added and ball-milled to form a YSZ slurry. The tape casting process was performed to prepare the porous YSZ scaffold. Both the NiO–YSZ support and the YSZ electrolyte were prepared by the same process. The YSZ electrolyte was sandwiched between the other two layers under vacuum conditions by a hot isostatic pressing process. The resulting tri-layer structure was calcined at 1300 °C *via* a one-step sintering method to obtain the YSZ scaffold-contained cell substrate.

Joseph Hartvigsen asked: In the processing slide showing a hot-press step, do you use hot isostatic pressing at high temperature as is required for sintering, or warm isostatic pressing as is used for tapes to help the binders attach?

Minfang Han replied: In the experimental section, the tape casting and thermally isostatic pressing technologies were used for the cell substrate preparation. Actually, the thermally isostatic pressing was carried out at 70 °C under vacuum conditions for the attachment of the tapes, thus warm isostatic pressing was preferable.

Scott Barnett commented: Our experience has been that the better attachment between the oxygen electrode scaffold and the electrolyte, expected in an infiltrated electrode, didn't yield a significant improvement in the electrode delamination, relative to a conventional printed electrode.

John Vohs remarked: Based on the SEM image in your paper, it is not clear to me that it shows detachment of the electrode. Do you have any additional images, perhaps for a larger length scale, that provide additional evidence for detachment?

Minfang Han answered: A higher magnification of the post-test interface between YSZ electrolyte and the SSC–YSZ electrode is provided in Fig. 1 below. A gap was visually formed between the YSZ and SSC–YSZ after long-term steam electrolysis operation.

Ann Call questioned: In reference to Fig. 8 in the paper, can you comment further on the presence of what appears to be a fibrous material between the scaffold/electrolyte and the infiltrated material interface? Do you know its composition or origin? It almost appears as though this material is pushing the infiltrated material away from the interface and may be contributing to the delamination observed.

Minfang Han responded: As shown in Fig. 8a, the delamination of the oxygen electrode from the electrolyte was observed and some fibrous (filament-like)

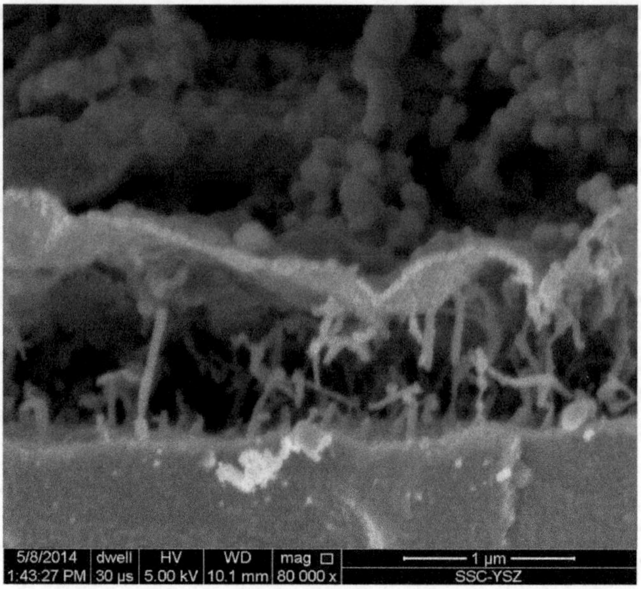

Fig. 1

material appeared at the interface. By employing EDS measurements at the selected gap, a cobalt-rich composite was detected, which could possibly be ascribed to the segregation of $Sm(Sr)CoO_3$. Further experiments and characterization may be needed to demonstrate the specific composition of the material.

Scott Barnett asked: For the EDS spectrum, over what area in the SEM image was the composition measured? Looking at your composition results, why was there zero Sr? Also, did you observe the formation of a Sr zirconate at the zirconia SSC interface?

Minfang Han answered: As shown in the SEM image, the EDS measurement was obtained within the area of the red lines in Fig. 8b (diameter of 1 μm). EDS was used to characterize the element distribution in the gap area between the LSCF–YSZ electrode and the YSZ electrolyte. In particular, the filament-like deposits were measured to be rich in cobalt. Strontium wasn't contained in the selected area *via* the EDS measurement, indicating that no Sr was deposited on the segregated interface. At the delaminated interface after long-term electrolysis operation, no Sr was detected. Moreover, further work will be required to investigate whether or not strontium zirconate was formed at the unseparated interface under different operation modes/conditions.

John Irvine queried: Do the cobalt rich particles originate from the loss of Sr to form zirconate at the interface?

Minfang Han replied: EDS characterization was used to measure the elemental distribution at a selected area of the gap between the SSC–YSZ electrode and the

YSZ electrolyte. In particular, a relatively large gap was found and the composition of the particles at the interface was investigated. No Sr was detected in the separated area *via* EDS measurements, indicating that strontium zirconate wasn't formed at the delaminated interface. Nevertheless, Sr was lost at the interface, which may be one of the causes of the cobalt enhancement.

Tatsumi Ishihara said: Did you measure the stability of the electrode at different temperatures or different current densities? Is there any condition which prevents delamination of the electrode? Maybe under some conditions $Sm(Sr)CoO_3$ becomes more stable and you can operate for a longer period.

Minfang Han responded: Yes, we have measured the electrode stability at lower temperatures (<700 °C) under power generation mode. As a result, the cell experienced a small current density change at 0.7 V, indicating relatively stable cell performance. Further work will be required to demonstrate conditions for preventing the electrode delamination or improving the stability of the $Sm(Sr)CoO_3$ electrode under steam electrolysis operation.

Raymond Gorte opened discussion of the paper by Lan Hu: Please describe your experimental set up and how you measure overpotentials.

Lan Hu answered: 1. Experimental set-up: The cell consists of a porous Ni–Cr alloy as a H_2-electrode and a porous Ni, oxidized and lithiated *in situ*, as an O_2-electrode. The electrolyte is a mixture of 62/38 mol% Li_2CO_3/K_2CO_3. Two reference electrodes, consisting of Au wires in equilibrium with a gas mixture of 33/67% O_2/CO_2, are placed in separate chambers filled with the same electrolyte as that in the cell. 2. Measuring overpotentials: The potential was applied on the cell to simulate the electrolysis cell mode, and the overpotentials were measured in potentiostatic steps. The polarization curves of the O_2-electrode were measured with the reference electrode.

Scott Barnett enquired: Regarding the experimental part of your presentation, when your cell was converting steam and CO_2, did you observe any methanation in addition to hydrogen production? This would be expected thermodynamically at the relatively low temperatures here, but it may not be kinetically favorable.

Lan Hu replied: In my opinion, it is possible to produce methane, since the reverse water–gas shift reaction could produce CO and nickel could be one of the catalysts for methanation. But I have not examined the outlet gas compositions yet, this will be performed in future work.

Scott Barnett asked: For the equation you use for the cell resistance, can you describe the assumptions? Is it a combination of mass transport and charge transfer processes?

Lan Hu responded: It is assumed that the NiO particles are formed as agglomerates, and the electrode reaction mainly takes place at the exterior surface area of the agglomerates. This equation only takes into account the charge-transfer process. We analyze the polarization data at low overpotential, assuming that in this range the electrode is under kinetic control where the mass-transfer limitations could be neglected.

Mogens Mogensen queried: How stable are your reference electrodes? These are supposed to be some of the most inert materials — are they quite stable? Are they made of gold?

Lan Hu answered: In the molten carbonate electrolysis cell the reference electrode is gold in equilibrium with 33/67% O_2/CO_2, and the reference electrode is stable.

John Irvine commented: Gold is about the only reference electrode possible as everything else dissolves in molten lithium carbonate, especially platinum.

Chengsheng Ni responded: It's very true that gold is more stable than platinum in molten carbonate, as shown in **ref. 1** below and in other literature. Actually for the reference in molten carbonate, the redox couple is the equilibration of oxygen species and carbonate, which is affected by the CO_2 and oxygen partial pressure. As long as the reference is in equilibrium, the effect of the metal is minor, but the dissolution of platinum should change the reference.

1 G. J. Janz, A. Conte and E. Neuenschwander, *Corrosion*, 1963, **19**, 292t–294t.

John Irvine asked: How would you see this as a whole system, especially in terms of CO_2 management as both electrodes will require CO_2? As this is for doing hydrogen production, how do you propose to manage the CO_2, including its separation from O_2?

Lan Hu responded: Since this is under molten carbonate conditions, the CO_2 is required to form the electrolyte ions (CO_3^{2-}). For the MCEC operation, it is not purely H_2 production. And it is still possible to form CO from the reverse water-gas shift reaction and/or CO_2 electrolysis. In general, it can use O_2/CO_2 recycle combustion to separate CO_2 from O_2.

Viola Birss questioned: Do you see any differences in the "wettabilities" of the Ni and NiO electrodes during fuel cell or electrolysis modes? As one is a metal and one is an oxide, they should interact differently with the molten carbonate medium. Related to this, are there no issues with bubble removal during oxygen evolution?

Lan Hu replied: In both MCFC and MCEC modes, the wetting property of the Ni electrode is better than that of the NiO electrode. This is one reason why the ohmic loss of the Ni electrode is relatively lower. Until now I have not taken into account the issue of bubble removal during oxygen evolution. We will consider this and investigate in future work. However, the SEM micrograph of the NiO electrode after the MCEC operation does not show significant differences when compared to the NiO electrode in the MCFC operation.

Viola Birss communicated: In the MCEC mode, is CO_2 reduced, along with steam, forming CO + H_2, similar to what is seen during SOEC operation under co-electrolysis conditions (steam + CO_2)? Only the formation of H_2 from steam was alluded to in the talk as occurring in the cathodic (reduction) reaction during electrolysis.

Lan Hu communicated in reply: Yes, in the MCEC mode, CO_2 is reduced along with steam, forming CO + H_2, similarly to the co-electrolysis (steam + CO_2) in the SOEC operation. It seems that the formation of CO is mainly from the reverse water–gas shift reaction. But there is still the possibility of CO_2 electrolysis forming CO. The formation of H_2 is both from electrolysis of steam and from the reverse water–gas shift reaction. During the electrolysis, the cathodic reaction may include CO_2 electrolysis (CO formation).

Faraday Discussions

PAPER

Solid oxide electrolysis: Concluding remarks

Areum Jun, Young-Wan Ju and Guntae Kim*

Received 19th August 2015, Accepted 19th August 2015
DOI: 10.1039/c5fd90072g

Renewable energy resources such as solar energy, wind energy, hydropower or geothermal energy have attracted significant attention in recent years. Renewable energy sources have to match supply with demand, therefore it is essential that energy storage devices (*e.g.*, secondary batteries) are developed. However, secondary batteries are accompanied with critical problems such as high cost for the limited energy storage capacity and loss of charge over time. Energy storage in the form of chemical species, such as H_2 or CO_2, have no constraints on energy storage capacity and will also be essential. When plentiful renewable energy exists, for example, it could be used to convert H_2O into hydrogen *via* water electrolysis. Also, renewable energy resources could be used to reduce CO_2 into CO and recycle CO_2 and H_2O into sustainable hydrocarbon fuels in solid oxide electrolysis (SOE).

Introduction

High temperature solid oxide electrolysis (HT-SOE) is a promising candidate technology that could play a key role in the transition to sustainable energy. The high temperature operation (heat energy) maximizes its efficiency and high reaction rate, offering thermodynamic and kinetic advantages. However, there are still some drawbacks when the cell is operated in SOE mode. One of the most important issues in SOE mode is the oxygen electrode degradation, which is attributed to the formation of a secondary phase and delamination of the oxygen electrode. For realization of a commercial SOE technology, it is necessary to develop robust and efficient electrodes and study more systematic fundamental physicochemical and electrochemical properties including surface chemistry and defect chemical behavior.

Recently, Graves *et al.* reported that operating a SOC in a reversible cycling mode between electrolysis and fuel-cell modes, similar to a rechargeable battery, inhibits the microstructural deterioration and long-term degradation compared with constant electrolysis operation.[1] SOCs (solid oxide cells) can both store renewable electricity as fuels (electrolysis mode) and convert fuels to electricity

Department of Energy Engineering, Ulsan National Institute of Science and Technology (UNIST), Ulsan, Republic of Korea. E-mail: gtkim@unist.ac.kr; Fax: +82 52 217 2909

(fuel cell mode). Incorporating SOECs and SOFCs into a single device that can work reversibly with high and comparable efficiency will provide significant overall cost benefits.

This discussion held in York brought together researchers from a wide range of disciplines, *i.e.*, fundamental electrochemistry, materials development, system, durability and oxygen electrodes, within the excellent traditional *Faraday Discussions* format, to discuss advances in areas relevant to the main theme of the meeting.

Fundamental electrochemistry

Kee *et al.* (DOI: 10.1039/c5fd00012b) developed new experimental and modeling approaches to establish thermodynamics (ΔH° and ΔS°) and transport properties (diffusion and surface kinetics) in a multiple charged defect conducting (MCDC) ceramic, yttrium-doped barium zirconate (BZY10). Because widely-practiced models, such as ambipolar diffusion theory, are not directly applicable to MCDCs, a more complex formulation based on the Nernst–Planck–Poisson was developed. The measured enthalpy and entropy changes and kinetics representing the incorporation of H_2, O_2, and H_2O may give insight into efficient hydrogen production using SOEs based on proton conducting materials.

Graves *et al.* (DOI: 10.1039/c5fd00048c) discussed both model and porous electrodes to obtain useful information for a more realistic technological electrode in SOCs. The conventional fuel-electrode material, a composite of nickel and yttria/scandia stabilized zirconia (Ni–SZ), is known to be more active for H_2/H_2O than for CO/CO_2 reactions. The ratio of electrode polarization resistance in CO/CO_2 *vs.* H_2/H_2O decreases from 33 to 2. Besides showing higher activity for H_2/H_2O reactions than CO/CO_2 reactions, Ni–SZ is more active for oxidation than reduction. Graves *et al.* found the opposite behavior in both cases for CGOn/STN model electrodes, thus reflecting the unprecedented observation of a higher electrocatalytic activity of CGO nanoparticles for CO/CO_2 than for H_2/H_2O reactions in the absence of a gas diffusion limitation.

In order to better understand the nature of the electrocatalytic activities observed for different electrode materials and geometries, which vary considerably for these reactions, they measured and compared AC impedance and DC current-overpotential for Ni–SZ, 3 phase boundary (3PB), and CGO nanoparticles on a porous mixed ionic conductor such as $Sr_{0.94}Ti_{0.9}Nb_{0.1}O_3$ (STN) backbones, 2 phase boundary (2PB). The enhanced surface reduction of nano CGO on STN results in a higher ratio of CO/CO_2 *vs.* H_2/H_2O and more activity on CO_2 reduction than H_2O reduction *via* increased surface concentrations of oxygen vacancies and polarons (Ce^{3+}). They also pointed out that a Pt current collector may selectively enhance the activity of the H_2/H_2O reaction.

Cumming *et al.* (DOI: 10.1039/c5fd00030k) at the University of Sheffield developed and designed an analytical tool to fully understand and characterize co-electrolysis of carbon dioxide and steam to produce syngas, *in situ* monitoring of the reactions, products, and SOC. They emphasized the importance of *in situ* or *operando* analysis of fuel cells and electrolysers. This *in situ* spectroscopy could provide invaluable information about the surfaces and species involved in the overall macroscopic reaction, along with information on the complex interplay of reactions occurring during co-electrolysis. Diffuse reflectance infrared Fourier transform spectroscopy (DRIFT) is ideal for the detection of CO and CO_2 species, and has previously not been used for this purpose. The designed rig has been used for symmetric cell testing at temperatures from 450 to 600 °C. In contrast with *ex situ* experiments, *in situ* DRIFT gives significant changes related to a combination of CO oxidation, the water gas shift reaction, and carbonate formation and decomposition processes, with the dominant process being both potential and temperature dependent. Further development is expected to allow the use of asymmetric cells in the presence of different atmospheres.

What is a real surface? Is the surface oxygen exchange rate at the true outer surface linked to bulk ion diffusivity in mixed conducting materials, *e.g.* Ruddlesden–Popper (RP) phases? McIntosh *et al.* (DOI: 10.1039/c5fd00014a) tried to determine a correlation between bulk ion mobility and the surface exchange rate between materials. They measured each rate with separate techniques, *in situ* neutron diffraction and pulsed isotopic surface exchange, respectively, for three RP phases, such as the general form $A_{n-1}A'_2B_nO_{3n+1}$ ($n = 1$, 2 and 3). The surface exchange rates vary by more than one order of magnitude with high anion mobility in the bulk of an oxygen vacancy-rich $n = 2$ RP material, correlating with rapid oxygen exchange. Even though it was concluded that surface oxygen incorporation is not rate limiting in these materials, the challenge of determining the true outer surface oxygen exchange rate experimentally remains.

Proton conducting solid oxide fuel cells (PCFCs) have attracted significant attention in recent years due to their advantages, such as higher ionic conductivity of electrolytes compared to oxide ion conductors, no dilution of the fuel with water vapor, faster start-up, and less degradation.[2] The reduction of the PCFC operating temperature, however, causes a drop of the electrocatalytic activity, which occurs especially in the cathode process.[3] In this meeting, Merkle *et al.* (DOI: 10.1039/c5fd00013k) presented thermogravimetric properties which changed to conditions of lower and higher p_{O_2} and rationalized the observed changes in the defect chemical behavior. By considering the defect relationship between three mobile carriers (*i.e.* $V_O^{\bullet\bullet}$, $OH_O^{\bullet}$, and holes), the concentrations of oxygen vacancies, protons, and holes can be calculated numerically as a function of $K_O p_{O_2}{}^{1/2}$ and $K_W p_{H_2O}{}^{1/2}$. From the calculated results, one can recognize that the proton uptake of materials having high redox-activity is mainly compensated by hole consumption (*i.e.* hydrogenation), while the proton uptake of electrolyte materials is mainly compensated by $V_O^{\bullet\bullet}$ (*i.e.* hydration). In addition, the thermogravimetric investigation of $Ba_{0.5}Sr_{0.5}Fe_{0.8}Zn_{0.2}O_{3-\delta}$ (BSFZ) under various p_{O_2} atmospheres shows that BSFZ incorporates a non-negligible amount of protons and the mode of proton incorporation changes from hydration (at low p_{O_2}) to hydrogenation (at high p_{O_2}), indicating that the presence of oxygen vacancies is not necessarily required for the proton uptake.

The surface and near-surface chemical composition of electroceramic materials often shows significant deviations from that of the bulk. The oxygen transport properties and their relationship with the surface chemistry in cation-ordered $GdBaCo_2O_{5+\delta}$ (GBCO) perovskite were investigated by Téllez *et al.* (DOI: 10.1039/c5fd00027k) using Low Energy Ion Scattering (LEIS) spectroscopy and ^{18}O tracer exchange coupled with Secondary Ion mass Spectrometry (SIMS). The LEIS showed that the surface and near-surface composition is very dynamic, and significant compositional changes can occur at relatively low temperatures and short time periods with accompanying structural rearrangements. In particular,

GBCO exhibited rapid segregation of the divalent Ba cation to the surface to form a BaO terminated layer, suggesting that a multiphase material is formed at the near-surface. Despite the different surface compositions and the extension of the Ba segregation profile, the surface exchange coefficients (k^*) appeared to be less sensitive to the surface composition than anticipated, even though the cation segregation led to a significantly different surface in terms of the Ba surface coverage. From these results, it appears that the near-surface disorder introduced by the cation segregation might have a more powerful influence on the observed surface exchange kinetics than the surface termination.

A reversible SOFC (RSOFC) with mixed ion-electron conducting (MIEC) with a simplified cell design and lower manufacturing costs was studied with $La_{0.3}Sr_{0.7}Fe_{0.7}Cr_{0.3}O_{3-\delta}$ (LSFCr-3) by Birss *et al.* (DOI: 10.1039/c5fd00029g). RSOFCs can run in the electrolysis mode (SOEC) to store energy by converting H_2O to H_2 and O_2, or co-electrolyze CO_2 and H_2O to syngas and O_2. The fuel cell mode (SOFC) can convert H_2 or syngas, and oxygen, to electricity and heat when electricity is needed with available excess electricity. Their careful electrochemical studies with 3 electrode half-cell impedance allowed them to isolate the reactions occurring in the fuel cell and electrolysis modes and determine the catalytic activity of the material towards each reaction separately. Interestingly, LSFCr-3 is a better catalyst for oxygen evolution and CO_2 reduction than for oxygen reduction or CO oxidation, indicating overall better performance in the electrolysis *vs.* fuel cell mode.

Material development

Various types of materials such as simple perovskite oxides and layered perovskite oxides have been investigated as electrodes for solid oxide electrolysis cells. Among them, pure oxide fuel electrodes have been attracting much interest because commercial cermet fuel electrodes require permanent circulation of a reducing gas (*e.g.* H_2 or CO, called a safe gas) to avoid oxidation of the metallic phase. Torrell *et al.* investigated steam/CO_2 co-electrolysis with a symmetrical solid oxide electrolysis cell consisting of reversible $La_{0.75}Sr_{0.25}Cr_{0.5}Mn_{0.5}O_{3-\delta}$ (LSCM) electrodes and an ytterbium scandium stabilized zirconia ($(ZrO_2)_{0.9}$-$(Yb_2O_3)_{0.06}(Sc_2O_3)_{0.04}$, YbScSZ) electrolyte (DOI: 10.1039/c5fd00018a). They demonstrated that no safe gas is required when using symmetrical full ceramic SOEC cells and ratios of $H_2 : CO$ of about 2 were obtained, meeting the requirement for the synthesis of standard hydrocarbon fuels or alcohols. However, significant degradation was also observed under galvanostatic operation conditions, caused by delamination of the oxygen–electrolyte interface. This may be associated with an increase of the internal oxygen pressure close to the surface of the oxygen electrode.

A new layered perovskite oxide transformed from a simple perovskite under a reducing atmosphere was introduced earlier this year.[4] The layered perovskite ($PrBaMn_2O_{5+\delta}$) oxide exhibits high electrical conductivity, excellent redox tolerance and coking tolerance. It is a promising cathode material for CO_2 electrolysis (DOI: 10.1039/c5fd00025d). By an *in situ* high temperature X-ray diffraction study, a reversible phase change between $PrBaMn_2O_{5+\delta}$ (layered perovskite) and $PrBaMn_2O_{6-\delta}$ (simple perovskite) during redox cycles was confirmed. An electrolysis cell using a layered perovskite oxide ($PrBaMn_2O_{5+\delta}$) exhibited a fairly high performance.

The remarkable electrolysis cell performance originates from the highly mixed ionic and electronic conductivity of the pure oxide fuel electrode. Mixed ionic and electronic conducting (MIEC) materials have been used not only as electrodes but also as membranes for air separation and for production of syngas (a mixture of CO and H_2). Yu *et al.* investigated oxygen permeation fluxes in mixed ionic and electronic conducting (MIEC) membranes based on composites of 40 vol% $La_{0.8}Sr_{0.2}CrO_3$ (LSCr) and 60 vol% yttria stabilized zirconia (YSZ). Oxygen fluxes in MIEC membranes consisting of LSCr and YSZ are often limited by surface reactions and can be improved by modifying the surface morphology with inflation of the MIEC (DOI: 10.1039/c5fd00007f).

Tsvetkov *et al.* (DOI: 10.1039/c5fd00023h) emphasized that surface modification can improve the surface oxygen kinetics and stability through suppression of Sr segregation and phase separation of dopant cations from the A-site, driven by elastic and electrostatic energy minimization and kinetic demixing. Careful surface modification was performed by adding the optimum amount of Ti (10%) into $La_{0.8}Sr_{0.2}CoO_3$ (LSC) by dipping the latter into a chemical bath of TiO_4 solution. Interestingly, the oxygen exchange kinetics on LSC-T cathodes was found to be up to 8 times higher than on bare LSC cathodes and their stability was retained. This kind of simple surface modification can open up a new route for stabilization and better electrochemical performance for perovskite oxide electrodes for high temperature electro- and thermo-chemical applications.

Composites consisting of a perovskite-based electronic or mixed conductor with a fluorite-structured ionic conductor are often used as electrodes in solid oxide electrochemical energy conversion devices. For fabricating the composite electrode, the composite materials are usually sintered at high temperature. However, after sintering the materials, inter-reactions between the two phases, or inter-diffusion of cations or impurities between the two phases, are often observed. Durce *et al.* studied the ^{18}O exchange properties and surface composition of a composite consisting of $Ce_{0.9}Gd_{0.1}O_{1.95}$ (CGO) and $La_{0.6}Sr_{0.4}Co_{0.2}Fe_{0.8}O_{3-\delta}$ (LSCF) in a 50 : 50 ratio by a high resolution time of flight-secondary ion mass spectrometry (ToF-SIMS) analysis and LEIS spectroscopy (DOI: 10.1039/c5fd00028a). High resolution ToF-SIMS mapping revealed that the ^{18}O fraction of the composite is much higher than expected from the diffusivity (D^*) and surface exchange coefficient (k^*) values for the single-phase parent material. By a surface compositional analysis with ToF-SIMS and LEIS spectroscopy, they confirmed that the surfaces of the CGO grains in the composite do not show impurities, which typically segregate into the surface in single-phase CGO. In order to explain this phenomenon, they considered three mechanisms: (1) catalytic spill-over; (2) activation of the CGO surface by incorporation of transition metals; and (3) cleaning of the CGO surface by the dissolution of impurities into the LSCF. Due to the absence of evidence to support mechanism (2), they concluded the apparent enhanced surface exchange for CGO in these composites may be related with mechanisms (1) and (3), or a combination of the two.

Novel materials compatible with YSZ should be developed for better performance as the oxygen electrode of RSOFCs at temperatures below 800 °C for the purpose of using metallic interconnects, enhancing stability, and reducing maintenance expenses. Ni *et al.* (DOI: 10.1039/c5fd00026b) introduced

$CaMn_{0.9}Nb_{0.1}O_{3-\delta}$ (CMN) with Nb doping as a novel oxygen electrode through an investigation of crystal structure, electrochemical characterization, physical/chemical property variations, and the difference in electrochemical performance of the oxygen electrode from $CaMnO_{3-\delta}$ (CM) and $CaMn_{0.9}Nb_{0.1}O_{3-\delta}$ (CMN) materials under SOE and SOFC operating conditions. The crystal structure of CM and CMN was determined to be an orthorhombic structure. The δ values of CM and CMN were analyzed by TGA and changed to 0.099 and 0.025 at 900 °C in air, respectively. The changes of oxygen content caused by the valence variation of the transition metal were anticipated to be beneficial to the electrochemical performance of the oxygen electrode.[5] At 700 °C, CM, as a cathode for SOFCs, achieved a polarization resistance of less than 0.1 Ω cm^2, while CMN showed superior performance (R_p < 0.1 Ω cm^2 at 0.4 V bias) as an anode for SOECs. A possible candidate material with a Nb content of less than 10 mol% doping on the Mn site would lead to better oxygen electrode performance in both SOFC and SOE modes.

System studies

For commercialization of a solid oxide electrolysis cell, in addition to fundamental issues, systemic issues should also be discussed. The study of high temperature (HT) steam/CO_2 co-electrolysis is much more complicated than that of HT steam or CO_2 electrolysis because (1) apart from electrochemical reactions, the chemical reverse water–gas shift reaction (RWGS) also plays an important part in the formation of CO, (2) the equilibrium of the RWGS is highly sensitive to the outlet gas temperature from the electrolysis cell, and (3) the performance and syngas composition change significantly with the microstructure of the SOECs and the operating conditions.[6–8] Wang *et al.* investigated reaction characteristics and the outlet syngas composition of high temperature (HT) steam/CO_2 coelectrolysis under different operating conditions with a single cell (5 × 5 cm^2) at 800 °C (DOI: 10.1039/c5fd00017c). Through a comparison of the area specific resistance (ASR) under different fuel conditions, it was confirmed that the RWGS more significantly contributes to the formation of CO than the electrochemical reduction reaction. The outlet syngas composition from HT steam/CO_2 coelectrolysis is very sensitive to the operating conditions.

Previously, performance maps based on closed form isothermal parametric models for both hydrogen and natural gas fuelled SOFC stacks have been demonstrated.[9] By applying this approach, Milobar *et al.* determined the power and the efficiency of a solid oxide electrolysis stack over a wide range of operating conditions (DOI: 10.1039/c5fd00015g). The determined values were then confirmed with estimated costs of principal components and sized for specific system design criteria to obtain a cost/performance model. The model was used to map the performance of a stack and the cost elements for a given set of performance parameters across the operating space in voltage and specific steam flow rate. Component sizing and cost estimates were used to predict the total system capital cost. A cost analysis framework is a useful tool to reveal opportunities for system optimization and cost minimization.

Tao *et al.* introduced a new application of solid oxide electrolysis cell. They directly synthesized ammonia from wet air at 400 °C using a solid oxide electrolysis cell consisting of a $Sm_{0.6}Ba_{0.4}Fe_{0.8}Cu_{0.2}O_{3-\delta}$ electrode and a $Ce_{0.8}Gd_{0.2}O_{2-\delta}$ and ternary carbonate mixture $(Li, Na, K)_2CO_3$ (32.1 wt% Li_2CO_3;

34.5 wt% K_2CO_3; 33.4 wt% Na_2CO_3) electrolyte (DOI: 10.1039/c5fd00033e). The ammonia formation rate was more than two orders of magnitude higher than that when synthesized from N_2 and H_2O at 600 °C.[10] The electrode material was also inexpensive compared to commercial catalysts for ammonia synthesis from air and water.

Understanding durability

Recently, a higher-efficiency storage method utilizing a reversible solid oxide cell (ReSOC) has been proposed.[11] One key question regarding ReSOCs is their long-term durability. Hughes *et al.* (DOI: 10.1039/c5fd00020c) presented the degradation of solid oxide cell $(La_{0.8}Sr_{0.2})_{0.98}MnO_{3-\delta}$–$Zr_{0.84}Y_{0.16}O_{2-\gamma}$ (LSM–YSZ) composite oxygen electrodes during reversing current operation. Degradation was characterized by impedance spectroscopy in symmetrical LSM–YSZ/YSZ/LSM–YSZ cells tested at 800 °C in air with a symmetric current cycle with a period of 12 hours. No change in cell resistance could be detected in 1000-h tests with a sensitivity of ~1% per kh, at a current density of 0.5 A cm^{-2} corresponding to an overpotential of 0.18 V, while higher current/overpotential values at a current density of 0.6 A cm^{-2} (0.33 V overpotential) led to faster degradation. As the authors indicated, the polarization degradation, mainly observed at higher current density, is associated with the resistance of YSZ grain boundaries within the electrode from an increase in LSM particle size and a decrease in the amount of YSZ and LSM at the electrode/electrolyte interface.

Similar to the problems facing solid oxide fuel cell (SOFC) technology, a notable issue is poisoning of the air electrode by impurities such as Cr released from metallic interconnects or balance-of-plant components.[12] Egger *et al.* (DOI: 10.1039/c5fd00021a) investigated $La_2NiO_{4+\delta}$ (LNO) as a potential anode material for high temperature solid oxide electrolyzer cells (SOECs). Long-term durability was tested at 800 °C for 140 days (~3300 hours), applying current densities up to −410 mA cm^{-2} under a pure Ar/O_2 atmosphere as well as in the presence of chromium and humidity. After 46 days (1100 h) in a pure 20% O_2/Ar atmosphere, Cr was added under dry conditions and after another 41 days (1000 h) 30% relative humidity was introduced into the gas stream. A total increase in the anode ASR of 350% from 0.22 to 1.0 Ω cm^2 was observed. A clear effect of Cr in dry/humid conditions on the degradation behavior unfortunately could not be revealed from cell measurements. Cr was not detected by EDX in the SOEC anode after long-term operation, indicating that Cr-poisoning may have an imperceptible influence on LNO electrodes under anodic polarization. Poor electrode-electrolyte contact or partial delamination is probably responsible for the observed performance degradation of the LNO anode.

Generally, the state-of-the-art commercial solid oxide cell (SOC) uses Ni–YSZ as the fuel electrode material. Operation of a Ni–YSZ electrode supported Solid Oxide Cell (SOC) was investigated by Ebbesen *et al.* (DOI: 10.1039/c5fd00032g), who showed the operation of a SOC in both fuel cell mode (FC-mode) and electrolysis cell mode (EC-mode) in mixtures of H_2O/H_2, CO_2/CO, and $H_2O/H_2O/CO_2$/CO at 750 °C, 800 °C, and 850 °C. The observed area specific resistance (ASR) values in both H_2O/H_2 and CO_2/CO mixtures are lower for oxidation than for reduction. Further, the ASR (both $ASR_{fuel\ cell}$ and $ASR_{electrolysis}$) values were higher for CO_2/CO mixtures compared to the ASR values for H_2O/H_2 mixtures due to lower activity for

CO oxidation/CO_2 reduction compared to H_2 oxidation/H_2O reduction. The slower kinetics for reduction was found to be partly related to the polarization heating and the entropy change. Due to the exothermic reaction in FC-mode and the endothermic reaction in EC-mode, both the ohmic resistance and the electrochemical electrode resistance are higher in EC-mode as compared to FC-mode. The effect of changing the porosity was also discussed by Ebbesen *et al.* during the meeting. The polarization in EC-mode at decreasing porosity increased, indicating that diffusion limitations cannot be neglected for support structures with porosity below 30% (with a support thickness of 300 μm). These diffusion limitations through the support and active electrode structure create an increased reducing atmosphere at the interface, which may be related to the degradation of the cells.

Fabrication routes for new SOEC mesoporous nanocomposite materials as electrodes were presented by Torrell *et al.* (DOI: 10.1039/c5fd00035a). Ceramic–metal (cermet) mesoporous composites of Ni–CGO obtained after reduction of mesoporous NiO impregnated with CGO were employed as a fuel electrode while samarium strontium cobaltite ($Sm_{0.5}Sr_{0.5}CoO_{3-\delta}$/SSC) impregnated scaffolds of mesoporous samarium doped ceria ($Ce_{0.8}Sm_{0.2}O_{1.9}$/SDC) were used as the oxygen electrode. The maximum value of injected current density at 1.7 V was 330 mA cm^{-2} with significant ohmic resistance at a temperature of 850 °C. This is likely to be due to insufficient adhesion of the electrode to the electrolyte caused by the high thermal stability of the mesoporous materials. Therefore, optimization of the electrode attachment temperatures should be pursued in future studies, especially when high thermal stability materials such as this are used.

Oxygen electrodes

Chen *et al.* (DOI: 10.1039/c5fd00010f) investigated the activity and stability of $La_{0.8}Sr_{0.2}MnO_3$ (LSM) oxygen electrodes for the first time under solid oxide electrolysis cell (SOEC) operating conditions at 800 °C in the presence of an Fe–Cr interconnect. Cr deposition occurs in the LSM electrode bulk and in particular on the YSZ electrolyte surface and on the LSM electrode inner surface close to the electrode/electrolyte interface. The results showed the formation of Cr, $CrO_{0.87}$, Cr_2O_3, Cr_2O_5, and $SrCrO_4$ on the YSZ electrolyte surface, within the contact rings and on the LSM inner surface. The observation of $SrCrO_4$ formation on the LSM electrode under SOEC operating conditions is attributable to anodic polarization promoting Sr segregation. The segregated Sr species could be highly mobile and readily spread to the YSZ electrolyte surface because of the high activity of gaseous Cr species with segregated Sr species. Chen *et al.* also presented clear evidence that the delamination or failure of LSM oxygen electrodes is primarily due to the disintegration or breaking of LSM particles attached to the YSZ electrolyte surface. The excess segregation of Sr could also result in partial decomposition of LSM, and this appears to be supported by the shift of $Mn2p_{3/2}$ to a lower binding energy (BE) of 639.5 eV under the rib of the interconnect.[13] This may indicate partial decomposition of the LSM perovskite phase on the YSZ surface under the rib of the interconnect.

Nano-sized composite electrodes were fabricated by infiltration technology and subsequently sintered at relatively low temperatures (700–900 °C),[14] which effectively hinders the formation of insulating phases. Nano-sized SSC–YSZ oxygen electrodes prepared by an infiltration process for SOFCs/SOECs were investigated by Fan *et al.* (DOI: 10.1039/c5fd00022j). The electrochemical performance of the cell was determined under both fuel cell and steam electrolysis modes using polarization curves and electrochemical impedance spectroscopy (EIS). The current densities of a Ni–YSZ/YSZ/SSC–YSZ cell at an electrolysis voltage of 1.3 V are 0.91 mA cm^{-2}, 0.70 mA cm^{-2}, 0.45 mA cm^{-2}, and 0.27 mA cm^{-2} at 800 °C, 750 °C, 700 °C, and 650 °C, respectively. The Ni–YSZ/YSZ/SSC–YSZ cell showed gradual degradation during electrolysis over 100 h in 50% H_2O, 25% H_2, and 25% N_2 due to delamination of the oxygen electrode from the electrolyte and agglomeration of SSC particles. Moreover, segregation of cobalt-enriched particles (particularly cobalt oxides) at the interface is likely related to cell degradation under the steam electrolysis mode.

The performance of a molten carbonate electrolysis cell (MCEC) is, to a great extent, determined by the anode, *i.e.* the oxygen production reaction at the porous NiO electrode.[15] The stationary polarization curves for the NiO electrode under varying gas compositions and temperatures were analyzed by Hu *et al.* (DOI: 10.1039/c5fd00011d). The activation energy of the NiO porous electrode for oxygen production decreased between 278 and 211 kJ mol^{-1} when increasing the O_2 concentration from 1.5 to 30%. In contrast, the value increased from 124 to 229 kJ mol^{-1} with increasing CO_2 content from 2.5 to 40%. Gas compositions of low O_2 partial pressure, *e.g.* 1.5–10% O_2, exhibit much larger activation energies than those with low CO_2 partial pressure, which means that oxygen appears to have a stronger effect on charge-transfer limitation. This indicates that the NiO electrode for oxygen production is generally under a charge-transfer limitation, except at very low CO_2 partial pressure where it is under mixed control. As the activation energy is evident in determining the rate-determining process,[16] it is probable that the NiO electrode is under mixed control at quite low CO_2 concentrations. Also, the reaction $CO_3^{2-} \leftrightharpoons O^{2-} + CO_2$ is considered to play an important role at low CO_2 partial pressure in oxygen production in MCEC mode.

Final comments

This meeting beginning with the possibility that a SOE would be the best energy conversion and storage technology to realize sustainable energy was held in the quaint city of York. The SOEC is only a reverse reaction of a SOFC, but there are some difficulties to solve as it has a more complicated electrochemical reaction and system than those of the SOFC, which is the common view of the participants. The newly developed technologies and prospects were discussed throughout this meeting, especially including fundamental electrochemistry, the overall results of new materials, systems, and durability in the traditional manner of *Faraday Discussions*. Some researchers introduced various types of materials such as simple perovskite oxides and layered perovskite oxides as electrodes with high electrolysis performance for solid oxide electrolysis cells. In particular, Professor Mogensen asked 'the most frequent questions' to participants and his enthusiasm made the meeting more effective. I believe these enthusiastic efforts and vision will realize the industrialization of SOE on a worldwide scale in the near future. I would like to close my comments with an expression of my heartfelt gratitude to the organizing committee members, including John T. S. Irvine, Raymond J. Gorte, Tatsumi Ishihara, Mogens Mogensen, Truls Norby and John Varcoe.

References

1 C. Graves, S. D. Ebbesen, S. H. Jensen, S. B. Simonsen and M. B. Mogensen, *Nat. Mater.*, 2014, **14**, 239–244.

2 K. D. Kreuer, *Annu. Rev. Mater. Res.*, 2003, **33**, 333–359.

3 A. Grimaud, F. Mauvy, J. Marc Bassat, S. Fourcade, M. Marrony and J. Claude Grenier, *J. Mater. Chem.*, 2012, **22**, 16017.

4 S. Sengodan, S. Choi, A. Jun, T. H. Shin, Y.-W. Ju, H. Y. Jeong, J. Shin, J. T. S. Irvine and G. Kim, *Nat. Mater.*, 2014, **14**, 205–209.

5 E. Konysheva and J. T. S. Irvine, *J. Mater. Chem.*, 2008, **18**, 5147–5154.

6 C. M. Stoots, *High-temperature co-electrolysis of H_2O and CO_2 for syngas production*, 2006, p. 12.

7 V. Menon, Q. Fu, V. M. Janardhanan and O. Deutschmann, *J. Power Sources*, 2015, **274**, 768–781.

8 Y. Xie and X. Xue, *Solid State Ionics*, 2012, **224**, 64–73.

9 J. Hartvigsen, S. Elangovan and A. Khandkar, *Solid Oxide Fuel Cells*, Pennington, NJ, 1993, p. 878.

10 A. Skodra and M. Stoukides, *Solid State Ionics*, 2009, **180**, 1332–1336.

11 D. M. Bierschenk, J. R. Wilson and S. A. Barnett, *Energy Environ. Sci.*, 2011, **4**, 944–951.

12 J. R. Mawdsley, J. D. Carter, A. J. Kropf, B. Yildiz and V. A. Maroni, *Int. J. Hydrogen Energy*, 2009, **34**, 4198–4207.

13 Y. Wang and Y. Ohtsuka, *Appl. Catal., A*, 2001, **219**, 183–193.

14 J. M. Vohs and R. J. Gorte, *Adv. Mater.*, 2009, **21**, 943–956.

15 L. Hu, I. Rexed, G. Lindbergh and C. Lagergren, *Int. J. Hydrogen Energy*, 2014, **39**, 12323–12329.

16 C. Y. Yuh and J. R. Selman, *J. Electrochem. Soc.*, 1991, **138**, 3642–3648.

Poster titles

Single-step, infiltrated cathodes for SOFC, **T.-S. Oh, Y. Chen, R. J. Gorte and J. M. Vohs**, *University of Pennsylvania, USA*

Thermodynamics and stoichiometry relaxation kinetics in materials with three carriers: analytic relations and numerical simulations, **R. Merkle, D. Poetzsch and J. Maier**, *Max Planck Institute for Solid State Research, Germany*

Mechanistic studies of RSOFCs by infiltration into porous MIEC electrode structures, **B. Molero-Sánchez, P. Addo, A. Buyukaksoy and V. Birss**, *University of Calgary, Canada*

Nickel-yttria stabilized zirconia fuel electrodes infiltrated with gadolinium doped ceria, **T. L. Skafte, C. Graves, P. Blennow and J. Hjelm**, *Haldor Topsøe A/S, Denmark*

Modular Fischer-Tropsch system for conversion of syngas from co-electrolysis of CO_2 and steam, **J. Hartvigsen, S. Elangovan, L. Frost, J. Elwell, P. Czernichowski, A. Gomez, M. Bokil and D. Claus**, *Ceramatec, Inc., USA*

Nonlinear degradation model for SOEC, **J. Hartvigsen and S. Elangovan**, *Ceramatec, Inc., USA*

Highly efficient performance and stability of layered perovskite electrodes for solid oxide electrolysis cells, **G. Kim and A. Jun**, *UNIST, South Korea*

Determination of solid oxide cell processes by the representation of electrochemical impedance spectroscopy data in different formalisms, **A. R. Thompson, A. Call, D. J. Cumming, D. C. Sinclair and R. H. Elder**, *University of Sheffield, United Kingdom*

Why the solid oxide cells can be reversibly operated under solid oxide electrolysis cell and fuel cell modes?, **K. Chen, S.-S. Liu, N. Ai, M. Koyama and S. P. Jiang**, *Curtin University, Australia*

Microchanneled Ni/YSZ cathode supports of SOECs for CO_2 electrolysis, **S. Xu, X. Shao, D. Dong, G. Parkinson, L. Hucker, J. Marin, T. Pham and C.-Z. Li**, *Curtin University of Technology, Australia*

High temperature co-electrolysis of steam/CO_2 mixtures for syngas production using Pd-infiltrated solid oxide cells, **S.-W. Kim, H. Kim, K. J. Yoon, J.-W. Son, J.-H. Lee, B.-K. Kim, J.-H. Lee and J. Hong**, *Korea Institute of Science and Technology, South Korea*

In situ Raman spectroscopy monitoring for solid oxide cells, **J. Manerova, K. W. Reeping, A. V. Call, D. C. Sinclair, R. A. Walker and R. H. Elder**, *University of Sheffield, United Kingdom*

Characterization of $BaZr_{0.85}Y_{0.15}O_{3-\delta}$ thin films for proton ceramic electrolyser cells by pulsed laser deposition, **E. Stefan, M. Stange, C. Denonville, Y. Larring, R. Dahl-Hansen, P. M. Rørvik, T. Norby and R. Haugsrud**, *University of Oslo, Norway*

Understanding CO_2 electrochemical reduction reaction in high temperature solid oxide electrolysis cells, **X. Yue and J. T. S. Irvine**, *University of St Andrews, United Kingdom*

The Skinner Prize for the best poster was awarded to Beatriz Molero-Sánchez of University of Calgary, Canada, for her poster on mechanistic studies of RSOFCs by infiltration into porous MIEC electrode structures.

List of participants

Mr Amane Abdoun, *University of St Andrews, United Kingdom*
Professor Scott Barnett, *Northwestern University, USA*
Dr Viola Birss, *University of Calgary, Canada*
Dr Ann Call, *University of Sheffield, United Kingdom*
Dr Denis Cumming, *University of Sheffield, United Kingdom*
Dr Dehua Dong, *Curtin University of Technology, Australia*
Dr John Druce, *Kyushu University, Japan*
Dr Sune Dalgaard Ebbesen, *Technical University of Denmark, Denmark*
Dr Andreas Egger, *Montanuniversitaet Leoben, Austria*
Dr Rachael Elder, *University of Sheffield, United Kingdom*
Miss Jie Feng, *University of St Andrews, United Kingdom*
Dr Raymond Gorte, *University of Pennsylvania, USA*
Dr Christopher Graves, *Technical University of Denmark, Denmark*
Miss Abigail Hallowes, *Royal Society of Chemistry, United Kingdom*
Professor Minfang Han, *China University of Mining and Technology, China*
Mr John Bøgild Hansen, *Haldor Topsøe A/S, Denmark*
Mr Joseph Hartvigsen, *Ceramatec, Inc., USA*
Ms Isobel Hogg, *Royal Society of Chemistry, United Kingdom*
Dr Joanne Holmes, *Metrohm, United Kingdom*
Dr Jongsup Hong, *Korea Institute of Science and Technology, South Korea*
Ms Lan Hu, *KTH Royal Institute of Technology, Sweden*
Mr Jianing Hui, *University of St Andrews, United Kingdom*
Professor John Irvine, *University of St Andrews, United Kingdom*
Professor Tatsumi Ishihara, *Kyushu University, Japan*
Professor San Ping Jiang, *Curtin University, Australia*
Professor Robert Kee, *Colorado School of Mines, USA*
Professor Guntae Kim, *UNIST, South Korea*
Miss Lisa Kleiminger, *Imperial College London, United Kingdom*
Dr Ming Li, *University of Nottingham, United Kingdom*
Professor Göran Lindbergh, *KTH Royal Institute of Technology, Sweden*
Dr Yingjun Liu, *University of Cambridge, United Kingdom*
Ms Kirsty Lunnon, *Royal Society of Chemistry, United Kingdom*
Miss Jen Manerova, *University of Sheffield, United Kingdom*
Professor Steven McIntosh, *Lehigh University, USA*
Mr Ian McPherson, *University of Oxford, United Kingdom*
Dr Rotraut Merkle, *MPI for Solid State Research, Germany*
Professor Ian Metcalfe, *Newcastle University, United Kingdom*
Professor Mogens Mogensen, *Technical University of Denmark, Denmark*
Ms Beatriz Molero-Sánchez, *University of Calgary, Canada*
Dr Alex Morata, *Catalonia Institute for Energy Research, Spain*
Dr Chengsheng Ni, *University of St Andrews, United Kingdom*
Dr Andrew Osborne, *Metrohm, United Kingdom*
Mr Robert Price, *University of St Andrews, United Kingdom*

Ms Chloë Pritchard, *Royal Society of Chemistry, United Kingdom*
Dr Victoria Richards, *Royal Society of Chemistry, United Kingdom*
Mr Giorgio Rinaldi, *École Polytechnique Fédérale de Lausanne, Switzerland*
Dr Guenter Schiller, *German Aerospace Center, Germany*
Dr Tae Ho Shin, *Korea Institute of Ceramic Engineering Technology, South Korea*
Mr Theis L. Skafte, *Haldor Topsøe A/S, Denmark*
Dr Elena Stefan, *University of Oslo, Norway*
Professor Shanwen Tao, *University of Warwick, United Kingdom*
Miss Alison Thompson, *University of Sheffield, United Kingdom*
Dr Marc Torrell, *Catalonia Institute for Energy Research, Spain*
Mr Christopher Tumilson, *Queen's University Belfast, United Kingdom*
Dr Helena Téllez, *Kyushu University, Japan*
Professor John Varcoe, *University of Surrey, United Kingdom*
Professor John Vohs, *University of Pennsylvania, USA*
Dr Bo Wei, *RWTH Aachen University, Germany*
Mr Chen Xinbing, *Shanghai Institute of Applied Physics, Chinese Academy of Sciences, China*
Professor Bilge Yildiz, *Massachusetts Institute of Technology, USA*
Dr Xiangling Yue, *University of St Andrews, United Kingdom*

